HYDROGEN: ITS TECHNOLOGY AND IMPLICATIONS

Editors

Kenneth E. Cox
Project Manager
Thermochemical Hydrogen
Los Alamos Scientific Laboratory
University of California
Los Alamos, New Mexico

K. D. Williamson, Jr.
Assistant Division Leader
Los Alamos Scientific Laboratory
University of California
Los Alamos, New Mexico

SERIES OUTLINE

Hydrogen: Its Technology and Implications

Hydrogen Properties
Volume III

Author:

Robert D. McCarty

Cryogenics Division
Institute for Basic Standards
National Bureau of Standards
Boulder, Colorado

Editors:

Kenneth E. Cox
Project Manager, Thermochemical Hydrogen
Los Alamos Scientific Laboratory
University of California
Los Alamos, New Mexico

K. D. Williamson, Jr.
Assistant Division Leader
Los Alamos Scientific Laboratory
University of California
Los Alamos, New Mexico

Based on National Aeronautics and Space Administration Special Publication 3089 (1975), prepared for Aerospace Safety Research and Data Institute, NASA Lewis Research Center, Cleveland, Ohio

CRC Press
Taylor & Francis Group
Boca Raton London New York

CRC Press is an imprint of the
Taylor & Francis Group, an informa business

First published 1975 by CRC Press
Taylor & Francis Group
6000 Broken Sound Parkway NW, Suite 300
Boca Raton, FL 33487-2742
Reissued 2018 by CRC Press

ISBN 13: 978-1-138-59704-4 (hbk)
ISBN 13: 978-0-429-48717-0 (ebk)

Visit the Taylor & Francis Web site at http://www.taylorandfrancis.com and the
CRC Press Web site at http://www.crcpress.com

PREFACE TO HYDROGEN: ITS TECHNOLOGY AND IMPLICATIONS

The United States, Western Europe, Japan, and several other countries are presently faced with an energy shortage due largely to an imbalance of energy consumption over fossil energy production. This problem was dramatized in October 1973 during the Arab embargo on the shipment of oil to the United States and the resultant large increases in the price of crude oil. This shortage in energy supply was then termed the "energy crisis," and it was a clear demonstration of the nation's dependence on imported petroleum and its vulnerability on both political and economic grounds. It is clear that the above problems would worsen in the future unless more attention and effort are directed toward increasing domestic energy production from both depletable and nondepletable sources and reducing energy consumption.

In the short-term, till the year 2000, coal and nuclear energy are expected to play dominant roles in meeting the energy shortage despite the environmental restrictions that hamper the production and consumption of high-sulfur coal and similar difficulties (siting and radioactive waste disposal) that have slowed the development of nuclear energy. In the long-term, beyond the year 2000, it is imperative that all forms of renewable energy be developed. These include solar energy, in such forms as wind, ocean thermal gradients, and biomass; geothermal energy; and fusion.

A major problem with several of the renewable energy sources is that they are intermittent and their energy density is low; thus, there is a need for an energy carrier that can act as both a storage and transportation medium to connect the energy source to the energy consumer. Many of the renewable energy forms together with coal and fission exhibit their energy in the form of heat release. It is necessary to develop an energy carrier, other than electricity, to supply the transportation sector as well as overcome the problems of electrical storage.

Hydrogen, the lightest element, has been suggested as the energy carrier of the future. In itself, it is not a primary energy source but rather serves as a medium through which a primary energy source (such as nuclear or solar energy) can be stored, transmitted, and utilized to fulfill our energy needs. There are several distinct advantages to the use of hydrogen as an energy medium. It can be made from water, an inexhaustible resource. On combustion, water is the main product; thus, hydrogen can be regarded as a clean, nonpolluting fuel. Indications from current research efforts suggest that hydrogen may be produced from high-temperature heat sources at an efficiency greater than that of electrical generation, thereby making hydrogen a more economical energy source than electricity. Technology has already been developed for storing hydrogen as a pressurized gas, a cryogenic liquid, or in the form of a metal hydride. Systems for transporting hydrogen as a gas or a liquid have been developed with liquid hydrogen playing a major part in NASA's putting a man on the moon. Finally, hydrogen is of value as a chemical intermediate, being used in fertilizer manufacture, methanol synthesis, and petroleum treatment. This area of hydrogen utilization represents 3% of today's energy consumption and is expected to grow by a factor of five by the year 2000. The above concept of using hydrogen is termed the "hydrogen energy economy" and has been receiving an increasing amount of attention from energy scientists and engineers in the United States and abroad.

This series in five volumes represents a serious attempt at providing information on all aspects of hydrogen at the postgraduate and professional level. It discusses recent developments in the science and technology of hydrogen production; hydrogen transmission and storage; hydrogen utilization; and the social, legal, political, environmental, and economic implications of hydrogen's adoption as an energy medium. Although there are several reports of selected studies on hydrogen as a fuel, this is the first comprehensive reference book that covers a wide range of topics of notable interest and timely importance.

Volume I of the series discusses such topics as hydrogen production from fossil fuels, nuclear energy, and solar energy. Hydrogen production technology from water by traditional methods such as water electrolysis and newer attempts to split water thermochemically are included with details of current research efforts and future directions.

Volume II provides detailed design information on systems necessary for the storage, transfer, and transmission of gaseous and liquid hydrogen. Cost factors, technical aspects, and models of hydrogen pipeline systems are included together with a discussion of materials for hydrogen service. Metallic hydride gaseous storage systems for the utility and transportation industry are covered in detail, and the design of Dewars and liquid hydrogen transfer systems are examined.

Volume III focuses on hydrogen's properties and provides in one location all of the hydrogen data measured and compiled by the National Bureau of Standards, Cryogenic Division. The properties are individually discussed, and tables of data are provided. The properties of slush hydrogen are also included.

Volume IV covers the present and future uses of hydrogen. Hydrogen has been suggested as a prime candidate for both air and surface transportation. In the utility industry, hydrogen systems for peak shaving promise to play an important future role. Both present and future domestic and industrial applications of hydrogen are surveyed. These include present uses in ammonia and methanol synthesis and future uses in the direct hydrogasification of coal to synthetic natural gas. Important to all of these applications are the safety considerations in the use of hydrogen to allow for public acceptance of hydrogen's role as an energy medium.

Volume V is primarily concerned with the nontechnical aspects of hydrogen. Economics of hydrogen energy systems will play a major part in determining the time frame for hydrogen's adoption. Cost analyses of such systems with return on investment considerations are surveyed from the point of view of production, transmission, and storage of hydrogen. The environmental, political, social, and legal implications of new secondary energy forms such as hydrogen are discussed with reference to governmental energy policy, the social costs of energy production and use, and the public's acceptance of a hydrogen energy medium.

The unusually broad nature of hydrogen demands the expertise that could only be provided by a wide authorship; thus, some of the authors are the original authorities in their respective fields. Although the subject matter treated in each chapter is, in general, the author's research work and his critical review of the state-of-the-art, the authors have had complete freedom in choosing the particular important areas to be emphasized. As a result, some chapters treat the subject matter in more detail than others with a greater emphasis on the engineering or design aspects of a particular system. Therefore, each chapter possesses its own special feature and appealing points. Due to the limited space in the series, the editors have encouraged each author to supply an extensive list of references at the end of his chapter for the benefit of interested readers. Detailed author and subject indexes have been provided at the end of each volume.

The editors, while striving to avoid duplication, have allowed some degree of overlap in certain of the chapters for the sake of continuity and allowing the reader to view a particular topic from two or more points of view. Further volumes on the topic of hydrogen are planned, and we wish to hear from our readers as to areas that might have been neglected or deserve a special chapter on their own.

We would like to express our sincere thanks to these authors and the staff of CRC Press, Inc., in particular, Mrs. Gayle Tavens and Miss Sandy Pearlman, for their efforts in making these volumes possible. Lastly, we would like to thank our wives, Patricia R. Cox and Ruth S. Williamson, for their encouragement and help during the time it took to edit these five volumes.

K. E. Cox
Albuquerque, New Mexico
K. D. Williamson, Jr.
Los Alamos, New Mexico
March 1975

PREFACE TO VOLUME III: PROPERTIES OF HYDROGEN

An important part of any treatise on a subject such as hydrogen is its thermophysical properties. To a large extent, these determine how hydrogen can be stored, transmitted, and then utilized as a secondary energy source, an energy storage system, or a building block for more complex molecules.

The National Bureau of Standards, Cryogenics Laboratory in Boulder, Colorado, has played a leading role in experimental programs to obtain hydrogen properties. The Cryogenic Data Center has been collecting data and documents pertinent to the field for over 15 years. The basis of this volume is Robert D. McCarty's *Hydrogen Technological Survey – Thermophysical Properties* issued in 1975 as National Aeronautics and Space Administration Report NASA SP-3089. This publication was part of a major hydrogen safety review by the NASA Aerospace Safety Research and Data Institute (ASRDI).

In editing this publication for use in the current CRC Press Uniscience Series, several changes in format were made. All of the properties are presented in SI units with conversion tables at the end of the volume to aid in conversion to the English system.

It has been a great pleasure working with Bob and his colleagues at NBS, and we all hope this volume will provide a readily accessible source of data for scientists and engineers new and old to the field.

K. D. Williamson, Jr.
Los Alamos, New Mexico
March 1975

THE EDITORS

Kenneth E. Cox, Ph.D., is a Staff Member in the High-Temperature Chemistry Group at the Los Alamos Scientific Laboratory, Los Alamos, New Mexico. He is currently performing research in developing practical thermochemical cycles that produce hydrogen by water-splitting. In his previous position as Professor of Chemical Engineering at the University of New Mexico, Albuquerque, he pioneered the development of photovoltaic-electrolytic methods to produce hydrogen from solar energy and water.

Dr. Cox graduated in 1956 from the Imperial College of Science and Technology of the University of London with B.Sc. and A.C.G.I. degrees in chemical engineering. He received his M.A.Sc. from the University of British Columbia in 1959 and his Ph.D. from Montana State University in 1962, also in chemical engineering.

Dr. Cox is a member of the American Institute of Chemical Engineers, the American Chemical Society, and Sigma Xi, the Scientific Research Society. He is also a member of the International Association for Hydrogen Energy and has contributed both as an author and reviewer to the *International Journal of Hydrogen Energy*.

Dr. Cox has published numerous research papers and has given over 100 presentations. His current research interest include: thermochemical generation of hydrogen from water, use of solar energy for hydrogen generation, techno-economic energy evaluations as well as traditional chemical engineering, e.g., thermodynamics of separation processes, process design, and dispersion phenomena in fluids.

K. D. Williamson Jr., Ph.D. is Assistant Division Leader of the Systems Analysis and Assessment Division, Los Alamos Scientific Laboratory, Los Alamos, New Mexico.

Dr. Williamson received his B.S., M.S., and Ph.D. degrees in chemical engineering from Pennsylvania State University in 1957, 1959, and 1961, respectively.

Dr. Williamson is a member and lecturer for the New Mexico Academy of Sciences and has served as an invited lecturer at the Univversity of Tennessee and New Mexico Highlands University.

Dr. Williamson has published more than 35 research papers. His current interests include technology assessments and systems analyses of energy-related topics in support of research and development decision makers at Los Alamos and the State and National government levels.

TABLE OF CONTENTS

Chapter 1

Introduction

Chapter 1

INTRODUCTION

This document gathers together the available thermophysical properties of hydrogen. The work was performed at the Cryogenic Data Center of the National Bureau of Standards in Boulder, Colorado. The Cryogenic Data Center has been collecting data and documents pertinent to the field for 15 years and presently has about 100,000 coded entries which may be computer searched to give bibliographies on the physical properties of materials at low temperatures. A computer search of the data center holdings produced a bibliography for hydrogen of about 2,500 references. All of these references were considered; some of them do not appear here as a result of the review and editing process.

1.1. SCOPE AND PHILOSOPHY

The scope of this volume is defined as all physical properties of fluid hydrogen (chemical properties were excluded) with no limits on pressure or temperature. Although the solid phase was not included in the original scope, articles dealing with the properties of solid which were encountered during the review of the literature have been noted on the description sheets, and values of properties for the solid at the triple point have been included where possible. In addition, short descriptions of slush and metallic hydrogen are given. Properties of the isotopes of hydrogen have not been included, except in a few instances; the reader is referred to Reference 14.

Papers on engineering processes utilizing hydrogen were specifically excluded in the bibliographic search and subsequent review. The subject of hydrogen embrittlement of materials is not treated here (see Reference 276).

The level of review of the properties of hydrogen varies from the extensive tabulations of "critically evaluated" thermodynamic and transport properties to the selection of a few representative values from the literature by persons knowledgeable in the field. In the latter case the data selected were judged to be the most useful to the largest number of potential users of this volume. For those who need to pursue the subject in greater detail, a list of references is provided. References to articles that treat areas of limited interest in which only a few articles exist have also been included.

1.2. DESCRIPTIVE SHEETS

These sheets serve as the starting point for the use of this volume. For a given property each sheet usually contains the following:

A. A brief description or definition of the property to distinguish between similar sounding terms.

B. A locator of tabular values of the property. The range and units of the tables are given for convenience.

C. A locator of graphs and charts of the property.

D. Mathematical equations which represent the property provided that the functional form is simple.

E. One or more values of the property, usually at fixed points to illustrate the range of the property.

F. An estimate of uncertainty is usually given. It is made on a 2 σ level of probability. This means that there is a 95% chance that new experimental measurements (of equal or greater accuracy) of the property will fall within a 2 σ band of the quoted numbers. The estimate of the uncertainty of a property is seldom straightforward and is often made on the basis of the accuracy of the instrumentation used rather than the random scatter of a given set of data. In the case of a property that is tabulated over a wide range of pressure and temperature, the stated uncertainty is for the most uncertain value, and the uncertainty of most of the tabular entries is much less. Exceptions to the above convention are noted individually on the descriptive sheets. The most common exception is the critical region where the uncertainty of most properties is greater than the stated value. In no case should the quoted uncertainty be associated with values from other than recommended sources or values extrapolated from the tables.

G. A list of references is given with each descriptive sheet.

1.3. FIGURES

Chapter 3 is a collection of property charts and diagrams. These figures were taken from the literature in the original form of presentation; consequently, the units are a complete mix of SI, metric, and engineering. In most cases the prime purpose of the figures is to illustrate the behavior of a property over a wide range of pressure and temperature. Although the figures will provide adequate values for rough calculations, the tabular values are recommended for precise calculations.

1.4. TABLES

The tables of properties are collected in Chapter 4. The extensive tabulations of thermodynamic and transport properties cover the range from the triple point to 3000 K, with pressures to 100 MPa. They are presented in SI units for both para and normal hydrogen.

1.5. COMPUTER PROGRAMS AND EXTRAPOLATION

In many instances a computer program is a convenient source of property data. In other instances, particularly where a large number of values is needed in a short time, a computer program to furnish these values is a necessity. For these and other reasons, the use of computer programs to furnish fluid property data has steadily increased in recent years. This increased usage has resulted in a variety of computer programs. At the present time there does not seem to be a single program that will satisfy every need, and a choice must often be made on the basis of the particular job requirements. Table 1 lists the more important fluid property programs for hydrogen and some of the pertinent information needed to select the proper program for a particular job. A column-by-column description of Table 1 is given below.

A. Column 1 lists four general methods used to computerize the thermodynamic properties of fluids. The linear interpolation method is, as the name implies, tabular interpolation of property values stored in the computer. This method is fast but not very accurate and requires more computer core storage than the other methods. The most accurate method, if programmed properly, is polynomial interpolation. The disadvantages of the polynomial method are a lack of versatility and slow computations relative to the other methods. A good compromise and a widely used method is the equation-of-state

Table 1
MAJOR HYDROGEN PROPERTIES COMPUTER PROGRAMS

Program name	1[a] Type	2 P–T range	3 List of properties	4 Accuracy	5 Availability	6 Continuous	7 Modifications	8 Input
Tab Code[99]	Linear interpolation	0–35 MPa 14–3300 K	P, T, ρ, H, S, λ, η C_p, C_v	1% average	NBS Cryogenics Division	No	Para and equilibrium	P, T and P, H
H_2 Props[286]	33-term MBWR	0–83 MPa 14–1100 K	P, T, ρ, H, S C_p, C_v, ω	0.1% average	NBS Cryogenics Division	Yes	Para, ortho, equilibrium, and normal	P, T or D, T
H_2 Gen[b,1]	Polynomial interpolation	0–103 MPa 14–3300 K	P, T, ρ, H, S, λ, η, ω C_p, C_v, E, γ, RI, ϵ	Better than 0.1% average	NBS Cryogenics Division	No	Para	P, T
GASP[287] MBWR	20-term 14–1100 K	0–83 MPa C_p, C_v, ω, $\frac{\partial P}{\partial \rho}$, $\frac{\partial P}{\partial T}$	P, T, ρ, H, S, γ, η	0.1% average	NASA	Yes	Para (T, ρ) (H, P)	(P, T) (ρ, T) (S, P)

[a] Numbers refer to explanations in text.
[b] This program used to calculate tables in Chapter 4.

approach using a modified Benedict Webb Rubin (MBWR) equation. This method offers accuracies only slightly worse than the polynomial interpolation method and has the advantages of being faster and more versatile and giving a continuous, thermodynamically consistent surface. The main disadvantage of the MBWR is that it is functionally incorrect in the critical region, i.e.,

$$\rho = \rho_c \pm 20\% \text{ and } T = T_c \pm 5\%$$

The nonanalytical equation-of-state approach (no programs using this method are listed in Table 1) gives theoretically correct results in the critical region, but these equations are usually difficult to use.

B. Column 2 gives the range of validity of a program in terms of pressure and temperature limits. Extrapolation of these programs beyond the stated limits is not recommended. If properties are needed beyond the stated range of validity, the list of references for that property in Chapter 2 should first be scanned, as references to properties at extreme temperatures and pressures have been noted there. If extrapolation of one of the programs is necessary, the modified BWR type is the most convenient and least likely to yield large errors.

C. The symbols used in column 3 are defined as follows:

P = pressure
T = temperature
ρ = density
H = enthalpy
s = entropy
λ = thermal conductivity
η = viscosity
c_p = specific heat capacity at constant pressure
c_v = specific heat capacity at constant volume
ϵ = dielectric constant
γ = surface tension
RI = refractive index
ω = speed of sound

The units of the properties vary from program to program.

D. The accuracy of the properties calculated from these programs varies from property to property within a given program. The accuracies quoted in column 4 are for densities calculated from a P-T input. Since they are an estimate of the *average* accuracies, they are most useful as a basis of comparison with the other programs.

E. Column 5 is self-explanatory. Copies or use of the program may be obtained by contacting the agency listed.

F. For certain applications the mathematical continuity of the properties is important. In other words, the numbers coming from the program must not contain abrupt changes in the property or derivatives. Such requirements often occur in iterative calculations. If column 6 has a "no" in it, discontinuities may be expected.

G. The modification in column 7 refers to the relative orientation of the nuclear spin of the two hydrogen atoms in the molecule (see the explanation in Chapter 2 under ortho-para modifications).

H. The input to the various programs is given in column 8. The units of both input and output of the programs are not mentioned because in some cases a choice of units is available and in most cases the user converts the units to suit his particular need. An explanation of the input and output is furnished when the program is requested.

1.6. EVALUATION OF THE LITERATURE

The preparation of this review involved the evaluation of each reference to hydrogen data found in the literature. The Cryogenic Data Center's storage and retrieval system was queried for references pertaining to the physical properties of hydrogen (chemical properties excluded); this search yielded 2543 references. The resulting bibliography from the Cryogenic Data Center was further cross referenced by property, which greatly aided the evaluation process.

In many cases preliminary examination of the bibliography for a given property was sufficient to determine whether or not an article should be included. In the final evaluation process a copy of each remaining article was then obtained, and evaluation was made on the basis of content of the article.

1.7. CRITICAL SELECTION OF "BEST VALUES"

When selecting so-called "best values" many factors must be considered. The first tendency is to equate "best" with most accurate; indeed, in some cases this is true, but in other cases it is not. For example, in many engineering problems a continuous and thermodynamically consistent set of properties is of utmost importance. This is not to say that accuracy is not important. In the selection process both factors were considered. Selections were made to provide the most accurate and consistent set of properties over the widest possible range of pressure and temperature.

In that regard, Tables 4.1 and 4.2 are thermodynamically consistent and mathematically continuous. The properties tabulated in Tables 4.1 and 4.2 were calculated by means of a computer program (see Section 1.5) which is available from the Cryogenic Data Center.

1.8. ACKNOWLEDGMENT

The author is indebted to many colleagues for their assistance in preparing this volume; however, special contributions were made by V. J. Johnson, who supervised the preparation of the charts, and W. R. Parrish, who wrote the section on mixtures. Special thanks go to Phil Angerhofer for his very able assistance in securing the required articles from the literature and to Mildred Birchfield and Salli Schiffmacher for the many hours of typing manuscripts. Paul Ordin and George Mandel of the Lewis Research Center were the Project Manager and Technical Advisor, respectively, for the Aerospace Safety Research and Data Institute.

Chapter 2

Discussion of the Properties

Chapter 2
DISCUSSION OF THE PROPERTIES

2.1. GENERAL INFORMATION

Hydrogen is the first element in the periodic table with the atomic number 1; therefore, it is the lightest of all gases. On the earth hydrogen occurs chiefly in combination with oxygen as water. It is the most abundant element in the universe. It is a colorless, tasteless, and odorless gas and liquefies at atmospheric pressure at 20.27 K.

Uses – In recent years hydrogen has been used extensively as a rocket fuel in the space effort. It is now being seriously considered as a renewable fuel substitute for petroleum. The utilization of hydrogen is described in detail in Volume IV of this series.

Preparation – Methods of preparation include steam on heated carbon, decomposition of certain hydrocarbons by heat, electrolysis of water, and displacement from acids by certain metals. Details are described in Volume I of this series.

2.2. COMPRESSIBILITY COEFFICIENT[1-3]

Definition – The compressibility coefficient is defined as

$$-\frac{P}{V}\left(\frac{\partial V}{\partial P}\right)_T$$

Tables of Values – This quantity is not tabulated here, but it may be easily obtained by forming the product of P times the isothermal compressibility (see Section 2.5). The reciprocal of the isothermal compressibility, $V(\partial P/\partial V)_T$, is tabulated in Tables 4.1, 4.2, 4.12, and 4.13.

Units	Range of table		Table location
Dimensionless	13.8–3000 K	0.01–100 MPa	4.1, 4.12
Dimensionless	Saturation boundary		4.2, 4.13

Note: The tables reflect no differences for this property due to ortho-para modifications.

Figure – None
Equation – See Section 2.12
Range of Values – Para and normal

	Triple point		Boiling point			
Units	Liquid	Vapor	Liquid	Vapor	Critical point	300 K, 1 atm
Dimensionless	7.79×10^{-5}	1.02	2.02×10^{-3}	1.12	∞	1.0

Uncertainty – For values derived from tables listed above, the uncertainty varies from 0% in the low density limit to about 3 to 4% at 35 MPa and 10% at 100 MPa. In the critical region (T_c ± 5% and ρ_c ± 20%) the uncertainties are greater and very difficult to estimate as the property diverges at the critical point.

2.3. ACCOMMODATION COEFFICIENT[4]

Definition – The accommodation coefficient, a, is defined as the ratio $a = (E_i - E_r)/(E_i - E_r')$, where $(E_i - E_r)$ is the mean energy change of molecules colliding with a wall and $(E_i - E_r')$ is the mean energy change if the molecules come into thermal equilibrium with the wall.

Discussion – The accommodation coefficients of gases on solids are used to estimate heat conduction by gases at low pressures ($p \leqslant 0.1$ mm Hg). The accommodation coefficient is a function of the gas, temperature, and characteristics of the solid surface, especially the smoothness and cleanliness. In general, the lighter the gas, the higher the temperature and the smoother and cleaner the surface, the smaller is the value of the accommodation coefficient. The coefficient is independent of the pressure if the pressure is sufficiently high ($p \geqslant 0.1$ mm Hg) to insure a monomolecular gas film on the solid surface.

Tables of Values – None

Figure – None

Equation – None

Range of Values –

Units	At room temperature, $p \approx 0.1$ mm Hg
Dimensionless	a = 0.311 on platinum[4]

Uncertainty – The uncertainty is estimated to be ±1%

2.4. COMPRESSIBILITY FACTOR[1-3]

Definition – The compressibility factor is defined as $Z = PV/RT$.

Tables of Values – This quantity is not tabulated but may be simply obtained by forming PV/RT from Tables 4.1, 4.2, 4.12, and 4.13.

Figure – 3.1

Equation – See Section 2.12

Range of Values –

Units	Triple point		Boiling point		Critical point	300 K, 1 atm
	Liquid	Vapor	Liquid	Vapor		
Dimensionless	0.00161	0.9852	0.01712	0.9060	0.3024	1.0006

Uncertainty – The uncertainty of Z derived from the tables in Chapter 4 will correspond to the uncertainty of the PVT in the above tables. McCarty and Weber[2] estimated these uncertainties to be

Temperature range	Pressure range	Uncertainty in density
700–3000 K	1–100 MPa	1% (extrapolation)
300–700 K	1–35 MPa	0.5%
300–700 K	35–100 MPa	
13–300 K	1–100 MPa	0.1% (except critical region)
Critical region	$T_c \pm 5\%$, $\rho_c \pm 20\%$	6%

The uncertainties quoted here are for parahydrogen (see Introduction). The tables in

Chapter 4 reflect no difference in Z between the normal and para modifications. To estimate the uncertainties for the normal hydrogen tables, add 0.1% for temperatures above 100 K, 0.2% for temperatures between critical and 100 K, and 0.4% for the compressed liquid region.

Ortho-para Differences[1,2] – There is probably a very small difference in densities (less than 0.1%) between the ortho and para modifications. An exception is the saturated vapor and liquid densities where the difference is larger due to a difference in the vapor pressure curve (see Sections 2.49 and 3.33).

2.5. ISOTHERMAL COMPRESSIBILITY[1-3]

Definition – The isothermal compressibility is defined as

$$\beta_T = \frac{1}{V}\left(\frac{\partial V}{\partial P}\right)_T$$

Tables of Values – The reciprocal of the isothermal compressibility is given in Tables 4.1, 4.2, 4.12, and 4.13.

Units	Range of table		Table location[a]
MPa^{-1}	13.8–3000 K	0.01–100 MPa	4.1, 4.12
MPa^{-1}	Saturation boundary		4.2, 4.13

[a] To make the table entries positive, $-1/\beta_T$ is tabulated. The isothermal compressibility is a negative quantity.

Figure – None
Equation – See Section 2.12
Range of Values –

	Triple point		Boiling point		Critical	300 K,
Units	Liquid	Vapor	Liquid	Vapor	point	1 atm
MPa^{-1}	-1.10×10^{-2}	−145	-1.99×10^{-2}	−11.1	∞	−10

Note: The tables in Chapter 4 reflect no differences in this property for the normal and para modifications.

Uncertainty – The uncertainty for values obtained from the tables in Chapter 4 varies from 0% in the low density limit to about 3 to 4% at 35 MPa and 10% at 100 MPa. In the critical region (T_c ± 5% and ρ_c ± 20%) the uncertainties are much greater and very difficult to estimate as the property diverges at the critical point.

2.6. DIELECTRIC CONSTANT[5-14]

Definition – The dielectric constant of a material may be defined as $\epsilon = C(P,T)/C(0,T)$, where C(P,T) is the capacitance of a capacitor whose plates are separated by the material at pressure P and temperature T, and C(0,T) refers to the capacitance of the same capacitor in a vacuum. In practical situations corrections are often applied for pressure distortion of the capacitor. The equation assumes no stray capacitances, i.e., the electric

field is affected only by the dielectric material. *Note:* The dielectric constant is also known as specific inductive capacity.
Tables of Values – This quantity is tabulated in the tables in Chapter 4.

Units	Range of table		Table location
Dimensionless	0.01–100 MPa	13.8–3000 K	4.1
Dimensionless	Saturation boundary		4.2, 4.13

Figure – 3.2, 3.3
Equation – The dielectric constant for the fluid parahydrogen may be calculated from

$$\frac{\epsilon - 1}{\epsilon + 2} = A\rho + B\rho^2 + C\rho^3$$

where A = 0.99575, B = -0.09069, C = 1.1227, and ϵ is the dielectric constant. The density, ρ, must be in the units of grams per cubic centimeter. The equation is valid over the range of the tables and may be extrapolated in the fluid phase with reasonable results.
Range of Values –

	Triple point			Boiling point			
Units	Liquid	Vapor	Solid	Liquid	Vapor	Critical point	300 K, 1 atm
Dimensionless	1.252	1.0004	1.285	1.229	1.004	1.0980	1.00025

Uncertainty – The uncertainty of $\epsilon - 1$ is estimated to be no greater than 0.1% for the fluid phase and 0.2% for the solid phase.

2.7. DIFFUSION COEFFICIENT

Definition – The diffusion coefficient is defined as the coefficient relating the flux of a given species in a mixture to the concentration gradient of the species under isothermal conditions. For a binary mixture of species i and j: $\mathbf{J}_i = -D_{ij}$ grad C_i, where $\mathbf{J}_i$ is the mass flux of i, C_i is the gradient of concentration, and D_{ij} is the diffusion coefficient. Diffusion is usually thought of as a property of mixtures; however, the term self-diffusion, as the name implies, refers to a diffusion of a single species. There is, of course, no true means of measuring such a quantity, so the process is simulated using different isotopes of the same species. A third type of diffusion appears in the literature for hydrogen where the binary mixture is that of molecules with different nuclear spin orientation, i.e., ortho-para mixtures.

Thermal Diffusion

Definition – In a binary system where a temperature gradient exists, the above equation should be extended to $\mathbf{J}_i = -D_{ij}\nabla_{ci} - D_{ij}^T \nabla \ln T$, where D_{ij}^T is the thermal diffusion coefficient. The same extension is made to isotopic and ortho-para mixtures.

The information on the diffusion of hydrogen has been categorized according to the above definitions; references are listed below.
Equations – From kinetic theory, for a dilute gas $D_{ij}\ \alpha\ T^{3/2}$, and at room temperature

D_{ij} for hydrogen systems is about 1 cm^2/sec; T is temperature. For liquid or gas dissolved in a liquid

$$\frac{D_{ij}\eta}{T} = \text{constant}$$

where η is the viscosity of the mixture and T is temperature. D_{ij} for H_2 in a liquid is about 10^{-4} cm^2/sec.

References –

A. Diffusion
  1. Isothermal[15–19]
  2. Thermal
     a. Argon[20–22,48]
     b. Krypton[18,20,22,48]
     c. Xenon[22–27]
     d. CO_2[21,28–30]
     e. SO_2[17]
     f. H_2O[17,31]
     g. N_2O[32]
     h. O_2[33,34]
     i. CO[34]
     j. N_2[21,22,27,29,30,33–36]
     k. Ne[37,38]
     l. He^4 [22,27,33,34,39–47]
     m. He^3 [48]

B. Self-diffusion
  1. Isothermal[19,42,43,49–53]
     a. High temperatures[54,55]
     b. Solids[56]
  2. Thermal[21,39–41,43,57–65]
  3. Ortho-para[47,63,64,66]

2.8. THERMAL DIFFUSIVITY[1–3]

Definition – The thermal diffusivity, α, is defined as

$$\alpha = \frac{\lambda}{\rho C_p}$$

where λ is the thermal conductivity, ρ is the density, and C_p is the specific heat capacity at constant pressure.

Tables of Values – This quantity is tabulated in the tables in Chapter 4.

Units	Range of table		Table location
m^2/hr	13.8–3000 K	0.01–100 MPa	4.1, 4.12
m^2/hr	Saturation boundary		4.2, 4.13

Figure – None

Equation – None, except above and the equation of state (see Section 2.12).

Range of Values –

Units	Triple point Liquid	Triple point Vapor	Boiling point Liquid	Boiling point Vapor	Critical point	1 atm, 300 K
m^2/hr	0.00053	0.034	0.00052	0.0038	0	0.52

Note: Values for normal and para are the same at these state points but differ at other state points.

Uncertainty – The uncertainty of the values tabulated in Chapter 4 is estimated to be ±10%, except in the critical region (T_c ±3% and ρ_c± 30%) and for $T \leqslant 180$ R and $P > 35$ MPa.

2.9. DIELECTRIC BREAKDOWN

Liquid[67–76] – Below some critical electric field gradient, called the "breakdown voltage," hydrogen is an insulator. In recent years there have been a few experiments to determine this breakdown voltage for liquid hydrogen. Two things seem evident from these experiments. First, there is a fairly large dispersion in the measurements (reproducibility in a single experiment commonly has a 25% scatter); second, the breakdown voltage is quite sensitive to pressure changes, i.e., an increase from 300 to 690 kV/cm was observed when the pressure was increased from 1 to 5 atm at 20 K.[70]

Gas[67,68,70,75,77–91] – Hydrogen in the gaseous phase is, like the liquid, an insulator below some critical field gradient. When the critical voltage is reached, the gas ionizes, a corona results, and the conduction takes place. This breakdown voltage follows Paschen's law, which states the breakdown voltage, V, does not change as long as ρd is a constant and it is proportional to ρd. $V \propto \rho d$, where ρ is the density and d is the electrode spacing. This relation is surprisingly independent of temperature, although some investigators claim to have observed temperature dependence at low temperatures. Also, a minimum breakdown voltage for hydrogen has been observed at about 250 to 300 V at a ρd of 10^{-7} g/cm^2.

Electrical Conductivity[76,92–97]

Hydrogen is essentially an insulator with a negligible electrical conductance except when voltages in excess of the breakdown voltages are applied. At elevated temperatures where ionization occurs, the gas becomes an electrical conductor, and conductivities for hydrogen to 30,000 and 40,000 K have been calculated.[83,84]

2.10. ENTHALPY[1–3]

Definition – Enthalpy is defined by the equation $H = U + PV$, where U is the internal energy, H is the enthalpy, and PV is the product of pressure and volume. The change in enthalpy is a measure of the heat absorbed by a system in a constant pressure process. The enthalpy of a substance is a derived thermodynamic property and may be calculated by means of a thermodynamic relationship which requires the PVT surface (see Section 2.12) and the specific heat capacity (see Sections 2.16 through 2.19).

Tables of Values – Tabulated in Tables 4.1, 4.2, 4.12, and 4.13

Units	Range of table		Table location
J/g	13.8–3000 K	0.01–100 MPa	4.1, 4.12
J/g	Saturation boundary		4.2, 4.13

Figure – 3.20, T-S chart; 3.21, T-S chart; 3.22, H-S chart.
Equation – None (see above)
Range of Values – The reference state for enthalpy is zero for the ideal gas at zero absolute temperature (see Section 2.20).

Units	Triple point		Boiling point		300 K,
	Liquid	Vapor	Liquid	Vapor	1 atm
Para					
J/g	-308.8	140.3	-256.3	189.3	4199.4
Normal					
J/g	218.3	667.4	270.9	716.5	4226.9

Uncertainty – The uncertainties of the enthalpy values tabulated here vary somewhat with pressure and temperature. In the region of $T < 300$ K uncertainty is estimated to range from 1.2 J/g at low densities to about 5.0 J/g in the liquid. For all densities at $300 < T < 500$ K the uncertainty is estimated to vary from 1 J/g at low pressures to about 15 J/g at the highest pressures. For temperatures above 500 K the uncertainty of the enthalpy at the lower densities could be much greater, especially for temperatures above about 1800 K where disassociation is a large factor.

2.11. ENTROPY[1-3]

Definition – Entropy is defined by the equation $dS = dQ/T$, or in any reversible process the change in entropy of a system is equal to the heat which it absorbs, divided by the absolute temperature. The entropy of a substance is a derived thermodynamic quantity, and many thermodynamic equations exist which relate the PVT surface of a substance to the changes in entropy. In addition to the PVT surface, or an equation of state, the specific heat capacity of a substance is needed to complete the calculation.
Tables of Values – This quantity is tabulated in Tables 4.1, 4.2, 4.12, and 4.13

Units	Range of table		Table location
J/g-K	13.8–3000 K	0–100 MPa	4.1, 4.12
J/g-K	Saturation boundary		4.2, 4.13

Figure – 3.20, T-S chart; 3.21, T-S chart; 3.22, H-S chart.
Equation – None (see definition)
Range of Values – The reference state for entropy is the ideal gas at 1 atm pressure and zero absolute temperature (see Section 2.20).

Units	Triple point		Boiling point		300 K,
	Liquid	Vapor	Liquid	Vapor	1 atm
Para					
J/g-K	4.964	37.52	7.977	29.97	64.77
Normal					
J/g-K	14.08	46.64	17.09	39.08	70.58

Uncertainty – The uncertainty of the entropy values tabulated here varies somewhat with pressure and temperature. In the region $T < 300$ K, the uncertainty is estimated to range from 0.04 J/g-K at low pressures to 0.17 J/g-K at the higher pressures. For all P at 300 <

T < 500 K the uncertainty is estimated to vary from 0.05 J/g-K at low pressures to about 1.0 J/g-K at the highest pressures. For temperatures above 500 K the region is again one of extrapolation and the uncertainties are greater, especially for the low pressures at temperatures above 1800 K where disassociation becomes significant.

2.12. EQUATION OF STATE[1-3,98-101]

Definition – The relationship between the pressure, temperature, and density of a substance is called the equation of state. This section will be concerned with the mathematical model of the equation of state. Because of the speed and availability of today's digital computers, the mathematical equation of state is an extremely useful tool. There are literally hundreds of these to be found in the literature. To recommend an equation of state, the detailed requirements must be known, or, to put it another way, there is no single mathematical model of the equation of state that will be the best choice for every purpose.

Equations – For the equations used to generate Tables 4.1, 4.2, 4.11, and 4.12, see References 1, 2, and 100. For a more versatile but not quite as accurate equation of state, see Reference 99. For an equation of state designed for extremely high-speed computation but less accuracy, see Reference 98. See also the main text on computer programs and Section 2.13.

2.13. VIRIAL COEFFICIENTS[1-3]

Definition – The virial coefficients are commonly defined in two ways, as follows: $P = RT\rho[1 + B(T)\rho + C(T)\rho^2 + \ldots]$ and $PV = RT + B'(T)P + C'(T)P^2$. The first two terms of these series may be inverted by the following:

$$B(T) = B'(T) \text{ and } C'(T) = \frac{C(T) - B(T)^2}{RT}$$

where P is pressure; T is temperature; ρ is density; V is $1/\rho$; R is the gas constant; and B(T), $B'(T)$ and C(T), $C'(T)$ are virial coefficients of a power series expansion in density and pressure, respectively. Both series are theoretically infinite in length; however, the coefficients beyond the first two [B(T) and C(T)] are of less interest because of their complexity. The temperature at which B(T) = 0 is called the Boyle temperature. Either of the above two equations is adequate to describe the PVT surface for densities up to about one half critical. The coefficients B(T) and C(T) for hydrogen for the density expansion are given in Table 4.3. Most equations of state may be algebraically arranged so that they are in a virial expansion form; therefore, these equations of state may be used to calculate the virial coefficients (see Section 2.12 for equations of state).

Tables of Values –

Units	Range of table	Table location
$B(cm^3)/mol$), $C(cm^3/mol)^2$	13–500 K	4.3

Figure – None

Equation – The values in Table 4.3 have been calculated from the following expressions. For temperatures below 100 K,

$$B(T) = (b_1T + b_2 + b_3/T + b_4/T^2)/RT$$

where b_1 = 1.9397741 X 10^3, b_2 = -1.9279522 X 10^5, b_3 = -2.2890051 X 10^6, and b_4 = 1.1094088 X 10^7.

For T < 55 K,

$$C(T) = (C_1T^2 + C_2T + C_3 + C_4/T + C_5/T^2 + C_6/T^3)/RT$$

There are two sets of C_i coefficients: one for temperatures below T_c = 32.957,

$$C_1 = 1.0541776 \times 10^5$$
$$C_2 = -1.6597141 \times 10^7$$
$$C_3 = 1.0431411 \times 10^9$$
$$C_4 = -3.2538718 \times 10^{10}$$
$$C_5 = 5.1405848 \times 10^{11}$$
$$C_6 = -3.3123453 \times 10^{12}$$

and a second set for temperatures between T_c and 55 K,

$$C_1 = 1.6971294 \times 10^3$$
$$C_2 = -5.0854223 \times 10^5$$
$$C_3 = 6.7284118 \times 10^7$$
$$C_4 = -3.8045171 \times 10^9$$
$$C_5 = 1.0789413 \times 10^{11}$$
$$C_6 = -1.1515642 \times 10^{12}$$

For 55 < T < 100,

$$C(T) = a_1 e^{a_2/T} [1 - e^{a_3 \{1-(T/a_4)^{a_5}\}}]$$

where a_1 = 388.682, a_2 = 45.5, a_3 = 0.6, a_4 = 20.0, and a_5 = 4.0.

For temperatures above 100 K,

$$B(T) = \sum_{i=1}^{4} b_i x_1^{(2i-4)/4}$$

and

$$C(T) = C_0 x_2^{1/2} [1 + Cx_2^3] [1 - \exp(1 - x_2^{-3})]$$

where b_1 = 42.464, b_2 = -37.1172, b_3 = -2.2982, b_4 = -3.0484, x_1 = 109.781 K/T, C_0 = 1310.5, C = 2.1486, and x_2 = 20.615 K/T.

Range of Values –

Units	14 K	20 K	100 K	200 K	300 K
B(T)cm^3/mol	-237.2	146.7	-2.51	10.73	14.38
C(T) $(cm^3/mol)^2$		-1503	608.6	421.7	343.8

Boyle Point (B = 0) – 112.4 K

Uncertainty – The uncertainty of B is probably a maximum of 5% at the highest and lowest temperatures. The uncertainty of C is a minimum of 5% between 55 and 100 K and as much as 20% for temperatures below critical temperature.

2.14. INTERMOLECULAR POTENTIAL FUNCTION[102-105]

Definition – The intermolecular potential function is the potential energy of interaction that exists between pairs of molecules. Mathematical models of this function are used to calculate macroscopic properties with kinetic theory and statistical mechanics.
General Comments – The potential energy of interaction between unlike molecules, i.e., hydrogen and some other fluid, has not been considered here. Only those papers with some relevance to $H_2 - H_2$ interactions have been used.
The Lennard-Jones Potential – The most common potential in the literature is the Lennard-Jones or 12-6 potential

$$\varphi(r) = 4\epsilon\left[\left(\frac{\sigma}{r}\right)^{12} - \left(\frac{\sigma}{r}\right)^{6}\right]$$

where $\varphi(r)$ is the potential energy of interaction, r is the intermolecular separation, ϵ is the maximum energy of attraction, and σ is the intermolecular separation for which $\varphi(r) = 0$. σ is sometimes referred to as the molecular diameter. Typical values for hydrogen[102] are $N\sigma = 15.6\ cm^3/mol$ and $\epsilon/K = 36.7$ where N is Avogadro's number. The above parameters for the Lennard-Jones potential have been shown[103] to be temperature and property dependent. A new potential model has been developed[104,105] which has shown significant improvement over the Lennard-Jones; unfortunately, parameters for hydrogen are not yet available.

2.15. FIXED POINTS[1,14,106-108]

Discussion – Below is a summary of the P-ρ-T data at selected fixed points. To be consistent with the property tables in Chapter 4, the temperatures in the para-equilibrium column are on the NBS-55 scale. The temperatures in the column labeled "normal" are also on the NBS scale. For relationships between the IPTS-68 scale and the NBS-55 scale, see Reference 108.

Para or equilibrium	Normal
Critical point	
T = 32.976 ± 0.05 K	33.19 K
P = 12.759 atm (1.2928 MPa)	12.98 atm (1.315 MPa)
ρ = 15.59 mol/l (31.43 kg/m³)	14.94 mol/l (30.12 kg/m³)
Normal boiling point	
T = 20.268 K	20.39 K
P = 1 atm (0.101325 MPa)	1 atm (0.101325 MPa)
ρ(liquid) = 35.11 mol/l (70.78 kg/m³)	35.2 mol/l (71.0 kg/m³)
ρ(vapor) = 0.6636 mol/l (1.338 kg/m³)	0.6604 mol/l (1.331 kg/m³)
Triple point	
T = 13.803 K	13.957 K
P = 0.0695 atm (0.00704 MPa)	0.0711 atm (0.00720 MPa)
ρ(solid) = 42.91 mol/l (86.50 kg/m³)	43.01 mol/l (86.71 kg/m³)
ρ(liquid) = 38.21 mol/l (77.03 kg/m³)	38.3 mol/l (77.2 kg/m³)
ρ(vapor) = 0.0623 mol/l (0.126 kg/m³)	0.0644 mol/l (0.130 kg/m³)

Solid-solid Transition – Solid hydrogen does undergo such transitions (see, for example, Reference 14).
Molecular Weight – 2.01594 on the C^{12} scale.[107]
Other Properties at the Fixed Points – See Tables 4.1, 4.2, 4.4, 4.11, 4.12, and 4.13.

2.16. HEAT CAPACITY (SPECIFIC HEAT) AT CONSTANT PRESSURE[1-3]

Definition – The heat capacity at constant pressure is defined as $C_p = (\partial H/\partial T)_p$, where H is enthalpy.

Tables of Values –

Units	Range of values		Table location
J/g-K	13.8–3000 K	0.1–100 MPa	4.1, 4.2, 4.12, 4.13

Figure – 3.5

Equation – None

Range of Values –

	Triple point		Boiling point		Critical	300 K,
Units	Liquid	Vapor	Liquid	Vapor	point	1 atm
Para J/g-K	6.36	10.52	9.66	12.15	∞	14.85
Normal J/g-K	6.36	10.52	9.66	12.15	∞	14.31

Uncertainty – The uncertainty is estimated to vary from 0.02% in the low density limit to 3% at 35 MPa and 8% at 100 MPa. In the critical region ($T_c \pm 5\%$, $\rho_c \pm 20\%$) the uncertainties are greater but difficult to estimate.

2.17. HEAT CAPACITY (SPECIFIC HEAT) AT CONSTANT VOLUME[1-3]

Definition – The heat capacity at constant volume is $C_v = (\partial U/\partial T)_v$, where U is internal energy, T is temperature, and v is specific volume.

Tables of Values –

Units	Range of tables		Table location
J/g-K	13.8–3000 K	0.01–100 MPa	4.1, 4.2, 4.12, 4.13

Figure – 3.6

Equation – None

Range of Values –

	Triple point		Boiling point		Critical	300 K,
Units	Liquid	Vapor	Liquid	Vapor	point	1 atm
Para J/g-K	4.67	6.21	5.78	6.50	∞	10.72
Normal J/g-K	4.67	6.21	5.78	6.50	∞	10.18

Uncertainty – The uncertainty is estimated to vary from 0.02% in the low density limit to 3% at 35 MPa and 8% at 100 MPa. In the near critical region ($T_c \pm 5\%$, $\rho_c \pm 20\%$) the uncertainties are greater but difficult to estimate.

2.18. HEAT CAPACITY (SPECIFIC HEAT) RATIO[1-3]

Definition – $\gamma = C_p/C_v$.
Tables of Values – The specific heat ratio is not tabulated in tables; however, C_p and C_v are tabulated and should be used to form the ratio as defined above.
Figure – None
Equation – None
Range of Values –

Units	Triple point Liquid	Triple point Vapor	Boiling point Liquid	Boiling point Vapor	Critical point	300 K, 1 atm
Para						
Dimensionless	1.38	1.69	1.69	1.88	∞	1.38
Normal						
Dimensionless	1.38	1.69	1.69	1.88	∞	1.41

Uncertainty – The uncertainty is estimated to vary from 0.02% in the low density limit to 3% at 35 MPa and 8% at 100 MPa. In the near critical region ($T_c \pm 5\%$, $\rho_c \pm 20\%$) the uncertainties are greater but difficult to estimate.

2.19. HEAT CAPACITY (SPECIFIC HEAT) OF THE SATURATED LIQUID AT CONSTANT SATURATION[109]

Definition – $C_\sigma = T(\partial S/\partial T)_\sigma = (dH/dT)_\sigma - V(dP/dT)_\sigma$ where the subscript σ denotes constant saturation.
Tabulated Values – None
Figure – None
Equation –

$$C_\sigma = \frac{AT}{(T_c - T)^n} + B + CT + DT^2 + ET^3 + FT^4 + GT^5$$

where T is temperature in Kelvins, n = 0.10, A = 1.6815742, B = -32.80279, C = 6.8169871, D = -0.73194341, E = 0.033574357, F = -0.0007682974, G = 0.0000069029224, and $T_c = 32.984$. C_σ is cal/mol·K.

Units	Triple point	Boiling point	Critical point
Para only			
J/gK	6.33	9.59	∞

Uncertainty – The uncertainty is estimated by Younglove and Diller[109] to be 0.3% except near the critical point.

2.20. IDEAL GAS PROPERTIES, THERMODYNAMIC[106]

Discussion – All thermodynamic properties tabulated here are based on thermodynamic properties of the ideal gas. These ideal gas properties are derived from spectroscopic measurements and statistical mechanics; they are taken from Reference 106. The reference state for these properties is the ideal gas at 0 K where the free energy, E_o° is taken to be zero. By convention the values of S° and C° are presented for 1 atm pressure.

Tables of Values – For normal, equilibrium, and parahydrogen, see Reference 106.

Properties	Units	Range of table	Table location
C_p^o, S^o, H^o-E^o, $(F^o$-$E_0^o)/T$	cal mol, K	10–5000 K	Reference 106

Figure – None
Equation – None
Uncertainty – The uncertainty for C_p° is estimated to be 0.2% to 700 K increasing to 0.4% at 5000 K. The uncertainty for S° and E° is estimated to be an order of magnitude less.

2.21. INDEX OF REFRACTION[2,110]

Definition – The index of refraction n_λ of a medium is the ratio of the speed of light in a vacuum to the speed of light in the medium, $n_\lambda = C/C_m$.
Discussion – The refractive index of a nonpolar fluid depends on the wavelength of the incident light, the density of the fluid, and, to a lesser extent, the temperature. The dependence on temperature is usually small enough to be neglected, as it has been here. The graph at location 3.8 gives an indication of the temperature dependence. Reference 2 gives a comparison between the index of refraction for normal and parahydrogen.
Table of Values –

Units	Range of values	Table location
Dimensionless	Saturated liquid	4.6

Figure – See 3.8.
Equation –

$$n_\lambda(\rho) = \left(\frac{2\rho r_\lambda(\rho) + 1}{1 - \rho r_\lambda(\rho)}\right)^{1/2}$$

where $r_\lambda(\rho) = 0.99575 - 0.09069\rho + 1.227\rho^2 + 0.7799569 \times 10^6/\lambda^2 + 0.495126 \times 10^{12}/\lambda^4$, the specific refraction $r_\lambda/(\rho)$ is in cubic centimeters per gram, densities are in grams per cubic centimeter, and the wavelengths are in angstroms.
Range of Values –

Units	Boiling point, liquid	300 K 1 atm	λ, A
Dimensionless	1.1137	1.0001	4358

Uncertainty – The uncertainty of the tabulated values is estimated to be less than 0.1%.

2.22. INTERNAL ENERGY[1-3]

Definition – Internal energy is defined by the equation $\Delta U = Q - P\Delta V$, where Q is the heat absorbed in a system, ΔU is the change in internal energy, P is the pressure, and ΔV is the change in volume. The internal energy of a system is a derived thermodynamic property and may be calculated by means of thermodynamic relationships which require the PVT surface (see Section 2.12) and the specific heat capacity (see Sections 2.16

through 2.19). The most common method of calculating the internal energy is by using H = U + PV where H is the enthalpy (see Section 2.10) and PV is the pressure volume product.

Tables of Values –

Units	Range of values		Table location
J/g	13.8–3000 K	0.1–100 MPa	4.1, 4.12
J/g	Saturation boundary		4.2, 4.13

Figure – None

Equation – None, see definition

Range of Values – The reference state for internal energy is zero for the ideal gas at zero absolute temperature (see Section 2.20).

	Triple point		Boiling point		300 K,
Units	Liquid	Vapor	Liquid	Vapor	1 atm
Para					
J/g	-309.0	84.2	-257.7	113.6	2961.38
Normal					
J/g	218.2	611.4	269.5	640.7	2989.0

Uncertainty – The uncertainty of the internal energy values tabulated here is estimated to be the same as the enthalpy (see Section 2.10).

2.23. JOULE-THOMSON COEFFICIENT[1-3,111]

Definition – The Joule-Thomson coefficient, μ, is defined as $\mu = -C_p^{-1}(\partial H/\partial P)_T = (\partial T/\partial P)_H$. The sign of μ indicates whether a gas expansion will cause an increase or decrease in the temperature. If μ is positive, the expanding gas will be cooled. The locus of points where $\mu = 0$ is called the Joule-Thomson inversion curve. These data are given in Table 4.7. Values of the Joule-Thomson coefficient are tabulated by Michels et al.[111] for pressures up to 2500 atm.

Tables of Values – This quantity is not tabulated here but it may be calculated from the property table in Chapter 4.

Figure – 3.9, the inversion curve, T vs. P.

Equation – None (may be calculated using an equation of state)

Range of Values –

Units	Pressure	40 K	110 K	180 K
K/atm	100 atm	-4.3×10^{-2}	2.5×10^{-2}	-7.6×10^{-3}

Uncertainty – The uncertainty is estimated to be 5%.

2.24. LATENT HEAT OF FUSION[14,112,113]

Definition – The latent heat of fusion is the amount of heat required to melt a unit mass of a substance at constant temperature and pressure at temperatures above the triple point.

Tables of Values – None
Figure – None
Equation – None
Range of Values – This quantity has been measured by several experimenters at the triple point. Reference 112 was selected for the triple point; the two values listed below at higher temperatures are found in Reference 113.

Units	Triple point	15.54 K	21.19 K
cal/mol	28.08	30.80	40.28
J/g	58.28	63.92	83.60

Note: Values are for parahydrogen.

Uncertainty – The estimate of uncertainty based on the existing data is about 0.25 cal/mol.

2.25. LATENT HEAT OF SUBLIMATION[14,114,115]

Definition – The latent heat of sublimation is the amount of heat per unit mass required to vaporize the solid phase.
Tables of Values – This quantity is tabulated in Table 4.8.
Figure – None
Equation – None
Range of Values –

Units	0 K	Triple point
Para		
cal/mol	181.5	244.5
J/g	376.7	507.4
Normal		
cal/mol	182.9	–
J/g	379.6	–

Uncertainty – The uncertainty is estimated to be about 2 cal/mol.

2.26. LATENT HEAT OF VAPORIZATION[1-3,14]

Definition – The latent heat of vaporization is the amount of heat required to convert a unit mass of a substance from the liquid to the vapor state at constant pressure.
Tables of Values – This quantity is not tabulated but values are easily obtained from Tables 4.2 and 4.13 by subtracting the liquid enthalpy from that of the vapor at the same temperature and pressure.
Figure – 3.10, heat of vaporization vs. T.
Equation – None
Range of Values –

Units	Triple point	Boiling point	Critical point
Para			
J/g	448.2	445.5	0
Normal			
J/g	449.1	445.6	0

Uncertainty – The uncertainty is estimated to be 2.5 J/mol for para and 5.0 J/mol for normal.

2.27. MELTING CURVE[2,3,14,116]

Definition – The melting curve is the boundary between the solid and liquid regions in a phase diagram.
Tables of Values – The density of the saturated liquid in coexistence with the solid phase is found on the first line of each isobar in Tables 4.1 and 4.12. The liquid and solid densities of three points are given here. For more extensive tabulations, the reader is referred to Reference 14.
Figure – 3.11 and 3.12
Equation – See Section 2.39 for vapor pressures
Range of Values –

	Triple point		100 atm		300 atm	
Units	Liquid	Solid	Fluid	Solid	Fluid	Solid
kg/m³	77.0	86.9	81.8	89.9	88.7	96.0

Note: These values are for parahydrogen. For an estimate of normal-para density differences, see Section 2.49 and Figure 3.34.

Uncertainty – The uncertainty is estimated to be 0.1% for the liquid phase and 0.5% for the solid phase.

2.28. MELTING PRESSURE[2,9,14,117,118]

Definition – The relationship between pressure and temperature along the solid-liquid boundary.
Tables of Values – The melting temperature for a given pressure is found on the first line of each isobar in Tables 4.1 and 4.12.

Units	Range of table	Table location
MPa	13.8–34.2 K	4.1, 4.12

Figure – 3.11, melting curve (T vs. P) for hydrogen, and 3.12.
Equation – See Reference 2
Range of Values –

Units	Triple point	20 K, 36 R	50 K, 90 R
MPa	0.00704	22.71	221.1

Uncertainty – The uncertainty of the pressures is about 0.1%.

2.29. PHASE DIAGRAM[1–3]

Definition – Phase diagrams indicate the boundaries between the solid, liquid, and/or gas

phases. The diagrams that were referenced here are only those which yield the PVT of the phases. For a diagram of the phases in other coordinates, consult the page on the particular property desired.
Figure – 3.14 and 3.15
Uncertainty – The uncertainty of values taken from the above (or any other) diagrams is determined by the resolution of the individual chart.

2.30. PRANDTL NUMBER[2]

Definition – The Prandtl number, Pr, is defined by the relation $Pr = C_p\eta/\lambda$, where C_p is the specific heat at constant pressure, η is the viscosity, and λ the thermal conductivity.
Tables of Values –

Units	Range of table		Table location
Dimensionless	13.8–3000 K	0.01–100 MPa	4.1, 4.12
Dimensionless	Saturation boundary		4.2, 4.13

Figure – 3.13
Equation – None other than above defining relationship
Range of Values –

	Triple point		Boiling point		Critical	300 K,
Units	Liquid	Vapor	Liquid	Vapor	point	1 atm
Dimensionless	2.24	0.623	1.29	0.804	∞	0.687

Uncertainty – The uncertainty is estimated to be 15% except in the critical region, where it is greater.

2.31. PVT[1-3,106]

Definition – PVT stands for pressure, volume, and temperature, the fundamental variables of the state of a pure fluid at equilibrium. Any two of three variables may be specified; the third is then a unique value for a given fluid, except where two or more phases coexist, in which case the pressure is a unique function of the temperature, but the volumes of the phases differ.
Tables of Values – The properties are tabulated in Tables 4.1, 4.2, 4.12, and 4.13.

Figure – 3.1, 3.14, and 3.15
Equation – See Sections 2.12, 2.13, 2.28, and 2.39
Range of Values –

Property	Units	Range of tables	Table location
Pressure	MPa	0.01–100	4.1, 4.12
Volume	m^3/kg	0.0097–1538	4.1, 4.12
Temperature	K	13.8–3000	4.1, 4.12

Note: The tables for normal and parahydrogen do not reflect any difference for PVT (see Section 2.49 for a more detailed discussion of ortho-para differences).

Uncertainty – The uncertainty is estimated to be

Temperature range	Pressure range	Uncertainty in density, %
700–3000 K	0.1–100 MPa	1
300–700 K	0.1–100 MPa	0.5
13–300 K	50–100 MPa	0.2
13–300 K	0.1–50 MPa	0.1
Critical region	$T_c \pm 1\%$, $\rho_c \pm 20\%$	6
Solid	340–700 atm	1

2.32. RADIATIVE PROPERTIES

No references to the emissivity of hydrogen were found in the literature; see Section 2.43 for spectral absorption.

2.33. SATURATION PROPERTIES

Definition – The term saturation refers to a state condition where two or more phases coexist in equilibrium.

Tables of Values – Saturation properties of the coexisting liquid and vapor phases are tabulated from the triple point to the critical point in Tables 4.2 and 4.12. The individual properties which are tabulated in these tables are listed below along with the section in which they are discussed in more detail.

Property	Section	Property	Section
Temperature	2.31	Constant pressure heat capacity	2.16
Pressure	2.31	Velocity of sound	2.34
Volume (or density)	2.31	Surface tension	2.36
Internal energy	2.22	Thermal conductivity	2.37
Enthalpy	2.10	Viscosity	2.40
Entropy	2.11	Thermal diffusivity	2.8
Constant volume heat capacity	2.17	Dielectric constant	2.6
		Prandtl number	2.30

For tables of PVT below the triple point, see Reference 14.

Figure – See pages listed above

2.34. SOUND VELOCITY[1-3,14]

Definition – The thermodynamic sound velocity, W, is defined by

$$W^2 = -V\left(\frac{\partial P}{\partial V}\right)_S$$

where P is pressure, V is specific volume, and S is entropy; since entropy is not usually a variable in the mathematical models for the equation of state, calculations are more easily made using

$$W = \left[\frac{C_p}{C_v}\left(\frac{\partial P}{\partial \rho}\right)_T\right]^{1/2}$$

which is thermodynamically equivalent to the above equation. C_p and C_v are the specific heat capacity at constant pressure and constant volume, respectively; P is again pressure, and ρ is the density. Experimentally measured sound velocity is a very precise and accurate measurement and is equivalent to the thermodynamic sound velocity defined above.

Tables of Values –

Units	Range of table		Table location
m/s	13.8–3000 K	0.01–100 MPa	4.1, 4.2, 4.12, 4.13
m/s	0–12 K (para)	Saturated solid	Reference 14
m/s	4.2–13.96 (normal)	Saturated solid	Reference 14

Figure – 3.16
Equation – None
Range of Values –

	Triple point		Boiling point		Critical	300 K,
Units	Liquid	Vapor	Liquid	Vapor	point	1 atm
Para						
m/s	1264	305.7	1089	355	0[a]	1309

[a] Theoretically the velocity of sound should go to 0 at the critical point. Experimentally only very low velocities have been measured in the critical region.

Uncertainty – The uncertainty of the tabulated values has been estimated by comparing tabulated and experimental sound velocities. The uncertainty varies with pressure from 1 to 2% in the liquid and is about 0.5% in the vapor or above the critical temperature. In the critical region ($T_c \pm 5\%$, $\rho_c \pm 20\%$) the uncertainties are larger but not estimable.

2.35. SOUND ABSORPTION[119,120]

Definition – The attenuation α is defined by the relation

$$P = P_o e^{-\alpha \exp x}$$

where x is the distance between the point where the peak pressure due to the sound wave is P_o and the point where it is P. The so-called "classical" sound absorption α_{cal} is often calculated from

$$\alpha_{cal} = \frac{2\pi^2}{\gamma V}\left[\frac{4}{3}\eta + \frac{\gamma - 1}{C_p}\lambda\right] f^2/P$$

where γ is $\frac{C_p}{C_v}$, V is velocity of sound, f is frequency, P is the pressure, η is the shear viscosity, and λ is the thermal conductivity.

Papers found in the literature report sound absorption in many different ways; some of these are as follows:

$$\alpha_{excess} = \alpha_{exp} - \alpha_{cal}$$
$$\text{nepers} = \ln (P/P_o)$$
$$\alpha_{exp}/\alpha_{cal}$$
$$\alpha_{exp}/f$$
$$\alpha_{exp}\lambda \text{ (where } \lambda \text{ is wavelength)}$$

Liquid – α_{exp} for the saturated liquid[119] at 17 ± 1 K is reported to be 0.11 cm^{-1} ± 10% for the frequency of 44.4 MHz.
Gas – Values of α_{exp}/f for both normal and parahydrogen at temperatures of 293, 90, and 77 K are given as a function of f/P in Reference 120.
Range of Values –

Temperature, K	f/P MHz/atm	α_{exp}/f (normal) $(cm\ MHz)^{-1}$	α_{exp}/f (para)
293	1.48	0.491	
293	1.53		0.483
77	1.60		0.677
77	2.10	0.314	

Uncertainty – The uncertainty of α is 10% for the liquid value and 5% for the gaseous data.

2.36. SURFACE TENSION[2,121,122]

Definition – Surface tension is defined as the amount of work required to increase the surface area of a liquid by one unit of area. Note that this property is defined only for liquid in coexistence with the vapor phase.
Tables of Values –

Units	Range of table	Table location
dyne/cm	13.8–32.976 K	4.2, 4.13

Figure – 3.17
Equation – $\gamma = \gamma_o (1 - T/T_c)^{1.065}$ (see discussion below), where γ is surface tension in dyn/cm, γ_o is 5.328 for parahydrogen and 5.369 for normal hydrogen, and T_c is 32.976 K for parahydrogen and 33.18 K for normal hydrogen.
Range of Values –

Units	Triple point	NBP	Critical point	300 K
Para				
dyne/cm	2.99	1.93	0	Not defined
Normal				
dyne/cm	3.00	1.95	0	Not defined

Discussion – Since Corruccini[121] correlated the older experimental data, a new set of experimental data for normal hydrogen[122] has appeared in the literature. These new data are consistently about 0.07 dyn/cm lower at all temperatures below 31.5 K than those calculated from the above equation. These differences cannot be explained on the basis of differences in temperatures or densities used in the calculations, but the experimental

apparatus used by Blagoi and Pashkou[122] does not inspire confidence in their data. The capillary rise technique was used, which is the most common technique but is beset with difficulties because of the corrections that must be applied. In the older hydrogen surface tension experiments, these corrections amounted to about 10% of the measured capillary rise. Because of the size of the capillaries used by Blagoi and Pashkou[122] the measured capillary rise must be corrected by about 40 to 50%.
Uncertainty – The uncertainty of the tabulated values is estimated to be 1% for temperatures below the normal boiling point and increasing to about 5% for temperatures above 0.8 T_c.

2.37. THERMAL CONDUCTIVITY[2,14,123,124]

Definition – The thermal conductivity coefficient relates the transfer of heat through a material via molecular interaction due to a temperature gradient across the material. $q = -\lambda \text{ grad } T$, where q is the heat flux (heat per unit area per unit time), grad T is the temperature gradient, and λ is the thermal conductivity coefficient.
Discussion – The literature search revealed no experimental data outside the P-T range of these tabulations. The tabulations include extrapolations to higher densities and temperatures of experimental data by Roder and Diller.[123] An alternative method of calculating thermal conductivity is provided by the modified Enskog theory.[124] The most popular method of correlating thermal conductivity data is to separate the property into a number of additive parts: $\lambda = \lambda_o(T) + \lambda_E(\rho,T) + \lambda_c(\rho,T)$, where $\lambda_o(T)$ is the dilute gas contribution (a function of T only), $\lambda_E(\rho,T)$ is the so-called excess function, and $\lambda_c(\rho,T)$ gives the enhancement due to the critical point behavior. $\lambda_E(\rho,T)$ and $\lambda_c(\rho,T)$ are taken to be the same for the para and normal modifications but $\lambda_o(T)$ shows marked differences for the para and normal modifications. A survey of the thermal conductivity of the solid is given in Reference 14 from which Figure 3.18 was taken.
Tables of Values – 4.1, 4.2, 4.12, and 4.13

Units	Range of table		Table location
Para and normal mW/cm-K	13.8–3000 K	0.01–100 MPa	4.1, 4.2, 4.12, 4.13

Figure – 3.18
Equation – None
Range of Values –

Units	Triple point		Boiling point		Critical point	300 K, 1 atm
	Liquid	Vapor	Liquid	Vapor		
Para mW/cm-K	0.7259	0.1243	0.9892	0.1694	∞	1.76
Normal mW/cm-K	0.7259	0.1243	0.9892	0.1694	∞	1.70

Uncertainty – The uncertainty of the tabulated data is estimated to be 3% except in the extrapolated regions ($T < 100$ K, $P > 350$ atm; $T > 100$ K, $P > 100$ atm; and $T > 400$ K) where the uncertainty is estimated at 10% and in the critical region where it is undoubtedly larger.

2.38. THERMODYNAMIC DIAGRAMS[1,2]

Discussion – The thermodynamic diagrams listed below are based on the PVT of Roder et al.[1] and McCarty and Weber.[2] These diagrams have been prepared very carefully and are accurate to the precision to which they may be read. For other figures the primary consideration in many cases was availability and in others the range of P to T covered by the chart.
Figure – 3.1, compressibility factor chart; 3.19, temperature entropy chart (para); 3.20, temperature entropy chart (4 to 22 K) (para); 3.21, Mollier diagram (para)

2.39. VAPOR PRESSURE[14,125,126]

Definition – The vapor pressure is the pressure, as a function of temperature, of a liquid in equilibrium with its own vapor.
Tables of Values –

Units	Range of table	Table location
MPa	13.8–32.938 K	4.2, 4.13
mm Hg	1–13.8 K	4.8

Figure – 3.23 and 3.24
Equation – Parahydrogen:[126] $\ln(P/P_T) = n_1 x + n_2 x^2 + n_3 x^3 + n_4 (1 - x)^{n_5}$, where n_1 = 3.05300134164, n_2 = 2.80810925813, n_3 = -0.655461216567, n_4 = 1.59514439374, n_5 = 1.5814454428, $x = (1 - T_t/T)/(1 - T_t/T_c)$, T_t = 13.8 K, T_c = 32.938 K, and P_t = 0.007 to 42 MPa.

Normal hydrogen:[125] $\ln P(\text{kg/cm}^2) = A + B/T + C\ln T$, where A = -1.55447, B = -31.875, and C = 2.39188.

Sublimation Pressures

Vapor pressures at temperatures below the triple point (13.8 K) are often called sublimation pressures. The available data are summarized in Reference 14, and a table of values is given in Table 4.9.

2.40. VISCOSITY

Dynamic[2,127]

Definition – The viscosity coefficient of a fluid relates the flux of momentum in a fluid to the traceless symmetric part of the velocity gradient: $J = -2\eta D$, where J is the flux of momentum, D is the traceless symmetric part of the velocity gradient, and η is the viscosity coefficient. *Note*: The coefficient η as defined is sometimes called the shear or dynamic viscosity.
Discussion – The viscosity of a fluid may be separated into two additive parts: $\eta = \eta_0(T) + \eta_E(\rho,T)$, where $\eta_0(T)$ is the dilute gas contribution and is a function of temperature only and $\eta_E(\rho,T)$ is called the "excess" viscosity and is a strong function of density. When compared on a density basis, there does not seem to be a dependence on ortho-para composition.

Tables of Values –

Units	Range of table		Table location
kg/m-s	13.8–3000 K	0.01–100 MPa	4.1, 4.12
kg/m-s	Saturation		4.2, 4.13

Figure – 3.38
Equation – Equations are presented in Reference 2.
Range of Values –

	Triple point		Boiling point			
Units	Liquid	Vapor	Liquid	Vapor	Critical point	300 K, 1 atm
Para kg/m-s	255.1×10^{-7}	7.36×10^{-7}	132.0×10^{-7}	11.28×10^{-7}	$35{:}43 \times 10^{-7}$	81.40×10^{-7}

Uncertainty – The uncertainty of the tabulations is estimated as follows: for $T < 100$ K and $P < 350$ atm the uncertainty is estimated to 1%; for $T < 100$ K and $P > 350$ atm the uncertainty is estimated to be 10%; for $T > 100$ K and $P < 100$ atm the uncertainty is estimated to be 5%; for $T > 100$ K and $P > 100$ atm the uncertainty is estimated to be 15%.

Bulk

The shear viscosity should not be confused with the bulk viscosity. The bulk viscosity relates the flow of momentum in a fluid to the divergence of the velocity: $J = 2\eta D - \eta_v I$ div u, where I is the unit tensor, div u is the divergence of the velocity of the fluid, and η_v the coefficient of bulk viscosity.

Kinematic

Definition – The kinematic viscosity, γ, is defined as $\gamma = \eta/\rho$, where η is the shear viscosity and ρ is the density.

2.41. ADIABATIC COMPRESSIBILITY[1-3]

Definition – Adiabatic compressibility is defined as

$$\beta_S = -\frac{1}{V}\left(\frac{\partial V}{\partial P}\right)_S$$

A useful relation is

$$\beta_S = \frac{C_v}{C_p}\beta_T$$

where β_T is the isothermal compressibility.
Tables of Values – This quantity is not tabulated, but it may be obtained by means of the relationship $\beta_S = 1/\rho W^2$, where ρ is density and W is sound velocity, from the following tables.

Units	Range of table		Table location
MPa^{-1}	13.8–3000 K	0.01–100 MPa	4.1, 4.12
MPa^{-1}	Saturation boundary		4.2, 4.13

Figure – None
Equation – None
Range of Values –

Units	Triple point		Boiling point		Critical point	300 K, 1 atm
	Liquid	Vapor	Liquid	Vapor		
Para						
MPa^{-1}	0.00807	8.56	0.0119	5.93	∞	7.29

Uncertainty – The uncertainty in the compressed liquid varies between 2% at low pressures to 4% at high pressures; in the gas, well above the critical temperature, it is 1%. In the critical region (T_c ± 5%, ρ_c ± 20%) the uncertainties are undoubtedly larger; however, they are not experimentally defined as the property diverges at the critical point.

2.42. MIXTURE PROPERTIES[128-136]

Thermodynamic Properties – A review by Hiza et al.[130] lists experimental data for fluid mixtures properties of cryogenic interest; included are data for hydrogen systems. In addition to the above review, two reviews[131,132] give experimental data and some theoretical discussion of solid-vapor equilibria. For liquid-vapor equilibria, a bibliography by Oellrich et al.[133] gives references for both experimental and theoretical studies; this review includes all components with normal boiling points below 350 K. Also, the liquid-vapor section of the Hiza et al. review has been published.[134] Figure 3.25 can be used to obtain a rough estimate of the solubility of hydrogen in various solvents as a function of temperature and pressure. To estimate hydrogen solubility use

$$X(P, T) = \frac{X(100, T)}{100 - p_0(T)} \times P$$

where X(P,T) is the mole percent H_2 in the solvent at the desired pressure P, in atmospheres, and temperature T, in Kelvin; X(100,T) is the mole percent hydrogen in the solvent at 100 atm and T which is obtained from Figure 3.25; $p_0(T)$ is the solvent's vapor pressure at T. It is recommended that the above equation be used only for pressures of 100 atm or less and that the curves not be extrapolated to higher temperature; one should refer to the original data for higher temperatures and/or pressures. To estimate the vapor phase composition one can use $y = 100 \times p_0(T)/P$, where y is the mole percent of the solvent in the vapor phase; however, this estimate will always be too low especially if the pressure is more than 10 atm above the vapor pressure. For additional information on thermodynamic properties of mixtures, see References 128 to 136. Solubility references differ from liquid-vapor equilibria references in that they consider only liquid phase compositions, whereas the liquid-vapor references consider both phases.

Transport Properties of Mixtures – This section considers only viscosity and thermal conductivity; Section 2.27 considers diffusion. For dilute gas mixtures, Wilke's correlation can be used to estimate the viscosity[135] and thermal conductivity.[136] To our knowledge, no reliable technique exists for estimating the effect of hydrogen on the viscosity or thermal conductivity in the liquid phase.

2.43. INFRARED ABSORPTION[137-140]

Definition – The absorption coefficient, k, of a material is a measure of the loss in intensity, I, of radiation which is passed through a unit length of a homogeneous region of the material, i.e., if Z is the axis of propagation, then $dI/I = -kdZ$ or in integral form $I = I(0)e^{(-kZ)}$.

Absorption Bands – k is for most gases a strong function of frequency, being zero over large ranges of frequency and sharply peaked in certain other regions. It is difficult to summarize this subject because so many transitions are possible for hydrogen, i.e., rotational, vibrational, Raman, ultraviolet, infrared, and far infrared spectra have all been investigated. As with other gases, the absorption coefficient k is density dependent, varying roughly as the density squared (see Figure 3.29); in the particular case of hydrogen, k is significantly dependent upon the ortho-para composition (see Figures 3.26 and 3.27). For an estimate of importance to heat transfer problems, see Reference 157.

Figure – 3.26, far infrared (rotational);[137] 3.27, far infrared (translational);[137] 3.28, far infrared (rotational);[138] 3.29, Raman;[139] 3.30, ultraviolet.[140]

Uncertainty – The uncertainty varies with frequency but a typical conservative estimate is 5%. *Note:* Differences of 30 to 40% for the peak absorption between Jones[137,158] and others have not been explained.

2.44. HEAT TRANSFER AND PRESSURIZATION PARAMETERS[1-3,141]

A number of thermodynamic expressions have become convenient for the engineer to use in dealing with the storage and transfer of liquefied gases. Tabulations for four such parameters for hydrogen have been included in this Handbook.

Specific heat input (commonly symbolized by θ)	$V\left(\frac{\partial H}{\partial V}\right)_P = \rho C_P \left[\left(\frac{\partial P}{\partial \rho}\right)_T / \left(\frac{\partial P}{\partial T}\right)_V\right]$
Energy derivative (commonly symbolized by Φ)	$V\left(\frac{\partial P}{\partial U}\right)_V = \frac{V}{C_v}\left(\frac{\partial P}{\partial T}\right)_V$
Isothermal bulk modulus (commonly symbolized by α)	$V\left(\frac{\partial P}{\partial V}\right)_T = -\rho\left(\frac{\partial P}{\partial \rho}\right)_T$
Volume expansivity (commonly symbolized by β)	$\frac{1}{V}\left(\frac{\partial V}{\partial T}\right)_P = \frac{1}{\rho}\left(\frac{\partial P}{\partial T}\right)_\rho / \left(\frac{\partial P}{\partial \rho}\right)_T$

Specific Heat Input, θ – This is sometimes referred to as dQ/dM, heat input per increment of mass. It is frequently expressed in terms of density instead of specific volume, i.e., $\rho(\partial H/\partial \rho)_P$. It is the amount of heat required to expel a unit mass of fluid from a storage vessel. The units are the same as those for enthalpy: J/mol.

Energy Derivative, Φ – This is also frequently expressed in density by substituting $1/\rho$ for V. Typical units are atm-cm^3/J. This parameter is often used in conjunction with the specific heat input, θ, to determined expulsion rates from rigid containers.

Isothermal Bulk Modulus, α – Likewise, this is frequently expressed in density: $-\rho(\partial P/\partial \rho)_T$ with the usual units of psia or atm. It is the reciprocal of isothermal compressibility $-(\partial V/\partial P)_T/V$ and is a convenient factor to use in certain engineering calculations.

Volume Expansivity, β – Often called simply expansivity or constant pressure expansion, it is separately described in Section 2.48. Units are K^{-1}. A related expression is the constant pressure expansion coefficient, $T(\partial V/\partial T)_P/V$, but values for this latter expression have not been tabulated and are not as widely used (see Sections 2.3 and 2.5).

Tables of Values – These parameters are tabulated in Tables 4.1, 4.2, 4.11, and 4.12 as follows:

Parameter	Units	Range of table		Table location
$\theta = V(\partial H/\partial V)_P$	J/g			
$\phi = V(\partial P/\partial U)_V$	MPa-m³/kJ	13.8–3000 K Saturation boundary	0.01–70 MPa	4.1, 4.2, 4.12, 4.13
$\alpha = -V(\partial P/\partial V)_T$	MPa			
$\beta = (\partial V/\partial T)_P/V$	K^{-1}			

Figure – 3.31

Equations – None except as shown above.

Range of Values –

Parameter	Units	Triple point		Boiling point		Critical point	STP
		Liquid	Vapor	Liquid	Vapor		
Para							
θ	J/g	a	a	a	a		4455
ϕ	MPa-m³/kJ	a	a	a	a	0[b]	0.0385
α	MPa	a	a	a	a	0	
β	K^{-1}	a	a	a	a	∞	
Normal							
θ	J/g	622.6	139.8	587.9	189.1	199.0	4295
ϕ	MPa-m³/kJ × 10	0.2565	0.06684	0.2018	0.0667	0[b]	0.0405
α	MPa	90.34	0.0069	50.21	0.0903	0	0.101
β	K^{-1}	0.01022	0.07523	0.0164	0.0642	∞	0.00333

[a] Values the same as normal table.
[b] Not conclusive; may be finite but small value instead of 0.

Uncertainty – The uncertainties for each parameter are estimated as follows:

Parameter	Compressed liquid, %	Gas and supercritical, %
θ	3–5	3
ϕ	3	2
α	3–5	2
β	3–5	2

2.45. THERMAL TRANSPIRATION[142]

Definition – Thermal transpiration is the phenomenon of the flow of a gas at very low pressures through a capillary due to a temperature gradient. In a closed system, with a constant temperature gradient, a steady state pressure difference will occur between the ends of the capillary, provided the mean free path of the gas is about equal to, or larger than, the diameter of the tube.

Thermomolecular Pressure Difference – If, at the steady state, the pressure at the ends of the tube are P_H and P_C, corresponding to temperatures of T_H and T_C, then

$$\frac{P_H}{P_C} = \frac{T_H}{T_C}$$

where the subscripts H and C refer to hot and cold, respectively. The equation is only valid in the limit of $P \to 0$. At higher pressures, correction terms must be added. However, it must be remembered that thermal transpiration is an *apparatus effect*; therefore, correction terms to the equation should not be added for a particular gas but rather for a particular apparatus.

2.46. FREE ENERGY[1,2]

Definition – The Gibbs free energy, G, is defined as G = H - TS, where H is enthalpy, T is absolute temperature, and S is entropy. Similarly, the Helmholtz free energy, F, is F = U - TS, where U is internal energy.
Tables of Values – These quantities are not tabulated but may be simply obtained, using the formula above, from the following tables.

Units	Range of table		Table location
J/g	13.8–3000 K	0.01–100 MPa	4.1, 4.12
J/g	Saturation boundary		4.2, 4.13

Figure – None
Equation – None
Range of Values – (Gibbs free energy)

	Triple point		Boiling point			
Units	Liquid	Vapor	Liquid	Vapor	Critical point	300 K, 1 atm
Para						
J/g	-377	-377	-420	-420	-542	-15,233
Normal						
J/g	-21.9	17.0	-66.5	-80.3	-138	-16,945

Uncertainty – The uncertainty varies from about 0.5 J/g in the low density limit to about 15 J/g in the liquid.

2.47. HEATS OF TRANSITION[14]

Definition – A heat of transition is the energy associated with a phase change. Besides the solid-liquid and liquid-gas transition, hydrogen has three known solid-solid transitions.[14] The energy changes due to these solid transitions are not known but are thought to be small. For values of the latent heats of fusion-sublimation and vaporization, see Sections 2.24 to 2.26.

2.48. VOLUME EXPANSIVITY[1–3]

Definition – The volume expansivity is defined as

$$\beta = \frac{1}{V}\left(\frac{\partial V}{\partial T}\right)_P$$

Volume expansivity is often simply called *expansivity* or *coefficient of volume expansion.* The property is one of the compressibilities of a gas, where here the path (condition) of the compression is at constant pressure.

Tables of Values – This quantity is tabulated in Tables 4.1, 4.2, 4.12, and 4.13.

Units	Range of table		Table location
K^{-1}	13.8–3000 K	0.01–100 MPa	4.1, 4.12
K^{-1}	Saturation boundary		4.2, 4.13

Figure – None
Equation – None
Range of Values –

	Triple point		Boiling point			
Units	Liquid	Vapor	Liquid	Vapor	Critical point	300 K, 1 atm
Para and normal K^{-1}	0.0102	0.0752	0.0164	0.0642	∞	0.00333

Uncertainty – The uncertainty in the compressed liquid is 2 to 3%; in the gas and supercritical fluid, 1 to 2%. In the critical region ($T_c \pm 5\%$, $\rho_c \pm 20\%$) the uncertainties are larger; however, they are not experimentally defined since the property diverges at the critical point.

2.49. ORTHO-PARA MODIFICATIONS[14,106,143]

Definition – The hydrogen atom is composed of a nucleus of one proton and an electron orbiting around the nucleus. The hydrogen molecule is composed of two hydrogen atoms; however, there are two modifications of hydrogen molecules called orthohydrogen and parahydrogen. The differentiating feature of these molecules is the relative orientation of the nuclear spin of the individual atoms. The nuclear spin is a rotational motion of the nucleus about an axis perpendicular to the molecular axis. It is therefore possible to have two relative orientations of these nuclear spins, i.e., the spins may be in the same direction or they may be in opposite directions. These orientations are commonly called parallel and antiparallel, respectively. The nuclear spin is a quantumized motion; therefore, only certain motions are allowed. Associated with this quantumization are quantum numbers for the spin and the corresponding energy levels. The molecules with antiparallel nuclear spins, called parahydrogen, have even quantum numbers and are in the lowest energy state. Conversely, the molecules with parallel nuclear spins, called orthohydrogen, have odd quantum numbers and are in a higher energy level.

Discussion – Fluid hydrogen may be thought of as a binary mixture of two different species of molecules differing from each other in physical properties. These differences are easily detectable on a macroscopic scale and in some cases are of great importance in design problems. The percentage of the ortho-para concentrations in the mixture is temperature dependent. The term equilibrium hydrogen is, as the name implies, the equilibrium concentration at a given temperature. For example, at a temperature of about

120 K, the equilibrium concentration of ortho and para is 50% for each. Near ambient temperature the composition is 75% ortho and 25% para. This 25-75 composition is called normal hydrogen.
Tables of Values – These tables are from Reference 106.

Units	Range of table	Table location
% Para	0–500 K	4.10

Figure – 3.33
Equation – None
Range of Values –

Units	20 K	300 K
% Para	99.821	25.072

Uncertainty – These values are estimated to be uncertain by less than 1%.

Heat of Conversion[106]

Definition – The conversion of orthohydrogen to parahydrogen is an exothermic process. The amount of heat given off in the conversion is temperature dependent.
Table – Values taken from Reference 106.

Units	Range of tables	Table location
cal/mol	10–300 K	4.11

Figure – 3.37
Equation – None
Range of Values –

Units	10 K	300 K
cal/mol	253.9865	13.28

Uncertainty – The uncertainty of the heats of conversion is estimated to be less than 0.1%.
Conversion Rates – Since the percentage of the ortho-para composition is temperature dependent, the rate of conversion is of interest in a variety of problems. The conversion of a nonequilibrium ortho-para composition to an equilibrium composition is a very slow process in the absence of a catalyst. This mode of conversion (no catalyst present) is called self-conversion. In general, the self-conversion rate is also a function of temperature. For example, at liquid air temperatures the half-life of the conversion is greater than 1 year, while at 923 K and 50 mm Hg pressure the half-life is on the order of 10 min. The "half-life period" is defined as the amount of time required to convert 1/2 of the excess ortho or para composition present at the starting time.
Tables of Values – None
Figure – None

Catalysis Conversion Rates[144]

The ortho-para conversion process is greatly enhanced by catalysis; since many substances as well as magnetic and radiation fields will speed up the conversion, only a typical example of conversion will be given here. Weitzel et al.,[12] using 1½ l of unsupported hydrous ferric oxide granules, converted liquid normal hydrogen to 90 to 95% equilibrium H_2 at an average rate of 235 l/hr.

Tables of Values – None

Figure – None

Equation – None

Property Differences – In general, the larger differences in the properties of ortho- and parahydrogen will occur in the derived properties where heat is important. Properties like enthalpy, specific heat capacity, and thermal conductivity show large differences. On the other hand, the density of ortho-para mixtures will vary very little. The difference in density between normal and pure parahydrogen is probably less than about 0.3%, except at the very low temperature liquid where it could be as much as 0.5 to 0.7%, the density of normal hydrogen having the larger density. For differences in a particular property, the reader should consult Chapter 2 for the property and the figures listed below.

Figure – 3.34, density differences between normal and parahydrogen; 3.35, specific heat capacity differences between normal and parahydrogen; 3.36, thermal conductivity differences between normal and parahydrogen.

2.50. ISOTOPES[14]

Definition – An isotope is an atom of an element having a different nuclear mass and atomic weight from other atoms of the same element.

Discussion – Hydrogen has two such isotopes, deuterium with an atomic weight of 2 and tritium with an atomic weight of 3. The three different kinds of hydrogen atoms may combine in six different ways to form a molecule, i.e., H_2, D_2, T_2, HD, HT, and DT. There are some data in the literature for D_2 but very little for any of the tritium-containing molecules. The isotopes are beyond the scope of this work, and the reader is referred to Reference 14 for data for the hydrogen isotopes.

2.51. METALLIC HYDROGEN[145-147]

Definition – Solid hydrogen under extreme pressure (about 2 to 3 Mbar) is thought to undergo a phase transition to the metallic state and may be superconducting. The subject of metallic hydrogen is beyond the scope of this work, but References 145 to 147 will serve as a starting point for those who wish to pursue the subject further.

2.52. SLUSH HYDROGEN[148,149]

Definition – Slush hydrogen is the name given to a homogenous mixture of the coexisting solid and liquid phases. Because of the potential application of slush hydrogen as a fuel, its properties were studied by NASA. The slush form of hydrogen as a fuel offers two distinct advantages over the pure liquid.[149] First, if the mixture is 0.6 solid by mass, the density will be 11.5% greater than the pure liquid at its normal boiling point; second, the heat capacity will increase by 86.2 J/g. The increased density is desirable because the available energy is increased per unit volume, and the increased heat capacity eases refrigeration requirements.

Preparation – There are a number of ways to produce slush hydrogen. The method that seems to be the best at the present time is the so-called freeze thaw method. To produce slush using this method, the ullage over the liquid is pumped and a solid layer forms at the surface. The pumping is then stopped, and the solid layer melts at the edge of the container and settles into the liquid. The cycle is repeated until the desired amount of solid has been formed. The solid is then mixed with the liquid by stirring to form a homogenous mixture of solid and liquid. The porosity of the solid formed in this manner is a function of the pumping rate, and the porosity is a critical factor in mixing the solid with the liquid.

Low pumping rates produce a dense solid which when broken forms large particles of solid and will not form a homogenous mixture with the liquid. Higher pumping rates produce a porous solid which is easily broken into much smaller particles and is easily mixed with the liquid to form a homogenous mixture which may be defined as slush. This method is capable of producing slush in large quantities. A disadvantage of the method is that the vapor over the liquid must be pumped down to very low pressures, which makes the possibility of a leak in the system a potential hazard since a leak would result in the possibility of air coming into the system. Because of this potential safety hazard, new methods of production are being investigated by NBS in Boulder, Colorado. New methods should be capable of operation at pressures above atmospheric and practical for large scale production.

Aging – Although the prime objective of research has been to investigate the preparation, storage, and transfer of slush hydrogen, some interesting aging affects have been observed.[148] Although the change in the particle size is insignificant, the structure changes dramatically during aging. The solid particles were observed to change from particles with irregular rough edges to particles with smooth edges. The smaller particles tend to combine with the larger particles to form a particle of higher density.

Transfer – Although some special techniques are required to handle slush hydrogen, it has the characteristics of a cryogenic fluid for most applications. Slush with solid content of up to 0.6 may be pumped with losses similar to that of hydrogen at its triple point. Slush with 0.3 or less solid was found to have transfer losses of 10% less than those of the triple point liquid when the Reynolds number was 7×10^5 or greater.

Problems – Although a good deal of experimental work has already been done on slush hydrogen, certain problems still exist. More data are needed on the heat transfer in slush, and good methods of measurement of the mass and the mass flow rates have not been developed. However, none of the experimental results in the slush hydrogen investigations to date have indicated that slush cannot or should not be used as a fuel; on the contrary, the use of slush hydrogen appears to be both practical and desirable.

2.53. HYDROGEN SAFETY[150-156]

There are many books, articles, and pamphlets that deal with the problem of safety where the presence of hydrogen causes a potential hazard. The purpose of this descriptive sheet is to call the attention of the reader to the problem and provide a few important references where more detailed information is available. Three of the best references are 151 to 153.

Discussion – Potential hazardous conditions due to the presence of hydrogen:

1. Hydrogen is colorless and odorless and therefore is not detectable by human senses.
2. Hydrogen gas can cause suffocation by dilution of the air. It is not toxic.
3. At temperatures above −418°F, hydrogen gas is lighter than air (at stp) and

tends to rise. Temperatures of less than −418°F may occur just after evaporation in which case the gas tends to fall.

4. Mixtures of hydrogen with either air or oxygen are highly flammable over a wide range of composition (see below).

5. Detonation of an explosion can occur with very low energy impact, only 1/10 that of a gasoline-air mixture.

6. At atmospheric pressure a temperature in excess of 1000°F is required for ignition, but at pressures of 2/10 to 5/10 of an atmosphere, ignition can occur at a temperature of 650°F.

7. Hydrogen flame is colorless.

Properties of Hydrogen Gas –

Gross heat of combustion, K-J/m^3	12,287
Gross heat of combustion, K-J/m^3	141.98
K-J/m^3 of stoichiometric gas/air mixture (F/A ratio, 0.418 vol, 0.029 wt; or 29.5 H_2 by vol)	3,073
m^3 of air required per m^3 of combustible	2.382
Kg air required per Kg combustible	34.226
Maximum flame temperature (F/A ratio 0.462 vol, 0.0313 wt; or 31.6% H_2 by vol)	2,403 K
Autoignition temperature in air	847 K
Autoignition temperature in oxygen	833 K
Flammability limits, % vol H_2 in air	4.0–75
Flammability limits, % vol H_2 in oxygen	4.5–94
Detonation limits, % vol H_2 in air	18.3–59
Detonation limits, % vol H_2 in oxygen	15.0–90
Nonflammable limits, air-hydrogen	Less than 8% H_2
Nonflammable limits, O_2-hydrogen	Less than 5% H_2

2.54. MISCELLANEOUS PROPERTIES[94,159-275]

Included in this section are properties that have not been discussed in the preceding pages and their corresponding references.

Property	Reference
Birefringence	159
Bulk modulus (solid)	160
Creep	161
Debye temperature	160, 162
Dispersion	163–169
Elastic constant	160
Electrical properties (miscellaneous)	94, 170
Electron diffraction	171, 172
Electron drift	173–176
Emissivity-reflectance	177–179
Hydrogen gel	180
Ionization potential	181–185
Irradiation	186–192
Isotopic phase separation	193, 194
Magnetic susceptibility	196–199
Magneto optics	195
Magneto resistance	200
Nuclear magnetic resonance (NMR)	201–222
Nuclear spin relaxation	208, 213, 218, 220–233

Property	Reference
Neutron diffraction	188, 234, 235
Neutron scattering	162, 236–244
Neutron spectra	237, 244–247
Opacity	248
Optical susceptibilities	249
Ortho-para absorption	250
Plastic deformation	251, 252
Relaxation time (electron spin)	253
Scattering	195, 237, 245, 247, 254–258
Shear strength (solid)	252, 259–261
Specific impulse	262
Surface energy	263–268
Tensile strength	252, 260
Thermomagnetism (Senftleben effect)	269, 270
X-ray diffraction	271–275
Young's modulus	252, 260

Chapter 3

Data Graphs

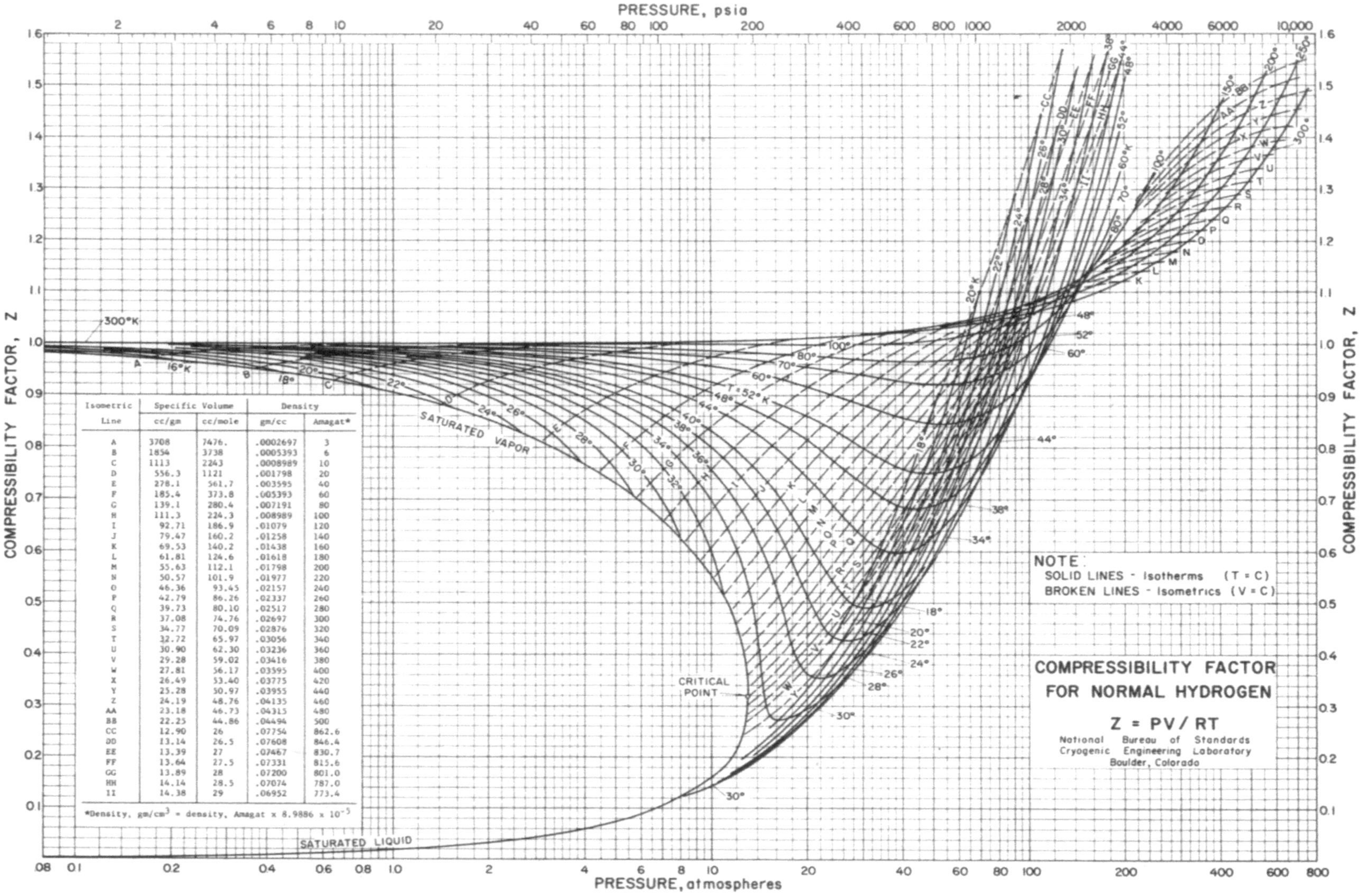

Isometric Line	Specific Volume		Density	
	cc/gm	cc/mole	gm/cc	Amagat*
A	3708	7476.	.0002697	3
B	1854	3738	.0005393	6
C	1113	2243	.0008989	10
D	556.3	1121	.001798	20
E	278.1	561.7	.003595	40
F	185.4	373.8	.005393	60
G	139.1	280.4	.007191	80
H	111.3	224.3	.008989	100
I	92.71	186.9	.01079	120
J	79.47	160.2	.01258	140
K	69.53	140.2	.01438	160
L	61.81	124.6	.01618	180
M	55.63	112.1	.01798	200
N	50.57	101.9	.01977	220
O	46.36	93.45	.02157	240
P	42.79	86.26	.02337	260
Q	39.73	80.10	.02517	280
R	37.08	74.76	.02697	300
S	34.77	70.09	.02876	320
T	32.72	65.97	.03056	340
U	30.90	62.30	.03236	360
V	29.28	59.02	.03416	380
W	27.81	56.17	.03595	400
X	26.49	53.40	.03775	420
Y	25.28	50.97	.03955	440
Z	24.19	48.76	.04135	460
AA	23.18	46.73	.04315	480
BB	22.25	44.86	.04494	500
CC	12.90	26	.07754	862.6
DD	13.14	26.5	.07608	846.4
EE	13.39	27	.07467	830.7
FF	13.64	27.5	.07331	815.6
GG	13.89	28	.07200	801.0
HH	14.14	28.5	.07074	787.0
II	14.38	29	.06952	773.4

*Density, gm/cm^3 = density, Amagat $\times 8.9886 \times 10^{-5}$

FIGURE 3.1. Compressibility factor for normal hydrogen

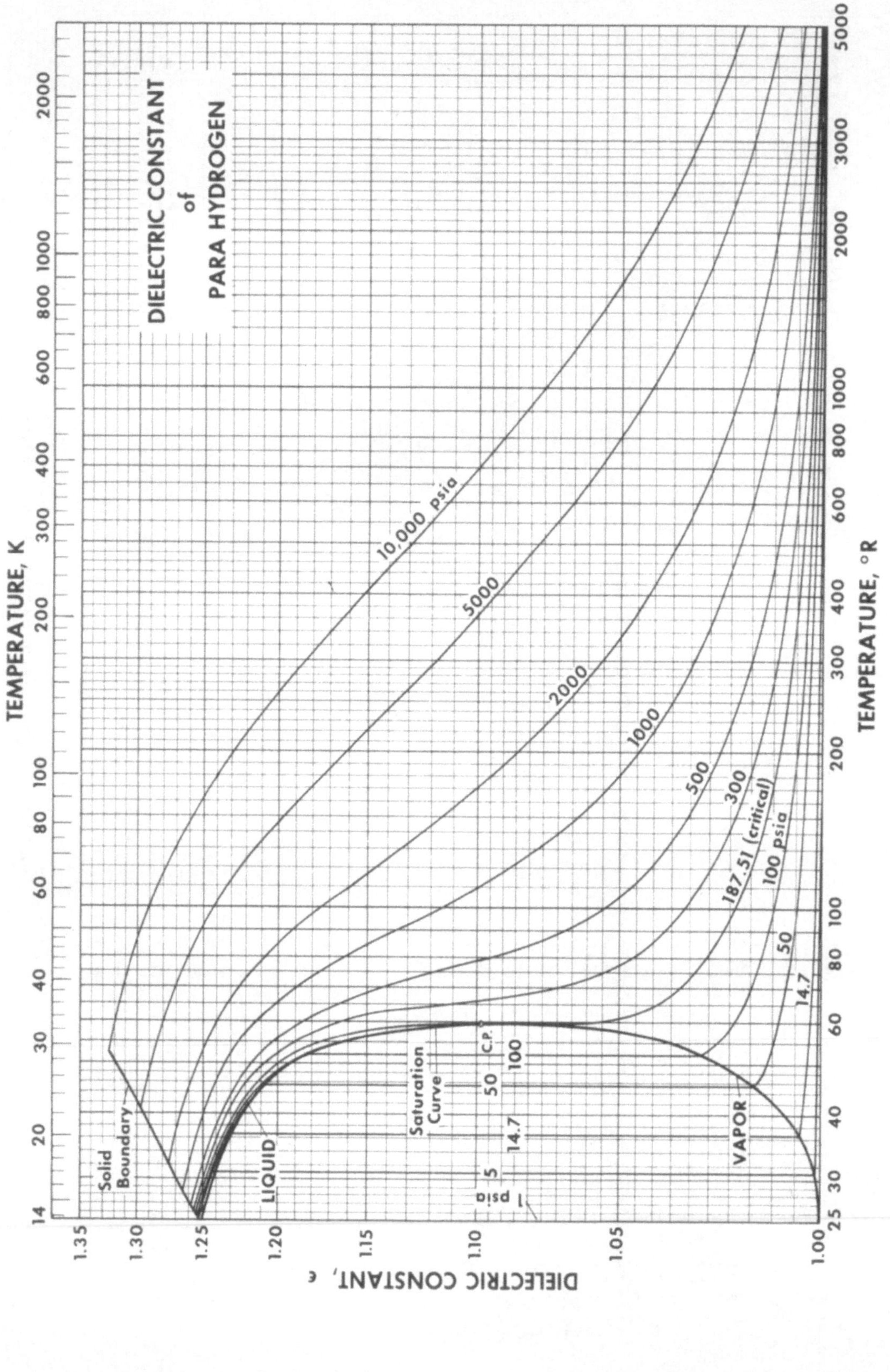

FIGURE 3.2. Dielectric constant of parahydrogen.

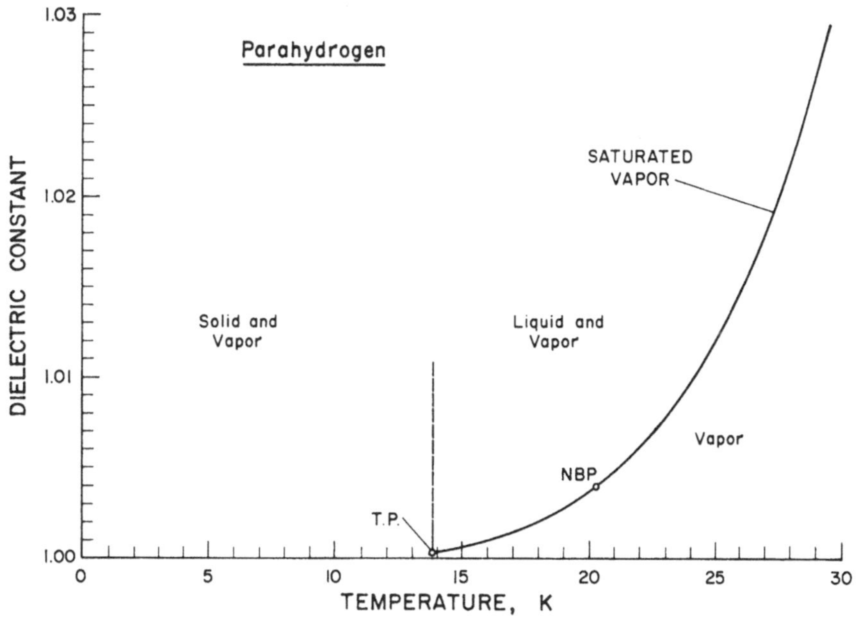

FIGURE 3.3. Dielectric constant of saturated vapor.

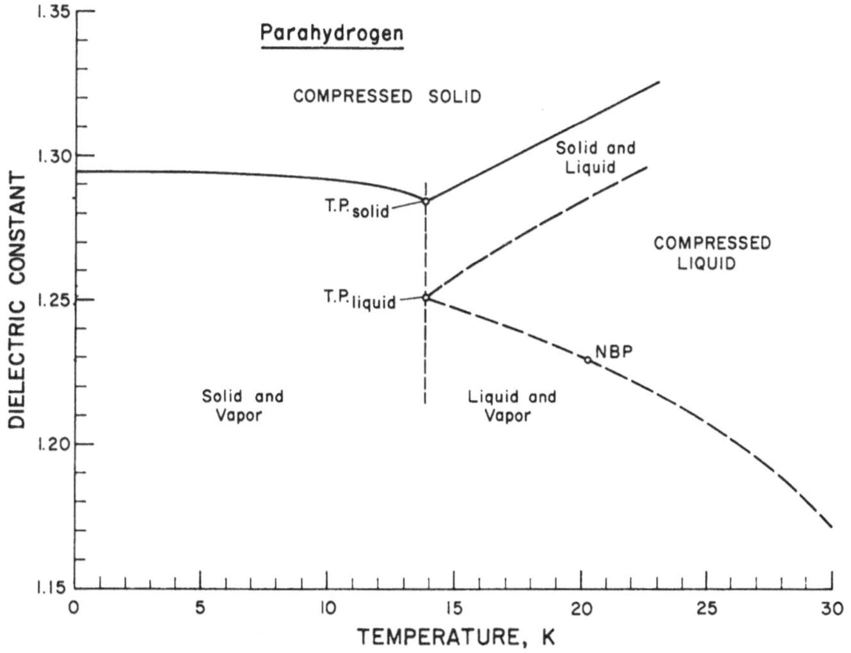

FIGURE 3.4. Dielectric constant of saturated liquid and solid.

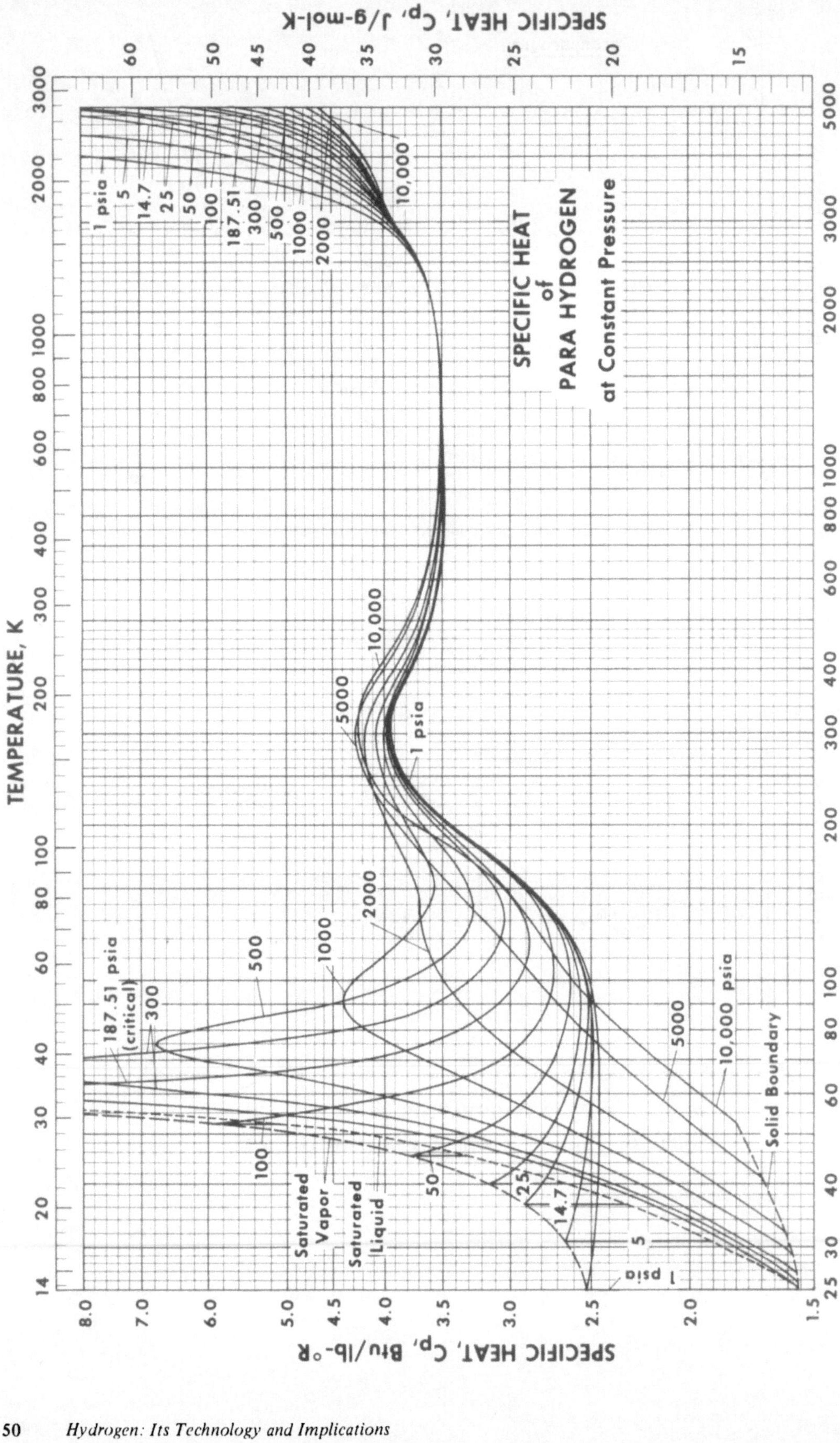

FIGURE 3.5. Specific heat of parahydrogen at constant pressure (heat capacity).

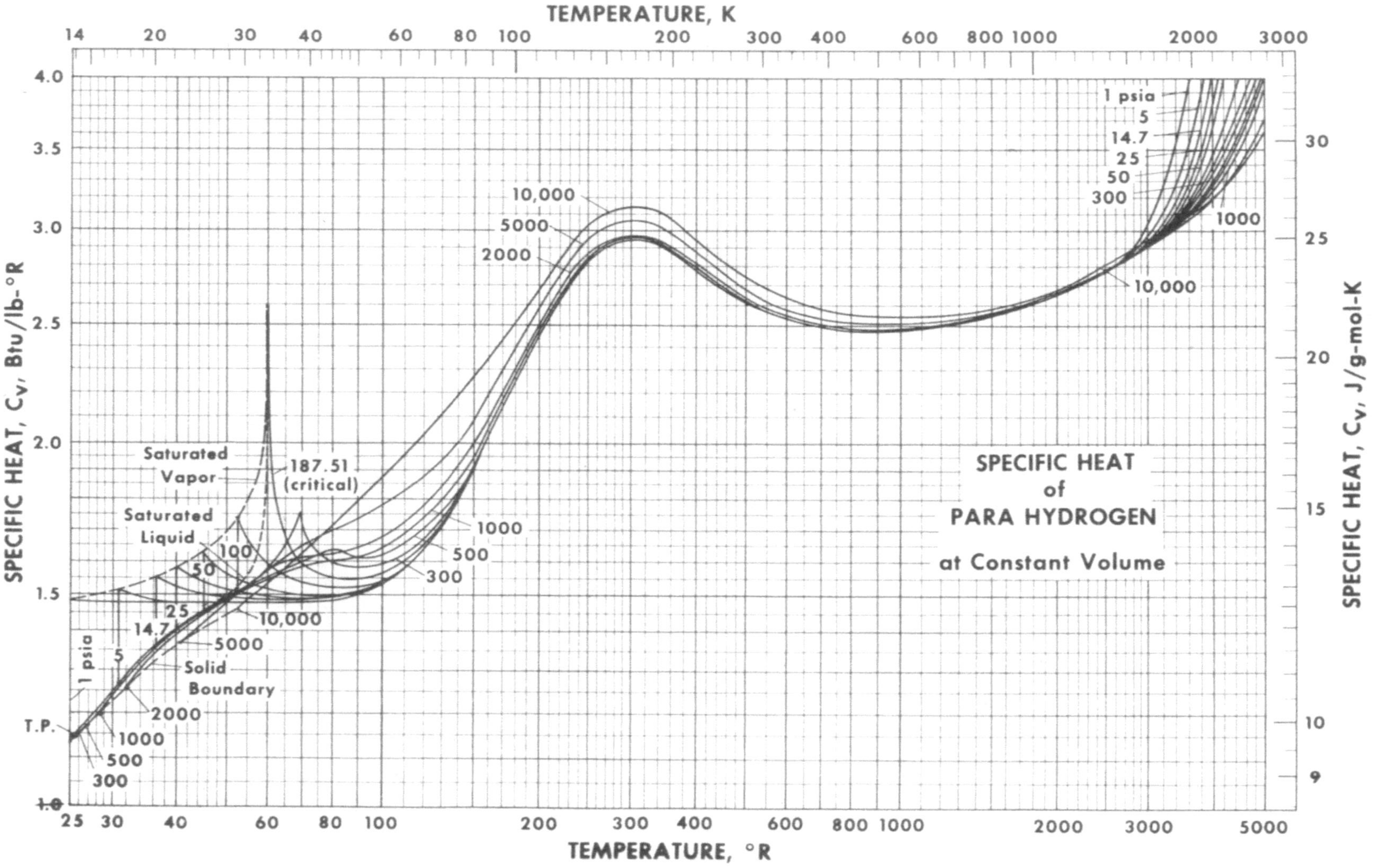

FIGURE 3.6. Specific heat of parahydrogen at constant volume (heat capacity).

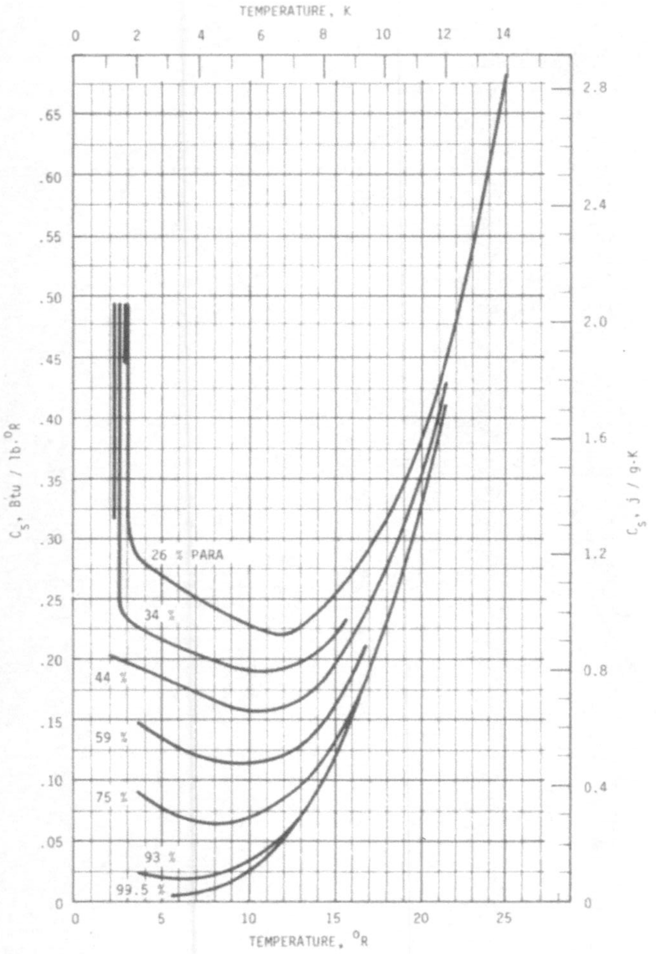

FIGURE 3.7. Specific heat of solid hydrogen (heat capacity).

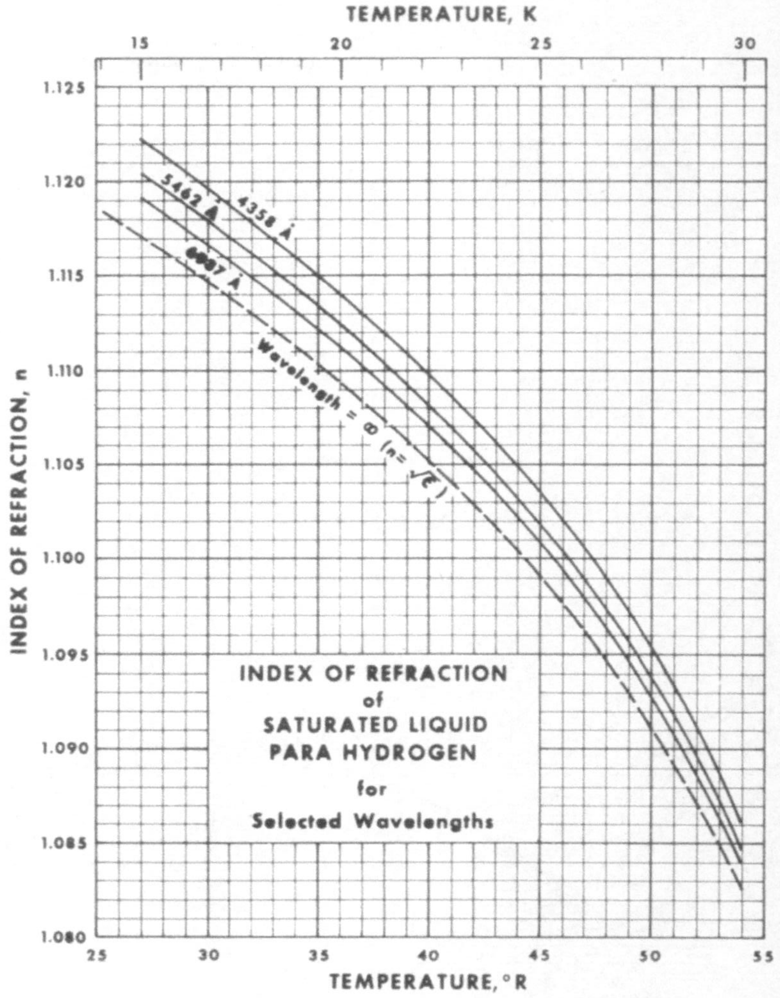

FIGURE 3.8. Index of refraction of saturated liquid.

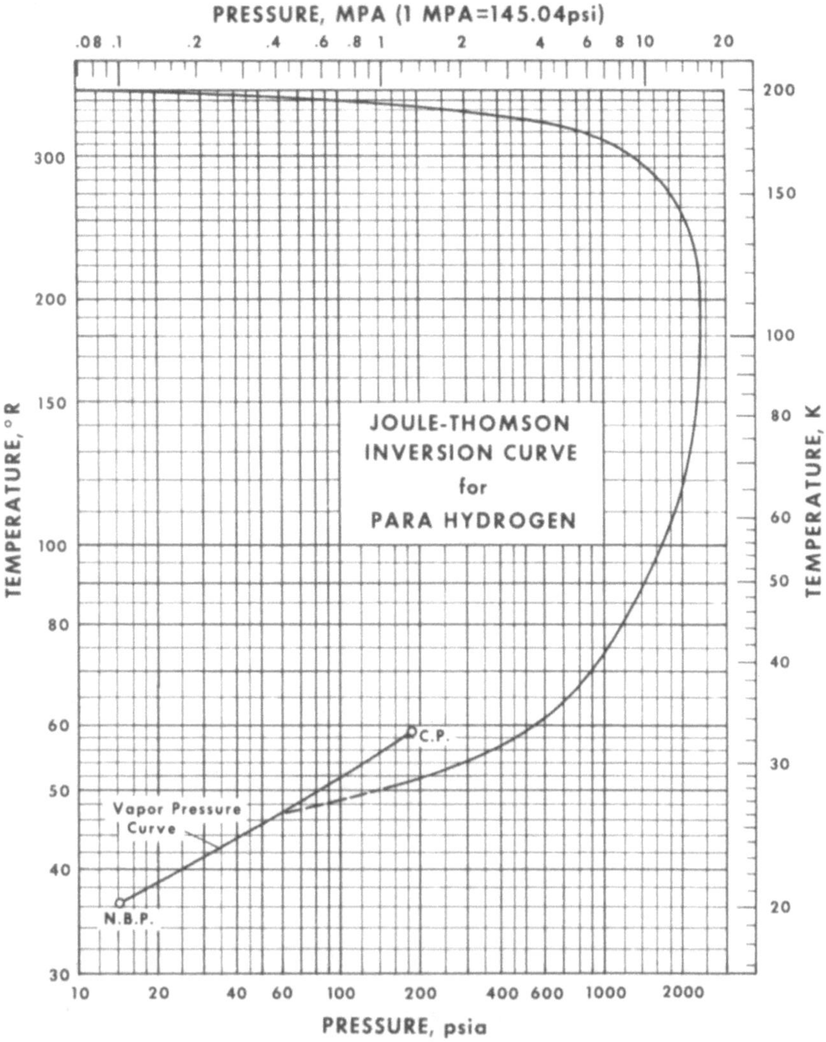

FIGURE 3.9. Joule-Thomson inversion curve.

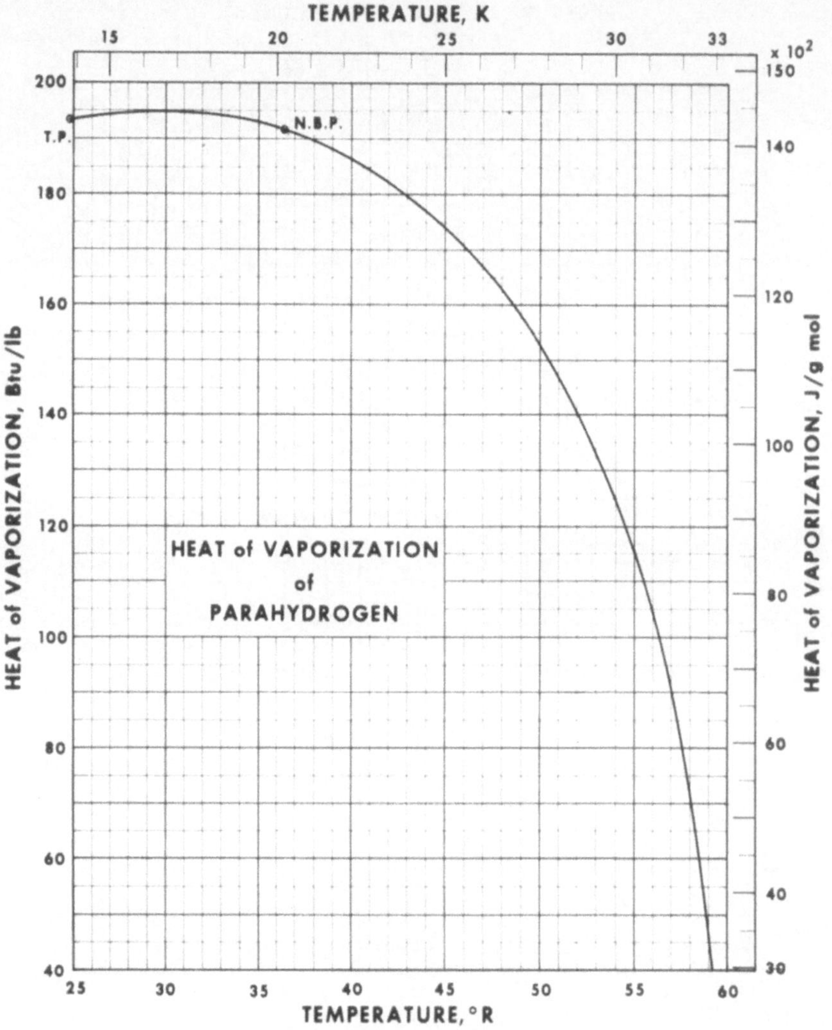

FIGURE 3.10. Heat of vaporization.

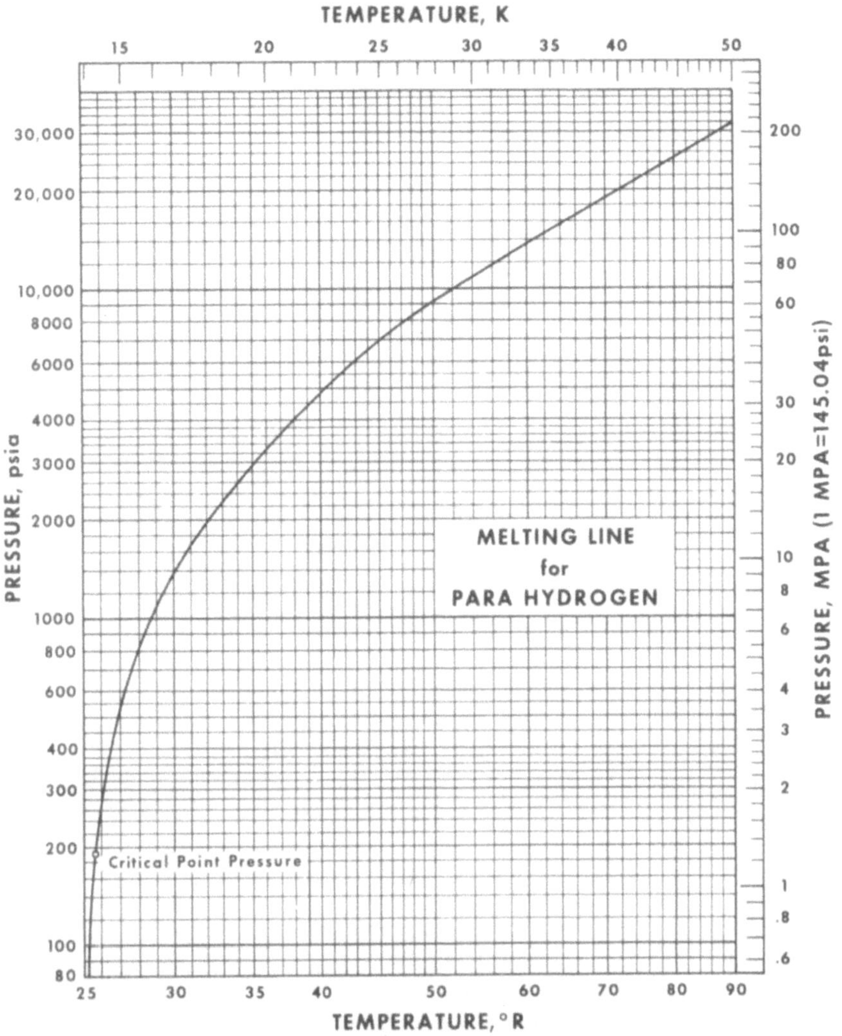

FIGURE 3.11. Melting line (80 to 30,000 psia).

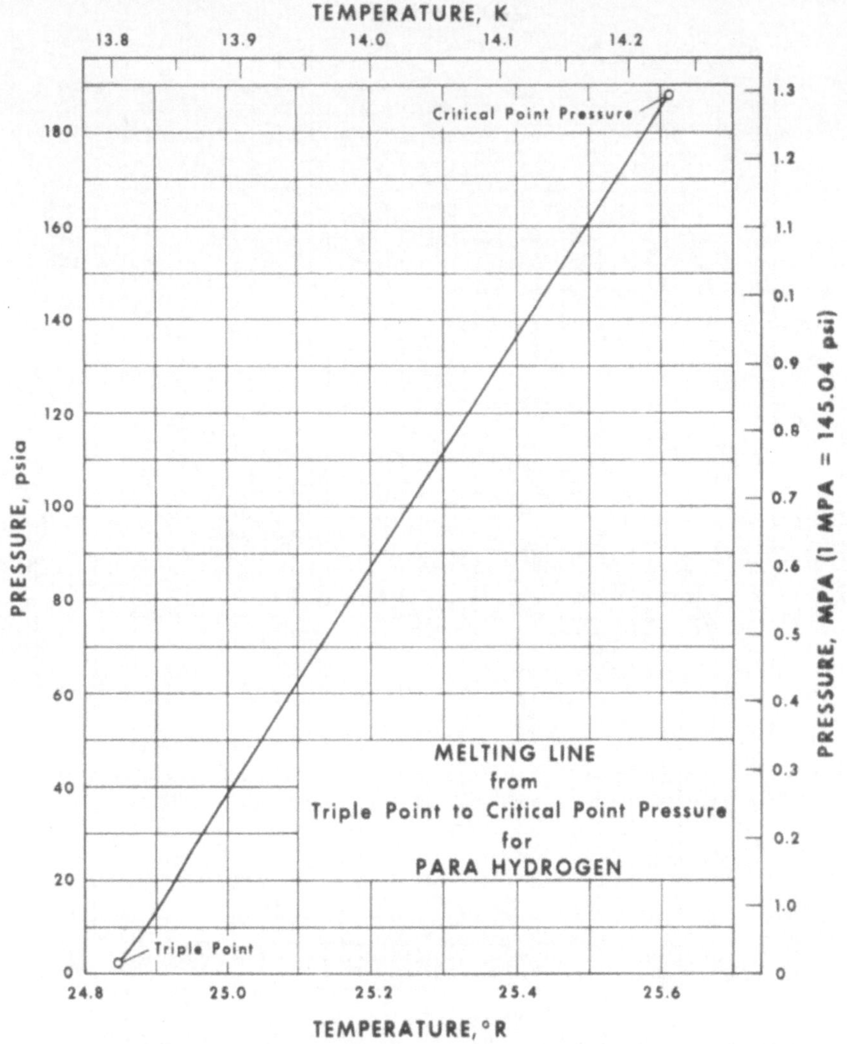

FIGURE 3.12. Melting line from triple point to critical point pressure.

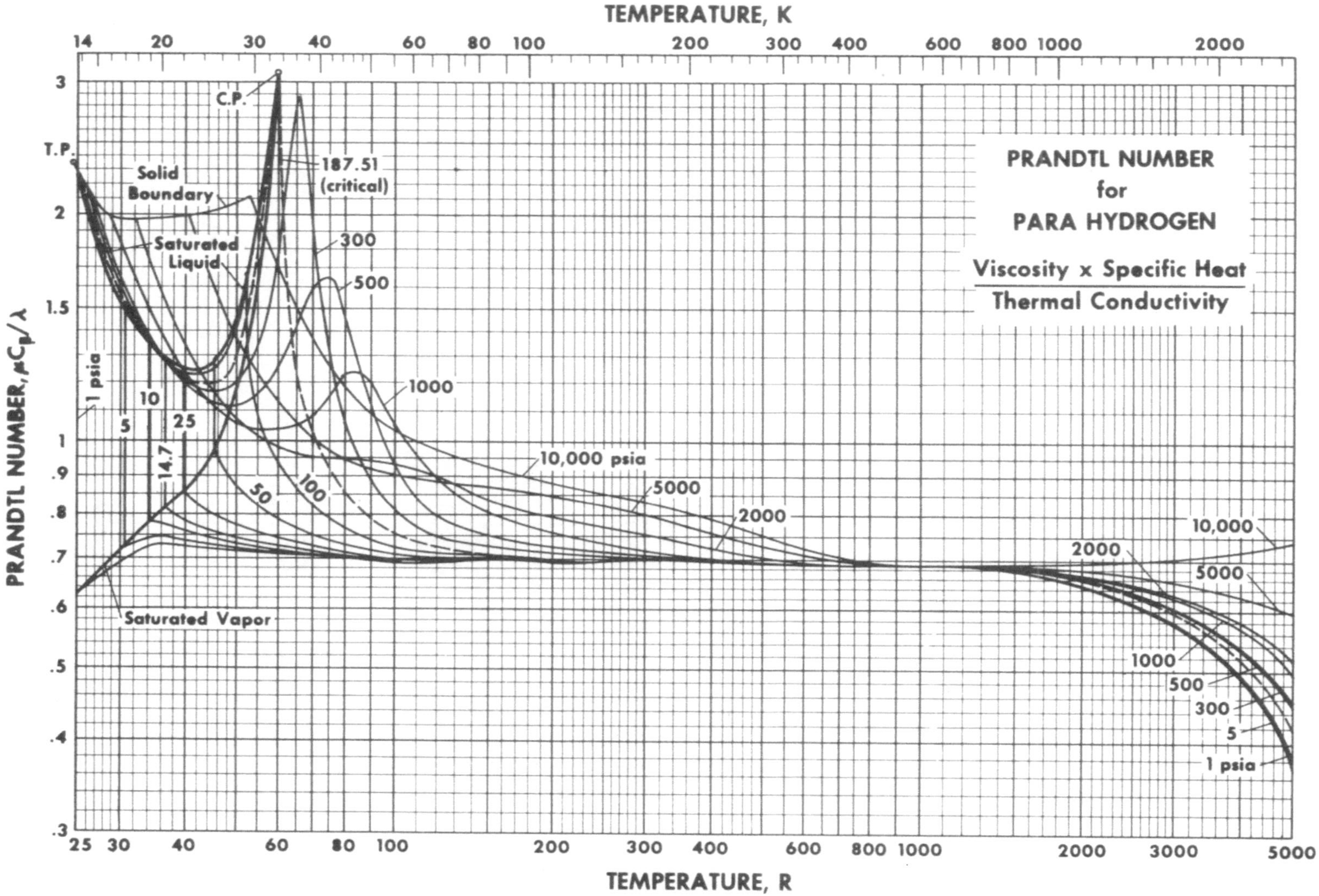

FIGURE 3.13. Prandtl number.

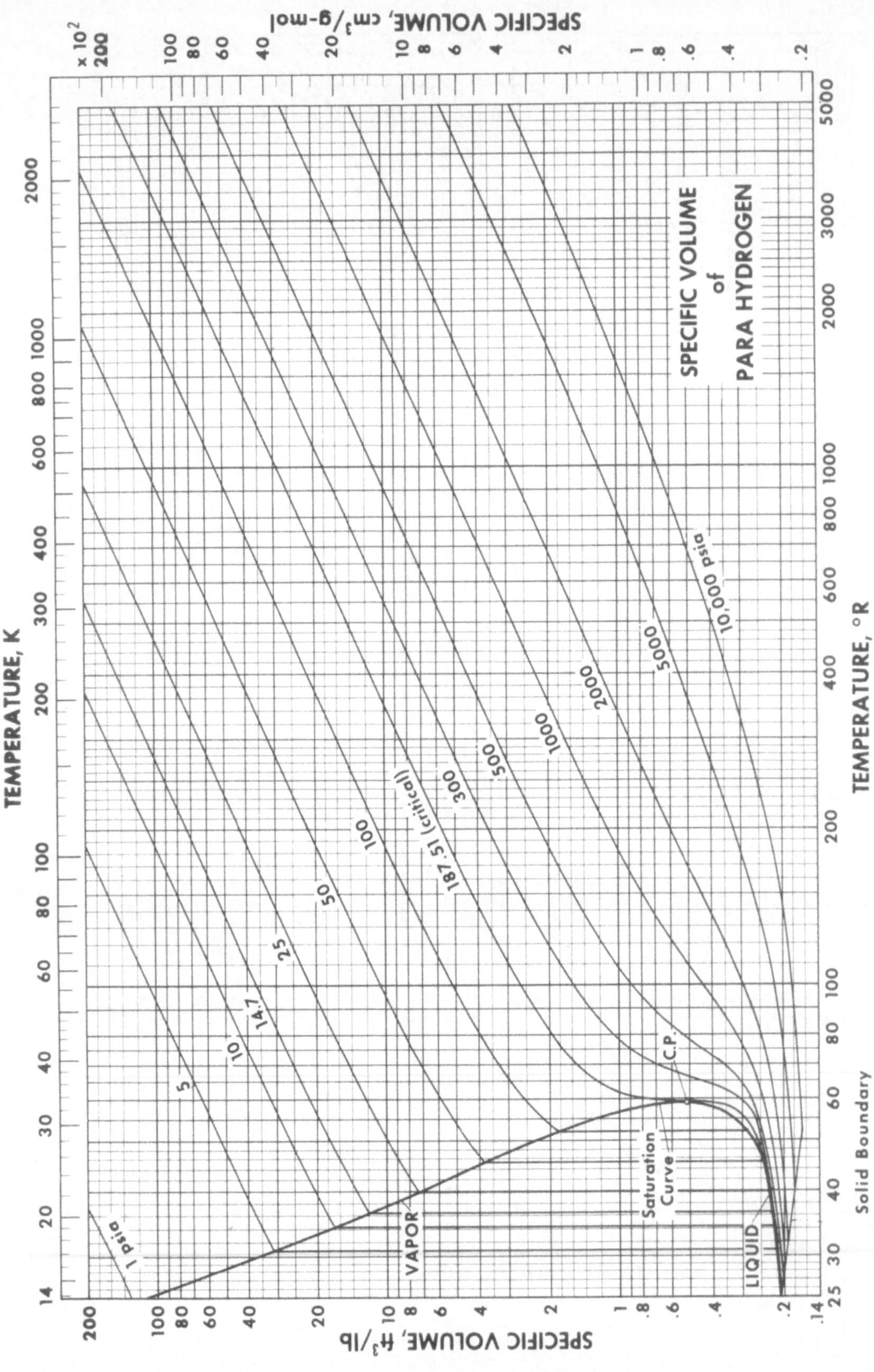

FIGURE 3.14. Specific volume (PVT).

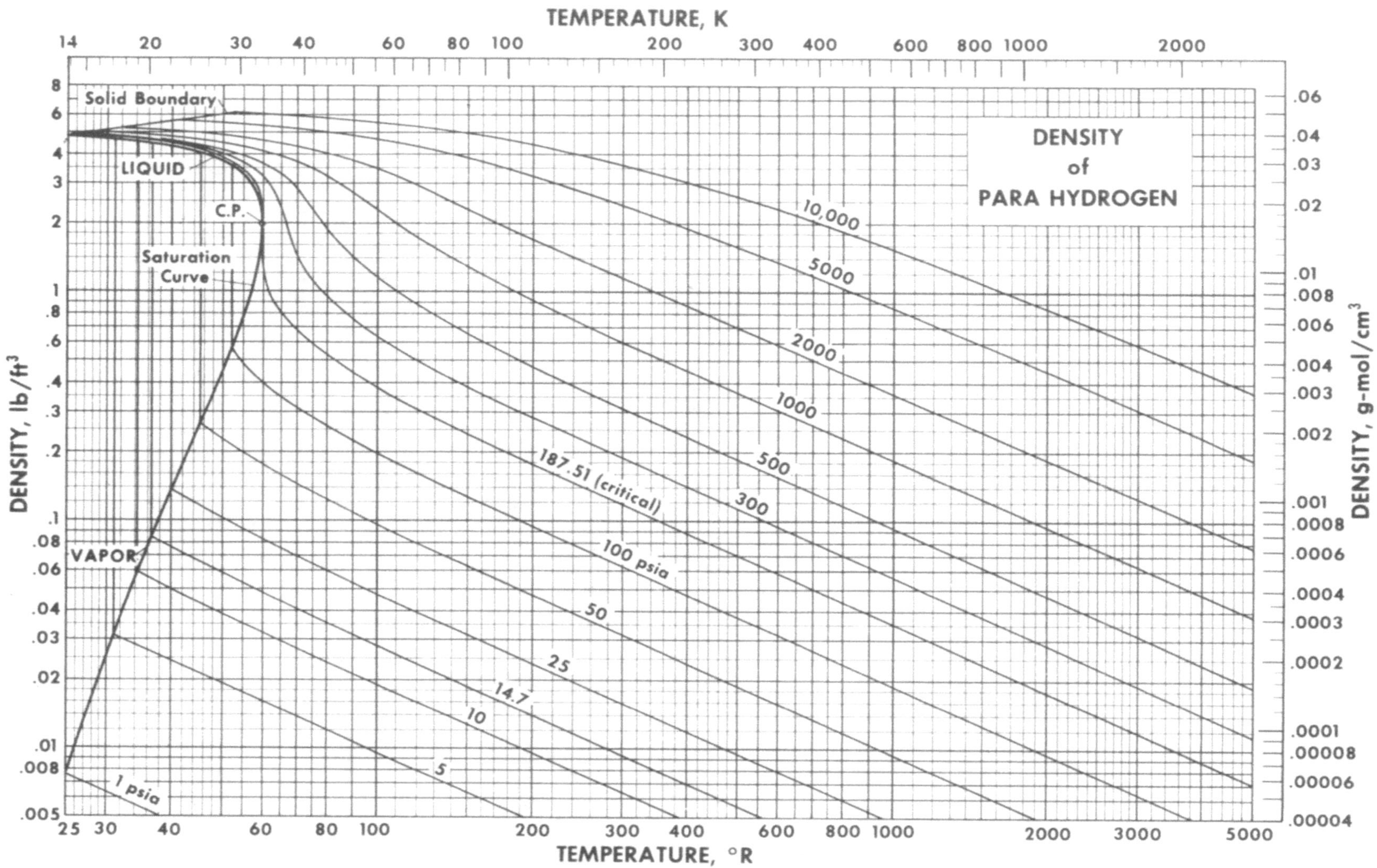

FIGURE 3.15. Density (P-ρ-T).

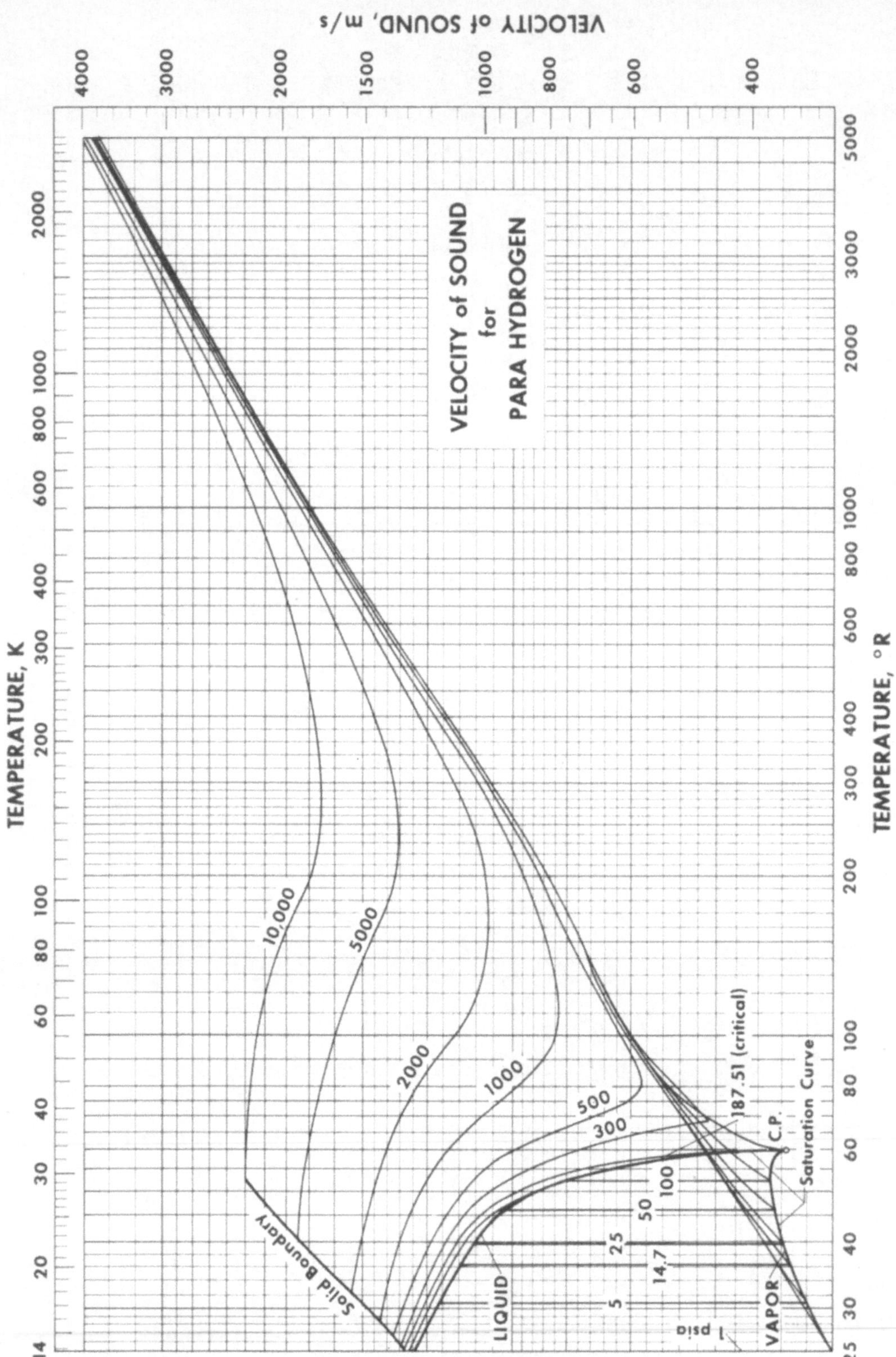

FIGURE 3.16. Velocity of sound.

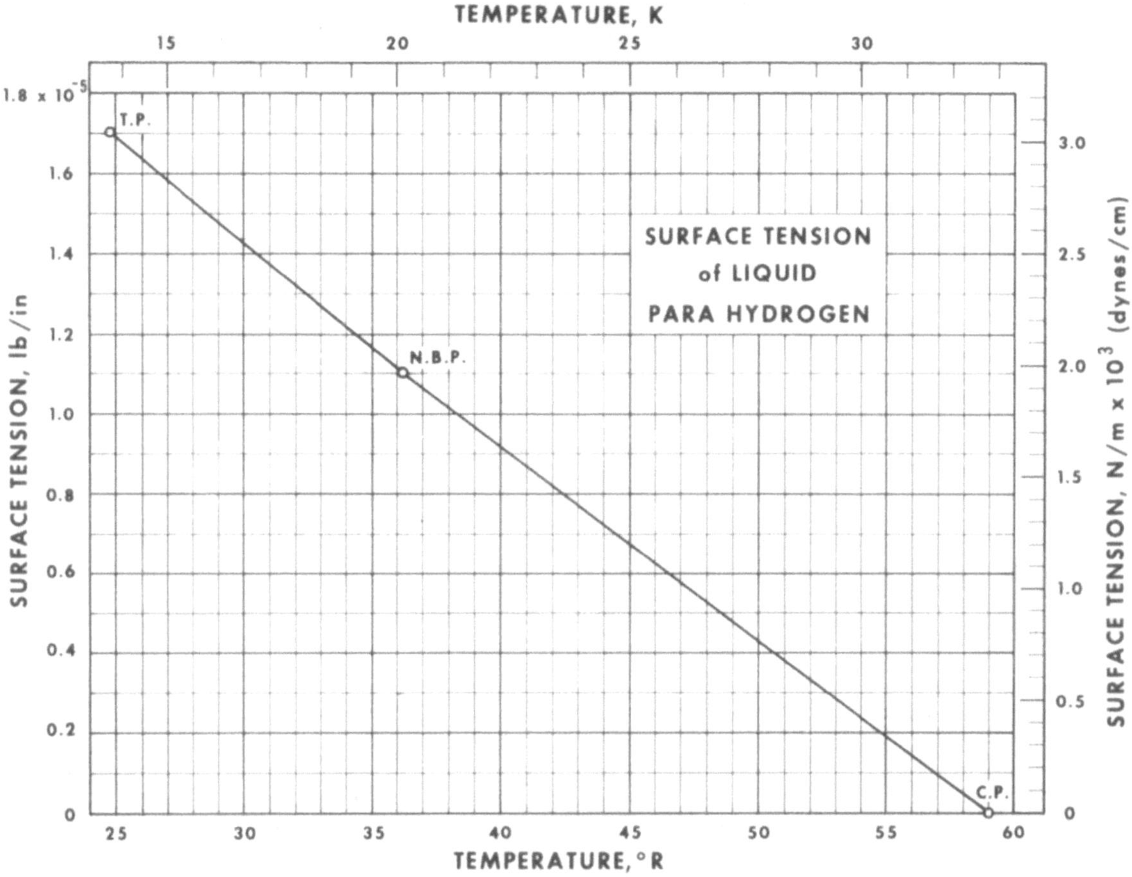

FIGURE 3.17. Surface tension.

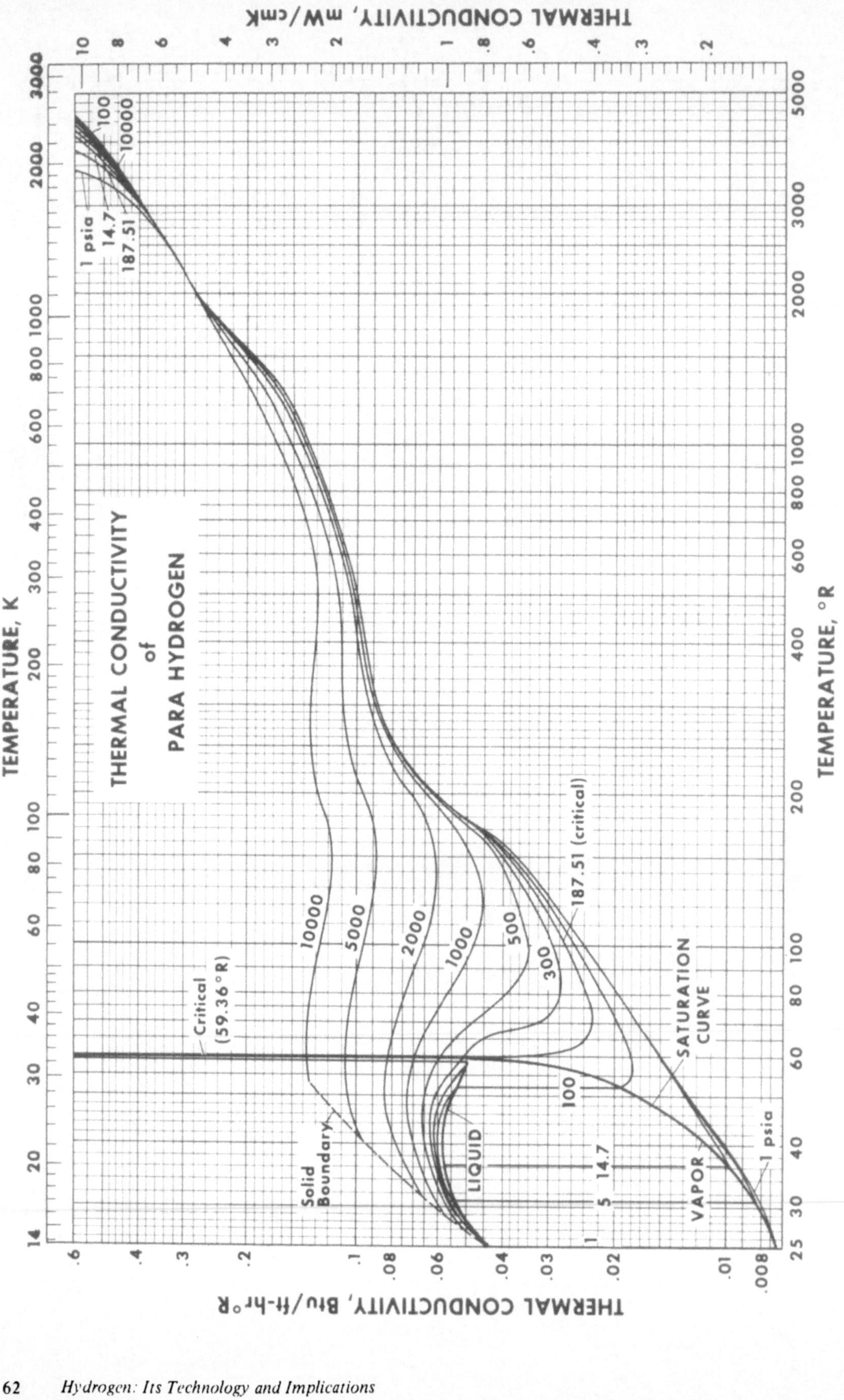

FIGURE 3.18. Thermal conductivity.

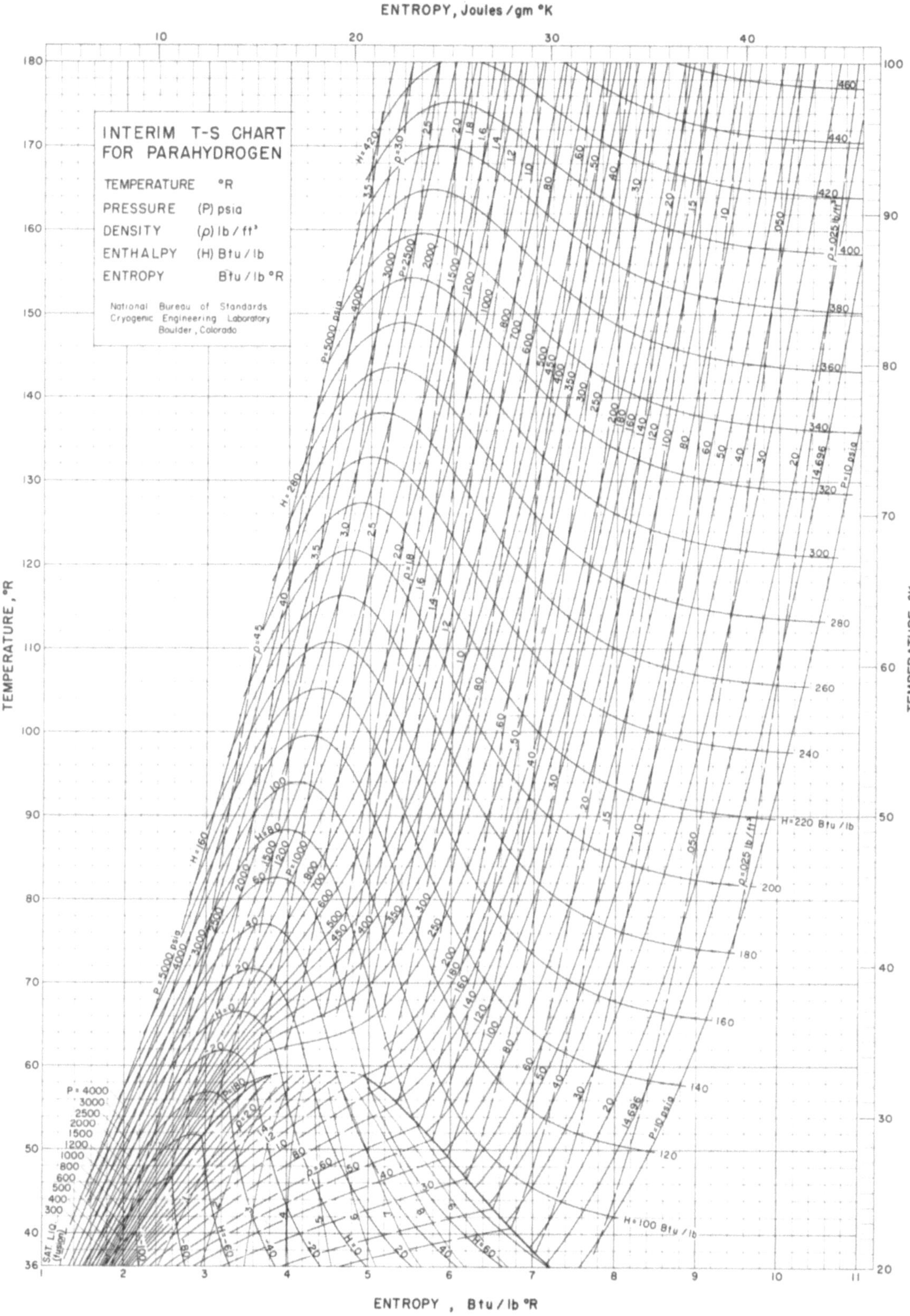

FIGURE 3.19. Temperature-entropy chart (36 to 180°R).

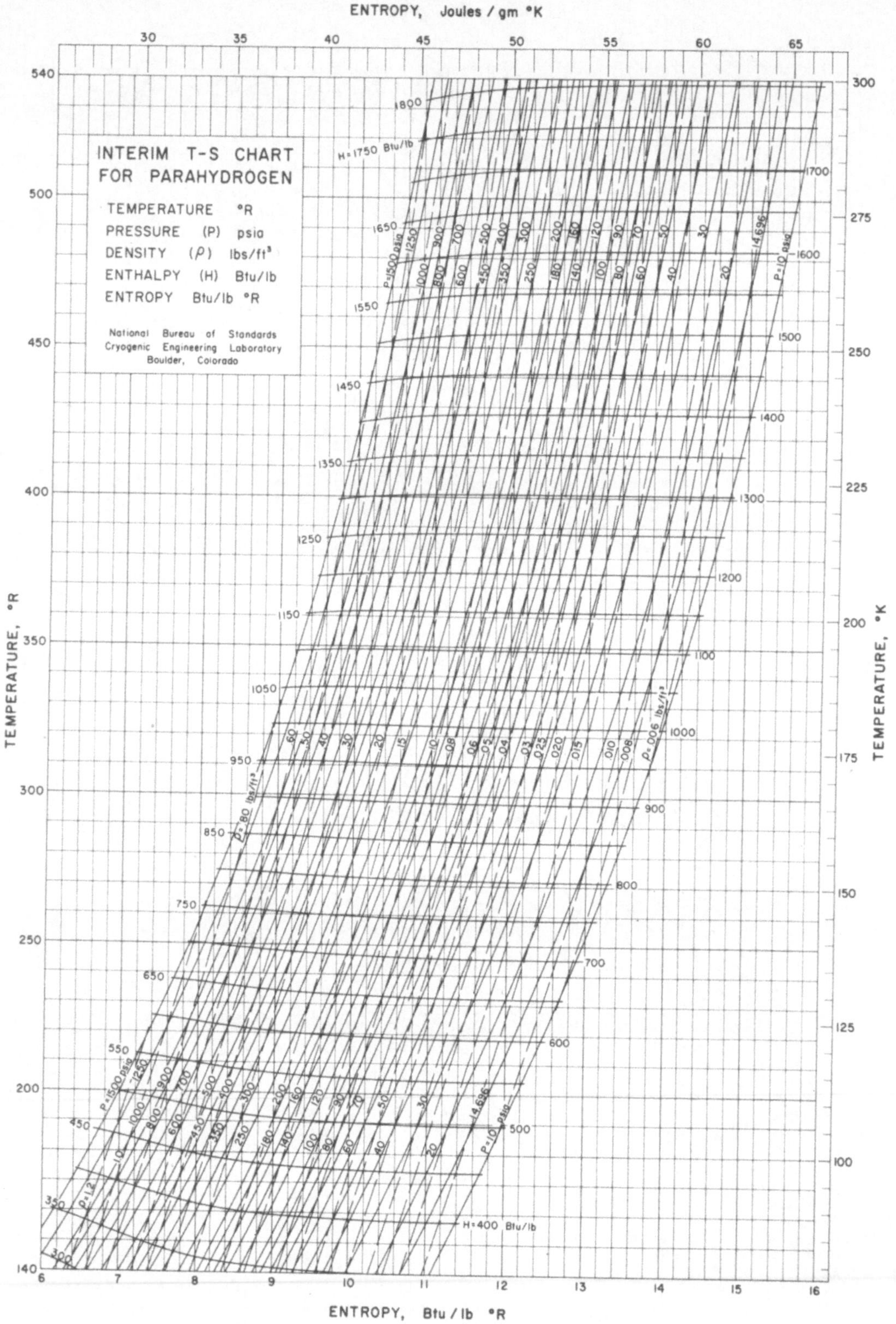

FIGURE 3.20. Temperature-entropy chart (140 to 540° R).

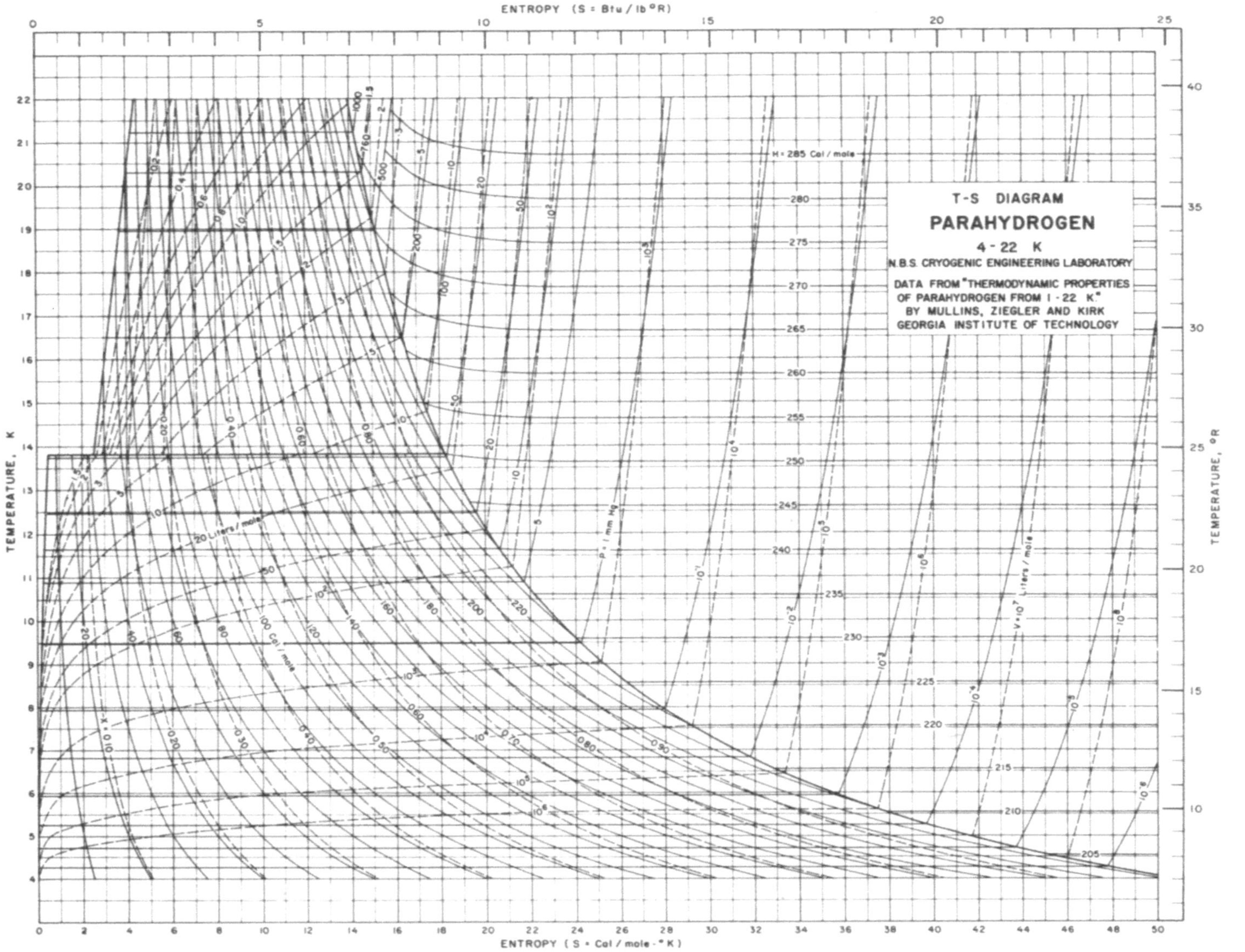

FIGURE 3.21. Temperature-entropy chart (4 to 22 K).

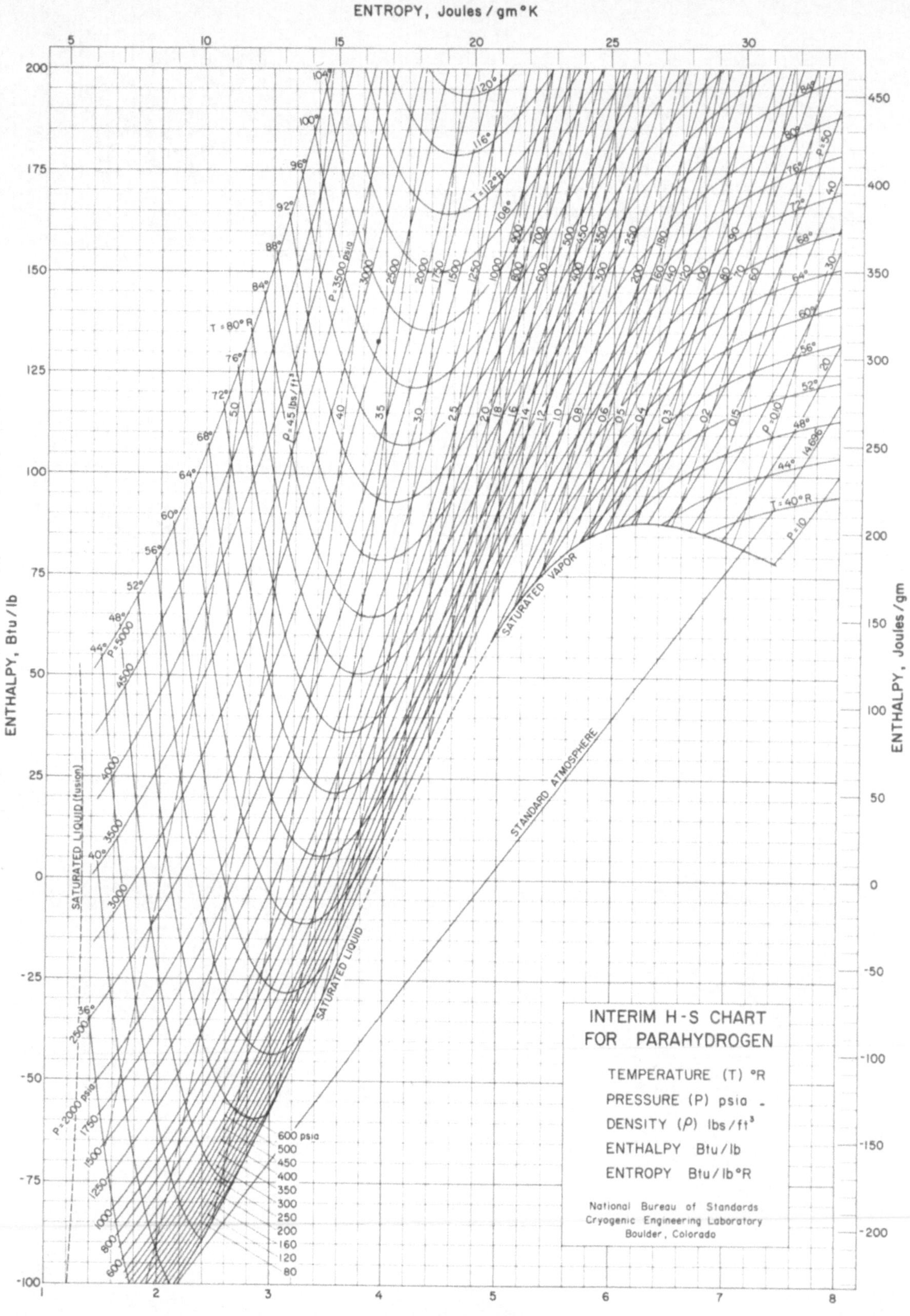

FIGURE 3.22. Enthalpy-entropy chart.

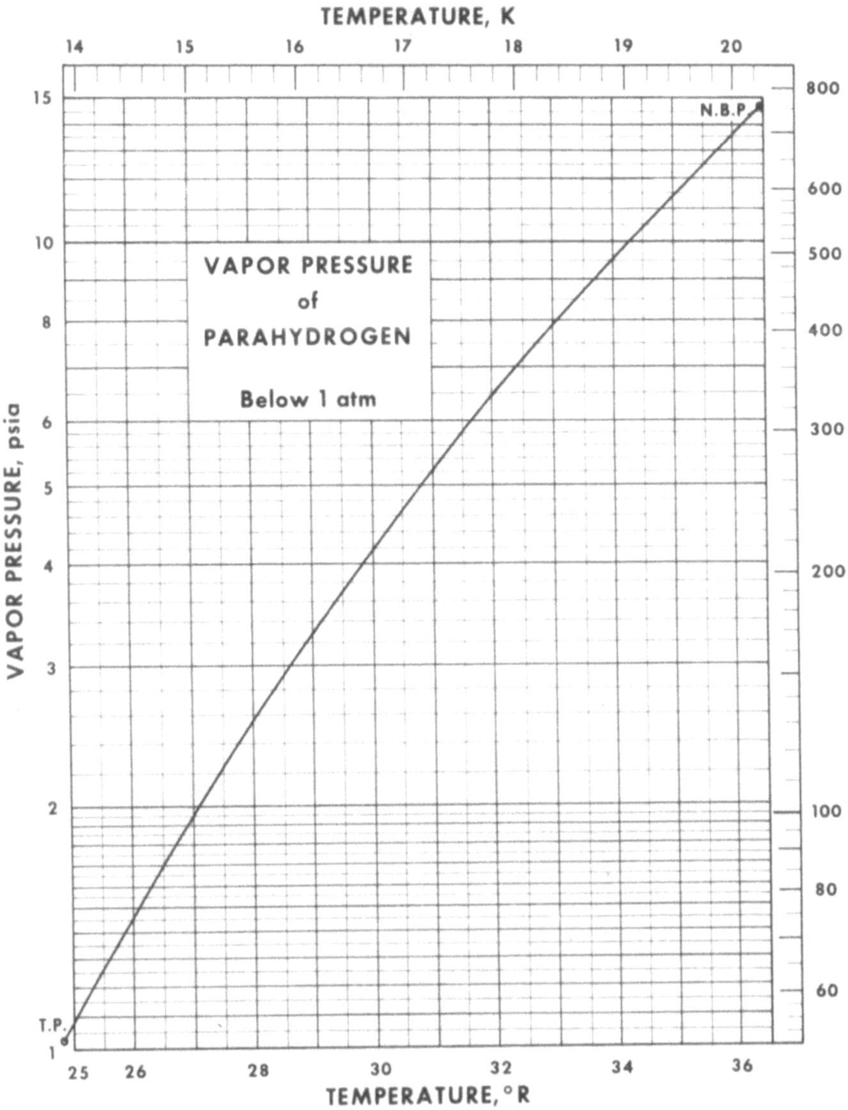

FIGURE 3.23. Vapor pressure (below 1 atm).

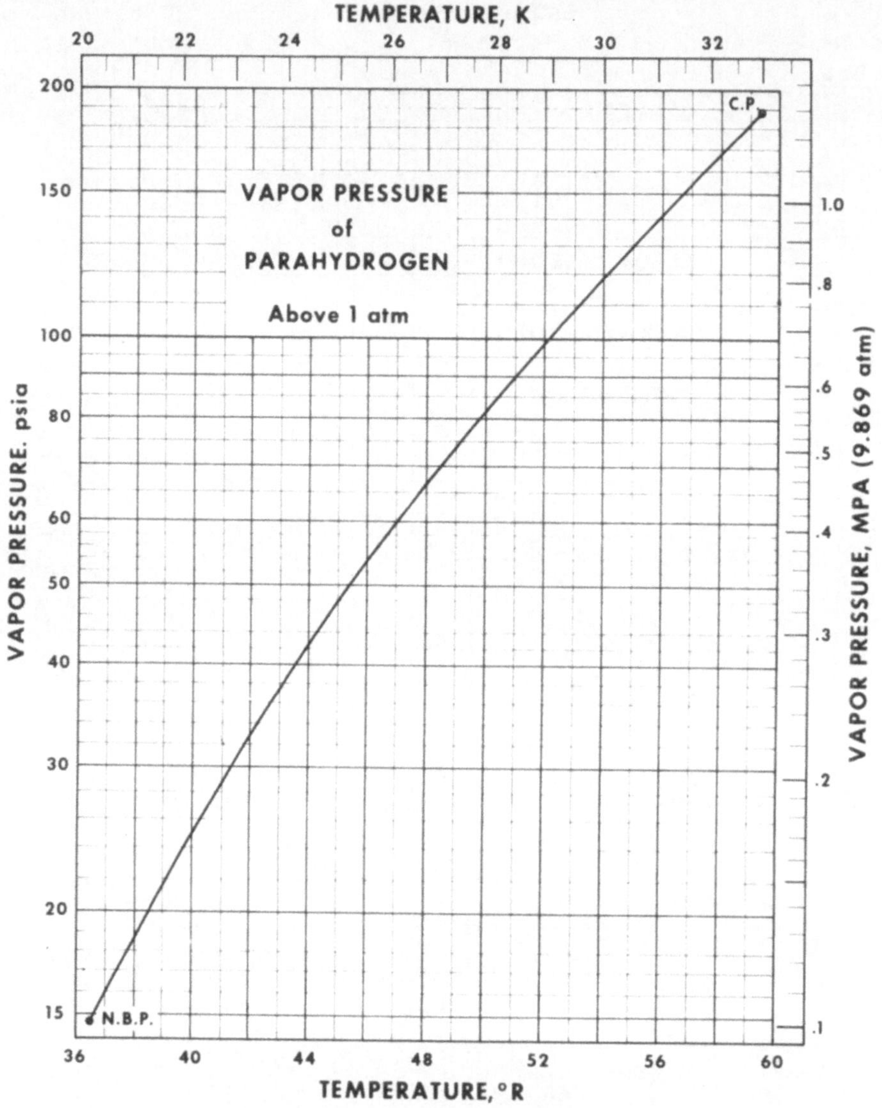

FIGURE 3.24. Vapor pressure (above 1 atm).

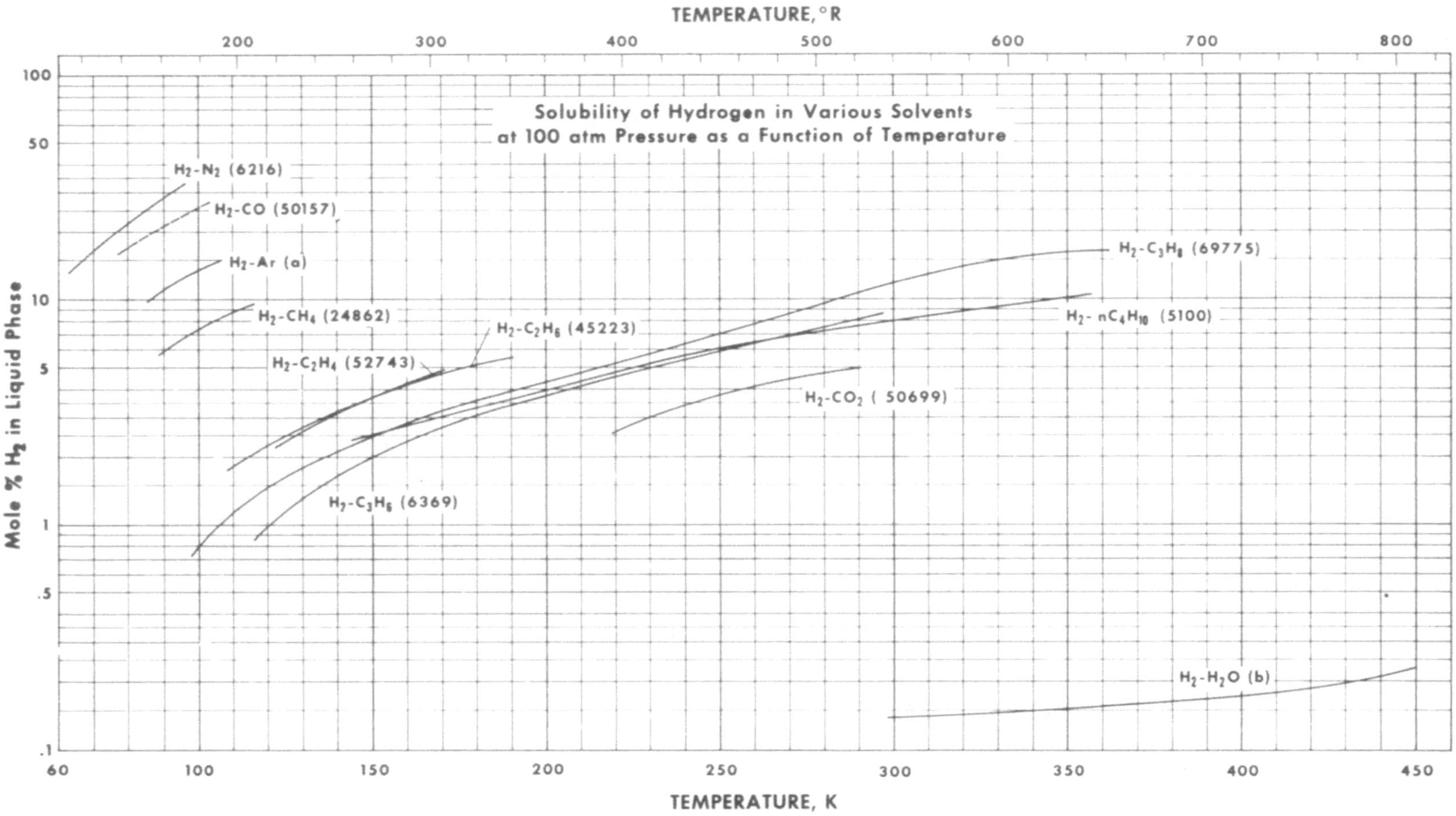

FIGURE 3.25. Solubility of hydrogen in various solvents at 100 atm.

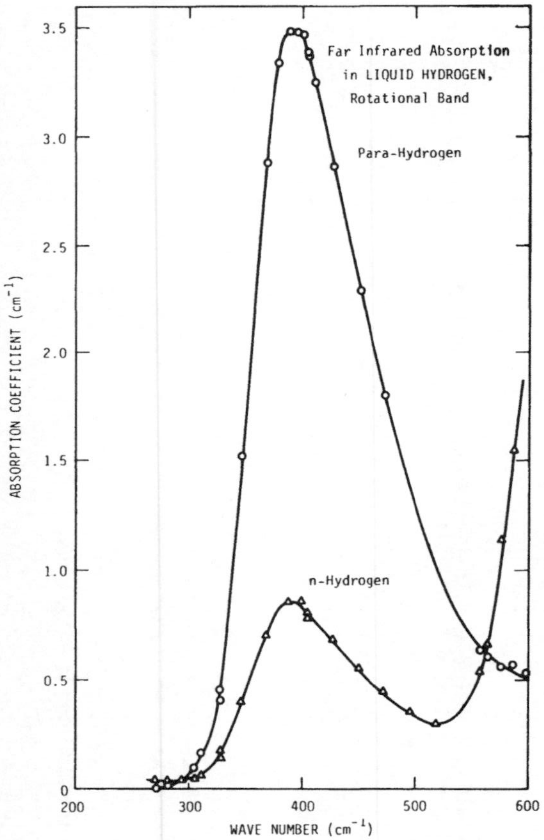

FIGURE 3.26. Far infrared absorption in liquid hydrogen (rotational band).

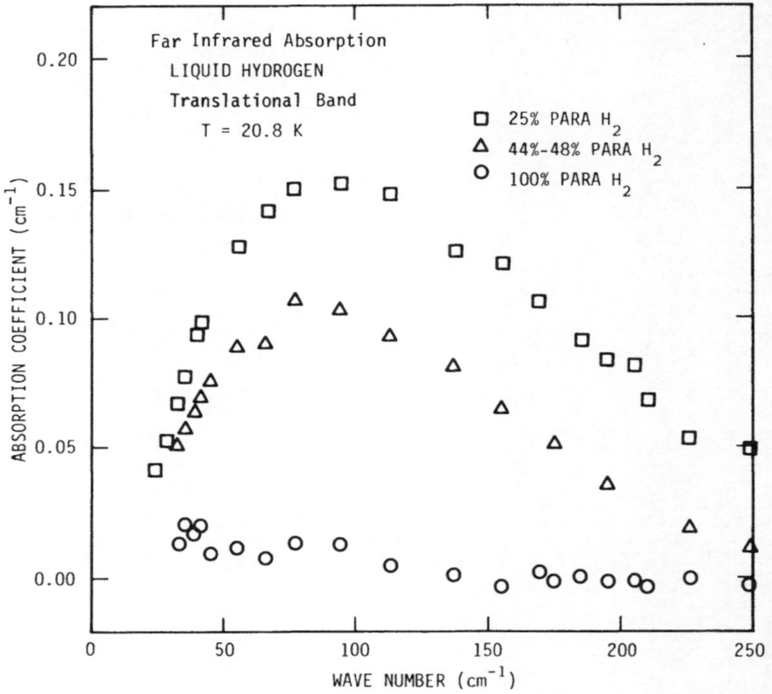

FIGURE 3.27. Far infrared absorption band in liquid hydrogen (translational band).

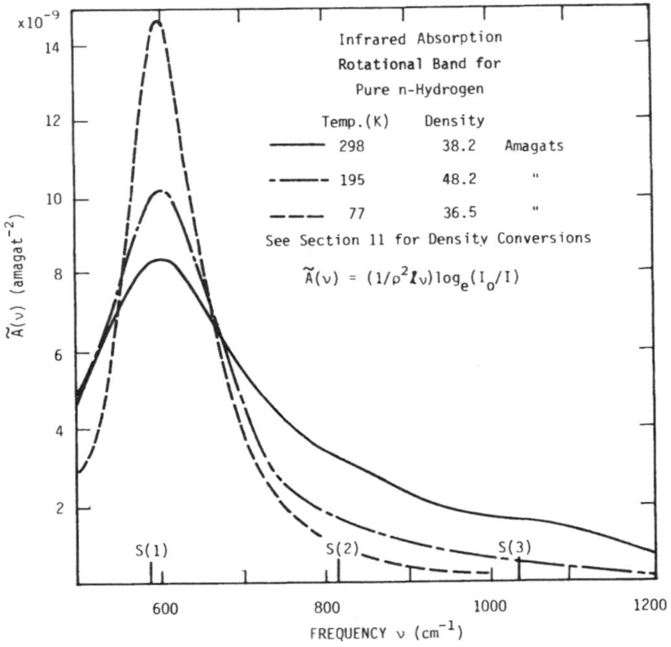

FIGURE 3.28. Infrared absorption (rotational band).

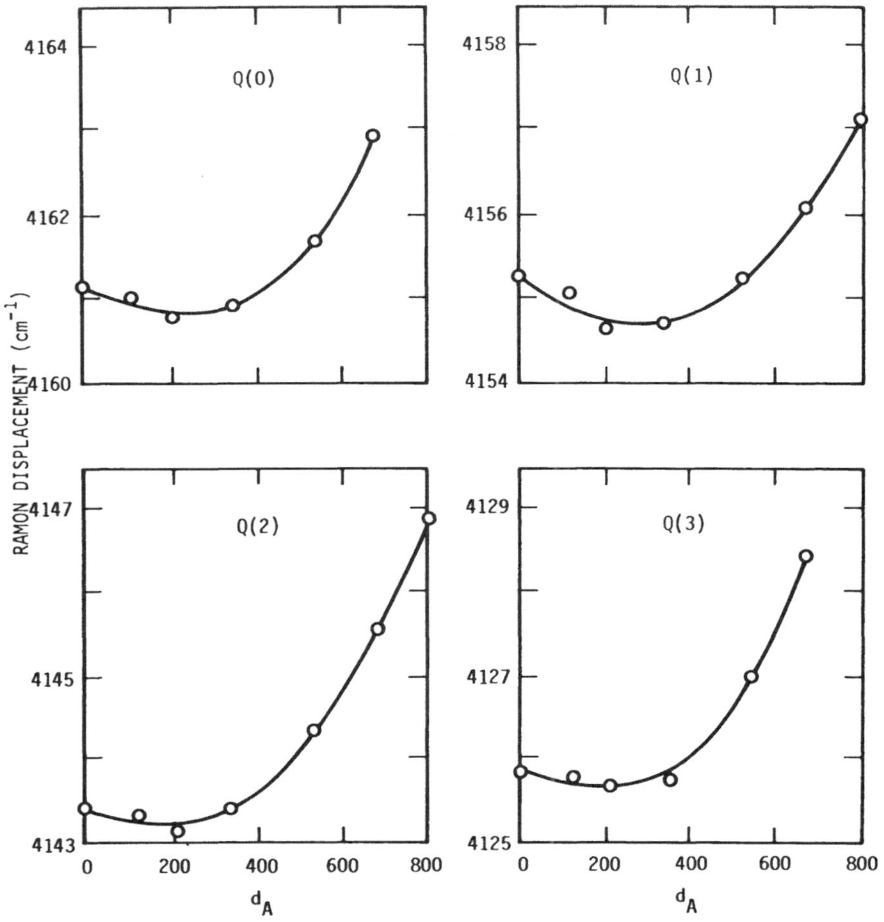

FIGURE 3.29. Ramon displacement, d_A = density Amagats. See Chapter 5 for density conversions.

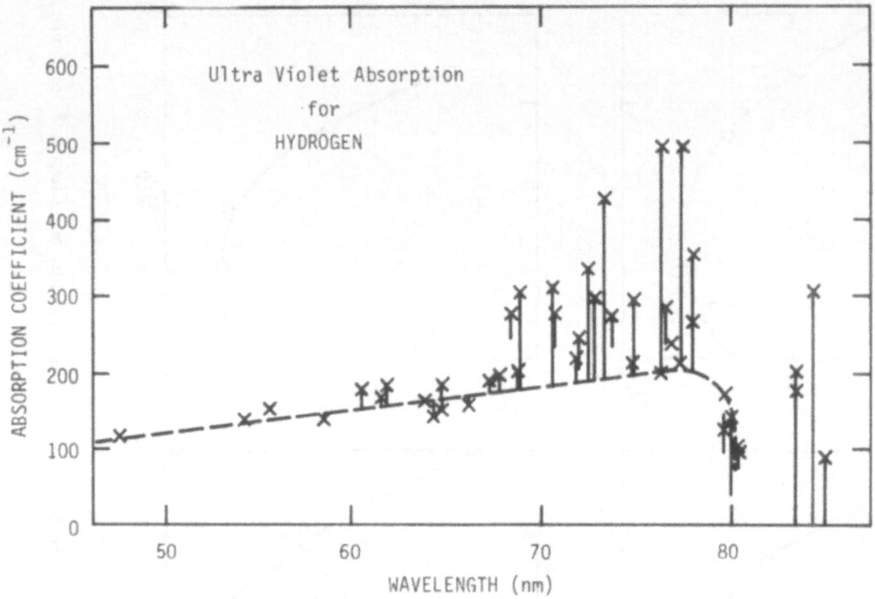

FIGURE 3.30. Ultraviolet absorption.

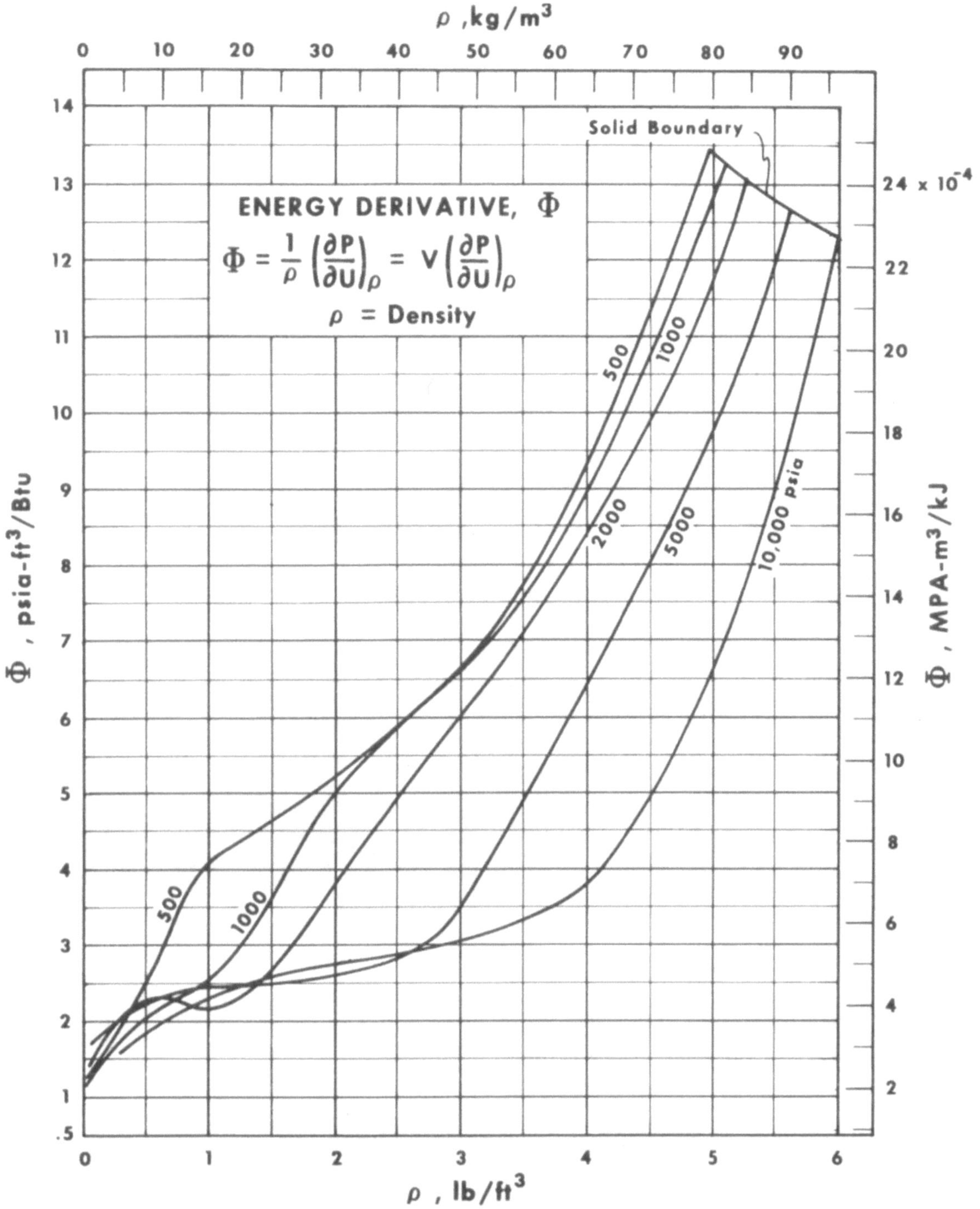

FIGURE 3.31. Energy derivative function.

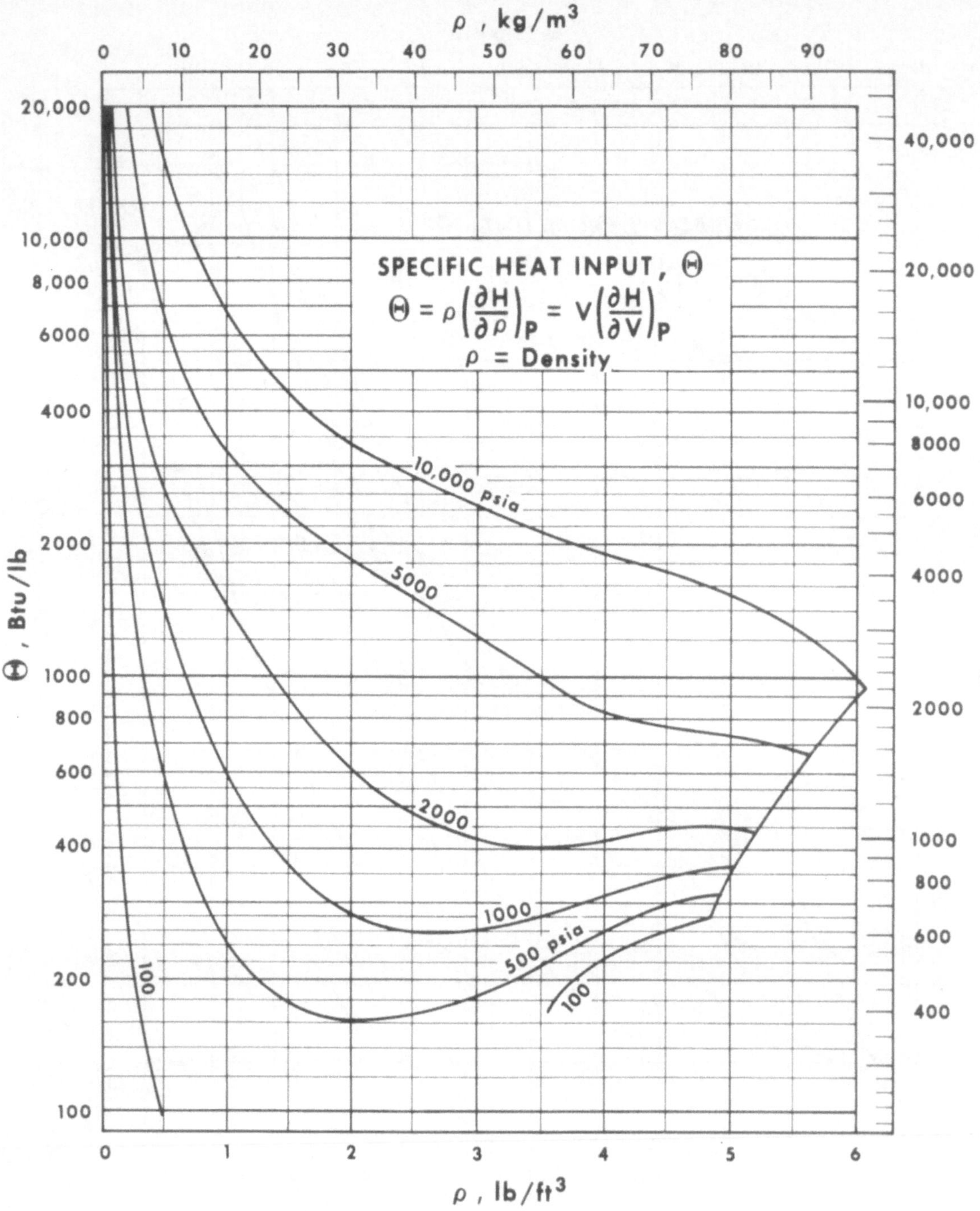

FIGURE 3.32. Specific heat input function.

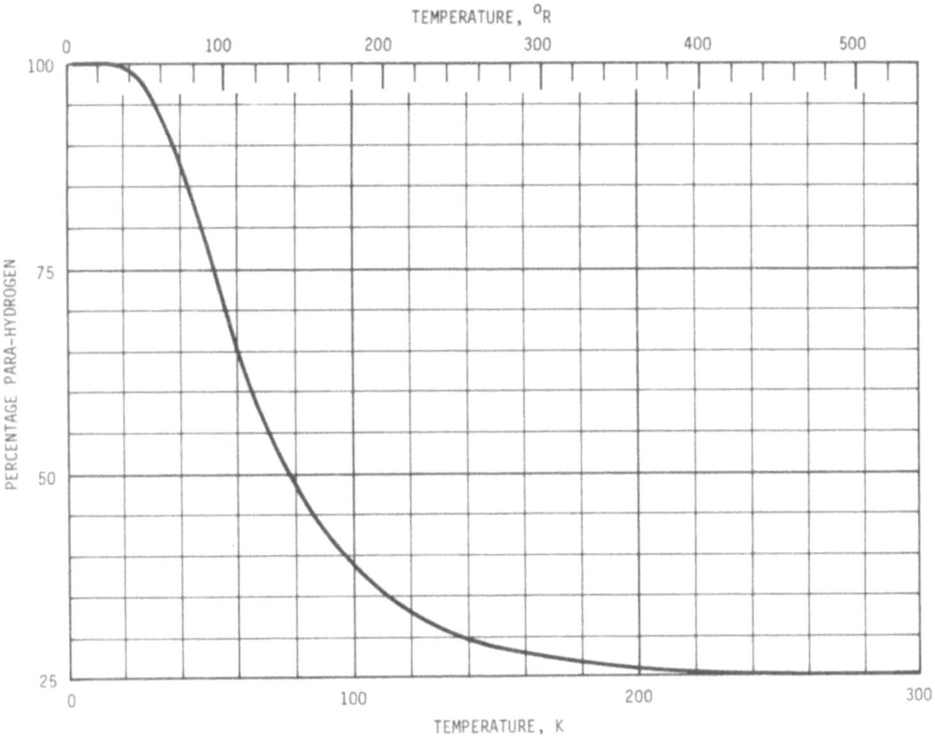

FIGURE 3.33. Equilibrium percentage of parahydrogen.

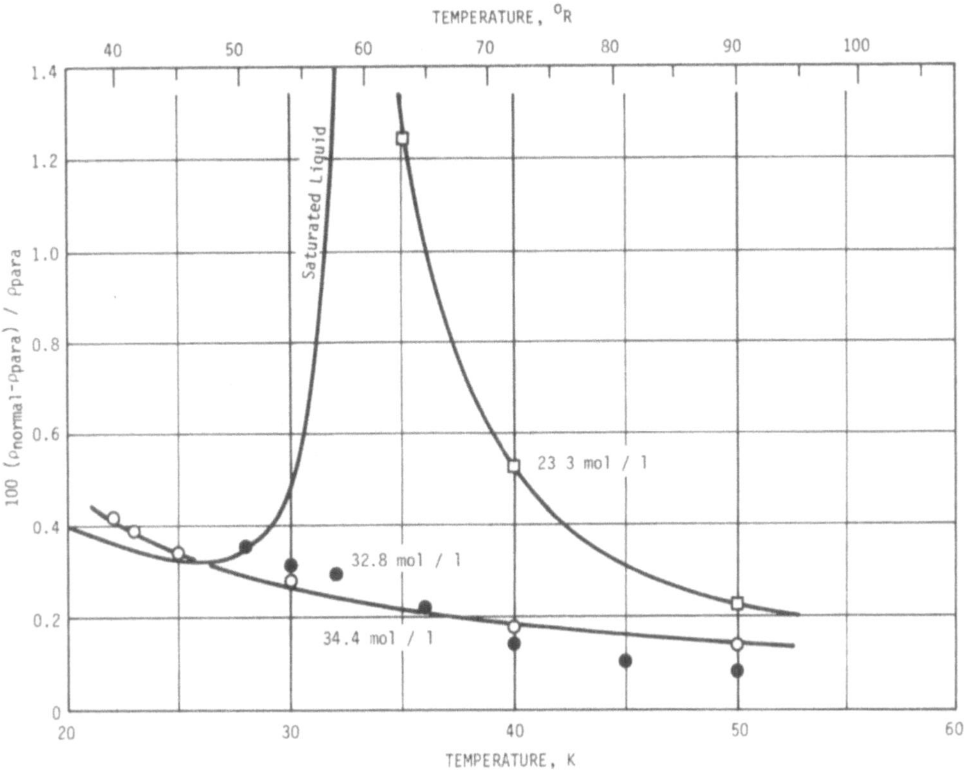

FIGURE 3.34. Density differences between normal and parahydrogen.

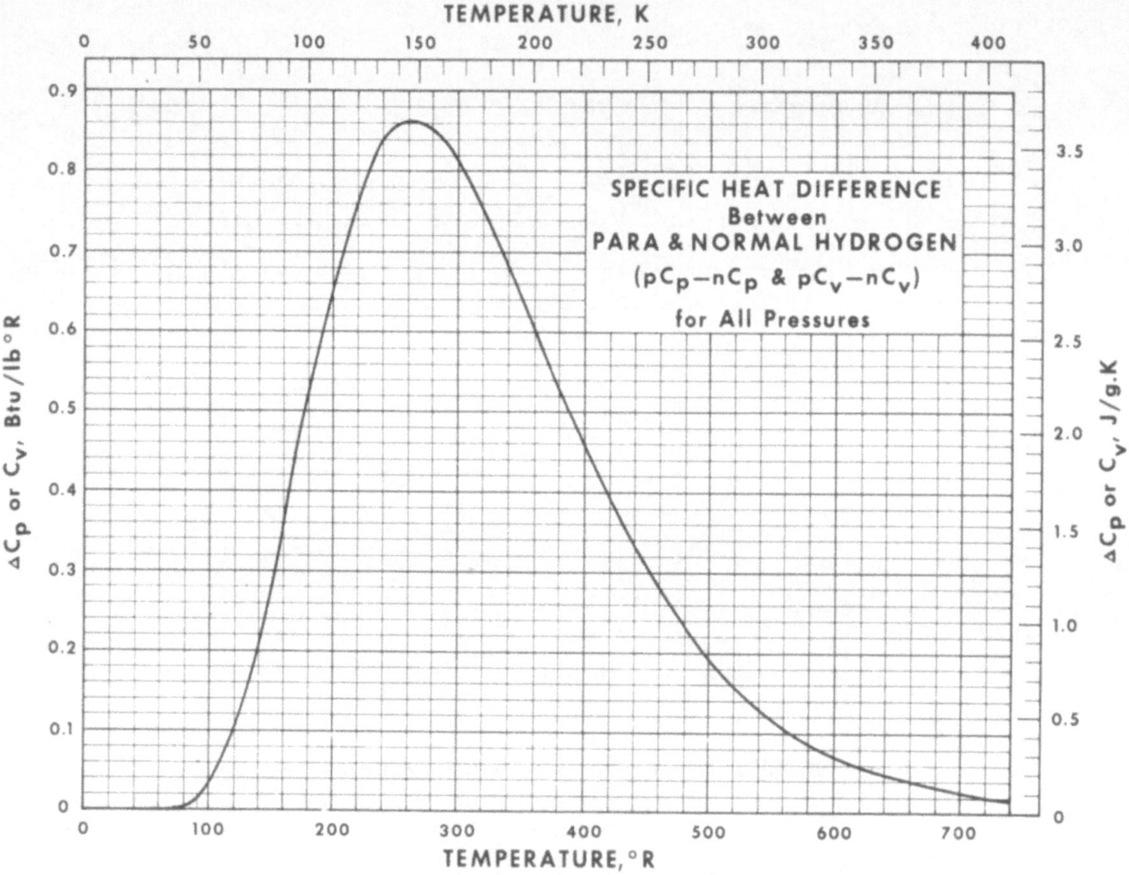

FIGURE 3.35. Specific heat difference between para and normal hydrogen.

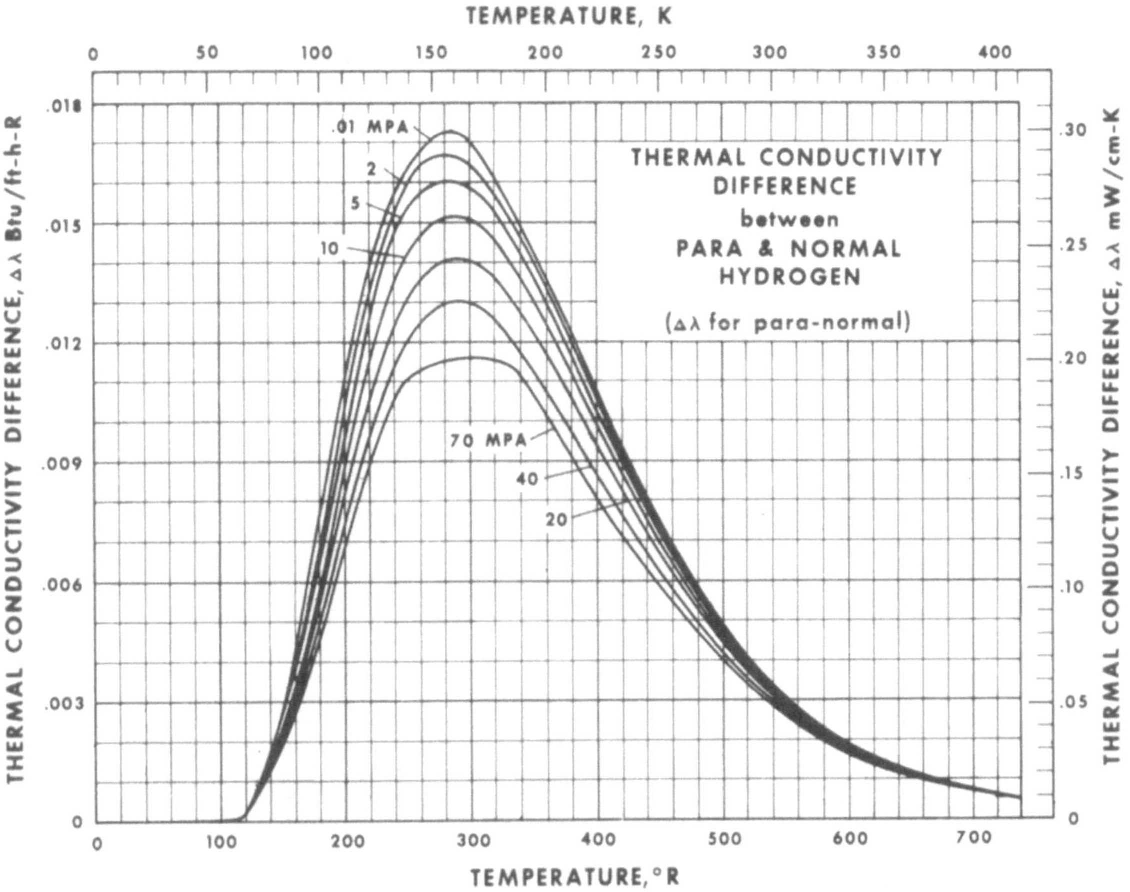

FIGURE 3.36. Thermal conductivity difference between para and normal hydrogen.

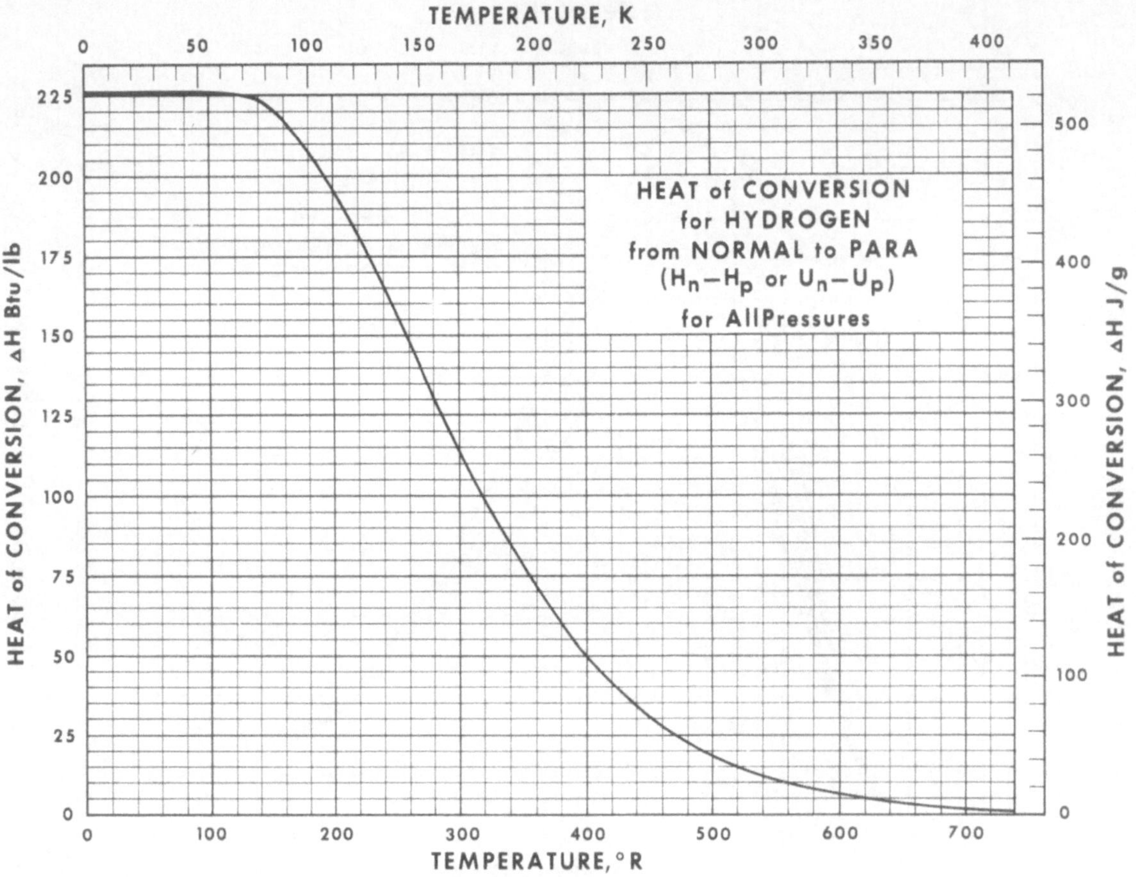

FIGURE 3.37. Heat of conversion for hydrogen from normal to para.

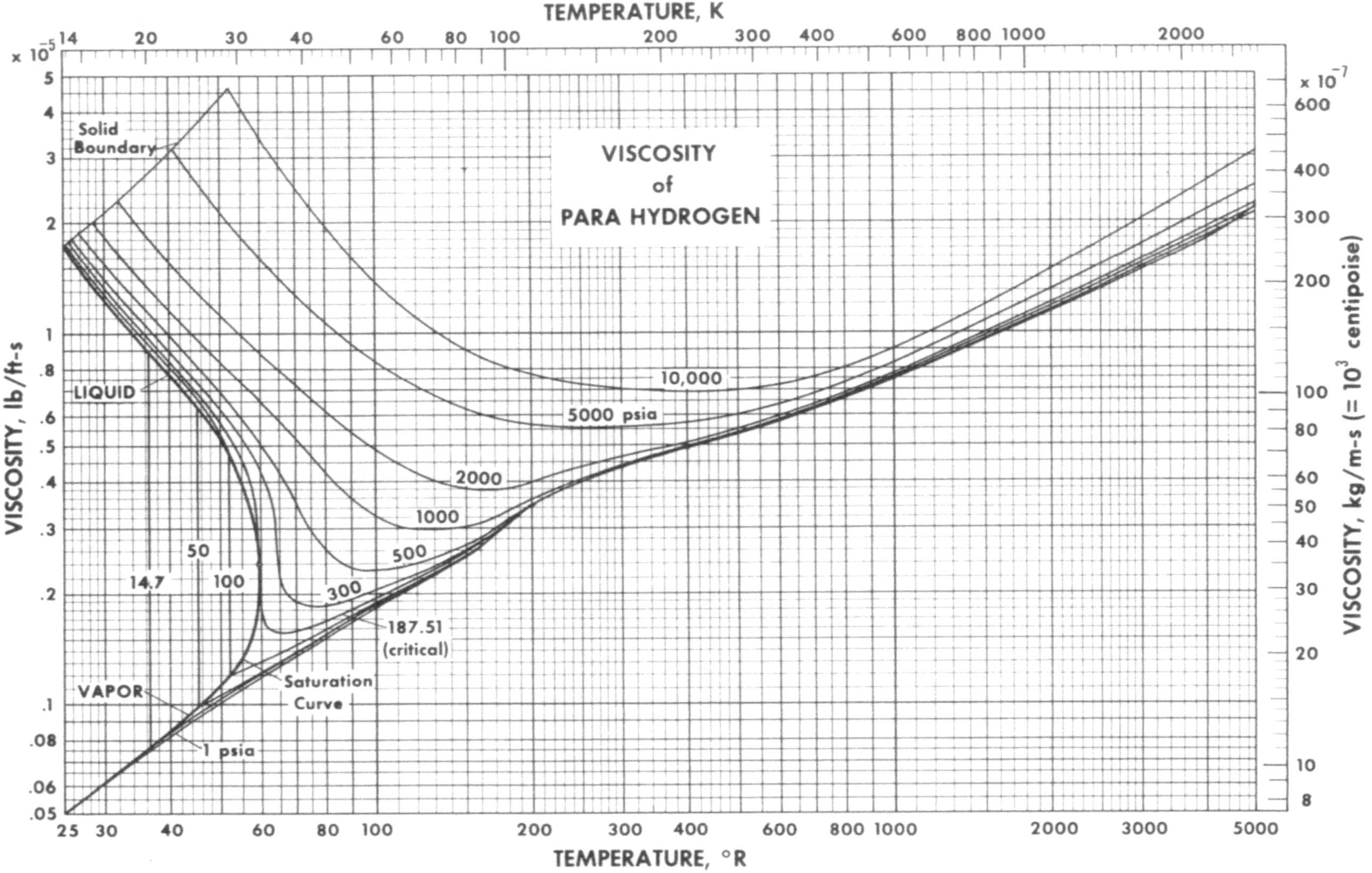

FIGURE 3.38. Viscosity of parahydrogen.

Chapter 4

Data Tables

Table 4.1
THERMODYNAMIC PROPERTIES OF PARAHYDROGEN (ISOBARS, SI UNITS)

0.01 MPA ISOBAR

TEMPERATURE DEG. K	VOLUME CU M/KG	ISOTHERM DERIVATIVE CU M-MPA/KG	ISOCHORE DERIVATIVE MPA/K	INTERNAL ENERGY KJ/KG-K	ENTHALPY KJ/KG	ENTROPY KJ/KG-K	CV KJ / KG-K	CP KJ / KG-K	VELOCITY OF SOUND M/S
* 13.804	0.01298	1.1628	0.9233	-308.9	-308.8	4.966	4.67	6.38	1260
14	0.01301	1.1302	0.9233	-307.6	-307.5	5.059	4.72	6.50	1248
* 14.430	0.01307	1.1173	0.9196	-304.8	-304.7	5.259	4.80	6.67	1245
* 14.430	5.83386	0.0571	0.0007	87.7	146.0	36.498	6.22	10.59	312
15	6.07554	0.0596	0.0007	91.3	152.1	36.907	6.22	10.56	318
16	6.49796	0.0640	0.0006	97.6	162.6	37.587	6.21	10.52	329
17	6.91887	0.0683	0.0006	103.9	173.1	38.224	6.20	10.49	340
18	7.33861	0.0725	0.0006	110.2	183.6	38.822	6.20	10.46	350
19	7.75743	0.0768	0.0005	116.4	194.0	39.388	6.20	10.44	360
20	8.17550	0.0810	0.0005	122.7	204.5	39.923	6.20	10.43	369
21	8.59295	0.0852	0.0005	128.9	214.9	40.431	6.20	10.42	379
22	9.00988	0.0895	0.0005	135.2	225.3	40.916	6.20	10.41	388
23	9.42636	0.0937	0.0004	141.4	235.7	41.378	6.20	10.40	396
24	9.84245	0.0979	0.0004	147.7	246.1	41.820	6.19	10.39	405
25	10.25822	0.1021	0.0004	153.9	256.5	42.244	6.19	10.38	414
26	10.67369	0.1062	0.0004	160.1	266.8	42.651	6.19	10.38	422
27	11.08891	0.1104	0.0004	166.3	277.2	43.043	6.19	10.37	430
28	11.50391	0.1146	0.0004	172.6	287.6	43.420	6.19	10.37	438
29	11.91870	0.1188	0.0003	178.8	298.0	43.784	6.19	10.36	446
30	12.33332	0.1229	0.0003	185.0	308.3	44.135	6.19	10.36	453
31	12.74778	0.1271	0.0003	191.2	318.7	44.475	6.19	10.36	461
32	13.16210	0.1313	0.0003	197.4	329.0	44.803	6.19	10.35	468
33	13.57629	0.1354	0.0003	203.6	339.4	45.122	6.19	10.35	476
34	13.99036	0.1396	0.0003	209.8	349.7	45.431	6.19	10.35	483
35	14.40432	0.1437	0.0003	216.0	360.1	45.731	6.19	10.35	490
36	14.81819	0.1479	0.0003	222.2	370.4	46.022	6.19	10.34	497
37	15.23198	0.1520	0.0003	228.4	380.8	46.306	6.19	10.34	504
38	15.64568	0.1562	0.0003	234.7	391.1	46.581	6.20	10.34	511
39	16.05931	0.1603	0.0003	240.9	401.5	46.850	6.20	10.34	517
40	16.47288	0.1645	0.0003	247.1	411.8	47.112	6.20	10.35	524
42	17.29983	0.1728	0.0002	259.5	432.5	47.617	6.21	10.35	537
44	18.12657	0.1811	0.0002	271.9	453.2	48.099	6.21	10.36	549
46	18.95313	0.1893	0.0002	284.4	473.9	48.559	6.23	10.37	561
48	19.77953	0.1976	0.0002	296.9	494.7	49.001	6.25	10.38	573
50	20.60580	0.2059	0.0002	309.4	515.5	49.425	6.27	10.41	585
52	21.43196	0.2142	0.0002	322.0	536.3	49.834	6.30	10.43	596
54	22.25800	0.2224	0.0002	334.6	557.2	50.228	6.33	10.47	606
56	23.08396	0.2307	0.0002	347.4	578.2	50.610	6.38	10.51	617
58	23.90983	0.2390	0.0002	360.2	599.3	50.979	6.43	10.56	627
60	24.73564	0.2472	0.0002	373.1	620.5	51.339	6.49	10.62	636
65	26.79987	0.2679	0.0002	406.0	674.0	52.196	6.68	10.82	658
70	28.86380	0.2886	0.0001	440.1	728.7	53.007	6.93	11.07	679
75	30.92747	0.3092	0.0001	475.5	784.8	53.780	7.24	11.37	697
80	32.99095	0.3299	0.0001	512.6	842.5	54.525	7.59	11.72	714
85	35.05427	0.3505	0.0001	551.5	902.1	55.247	7.99	12.11	729
90	37.11746	0.3711	0.0001	592.5	963.7	55.952	8.41	12.53	744
95	39.18054	0.3918	0.0001	635.6	1027.4	56.641	8.84	12.97	758
100	41.24353	0.4124	0.0001	680.9	1093.4	57.317	9.27	13.40	772
120	49.49480	0.4950	0.0001	882.6	1377.5	59.901	10.83	14.95	827
140	57.74534	0.5775	0.0001	1110.1	1687.5	62.288	11.82	15.94	883
160	65.99544	0.6600	0.0001	1351.2	2011.2	64.449	12.21	16.34	940
180	74.24525	0.7425	0.0001	1595.7	2338.2	66.375	12.19	16.31	997
200	82.49487	0.8250	0.0001	1837.2	2662.2	68.081	11.94	16.07	1053
220	90.74435	0.9075	0.0000	2073.0	2980.5	69.598	11.63	15.76	1109
240	98.99373	0.9900	0.0000	2302.5	3292.5	70.955	11.33	15.45	1162
260	107.24303	1.0725	0.0000	2526.5	3598.9	72.182	11.07	15.20	1213
280	115.49227	1.1550	0.0000	2745.8	3900.7	73.301	10.87	15.00	1262
300	123.73215	1.2374	0.0000	2961.7	4199.0	74.330	10.72	14.84	1309
350	144.35340	1.4436	0.0000	3491.6	4935.1	76.598	10.51	14.63	1418
400	164.97452	1.6498	0.0000	4014.6	5664.3	78.547	10.42	14.55	1517
450	185.59556	1.8560	0.0000	4534.8	6390.8	80.260	10.40	14.53	1610
500	206.21654	2.0622	0.0000	5055.1	7117.3	81.789	10.40	14.52	1697
550	226.83749	2.2685	0.0000	5575.1	7843.5	83.174	10.41	14.53	1780
600	247.45841	2.4747	0.0000	6096.1	8570.6	84.437	10.42	14.55	1858
700	288.70020	2.8871	0.0000	7141.0	10028.0	86.685	10.48	14.61	2006
800	329.94195	3.2995	0.0000	8193.4	11492.9	88.640	10.57	14.70	2142
900	371.18367	3.7119	0.0000	9256.8	12968.6	90.376	10.70	14.82	2268
1000	412.42538	4.1243	0.0000	10334.0	14458.2	91.949	10.86	14.98	2386
1200	494.90891	4.9492	0.0000	12544.2	17493.3	94.702	11.24	15.37	2601
1400	577.39658	5.7740	0.0000	14838.3	20612.3	97.110	11.70	15.83	2795
1600	659.93251	6.5988	0.0000	17233.3	23832.7	99.237	12.28	16.42	2970
1800	742.78144	7.4237	0.0000	19799.6	27227.4	101.197	13.46	17.66	3121
2000	826.97486	8.2485	0.0000	22778.4	31048.1	103.273	16.49	20.91	3235
2500	1072.23177	10.3106	0.0000	37496.2	48218.6	113.139	49.10	57.60	3476
3000	1538.60332	12.3727	0.0000	88375.8	103761.9	136.424	145.32	179.51	3904

* TWO-PHASE BOUNDARY

Table 4.1 (continued)
THERMODYNAMIC PROPERTIES OF PARAHYDROGEN (ISOBARS, SI UNITS)

0.01 MPA ISOBAR

TEMPERATURE DEG. K	DENSITY KG/CU M	V(DH/DV)$_P$ KJ/KG	V(DP/DU)$_V$ MPA-CU M/KJ X 10^2	-V(DP/DV)$_T$ MPA	-(DV/DT)$_P$/V 1/DEG. K	THERMAL CONDUCTIVITY W/K-M X 10^3	VISCOSITY KG/M-S X 10^7	THERMAL DIFFUSIVITY SQ M/HR	DIELECTRIC CONSTANT	PRANDTL NUMBER
* 13.804	77.0167	618.8	0.25650	89.5531	0.01331	72.63	254.83	0.00053	1.25158	2.2385
14	76.8435	611.8	0.25478	86.8501	0.01363	74.61	247.76	0.00054	1.25097	2.1598
* 14.430	76.4900	620.0	0.25021	85.4586	0.01076	78.59	234.09	0.00055	1.24972	1.9873
* 14.430	0.1714	145.5	0.06684	0.0098	0.07279	12.77	7.74	0.02532	1.00052	0.6420
15	0.1646	151.5	0.06684	0.0098	0.06968	13.01	8.07	0.02696	1.00050	0.6546
16	0.1539	162.1	0.06684	0.0098	0.06488	13.45	8.64	0.02992	1.00046	0.6756
17	0.1445	172.6	0.06683	0.0099	0.06074	13.89	9.21	0.03300	1.00044	0.6949
18	0.1363	183.1	0.06681	0.0099	0.05713	14.40	9.77	0.03635	1.00041	0.7099
19	0.1289	193.6	0.06680	0.0099	0.05394	14.90	10.32	0.03985	1.00039	0.7235
20	0.1223	204.1	0.06679	0.0099	0.05109	15.53	10.87	0.04382	1.00037	0.7303
21	0.1164	214.6	0.06678	0.0099	0.04855	16.32	11.41	0.04846	1.00035	0.7286
22	0.1110	225.0	0.06677	0.0099	0.04625	17.11	11.95	0.05334	1.00033	0.7267
23	0.1061	235.4	0.06676	0.0099	0.04416	17.91	12.48	0.05844	1.00032	0.7247
24	0.1016	245.9	0.06675	0.0099	0.04226	18.70	13.01	0.06378	1.00031	0.7225
25	0.0975	256.3	0.06675	0.0099	0.04051	19.49	13.52	0.06934	1.00029	0.7202
26	0.0937	266.7	0.06674	0.0100	0.03891	20.28	14.04	0.07510	1.00028	0.7181
27	0.0902	277.1	0.06674	0.0100	0.03743	21.07	14.54	0.08109	1.00027	0.7159
28	0.0869	287.4	0.06673	0.0100	0.03606	21.85	15.04	0.08730	1.00026	0.7136
29	0.0839	297.8	0.06673	0.0100	0.03479	22.64	15.54	0.09374	1.00025	0.7112
30	0.0811	308.2	0.06672	0.0100	0.03361	23.42	16.03	0.10040	1.00024	0.7088
31	0.0784	318.6	0.06672	0.0100	0.03251	24.12	16.51	0.10690	1.00024	0.7089
32	0.0760	328.9	0.06671	0.0100	0.03147	24.82	16.99	0.11359	1.00023	0.7088
33	0.0737	339.3	0.06671	0.0100	0.03050	25.51	17.47	0.12049	1.00022	0.7085
34	0.0715	349.7	0.06670	0.0100	0.02959	26.22	17.93	0.12761	1.00022	0.7078
35	0.0694	360.1	0.06669	0.0100	0.02873	26.92	18.40	0.13494	1.00021	0.7070
36	0.0675	370.4	0.06668	0.0100	0.02793	27.63	18.86	0.14247	1.00020	0.7061
37	0.0657	380.8	0.06667	0.0100	0.02716	28.33	19.31	0.15020	1.00020	0.7051
38	0.0639	391.2	0.06665	0.0100	0.02644	29.04	19.76	0.15812	1.00019	0.7041
39	0.0623	401.7	0.06663	0.0100	0.02575	29.74	20.21	0.16624	1.00019	0.7029
40	0.0607	412.1	0.06660	0.0100	0.02510	30.45	20.65	0.17454	1.00018	0.7017
42	0.0578	433.1	0.06653	0.0100	0.02390	31.85	21.53	0.19166	1.00017	0.6994
44	0.0552	454.2	0.06643	0.0100	0.02280	33.25	22.38	0.20950	1.00017	0.6971
46	0.0528	475.5	0.06628	0.0100	0.02180	34.65	23.22	0.22802	1.00016	0.6949
48	0.0506	497.1	0.06609	0.0100	0.02089	36.05	24.05	0.24720	1.00015	0.6927
50	0.0485	519.1	0.06585	0.0100	0.02005	37.45	24.86	0.26696	1.00015	0.6908
52	0.0467	541.4	0.06554	0.0100	0.01927	38.85	25.66	0.28725	1.00014	0.6893
54	0.0449	564.2	0.06516	0.0100	0.01855	40.25	26.45	0.30803	1.00014	0.6880
56	0.0433	587.7	0.06471	0.0100	0.01789	41.65	27.22	0.32922	1.00013	0.6872
58	0.0418	611.7	0.06418	0.0100	0.01727	43.05	27.99	0.35074	1.00013	0.6869
60	0.0404	636.5	0.06357	0.0100	0.01669	44.51	28.74	0.37306	1.00012	0.6861
65	0.0373	702.2	0.06174	0.0100	0.01540	48.38	30.58	0.43162	1.00011	0.6836
70	0.0346	773.8	0.05950	0.0100	0.01430	52.26	32.36	0.49076	1.00010	0.6852
75	0.0323	852.1	0.05699	0.0100	0.01334	56.13	34.09	0.54968	1.00010	0.6905
80	0.0303	937.2	0.05433	0.0100	0.01251	60.04	35.77	0.60829	1.00009	0.6985
85	0.0285	1029.2	0.05166	0.0100	0.01177	65.00	37.41	0.67708	1.00009	0.6973
90	0.0269	1127.4	0.04909	0.0100	0.01112	69.96	39.01	0.74586	1.00008	0.6988
95	0.0255	1231.2	0.04668	0.0100	0.01053	74.91	40.57	0.81499	1.00008	0.7021
100	0.0242	1339.5	0.04450	0.0100	0.01000	89.50	45.74	0.99173	1.00007	0.6847
120	0.0202	1793.9	0.03811	0.0100	0.00833	117.12	54.09	1.39567	1.00006	0.6905
140	0.0173	2231.9	0.03491	0.0100	0.00714	136.14	59.20	1.77511	1.00005	0.6933
160	0.0152	2613.6	0.03378	0.0100	0.00625	148.29	63.00	2.15677	1.00005	0.6940
180	0.0135	2935.8	0.03385	0.0100	0.00556	155.64	66.17	2.55059	1.00004	0.6934
200	0.0121	3213.9	0.03453	0.0100	0.00500	160.18	69.01	2.96043	1.00004	0.6923
220	0.0110	3466.4	0.03546	0.0100	0.00455	163.39	71.65	3.38766	1.00003	0.6910
240	0.0101	3708.8	0.03641	0.0100	0.00417	166.20	74.18	3.83294	1.00003	0.6897
260	0.0093	3951.5	0.03725	0.0100	0.00385	169.11	76.63	4.29619	1.00003	0.6886
280	0.0087	4199.1	0.03794	0.0100	0.00357	172.33	79.03	4.77794	1.00003	0.6877
300	0.0081	4453.3	0.03848	0.0100	0.00333	175.87	81.39	5.27759	1.00002	0.6870
350	0.0069	5121.4	0.03925	0.0100	0.00286	186.08	87.21	6.60883	1.00002	0.6858
400	0.0061	5819.8	0.03956	0.0100	0.00250	197.39	92.95	8.05766	1.00002	0.6851
450	0.0054	6538.4	0.03964	0.0100	0.00222	209.27	98.62	9.62376	1.00002	0.6847
500	0.0048	7262.3	0.03966	0.0100	0.00200	221.21	104.23	11.30685	1.00001	0.6844
550	0.0044	7993.3	0.03962	0.0100	0.00182	233.22	109.79	13.10472	1.00001	0.6841
600	0.0040	8729.7	0.03956	0.0100	0.00167	245.22	115.28	15.01509	1.00001	0.6839
700	0.0035	10223.8	0.03935	0.0100	0.00143	269.30	126.07	19.16374	1.00001	0.6837
800	0.0030	11757.5	0.03901	0.0100	0.00125	293.66	136.61	23.73407	1.00001	0.6837
900	0.0027	13341.4	0.03855	0.0100	0.00111	318.47	146.90	28.70857	1.00001	0.6838
1000	0.0024	14983.1	0.03798	0.0100	0.00100	440.56	156.95	43.65775	1.00001	0.5338
1200	0.0020	18442.2	0.03668	0.0100	0.00083	511.19	176.42	59.26340	1.00001	0.5304
1400	0.0017	22156.4	0.03525	0.0100	0.00071	584.62	195.11	76.78658	1.00001	0.5282
1600	0.0015	26269.7	0.03358	0.0100	0.00063	668.03	213.14	96.65688	1.00000	0.5239
1800	0.0013	31767.5	0.03065	0.0100	0.00056	800.59	230.62	121.23303	1.00000	0.5087
2000	0.0012	41718.6	0.02508	0.0100	0.00050	1099.16	247.74	156.47369	1.00000	0.4713
2500	0.0009	138471.2	0.00874	0.0096	0.00042	4536.70	292.93	304.02305	1.00000	0.3719
3000	0.0006	433048.8	0.00353	0.0080	0.00041	15311.58	346.27	472.46613	1.00000	0.4059

* TWO-PHASE BOUNDARY

Table 4.1 (continued)
THERMODYNAMIC PROPERTIES OF PARAHYDROGEN (ISOBARS, SI UNITS)

0.02 MPA ISOBAR

TEMPERATURE DEG. K	VOLUME CU M/KG	ISOTHERM DERIVATIVE CU M-MPA/KG	ISOCHORE DERIVATIVE MPA/K	INTERNAL ENERGY KJ/KG-K	ENTHALPY KJ/KG	ENTROPY KJ/KG-K	CV KJ / KG-K	CP KJ / KG-K	VELOCITY OF SOUND M/S
* 13.807	0.01298	1.1634	0.9234	-308.9	-308.7	4.966	4.67	6.38	1260
14	0.01301	1.1318	0.9234	-307.6	-307.4	5.058	4.72	6.50	1249
15	0.01315	1.0325	0.9076	-300.9	-300.7	5.520	4.92	6.99	1211
* 15.832	0.01329	1.0154	0.8919	-294.9	-294.7	5.911	5.09	7.28	1205
* 15.832	3.15931	0.0610	0.0013	95.0	158.2	34.518	6.27	10.81	325
16	3.19578	0.0618	0.0013	96.1	160.0	34.632	6.26	10.80	326
17	3.41127	0.0663	0.0012	102.5	170.8	35.284	6.24	10.71	337
18	3.62529	0.0708	0.0011	108.9	181.4	35.894	6.23	10.64	348
19	3.83822	0.0752	0.0011	115.3	192.1	36.468	6.22	10.59	358
20	4.05030	0.0795	0.0010	121.6	202.6	37.010	6.21	10.56	368
21	4.26168	0.0839	0.0010	128.0	213.2	37.525	6.21	10.53	377
22	4.47250	0.0882	0.0009	134.3	223.7	38.014	6.21	10.51	386
23	4.68285	0.0925	0.0009	140.5	234.2	38.481	6.21	10.49	395
24	4.89279	0.0967	0.0008	146.8	244.7	38.927	6.20	10.47	404
25	5.10238	0.1010	0.0008	153.1	255.1	39.354	6.20	10.46	413
26	5.31167	0.1052	0.0008	159.4	265.6	39.764	6.20	10.44	421
27	5.52070	0.1095	0.0008	165.6	276.0	40.157	6.20	10.43	429
28	5.72950	0.1137	0.0007	171.9	286.5	40.537	6.20	10.42	437
29	5.93809	0.1179	0.0007	178.1	296.9	40.902	6.20	10.41	445
30	6.14651	0.1221	0.0007	184.4	307.3	41.255	6.20	10.41	453
31	6.35476	0.1263	0.0007	190.6	317.7	41.596	6.20	10.40	460
32	6.56286	0.1305	0.0006	196.8	328.1	41.926	6.20	10.39	468
33	6.77084	0.1347	0.0006	203.1	338.5	42.246	6.20	10.39	475
34	6.97869	0.1389	0.0006	209.3	348.9	42.556	6.20	10.38	482
35	7.18644	0.1431	0.0006	215.5	359.2	42.857	6.20	10.38	490
36	7.39410	0.1473	0.0006	221.7	369.6	43.150	6.20	10.38	497
37	7.60166	0.1515	0.0005	228.0	380.0	43.434	6.20	10.37	504
38	7.80915	0.1556	0.0005	234.2	390.4	43.710	6.20	10.37	510
39	8.01656	0.1598	0.0005	240.4	400.7	43.980	6.20	10.37	517
40	8.22390	0.1640	0.0005	246.6	411.1	44.242	6.20	10.37	524
42	8.63841	0.1723	0.0005	259.1	431.9	44.748	6.21	10.37	537
44	9.05271	0.1806	0.0005	271.6	452.6	45.231	6.22	10.38	549
46	9.46683	0.1889	0.0004	284.0	473.4	45.692	6.23	10.39	561
48	9.88078	0.1973	0.0004	296.5	494.2	46.135	6.25	10.40	573
50	10.29461	0.2056	0.0004	309.1	515.0	46.560	6.27	10.42	584
52	10.70831	0.2139	0.0004	321.7	535.8	46.969	6.30	10.45	596
54	11.12192	0.2222	0.0004	334.3	556.8	47.364	6.34	10.48	606
56	11.53542	0.2304	0.0004	347.1	577.8	47.746	6.38	10.52	617
58	11.94885	0.2387	0.0003	359.9	598.9	48.116	6.43	10.57	627
60	12.36221	0.2470	0.0003	372.8	620.1	48.475	6.49	10.63	636
65	13.39533	0.2677	0.0003	405.8	673.7	49.334	6.68	10.82	658
70	14.42813	0.2884	0.0003	439.9	728.4	50.145	6.94	11.07	679
75	15.46069	0.3091	0.0003	475.3	784.5	50.919	7.24	11.38	697
80	16.49305	0.3298	0.0003	512.4	842.3	51.664	7.59	11.73	714
85	17.52525	0.3504	0.0002	551.4	901.9	52.386	7.99	12.12	729
90	18.55732	0.3711	0.0002	592.4	963.5	53.091	8.41	12.54	744
95	19.58928	0.3917	0.0002	635.5	1027.3	53.780	8.84	12.97	758
100	20.62115	0.4124	0.0002	680.8	1093.2	54.457	9.27	13.40	772
120	24.74793	0.4950	0.0002	882.4	1377.4	57.041	10.83	14.95	827
140	28.87400	0.5775	0.0001	1110.0	1687.4	59.428	11.82	15.94	883
160	32.99962	0.6601	0.0001	1351.2	2011.1	61.589	12.21	16.34	940
180	37.12496	0.7426	0.0001	1595.7	2338.2	63.516	12.19	16.31	997
200	41.25010	0.8251	0.0001	1837.2	2662.2	65.222	11.94	16.07	1054
220	45.37510	0.9076	0.0001	2073.0	2980.5	66.739	11.63	15.76	1109
240	49.50000	0.9901	0.0001	2302.5	3292.5	68.096	11.33	15.45	1162
260	53.62482	1.0726	0.0001	2526.4	3598.9	69.323	11.07	15.20	1213
280	57.74958	1.1551	0.0001	2745.8	3900.8	70.442	10.87	15.00	1262
300	61.86966	1.2375	0.0001	2961.7	4199.0	71.471	10.72	14.84	1309
350	72.18050	1.4438	0.0001	3491.6	4935.2	73.740	10.51	14.63	1418
400	82.49121	1.6500	0.0001	4014.6	5664.4	75.689	10.42	14.55	1517
450	92.80183	1.8562	0.0000	4534.8	6390.8	77.401	10.40	14.53	1610
500	103.11240	2.0624	0.0000	5055.1	7117.4	78.930	10.40	14.52	1697
550	113.42293	2.2686	0.0000	5575.1	7843.5	80.315	10.41	14.53	1780
600	123.73343	2.4748	0.0000	6096.0	8570.7	81.579	10.42	14.55	1858
700	144.35440	2.8873	0.0000	7141.0	10028.0	83.826	10.48	14.61	2006
800	164.97531	3.2997	0.0000	8193.4	11492.9	85.781	10.57	14.70	2142
900	185.59621	3.7121	0.0000	9256.8	12968.7	87.518	10.70	14.82	2268
1000	206.21708	4.1245	0.0000	10333.9	14458.3	89.090	10.86	14.98	2386
1200	247.45886	4.9494	0.0000	12544.2	17493.4	91.843	11.24	15.37	2601
1400	288.70208	5.7742	0.0000	14837.9	20611.9	94.251	11.69	15.82	2795
1600	329.96236	6.5990	0.0000	17227.8	23827.1	96.374	12.23	16.36	2971
1800	371.33330	7.4239	0.0000	19763.5	27190.1	98.316	13.17	17.34	3127
2000	413.17959	8.2487	0.0000	22614.2	30877.7	100.323	15.41	19.75	3251
2500	530.09558	10.3108	0.0000	34967.0	45568.9	109.124	38.86	45.95	3490
3000	726.79296	12.3729	0.0000	73552.5	88088.4	127.747	113.04	137.57	3875

* TWO-PHASE BOUNDARY

Table 4.1 (continued)
THERMODYNAMIC PROPERTIES OF PARAHYDROGEN (ISOBARS, SI UNITS)

0.02 MPA ISOBAR

TEMPERATURE DEG. K	DENSITY KG/CU M	$V(DH/DV)_P$ KJ/KG	$V(DP/DU)_V$ MPA-CU M/KJ X 10^2	$-V(DP/DV)_T$ MPA	$-(DV/DT)_P/V$ 1/DEG. K	THERMAL CONDUCTIVITY W/K-M X 10^3	VISCOSITY KG/M-S X 10^7	THERMAL DIFFUSIVITY SQ M/HR	DIELECTRIC CONSTANT	PRANDTL NUMBER
* 13.807	77.0224	619.1	0.25649	89.6087	0.01031	72.67	254.89	0.00053	1.25160	2.2378
14	76.8522	612.4	0.25479	86.9799	0.01062	74.62	247.94	0.00054	1.25100	2.1602
15	76.0247	604.6	0.24263	78.4985	0.01156	83.06	218.31	0.00056	1.24809	1.8373
* 15.832	75.2560	623.4	0.23300	76.4147	0.01167	88.03	197.78	0.00058	1.24539	1.6349
* 15.832	0.3165	157.7	0.06680	0.0193	0.06860	13.56	8.58	0.01426	1.00095	0.6841
16	0.3129	159.4	0.06682	0.0193	0.06771	13.63	8.67	0.01452	1.00094	0.6870
17	0.2931	170.1	0.06687	0.0194	0.06293	14.04	9.24	0.01611	1.00088	0.7043
18	0.2758	180.8	0.06689	0.0195	0.05887	14.52	9.80	0.01781	1.00083	0.7178
19	0.2605	191.4	0.06689	0.0196	0.05535	15.02	10.35	0.01958	1.00079	0.7304
20	0.2469	202.0	0.06689	0.0196	0.05227	15.63	10.90	0.02158	1.00074	0.7363
21	0.2346	212.6	0.06688	0.0197	0.04953	16.41	11.44	0.02391	1.00071	0.7339
22	0.2236	223.2	0.06686	0.0197	0.04708	17.20	11.98	0.02636	1.00067	0.7315
23	0.2135	233.7	0.06685	0.0197	0.04487	17.99	12.50	0.02892	1.00064	0.7290
24	0.2044	244.2	0.06684	0.0198	0.04287	18.78	13.03	0.03159	1.00062	0.7264
25	0.1960	254.7	0.06683	0.0198	0.04105	19.57	13.55	0.03438	1.00059	0.7238
26	0.1883	265.2	0.06682	0.0198	0.03938	20.35	14.06	0.03726	1.00057	0.7214
27	0.1811	275.7	0.06681	0.0198	0.03784	21.13	14.56	0.04026	1.00055	0.7189
28	0.1745	286.2	0.06680	0.0198	0.03642	21.91	15.06	0.04337	1.00053	0.7164
29	0.1684	296.6	0.06679	0.0199	0.03511	22.70	15.56	0.04659	1.00051	0.7138
30	0.1627	307.0	0.06679	0.0199	0.03389	23.48	16.05	0.04992	1.00049	0.7112
31	0.1574	317.5	0.06678	0.0199	0.03276	24.18	16.53	0.05318	1.00047	0.7111
32	0.1524	327.9	0.06677	0.0199	0.03170	24.87	17.01	0.05654	1.00046	0.7108
33	0.1477	338.3	0.06676	0.0199	0.03071	25.57	17.48	0.05999	1.00045	0.7104
34	0.1433	348.7	0.06675	0.0199	0.02978	26.27	17.95	0.06356	1.00043	0.7096
35	0.1392	359.2	0.06674	0.0199	0.02890	26.98	18.42	0.06724	1.00042	0.7086
36	0.1352	369.6	0.06673	0.0199	0.02808	27.68	18.88	0.07101	1.00041	0.7076
37	0.1316	380.0	0.06672	0.0199	0.02730	28.38	19.33	0.07487	1.00040	0.7065
38	0.1281	390.4	0.06670	0.0199	0.02656	29.09	19.78	0.07884	1.00039	0.7053
39	0.1247	400.9	0.06668	0.0199	0.02587	29.79	20.23	0.08291	1.00038	0.7041
40	0.1216	411.4	0.06665	0.0199	0.02521	30.50	20.67	0.08706	1.00037	0.7028
42	0.1158	432.4	0.06657	0.0199	0.02398	31.89	21.54	0.09563	1.00035	0.7004
44	0.1105	453.6	0.06647	0.0200	0.02288	33.29	22.40	0.10456	1.00033	0.6980
46	0.1056	474.9	0.06632	0.0200	0.02187	34.69	23.24	0.11384	1.00032	0.6956
48	0.1012	496.6	0.06613	0.0200	0.02094	36.09	24.06	0.12343	1.00030	0.6934
50	0.0971	518.6	0.06588	0.0200	0.02010	37.49	24.87	0.13332	1.00029	0.6915
52	0.0934	540.9	0.06557	0.0200	0.01931	38.88	25.67	0.14347	1.00028	0.6898
54	0.0899	563.8	0.06519	0.0200	0.01859	40.28	26.46	0.15388	1.00027	0.6885
56	0.0867	587.2	0.06474	0.0200	0.01792	41.68	27.24	0.16448	1.00026	0.6877
58	0.0837	611.3	0.06421	0.0200	0.01730	43.08	28.00	0.17524	1.00025	0.6873
60	0.0809	636.1	0.06360	0.0200	0.01672	44.54	28.75	0.18641	1.00024	0.6865
65	0.0747	701.8	0.06176	0.0200	0.01542	48.42	30.59	0.21571	1.00022	0.6839
70	0.0693	773.5	0.05953	0.0200	0.01431	52.29	32.37	0.24530	1.00021	0.6855
75	0.0647	851.8	0.05701	0.0200	0.01336	56.16	34.10	0.27477	1.00019	0.6908
80	0.0606	937.0	0.05435	0.0200	0.01252	60.07	35.78	0.30409	1.00018	0.6986
85	0.0571	1028.9	0.05168	0.0200	0.01178	65.02	37.42	0.33850	1.00017	0.6974
90	0.0539	1127.2	0.04910	0.0200	0.01112	69.98	39.02	0.37291	1.00016	0.6990
95	0.0510	1231.0	0.04670	0.0200	0.01054	74.94	40.58	0.40749	1.00015	0.7022
100	0.0485	1339.3	0.04451	0.0200	0.01001	89.50	45.74	0.49576	1.00015	0.6849
120	0.0404	1793.8	0.03812	0.0200	0.00834	117.11	54.09	0.69775	1.00012	0.6906
140	0.0346	2231.8	0.03491	0.0200	0.00714	136.14	59.20	0.88750	1.00010	0.6934
160	0.0303	2613.6	0.03379	0.0200	0.00625	148.29	63.00	1.07837	1.00009	0.6940
180	0.0269	2935.9	0.03386	0.0200	0.00556	155.64	66.17	1.27531	1.00008	0.6934
200	0.0242	3214.0	0.03454	0.0200	0.00500	160.18	69.01	1.48026	1.00007	0.6923
220	0.0220	3466.5	0.03547	0.0200	0.00455	163.39	71.65	1.69391	1.00007	0.6910
240	0.0202	3709.0	0.03642	0.0200	0.00417	166.20	74.18	1.91657	1.00006	0.6897
260	0.0186	3951.7	0.03726	0.0200	0.00385	169.12	76.63	2.14822	1.00006	0.6886
280	0.0173	4199.3	0.03794	0.0200	0.00357	172.33	79.03	2.38912	1.00005	0.6877
300	0.0162	4453.6	0.03848	0.0200	0.00333	175.88	81.39	2.63897	1.00005	0.6870
350	0.0139	5121.7	0.03925	0.0200	0.00286	186.09	87.21	3.30464	1.00004	0.6858
400	0.0121	5820.1	0.03957	0.0200	0.00250	197.40	92.95	4.02912	1.00004	0.6851
450	0.0108	6538.6	0.03964	0.0200	0.00222	209.28	98.62	4.81222	1.00003	0.6847
500	0.0097	7262.6	0.03966	0.0200	0.00200	221.22	104.24	5.65381	1.00003	0.6844
550	0.0088	7993.6	0.03963	0.0200	0.00182	233.23	109.79	6.55281	1.00003	0.6841
600	0.0081	8730.0	0.03956	0.0200	0.00167	245.23	115.28	7.50806	1.00002	0.6839
700	0.0069	10224.1	0.03935	0.0200	0.00143	269.31	126.07	9.58252	1.00002	0.6837
800	0.0061	11757.8	0.03901	0.0200	0.00125	293.67	136.61	11.86783	1.00002	0.6837
900	0.0054	13341.7	0.03855	0.0200	0.00111	318.48	146.90	14.35521	1.00002	0.6838
1000	0.0048	14983.4	0.03798	0.0200	0.00100	440.56	156.96	21.82932	1.00001	0.5338
1200	0.0040	18442.1	0.03668	0.0200	0.00083	511.17	176.43	29.63186	1.00001	0.5304
1400	0.0035	22147.0	0.03527	0.0200	0.00071	584.23	195.12	38.38501	1.00001	0.5283
1600	0.0030	26174.7	0.03373	0.0200	0.00063	664.26	213.15	48.23109	1.00001	0.5250
1800	0.0027	31207.2	0.03133	0.0200	0.00056	778.71	230.62	60.01969	1.00001	0.5136
2000	0.0024	39423.8	0.02681	0.0200	0.00050	1010.58	247.68	76.12031	1.00001	0.4840
2500	0.0019	111728.4	0.01091	0.0195	0.00041	3524.07	291.64	146.34712	1.00001	0.3803
3000	0.0014	351285.0	0.00429	0.0170	0.00039	12245.90	345.55	232.91357	1.00000	0.3882

* TWO-PHASE BOUNDARY

Table 4.1 (continued)
THERMODYNAMIC PROPERTIES OF PARAHYDROGEN (ISOBARS, SI UNITS)

0.04 MPA ISOBAR

TEMPERATURE DEG. K	VOLUME CU M/KG	ISOTHERM DERIVATIVE CU M-MPA/KG	ISOCHORE DERIVATIVE MPA/K	INTERNAL ENERGY KJ/KG-K	ENTHALPY KJ/KG	ENTROPY KJ/KG-K	CV KJ / KG-K	CP KJ / KG-K	VELOCITY OF SOUND M/S
* 13.814	0.01298	1.1647	0.9237	-308.9	-308.4	4.967	4.68	6.38	1261
14	0.01301	1.1349	0.9236	-307.7	-307.2	5.055	4.72	6.50	1250
15	0.01315	1.0346	0.9083	-301.0	-300.5	5.517	4.92	6.99	1212
16	0.01331	0.9689	0.8905	-293.8	-293.2	5.984	5.12	7.44	1187
17	0.01348	0.9053	0.8754	-286.1	-285.5	6.450	5.30	7.92	1163
* 17.497	0.01357	0.8806	0.8682	-282.1	-281.5	6.683	5.39	8.15	1154
* 17.497	1.71035	0.0646	0.0025	102.9	171.3	32.566	6.34	11.20	338
18	1.76648	0.0670	0.0024	106.3	176.9	32.882	6.32	11.10	343
19	1.87705	0.0718	0.0022	112.9	188.0	33.478	6.28	10.96	354
20	1.98652	0.0764	0.0021	119.4	198.9	34.038	6.26	10.86	364
21	2.09514	0.0810	0.0020	125.9	209.7	34.566	6.25	10.79	374
22	2.20310	0.0855	0.0019	132.3	220.4	35.066	6.24	10.73	384
23	2.31052	0.0900	0.0018	138.7	231.2	35.542	6.23	10.68	393
24	2.41749	0.0944	0.0017	145.1	241.8	35.996	6.22	10.64	402
25	2.52409	0.0988	0.0017	151.5	252.4	36.430	6.22	10.61	411
26	2.63035	0.1032	0.0016	157.8	263.0	36.845	6.22	10.58	419
27	2.73634	0.1076	0.0015	164.2	273.6	37.244	6.21	10.56	427
28	2.84208	0.1119	0.0015	170.5	284.2	37.628	6.21	10.54	436
29	2.94761	0.1162	0.0014	176.8	294.7	37.998	6.21	10.52	444
30	3.05295	0.1205	0.0014	183.1	305.2	38.354	6.21	10.51	452
31	3.15812	0.1248	0.0013	189.4	315.7	38.698	6.21	10.49	459
32	3.26314	0.1291	0.0013	195.7	326.2	39.031	6.21	10.48	467
33	3.36802	0.1333	0.0012	201.9	336.7	39.353	6.21	10.47	474
34	3.47278	0.1376	0.0012	208.2	347.1	39.666	6.20	10.46	482
35	3.57744	0.1418	0.0012	214.5	357.6	39.969	6.20	10.45	489
36	3.68199	0.1461	0.0011	220.7	368.0	40.263	6.20	10.44	496
37	3.78646	0.1503	0.0011	227.0	378.5	40.549	6.20	10.43	503
38	3.89084	0.1545	0.0011	233.3	388.9	40.827	6.20	10.43	510
39	3.99515	0.1588	0.0010	239.5	399.3	41.098	6.21	10.42	516
40	4.09939	0.1630	0.0010	245.8	409.7	41.362	6.21	10.42	523
42	4.30768	0.1714	0.0010	258.3	430.6	41.870	6.21	10.42	536
44	4.51576	0.1798	0.0009	270.8	451.4	42.355	6.22	10.42	549
46	4.72366	0.1882	0.0009	283.3	472.2	42.818	6.23	10.42	561
48	4.93140	0.1965	0.0008	295.8	493.1	43.262	6.25	10.43	573
50	5.13901	0.2049	0.0008	308.4	514.0	43.688	6.27	10.45	584
52	5.34649	0.2132	0.0008	321.0	534.9	44.098	6.30	10.47	595
54	5.55387	0.2216	0.0007	333.7	555.9	44.494	6.34	10.51	606
56	5.76116	0.2299	0.0007	346.5	576.9	44.877	6.38	10.55	616
58	5.96837	0.2383	0.0007	359.3	598.1	45.248	6.43	10.60	626
60	6.17550	0.2466	0.0007	372.3	61[illegible].3	45.608	6.49	10.65	636
65	6.69306	0.2674	0.0006	405.3	673.0	46.468	6.69	10.84	658
70	7.21030	0.2881	0.0006	439.4	727.8	47.280	6.94	11.09	679
75	7.72730	0.3088	0.0005	474.9	784.0	48.054	7.24	11.39	697
80	8.24410	0.3296	0.0005	512.0	841.8	48.800	7.60	11.74	714
85	8.76074	0.3503	0.0005	551.0	901.5	49.524	7.99	12.13	729
90	9.27725	0.3710	0.0004	592.0	963.1	50.229	8.41	12.54	744
95	9.79365	0.3917	0.0004	635.2	1026.9	50.918	8.84	12.98	758
100	10.30996	0.4123	0.0004	680.5	1092.9	51.595	9.27	13.41	772
120	12.37451	0.4950	0.0003	882.2	1377.2	54.180	10.83	14.96	827
140	14.43833	0.5776	0.0003	1109.8	1687.3	56.568	11.82	15.95	883
160	16.50171	0.6602	0.0003	1351.0	2011.1	58.730	12.21	16.34	940
180	18.56481	0.7428	0.0002	1595.5	2338.1	60.656	12.19	16.31	997
200	20.62771	0.8253	0.0002	1837.1	2662.2	62.362	11.94	16.07	1054
220	22.69048	0.9079	0.0002	2072.9	2980.5	63.880	11.63	15.76	1109
240	24.75314	0.9904	0.0002	2302.4	3292.5	65.237	11.33	15.45	1162
260	26.81572	1.0729	0.0002	2526.3	3599.0	66.464	11.07	15.20	1213
280	28.87824	1.1554	0.0001	2745.7	3900.8	67.583	10.87	15.00	1262
300	30.93842	1.2378	0.0001	2961.6	4199.1	68.612	10.72	14.84	1309
350	36.09405	1.4441	0.0001	3491.5	4935.3	70.881	10.51	14.63	1418
400	41.24955	1.6503	0.0001	4014.5	5664.5	72.830	10.42	14.55	1518
450	46.40496	1.8565	0.0001	4534.8	6391.0	74.542	10.40	14.53	1610
500	51.56032	2.0627	0.0001	5055.1	7117.5	76.072	10.40	14.52	1697
550	56.71565	2.2690	0.0001	5575.1	7843.7	77.456	10.41	14.53	1780
600	61.87095	2.4752	0.0001	6096.0	8570.9	78.720	10.42	14.55	1859
700	72.18149	2.8876	0.0001	7140.9	10028.2	80.968	10.48	14.61	2006
800	82.49200	3.3000	0.0001	8193.4	11493.1	82.922	10.57	14.70	2142
900	92.80248	3.7124	0.0000	9256.7	12968.8	84.659	10.70	14.82	2268
1000	103.11294	4.1249	0.0000	10333.9	14458.5	86.231	10.86	14.98	2386
1200	123.73385	4.9497	0.0000	12544.2	17493.5	88.985	11.24	15.37	2601
1400	144.35526	5.7745	0.0000	14837.5	20611.7	91.392	11.69	15.81	2795
1600	164.98269	6.5994	0.0000	17223.9	23823.3	93.513	12.19	16.32	2972
1800	185.64924	7.4242	0.0000	19737.9	27163.9	95.442	12.96	17.12	3132
2000	206.48386	8.2490	0.0000	22498.0	30757.4	97.400	14.65	18.92	3264
2500	262.92037	10.3111	0.0000	33176.9	43693.8	105.447	31.54	37.70	3509
3000	347.85821	12.3732	0.0000	62713.5	76627.9	120.644	86.94	104.62	3853

* TWO-PHASE BOUNDARY

Table 4.1 (continued)
THERMODYNAMIC PROPERTIES OF PARAHYDROGEN (ISOBARS, SI UNITS)

0.04 MPA ISOBAR

TEMPERATURE DEG. K	DENSITY KG/CU M	V(DH/DV)$_P$ KJ/KG	V(DP/DU)$_V$ MPA-CU M/KJ X 10^2	-V(DP/DV)$_T$ MPA	-(DV/DT)$_P$/V 1/DEG. K	THERMAL CONDUCTIVITY W/K-M X 10^3	VISCOSITY KG/M-S X 10^7	THERMAL DIFFUSIVITY SQ M/HR	DIELECTRIC CONSTANT	PRANDTL NUMBER
* 13.814	77.0340	619.8	0.25646	89.7200	0.01029	72.76	255.02	0.00053	1.25164	2.2364
14	76.8696	613.6	0.25483	87.2391	0.01059	74.63	248.30	0.00054	1.25106	2.1611
15	76.0433	605.3	0.24280	78.6773	0.01154	83.08	218.62	0.00056	1.24816	1.8387
16	75.1384	608.0	0.23158	72.8003	0.01223	88.91	194.63	0.00057	1.24498	1.6280
17	74.1849	607.4	0.22254	67.1585	0.01304	92.73	175.22	0.00057	1.24164	1.4961
* 17.497	73.6915	608.9	0.21861	64.8894	0.01338	94.22	166.89	0.00056	1.23991	1.4431
* 17.497	0.5847	170.9	0.06672	0.0378	0.06552	14.61	9.58	0.00803	1.00176	0.7338
18	0.5661	176.3	0.06682	0.0379	0.06297	14.83	9.86	0.00849	1.00171	0.7381
19	0.5328	187.1	0.06694	0.0382	0.05858	15.27	10.41	0.00941	1.00161	0.7470
20	0.5034	197.9	0.06700	0.0385	0.05487	15.86	10.95	0.01044	1.00152	0.7503
21	0.4773	208.7	0.06702	0.0387	0.05167	16.62	11.49	0.01162	1.00144	0.7460
22	0.4539	219.5	0.06702	0.0388	0.04887	17.39	12.03	0.01285	1.00137	0.7420
23	0.4328	230.3	0.06701	0.0389	0.04639	18.16	12.55	0.01414	1.00130	0.7393
24	0.4137	241.0	0.06700	0.0391	0.04416	18.94	13.07	0.01549	1.00125	0.7348
25	0.3962	251.7	0.06698	0.0392	0.04216	19.72	13.59	0.01689	1.00119	0.7313
26	0.3802	262.4	0.06697	0.0392	0.04034	20.49	14.10	0.01834	1.00115	0.7282
27	0.3655	273.0	0.06695	0.0393	0.03869	21.27	14.61	0.01984	1.00110	0.7252
28	0.3519	283.6	0.06694	0.0394	0.03717	22.04	15.10	0.02140	1.00106	0.7222
29	0.3393	294.2	0.06692	0.0394	0.03577	22.82	15.60	0.02301	1.00102	0.7192
30	0.3276	304.7	0.06691	0.0395	0.03447	23.59	16.09	0.02468	1.00099	0.7163
31	0.3166	315.3	0.06690	0.0395	0.03328	24.29	16.57	0.02632	1.00095	0.7157
32	0.3065	325.8	0.06688	0.0396	0.03216	24.98	17.05	0.02801	1.00092	0.7150
33	0.2969	336.3	0.06687	0.0396	0.03112	25.68	17.52	0.02974	1.00089	0.7142
34	0.2880	346.8	0.06686	0.0396	0.03015	26.38	17.99	0.03154	1.00087	0.7131
35	0.2795	357.3	0.06685	0.0396	0.02924	27.08	18.45	0.03338	1.00084	0.7119
36	0.2716	367.8	0.06683	0.0397	0.02838	27.78	18.91	0.03527	1.00082	0.7106
37	0.2641	378.3	0.06681	0.0397	0.02758	28.48	19.36	0.03721	1.00080	0.7093
38	0.2570	388.9	0.06679	0.0397	0.02682	29.19	19.81	0.03920	1.00077	0.7079
39	0.2503	399.4	0.06677	0.0397	0.02610	29.89	20.26	0.04124	1.00075	0.7065
40	0.2439	409.9	0.06674	0.0398	0.02542	30.59	20.70	0.04332	1.00074	0.7051
42	0.2321	431.1	0.06666	0.0398	0.02416	31.98	21.57	0.04762	1.00070	0.7024
44	0.2214	452.3	0.06655	0.0398	0.02303	33.38	22.42	0.05209	1.00067	0.6998
46	0.2117	473.8	0.06640	0.0398	0.02200	34.77	23.26	0.05674	1.00064	0.6972
48	0.2028	495.5	0.06620	0.0399	0.02106	36.17	24.09	0.06155	1.00061	0.6948
50	0.1946	517.5	0.06595	0.0399	0.02019	37.56	24.90	0.06650	1.00059	0.6927
52	0.1870	540.0	0.06564	0.0399	0.01940	38.96	25.70	0.07159	1.00056	0.6909
54	0.1801	562.9	0.06526	0.0399	0.01866	40.36	26.48	0.07680	1.00054	0.6895
56	0.1736	586.4	0.06480	0.0399	0.01799	41.75	27.26	0.08211	1.00052	0.6886
58	0.1676	610.5	0.06427	0.0399	0.01735	43.15	28.02	0.08750	1.00050	0.6881
60	0.1619	635.4	0.06366	0.0399	0.01677	44.61	28.78	0.09309	1.00049	0.6872
65	0.1494	701.2	0.06182	0.0399	0.01546	48.48	30.61	0.10776	1.00045	0.6845
70	0.1387	772.9	0.05957	0.0400	0.01434	52.35	32.39	0.12257	1.00042	0.6860
75	0.1294	851.2	0.05705	0.0400	0.01338	56.21	34.12	0.13732	1.00039	0.6912
80	0.1213	936.5	0.05439	0.0400	0.01253	60.12	35.80	0.15200	1.00037	0.6990
85	0.1141	1028.5	0.05171	0.0400	0.01179	65.07	37.43	0.16922	1.00034	0.6977
90	0.1078	1126.8	0.04913	0.0400	0.01113	70.03	39.03	0.18643	1.00032	0.6992
95	0.1021	1230.6	0.04673	0.0400	0.01054	74.98	40.59	0.20374	1.00031	0.7024
100	0.0970	1338.9	0.04453	0.0400	0.01001	89.51	45.74	0.24777	1.00029	0.6851
120	0.0808	1793.6	0.03813	0.0400	0.00834	117.11	54.08	0.34878	1.00024	0.6908
140	0.0693	2231.8	0.03493	0.0400	0.00715	136.14	59.20	0.44370	1.00021	0.6935
160	0.0606	2613.6	0.03380	0.0400	0.00625	148.29	63.00	0.53917	1.00018	0.6941
180	0.0539	2936.1	0.03387	0.0400	0.00556	155.64	66.17	0.63767	1.00016	0.6935
200	0.0485	3214.3	0.03455	0.0400	0.00500	160.19	69.01	0.74019	1.00015	0.6924
220	0.0441	3466.8	0.03548	0.0400	0.00455	163.39	71.66	0.84703	1.00013	0.6910
240	0.0404	3709.3	0.03642	0.0400	0.00417	166.20	74.18	0.95838	1.00012	0.6898
260	0.0373	3952.1	0.03726	0.0400	0.00385	169.12	76.63	1.07423	1.00011	0.6886
280	0.0346	4199.7	0.03795	0.0400	0.00357	172.34	79.03	1.19471	1.00010	0.6877
300	0.0323	4454.0	0.03849	0.0400	0.00333	175.89	81.40	1.31966	1.00010	0.6870
350	0.0277	5122.2	0.03926	0.0400	0.00286	186.10	87.22	1.65255	1.00008	0.6858
400	0.0242	5820.6	0.03957	0.0400	0.00250	197.41	92.96	2.01484	1.00007	0.6851
450	0.0215	6539.2	0.03964	0.0400	0.00222	209.30	98.63	2.40645	1.00006	0.6847
500	0.0194	7263.1	0.03966	0.0400	0.00200	221.23	104.24	2.82730	1.00006	0.6844
550	0.0176	7994.2	0.03963	0.0400	0.00182	233.24	109.80	3.27686	1.00005	0.6841
600	0.0162	8730.5	0.03957	0.0400	0.00167	245.25	115.29	3.75455	1.00005	0.6839
700	0.0139	10224.7	0.03935	0.0400	0.00143	269.33	126.08	4.79190	1.00004	0.6837
800	0.0121	11758.4	0.03901	0.0400	0.00125	293.70	136.62	5.93471	1.00004	0.6837
900	0.0108	13342.3	0.03855	0.0400	0.00111	318.51	146.92	7.17853	1.00003	0.6838
1000	0.0097	14984.0	0.03798	0.0400	0.00100	440.56	156.97	10.91511	1.00003	0.5338
1200	0.0081	18442.4	0.03668	0.0400	0.00083	511.16	176.44	14.81632	1.00002	0.5305
1400	0.0069	22140.9	0.03528	0.0400	0.00071	583.96	195.14	19.18995	1.00002	0.5285
1600	0.0061	26108.0	0.03384	0.0400	0.00062	661.59	213.17	24.08132	1.00002	0.5258
1800	0.0054	30811.3	0.03183	0.0400	0.00056	763.23	230.63	29.79266	1.00002	0.5174
2000	0.0048	37799.5	0.02819	0.0400	0.00050	947.86	247.66	37.23361	1.00001	0.4944
2500	0.0038	92396.5	0.01334	0.0392	0.00041	2791.89	290.67	70.10222	1.00001	0.3925
3000	0.0029	279107.0	0.00533	0.0356	0.00037	9541.08	342.92	114.20189	1.00001	0.3760

* TWO-PHASE BOUNDARY

Table 4.1 (continued)
THERMODYNAMIC PROPERTIES OF PARAHYDROGEN (ISOBARS, SI UNITS)

0.06 MPA ISOBAR

TEMPERATURE DEG. K	VOLUME CU M/KG	ISOTHERM DERIVATIVE CU M-MPA/KG	ISOCHORE DERIVATIVE MPA/K	INTERNAL ENERGY KJ/KG-K	ENTHALPY KJ/KG	ENTROPY KJ/KG-K	CV KJ / KG-K	CP KJ / KG-K	VELOCITY OF SOUND M/S
* 13.821	0.01298	1.1660	0.9239	-308.9	-308.1	4.967	4.68	6.38	1261
14	0.01301	1.1380	0.9239	-307.7	-306.9	5.052	4.71	6.49	1252
15	0.01315	1.0367	0.9090	-301.0	-300.2	5.514	4.92	6.99	1213
16	0.01331	0.9709	0.8912	-293.8	-293.0	5.980	5.12	7.43	1188
17	0.01348	0.9073	0.8761	-286.1	-285.3	6.447	5.30	7.91	1164
18	0.01366	0.8460	0.8616	-278.0	-277.2	6.914	5.47	8.42	1141
* 18.621	0.01378	0.8147	0.8524	-272.6	-271.8	7.205	5.56	8.72	1130
* 18.621	1.19270	0.0663	0.0036	107.6	179.2	31.430	6.41	11.53	345
19	1.22174	0.0682	0.0035	110.2	183.6	31.662	6.38	11.43	350
20	1.29743	0.0732	0.0033	117.0	194.9	32.243	6.33	11.23	360
21	1.37208	0.0780	0.0031	123.7	206.0	32.787	6.29	11.09	371
22	1.44595	0.0828	0.0029	130.3	217.1	33.300	6.27	10.98	381
23	1.51921	0.0875	0.0028	136.9	228.0	33.787	6.26	10.90	390
24	1.59196	0.0921	0.0026	143.4	238.9	34.249	6.25	10.83	400
25	1.66430	0.0966	0.0025	149.8	249.7	34.690	6.24	10.78	409
26	1.73629	0.1011	0.0024	156.3	260.4	35.112	6.23	10.73	417
27	1.80798	0.1056	0.0023	162.7	271.2	35.516	6.23	10.70	426
28	1.87941	0.1101	0.0022	169.1	281.8	35.905	6.23	10.66	434
29	1.95062	0.1145	0.0021	175.4	292.5	36.278	6.22	10.63	442
30	2.02162	0.1189	0.0021	181.8	303.1	36.639	6.22	10.61	450
31	2.09245	0.1232	0.0020	188.2	313.7	36.986	6.22	10.59	458
32	2.16313	0.1276	0.0019	194.5	324.3	37.322	6.22	10.57	466
33	2.23367	0.1319	0.0019	200.8	334.8	37.647	6.21	10.55	473
34	2.30408	0.1363	0.0018	207.1	345.4	37.961	6.21	10.53	481
35	2.37437	0.1406	0.0018	213.4	355.9	38.266	6.21	10.52	488
36	2.44457	0.1449	0.0017	219.7	366.4	38.563	6.21	10.51	495
37	2.51468	0.1492	0.0017	226.0	376.9	38.850	6.21	10.49	502
38	2.58470	0.1534	0.0016	232.3	387.4	39.130	6.21	10.49	509
39	2.65465	0.1577	0.0016	238.6	397.9	39.402	6.21	10.48	516
40	2.72452	0.1620	0.0015	244.9	408.4	39.668	6.21	10.47	522
42	2.86409	0.1705	0.0014	257.4	429.3	40.178	6.22	10.46	536
44	3.00344	0.1789	0.0014	270.0	450.2	40.665	6.23	10.46	548
46	3.14260	0.1874	0.0013	282.6	471.1	41.129	6.24	10.46	561
48	3.28161	0.1958	0.0013	295.1	492.0	41.575	6.25	10.47	572
50	3.42047	0.2042	0.0012	307.8	513.0	42.002	6.28	10.48	584
52	3.55922	0.2126	0.0012	320.4	534.0	42.414	6.30	10.50	595
54	3.69786	0.2210	0.0011	333.1	555.0	42.811	6.34	10.53	606
56	3.83641	0.2294	0.0011	345.9	576.1	43.194	6.38	10.57	616
58	3.97487	0.2378	0.0010	358.8	597.3	43.566	6.44	10.62	626
60	4.11327	0.2461	0.0010	371.8	618.6	43.927	6.50	10.67	636
65	4.45897	0.2670	0.0009	404.8	672.4	44.788	6.69	10.86	658
70	4.80437	0.2878	0.0009	439.0	727.3	45.601	6.94	11.10	679
75	5.14951	0.3086	0.0008	474.5	783.5	46.377	7.24	11.40	697
80	5.49446	0.3294	0.0008	511.7	841.3	47.124	7.60	11.75	714
85	5.83924	0.3501	0.0007	550.7	901.0	47.847	7.99	12.14	729
90	6.18390	0.3709	0.0007	591.7	962.8	48.553	8.41	12.55	744
95	6.52845	0.3916	0.0006	634.9	1026.6	49.243	8.84	12.98	758
100	6.87291	0.4123	0.0006	680.2	1092.6	49.920	9.27	13.41	772
120	8.25003	0.4951	0.0005	882.0	1377.0	52.506	10.83	14.96	827
140	9.62644	0.5777	0.0004	1109.6	1687.2	54.895	11.82	15.95	883
160	11.00241	0.6604	0.0004	1350.8	2011.0	57.056	12.21	16.34	940
180	12.37810	0.7430	0.0003	1595.4	2338.1	58.983	12.19	16.31	997
200	13.75359	0.8255	0.0003	1837.0	2662.2	60.690	11.94	16.07	1054
220	15.12893	0.9081	0.0003	2072.8	2980.5	62.207	11.63	15.76	1109
240	16.50418	0.9906	0.0003	2302.3	3292.6	63.565	11.33	15.45	1163
260	17.87935	1.0732	0.0002	2526.3	3599.0	64.791	11.07	15.20	1214
280	19.25446	1.1557	0.0002	2745.6	3900.9	65.910	10.87	15.00	1263
300	20.62801	1.2381	0.0002	2961.5	4199.2	66.940	10.72	14.85	1309
350	24.06523	1.4444	0.0002	3491.5	4935.4	69.208	10.51	14.63	1418
400	27.50233	1.6506	0.0002	4014.5	5664.6	71.157	10.42	14.55	1518
450	30.93934	1.8568	0.0001	4534.7	6391.1	72.870	10.40	14.53	1610
500	34.37630	2.0631	0.0001	5055.1	7117.7	74.399	10.40	14.52	1697
550	37.81322	2.2693	0.0001	5575.0	7843.8	75.784	10.41	14.53	1780
600	41.25012	2.4755	0.0001	6096.0	8571.0	77.048	10.42	14.55	1859
700	48.12386	2.8879	0.0001	7140.9	10028.4	79.295	10.48	14.61	2006
800	54.99756	3.3004	0.0001	8193.4	11493.3	81.250	10.57	14.70	2142
900	61.87123	3.7128	0.0001	9256.7	12969.0	82.987	10.70	14.82	2268
1000	68.74489	4.1252	0.0001	10333.9	14458.6	84.559	10.86	14.98	2386
1200	82.49218	4.9501	0.0001	12544.2	17493.7	87.312	11.24	15.37	2601
1400	96.23973	5.7749	0.0000	14837.3	20611.7	89.720	11.69	15.81	2795
1600	109.99056	6.5997	0.0000	17222.2	23821.7	91.840	12.17	16.30	2973
1800	123.76268	7.4246	0.0000	19726.6	27152.3	93.763	12.87	17.02	3134
2000	137.62628	8.2494	0.0000	22446.6	30704.2	95.699	14.31	18.56	3271
2500	174.65337	10.3115	0.0000	32383.6	42862.8	103.412	28.28	34.03	3521
3000	227.24916	12.3736	0.0000	57839.5	71474.4	117.065	74.57	39.35	3845

* TWO-PHASE BOUNDARY

Table 4.1 (continued)
THERMODYNAMIC PROPERTIES OF PARAHYDROGEN (ISOBARS, SI UNITS)

0.06 MPA ISOBAR

TEMPERATURE DEG. K	DENSITY KG/CU M	$V(\partial H/\partial V)_P$ KJ/KG	$V(\partial P/\partial U)_V$ MPA-CU M/KJ X 10^2	$-V(\partial P/\partial V)_T$ MPA	$-(\partial V/\partial T)_P/V$ 1/DEG. K	THERMAL CONDUCTIVITY W/K-M X 10^3	VISCOSITY KG/M-S X 10^7	THERMAL DIFFUSIVITY SQ M/HR	DIELECTRIC CONSTANT	PRANDTL NUMBER
* 13.821	77.0455	620.4	0.25644	89.8314	0.01028	72.84	255.14	0.00053	1.25168	2.2350
14	76.8869	614.7	0.25486	87.4981	0.01056	74.65	248.65	0.00054	1.25112	2.1620
15	76.0618	606.0	0.24296	78.8559	0.01153	83.11	218.92	0.00056	1.24822	1.8401
16	75.1582	608.7	0.23173	72.9692	0.01221	88.95	194.90	0.00057	1.24505	1.6289
17	74.2067	608.1	0.22269	67.3243	0.01301	92.77	175.47	0.00057	1.24172	1.4968
18	73.2096	605.0	0.21517	61.9320	0.01391	95.46	159.47	0.00056	1.23823	1.4062
* 18.621	72.5645	604.7	0.21108	59.1157	0.01442	96.52	150.87	0.00055	1.23598	1.3628
* 18.621	0.8384	178.9	0.06666	0.0556	0.06446	15.43	10.26	0.00574	1.00253	0.7667
19	0.8185	183.1	0.06678	0.0558	0.06246	15.58	10.47	0.00599	1.00247	0.7681
20	0.7708	194.0	0.06699	0.0564	0.05790	16.11	11.01	0.00670	1.00232	0.7673
21	0.7288	205.0	0.06709	0.0569	0.05410	16.84	11.55	0.00750	1.00220	0.7600
22	0.6916	215.9	0.06714	0.0573	0.05086	17.59	12.08	0.00834	1.00209	0.7539
23	0.6582	226.9	0.06715	0.0576	0.04804	18.35	12.60	0.00921	1.00198	0.7486
24	0.6282	237.8	0.06714	0.0578	0.04556	19.11	13.12	0.01011	1.00189	0.7438
25	0.6009	248.6	0.06713	0.0581	0.04335	19.88	13.64	0.01105	1.00181	0.7394
26	0.5759	259.5	0.06711	0.0582	0.04137	20.65	14.15	0.01202	1.00174	0.7355
27	0.5531	270.3	0.06709	0.0584	0.03958	21.41	14.65	0.01303	1.00167	0.7318
28	0.5321	281.0	0.06707	0.0586	0.03794	22.18	15.15	0.01407	1.00160	0.7283
29	0.5127	291.7	0.06705	0.0587	0.03645	22.94	15.64	0.01515	1.00155	0.7249
30	0.4947	302.4	0.06703	0.0588	0.03508	23.71	16.13	0.01627	1.00149	0.7215
31	0.4779	313.1	0.06701	0.0589	0.03381	24.40	16.61	0.01737	1.00144	0.7204
32	0.4623	323.7	0.06700	0.0590	0.03264	25.10	17.08	0.01850	1.00139	0.7193
33	0.4477	334.3	0.06698	0.0591	0.03155	25.79	17.56	0.01966	1.00135	0.7181
34	0.4340	344.9	0.06697	0.0591	0.03053	26.49	18.02	0.02086	1.00131	0.7167
35	0.4212	355.5	0.06695	0.0592	0.02958	27.19	18.49	0.02209	1.00127	0.7152
36	0.4091	366.1	0.06693	0.0593	0.02870	27.89	18.94	0.02336	1.00123	0.7137
37	0.3977	376.7	0.06691	0.0593	0.02786	28.58	19.40	0.02466	1.00120	0.7122
38	0.3869	387.3	0.06689	0.0594	0.02707	29.28	19.85	0.02599	1.00117	0.7106
39	0.3757	397.9	0.06686	0.0594	0.02633	29.98	20.29	0.02735	1.00114	0.7090
40	0.3670	408.5	0.06683	0.0594	0.02563	30.68	20.73	0.02874	1.00111	0.7074
42	0.3492	429.7	0.06674	0.0595	0.02434	32.07	21.60	0.03161	1.00105	0.7044
44	0.3330	451.1	0.06663	0.0596	0.02318	33.47	22.45	0.03460	1.00100	0.7015
46	0.3182	472.6	0.06648	0.0596	0.02213	34.86	23.29	0.03771	1.00096	0.6988
48	0.3047	494.4	0.06628	0.0597	0.02117	36.25	24.12	0.04092	1.00092	0.6962
50	0.2924	516.5	0.06602	0.0597	0.02029	37.64	24.93	0.04423	1.00088	0.6940
52	0.2810	539.0	0.06571	0.0597	0.01948	39.03	25.72	0.04762	1.00085	0.6921
54	0.2704	562.0	0.06532	0.0598	0.01874	40.43	26.51	0.05110	1.00081	0.6906
56	0.2607	585.6	0.06486	0.0598	0.01805	41.82	27.28	0.05465	1.00079	0.6895
58	0.2516	609.7	0.06433	0.0598	0.01741	43.22	28.05	0.05825	1.00076	0.6890
60	0.2431	634.6	0.06372	0.0598	0.01682	44.68	28.80	0.06198	1.00073	0.6880
65	0.2243	700.5	0.06187	0.0599	0.01550	48.54	30.63	0.07177	1.00068	0.6851
70	0.2081	772.3	0.05962	0.0599	0.01437	52.41	32.41	0.08166	1.00063	0.6865
75	0.1942	850.7	0.05709	0.0599	0.01340	56.26	34.14	0.09150	1.00059	0.6916
80	0.1820	936.0	0.05443	0.0599	0.01255	60.17	35.82	0.10130	1.00055	0.6994
85	0.1713	1028.0	0.05175	0.0600	0.01181	65.12	37.45	0.11279	1.00052	0.6980
90	0.1617	1126.3	0.04916	0.0600	0.01114	70.07	39.05	0.12428	1.00049	0.6995
95	0.1532	1230.2	0.04675	0.0600	0.01055	75.03	40.61	0.13582	1.00046	0.7026
100	0.1455	1338.5	0.04456	0.0600	0.01002	89.52	45.74	0.16511	1.00044	0.6854
120	0.1212	1793.3	0.03815	0.0600	0.00834	117.11	54.08	0.23246	1.00037	0.6909
140	0.1039	2231.7	0.03494	0.0600	0.00715	136.13	59.20	0.29577	1.00031	0.6936
160	0.0909	2613.7	0.03381	0.0600	0.00625	148.29	63.00	0.35944	1.00027	0.6942
180	0.0808	2936.2	0.03388	0.0600	0.00556	155.64	66.17	0.42513	1.00024	0.6936
200	0.0727	3214.5	0.03456	0.0600	0.00500	160.19	69.01	0.49348	1.00022	0.6924
220	0.0661	3467.1	0.03548	0.0600	0.00455	163.40	71.66	0.56473	1.00020	0.6911
240	0.0606	3709.7	0.03643	0.0600	0.00417	166.21	74.18	0.63899	1.00018	0.6898
260	0.0559	3952.5	0.03727	0.0600	0.00385	169.13	76.63	0.71624	1.00017	0.6887
280	0.0519	4200.1	0.03796	0.0600	0.00357	172.35	79.04	0.79657	1.00016	0.6878
300	0.0485	4454.5	0.03849	0.0600	0.00333	175.90	81.40	0.87989	1.00015	0.6870
350	0.0416	5122.7	0.03926	0.0600	0.00286	186.11	87.22	1.10185	1.00013	0.6858
400	0.0364	5821.2	0.03957	0.0600	0.00250	197.42	92.96	1.34342	1.00011	0.6851
450	0.0323	6539.7	0.03965	0.0600	0.00222	209.31	98.63	1.60452	1.00010	0.6847
500	0.0291	7263.7	0.03967	0.0600	0.00200	221.25	104.25	1.88513	1.00009	0.6844
550	0.0264	7994.7	0.03963	0.0600	0.00182	233.26	109.80	2.18488	1.00008	0.6841
600	0.0242	8731.1	0.03957	0.0600	0.00167	245.27	115.30	2.50338	1.00007	0.6839
700	0.0208	10225.3	0.03936	0.0600	0.00143	269.35	126.09	3.19503	1.00006	0.6837
800	0.0182	11759.0	0.03902	0.0600	0.00125	293.72	136.64	3.95699	1.00005	0.6837
900	0.0162	13343.0	0.03855	0.0600	0.00111	318.54	146.93	4.78632	1.00005	0.6838
1000	0.0145	14984.6	0.03799	0.0600	0.00100	440.56	156.99	7.27704	1.00004	0.5339
1200	0.0121	18442.9	0.03669	0.0600	0.00083	511.15	176.46	9.87786	1.00004	0.5305
1400	0.0104	22138.5	0.03529	0.0600	0.00071	583.83	195.16	12.79276	1.00003	0.5285
1600	0.0091	26078.9	0.03389	0.0600	0.00062	660.41	213.19	16.04431	1.00003	0.5261
1800	0.0081	30636.3	0.03206	0.0600	0.00056	756.37	230.65	19.79668	1.00002	0.5191
2000	0.0073	37079.7	0.02885	0.0599	0.00050	920.05	247.66	24.56292	1.00002	0.4996
2500	0.0057	83721.9	0.01482	0.0590	0.00041	2463.29	290.23	45.50813	1.00002	0.4010
3000	0.0044	243245.4	0.00610	0.0544	0.00037	8196.95	341.07	75.05428	1.00001	0.3718

* TWO-PHASE BOUNDARY

Table 4.1 (continued)
THERMODYNAMIC PROPERTIES OF PARAHYDROGEN (ISOBARS, SI UNITS)

0.08 MPA ISOBAR

TEMPERATURE	VOLUME	ISOTHERM DERIVATIVE	ISOCHORE DERIVATIVE	INTERNAL ENERGY	ENTHALPY	ENTROPY	CV	CP	VELOCITY OF SOUND
DEG. K	CU M/KG	CU M-MPA/KG	MPA/K	KJ/KG-K	KJ/KG	KJ/KG-K	KJ /	KG-K	M/S
* 13.828	0.01298	1.1672	0.9241	-308.9	-307.9	4.968	4.68	6.38	1262
14	0.01300	1.1411	0.9241	-307.8	-306.7	5.050	4.71	6.49	1253
15	0.01314	1.0388	0.9097	-301.1	-300.0	5.511	4.92	6.98	1214
16	0.01330	0.9729	0.8919	-293.9	-292.8	5.977	5.12	7.43	1189
17	0.01347	0.9092	0.8768	-286.2	-285.1	6.444	5.30	7.91	1165
18	0.01366	0.8480	0.8623	-278.0	-276.9	6.911	5.47	8.41	1142
19	0.01385	0.7833	0.8470	-269.4	-268.3	7.379	5.62	8.96	1117
* 19.495	0.01396	0.7643	0.8384	-264.9	-263.8	7.614	5.69	9.18	1111
* 19.496	0.92223	0.0671	0.0047	110.9	184.7	30.623	6.46	11.84	351
20	0.95189	0.0698	0.0045	114.5	190.7	30.923	6.41	11.68	356
21	1.00981	0.0750	0.0042	121.4	202.2	31.487	6.35	11.44	367
22	1.06682	0.0800	0.0040	128.2	213.6	32.015	6.32	11.27	378
23	1.12311	0.0849	0.0038	134.9	224.8	32.513	6.29	11.14	388
24	1.17885	0.0897	0.0036	141.5	235.9	32.985	6.27	11.04	397
25	1.23413	0.0944	0.0034	148.1	246.9	33.434	6.26	10.96	406
26	1.28903	0.0990	0.0033	154.7	257.8	33.863	6.25	10.89	415
27	1.34361	0.1036	0.0031	161.2	268.7	34.273	6.25	10.84	424
28	1.39792	0.1082	0.0030	167.6	279.5	34.666	6.24	10.79	433
29	1.45199	0.1127	0.0029	174.1	290.2	35.044	6.24	10.75	441
30	1.50585	0.1172	0.0028	180.5	301.0	35.408	6.23	10.72	449
31	1.55953	0.1217	0.0027	186.9	311.7	35.759	6.23	10.68	457
32	1.61305	0.1261	0.0026	193.3	322.3	36.098	6.23	10.66	465
33	1.66642	0.1305	0.0025	199.7	333.0	36.425	6.22	10.63	472
34	1.71967	0.1349	0.0024	206.0	343.6	36.742	6.22	10.61	480
35	1.77280	0.1393	0.0024	212.4	354.2	37.050	6.22	10.59	487
36	1.82582	0.1437	0.0023	218.7	364.8	37.348	6.22	10.57	494
37	1.87875	0.1480	0.0022	225.1	375.4	37.637	6.22	10.56	501
38	1.93160	0.1523	0.0022	231.4	385.9	37.919	6.22	10.54	508
39	1.98437	0.1567	0.0021	237.7	396.4	38.192	6.22	10.53	515
40	2.03707	0.1610	0.0020	244.0	407.0	38.459	6.22	10.52	522
42	2.14228	0.1695	0.0019	256.6	428.0	38.972	6.22	10.51	535
44	2.24726	0.1781	0.0018	269.2	449.0	39.460	6.23	10.50	548
46	2.35206	0.1866	0.0018	281.8	470.0	39.927	6.24	10.49	560
48	2.45670	0.1951	0.0017	294.5	491.0	40.374	6.26	10.50	572
50	2.56120	0.2036	0.0016	307.1	512.0	40.803	6.28	10.51	584
52	2.66558	0.2120	0.0016	319.8	533.0	41.215	6.31	10.53	595
54	2.76985	0.2205	0.0015	332.5	554.1	41.613	6.34	10.56	606
56	2.87403	0.2289	0.0014	345.4	575.3	41.998	6.39	10.59	616
58	2.97813	0.2373	0.0014	358.3	596.5	42.370	6.44	10.64	626
60	3.08215	0.2457	0.0013	371.3	617.8	42.732	6.50	10.69	636
65	3.34194	0.2666	0.0012	404.4	671.7	43.594	6.69	10.87	658
70	3.60140	0.2875	0.0011	438.6	726.7	44.408	6.94	11.11	679
75	3.86062	0.3084	0.0011	474.1	783.0	45.185	7.24	11.41	697
80	4.11964	0.3292	0.0010	511.3	840.9	45.932	7.60	11.76	714
85	4.37850	0.3500	0.0009	550.3	900.6	46.657	7.99	12.15	729
90	4.63723	0.3708	0.0009	591.4	962.4	47.363	8.41	12.56	744
95	4.89585	0.3915	0.0008	634.6	1026.2	48.053	8.84	12.99	758
100	5.15438	0.4123	0.0008	679.9	1092.3	48.730	9.27	13.42	772
120	6.18780	0.4951	0.0007	881.8	1376.8	51.318	10.83	14.97	827
140	7.22050	0.5779	0.0006	1109.4	1687.0	53.707	11.82	15.95	883
160	8.25276	0.6605	0.0005	1350.7	2010.9	55.869	12.21	16.34	940
180	9.28474	0.7432	0.0004	1595.3	2338.0	57.796	12.19	16.32	997
200	10.31652	0.8257	0.0004	1836.8	2662.2	59.502	11.95	16.07	1054
220	11.34817	0.9083	0.0004	2072.7	2980.5	61.020	11.63	15.76	1109
240	12.37971	0.9909	0.0003	2302.2	3292.6	62.378	11.33	15.46	1163
260	13.41117	1.0734	0.0003	2526.2	3599.1	63.605	11.07	15.20	1214
280	14.44257	1.1560	0.0003	2745.6	3901.0	64.724	10.87	15.00	1263
300	15.47280	1.2384	0.0003	2961.5	4199.3	65.753	10.72	14.85	1310
350	18.05083	1.4447	0.0002	3491.4	4935.5	68.022	10.51	14.63	1418
400	20.62872	1.6509	0.0002	4014.5	5664.8	69.971	10.42	14.55	1518
450	23.20653	1.8572	0.0002	4534.7	6391.2	71.683	10.41	14.53	1610
500	25.78429	2.0634	0.0002	5055.1	7117.8	73.213	10.40	14.52	1698
550	28.36201	2.2696	0.0001	5575.0	7844.0	74.597	10.41	14.53	1780
600	30.93970	2.4759	0.0001	6096.0	8571.2	75.861	10.42	14.55	1859
700	36.09504	2.8883	0.0001	7140.9	10028.5	78.109	10.48	14.61	2006
800	41.25034	3.3007	0.0001	8193.4	11493.4	80.064	10.57	14.70	2142
900	46.40561	3.7131	0.0001	9256.7	12969.2	81.800	10.70	14.82	2268
1000	51.56086	4.1256	0.0001	10333.9	14458.8	83.372	10.86	14.98	2386
1200	61.87135	4.9504	0.0001	12544.2	17493.9	86.126	11.24	15.37	2601
1400	72.18199	5.7753	0.0001	14837.2	20611.8	88.533	11.69	15.81	2795
1600	82.49476	6.6001	0.0001	17221.2	23820.8	90.652	12.16	16.29	2973
1800	92.82135	7.4249	0.0000	19719.8	27145.5	92.572	12.81	16.96	3135
2000	103.20737	8.2498	0.0000	22415.9	30672.5	94.495	14.11	18.34	3275
2500	130.71063	10.3119	0.0000	31910.6	42367.5	102.009	26.33	31.85	3531
3000	168.34401	12.3739	0.0000	54917.2	68384.7	114.736	66.97	80.08	3841

* TWO-PHASE BOUNDARY

Table 4.1 (continued)
THERMODYNAMIC PROPERTIES OF PARAHYDROGEN (ISOBARS, SI UNITS)

0.08 MPA ISOBAR

TEMPERATURE DEG. K	DENSITY KG/CU M	$V(DH/DV)_P$ KJ/KG	$V(DP/DU)_V$ MPA-CU M/KJ X 10^2	$-V(DP/DV)_T$ MPA	$-(DV/DT)_P/V$ 1/DEG. K	THERMAL CONDUCTIVITY W/K-M X 10^3	VISCOSITY KG/M-S X 10^7	THERMAL DIFFUSIVITY SQ M/HR	DIELECTRIC CONSTANT	PRANDTL NUMBER
* 13.828	77.0570	621.0	0.25641	89.9428	0.01027	72.92	255.26	0.00053	1.25172	2.2336
14	76.9043	615.9	0.25489	87.7567	0.01053	74.67	249.01	0.00054	1.25118	2.1630
15	76.0803	606.6	0.24313	79.0342	0.01151	83.13	219.23	0.00056	1.24824	1.8415
16	75.1781	609.4	0.23189	73.1378	0.01219	88.98	195.17	0.00057	1.24512	1.6299
17	74.2283	608.8	0.22283	67.4898	0.01299	92.81	175.71	0.00057	1.24179	1.4974
18	73.2332	605.8	0.21531	62.1031	0.01388	95.51	159.70	0.00056	1.23831	1.4065
19	72.1879	598.0	0.20884	56.5441	0.01498	97.04	146.25	0.00054	1.23466	1.3500
* 19.496	71.6402	599.5	0.20583	54.7548	0.01531	97.43	140.27	0.00053	1.23276	1.3216
* 19.496	1.0843	184.6	0.06665	0.0727	0.06416	16.11	10.80	0.00452	1.00327	0.7938
20	1.0505	190.1	0.06684	0.0733	0.06144	16.41	11.07	0.00481	1.00317	0.7879
21	0.9903	201.2	0.06708	0.0742	0.05686	17.10	11.60	0.00543	1.00299	0.7764
22	0.9374	212.4	0.06720	0.0750	0.05307	17.81	12.13	0.00607	1.00283	0.7675
23	0.8904	223.5	0.06726	0.0756	0.04986	18.55	12.65	0.00673	1.00269	0.7601
24	0.8483	234.6	0.06727	0.0761	0.04707	19.30	13.17	0.00742	1.00256	0.7538
25	0.8103	245.6	0.06726	0.0765	0.04463	20.05	13.69	0.00813	1.00244	0.7481
26	0.7758	256.6	0.06725	0.0768	0.04246	20.80	14.19	0.00886	1.00234	0.7433
27	0.7443	267.5	0.06722	0.0771	0.04051	21.56	14.69	0.00962	1.00224	0.7388
28	0.7153	278.4	0.06720	0.0774	0.03876	22.31	15.19	0.01041	1.00216	0.7347
29	0.6887	289.3	0.06718	0.0776	0.03716	23.07	15.68	0.01122	1.00208	0.7307
30	0.6641	300.1	0.06715	0.0778	0.03570	23.83	16.17	0.01206	1.00200	0.7270
31	0.6412	310.9	0.06713	0.0780	0.03436	24.52	16.65	0.01289	1.00193	0.7253
32	0.6199	321.7	0.06711	0.0782	0.03313	25.21	17.12	0.01374	1.00187	0.7237
33	0.6001	332.4	0.06709	0.0783	0.03199	25.90	17.59	0.01462	1.00181	0.7222
34	0.5815	343.1	0.06707	0.0785	0.03093	26.60	18.06	0.01552	1.00175	0.7204
35	0.5641	353.7	0.06705	0.0786	0.02994	27.29	18.52	0.01645	1.00170	0.7186
36	0.5477	364.4	0.06703	0.0787	0.02901	27.99	18.98	0.01740	1.00165	0.7169
37	0.5323	375.0	0.06701	0.0788	0.02815	28.69	19.43	0.01839	1.00160	0.7151
38	0.5177	385.7	0.06698	0.0789	0.02734	29.38	19.88	0.01939	1.00156	0.7133
39	0.5039	396.3	0.06695	0.0789	0.02657	30.08	20.32	0.02040	1.00152	0.7115
40	0.4909	407.0	0.06692	0.0790	0.02585	30.78	20.76	0.02145	1.00148	0.7098
42	0.4668	428.4	0.06683	0.0791	0.02453	32.17	21.63	0.02361	1.00141	0.7065
44	0.4450	449.8	0.06671	0.0792	0.02334	33.55	22.48	0.02586	1.00134	0.7033
46	0.4252	471.5	0.06656	0.0793	0.02226	34.94	23.32	0.02819	1.00128	0.7004
48	0.4070	493.3	0.06635	0.0794	0.02128	36.33	24.14	0.03060	1.00123	0.6977
50	0.3904	515.5	0.06610	0.0795	0.02039	37.72	24.95	0.03309	1.00118	0.6952
52	0.3752	538.1	0.06578	0.0795	0.01957	39.11	25.75	0.03564	1.00113	0.6932
54	0.3610	561.1	0.06539	0.0796	0.01881	40.50	26.53	0.03826	1.00109	0.6916
56	0.3479	584.7	0.06493	0.0796	0.01812	41.89	27.31	0.04092	1.00105	0.6904
58	0.3358	608.9	0.06439	0.0797	0.01747	43.29	28.07	0.04362	1.00101	0.6898
60	0.3244	633.9	0.06378	0.0797	0.01687	44.74	28.82	0.04643	1.00098	0.6887
65	0.2992	699.8	0.06192	0.0798	0.01554	48.60	30.65	0.05378	1.00090	0.6857
70	0.2777	771.7	0.05967	0.0798	0.01440	52.47	32.43	0.06121	1.00084	0.6869
75	0.2590	850.1	0.05714	0.0799	0.01342	56.32	34.16	0.06859	1.00078	0.6921
80	0.2427	935.4	0.05446	0.0799	0.01257	60.22	35.83	0.07595	1.00073	0.6997
85	0.2284	1027.5	0.05178	0.0799	0.01182	65.17	37.47	0.08457	1.00069	0.6983
90	0.2156	1125.9	0.04920	0.0800	0.01116	70.12	39.06	0.09320	1.00065	0.6997
95	0.2043	1229.7	0.04678	0.0800	0.01056	75.07	40.62	0.10186	1.00062	0.7028
100	0.1940	1338.1	0.04459	0.0800	0.01003	89.53	45.74	0.12378	1.00058	0.6856
120	0.1616	1793.1	0.03817	0.0800	0.00835	117.11	54.08	0.17430	1.00049	0.6911
140	0.1385	2231.6	0.03495	0.0800	0.00715	136.13	59.20	0.22180	1.00042	0.6937
160	0.1212	2613.7	0.03382	0.0800	0.00625	148.29	63.00	0.26957	1.00037	0.6943
180	0.1077	2936.4	0.03389	0.0800	0.00556	155.65	66.17	0.31885	1.00032	0.6937
200	0.0969	3214.7	0.03456	0.0800	0.00500	160.19	69.01	0.37014	1.00029	0.6925
220	0.0881	3467.4	0.03549	0.0800	0.00455	163.40	71.66	0.42359	1.00027	0.6911
240	0.0808	3710.0	0.03644	0.0800	0.00417	166.22	74.19	0.47929	1.00024	0.6898
260	0.0746	3952.9	0.03728	0.0800	0.00385	169.14	76.64	0.53724	1.00022	0.6887
280	0.0692	4200.6	0.03796	0.0800	0.00357	172.36	79.04	0.59750	1.00021	0.6878
300	0.0646	4454.9	0.03850	0.0800	0.00333	175.90	81.40	0.66009	1.00019	0.6870
350	0.0554	5123.2	0.03927	0.0800	0.00286	186.12	87.23	0.82650	1.00017	0.6858
400	0.0485	5821.7	0.03958	0.0800	0.00250	197.43	92.97	1.00770	1.00015	0.6851
450	0.0431	6540.3	0.03965	0.0800	0.00222	209.32	98.64	1.20356	1.00013	0.6847
500	0.0388	7264.3	0.03967	0.0800	0.00200	221.26	104.26	1.41405	1.00012	0.6844
550	0.0353	7995.3	0.03964	0.0800	0.00182	233.28	109.81	1.63889	1.00011	0.6841
600	0.0323	8731.7	0.03957	0.0800	0.00167	245.28	115.31	1.87779	1.00010	0.6839
700	0.0277	10225.9	0.03936	0.0800	0.00143	269.37	126.10	2.39660	1.00009	0.6837
800	0.0242	11759.6	0.03902	0.0800	0.00125	293.75	136.65	2.96814	1.00007	0.6837
900	0.0215	13343.6	0.03855	0.0800	0.00111	318.57	146.94	3.59021	1.00006	0.6837
1000	0.0194	14985.3	0.03799	0.0800	0.00100	440.56	157.00	5.45800	1.00006	0.5339
1200	0.0162	18443.5	0.03669	0.0800	0.00083	511.15	176.47	7.40863	1.00005	0.5306
1400	0.0139	22137.4	0.03530	0.0800	0.00071	583.76	195.18	9.59445	1.00004	0.5286
1600	0.0121	26061.8	0.03392	0.0800	0.00062	659.70	213.21	12.02892	1.00004	0.5264
1800	0.0108	30532.2	0.03219	0.0800	0.00056	752.28	230.67	14.81827	1.00003	0.5202
2000	0.0097	36650.8	0.02926	0.0799	0.00050	903.47	247.67	18.30263	1.00003	0.5028
2500	0.0077	78519.0	0.01589	0.0789	0.00041	2266.17	289.97	33.48168	1.00002	0.4075
3000	0.0059	220733.0	0.00670	0.0735	0.00036	7352.97	339.76	55.64640	1.00002	0.3700

* TWO-PHASE BOUNDARY

Table 4.1 (continued)
THERMODYNAMIC PROPERTIES OF PARAHYDROGEN (ISOBARS, SI UNITS)

0.10 MPA ISOBAR

TEMPERATURE DEG. K	VOLUME CU M/KG	ISOTHERM DERIVATIVE CU M-MPA/KG	ISOCHORE DERIVATIVE MPA/K	INTERNAL ENERGY KJ/KG-K	ENTHALPY KJ/KG	ENTROPY KJ/KG-K	CV KJ / KG-K	CP KJ / KG-K	VELOCITY OF SOUND M/S
* 13.83	0.01298	1.1685	0.9244	-308.9	-307.6	4.968	4.68	6.38	1262
14	0.01300	1.1442	0.9243	-307.8	-306.5	5.047	4.71	6.48	1254
15	0.01314	1.0409	0.9104	-301.1	-299.8	5.509	4.92	6.98	1216
16	0.01330	0.9748	0.8925	-293.9	-292.6	5.974	5.12	7.43	1190
17	0.01347	0.9112	0.8774	-286.3	-284.9	6.440	5.30	7.91	1166
18	0.01365	0.8501	0.8630	-278.1	-276.7	6.907	5.47	8.41	1143
19	0.01385	0.7853	0.8476	-269.5	-268.1	7.375	5.62	8.95	1119
20	0.01406	0.7221	0.8304	-260.3	-258.9	7.848	5.75	9.53	1094
* 20.224	0.01412	0.7122	0.8261	-258.1	-256.7	7.956	5.77	9.64	1090
* 20.224	0.75454	0.0675	0.0057	113.4	188.9	29.994	6.50	12.14	355
21	0.79178	0.0717	0.0054	119.0	198.2	30.445	6.43	11.86	364
22	0.83883	0.0770	0.0051	126.0	209.9	30.991	6.37	11.60	375
23	0.88507	0.0822	0.0048	132.9	221.4	31.502	6.33	11.41	385
24	0.93068	0.0872	0.0046	139.7	232.8	31.985	6.30	11.27	395
25	0.97579	0.0921	0.0043	146.4	244.0	32.442	6.29	11.16	404
26	1.02048	0.0969	0.0041	153.0	255.1	32.878	6.27	11.07	413
27	1.06484	0.1016	0.0040	159.6	266.1	33.294	6.26	10.99	422
28	1.10890	0.1063	0.0038	166.2	277.1	33.693	6.26	10.93	431
29	1.15271	0.1110	0.0036	172.7	288.0	34.075	6.25	10.87	439
30	1.19630	0.1156	0.0035	179.2	298.8	34.443	6.25	10.83	448
31	1.23970	0.1201	0.0034	185.7	309.6	34.798	6.24	10.79	456
32	1.28294	0.1246	0.0033	192.1	320.4	35.140	6.24	10.75	463
33	1.32603	0.1291	0.0032	198.5	331.1	35.470	6.23	10.72	471
34	1.36898	0.1336	0.0031	204.9	341.8	35.789	6.23	10.69	479
35	1.41181	0.1380	0.0030	211.3	352.5	36.099	6.23	10.66	486
36	1.45454	0.1425	0.0029	217.7	363.2	36.399	6.23	10.64	493
37	1.49717	0.1469	0.0028	224.1	373.8	36.690	6.22	10.62	501
38	1.53972	0.1512	0.0027	230.4	384.4	36.973	6.22	10.60	508
39	1.58218	0.1556	0.0026	236.8	395.0	37.248	6.22	10.59	514
40	1.62458	0.1600	0.0026	243.1	405.6	37.516	6.22	10.57	521
42	1.70918	0.1686	0.0024	255.8	426.7	38.032	6.23	10.55	535
44	1.79355	0.1772	0.0023	268.4	447.8	38.522	6.23	10.54	547
46	1.87774	0.1858	0.0022	281.1	468.9	38.991	6.25	10.53	560
48	1.96176	0.1944	0.0021	293.8	489.9	39.439	6.26	10.53	572
50	2.04564	0.2029	0.0020	306.4	511.0	39.869	6.28	10.54	583
52	2.12940	0.2114	0.0020	319.2	532.1	40.283	6.31	10.56	595
54	2.21305	0.2199	0.0019	331.9	553.2	40.682	6.35	10.58	606
56	2.29661	0.2284	0.0018	344.8	574.4	41.067	6.39	10.62	616
58	2.38009	0.2368	0.0017	357.7	595.7	41.440	6.44	10.66	626
60	2.46349	0.2452	0.0017	370.7	617.1	41.803	6.50	10.71	636
65	2.67172	0.2663	0.0016	403.9	671.1	42.667	6.69	10.89	658
70	2.87963	0.2872	0.0014	438.1	726.1	43.482	6.94	11.13	679
75	3.08729	0.3081	0.0013	473.7	782.4	44.259	7.25	11.42	697
80	3.29475	0.3290	0.0013	510.9	840.4	45.007	7.60	11.77	714
85	3.50206	0.3498	0.0012	550.0	900.2	45.732	7.99	12.15	729
90	3.70923	0.3706	0.0011	591.1	962.0	46.439	8.41	12.57	744
95	3.91630	0.3914	0.0011	634.3	1025.9	47.130	8.84	13.00	759
100	4.12327	0.4122	0.0010	679.6	1092.0	47.807	9.27	13.43	773
120	4.95046	0.4952	0.0008	881.5	1376.6	50.396	10.83	14.97	827
140	5.77694	0.5780	0.0007	1109.2	1686.9	52.785	11.82	15.96	883
160	6.60298	0.6607	0.0006	1350.5	2010.8	54.947	12.21	16.35	940
180	7.42873	0.7433	0.0006	1595.1	2338.0	56.875	12.19	16.32	998
200	8.25429	0.8260	0.0005	1836.7	2662.1	58.582	11.95	16.08	1054
220	9.07970	0.9086	0.0005	2072.6	2980.5	60.099	11.63	15.76	1110
240	9.90502	0.9911	0.0004	2302.1	3292.6	61.457	11.33	15.46	1163
260	10.73026	1.0737	0.0004	2526.1	3599.1	62.684	11.07	15.20	1214
280	11.55544	1.1562	0.0004	2745.5	3901.0	63.803	10.87	15.00	1263
300	12.37968	1.2387	0.0003	2961.4	4199.4	64.832	10.72	14.85	1310
350	14.44218	1.4450	0.0003	3491.4	4935.6	67.101	10.51	14.63	1419
400	16.50455	1.6512	0.0003	4014.4	5664.9	69.050	10.43	14.55	1518
450	18.56684	1.8575	0.0002	4534.7	6391.4	70.763	10.41	14.53	1611
500	20.62908	2.0637	0.0002	5055.0	7117.9	72.292	10.40	14.52	1698
550	22.69128	2.2700	0.0002	5575.0	7844.1	73.677	10.41	14.53	1780
600	24.75345	2.4762	0.0002	6096.0	8571.3	74.941	10.42	14.55	1859
700	28.87775	2.8886	0.0001	7140.9	10028.7	77.189	10.48	14.61	2006
800	33.00201	3.3011	0.0001	8193.4	11493.6	79.143	10.57	14.70	2142
900	37.12624	3.7135	0.0001	9256.7	12969.3	80.880	10.70	14.82	2268
1000	41.25045	4.1259	0.0001	10333.9	14459.0	82.452	10.86	14.98	2386
1200	49.49885	4.9508	0.0001	12544.2	17494.0	85.206	11.24	15.37	2601
1400	57.74736	5.7756	0.0001	14837.2	20611.9	87.613	11.68	15.81	2795
1600	65.99738	6.6004	0.0001	17220.5	23820.2	89.732	12.15	16.28	2974
1800	74.25729	7.4253	0.0001	19715.2	27140.9	91.649	12.78	16.92	3136
2000	82.55973	8.2501	0.0001	22395.0	30651.0	93.563	13.97	18.19	3277
2500	104.41653	10.3122	0.0000	31587.8	42029.5	100.941	24.99	30.36	3538
3000	133.52968	12.3743	0.0000	52916.7	66269.7	113.034	61.69	73.70	3839

* TWO-PHASE BOUNDARY

Table 4.1 (continued)
THERMODYNAMIC PROPERTIES OF PARAHYDROGEN (ISOBARS, SI UNITS)

0.10 MPA ISOBAR

TEMPERATURE DEG. K	DENSITY KG/CU M	V(DH/DV)$_P$ KJ/KG	V(DP/DU)$_V$ MPA-CU M/KJ X 10^2	-V(DP/DV)$_T$ MPA	-(DV/DT)$_P$/V 1/DEG. K	THERMAL CONDUCTIVITY W/K-M X 10^3	VISCOSITY KG/M-S X 10^7	THERMAL DIFFUSIVITY SQ M/HR	DIELECTRIC CONSTANT	PRANDTL NUMBER
* 13.834	77.0685	621.7	0.25639	90.0543	0.01025	73.01	255.39	0.00053	1.25176	2.2323
14	76.9216	617.1	0.25492	88.0151	0.01050	74.68	249.37	0.00054	1.25124	2.1639
15	76.0988	607.3	0.24330	79.2124	0.01143	83.15	219.54	0.00056	1.24835	1.8428
16	75.1978	610.0	0.23204	73.3062	0.01218	89.01	195.44	0.00057	1.24519	1.6308
17	74.2500	609.5	0.22298	67.6550	0.01297	92.85	175.96	0.00057	1.24187	1.4981
18	73.2568	606.6	0.21545	62.2739	0.01386	95.55	159.92	0.00056	1.23840	1.4069
19	72.2131	598.8	0.20897	56.7088	0.01495	97.09	146.46	0.00054	1.23475	1.3501
20	71.1039	588.9	0.20322	51.3422	0.01617	98.44	134.88	0.00053	1.23089	1.3051
* 20.224	70.8360	588.4	0.20198	50.4463	0.01638	98.34	132.43	0.00052	1.22996	1.2911
* 20.224	1.3253	188.9	0.06666	0.0894	0.06427	16.89	11.25	0.00378	1.00400	0.8085
21	1.2630	197.5	0.06698	0.0906	0.06304	17.38	11.66	0.00418	1.00381	0.7956
22	1.1921	208.8	0.06722	0.0918	0.05555	18.06	12.19	0.00470	1.00360	0.7829
23	1.1299	220.1	0.06733	0.0929	0.05186	18.76	12.71	0.00524	1.00341	0.7729
24	1.0745	231.3	0.06738	0.0937	0.04872	19.49	13.22	0.00579	1.00324	0.7646
25	1.0248	242.5	0.06739	0.0944	0.04600	20.23	13.73	0.00637	1.00309	0.7576
26	0.9799	253.7	0.06738	0.0950	0.04362	20.97	14.24	0.00696	1.00296	0.7516
27	0.9391	264.8	0.06736	0.0955	0.04151	21.71	14.74	0.00757	1.00283	0.7463
28	0.9018	275.9	0.06733	0.0959	0.03961	22.46	15.23	0.00820	1.00272	0.7414
29	0.8675	286.9	0.06730	0.0963	0.03791	23.20	15.72	0.00885	1.00262	0.7369
30	0.8359	297.8	0.06727	0.0966	0.03636	23.95	16.21	0.00953	1.00252	0.7326
31	0.8066	308.7	0.06725	0.0969	0.03494	24.64	16.69	0.01020	1.00243	0.7304
32	0.7795	319.6	0.06722	0.0971	0.03364	25.33	17.16	0.01089	1.00235	0.7283
33	0.7541	330.4	0.06720	0.0974	0.03244	26.02	17.63	0.01159	1.00227	0.7264
34	0.7305	341.2	0.06718	0.0976	0.03133	26.71	18.10	0.01231	1.00220	0.7242
35	0.7083	351.9	0.06715	0.0978	0.03030	27.40	18.56	0.01305	1.00214	0.7221
36	0.6875	362.7	0.06713	0.0979	0.02934	28.10	19.01	0.01383	1.00207	0.7201
37	0.6679	373.4	0.06711	0.0981	0.02844	28.79	19.47	0.01461	1.00201	0.7180
38	0.6495	384.1	0.06708	0.0982	0.02760	29.49	19.91	0.01541	1.00196	0.7160
39	0.6320	394.8	0.06704	0.0983	0.02682	30.18	20.36	0.01624	1.00191	0.7141
40	0.6155	405.5	0.06701	0.0985	0.02607	30.88	20.79	0.01708	1.00186	0.7121
42	0.5851	427.0	0.06692	0.0987	0.02471	32.26	21.66	0.01881	1.00176	0.7086
44	0.5576	448.6	0.06679	0.0988	0.02349	33.64	22.51	0.02061	1.00168	0.7052
46	0.5326	470.3	0.06663	0.0990	0.02239	35.03	23.35	0.02248	1.00161	0.7020
48	0.5097	492.3	0.06643	0.0991	0.02140	36.41	24.17	0.02442	1.00154	0.6991
50	0.4888	514.5	0.06617	0.0992	0.02049	37.80	24.98	0.02641	1.00147	0.6965
52	0.4696	537.2	0.06585	0.0993	0.01965	39.19	25.77	0.02845	1.00142	0.6944
54	0.4519	560.2	0.06545	0.0994	0.01883	40.58	26.56	0.03055	1.00136	0.6926
56	0.4354	583.9	0.06499	0.0994	0.01818	41.96	27.33	0.03268	1.00131	0.6914
58	0.4202	608.2	0.06445	0.0995	0.01753	43.35	28.09	0.03485	1.00127	0.6907
60	0.4059	633.1	0.06384	0.0995	0.01692	44.81	28.84	0.03710	1.00122	0.6895
65	0.3743	699.2	0.06198	0.0997	0.01557	48.67	30.67	0.04299	1.00113	0.6863
70	0.3473	771.1	0.05972	0.0997	0.01443	52.52	32.45	0.04894	1.00105	0.6874
75	0.3239	849.6	0.05718	0.0998	0.01345	56.37	34.17	0.05485	1.00098	0.6925
80	0.3035	934.9	0.05450	0.0999	0.01259	60.27	35.85	0.06074	1.00091	0.7001
85	0.2855	1027.1	0.05182	0.0999	0.01183	65.22	37.48	0.06765	1.00086	0.6986
90	0.2696	1125.4	0.04923	0.0999	0.01117	70.17	39.08	0.07455	1.00081	0.6999
95	0.2553	1229.3	0.04681	0.1000	0.01057	75.12	40.63	0.08149	1.00077	0.7030
100	0.2425	1337.7	0.04461	0.1000	0.01004	89.54	45.74	0.09898	1.00073	0.6859
120	0.2020	1792.9	0.03819	0.1000	0.00835	117.11	54.08	0.13941	1.00061	0.6913
140	0.1731	2231.5	0.03497	0.1000	0.00715	136.13	59.19	0.17742	1.00052	0.6939
160	0.1514	2613.7	0.03383	0.1001	0.00625	148.29	63.00	0.21565	1.00046	0.6944
180	0.1346	2936.5	0.03390	0.1001	0.00556	155.65	66.17	0.25509	1.00041	0.6937
200	0.1211	3215.0	0.03457	0.1001	0.00500	160.20	69.01	0.29613	1.00037	0.6925
220	0.1101	3467.7	0.03550	0.1001	0.00454	163.41	71.66	0.33890	1.00033	0.6912
240	0.1010	3710.4	0.03645	0.1001	0.00417	166.22	74.19	0.38347	1.00030	0.6899
260	0.0932	3953.2	0.03728	0.1001	0.00385	169.14	76.64	0.42984	1.00029	0.6887
280	0.0865	4201.0	0.03797	0.1001	0.00357	172.36	79.04	0.47806	1.00026	0.6878
300	0.0808	4455.4	0.03850	0.1001	0.00333	175.31	81.41	0.52807	1.00024	0.6870
350	0.0692	5123.7	0.03927	0.1001	0.00286	186.13	87.23	0.66129	1.00021	0.6858
400	0.0606	5822.2	0.03958	0.1000	0.00250	197.45	92.97	0.80628	1.00018	0.6851
450	0.0539	6540.8	0.03966	0.1000	0.00222	209.34	98.65	0.96299	1.00016	0.6847
500	0.0485	7264.8	0.03967	0.1000	0.00200	221.28	104.26	1.13139	1.00015	0.6844
550	0.0441	7995.9	0.03964	0.1000	0.00182	233.29	109.82	1.31129	1.00013	0.6841
600	0.0404	8732.3	0.03958	0.1000	0.00167	245.30	115.31	1.50244	1.00012	0.6840
700	0.0346	10226.5	0.03936	0.1000	0.00143	269.39	126.11	1.91754	1.00010	0.6837
800	0.0303	11760.2	0.03902	0.1000	0.00125	293.77	136.66	2.37483	1.00009	0.6837
900	0.0269	13344.2	0.03856	0.1000	0.00111	318.59	146.95	2.87254	1.00008	0.6837
1000	0.0242	14985.9	0.03799	0.1000	0.00100	440.56	157.01	4.36658	1.00007	0.5340
1200	0.0202	18444.1	0.03669	0.1000	0.00083	511.15	176.49	5.92711	1.00006	0.5306
1400	0.0173	22136.8	0.03530	0.1000	0.00071	583.71	195.19	7.67556	1.00005	0.5267
1600	0.0152	26050.3	0.03395	0.1000	0.00062	659.22	213.23	9.62086	1.00005	0.5266
1800	0.0135	30461.4	0.03229	0.1000	0.00055	749.49	230.69	11.83865	1.00004	0.5209
2000	0.0121	36358.3	0.02954	0.0999	0.00050	892.15	247.69	14.57561	1.00004	0.5051
2500	0.0096	74954.9	0.01671	0.0988	0.00041	2131.13	289.81	26.38781	1.00003	0.4128
3000	0.0075	204885.4	0.00721	0.0927	0.00035	6758.73	338.77	44.08577	1.00002	0.3694

* TWO-PHASE BOUNDARY

Table 4.1 (continued)
THERMODYNAMIC PROPERTIES OF PARAHYDROGEN (ISOBARS, SI UNITS)

0.101325 MPA ISOBAR

TEMPERATURE DEG. K	VOLUME CU M/KG	ISOTHERM DERIVATIVE CU M-MPA/KG	ISOCHORE DERIVATIVE MPA/K	INTERNAL ENERGY KJ/KG-K	ENTHALPY KJ/KG	ENTROPY KJ/KG-K	CV KJ / KG-K	CP KJ / KG-K	VELOCITY OF SOUND M/S
* 13.835	0.01298	1.1686	0.9244	-308.9	-307.6	4.968	4.68	6.38	1263
14	0.01300	1.1444	0.9243	-307.8	-306.5	5.047	4.71	6.48	1254
15	0.01314	1.0411	0.9105	-301.1	-299.8	5.508	4.92	6.98	1216
16	0.01330	0.9750	0.8926	-293.9	-292.6	5.974	5.12	7.43	1190
17	0.01347	0.9113	0.8775	-286.3	-284.9	6.440	5.30	7.90	1166
18	0.01365	0.8502	0.8630	-278.1	-276.7	6.906	5.47	8.41	1143
19	0.01385	0.7854	0.8477	-269.5	-268.1	7.375	5.62	8.95	1119
20	0.01406	0.7222	0.8305	-260.3	-258.8	7.848	5.75	9.52	1094
* 20.268	0.01413	0.7093	0.8253	-257.7	-256.2	7.977	5.78	9.66	1089
* 20.268	0.74739	0.0675	0.0058	113.6	189.3	29.967	6.50	12.15	355
21	0.78035	0.0715	0.0055	118.8	197.9	30.383	6.44	11.89	363
22	0.82689	0.0768	0.0052	125.9	209.7	30.929	6.37	11.62	374
23	0.87260	0.0820	0.0049	132.8	221.2	31.442	6.33	11.43	385
24	0.91769	0.0870	0.0046	139.6	232.5	31.925	6.31	11.29	395
25	0.96226	0.0919	0.0044	146.3	243.8	32.383	6.29	11.17	404
26	1.00643	0.0968	0.0042	152.9	254.9	32.819	6.27	11.08	413
27	1.05025	0.1015	0.0040	159.5	265.9	33.236	6.26	11.00	422
28	1.09378	0.1062	0.0039	166.1	276.9	33.635	6.26	10.94	431
29	1.13705	0.1109	0.0037	172.6	287.8	34.018	6.25	10.88	439
30	1.18011	0.1154	0.0036	179.1	298.7	34.386	6.25	10.83	448
31	1.22297	0.1200	0.0034	185.6	309.5	34.741	6.24	10.79	456
32	1.26567	0.1245	0.0033	192.0	320.3	35.083	6.24	10.76	463
33	1.30822	0.1290	0.0032	198.4	331.0	35.413	6.23	10.72	471
34	1.35063	0.1335	0.0031	204.9	341.7	35.733	6.23	10.69	479
35	1.39293	0.1379	0.0030	211.3	352.4	36.043	6.23	10.67	486
36	1.43512	0.1424	0.0029	217.6	363.0	36.343	6.23	10.65	493
37	1.47721	0.1468	0.0028	224.0	373.7	36.634	6.23	10.62	501
38	1.51922	0.1512	0.0027	230.4	384.3	36.917	6.22	10.61	508
39	1.56115	0.1555	0.0027	236.7	394.9	37.193	6.22	10.59	514
40	1.60300	0.1599	0.0026	243.1	405.5	37.461	6.22	10.58	521
42	1.68652	0.1686	0.0025	255.7	426.6	37.976	6.23	10.56	535
44	1.76982	0.1772	0.0024	268.4	447.7	38.467	6.23	10.54	547
46	1.85292	0.1858	0.0022	281.0	468.8	38.935	6.25	10.53	560
48	1.93587	0.1943	0.0021	293.7	489.9	39.384	6.26	10.53	572
50	2.01867	0.2029	0.0021	306.4	510.9	39.814	6.28	10.54	583
52	2.10135	0.2114	0.0020	319.1	532.0	40.228	6.31	10.56	595
54	2.18393	0.2199	0.0019	331.9	553.2	40.626	6.35	10.58	606
56	2.26641	0.2283	0.0018	344.7	574.4	41.012	6.39	10.62	616
58	2.34880	0.2368	0.0018	357.7	595.7	41.385	6.44	10.66	626
60	2.43113	0.2452	0.0017	370.7	617.0	41.748	6.50	10.71	636
65	2.63666	0.2662	0.0016	403.9	671.0	42.612	6.69	10.89	658
70	2.84187	0.2872	0.0015	438.1	726.0	43.427	6.94	11.13	679
75	3.04684	0.3081	0.0014	473.7	782.4	44.205	7.25	11.42	697
80	3.25161	0.3290	0.0013	510.9	840.4	44.953	7.60	11.77	714
85	3.45621	0.3498	0.0012	550.0	900.2	45.678	7.99	12.16	730
90	3.66069	0.3706	0.0011	591.1	962.0	46.384	8.41	12.57	744
95	3.86506	0.3914	0.0011	634.3	1025.9	47.075	8.84	13.00	759
100	4.06933	0.4122	0.0010	679.6	1091.9	47.753	9.27	13.43	773
120	4.88574	0.4952	0.0008	881.5	1376.6	50.341	10.83	14.97	827
140	5.70143	0.5780	0.0007	1109.2	1686.9	52.731	11.82	15.96	883
160	6.51568	0.6607	0.0006	1350.5	2010.8	54.893	12.21	16.35	940
180	7.33165	0.7434	0.0006	1595.1	2338.0	56.820	12.19	16.32	998
200	8.14642	0.8260	0.0005	1836.7	2662.1	58.527	11.95	16.08	1054
220	8.96105	0.9086	0.0005	2072.5	2980.5	60.045	11.63	15.76	1110
240	9.77558	9911	0.0004	2302.1	3292.6	61.403	11.33	15.46	1163
260	10.59003	0737	0.0004	2526.1	3599.1	62.630	11.07	15.20	1214
280	11.40442	1563	0.0004	2745.5	3901.0	63.749	10.87	15.00	1263
300	12.21788	1.2387	0.0003	2961.4	4199.4	64.778	10.72	14.85	1310
350	14.25342	1.4450	0.0003	3491.4	4935.6	67.047	10.51	14.63	1419
400	16.28883	1.6513	0.0003	4014.4	5664.9	68.996	10.43	14.55	1518
450	18.32416	1.8575	0.0002	4534.7	6391.4	70.708	10.41	14.53	1611
500	20.35943	2.0638	0.0002	5055.0	7117.9	72.238	10.40	14.52	1698
550	22.39466	2.2700	0.0002	5575.0	7844.1	73.623	10.41	14.53	1780
600	24.42987	2.4762	0.0002	6096.0	8571.3	74.887	10.42	14.55	1859
700	28.50024	2.8887	0.0001	7140.9	10028.7	77.134	10.48	14.61	2006
800	32.57056	3.3011	0.0001	8193.4	11493.6	79.089	10.57	14.70	2142
900	36.64086	3.7135	0.0001	9256.7	12969.3	80.826	10.70	14.82	2268
1000	40.71114	4.1259	0.0001	10333.9	14459.0	82.398	10.86	14.98	2386
1200	48.85168	4.9508	0.0001	12544.2	17494.1	85.151	11.24	15.37	2601
1400	56.99233	5.7756	0.0001	14837.2	20611.9	87.558	11.68	15.81	2795
1600	65.13445	6.6065	0.0001	17220.5	23820.2	89.677	12.15	16.28	2974
1800	73.28628	7.4253	0.0001	197[illegible]4.9	27140.7	91.594	12.77	16.92	3136
2000	81.47980	8.2501	0.0001	22393.8	30649.8	93.508	13.97	18.18	3277
2500	103.04279	10.3122	0.0000	31569.9	42010.7	100.879	24.92	30.28	3539
3000	131.72066	12.3743	0.0000	52805.4	66152.0	112.936	61.40	73.34	3839

* TWO-PHASE BOUNDARY

Table 4.1 (continued)
THERMODYNAMIC PROPERTIES OF PARAHYDROGEN (ISOBARS, SI UNITS)

0.101325 MPA ISOBAR

TEMPERATURE DEG. K	DENSITY KG/CU M	$V(DH/DV)_P$ KJ/KG	$V(DP/DU)_V$ MPA-CU M/KJ X 10^2	$-V(DP/DV)_T$ MPA	$-(DV/DT)_P/V$ 1/DEG. K	THERMAL CONDUCTIVITY W/K-M X 10^3	VISCOSITY KG/M-S X 10^7	THERMAL DIFFUSIVITY SQ M/HR	DIELECTRIC CONSTANT	PRANDTL NUMBER
* 13.835	77.0693	621.7	0.25638	90.0617	0.01026	73.01	255.40	0.00053	1.25176	2.2322
14	76.9227	617.2	0.25493	88.0322	0.01050	74.68	249.39	0.00054	1.25125	2.1640
15	76.1000	607.3	0.24331	79.2241	0.01149	83.16	219.56	0.00056	1.24835	1.8429
16	75.1991	610.1	0.23205	73.3174	0.01217	89.01	195.46	0.00057	1.24519	1.6309
17	74.2514	609.6	0.22299	67.6660	0.01297	92.85	175.98	0.00057	1.24187	1.4981
18	73.2583	606.7	0.21546	62.2852	0.01386	95.56	159.94	0.00056	1.23840	1.4069
19	72.2148	598.9	0.20898	56.7197	0.01494	97.10	146.47	0.00054	1.23476	1.3501
20	71.1057	589.0	0.20322	51.3532	0.01617	98.44	134.89	0.00052	1.23089	1.3051
* 20.268	70.7859	587.9	0.20175	50.2056	0.01644	98.92	131.98	0.00052	1.22978	1.2893
* 20.268	1.3380	189.1	0.06667	0.0903	0.06424	16.94	11.28	0.00375	1.00404	0.8091
21	1.2815	197.3	0.06697	0.0916	0.06027	17.40	11.67	0.00411	1.00387	0.7970
22	1.2094	208.6	0.06721	0.0929	0.05573	18.08	12.19	0.00463	1.00365	0.7840
23	1.1460	219.8	0.06734	0.0940	0.05200	18.78	12.71	0.00516	1.00346	0.7738
24	1.0897	231.1	0.06739	0.0948	0.04883	19.50	13.23	0.00571	1.00329	0.7654
25	1.0392	242.3	0.06740	0.0955	0.04610	20.24	13.74	0.00628	1.00313	0.7582
26	0.9936	253.5	0.06739	0.0961	0.04370	20.98	14.24	0.00686	1.00300	0.7522
27	0.9522	264.6	0.06736	0.0967	0.04157	21.72	14.74	0.00746	1.00287	0.7468
28	0.9143	275.7	0.06734	0.0971	0.03967	22.47	15.24	0.00809	1.00276	0.7418
29	0.8795	286.7	0.06731	0.0975	0.03796	23.21	15.73	0.00873	1.00265	0.7373
30	0.8474	297.7	0.06728	0.0978	0.03640	23.96	16.21	0.00940	1.00256	0.7330
31	0.8177	308.6	0.06725	0.0981	0.03498	24.65	16.69	0.01006	1.00247	0.7308
32	0.7901	319.4	0.06723	0.0984	0.03367	25.34	17.16	0.01073	1.00238	0.7286
33	0.7644	330.3	0.06720	0.0986	0.03247	26.02	17.63	0.01143	1.00230	0.7266
34	0.7404	341.0	0.06718	0.0988	0.03136	26.72	18.10	0.01215	1.00223	0.7245
35	0.7179	351.8	0.06716	0.0990	0.03032	27.41	18.56	0.01288	1.00216	0.7224
36	0.6968	362.5	0.06714	0.0992	0.02936	28.10	19.02	0.01364	1.00210	0.7203
37	0.6770	373.3	0.06711	0.0994	0.02846	28.80	19.47	0.01441	1.00204	0.7182
38	0.6582	384.0	0.06708	0.0995	0.02762	29.49	19.91	0.01521	1.00198	0.7162
39	0.6406	394.7	0.06705	0.0996	0.02683	30.19	20.36	0.01602	1.00193	0.7142
40	0.6238	405.4	0.06701	0.0997	0.02609	30.88	20.80	0.01685	1.00188	0.7123
42	0.5929	426.9	0.06692	0.0999	0.02472	32.27	21.66	0.01856	1.00179	0.7087
44	0.5650	448.5	0.06680	0.1001	0.02350	33.65	22.51	0.02034	1.00170	0.7053
46	0.5397	470.2	0.06664	0.1003	0.02240	35.03	23.35	0.02218	1.00163	0.7021
48	0.5166	492.2	0.06643	0.1004	0.02140	36.42	24.17	0.02409	1.00156	0.6992
50	0.4954	514.5	0.06617	0.1005	0.02049	37.81	24.98	0.02606	1.00149	0.6966
52	0.4759	537.1	0.06585	0.1006	0.01966	39.19	25.78	0.02809	1.00143	0.6944
54	0.4579	560.2	0.06546	0.1007	0.01889	40.58	26.56	0.03015	1.00138	0.6927
56	0.4412	583.8	0.06499	0.1007	0.01819	41.97	27.33	0.03225	1.00133	0.6914
58	0.4257	608.1	0.06446	0.1008	0.01753	43.36	28.09	0.03439	1.00128	0.6907
60	0.4113	633.1	0.06384	0.1009	0.01692	44.81	28.84	0.03661	1.00124	0.6896
65	0.3793	699.1	0.06198	0.1010	0.01558	48.67	30.68	0.04242	1.00114	0.6864
70	0.3519	771.0	0.05972	0.1011	0.01443	52.53	32.45	0.04829	1.00106	0.6875
75	0.3282	849.5	0.05718	0.1011	0.01345	56.37	34.17	0.05413	1.00099	0.6925
80	0.3075	934.9	0.05451	0.1012	0.01259	60.27	35.85	0.05994	1.00093	0.7001
85	0.2893	1027.0	0.05182	0.1012	0.01184	65.22	37.48	0.06676	1.00087	0.6986
90	0.2732	1125.4	0.04923	0.1012	0.01117	70.17	39.08	0.07357	1.00082	0.7000
95	0.2587	1229.3	0.04681	0.1013	0.01057	75.12	40.64	0.08042	1.00078	0.7031
100	0.2457	1337.7	0.04461	0.1013	0.01004	89.54	45.74	0.09769	1.00074	0.6859
120	0.2047	1792.9	0.03819	0.1013	0.00835	117.11	54.08	0.13758	1.00062	0.6913
140	0.1754	2231.5	0.03497	0.1014	0.00715	136.13	59.19	0.17510	1.00053	0.6939
160	0.1535	2613.8	0.03384	0.1014	0.00625	148.29	63.00	0.21283	1.00046	0.6944
180	0.1364	2936.5	0.03390	0.1014	0.00556	155.65	66.17	0.25176	1.00041	0.6937
200	0.1228	3215.0	0.03457	0.1014	0.00500	160.20	69.01	0.29226	1.00037	0.6925
220	0.1116	3467.8	0.03550	0.1014	0.00454	163.41	71.66	0.33447	1.00034	0.6912
240	0.1023	3710.4	0.03645	0.1014	0.00417	166.22	74.19	0.37846	1.00031	0.6859
260	0.0944	3953.3	0.03728	0.1014	0.00385	169.14	76.64	0.42423	1.00028	0.6887
280	0.0877	4201.0	0.03797	0.1014	0.00357	172.36	79.04	0.47181	1.00026	0.6878
300	0.0818	4455.4	0.03850	0.1014	0.00333	175.91	81.41	0.52117	1.00025	0.6870
350	0.0702	5123.7	0.03927	0.1014	0.00286	186.13	87.23	0.65265	1.00021	0.6858
400	0.0614	5822.2	0.03958	0.1014	0.00250	197.45	92.97	0.79574	1.00018	0.6851
450	0.0546	6540.9	0.03966	0.1014	0.00222	209.34	98.65	0.95040	1.00016	0.6847
500	0.0491	7264.9	0.03967	0.1014	0.00200	221.28	104.26	1.11661	1.00015	0.6844
550	0.0447	7995.9	0.03964	0.1014	0.00182	233.29	109.82	1.29415	1.00013	0.6841
600	0.0409	8732.3	0.03958	0.1014	0.00167	245.30	115.31	1.48280	1.00012	0.6840
700	0.0351	10226.5	0.03936	0.1014	0.00143	269.40	126.11	1.89248	1.00011	0.6837
800	0.0307	11760.3	0.03902	0.1014	0.00125	293.77	136.66	2.34379	1.00009	0.6837
900	0.0273	13344.2	0.03856	0.1013	0.00111	318.59	146.95	2.83500	1.00008	0.6837
1000	0.0246	14985.9	0.03799	0.1013	0.00100	440.56	157.01	4.30949	1.00007	0.5340
1200	0.0205	18444.1	0.03669	0.1013	0.00083	511.15	176.49	5.84961	1.00006	0.5306
1400	0.0175	22136.8	0.03530	0.1013	0.00071	583.71	195.19	7.57520	1.00005	0.5287
1600	0.0154	26049.7	0.03395	0.1013	0.00062	659.20	213.23	9.49493	1.00005	0.5266
1800	0.0136	30457.5	0.03229	0.1013	0.00056	749.33	230.69	11.68297	1.00004	0.5210
2000	0.0123	36342.1	0.02956	0.1013	0.00050	891.52	247.69	14.38133	1.00004	0.3052
2500	0.0097	74756.5	0.01676	0.1001	0.00040	2123.61	289.80	26.01980	1.00003	0.4132
3000	0.0076	203993.2	0.00725	0.0939	0.00036	6725.27	338.71	43.48332	1.00002	0.3694

* TWO-PHASE BOUNDARY

Table 4.1 (continued)
THERMODYNAMIC PROPERTIES OF PARAHYDROGEN (ISOBARS, SI UNITS)

0.15 MPA ISOBAR

TEMPERATURE	VOLUME	ISOTHERM DERIVATIVE	ISOCHORE DERIVATIVE	INTERNAL ENERGY	ENTHALPY	ENTROPY	CV	CP	VELOCITY OF SOUND
DEG. K	CU M/KG	CU M-MPA/KG	MPA/K	KJ/KG-K	KJ/KG	KJ/KG-K	KJ /	KG-K	M/S
* 13.851	0.01297	1.1717	0.9250	-308.9	-306.9	4.969	4.68	6.38	1264
14	0.01299	1.1520	0.9249	-307.9	-305.9	5.040	4.71	6.47	1257
15	0.01313	1.0461	0.9122	-301.2	-299.3	5.501	4.92	6.97	1218
16	0.01329	0.9798	0.8942	-294.0	-292.1	5.966	5.11	7.42	1192
17	0.01346	0.9161	0.8791	-286.4	-284.4	6.431	5.30	7.89	1168
18	0.01364	0.8552	0.8646	-278.3	-276.2	6.897	5.46	8.39	1146
19	0.01384	0.7903	0.8493	-269.7	-267.6	7.365	5.61	8.93	1121
20	0.01405	0.7272	0.8321	-260.5	-258.4	7.837	5.74	9.50	1097
21	0.01429	0.6608	0.8128	-250.7	-248.5	8.316	5.86	10.14	1070
* 21.672	0.01447	0.6202	0.7984	-243.7	-241.6	8.644	5.92	10.58	1053
* 21.672	0.52255	0.0673	0.0084	117.6	196.0	28.842	6.59	12.84	362
22	0.53312	0.0692	0.0082	120.1	200.0	29.028	6.55	12.67	366
23	0.56539	0.0751	0.0077	127.5	212.5	29.582	6.46	12.25	378
24	0.59879	0.0807	0.0072	134.7	224.6	30.096	6.40	11.95	388
25	0.63055	0.0861	0.0068	141.8	236.4	30.579	6.36	11.73	399
26	0.66180	0.0914	0.0065	148.8	248.0	31.036	6.33	11.55	408
27	0.69263	0.0965	0.0062	155.6	259.5	31.469	6.31	11.42	418
28	0.72313	0.1015	0.0059	162.4	270.9	31.882	6.30	11.30	427
29	0.75333	0.1065	0.0056	169.1	282.1	32.277	6.29	11.21	436
30	0.78329	0.1113	0.0054	175.8	293.3	32.656	6.28	11.13	444
31	0.81304	0.1161	0.0052	182.4	304.4	33.020	6.27	11.06	453
32	0.84260	0.1209	0.0050	189.0	315.4	33.370	6.26	11.00	461
33	0.87200	0.1256	0.0048	195.6	326.4	33.708	6.26	10.94	469
34	0.90126	0.1302	0.0047	202.1	337.3	34.034	6.25	10.90	476
35	0.93039	0.1348	0.0045	208.6	348.2	34.349	6.25	10.86	484
36	0.95940	0.1394	0.0044	215.1	359.0	34.654	6.25	10.82	491
37	0.98832	0.1439	0.0043	221.6	369.8	34.950	6.24	10.79	499
38	1.01714	0.1485	0.0041	228.0	380.6	35.237	6.24	10.76	506
39	1.04588	0.1530	0.0040	234.5	391.3	35.516	6.24	10.73	513
40	1.07455	0.1574	0.0039	240.9	402.1	35.788	6.24	10.71	520
42	1.13168	0.1663	0.0037	253.7	423.4	36.309	6.24	10.67	533
44	1.18858	0.1751	0.0035	266.5	444.8	36.805	6.25	10.64	546
46	1.24528	0.1839	0.0034	279.2	466.0	37.278	6.25	10.63	559
48	1.30182	0.1926	0.0032	292.0	487.3	37.730	6.27	10.62	571
50	1.35822	0.2013	0.0031	304.8	508.5	38.163	6.29	10.62	583
52	1.41449	0.2099	0.0029	317.6	529.8	38.580	6.32	10.63	594
54	1.47065	0.2185	0.0028	330.4	551.0	38.981	6.35	10.65	605
56	1.52672	0.2271	0.0027	343.3	572.3	39.369	6.39	10.67	616
58	1.58270	0.2356	0.0026	356.3	593.7	39.744	6.45	10.71	626
60	1.63861	0.2441	0.0025	369.4	615.2	40.108	6.50	10.76	636
65	1.77810	0.2654	0.0023	402.7	669.4	40.976	6.69	10.93	658
70	1.91727	0.2865	0.0022	437.0	724.6	41.794	6.94	11.16	679
75	2.05619	0.3075	0.0020	472.7	781.1	42.574	7.25	11.45	697
80	2.19491	0.3285	0.0019	510.0	839.2	43.324	7.60	11.79	714
85	2.33348	0.3495	0.0018	549.1	899.2	44.050	7.99	12.18	730
90	2.47191	0.3704	0.0017	590.3	961.1	44.758	8.41	12.59	745
95	2.61023	0.3913	0.0016	633.5	1025.1	45.449	8.84	13.01	759
100	2.74846	0.4121	0.0015	678.9	1091.2	46.128	9.28	13.44	773
120	3.30069	0.4953	0.0013	881.0	1376.1	48.719	10.83	14.98	828
140	3.85219	0.5782	0.0011	1108.8	1686.6	51.110	11.82	15.96	884
160	4.40326	0.6611	0.0009	1350.1	2010.6	53.273	12.21	16.35	941
180	4.95405	0.7438	0.0008	1594.8	2337.9	55.201	12.19	16.32	998
200	5.50464	0.8265	0.0008	1836.4	2662.1	56.908	11.95	16.08	1055
220	6.05509	0.9091	0.0007	2072.3	2980.6	58.426	11.63	15.76	1110
240	6.60544	0.9918	0.0006	2301.9	3292.7	59.784	11.33	15.46	1163
260	7.15571	1.0744	0.0006	2525.9	3599.2	61.011	11.07	15.20	1214
280	7.70593	1.1569	0.0005	2745.3	3901.2	62.130	10.87	15.00	1263
300	8.25551	1.2394	0.0005	2961.2	4199.6	63.160	10.72	14.85	1310
350	9.63066	1.4457	0.0004	3491.3	4935.9	65.429	10.51	14.63	1419
400	11.00567	1.6520	0.0004	4014.3	5665.2	67.378	10.43	14.55	1518
450	12.38060	1.8583	0.0003	4534.6	6391.7	69.090	10.41	14.53	1611
500	13.75547	2.0646	0.0003	5055.0	7118.3	70.620	10.40	14.53	1698
550	15.13031	2.2708	0.0003	5574.9	7844.5	72.004	10.41	14.53	1781
600	16.50512	2.4770	0.0003	6095.9	8571.7	73.268	10.42	14.55	1859
700	19.25470	2.8895	0.0002	7140.9	10029.1	75.516	10.48	14.61	2007
800	22.00423	3.3019	0.0002	8193.4	11494.0	77.471	10.57	14.70	2142
900	24.75374	3.7144	0.0002	9256.7	12969.7	79.208	10.70	14.82	2268
1000	27.50323	4.1268	0.0002	10333.9	14459.4	80.780	10.86	14.98	2386
1200	33.00218	4.9517	0.0001	12544.1	17494.5	83.533	11.24	15.37	2602
1400	38.50118	5.7765	0.0001	14837.1	20612.3	85.940	11.68	15.81	2796
1600	44.00101	6.6013	0.0001	17219.4	23819.5	88.059	12.14	16.27	2974
1800	49.50621	7.4262	0.0001	19708.0	27133.9	89.972	12.72	16.86	3138
2000	55.03455	8.2510	0.0001	22362.5	30617.7	91.873	13.76	17.96	3282
2500	69.45464	10.3131	0.0001	31085.9	41504.1	99.039	22.91	28.04	3552
3000	87.83039	12.3752	0.0001	49798.0	62972.6	110.142	53.35	63.69	3839

* TWO-PHASE BOUNDARY

Table 4.1 (continued)
THERMODYNAMIC PROPERTIES OF PARAHYDROGEN (ISOBARS, SI UNITS)

0.15 MPA ISOBAR

TEMPERATURE DEG. K	DENSITY KG/CU M	$V(DH/DV)_P$ KJ/KG	$V(DP/DU)_V$ MPA-CU M/KJ x 10^2	$-V(DP/DV)_T$ MPA	$-(DV/DT)_P/V$ 1/DEG. K	THERMAL CONDUCTIVITY W/K-M x 10^3	VISCOSITY KG/M-S x 10^7	THERMAL DIFFUSIVITY SQ M/HR	DIELECTRIC CONSTANT	PRANDTL NUMBER
* 13.851	77.0972	623.3	0.25632	90.3332	0.01024	73.22	255.70	0.00054	1.25186	2.2289
14	76.9648	620.0	0.25501	88.6598	0.01043	74.72	250.26	0.00054	1.25139	2.1663
15	76.1448	609.0	0.24371	79.6566	0.01145	83.21	220.31	0.00056	1.24851	1.8463
16	75.2472	611.7	0.23242	73.7262	0.01213	89.09	196.11	0.00057	1.24536	1.6332
17	74.3038	611.3	0.22334	68.0668	0.01292	92.95	176.57	0.00057	1.24206	1.4998
18	73.3154	608.6	0.21579	62.6992	0.01379	95.67	160.49	0.00056	1.23860	1.4078
19	72.2750	600.8	0.20930	57.1192	0.01487	97.23	146.98	0.00054	1.23497	1.3505
20	71.1714	591.1	0.20352	51.7551	0.01608	98.59	135.36	0.00052	1.23112	1.3048
21	69.9823	577.2	0.19830	46.2470	0.01759	100.14	125.13	0.00051	1.22699	1.2675
* 21.672	69.1286	568.4	0.19499	42.8767	0.01962	100.73	118.87	0.00050	1.22403	1.2490
* 21.672	1.9137	196.1	0.06680	0.1287	0.06547	18.61	12.17	0.00273	1.00578	0.8398
22	1.8757	199.9	0.06699	0.1299	0.06337	18.79	12.34	0.00285	1.00566	0.8319
23	1.7656	211.6	0.06737	0.1326	0.05790	19.39	12.85	0.00323	1.00533	0.8119
24	1.6700	223.2	0.06757	0.1348	0.05353	20.04	13.36	0.00361	1.00504	0.7967
25	1.5859	234.9	0.06765	0.1366	0.04993	20.72	13.86	0.00401	1.00479	0.7847
26	1.5110	246.5	0.06768	0.1381	0.04688	21.41	14.36	0.00442	1.00456	0.7751
27	1.4438	258.0	0.06767	0.1394	0.04426	22.12	14.86	0.00483	1.00436	0.7669
28	1.3829	269.4	0.06764	0.1404	0.04196	22.83	15.35	0.00526	1.00417	0.7598
29	1.3274	280.8	0.06760	0.1413	0.03992	23.55	15.83	0.00570	1.00401	0.7535
30	1.2767	292.0	0.06757	0.1421	0.03810	24.28	16.31	0.00615	1.00385	0.7478
31	1.2300	303.2	0.06753	0.1428	0.03647	24.96	16.79	0.00661	1.00371	0.7439
32	1.1868	314.4	0.06749	0.1434	0.03499	25.64	17.26	0.00707	1.00358	0.7405
33	1.1468	325.5	0.06746	0.1440	0.03363	26.32	17.73	0.00755	1.00346	0.7373
34	1.1096	336.5	0.06744	0.1445	0.03239	27.00	18.19	0.00804	1.00335	0.7342
35	1.0748	347.4	0.06741	0.1449	0.03124	27.68	18.65	0.00854	1.00324	0.7312
36	1.0423	358.4	0.06738	0.1453	0.03019	28.37	19.10	0.00906	1.00314	0.7284
37	1.0118	369.3	0.06735	0.1456	0.02921	29.06	19.55	0.00959	1.00305	0.7257
38	0.9831	380.2	0.06732	0.1460	0.02829	29.75	20.00	0.01013	1.00297	0.7231
39	0.9561	391.0	0.06728	0.1462	0.02744	30.43	20.44	0.01068	1.00288	0.7206
40	0.9306	401.9	0.06724	0.1465	0.02664	31.12	20.88	0.01124	1.00281	0.7182
42	0.8836	423.7	0.06713	0.1470	0.02519	32.49	21.74	0.01241	1.00266	0.7139
44	0.8413	445.5	0.06700	0.1473	0.02389	33.87	22.58	0.01361	1.00254	0.7098
46	0.8030	467.4	0.06683	0.1477	0.02273	35.24	23.42	0.01487	1.00242	0.7061
48	0.7682	489.6	0.06662	0.1479	0.02169	36.62	24.24	0.01616	1.00232	0.7027
50	0.7363	512.0	0.06635	0.1482	0.02074	38.00	25.04	0.01750	1.00222	0.6997
52	0.7070	534.8	0.06602	0.1484	0.01987	39.38	25.84	0.01887	1.00213	0.6973
54	0.6800	558.1	0.06562	0.1486	0.01908	40.76	26.62	0.02027	1.00205	0.6952
56	0.6550	581.8	0.06515	0.1487	0.01835	42.14	27.39	0.02170	1.00197	0.6937
58	0.6318	606.2	0.06460	0.1489	0.01767	43.53	28.15	0.02315	1.00190	0.6928
60	0.6103	631.3	0.06398	0.1490	0.01705	44.98	28.90	0.02465	1.00184	0.6914
65	0.5624	697.5	0.06211	0.1492	0.01567	48.82	30.73	0.02859	1.00170	0.6879
70	0.5216	769.6	0.05984	0.1494	0.01450	52.67	32.50	0.03257	1.00157	0.6887
75	0.4863	848.2	0.05729	0.1496	0.01350	56.50	34.22	0.03652	1.00147	0.6936
80	0.4556	933.7	0.05460	0.1497	0.01263	60.39	35.89	0.04046	1.00137	0.7010
85	0.4285	1025.9	0.05190	0.1498	0.01187	65.34	37.52	0.04507	1.00129	0.6993
90	0.4045	1124.4	0.04931	0.1498	0.01120	70.28	39.12	0.04969	1.00122	0.7005
95	0.3831	1228.3	0.04688	0.1499	0.01059	75.23	40.67	0.05432	1.00115	0.7035
100	0.3638	1336.8	0.04468	0.1499	0.01006	89.56	45.74	0.06592	1.00110	0.6865
120	0.3030	1792.3	0.03823	0.1501	0.00836	117.10	54.07	0.09288	1.00091	0.6917
140	0.2596	2231.3	0.03500	0.1501	0.00715	136.12	59.19	0.11825	1.00078	0.6942
160	0.2271	2613.9	0.03386	0.1501	0.00626	148.28	63.00	0.14376	1.00068	0.6947
180	0.2019	2936.9	0.03392	0.1501	0.00556	155.65	66.17	0.17007	1.00061	0.6939
200	0.1817	3215.6	0.03460	0.1501	0.00500	160.21	69.02	0.19745	1.00055	0.6927
220	0.1652	3468.5	0.03552	0.1501	0.00454	163.42	71.67	0.22598	1.00050	0.6913
240	0.1514	3711.3	0.03646	0.1501	0.00417	166.24	74.20	0.25572	1.00046	0.6900
260	0.1397	3954.2	0.03730	0.1501	0.00384	169.16	76.65	0.28665	1.00042	0.6888
280	0.1298	4202.0	0.03799	0.1501	0.00357	172.38	79.05	0.31881	1.00039	0.6879
300	0.1211	4456.5	0.03852	0.1501	0.00333	175.94	81.42	0.35216	1.00036	0.6871
350	0.1038	5124.9	0.03929	0.1501	0.00286	186.16	87.24	0.44102	1.00031	0.6859
400	0.0909	5823.5	0.03960	0.1501	0.00250	197.48	92.98	0.53771	1.00027	0.6851
450	0.0808	6542.2	0.03967	0.1501	0.00222	209.37	98.66	0.64221	1.00024	0.6847
500	0.0727	7266.2	0.03968	0.1501	0.00200	221.32	104.28	0.75453	1.00022	0.6844
550	0.0661	7997.3	0.03965	0.1501	0.00182	233.34	109.84	0.87450	1.00020	0.6841
600	0.0606	8733.7	0.03958	0.1501	0.00167	245.35	115.33	1.00197	1.00018	0.6840
700	0.0519	10228.0	0.03937	0.1501	0.00143	269.45	126.14	1.27879	1.00016	0.6837
800	0.0454	11761.7	0.03903	0.1501	0.00125	293.83	136.68	1.58374	1.00014	0.6837
900	0.0404	13345.8	0.03856	0.1501	0.00111	318.66	146.98	1.91565	1.00012	0.6837
1000	0.0364	14987.5	0.03799	0.1500	0.00100	440.56	157.05	2.91135	1.00011	0.5341
1200	0.0303	18445.6	0.03669	0.1500	0.00083	511.14	176.53	3.95174	1.00009	0.5307
1400	0.0260	22136.6	0.03531	0.1500	0.00071	583.63	195.24	5.11720	1.00008	0.5288
1600	0.0227	26033.2	0.03398	0.1500	0.00062	658.47	213.28	6.41171	1.00007	0.5269
1800	0.0202	30352.1	0.03244	0.1500	0.00056	745.15	230.74	7.87596	1.00006	0.5221
2000	0.0182	35904.1	0.03000	0.1499	0.00050	874.55	247.74	9.64689	1.00005	0.5088
2500	0.0144	69392.8	0.01819	0.1485	0.00040	1920.33	289.57	17.12389	1.00004	0.4228
3000	0.0114	179472.6	0.00823	0.1409	0.00035	5805.59	337.08	29.82253	1.00003	0.3698

* TWO-PHASE BOUNDARY

THERMODYNAMIC PROPERTIES OF PARAHYDROGEN (ISOBARS, SI UNITS)

0.20 MPA ISOBAR

TEMPERATURE DEG. K	VOLUME CU M/KG	ISOTHERM DERIVATIVE CU M-MPA/KG	ISOCHORE DERIVATIVE MPA/K	INTERNAL ENERGY KJ/KG-K	ENTHALPY KJ/KG	ENTROPY KJ/KG-K	CV KJ / KG-K	CP KJ / KG-K	VELOCITY OF SOUND M/S
* 13.868	0.01297	1.1749	0.9255	-308.9	-306.3	4.971	4.68	6.38	1265
14	0.01299	1.1597	0.9255	-308.0	-305.4	5.034	4.71	6.46	1260
15	0.01312	1.0513	0.9140	-301.3	-298.7	5.494	4.91	6.97	1221
16	0.01328	0.9847	0.8959	-294.2	-291.5	5.959	5.11	7.41	1195
17	0.01345	0.9209	0.8807	-286.5	-283.9	6.423	5.29	7.88	1171
18	0.01363	0.8603	0.8662	-278.4	-275.7	6.888	5.46	8.38	1149
19	0.01382	0.7953	0.8509	-269.8	-267.1	7.355	5.61	8.92	1124
20	0.01404	0.7323	0.8337	-260.7	-257.9	7.826	5.74	9.48	1100
21	0.01427	0.6661	0.8146	-251.0	-248.1	8.304	5.85	10.12	1073
22	0.01454	0.5995	0.7928	-240.5	-237.6	8.790	5.95	10.82	1044
* 22.805	0.01478	0.5396	0.7723	-231.6	-228.6	9.192	6.01	11.52	1017
* 22.805	0.40014	0.0661	0.0112	119.9	199.9	27.996	6.66	13.55	367
23	0.40532	0.0674	0.0110	121.5	202.5	28.110	6.63	13.41	369
24	0.43157	0.0738	0.0102	129.3	215.6	28.668	6.52	12.84	381
25	0.45696	0.0799	0.0096	136.9	228.3	29.183	6.45	12.44	392
26	0.48170	0.0856	0.0090	144.2	240.6	29.665	6.40	12.15	403
27	0.50594	0.0912	0.0086	151.4	252.6	30.119	6.37	11.92	413
28	0.52976	0.0966	0.0081	158.5	264.4	30.549	6.35	11.74	423
29	0.55326	0.1018	0.0078	165.4	276.1	30.958	6.33	11.59	432
30	0.57647	0.1070	0.0074	172.3	287.6	31.349	6.32	11.46	441
31	0.59944	0.1121	0.0071	179.1	299.0	31.723	6.30	11.36	449
32	0.62222	0.1170	0.0069	185.9	310.3	32.083	6.29	11.27	458
33	0.64481	0.1219	0.0066	192.6	321.6	32.428	6.29	11.19	466
34	0.66725	0.1268	0.0064	199.3	332.7	32.761	6.28	11.12	474
35	0.68955	0.1316	0.0062	205.9	343.8	33.083	6.27	11.06	482
36	0.71173	0.1363	0.0060	212.5	354.8	33.393	6.26	11.01	489
37	0.73380	0.1410	0.0058	219.1	365.8	33.694	6.26	10.96	497
38	0.75578	0.1457	0.0056	225.6	376.7	33.986	6.26	10.92	504
39	0.77766	0.1503	0.0054	232.1	387.6	34.269	6.25	10.88	511
40	0.79948	0.1549	0.0053	238.6	398.5	34.544	6.25	10.85	518
42	0.84289	0.1640	0.0050	251.6	420.2	35.072	6.25	10.79	532
44	0.88607	0.1730	0.0047	264.5	441.7	35.573	6.26	10.75	545
46	0.92904	0.1819	0.0045	277.4	463.2	36.051	6.26	10.72	558
48	0.97185	0.1908	0.0043	290.2	484.6	36.506	6.28	10.70	570
50	1.01450	0.1996	0.0041	303.1	506.0	36.943	6.30	10.70	582
52	1.05704	0.2084	0.0040	316.0	527.4	37.363	6.32	10.70	594
54	1.09946	0.2171	0.0038	328.9	548.8	37.767	6.36	10.71	605
56	1.14178	0.2258	0.0037	341.9	570.2	38.157	6.40	10.73	615
58	1.18402	0.2344	0.0035	354.9	591.7	38.534	6.45	10.77	626
60	1.22618	0.2430	0.0034	368.1	613.3	38.900	6.51	10.81	635
65	1.33130	0.2645	0.0031	401.5	667.8	39.771	6.70	10.97	658
70	1.43610	0.2858	0.0029	435.9	723.2	40.592	6.95	11.20	679
75	1.54066	0.3069	0.0027	471.7	779.8	41.374	7.25	11.48	697
80	1.64501	0.3281	0.0025	509.1	838.1	42.126	7.60	11.82	714
85	1.74920	0.3491	0.0024	548.3	898.1	42.853	7.99	12.20	730
90	1.85326	0.3701	0.0022	589.5	960.1	43.562	8.41	12.61	745
95	1.95721	0.3911	0.0021	632.8	1024.2	44.255	8.84	13.03	759
100	2.06107	0.4120	0.0020	678.2	1090.4	44.934	9.28	13.46	773
120	2.47580	0.4954	0.0017	880.4	1375.6	47.528	10.83	14.99	828
140	2.88982	0.5785	0.0014	1108.3	1686.3	49.920	11.82	15.97	884
160	3.30341	0.6615	0.0013	1349.7	2010.4	52.084	12.21	16.36	941
180	3.71671	0.7443	0.0011	1594.4	2337.8	54.012	12.19	16.33	999
200	4.12982	0.8270	0.0010	1836.1	2662.1	55.720	11.95	16.08	1055
220	4.54279	0.9097	0.0009	2072.0	2980.6	57.238	11.63	15.77	1110
240	4.95565	0.9924	0.0008	2301.7	3292.8	58.596	11.33	15.46	1164
260	5.36844	1.0750	0.0008	2525.7	3599.4	59.824	11.07	15.20	1215
280	5.78117	1.1576	0.0007	2745.1	3901.4	60.943	10.87	15.00	1264
300	6.19343	1.2401	0.0007	2961.1	4199.8	61.973	10.72	14.85	1311
350	7.22489	1.4465	0.0006	3491.1	4936.1	64.242	10.51	14.64	1419
400	8.25622	1.6528	0.0005	4014.2	5665.5	66.191	10.43	14.55	1519
450	9.28747	1.8591	0.0004	4534.5	6392.0	67.904	10.41	14.53	1611
500	10.31867	2.0654	0.0004	5054.9	7118.6	69.433	10.40	14.53	1698
550	11.34982	2.2716	0.0004	5574.9	7844.9	70.818	10.41	14.53	1781
600	12.38096	2.4779	0.0003	6095.9	8572.1	72.082	10.42	14.55	1860
700	14.44317	2.8904	0.0003	7140.8	10029.5	74.330	10.48	14.61	2007
800	16.50534	3.3028	0.0003	8193.3	11494.4	76.285	10.57	14.70	2143
900	18.56749	3.7152	0.0002	9256.7	12970.2	78.021	10.70	14.82	2269
1000	20.62962	4.1277	0.0002	10333.9	14459.8	79.593	10.86	14.98	2386
1200	24.75385	4.9525	0.0002	12544.1	17494.9	82.347	11.24	15.37	2602
1400	28.87810	5.7774	0.0001	14837.0	20612.6	84.754	11.68	15.81	2796
1600	33.00288	6.6022	0.0001	17218.7	23819.3	86.872	12.13	16.26	2974
1800	37.13116	7.4271	0.0001	19703.7	27130.0	88.783	12.68	16.82	3139
2000	41.27447	8.2519	0.0001	22343.1	30598.0	90.676	13.63	17.82	3285
2500	52.02197	10.3140	0.0001	30786.7	41191.1	97.716	21.67	26.66	3561
3000	65.34074	12.3761	0.0001	47934.6	61002.7	108.228	48.29	57.68	3840

* TWO-PHASE BOUNDARY

Table 4.1 (continued)
THERMODYNAMIC PROPERTIES OF PARAHYDROGEN (ISOBARS, SI UNITS)

0.20 MPA ISOBAR

TEMPERATURE	DENSITY	$V(DH/DV)_P$	$V(DP/DU)_V$	$-V(DP/DV)_T$	$-(DV/DT)/V$	THERMAL CONDUCTIVITY	VISCOSITY	THERMAL DIFFUSIVITY	DIELECTRIC CONSTANT	PRANDTL NUMBER
DEG. K	KG/CU M	KJ/KG	MPA-CU M/KJ $\times 10^{2}$	MPA	1/DEG. K	W/K-M $\times 10^{3}$	KG/M-S $\times 10^{7}$	SQ M/HR		
* 13.868	77.1259	624.9	0.25626	90.6124	0.01021	73.43	256.01	0.00054	1.25196	2.2255
14	77.0078	622.9	0.25509	89.3027	0.01036	74.76	251.16	0.00054	1.25155	2.1687
15	76.1907	610.6	0.24412	80.0995	0.01141	83.27	221.08	0.00056	1.24867	1.8499
16	75.2962	613.4	0.23280	74.1448	0.01208	89.17	196.79	0.00058	1.24553	1.6356
17	74.3574	613.0	0.22370	68.4766	0.01286	93.05	177.19	0.00057	1.24224	1.5014
18	73.3736	610.5	0.21613	63.1221	0.01372	95.79	161.05	0.00056	1.23880	1.4087
19	72.3384	602.9	0.20963	57.5276	0.01479	97.36	147.49	0.00054	1.23519	1.3510
20	71.2386	593.3	0.20383	52.1662	0.01598	98.74	135.84	0.00053	1.23136	1.3046
21	70.0571	579.6	0.19860	46.6655	0.01746	100.31	125.59	0.00051	1.22725	1.2666
22	68.7873	563.1	0.19370	41.2384	0.01923	101.11	116.47	0.00049	1.22284	1.2469
* 22.805	67.6762	544.6	0.18973	36.5158	0.02115	101.20	109.69	0.00047	1.21900	1.2486
* 22.805	2.4991	200.5	0.06703	0.1652	0.06755	20.09	12.91	0.00214	1.00755	0.8709
23	2.4672	202.9	0.06715	0.1663	0.06610	20.18	13.01	0.00220	1.00745	0.8646
24	2.3171	215.0	0.06761	0.1710	0.05973	20.70	13.51	0.00250	1.00700	0.8380
25	2.1884	227.1	0.06784	0.1747	0.05478	21.29	14.00	0.00282	1.00661	0.8183
26	2.0760	239.1	0.06793	0.1778	0.05079	21.92	14.50	0.00313	1.00627	0.8032
27	1.9765	251.1	0.06795	0.1802	0.04747	22.58	14.98	0.00345	1.00597	0.7910
28	1.8876	262.9	0.06793	0.1823	0.04464	23.25	15.47	0.00378	1.00570	0.7808
29	1.8075	274.6	0.06790	0.1841	0.04219	23.94	15.95	0.00411	1.00546	0.7722
30	1.7347	286.3	0.06785	0.1856	0.04005	24.63	16.43	0.00446	1.00524	0.7646
31	1.6682	297.8	0.06781	0.1869	0.03815	25.30	16.90	0.00481	1.00504	0.7588
32	1.6072	309.2	0.06776	0.1881	0.03645	25.96	17.37	0.00516	1.00485	0.7537
33	1.5508	320.5	0.06772	0.1891	0.03491	26.63	17.83	0.00552	1.00468	0.7492
34	1.4987	331.8	0.06769	0.1900	0.03352	27.30	18.29	0.00590	1.00452	0.7449
35	1.4502	342.9	0.06767	0.1908	0.03225	27.98	18.74	0.00628	1.00438	0.7409
36	1.4050	354.1	0.06763	0.1915	0.03109	28.65	19.19	0.00667	1.00424	0.7372
37	1.3628	365.2	0.06760	0.1922	0.03001	29.33	19.64	0.00707	1.00411	0.7338
38	1.3231	376.2	0.06756	0.1927	0.02902	30.01	20.08	0.00748	1.00399	0.7305
39	1.2859	387.3	0.06752	0.1933	0.02809	30.69	20.52	0.00790	1.00388	0.7275
40	1.2508	398.3	0.06747	0.1937	0.02723	31.38	20.96	0.00832	1.00377	0.7246
42	1.1864	420.3	0.06735	0.1946	0.02568	32.74	21.82	0.00920	1.00359	0.7193
44	1.1286	442.4	0.06721	0.1953	0.02430	34.10	22.66	0.01012	1.00340	0.7146
46	1.0764	464.6	0.06703	0.1958	0.02308	35.46	23.49	0.01106	1.00325	0.7103
48	1.0290	486.9	0.06681	0.1963	0.02198	36.83	24.31	0.01204	1.00310	0.7064
50	0.9857	509.6	0.06653	0.1968	0.02099	38.20	25.11	0.01304	1.00297	0.7030
52	0.9460	532.5	0.06619	0.1971	0.02009	39.57	25.90	0.01409	1.00285	0.7002
54	0.9095	555.9	0.06579	0.1975	0.01927	40.95	26.68	0.01513	1.00274	0.6979
56	0.8758	579.8	0.06531	0.1977	0.01851	42.33	27.45	0.01621	1.00264	0.6961
58	0.8446	604.3	0.06476	0.1980	0.01782	43.70	28.21	0.01730	1.00255	0.6950
60	0.8155	629.5	0.06413	0.1982	0.01718	45.15	28.95	0.01843	1.00246	0.6934
65	0.7511	695.9	0.06224	0.1986	0.01577	48.98	30.78	0.02140	1.00226	0.6894
70	0.6963	768.2	0.05996	0.1990	0.01458	52.82	32.55	0.02439	1.00210	0.6899
75	0.6491	846.9	0.05740	0.1992	0.01356	56.63	34.26	0.02736	1.00196	0.6947
80	0.6079	932.5	0.05470	0.1994	0.01269	60.52	35.93	0.03032	1.00183	0.7019
85	0.5717	1024.7	0.05199	0.1996	0.01190	65.46	37.56	0.03379	1.00172	0.7001
90	0.5396	1123.3	0.04939	0.1997	0.01122	70.40	39.15	0.03726	1.00163	0.7011
95	0.5109	1227.3	0.04695	0.1998	0.01062	75.34	40.71	0.04074	1.00154	0.7040
100	0.4852	1335.9	0.04474	0.1999	0.01007	89.58	45.74	0.04939	1.00146	0.6871
120	0.4039	1791.8	0.03827	0.2001	0.00837	117.10	54.06	0.06962	1.00122	0.6922
140	0.3460	2231.1	0.03503	0.2002	0.00716	136.12	59.19	0.08866	1.00104	0.6945
160	0.3027	2614.0	0.03389	0.2002	0.00626	148.28	63.00	0.10781	1.00091	0.6949
180	0.2691	2937.3	0.03394	0.2003	0.00556	155.65	66.18	0.12756	1.00081	0.6941
200	0.2421	3216.2	0.03462	0.2003	0.00500	160.21	69.02	0.14811	1.00073	0.6928
220	0.2201	3469.3	0.03554	0.2003	0.00454	163.43	71.67	0.16953	1.00066	0.6914
240	0.2018	3712.1	0.03648	0.2003	0.00417	166.26	74.20	0.19184	1.00061	0.6901
260	0.1863	3955.2	0.03732	0.2002	0.00384	169.18	76.65	0.21505	1.00056	0.6889
280	0.1730	4203.1	0.03800	0.2002	0.00357	172.41	79.06	0.23918	1.00052	0.6879
300	0.1615	4457.7	0.03853	0.2002	0.00333	175.96	81.42	0.26421	1.00049	0.6871
350	0.1384	5126.2	0.03930	0.2002	0.00286	186.18	87.25	0.33088	1.00042	0.6859
400	0.1211	5824.8	0.03961	0.2002	0.00250	197.51	93.00	0.40342	1.00036	0.6852
450	0.1077	6543.5	0.03968	0.2002	0.00222	209.41	98.67	0.48183	1.00032	0.6847
500	0.0969	7267.6	0.03969	0.2002	0.00200	221.36	104.29	0.56609	1.00029	0.6844
550	0.0881	7998.7	0.03966	0.2001	0.00182	233.38	109.86	0.65610	1.00027	0.6841
600	0.0808	8735.2	0.03959	0.2001	0.00167	245.40	115.35	0.75173	1.00024	0.6840
700	0.0692	10229.5	0.03937	0.2001	0.00143	269.50	126.16	0.95942	1.00021	0.6837
800	0.0606	11763.3	0.03903	0.2001	0.00125	293.89	136.71	1.18820	1.00018	0.6837
900	0.0539	13347.3	0.03857	0.2001	0.00111	318.73	147.01	1.43721	1.00016	0.6837
1000	0.0485	14989.1	0.03800	0.2001	0.00100	440.56	157.08	2.18374	1.00015	0.5342
1200	0.0404	18447.2	0.03670	0.2001	0.00083	511.14	176.57	2.96405	1.00012	0.5309
1400	0.0346	22137.1	0.03531	0.2001	0.00071	583.59	195.28	3.83808	1.00010	0.5290
1600	0.0303	26023.6	0.03400	0.2001	0.00062	658.03	213.33	4.80793	1.00009	0.5272
1800	0.0269	30287.6	0.03253	0.2000	0.00056	742.56	230.80	5.89970	1.00008	0.5229
2000	0.0242	35634.1	0.03028	0.1999	0.00050	864.05	247.79	7.20329	1.00007	0.5111
2500	0.0192	66065.2	0.01921	0.1983	0.00040	1794.18	289.46	12.60473	1.00006	0.4301
3000	0.0153	163878.2	0.00902	0.1894	0.00035	5220.51	336.01	21.28967	1.00005	0.3713

* TWO-PHASE BOUNDARY

Table 4.1 (continued)
THERMODYNAMIC PROPERTIES OF PARAHYDROGEN (ISOBARS, SI UNITS)

0.25 MPA ISOBAR

TEMPERATURE DEG. K	VOLUME CU M/KG	ISOTHERM DERIVATIVE CU M-MPA/KG	ISOCHORE DERIVATIVE MPA/K	INTERNAL ENERGY KJ/KG-K	ENTHALPY KJ/KG	ENTROPY KJ/KG-K	CV KJ / KG-K	CP KJ / KG-K	VELOCITY OF SOUND M/S
* 13.885	0.01296	1.1781	0.9261	-308.8	-305.6	4.972	4.69	6.38	1267
14	0.01298	1.1673	0.9261	-308.1	-304.8	5.027	4.71	6.44	1264
15	0.01312	1.0565	0.9158	-301.4	-298.2	5.487	4.91	6.96	1224
16	0.01327	0.9896	0.8975	-294.3	-291.0	5.951	5.11	7.40	1197
17	0.01344	0.9257	0.8824	-286.7	-283.3	6.414	5.29	7.87	1174
18	0.01362	0.8653	0.8678	-278.6	-275.2	6.879	5.46	8.37	1151
19	0.01381	0.8002	0.8526	-270.0	-266.6	7.345	5.61	8.90	1127
20	0.01402	0.7373	0.8354	-260.9	-257.4	7.815	5.74	9.46	1103
21	0.01426	0.6713	0.8163	-251.2	-247.6	8.292	5.85	10.09	1076
22	0.01452	0.6051	0.7946	-240.8	-237.2	8.776	5.95	10.79	1048
23	0.01482	0.5340	0.7690	-229.7	-226.0	9.275	6.03	11.62	1015
* 23.751	0.01507	0.4836	0.7464	-220.8	-217.0	9.659	6.08	12.29	989
* 23.751	0.32422	0.0645	0.0140	121.2	202.2	27.322	6.73	14.28	370
24	0.32992	0.0663	0.0137	123.3	205.8	27.471	6.69	14.05	373
25	0.35185	0.0731	0.0127	131.5	219.5	28.029	6.56	13.36	386
26	0.37292	0.0795	0.0119	139.3	232.6	28.543	6.49	12.87	397
27	0.39336	0.0856	0.0112	146.9	245.2	29.022	6.44	12.52	408
28	0.41331	0.0914	0.0106	154.3	257.6	29.472	6.40	12.24	418
29	0.43286	0.0971	0.0100	161.5	269.7	29.898	6.38	12.02	428
30	0.45210	0.1026	0.0096	168.7	281.7	30.302	6.36	11.84	437
31	0.47106	0.1079	0.0092	175.7	293.4	30.688	6.34	11.69	446
32	0.48980	0.1131	0.0088	182.6	305.1	31.057	6.33	11.57	455
33	0.50834	0.1183	0.0084	189.5	316.6	31.412	6.32	11.46	463
34	0.52671	0.1233	0.0081	196.3	328.0	31.752	6.30	11.36	471
35	0.54494	0.1283	0.0078	203.1	339.3	32.080	6.29	11.28	479
36	0.56303	0.1332	0.0076	209.8	350.6	32.397	6.28	11.21	487
37	0.58101	0.1380	0.0073	216.5	361.7	32.703	6.28	11.14	495
38	0.59889	0.1429	0.0071	223.1	372.8	33.000	6.27	11.09	502
39	0.61668	0.1476	0.0069	229.7	383.9	33.287	6.27	11.04	510
40	0.63439	0.1523	0.0067	236.3	394.9	33.566	6.27	10.99	517
42	0.66959	0.1617	0.0063	249.4	416.8	34.100	6.26	10.92	531
44	0.70454	0.1709	0.0060	262.5	438.6	34.607	6.27	10.86	544
46	0.73928	0.1800	0.0057	275.5	460.3	35.089	6.27	10.82	557
48	0.77385	0.1890	0.0054	288.4	481.9	35.549	6.29	10.79	570
50	0.80827	0.1980	0.0052	301.4	503.5	35.989	6.31	10.78	582
52	0.84256	0.2069	0.0050	314.4	525.0	36.412	6.33	10.77	593
54	0.87674	0.2157	0.0048	327.4	546.6	36.818	6.36	10.78	604
56	0.91082	0.2245	0.0046	340.4	568.1	37.210	6.41	10.79	615
58	0.94481	0.2332	0.0044	353.6	589.8	37.590	6.46	10.82	625
60	0.97873	0.2419	0.0043	366.8	611.4	37.957	6.51	10.86	635
65	1.06323	0.2636	0.0039	400.3	666.1	38.832	6.70	11.01	658
70	1.14741	0.2850	0.0036	434.8	721.7	39.656	6.95	11.23	679
75	1.23134	0.3064	0.0034	470.7	778.5	40.440	7.25	11.51	697
80	1.31507	0.3276	0.0032	508.1	836.9	41.194	7.61	11.85	714
85	1.39864	0.3488	0.0030	547.4	897.1	41.923	8.00	12.22	730
90	1.48207	0.3699	0.0028	588.7	959.2	42.633	8.41	12.63	745
95	1.56540	0.3909	0.0027	632.0	1023.4	43.327	8.84	13.05	759
100	1.64864	0.4119	0.0025	677.5	1089.7	44.007	9.28	13.47	773
120	1.98088	0.4955	0.0021	879.8	1375.1	46.603	10.83	15.00	828
140	2.31241	0.5788	0.0018	1107.8	1685.9	48.996	11.82	15.98	884
160	2.64350	0.6618	0.0016	1349.3	2010.2	51.161	12.21	16.36	942
180	2.97431	0.7448	0.0014	1594.1	2337.7	53.090	12.19	16.33	999
200	3.30493	0.8276	0.0013	1835.8	2662.1	54.798	11.95	16.09	1056
220	3.63540	0.9103	0.0011	2071.8	2980.6	56.317	11.63	15.77	1111
240	3.96578	0.9930	0.0010	2301.4	3292.9	57.675	11.33	15.46	1164
260	4.29608	1.0757	0.0010	2525.5	3599.5	58.903	11.07	15.21	1215
280	4.62632	1.1583	0.0009	2744.9	3901.5	60.022	10.87	15.00	1264
300	4.95618	1.2408	0.0008	2960.9	4200.0	61.052	10.72	14.85	1311
350	5.78144	1.4473	0.0007	3491.0	4936.4	63.321	10.51	14.64	1420
400	6.60656	1.6536	0.0006	4014.1	5665.8	65.271	10.43	14.55	1519
450	7.43160	1.8599	0.0006	4534.4	6392.3	66.983	10.41	14.53	1612
500	8.25658	2.0662	0.0005	5054.8	7119.0	68.513	10.40	14.53	1699
550	9.08153	2.2725	0.0005	5574.8	7845.2	69.897	10.41	14.53	1781
600	9.90646	2.4787	0.0004	6095.8	8572.4	71.162	10.42	14.55	1860
700	11.55625	2.8912	0.0004	7140.8	10029.9	73.409	10.48	14.61	2007
800	13.20601	3.3037	0.0003	8193.3	11494.8	75.364	10.57	14.70	2143
900	14.85574	3.7161	0.0003	9256.6	12970.6	77.101	10.70	14.82	2269
1000	16.50545	4.1286	0.0003	10333.9	14460.2	78.673	10.86	14.98	2387
1200	19.80485	4.9534	0.0002	12544.1	17495.3	81.427	11.24	15.37	2602
1400	23.10426	5.7783	0.0002	14837.0	20613.0	83.833	11.68	15.81	2796
1600	26.40404	6.6031	0.0002	17218.3	23819.3	85.951	12.13	16.26	2975
1800	29.70632	7.4280	0.0001	19700.8	27127.4	87.861	12.66	16.80	3139
2000	33.01935	8.2528	0.0001	22329.8	30584.7	89.748	13.55	17.73	3287
2500	41.58048	10.3149	0.0001	30582.5	40977.6	96.702	20.82	25.71	3568
3000	51.98234	12.3770	0.0001	46661.3	59656.9	106.810	44.80	53.57	3842

* TWO-PHASE BOUNDARY

Table 4.1 (continued)
THERMODYNAMIC PROPERTIES OF PARAHYDROGEN (ISOBARS, SI UNITS)

0.25 MPA ISOBAR

TEMPERATURE DEG. K	DENSITY KG/CU M	$V(DH/DV)_P$ KJ/KG	$V(DP/DU)_V$ MPA-CU M/KJ X 10^2	$-V(DP/DV)_T$ MPA	$-(DV/DT)_P/V$ 1/DEG. K	THERMAL CONDUCTIVITY W/K-M X 10^3	VISCOSITY KG/M-S X 10^7	THERMAL DIFFUSIVITY SQ M/HR	DIELECTRIC CONSTANT	PRANDTL NUMBER
* 13.885	77.1545	626.5	0.25619	90.8919	0.01019	73.63	256.32	0.00054	1.25206	2.2222
14	77.0586	625.7	0.25517	89.9438	0.01030	74.79	252.05	0.00054	1.25170	2.1711
15	76.2363	612.3	0.24452	80.5409	0.01137	83.33	221.85	0.00057	1.24883	1.8534
16	75.3451	615.0	0.23317	74.5620	0.01204	89.24	197.46	0.00058	1.24571	1.6381
17	74.4187	614.8	0.22406	68.8847	0.01281	93.14	177.80	0.00057	1.24243	1.5032
18	73.4314	612.5	0.21646	63.5428	0.01366	95.90	161.61	0.00056	1.23901	1.4097
19	72.4005	604.9	0.20995	57.9339	0.01472	97.49	148.01	0.00054	1.23541	1.3514
20	71.3054	595.5	0.20413	52.5755	0.01589	98.89	136.31	0.00053	1.23159	1.3044
21	70.1314	582.0	0.19889	47.0814	0.01734	100.49	126.05	0.00051	1.22751	1.2657
22	68.8698	565.8	0.19399	41.6725	0.01907	101.30	116.92	0.00049	1.22313	1.2451
23	67.4908	544.5	0.18907	36.0382	0.02134	101.38	108.58	0.00047	1.21836	1.2443
* 23.751	66.3708	528.3	0.18510	32.0967	0.02325	100.99	102.75	0.00045	1.21449	1.2500
* 23.751	3.0843	203.3	0.06730	0.1988	0.07022	21.44	13.56	0.00175	1.00932	0.9028
24	3.0310	206.5	0.06747	0.2009	0.06806	21.53	13.68	0.00182	1.00916	0.8928
25	2.8422	219.1	0.06792	0.2078	0.06097	21.98	14.16	0.00208	1.00859	0.8607
26	2.6816	231.6	0.06813	0.2132	0.05558	22.51	14.64	0.00235	1.00810	0.8374
27	2.5422	244.1	0.06820	0.2176	0.05129	23.09	15.12	0.00261	1.00768	0.8195
28	2.4195	256.3	0.06821	0.2212	0.04776	23.71	15.60	0.00288	1.00731	0.8051
29	2.3102	268.5	0.06818	0.2243	0.04478	24.35	16.07	0.00316	1.00698	0.7934
30	2.2119	280.4	0.06813	0.2268	0.04223	25.01	16.54	0.00344	1.00668	0.7834
31	2.1229	292.3	0.06808	0.2290	0.04000	25.66	17.01	0.00372	1.00641	0.7752
32	2.0417	304.0	0.06803	0.2310	0.03804	26.31	17.47	0.00401	1.00616	0.7681
33	1.9672	315.6	0.06797	0.2326	0.03630	26.96	17.93	0.00431	1.00594	0.7620
34	1.8986	327.1	0.06795	0.2341	0.03473	27.62	18.39	0.00461	1.00573	0.7563
35	1.8351	338.5	0.06792	0.2354	0.03332	28.28	18.84	0.00492	1.00554	0.7512
36	1.7761	349.8	0.06789	0.2366	0.03203	28.95	19.29	0.00524	1.00536	0.7466
37	1.7211	361.1	0.06785	0.2376	0.03086	29.62	19.73	0.00556	1.00520	0.7423
38	1.6697	372.3	0.06780	0.2385	0.02978	30.29	20.17	0.00589	1.00504	0.7383
39	1.6216	383.5	0.06776	0.2394	0.02878	30.96	20.61	0.00623	1.00489	0.7346
40	1.5763	394.7	0.06770	0.2401	0.02785	31.64	21.04	0.00657	1.00476	0.7311
42	1.4935	417.0	0.06758	0.2415	0.02618	32.98	21.90	0.00728	1.00451	0.7250
44	1.4194	439.4	0.06743	0.2426	0.02473	34.33	22.74	0.00802	1.00428	0.7195
46	1.3527	461.8	0.06724	0.2435	0.02344	35.69	23.56	0.00878	1.00408	0.7146
48	1.2922	484.3	0.06700	0.2443	0.02228	37.04	24.38	0.00956	1.00390	0.7102
50	1.2372	507.1	0.06672	0.2450	0.02125	38.41	25.18	0.01037	1.00373	0.7064
52	1.1869	530.2	0.06637	0.2455	0.02031	39.77	25.97	0.01120	1.00358	0.7032
54	1.1406	553.8	0.06596	0.2460	0.01946	41.14	26.74	0.01205	1.00344	0.7006
56	1.0979	577.8	0.06547	0.2465	0.01868	42.51	27.51	0.01291	1.00331	0.6985
58	1.0584	602.4	0.06491	0.2469	0.01797	43.88	28.26	0.01379	1.00319	0.6971
60	1.0217	627.7	0.06428	0.2472	0.01731	45.32	29.01	0.01470	1.00308	0.6954
65	0.9405	694.3	0.06238	0.2479	0.01586	49.14	30.83	0.01708	1.00284	0.6909
70	0.8715	766.7	0.06008	0.2484	0.01465	52.97	32.59	0.01948	1.00263	0.6911
75	0.8121	845.6	0.05751	0.2488	0.01361	56.76	34.31	0.02186	1.00245	0.6958
80	0.7604	931.3	0.05480	0.2491	0.01272	60.64	35.98	0.02424	1.00229	0.7028
85	0.7150	1023.6	0.05208	0.2494	0.01194	65.58	37.60	0.02702	1.00216	0.7008
90	0.6747	1122.3	0.04946	0.2496	0.01125	70.52	39.19	0.02980	1.00203	0.7017
95	0.6388	1226.4	0.04702	0.2497	0.01064	75.46	40.74	0.03259	1.00193	0.7045
100	0.6066	1335.0	0.04481	0.2498	0.01009	89.60	45.74	0.03947	1.00183	0.6878
120	0.5048	1791.2	0.03832	0.2502	0.00838	117.10	54.06	0.05566	1.00152	0.6926
140	0.4324	2230.9	0.03507	0.2503	0.00716	136.11	59.19	0.07091	1.00130	0.6948
160	0.3783	2614.1	0.03392	0.2504	0.00626	148.28	63.00	0.08624	1.00114	0.6951
180	0.3362	2937.7	0.03397	0.2504	0.00556	155.66	66.18	0.10206	1.00101	0.6943
200	0.3026	3216.8	0.03464	0.2504	0.00500	160.22	69.03	0.11851	1.00091	0.6930
220	0.2751	3470.0	0.03556	0.2504	0.00454	163.44	71.68	0.13565	1.00083	0.6915
240	0.2522	3713.0	0.03650	0.2504	0.00416	166.27	74.21	0.15351	1.00076	0.6901
260	0.2328	3956.2	0.03733	0.2504	0.00384	169.20	76.66	0.17209	1.00070	0.6890
280	0.2162	4204.2	0.03802	0.2504	0.00357	172.43	79.07	0.19140	1.00065	0.6880
300	0.2018	4458.8	0.03855	0.2504	0.00333	175.98	81.43	0.21144	1.00061	0.6872
350	0.1730	5127.4	0.03931	0.2503	0.00285	186.21	87.26	0.26479	1.00052	0.6859
400	0.1514	5826.1	0.03962	0.2503	0.00250	197.54	93.01	0.32285	1.00046	0.6852
450	0.1346	6544.9	0.03969	0.2503	0.00222	209.44	98.69	0.38560	1.00041	0.6847
500	0.1211	7269.0	0.03970	0.2503	0.00200	221.39	104.31	0.45303	1.00036	0.6844
550	0.1101	8000.2	0.03966	0.2502	0.00182	233.42	109.87	0.52506	1.00033	0.6841
600	0.1009	8736.6	0.03960	0.2502	0.00167	245.44	115.37	0.60159	1.00030	0.6840
700	0.0865	10231.0	0.03938	0.2502	0.00143	269.55	126.18	0.76779	1.00026	0.6837
800	0.0757	11764.8	0.03904	0.2502	0.00125	293.95	136.74	0.95088	1.00023	0.6837
900	0.0673	13348.9	0.03857	0.2501	0.00111	318.80	147.04	1.15014	1.00020	0.6837
1000	0.0606	14990.6	0.03800	0.2501	0.00100	440.56	157.11	1.74717	1.00018	0.5343
1200	0.0505	18448.8	0.03670	0.2501	0.00083	511.14	176.61	2.37145	1.00015	0.5310
1400	0.0433	22138.0	0.03532	0.2501	0.00071	583.56	195.33	3.07064	1.00013	0.5291
1600	0.0379	26017.7	0.03401	0.2501	0.00062	657.72	213.38	3.84596	1.00011	0.5274
1800	0.0337	30244.2	0.03259	0.2500	0.00056	740.80	230.85	4.71587	1.00010	0.5235
2000	0.0303	35450.3	0.03047	0.2499	0.00050	856.89	247.84	5.74509	1.00009	0.5128
2500	0.0240	63789.6	0.01997	0.2481	0.00040	1707.89	289.40	9.94206	1.00007	0.4357
3000	0.0192	153046.9	0.00967	0.2381	0.00035	4814.04	335.26	16.81836	1.00006	0.3730

* TWO-PHASE BOUNDARY

Table 4.1 (continued)
THERMODYNAMIC PROPERTIES OF PARAHYDROGEN (ISOBARS, SI UNITS)

0.30 MPA ISOBAR

TEMPERATURE DEG. K	VOLUME CU M/KG	ISOTHERM DERIVATIVE CU M-MPA/KG	ISOCHORE DERIVATIVE MPA/K	INTERNAL ENERGY KJ/KG-K	ENTHALPY KJ/KG	ENTROPY KJ/KG-K	CV KJ / KG-K	CP KJ / KG-K	VELOCITY OF SOUND M/S
* 13.901	0.01296	1.1812	0.9267	-308.8	-304.9	4.973	4.69	6.38	1268
14	0.01297	1.1750	0.9266	-308.2	-304.3	5.020	4.71	6.43	1267
15	0.01311	1.0616	0.9176	-301.5	-297.6	5.480	4.91	6.96	1226
16	0.01326	0.9945	0.8992	-294.4	-290.4	5.943	5.11	7.40	1200
17	0.01343	0.9305	0.8840	-286.8	-282.8	6.406	5.29	7.86	1176
18	0.01361	0.8704	0.8694	-278.8	-274.7	6.869	5.46	8.35	1154
19	0.01380	0.8051	0.8542	-270.2	-266.1	7.335	5.61	8.89	1130
20	0.01401	0.7424	0.8370	-261.1	-256.9	7.804	5.74	9.44	1105
21	0.01424	0.6765	0.8180	-251.5	-247.2	8.279	5.85	10.06	1079
22	0.01450	0.6106	0.7964	-241.1	-236.8	8.763	5.95	10.75	1051
23	0.01480	0.5414	0.7710	-230.1	-225.6	9.259	6.02	11.55	1019
24	0.01513	0.4708	0.7406	-218.2	-213.6	9.770	6.09	12.49	983
* 24.571	0.01535	0.4345	0.7207	-210.9	-206.3	10.071	6.12	13.04	962
* 24.571	0.27220	0.0625	0.0168	121.7	203.3	26.754	6.78	15.05	372
25	0.28074	0.0658	0.0162	125.5	209.7	27.012	6.71	14.59	378
26	0.29965	0.0730	0.0150	134.0	223.9	27.568	6.59	13.80	391
27	0.31775	0.0797	0.0140	142.1	237.4	28.078	6.52	13.25	403
28	0.33525	0.0861	0.0132	149.9	250.5	28.552	6.47	12.84	413
29	0.35227	0.0921	0.0125	157.4	263.1	28.997	6.43	12.52	424
30	0.36892	0.0980	0.0119	164.8	275.5	29.417	6.40	12.27	433
31	0.38526	0.1036	0.0113	172.1	287.7	29.816	6.38	12.07	443
32	0.40135	0.1091	0.0108	179.3	299.7	30.196	6.36	11.89	452
33	0.41722	0.1145	0.0104	186.3	311.5	30.560	6.35	11.75	460
34	0.43291	0.1198	0.0100	193.3	323.2	30.909	6.33	11.62	469
35	0.44844	0.1250	0.0096	200.2	334.7	31.244	6.32	11.51	477
36	0.46382	0.1300	0.0093	207.1	346.2	31.567	6.31	11.42	485
37	0.47909	0.1351	0.0090	213.8	357.6	31.879	6.30	11.33	493
38	0.49425	0.1400	0.0087	220.6	368.9	32.180	6.29	11.26	501
39	0.50932	0.1449	0.0084	227.3	380.1	32.472	6.28	11.20	508
40	0.52429	0.1498	0.0081	234.0	391.3	32.755	6.28	11.14	515
42	0.55403	0.1593	0.0077	247.3	413.5	33.296	6.28	11.05	530
44	0.58350	0.1688	0.0073	260.4	435.5	33.809	6.28	10.98	543
46	0.61277	0.1781	0.0069	273.6	457.4	34.295	6.28	10.92	556
48	0.64185	0.1873	0.0066	286.6	479.2	34.759	6.30	10.88	569
50	0.67078	0.1964	0.0063	299.7	500.9	35.203	6.31	10.86	581
52	0.69958	0.2054	0.0060	312.8	522.6	35.629	6.34	10.84	593
54	0.72827	0.2143	0.0058	325.9	544.3	36.038	6.37	10.84	604
56	0.75685	0.2232	0.0056	339.0	566.0	36.433	6.41	10.86	615
58	0.78535	0.2321	0.0054	352.2	587.8	36.814	6.46	10.88	625
60	0.81376	0.2409	0.0052	365.4	609.6	37.183	6.52	10.91	635
65	0.88453	0.2627	0.0047	399.1	664.5	38.062	6.71	11.06	658
70	0.95496	0.2843	0.0044	433.7	720.2	38.889	6.95	11.27	679
75	1.02514	0.3058	0.0041	469.7	777.2	39.675	7.26	11.54	697
80	1.09512	0.3272	0.0038	507.2	835.8	40.430	7.61	11.87	715
85	1.16494	0.3484	0.0036	546.5	896.0	41.161	8.00	12.24	730
90	1.23463	0.3696	0.0034	587.9	958.2	41.872	8.41	12.65	745
95	1.30420	0.3907	0.0032	631.3	1022.5	42.567	8.85	13.07	760
100	1.37369	0.4118	0.0030	676.8	1088.9	43.248	9.28	13.49	774
120	1.65094	0.4956	0.0025	879.3	1374.6	45.846	10.83	15.01	829
140	1.92747	0.5791	0.0022	1107.4	1685.6	48.241	11.82	15.99	885
160	2.20356	0.6622	0.0019	1349.0	2010.0	50.407	12.21	16.37	942
180	2.47938	0.7452	0.0017	1593.8	2337.6	52.336	12.19	16.34	999
200	2.75500	0.8281	0.0015	1835.5	2662.0	54.045	11.95	16.09	1056
220	3.03048	0.9109	0.0014	2071.5	2980.7	55.564	11.63	15.77	1111
240	3.30587	0.9937	0.0013	2301.2	3293.0	56.922	11.33	15.47	1165
260	3.58117	1.0763	0.0012	2525.3	3599.6	58.150	11.07	15.21	1216
280	3.85642	1.1590	0.0011	2744.8	3901.7	59.269	10.87	15.01	1265
300	4.13135	1.2416	0.0010	2960.8	4200.2	60.239	10.72	14.95	1311
350	4.81913	1.4480	0.0009	3490.9	4936.6	62.569	10.51	14.64	1420
400	5.50678	1.6544	0.0008	4014.0	5566.1	64.518	10.43	14.55	1520
450	6.19435	1.8607	0.0007	4534.4	6392.7	66.231	10.41	14.53	1612
500	6.88186	2.0670	0.0006	5054.8	7119.3	67.761	10.40	14.53	1699
550	7.56934	2.2733	0.0005	5574.8	7845.6	69.145	10.41	14.53	1782
600	8.25679	2.4796	0.0005	6095.8	8572.8	70.410	10.43	14.55	1860
700	9.63164	2.8921	0.0004	7140.8	10030.2	72.657	10.48	14.61	2008
800	11.00645	3.3045	0.0004	8193.3	11495.2	74.612	10.57	14.70	2143
900	12.38124	3.7170	0.0003	9256.6	12971.0	76.349	10.70	14.82	2269
1000	13.75601	4.1294	0.0003	10333.8	14460.6	77.921	10.86	14.98	2387
1200	16.50551	4.9543	0.0003	12544.1	17495.7	80.675	11.24	15.37	2602
1400	19.25502	5.7792	0.0002	14836.9	20613.4	83.081	11.68	15.81	2796
1600	22.00482	6.6040	0.0002	17217.9	23813.4	85.199	12.13	16.25	2975
1800	24.75651	7.4289	0.0002	19698.7	27125.6	87.108	12.64	16.78	3140
2000	27.51638	9.2537	0.0002	22320.1	30575.0	88.990	13.48	17.66	3288
2500	34.62797	10.3158	0.0001	30431.8	40820.2	95.892	20.19	25.02	3574
3000	43.14034	12.3779	0.0001	45720.7	58662.8	105.690	42.21	30.52	3844

* TWO-PHASE BOUNDARY

Table 4.1 (continued)
THERMODYNAMIC PROPERTIES OF PARAHYDROGEN (ISOBARS, SI UNITS)

0.30 MPA ISOBAR

TEMPERATURE DEG. K	DENSITY KG/CU M	$V(\partial H/\partial V)_P$ KJ/KG	$V(\partial P/\partial U)_V$ MPA-CU M/KJ $\times 10^2$	$-V(\partial P/\partial V)_T$ MPA	$-(\partial V/\partial T)_P/V$ 1/DEG. K	THERMAL CONDUCTIVITY W/K-M $\times 10^3$	VISCOSITY KG/M-S $\times 10^7$	THERMAL DIFFUSIVITY SQ M/HR	DIELECTRIC CONSTANT	PRANDTL NUMBER
* 13.901	77.1830	628.1	0.25612	91.1717	0.01016	73.84	256.63	0.00054	1.25216	2.2189
14	77.0933	628.6	0.25525	90.5832	0.01023	74.83	252.95	0.00054	1.25185	2.1737
15	76.2817	613.9	0.24492	80.9809	0.01133	83.39	222.62	0.00057	1.24899	1.8570
16	75.3938	616.7	0.23354	74.9777	0.01199	89.32	198.14	0.00058	1.24588	1.6405
17	74.4636	616.5	0.22441	69.2910	0.01276	93.24	178.41	0.00057	1.24262	1.5049
18	73.4888	614.4	0.21680	63.9613	0.01359	96.02	162.17	0.00056	1.23921	1.4107
19	72.4621	606.9	0.21028	58.3382	0.01464	97.62	148.52	0.00055	1.23562	1.3519
20	71.3718	597.7	0.20443	52.9831	0.01580	99.03	136.79	0.00053	1.23182	1.3042
21	70.2851	584.3	0.19917	47.4947	0.01722	100.66	126.51	0.00051	1.22776	1.2649
22	68.9515	568.4	0.19426	42.1036	0.01892	101.50	117.36	0.00049	1.22341	1.2434
23	67.5828	548.3	0.18937	36.5865	0.02107	101.60	109.02	0.00047	1.21868	1.2398
24	66.0882	524.8	0.18403	31.1119	0.02380	101.03	101.33	0.00044	1.21352	1.2529
* 24.571	65.1597	512.3	0.18067	28.3138	0.02545	100.39	97.14	0.00043	1.21032	1.2616
* 24.571	3.6738	285.1	0.06760	0.2296	0.07336	22.71	14.13	0.00148	1.01111	0.9363
25	3.5621	210.8	0.06787	0.2344	0.06921	22.83	14.33	0.00158	1.01077	0.9161
26	3.3372	224.0	0.06825	0.2437	0.06162	23.20	14.80	0.00181	1.01009	0.8801
27	3.1471	236.9	0.06842	0.2509	0.05592	23.69	15.27	0.00205	1.00951	0.8537
28	2.9829	249.7	0.06846	0.2568	0.05142	24.23	15.74	0.00228	1.00902	0.8336
29	2.8387	262.2	0.06844	0.2616	0.04775	24.81	16.20	0.00251	1.00858	0.8177
30	2.7106	274.6	0.06840	0.2656	0.04469	25.42	16.67	0.00275	1.00819	0.8045
31	2.5956	286.8	0.06834	0.2690	0.04207	26.05	17.13	0.00299	1.00784	0.7934
32	2.4916	298.8	0.06828	0.2719	0.03980	26.68	17.59	0.00324	1.00753	0.7840
33	2.3968	310.7	0.06822	0.2745	0.03781	27.31	18.04	0.00349	1.00724	0.7760
34	2.3100	322.4	0.06820	0.2767	0.03604	27.96	18.49	0.00375	1.00698	0.7687
35	2.2300	334.0	0.06818	0.2786	0.03446	28.61	18.94	0.00401	1.00673	0.7622
36	2.1560	345.5	0.06814	0.2804	0.03304	29.26	19.39	0.00428	1.00651	0.7564
37	2.0873	357.0	0.06810	0.2819	0.03175	29.92	19.83	0.00455	1.00630	0.7512
38	2.0233	368.4	0.06805	0.2833	0.03057	30.58	20.26	0.00483	1.00611	0.7465
39	1.9634	379.8	0.06800	0.2846	0.02949	31.24	20.70	0.00511	1.00593	0.7421
40	1.9073	391.1	0.06794	0.2857	0.02849	31.91	21.13	0.00540	1.00576	0.7380
42	1.8050	413.8	0.06781	0.2876	0.02671	33.23	21.98	0.00600	1.00545	0.7308
44	1.7138	436.3	0.06764	0.2892	0.02516	34.57	22.81	0.00661	1.00517	0.7245
46	1.6319	459.0	0.06744	0.2906	0.02380	35.91	23.64	0.00725	1.00493	0.7190
48	1.5580	481.7	0.06720	0.2917	0.02259	37.26	24.45	0.00791	1.00470	0.7141
50	1.4908	504.7	0.06690	0.2927	0.02151	38.62	25.25	0.00859	1.00450	0.7098
52	1.4294	528.0	0.06655	0.2936	0.02054	39.97	26.03	0.00928	1.00431	0.7063
54	1.3731	551.6	0.06613	0.2943	0.01966	41.33	26.81	0.00999	1.00414	0.7033
56	1.3213	575.8	0.06563	0.2949	0.01885	42.69	27.57	0.01072	1.00399	0.7010
58	1.2733	600.5	0.06507	0.2955	0.01812	44.06	28.32	0.01145	1.00384	0.6993
60	1.2289	625.9	0.06443	0.2960	0.01744	45.49	29.07	0.01221	1.00371	0.6974
65	1.1305	692.7	0.06251	0.2970	0.01596	49.30	30.88	0.01420	1.00341	0.6925
70	1.0472	765.3	0.06020	0.2977	0.01472	53.12	32.64	0.01621	1.00316	0.6924
75	0.9755	844.3	0.05761	0.2983	0.01367	56.90	34.36	0.01819	1.00294	0.6969
80	0.9131	930.1	0.05489	0.2987	0.01276	60.77	36.02	0.02018	1.00275	0.7037
85	0.8584	1022.5	0.05217	0.2991	0.01197	65.70	37.64	0.02250	1.00259	0.7015
90	0.8100	1121.2	0.04954	0.2994	0.01128	70.64	39.23	0.02483	1.00244	0.7023
95	0.7668	1225.4	0.04710	0.2996	0.01066	75.57	40.78	0.02716	1.00231	0.7050
100	0.7280	1334.1	0.04487	0.2998	0.01011	89.62	45.74	0.03286	1.00219	0.6884
120	0.6057	1790.7	0.03836	0.3002	0.00838	117.09	54.05	0.04636	1.00183	0.6930
140	0.5188	2230.7	0.03510	0.3004	0.00717	136.10	59.18	0.05908	1.00156	0.6952
160	0.4538	2614.2	0.03395	0.3005	0.00626	148.28	63.00	0.07187	1.00137	0.6954
180	0.4033	2938.1	0.03399	0.3006	0.00556	155.66	66.18	0.08506	1.00122	0.6945
200	0.3630	3217.4	0.03466	0.3006	0.00500	160.23	69.03	0.09877	1.00109	0.6931
220	0.3300	3470.8	0.03558	0.3006	0.00454	163.46	71.68	0.11307	1.00099	0.6916
240	0.3825	3713.9	0.03652	0.3006	0.00416	166.29	74.21	0.12796	1.00091	0.6902
260	0.2792	3957.2	0.03735	0.3006	0.00384	169.22	76.67	0.14345	1.00084	0.6890
280	0.2593	4205.2	0.03803	0.3005	0.00357	172.45	79.07	0.15955	1.00078	0.6880
300	0.2421	4459.9	0.03856	0.3005	0.00333	176.00	81.44	0.17626	1.00073	0.6872
350	0.2875	5128.6	0.03933	0.3005	0.00285	186.24	87.27	0.22074	1.00063	0.6859
400	0.1816	5827.4	0.03963	0.3004	0.00250	197.57	93.02	0.26914	1.00055	0.6852
450	0.1614	6546.3	0.03970	0.3004	0.00222	209.48	98.70	0.32145	1.00049	0.6847
500	0.1453	7270.4	0.03971	0.3004	0.00200	221.43	104.33	0.37766	1.00044	0.6844
550	0.1321	8001.6	0.03967	0.3003	0.00182	233.46	109.89	0.43770	1.00040	0.6841
600	0.1211	8738.1	0.03961	0.3003	0.00167	245.49	115.39	0.50150	1.00036	0.6840
700	0.1038	10232.4	0.03939	0.3003	0.00143	269.61	126.21	0.64004	1.00031	0.6837
800	0.0909	11766.3	0.03904	0.3002	0.00125	294.01	136.77	0.79266	1.00027	0.6837
900	0.0808	13350.4	0.03857	0.3002	0.00111	318.87	147.07	0.95876	1.00024	0.6837
1000	0.0727	14992.2	0.03801	0.3002	0.00100	440.56	157.15	1.45613	1.00022	0.5344
1200	0.0606	18450.4	0.03670	0.3002	0.00083	511.14	176.65	1.97638	1.00018	0.5311
1400	0.0519	22139.1	0.03532	0.3001	0.00071	583.53	195.38	2.55903	1.00016	0.5292
1600	0.0454	26013.8	0.03403	0.3001	0.00062	657.50	213.43	3.20478	1.00014	0.5276
1800	0.0404	30212.6	0.03264	0.3001	0.00056	739.50	230.91	3.92755	1.00012	0.5240
2000	0.0363	35315.2	0.03062	0.3000	0.00050	851.60	247.90	4.77675	1.00011	0.5141
2500	0.0289	62107.6	0.02058	0.2979	0.00040	1644.08	289.38	8.19221	1.00009	0.4403
3000	0.0232	144953.9	0.01022	0.2869	0.00035	4510.27	334.70	13.86503	1.00007	0.3749

* TWO-PHASE BOUNDARY

Table 4.1 (continued)
THERMODYNAMIC PROPERTIES OF PARAHYDROGEN (ISOBARS, SI UNITS)

0.35 MPA ISOBAR

TEMPERATURE DEG. K	VOLUME CU M/KG	ISOTHERM DERIVATIVE CU M-MPA/KG	ISOCHORE DERIVATIVE MPA/K	INTERNAL ENERGY KJ/KG-K	ENTHALPY KJ/KG	ENTROPY KJ/KG-K	CV KJ /	CP KG-K	VELOCITY OF SOUND M/S
* 13.919	0.01295	1.1844	0.9273	-308.8	-304.3	4.974	4.69	6.39	1270
14	0.01296	1.1826	0.9272	-308.3	-303.7	5.014	4.71	6.42	1270
15	0.01310	1.0667	0.9194	-301.6	-297.1	5.473	4.91	6.95	1229
16	0.01326	0.9993	0.9009	-294.5	-289.9	5.936	5.10	7.39	1203
17	0.01342	0.9353	0.8856	-287.0	-282.3	6.398	5.29	7.86	1179
18	0.01360	0.8753	0.8710	-278.9	-274.2	6.860	5.45	8.34	1157
19	0.01379	0.8100	0.8559	-270.4	-265.6	7.325	5.60	8.87	1132
20	0.01400	0.7473	0.8386	-261.3	-256.4	7.793	5.73	9.42	1108
21	0.01423	0.6817	0.8197	-251.7	-246.7	8.267	5.85	10.04	1082
22	0.01449	0.6161	0.7982	-241.4	-236.4	8.749	5.94	10.72	1054
23	0.01478	0.5473	0.7730	-230.4	-225.2	9.244	6.02	11.51	1023
24	0.01511	0.4770	0.7430	-218.6	-213.3	9.752	6.09	12.43	987
25	0.01549	0.4056	0.7075	-205.8	-200.4	10.280	6.14	13.55	946
* 25.299	0.01562	0.3858	0.6949	-201.7	-196.2	10.445	6.16	13.89	933
* 25.299	0.23414	0.0603	0.0198	121.6	203.6	26.258	6.84	15.88	374
26	0.24651	0.0660	0.0186	128.2	214.4	26.682	6.72	15.02	384
27	0.26317	0.0735	0.0172	136.9	229.0	27.231	6.61	14.16	397
28	0.27906	0.0805	0.0161	145.2	242.8	27.735	6.54	13.56	408
29	0.29438	0.0870	0.0151	153.1	256.2	28.202	6.49	13.11	419
30	0.30925	0.0933	0.0143	160.9	269.1	28.641	6.45	12.76	429
31	0.32378	0.0993	0.0136	168.4	281.7	29.054	6.42	12.48	439
32	0.33801	0.1051	0.0130	175.8	294.1	29.447	6.40	12.26	449
33	0.35201	0.1107	0.0124	183.0	306.3	29.821	6.38	12.07	458
34	0.36580	0.1162	0.0119	190.2	318.2	30.179	6.36	11.90	466
35	0.37942	0.1216	0.0114	197.3	330.1	30.522	6.34	11.76	475
36	0.39289	0.1269	0.0110	204.3	341.8	30.852	6.33	11.64	483
37	0.40623	0.1321	0.0106	211.2	353.4	31.169	6.32	11.54	491
38	0.41946	0.1372	0.0103	218.0	364.8	31.476	6.31	11.45	499
39	0.43259	0.1422	0.0099	224.9	376.3	31.772	6.30	11.37	507
40	0.44562	0.1472	0.0096	231.6	387.6	32.059	6.30	11.30	514
42	0.47146	0.1570	0.0091	245.1	410.1	32.608	6.29	11.19	528
44	0.49704	0.1666	0.0086	258.4	432.4	33.126	6.29	11.10	542
46	0.52239	0.1761	0.0082	271.6	454.5	33.618	6.29	11.03	556
48	0.54756	0.1855	0.0078	284.8	476.5	34.086	6.30	10.98	568
50	0.57257	0.1947	0.0074	298.0	498.4	34.533	6.32	10.94	581
52	0.59745	0.2039	0.0071	311.2	520.3	34.962	6.35	10.92	592
54	0.62222	0.2130	0.0068	324.3	542.1	35.374	6.38	10.91	604
56	0.64688	0.2220	0.0065	337.5	563.9	35.771	6.42	10.92	614
58	0.67145	0.2309	0.0063	350.8	585.8	36.154	6.47	10.94	625
60	0.69594	0.2398	0.0061	364.1	607.7	36.525	6.52	10.97	635
65	0.75689	0.2618	0.0056	397.9	662.8	37.408	6.71	11.10	658
70	0.81750	0.2836	0.0051	432.7	718.8	38.237	6.96	11.30	679
75	0.87786	0.3052	0.0048	468.7	775.9	39.026	7.26	11.57	698
80	0.93802	0.3267	0.0045	506.3	834.6	39.783	7.61	11.90	715
85	0.99802	0.3481	0.0042	545.7	895.0	40.515	8.00	12.27	731
90	1.05788	0.3694	0.0039	587.1	957.3	41.227	8.42	12.67	746
95	1.11764	0.3906	0.0037	630.5	1021.7	41.923	8.85	13.08	760
100	1.17730	0.4117	0.0035	676.1	1088.1	42.605	9.28	13.50	774
120	1.41526	0.4958	0.0029	878.7	1374.1	45.205	10.83	15.02	829
140	1.65251	0.5794	0.0025	1106.9	1685.3	47.602	11.82	15.99	885
160	1.88933	0.6626	0.0022	1348.6	2009.8	49.768	12.21	16.37	942
180	2.12586	0.7457	0.0020	1593.4	2337.5	51.699	12.19	16.34	1000
200	2.36220	0.8287	0.0018	1835.3	2662.0	53.407	11.95	16.09	1056
220	2.59840	0.9115	0.0016	2071.3	2980.7	54.927	11.63	15.77	1112
240	2.83450	0.9943	0.0015	2301.0	3293.1	56.285	11.33	15.47	1165
260	3.07053	1.0770	0.0013	2525.1	3599.8	57.513	11.08	15.21	1216
280	3.30649	1.1597	0.0013	2744.6	3901.9	58.633	10.87	15.01	1265
300	3.54219	1.2423	0.0012	2960.6	4200.4	59.663	10.72	14.85	1312
350	4.13177	1.4488	0.0010	3490.8	4936.9	61.933	10.51	14.64	1421
400	4.72123	1.6552	0.0009	4013.9	5666.4	63.882	10.43	14.55	1520
450	5.31060	1.8615	0.0008	4534.3	6393.0	65.595	10.41	14.53	1612
500	5.89992	2.0679	0.0007	5054.7	7119.7	67.125	10.40	14.53	1699
550	6.48920	2.2742	0.0006	5574.7	7845.9	68.510	10.41	14.53	1782
600	7.07846	2.4804	0.0006	6095.7	8573.2	69.774	10.43	14.55	1861
700	8.25692	2.8929	0.0005	7140.7	10030.6	72.022	10.48	14.61	2008
800	9.43534	3.3054	0.0004	8193.2	11495.6	73.976	10.57	14.70	2144
900	10.61374	3.7179	0.0004	9256.6	12971.4	75.713	10.70	14.82	2270
1000	11.79212	4.1303	0.0004	10333.8	14461.1	77.285	10.86	14.98	2387
1200	14.14885	4.9552	0.0003	12544.1	17496.2	80.039	11.24	15.37	2602
1400	16.50557	5.7801	0.0003	14836.9	20613.8	82.446	11.68	15.81	2797
1600	18.86252	6.6049	0.0002	17217.7	23819.6	84.563	12.12	16.25	2975
1800	21.22097	7.4298	0.0002	19697.0	27124.3	86.471	12.63	16.77	3141
2000	23.58592	8.2546	0.0002	22312.5	30567.5	88.350	13.43	17.61	3289
2500	29.66646	10.3167	0.0001	30314.6	40697.9	95.192	19.71	24.48	3579
3000	36.85890	12.3788	0.0001	44989.2	57889.9	104.769	40.19	48.15	3846

* TWO-PHASE BOUNDARY

Table 4.1 (continued)

THERMODYNAMIC PROPERTIES OF PARAHYDROGEN (ISOBARS, SI UNITS)

0.35 MPA ISOBAR

TEMPERATURE DEG. K	DENSITY KG/CU M	$V(DH/DV)_P$ KJ/KG	$V(DP/DU)_V$ MPA-CU M/KJ X 10^2	$-V(DP/DV)_T$ MPA	$-(DV/DT)_P/V$ 1/DEG. K	THERMAL CONDUCTIVITY W/K-M X 10^3	VISCOSITY KG/M-S X 10^7	THERMAL DIFFUSIVITY SQ M/HR	DIELECTRIC CONSTANT	PRANDTL NUMBER
* 13.918	77.2115	629.7	0.25606	91.4518	0.01014	74.05	256.94	0.00054	1.25226	2.2156
14	77.1358	631.5	0.25533	91.2209	0.01016	74.87	253.85	0.00054	1.25200	2.1762
15	76.3278	615.5	0.24531	81.4195	0.01129	83.45	223.39	0.00057	1.24915	1.8606
16	75.4422	618.3	0.23391	75.3921	0.01195	89.40	198.82	0.00058	1.24605	1.6430
17	74.5163	618.2	0.22476	69.6955	0.01271	93.34	179.02	0.00057	1.24280	1.5066
18	73.5458	616.3	0.21712	64.3776	0.01353	96.13	162.73	0.00056	1.23941	1.4117
19	72.5233	608.8	0.21059	58.7405	0.01457	97.75	149.03	0.00055	1.23583	1.3524
20	71.4378	599.8	0.20472	53.3890	0.01571	99.18	137.27	0.00053	1.23205	1.3041
21	70.2782	586.7	0.19946	47.9054	0.01711	100.82	126.97	0.00051	1.22802	1.2641
22	69.0324	571.1	0.19454	42.5317	0.01877	101.69	117.81	0.00049	1.22369	1.2417
23	67.6752	551.3	0.18966	37.0390	0.02087	101.82	109.47	0.00047	1.21899	1.2370
24	66.1935	528.1	0.18438	31.5729	0.02353	101.28	101.78	0.00044	1.21388	1.2489
25	64.5558	501.3	0.17842	26.1848	0.02702	100.08	94.56	0.00041	1.20824	1.2799
* 25.299	64.0123	493.5	0.17627	24.6966	0.02814	99.54	92.41	0.00040	1.20637	1.2891
* 25.299	4.2710	206.2	0.06790	0.2575	0.07699	23.94	14.66	0.00127	1.01293	0.9724
26	4.0566	216.0	0.06830	0.2678	0.06953	24.06	14.98	0.00142	1.01228	0.9349
27	3.7998	229.6	0.06858	0.2793	0.06168	24.39	15.43	0.00163	1.01150	0.8958
28	3.5834	242.9	0.06868	0.2883	0.05581	24.83	15.88	0.00184	1.01084	0.8674
29	3.3978	255.9	0.06869	0.2956	0.05122	25.33	16.34	0.00205	1.01027	0.8457
30	3.2336	268.7	0.06865	0.3016	0.04749	25.88	16.80	0.00226	1.00978	0.8284
31	3.0885	281.3	0.06859	0.3066	0.04438	26.47	17.25	0.00247	1.00934	0.8137
32	2.9585	293.6	0.06853	0.3108	0.04174	27.08	17.70	0.00269	1.00894	0.8014
33	2.8408	305.8	0.06846	0.3145	0.03945	27.69	18.15	0.00291	1.00859	0.7911
34	2.7337	317.7	0.06845	0.3177	0.03745	28.31	18.60	0.00313	1.00826	0.7819
35	2.6356	329.6	0.06843	0.3205	0.03569	28.94	19.05	0.00336	1.00796	0.7740
36	2.5452	341.3	0.06840	0.3229	0.03411	29.58	19.49	0.00359	1.00769	0.7669
37	2.4616	352.9	0.06836	0.3251	0.03269	30.22	19.92	0.00383	1.00744	0.7607
38	2.3840	364.5	0.06830	0.3270	0.03141	30.87	20.36	0.00407	1.00720	0.7550
39	2.3117	376.1	0.06825	0.3288	0.03023	31.52	20.79	0.00432	1.00698	0.7499
40	2.2441	387.6	0.06818	0.3303	0.02916	32.18	21.22	0.00457	1.00678	0.7451
42	2.1211	410.5	0.06804	0.3330	0.02725	33.49	22.06	0.00508	1.00641	0.7369
44	2.0119	433.3	0.06786	0.3353	0.02561	34.82	22.89	0.00561	1.00607	0.7297
46	1.9143	456.2	0.06765	0.3372	0.02417	36.15	23.71	0.00616	1.00578	0.7235
48	1.8263	479.2	0.06740	0.3387	0.02291	37.48	24.52	0.00673	1.00551	0.7180
50	1.7465	502.3	0.06709	0.3401	0.02178	38.83	25.31	0.00732	1.00527	0.7133
52	1.6738	525.7	0.06673	0.3413	0.02077	40.17	26.10	0.00791	1.00505	0.7093
54	1.6072	549.5	0.06630	0.3423	0.01986	41.53	26.87	0.00852	1.00485	0.7068
56	1.5459	573.8	0.06580	0.3431	0.01902	42.88	27.63	0.00915	1.00467	0.7034
58	1.4893	598.6	0.06522	0.3439	0.01827	44.24	28.38	0.00978	1.00449	0.7016
60	1.4369	624.1	0.06458	0.3446	0.01757	45.67	29.12	0.01043	1.00434	0.6994
65	1.3212	691.2	0.06265	0.3459	0.01606	49.47	30.94	0.01214	1.00399	0.6941
70	1.2232	763.9	0.06032	0.3469	0.01480	53.27	32.69	0.01387	1.00369	0.6936
75	1.1391	843.0	0.05772	0.3477	0.01373	57.03	34.40	0.01558	1.00344	0.6980
80	1.0661	928.9	0.05499	0.3483	0.01281	60.90	36.06	0.01728	1.00322	0.7046
85	1.0020	1021.4	0.05226	0.3488	0.01201	65.82	37.69	0.01928	1.00302	0.7022
90	0.9453	1120.2	0.04962	0.3492	0.01131	70.75	39.27	0.02127	1.00285	0.7038
95	0.8947	1224.5	0.04717	0.3495	0.01068	75.68	40.82	0.02328	1.00270	0.7055
100	0.8494	1333.2	0.04494	0.3497	0.01013	89.65	45.74	0.02814	1.00256	0.6890
120	0.7066	1790.2	0.03840	0.3503	0.00839	117.09	54.05	0.03971	1.00213	0.6935
140	0.6051	2230.5	0.03514	0.3506	0.00717	136.10	59.18	0.05062	1.00182	0.6955
160	0.5293	2614.4	0.03397	0.3507	0.00626	148.28	63.00	0.06160	1.00160	0.6956
180	0.4704	2938.5	0.03402	0.3508	0.00556	155.67	66.18	0.07291	1.00142	0.6947
200	0.4233	3218.0	0.03468	0.3508	0.00500	160.24	69.03	0.08468	1.00128	0.6933
220	0.3849	3471.6	0.03560	0.3508	0.00454	163.47	71.69	0.09694	1.00116	0.6917
240	0.3528	3714.8	0.03654	0.3508	0.00416	166.30	74.22	0.10971	1.00106	0.6903
260	0.3257	3958.2	0.03737	0.3508	0.00384	169.24	76.67	0.12299	1.00098	0.6891
280	0.3024	4206.3	0.03805	0.3507	0.00357	172.47	79.08	0.13680	1.00091	0.6881
300	0.2823	4461.1	0.03858	0.3507	0.00333	176.03	81.45	0.15113	1.00085	0.6873
350	0.2420	5129.9	0.03934	0.3506	0.00285	186.26	87.28	0.18927	1.00073	0.6860
400	0.2118	5828.8	0.03964	0.3506	0.00250	197.60	93.03	0.23077	1.00064	0.6852
450	0.1883	6547.6	0.03971	0.3505	0.00222	209.51	98.72	0.27562	1.00057	0.6847
500	0.1695	7271.8	0.03972	0.3505	0.00200	221.47	104.34	0.32382	1.00051	0.6844
550	0.1541	8003.0	0.03968	0.3505	0.00182	233.50	109.91	0.37530	1.00046	0.6841
600	0.1413	8739.6	0.03961	0.3504	0.00166	245.53	115.41	0.43001	1.00043	0.6840
700	0.1211	10233.9	0.03939	0.3504	0.00143	269.66	126.23	0.54879	1.00036	0.6837
800	0.1060	11767.8	0.03905	0.3503	0.00125	294.07	136.79	0.67965	1.00032	0.6837
900	0.0942	13352.0	0.03858	0.3503	0.00111	318.93	147.11	0.82207	1.00028	0.6837
1000	0.0848	14993.8	0.03801	0.3503	0.00100	440.56	157.18	1.24824	1.00026	0.5346
1200	0.0707	18452.0	0.03671	0.3502	0.00083	511.14	176.69	1.69419	1.00021	0.5312
1400	0.0606	22140.4	0.03532	0.3502	0.00071	583.52	195.42	2.19359	1.00018	0.5294
1600	0.0530	25011.1	0.03404	0.3502	0.00062	657.33	213.48	2.74687	1.00016	0.5277
1800	0.0471	30188.4	0.03267	0.3501	0.00056	738.48	230.97	3.36495	1.00014	0.5244
2000	0.0424	35210.5	0.03073	0.3500	0.00050	847.49	247.96	4.08714	1.00013	0.5151
2500	0.0337	60799.1	0.02108	0.3478	0.00040	1594.42	289.38	6.95696	1.00010	0.4442
3000	0.0271	138607.2	0.01070	0.3358	0.00035	4271.99	334.27	11.77275	1.00008	0.3768

* TWO-PHASE BOUNDARY

Table 4.1 (continued)
THERMODYNAMIC PROPERTIES OF PARAHYDROGEN (ISOBARS, SI UNITS)

0.40 MPA ISOBAR

TEMPERATURE DEG. K	VOLUME CU M/KG	ISOTHERM DERIVATIVE CU M-MPA/KG	ISOCHORE DERIVATIVE MPA/K	INTERNAL ENERGY KJ/KG-K	ENTHALPY KJ/KG	ENTROPY KJ/KG-K	CV KJ / KG-K	CP KJ / KG-K	VELOCITY OF SOUND M/S
* 13.935	0.01295	1.1876	0.9279	-308.8	-303.6	4.975	4.69	6.39	1271
14	0.01296	1.1902	0.9278	-308.3	-303.2	5.007	4.71	6.41	1273
15	0.01309	1.0718	0.9211	-301.8	-296.5	5.465	4.91	6.94	1231
16	0.01325	1.0042	0.9025	-294.6	-289.3	5.928	5.10	7.38	1205
17	0.01341	0.9401	0.8872	-287.1	-281.7	6.389	5.29	7.85	1181
18	0.01358	0.8826	0.8729	-279.1	-273.7	6.850	5.45	8.32	1160
19	0.01378	0.8191	0.8576	-270.6	-265.1	7.314	5.60	8.84	1137
20	0.01399	0.7523	0.8402	-261.6	-256.0	7.782	5.73	9.40	1111
21	0.01421	0.6868	0.8213	-252.0	-246.3	8.255	5.84	10.01	1085
22	0.01447	0.6215	0.8000	-241.7	-235.9	8.736	5.94	10.68	1057
23	0.01476	0.5532	0.7751	-230.8	-224.9	9.228	6.02	11.46	1026
24	0.01508	0.4831	0.7454	-219.0	-213.0	9.734	6.09	12.37	991
25	0.01546	0.4122	0.7103	-206.3	-200.1	10.259	6.14	13.46	950
* 25.957	0.01590	0.3437	0.6694	-192.9	-186.6	10.790	6.19	14.74	905
* 25.957	0.20496	0.0579	0.0229	121.1	203.1	25.813	6.89	16.79	376
26	0.20572	0.0583	0.0228	121.6	203.9	25.842	6.88	16.71	376
27	0.22161	0.0668	0.0208	131.2	219.9	26.445	6.72	15.34	390
28	0.23648	0.0745	0.0193	140.1	234.7	26.985	6.62	14.44	403
29	0.25063	0.0817	0.0180	148.6	248.8	27.480	6.56	13.80	415
30	0.26426	0.0884	0.0170	156.7	262.4	27.940	6.51	13.33	425
31	0.27747	0.0948	0.0160	164.5	275.5	28.371	6.47	12.96	436
32	0.29036	0.1009	0.0153	172.2	288.3	28.777	6.44	12.66	445
33	0.30298	0.1068	0.0145	179.7	300.9	29.163	6.42	12.41	455
34	0.31537	0.1126	0.0139	187.0	313.2	29.531	6.39	12.21	464
35	0.32758	0.1182	0.0133	194.2	325.3	29.882	6.37	12.03	473
36	0.33963	0.1237	0.0128	201.4	337.2	30.219	6.35	11.88	481
37	0.35154	0.1290	0.0124	208.4	349.1	30.543	6.34	11.76	489
38	0.36332	0.1343	0.0119	215.4	360.8	30.855	6.33	11.65	497
39	0.37500	0.1395	0.0115	222.4	372.4	31.156	6.32	11.55	505
40	0.38659	0.1446	0.0112	229.2	383.9	31.447	6.31	11.47	513
42	0.40952	0.1547	0.0105	242.8	406.7	32.003	6.30	11.33	527
44	0.43217	0.1645	0.0099	256.3	429.2	32.528	6.30	11.22	541
46	0.45460	0.1742	0.0094	269.7	451.5	33.024	6.30	11.14	555
48	0.47684	0.1837	0.0089	283.0	473.8	33.497	6.31	11.07	568
50	0.49892	0.1931	0.0085	296.3	495.8	33.948	6.33	11.03	580
52	0.52086	0.2024	0.0082	309.5	517.9	34.380	6.35	11.00	592
54	0.54268	0.2116	0.0078	322.8	539.8	34.794	6.38	10.98	603
56	0.56440	0.2207	0.0075	336.0	561.8	35.194	6.42	10.98	614
58	0.58603	0.2297	0.0072	349.4	583.8	35.579	6.47	10.99	625
60	0.60758	0.2387	0.0070	362.8	605.8	35.952	6.53	11.02	635
65	0.66116	0.2610	0.0064	396.7	661.1	36.839	6.71	11.14	658
70	0.71441	0.2829	0.0059	431.6	717.3	37.671	6.96	11.34	679
75	0.76741	0.3047	0.0055	467.7	774.6	38.462	7.26	11.60	698
80	0.82020	0.3263	0.0051	505.4	833.4	39.221	7.61	11.92	715
85	0.87283	0.3478	0.0048	544.8	894.0	39.954	8.00	12.29	731
90	0.92533	0.3691	0.0045	586.2	956.4	40.668	8.42	12.69	746
95	0.97772	0.3904	0.0043	629.8	1020.8	41.365	8.85	13.10	760
100	1.03002	0.4116	0.0041	675.4	1087.4	42.047	9.28	13.52	774
120	1.23851	0.4959	0.0034	878.2	1373.6	44.650	10.83	15.03	830
140	1.44630	0.5796	0.0029	1106.4	1685.0	47.048	11.82	16.00	886
160	1.65365	0.6630	0.0025	1348.2	2009.6	49.215	12.22	16.38	943
180	1.86072	0.7462	0.0022	1593.1	2337.4	51.146	12.19	16.34	1000
200	2.06760	0.8292	0.0020	1835.0	2662.0	52.855	11.95	16.10	1057
220	2.27434	0.9121	0.0018	2071.0	2980.8	54.375	11.63	15.78	1112
240	2.48098	0.9949	0.0017	2300.8	3293.2	55.734	11.33	15.47	1165
260	2.68754	1.0777	0.0015	2524.9	3599.9	56.962	11.08	15.21	1217
280	2.89405	1.1604	0.0014	2744.4	3902.0	58.081	10.87	15.01	1266
300	3.10031	1.2430	0.0013	2960.5	4200.6	59.112	10.72	14.85	1312
350	3.61625	1.4496	0.0011	3490.7	4937.2	61.382	10.51	14.64	1421
400	4.13206	1.6560	0.0010	4013.8	5666.7	63.331	10.43	14.55	1520
450	4.64779	1.8624	0.0009	4534.2	6393.3	65.044	10.41	14.53	1613
500	5.16346	2.0687	0.0008	5054.6	7120.0	66.574	10.40	14.53	1700
550	5.67910	2.2750	0.0007	5574.7	7846.3	67.959	10.41	14.53	1782
600	6.19471	2.4813	0.0007	6095.7	8573.6	69.223	10.43	14.55	1861
700	7.22588	2.8938	0.0006	7140.7	10031.0	71.471	10.48	14.61	2008
800	8.25701	3.3063	0.0005	8193.2	11496.0	73.426	10.57	14.70	2144
900	9.28811	3.7187	0.0004	9256.6	12971.8	75.162	10.70	14.82	2270
1000	10.31920	4.1312	0.0004	10333.8	14461.5	76.734	10.86	14.98	2388
1200	12.38135	4.9561	0.0003	12544.1	17496.6	79.488	11.24	15.37	2603
1400	14.44349	5.7809	0.0003	14836.8	20614.2	81.895	11.68	15.81	2797
1600	16.50580	6.6058	0.0003	17217.5	23819.8	84.012	12.12	16.25	2976
1800	18.56935	7.4306	0.0002	19695.6	27123.4	85.919	12.62	16.75	3141
2000	20.63822	8.2555	0.0002	22306.3	30561.6	87.796	13.39	17.56	3290
2500	25.94803	10.3176	0.0002	30220.2	40599.4	94.598	19.31	24.04	3583
3000	32.16812	12.3797	0.0001	44399.4	57266.7	103.987	38.56	46.24	3848

* TWO-PHASE BOUNDARY

Table 4.1 (continued)
THERMODYNAMIC PROPERTIES OF PARAHYDROGEN (ISOBARS, SI UNITS)

0.40 MPA ISOBAR

TEMPERATURE DEG. K	DENSITY KG/CU M	V(DH/DV)$_P$ KJ/KG	V(DP/DU)$_V$ MPA-CU M/KJ X 10^2	-V(DP/DV)$_T$ MPA	-(DV/DT)$_P$/V 1/DEG. K	THERMAL CONDUCTIVITY W/K-M X 10^3	VISCOSITY KG/M-S X 10^7	THERMAL DIFFUSIVITY SQ M/HR	DIELECTRIC CONSTANT	PRANDTL NUMBER
* 13.935	77.2399	631.3	0.25599	91.7322	0.01011	74.25	257.26	0.00054	1.25236	2.2124
14	77.1782	634.3	0.25541	91.8568	0.01010	74.91	254.75	0.00055	1.25215	2.1789
15	76.3720	617.1	0.24570	81.8567	0.01125	83.51	224.16	0.00057	1.24931	1.8642
16	75.4984	619.9	0.23427	75.8051	0.01191	89.48	199.49	0.00058	1.24621	1.6455
17	74.5686	619.9	0.22511	70.0983	0.01266	93.43	179.63	0.00057	1.24298	1.5084
18	73.6116	619.2	0.21750	64.9667	0.01344	96.26	163.38	0.00057	1.23964	1.4120
19	72.5911	612.8	0.21094	59.4563	0.01442	97.89	149.60	0.00055	1.23607	1.3508
20	71.5035	602.0	0.20502	53.7931	0.01562	99.33	137.75	0.00053	1.23228	1.3040
21	70.3507	589.0	0.19974	48.3137	0.01700	100.99	127.43	0.00052	1.22827	1.2634
22	69.1127	573.7	0.19481	42.9569	0.01862	101.88	118.25	0.00050	1.22397	1.2401
23	67.7665	554.3	0.18995	37.4872	0.02068	102.04	109.91	0.00047	1.21931	1.2344
24	66.2974	531.4	0.18472	32.0293	0.02327	101.53	102.22	0.00045	1.21424	1.2450
25	64.6759	505.0	0.17883	26.6581	0.02664	100.37	95.02	0.00042	1.20865	1.2739
* 25.957	62.9088	476.2	0.17182	21.6193	0.03096	98.50	88.30	0.00038	1.20259	1.3216
* 25.957	4.8798	206.9	0.06828	0.2826	0.08113	25.16	15.16	0.00111	1.01478	1.0116
26	4.8610	207.6	0.06823	0.2835	0.08050	25.15	15.18	0.00111	1.01472	1.0083
27	4.5124	222.0	0.06869	0.3016	0.06910	25.23	15.61	0.00131	1.01366	0.9489
28	4.2287	235.9	0.06888	0.3152	0.06120	25.51	16.05	0.00150	1.01280	0.9083
29	3.9899	249.5	0.06892	0.3258	0.05532	25.91	16.49	0.00169	1.01207	0.8787
30	3.7842	262.8	0.06889	0.3344	0.05072	26.38	16.94	0.00188	1.01145	0.8557
31	3.6040	275.7	0.06883	0.3415	0.04699	26.93	17.38	0.00208	1.01090	0.8364
32	3.4448	288.4	0.06877	0.3475	0.04389	27.50	17.83	0.00227	1.01042	0.8207
33	3.3005	301.0	0.06869	0.3526	0.04125	28.09	18.27	0.00247	1.00998	0.8077
34	3.1788	313.1	0.06870	0.3570	0.03898	28.68	18.71	0.00267	1.00959	0.7963
35	3.0527	325.1	0.06869	0.3608	0.03700	29.29	19.15	0.00287	1.00923	0.7866
36	2.9444	337.0	0.06866	0.3641	0.03525	29.91	19.59	0.00308	1.00890	0.7781
37	2.8447	348.9	0.06861	0.3670	0.03369	30.54	20.02	0.00329	1.00860	0.7707
38	2.7524	360.7	0.06856	0.3697	0.03229	31.18	20.45	0.00350	1.00832	0.7640
39	2.6666	372.4	0.06858	0.3720	0.03102	31.82	20.88	0.00372	1.00806	0.7580
40	2.5867	384.1	0.06843	0.3741	0.02986	32.47	21.31	0.00394	1.00782	0.7525
42	2.4419	407.3	0.06827	0.3777	0.02781	33.76	22.15	0.00439	1.00738	0.7431
44	2.3139	430.4	0.06809	0.3807	0.02607	35.07	22.97	0.00486	1.00699	0.7351
46	2.1997	453.4	0.06786	0.3832	0.02456	36.38	23.79	0.00535	1.00664	0.7281
48	2.0971	476.6	0.06760	0.3853	0.02323	37.71	24.59	0.00585	1.00633	0.7220
50	2.0043	499.9	0.06728	0.3871	0.02205	39.04	25.38	0.00636	1.00605	0.7169
52	1.9199	523.5	0.06691	0.3886	0.02100	40.38	26.16	0.00689	1.00580	0.7125
54	1.8427	547.5	0.06647	0.3899	0.02006	41.72	26.93	0.00742	1.00556	0.7088
56	1.7718	571.9	0.06596	0.3910	0.01920	43.07	27.69	0.00797	1.00535	0.7059
58	1.7064	596.8	0.06538	0.3920	0.01842	44.42	28.44	0.00853	1.00515	0.7038
60	1.6459	622.4	0.06473	0.3929	0.01770	45.84	29.18	0.00910	1.00497	0.7014
65	1.5125	689.7	0.06278	0.3947	0.01615	49.63	30.99	0.01060	1.00456	0.6956
70	1.3998	762.5	0.06044	0.3960	0.01487	53.43	32.74	0.01212	1.00422	0.6948
75	1.3031	841.8	0.05783	0.3970	0.01378	57.17	34.45	0.01361	1.00393	0.6991
80	1.2192	927.7	0.05509	0.3978	0.01285	61.02	36.11	0.01511	1.00368	0.7055
85	1.1457	1020.4	0.05234	0.3984	0.01204	65.95	37.73	0.01686	1.00346	0.7030
90	1.0807	1119.2	0.04970	0.3989	0.01133	70.87	39.31	0.01861	1.00326	0.7036
95	1.0228	1223.5	0.04724	0.3993	0.01071	75.80	40.85	0.02037	1.00308	0.7060
100	0.9709	1332.4	0.04500	0.3996	0.01015	89.67	45.75	0.02460	1.00293	0.6897
120	0.8074	1789.7	0.03845	0.4004	0.00840	117.09	54.05	0.03473	1.00243	0.6939
140	0.6914	2230.4	0.03517	0.4008	0.00717	136.10	59.18	0.04428	1.00208	0.6958
160	0.6047	2614.5	0.03400	0.4010	0.00626	148.28	63.00	0.05389	1.00182	0.6959
180	0.5374	2938.9	0.03404	0.4010	0.00556	155.67	66.19	0.06380	1.00162	0.6949
200	0.4837	3218.6	0.03471	0.4010	0.00500	160.25	69.04	0.07411	1.00146	0.6934
220	0.4397	3472.3	0.03562	0.4010	0.00454	163.48	71.69	0.08484	1.00133	0.6919
240	0.4031	3715.7	0.03656	0.4010	0.00416	166.32	74.23	0.09602	1.00121	0.6904
260	0.3721	3959.2	0.03739	0.4010	0.00384	169.26	76.68	0.10765	1.00112	0.6892
280	0.3455	4207.3	0.03807	0.4010	0.00357	172.49	79.09	0.11974	1.00104	0.6882
300	0.3225	4462.2	0.03859	0.4009	0.00333	176.05	81.46	0.13228	1.00097	0.6873
350	0.2765	5131.1	0.03935	0.4008	0.00285	186.29	87.29	0.16567	1.00083	0.6860
400	0.2420	5830.1	0.03965	0.4008	0.00250	197.63	93.05	0.20199	1.00073	0.6852
450	0.2152	6549.0	0.03972	0.4007	0.00222	209.54	98.73	0.24125	1.00065	0.6848
500	0.1937	7273.2	0.03973	0.4006	0.00200	221.51	104.36	0.28344	1.00058	0.6844
550	0.1761	8004.5	0.03969	0.4006	0.00182	233.55	109.93	0.32850	1.00053	0.6842
600	0.1614	8741.0	0.03962	0.4005	0.00166	245.58	115.43	0.37638	1.00049	0.6840
700	0.1384	10235.4	0.03940	0.4005	0.00143	269.71	126.25	0.48035	1.00042	0.6837
800	0.1211	11769.3	0.03905	0.4004	0.00125	294.13	136.82	0.59489	1.00036	0.6837
900	0.1077	13353.5	0.03858	0.4004	0.00111	319.00	147.14	0.71954	1.00032	0.6837
1000	0.0969	14995.4	0.03601	0.4003	0.00100	440.56	157.21	1.09232	1.00029	0.5347
1200	0.0808	18453.6	0.03671	0.4003	0.00083	511.14	176.73	1.48254	1.00024	0.5313
1400	0.0692	22141.7	0.03533	0.4002	0.00071	583.50	195.47	1.91952	1.00021	0.5295
1600	0.0606	26009.3	0.03404	0.4002	0.00062	657.19	213.53	2.40349	1.00018	0.5279
1800	0.0539	30169.3	0.03270	0.4002	0.00056	737.67	231.02	2.94329	1.00016	0.5247
2000	0.0485	35126.5	0.03082	0.4000	0.00050	844.18	248.02	3.57117	1.00015	0.5160
2500	0.0385	59743.7	0.02150	0.3976	0.00040	1554.35	289.39	6.03972	1.00012	0.4476
3000	0.0311	133455.3	0.01112	0.3848	0.00035	4078.54	333.93	10.21507	1.00009	0.3786

* TWO-PHASE BOUNDARY

Table 4.1 (continued)

THERMODYNAMIC PROPERTIES OF PARAHYDROGEN (ISOBARS, SI UNITS)

0.50 MPA ISOBAR

	TEMPERATJRE	VOLUME	ISOTHERM DERIVATIVE	ISOCHORE DERIVATIVE	INTERNAL ENERGY	ENTHALPY	ENTROPY	CV	CP	VELOCITY OF SOUND
	DEG. K	CU M/KG	CU M-MPA/KG	MPA/K	KJ/KG-K	KJ/KG	KJ/KG-K	KJ /	KG-K	M/S
*	13.968	0.01294	1.1940	0.9290	-308.7	-302.3	4.978	4.70	6.39	1274
	14	0.01294	1.2053	0.9289	-308.5	-302.0	4.994	4.70	6.38	1279
	15	0.01308	1.0819	0.9246	-302.0	-295.4	5.451	4.91	6.93	1237
	16	0.01323	1.0138	0.9058	-294.9	-288.3	5.913	5.10	7.37	1210
	17	0.01339	0.9559	0.8908	-287.4	-280.7	6.370	5.28	7.81	1189
	18	0.01356	0.8927	0.8759	-279.4	-272.7	6.832	5.45	8.29	1166
	19	0.01375	0.8295	0.8607	-271.0	-264.1	7.294	5.60	8.81	1142
	20	0.01396	0.7653	0.8436	-262.0	-255.0	7.760	5.73	9.35	1118
	21	0.01419	0.6968	0.8246	-252.4	-245.3	8.231	5.84	9.96	1090
	22	0.01444	0.6323	0.8035	-242.3	-235.1	8.709	5.94	10.62	1063
	23	0.01472	0.5647	0.7790	-231.4	-224.1	9.198	6.02	11.37	1033
	24	0.01504	0.4952	0.7500	-219.8	-212.3	9.700	6.08	12.25	998
	25	0.01541	0.4251	0.7158	-207.3	-199.6	10.219	6.14	13.29	959
	26	0.01584	0.3527	0.6746	-193.5	-185.6	10.766	6.19	14.61	912
	27	0.01638	0.2793	0.6263	-178.4	-170.2	11.348	6.25	16.43	856
*	27.115	0.01645	0.2724	0.6200	-176.4	-168.2	11.421	6.26	16.62	850
*	27.116	0.16290	0.0528	0.0296	119.0	200.5	25.025	7.00	18.92	378
	28	0.17542	0.0614	0.0270	128.6	216.3	25.600	6.84	17.03	391
	29	0.18839	0.0701	0.0247	138.4	232.6	26.172	6.72	15.69	405
	30	0.20054	0.0780	0.0229	147.5	247.8	26.688	6.64	14.78	417
	31	0.21210	0.0854	0.0215	156.2	262.2	27.161	6.58	14.12	428
	32	0.22323	0.0923	0.0203	164.5	276.1	27.601	6.53	13.62	439
	33	0.23401	0.0988	0.0192	172.5	289.6	28.015	6.50	13.23	449
	34	0.24452	0.1051	0.0183	180.4	302.6	28.405	6.45	12.90	458
	35	0.25480	0.1112	0.0174	188.0	315.4	28.775	6.42	12.64	468
	36	0.26489	0.1171	0.0167	195.5	327.9	29.128	6.40	12.42	477
	37	0.27482	0.1229	0.0160	202.8	340.2	29.465	6.38	12.23	485
	38	0.28462	0.1285	0.0154	210.1	352.4	29.789	6.36	12.07	494
	39	0.29430	0.1340	0.0149	217.2	364.4	30.101	6.35	11.94	502
	40	0.30387	0.1394	0.0144	224.3	376.2	30.402	6.34	11.82	510
	42	0.32275	0.1500	0.0135	238.3	399.7	30.973	6.33	11.63	525
	44	0.34133	0.1603	0.0127	252.1	422.8	31.511	6.32	11.47	539
	46	0.35967	0.1703	0.0120	265.8	445.6	32.018	6.32	11.36	553
	48	0.37782	0.1802	0.0114	279.3	468.2	32.499	6.33	11.27	566
	50	0.39579	0.1899	0.0108	292.8	490.7	32.958	6.34	11.20	579
	52	0.41363	0.1994	0.0104	306.2	513.0	33.396	6.37	11.15	591
	54	0.43134	0.2089	0.0099	319.6	535.3	33.816	6.40	11.12	603
	56	0.44894	0.2182	0.0095	333.1	557.5	34.221	6.44	11.11	614
	58	0.46645	0.2275	0.0091	346.5	579.8	34.610	6.48	11.11	624
	60	0.48388	0.2366	0.0088	360.1	602.0	34.987	6.54	11.12	634
	65	0.52716	0.2593	0.0080	394.3	657.8	35.881	6.72	11.23	658
	70	0.57010	0.2816	0.0074	429.4	714.4	36.720	6.97	11.41	679
	75	0.61279	0.3036	0.0069	465.7	772.1	37.515	7.27	11.66	698
	80	0.65527	0.3254	0.0064	503.5	831.1	38.277	7.62	11.97	715
	85	0.69759	0.3471	0.0060	543.1	891.9	39.014	8.00	12.33	731
	90	0.73977	0.3687	0.0057	584.6	954.5	39.730	8.42	12.72	746
	95	0.78185	0.3901	0.0054	628.2	1019.2	40.429	8.85	13.13	761
	100	0.82383	0.4115	0.0051	674.0	1085.9	41.113	9.28	13.55	775
	120	0.99107	0.4962	0.0042	877.0	1372.6	43.721	10.84	15.05	830
	140	1.15760	0.5802	0.0036	1105.5	1684.3	46.121	11.83	16.02	886
	160	1.32371	0.6638	0.0031	1347.4	2009.2	48.290	12.22	16.39	944
	180	1.48953	0.7472	0.0028	1592.4	2337.2	50.222	12.19	16.35	1001
	200	1.65516	0.8303	0.0025	1834.4	2662.0	51.932	11.95	16.10	1058
	220	1.82065	0.9133	0.0023	2070.5	2980.8	53.452	11.64	15.78	1113
	240	1.98605	0.9962	0.0021	2300.3	3293.3	54.812	11.33	15.47	1166
	260	2.15136	1.0790	0.0019	2524.5	3600.2	56.040	11.08	15.22	1217
	280	2.31662	1.1618	0.0018	2744.0	3902.4	57.160	10.87	15.01	1266
	300	2.48169	1.2445	0.0017	2960.2	4201.0	58.191	10.72	14.86	1313
	350	2.89453	1.4511	0.0014	3490.4	4937.7	60.461	10.51	14.64	1422
	400	3.30723	1.6576	0.0013	4013.6	5667.3	62.411	10.43	14.56	1521
	450	3.71985	1.8640	0.0011	4534.1	6394.0	64.124	10.41	14.53	1613
	500	4.13242	2.0703	0.0010	5054.5	7120.7	65.654	10.40	14.53	1701
	550	4.54495	2.2767	0.0009	5574.6	7847.0	67.038	10.41	14.54	1783
	600	4.95746	2.4830	0.0008	6095.6	8574.3	68.302	10.43	14.55	1862
	700	5.78242	2.8955	0.0007	7140.6	10031.8	70.550	10.48	14.61	2009
	800	6.60734	3.3080	0.0006	8193.1	11496.8	72.505	10.57	14.70	2144
	900	7.43224	3.7205	0.0006	9256.5	12972.6	74.242	10.70	14.82	2270
	1000	8.25712	4.1330	0.0005	10333.8	14462.3	75.814	10.86	14.98	2388
	1200	9.90685	4.9578	0.0004	12544.0	17497.5	78.568	11.24	15.37	2603
	1400	11.55656	5.7827	0.0004	14836.8	20615.1	80.975	11.68	15.81	2797
	1600	13.20641	6.6076	0.0003	17217.1	23820.3	83.092	12.12	16.24	2976
	1800	14.85713	7.4324	0.0003	19693.5	27122.1	84.998	12.60	16.74	3142
	2000	16.51165	8.2573	0.0003	22297.0	30552.8	86.871	13.33	17.50	3292
	2500	20.74647	10.3194	0.0002	30075.8	40449.0	93.612	18.71	23.37	3590
	3000	25.63286	12.3815	0.0002	43497.1	56313.6	102.715	36.04	43.31	3853

* TWO-PHASE BOUNDARY

Table 4.1 (continued)
THERMODYNAMIC PROPERTIES OF PARAHYDROGEN (ISOBARS, SI UNITS)

0.50 MPA ISOBAR

TEMPERATURE DEG. K	DENSITY KG/CU M	V(DH/DV)$_P$ KJ/KG	V(DP/DU)$_V$ MPA-CU M/KJ x 10^2	-V(DP/DV)$_T$ MPA	-(DV/DT)$_P$/V 1/DEG. K	THERMAL CONDUCTIVITY W/K-M x 10^3	VISCOSITY KG/M-S x 10^7	THERMAL DIFFUSIVITY SQ M/HR	DIELECTRIC CONSTANT	PRANDTL NUMBER
* 13.968	77.2965	634.5	0.25586	92.2941	0.01007	74.66	257.88	0.00054	1.25256	2.2061
14	77.2625	639.9	0.25557	93.1235	0.00998	74.98	256.56	0.00055	1.25244	2.1842
15	76.4616	620.4	0.24647	82.7270	0.01118	83.62	225.71	0.00057	1.24962	1.8716
16	75.5862	623.1	0.23499	76.6271	0.01182	89.63	200.85	0.00058	1.24655	1.6506
17	74.6876	626.0	0.22585	71.3910	0.01248	93.65	181.04	0.00058	1.24340	1.5099
18	73.7238	623.1	0.21812	65.8110	0.01331	96.49	164.50	0.00057	1.24003	1.4140
19	72.7116	617.0	0.21153	60.3138	0.01427	98.15	150.63	0.00055	1.23649	1.3514
20	71.6412	607.7	0.20563	54.8242	0.01539	99.63	138.76	0.00054	1.23276	1.3022
21	70.4941	593.6	0.20030	49.1229	0.01679	101.32	128.35	0.00052	1.22877	1.2621
22	69.2711	578.8	0.19535	43.7989	0.01834	102.25	119.14	0.00050	1.22452	1.2372
23	67.9460	560.1	0.19052	38.3717	0.02030	102.47	110.79	0.00048	1.21993	1.2294
24	66.5813	537.8	0.18538	32.9286	0.02278	102.02	103.11	0.00045	1.21494	1.2379
25	64.9112	512.3	0.17963	27.5905	0.02594	100.93	95.91	0.00042	1.20946	1.2629
26	63.1156	482.2	0.17263	22.2624	0.03030	99.09	88.98	0.00039	1.20330	1.3120
27	61.0558	447.2	0.16402	17.0512	0.03673	96.50	82.17	0.00035	1.19625	1.3988
* 27.116	60.7811	443.9	0.16287	16.5596	0.03744	96.11	81.33	0.00034	1.19531	1.4064
* 27.116	6.1386	207.3	0.06874	0.3240	0.09124	27.63	16.10	0.00086	1.01862	1.1024
28	5.7006	221.3	0.06912	0.3503	0.07695	27.33	16.44	0.00101	1.01728	1.0239
29	5.3082	236.3	0.06929	0.3723	0.06639	27.35	16.84	0.00118	1.01609	0.9656
30	4.9866	250.6	0.06931	0.3891	0.05897	27.58	17.25	0.00135	1.01511	0.9246
31	4.7147	264.5	0.06927	0.4024	0.05340	27.99	17.67	0.00151	1.01428	0.8918
32	4.4797	278.0	0.06922	0.4133	0.04901	28.46	18.10	0.00168	1.01356	0.8664
33	4.2733	291.2	0.06912	0.4223	0.04544	28.97	18.53	0.00184	1.01293	0.8461
34	4.0897	303.8	0.06918	0.4300	0.04247	29.50	18.96	0.00201	1.01238	0.8290
35	3.9247	316.3	0.06919	0.4365	0.03995	30.06	19.38	0.00218	1.01188	0.8148
36	3.7752	328.6	0.06918	0.4422	0.03778	30.63	19.81	0.00235	1.01142	0.8028
37	3.6387	340.9	0.06914	0.4471	0.03588	31.22	20.23	0.00253	1.01101	0.7925
38	3.5135	353.0	0.06908	0.4515	0.03420	31.83	20.66	0.00270	1.01063	0.7835
39	3.3979	365.1	0.06901	0.4554	0.03270	32.44	21.08	0.00288	1.01028	0.7755
40	3.2909	377.1	0.06894	0.4588	0.03134	33.06	21.49	0.00306	1.00995	0.7684
42	3.0984	400.8	0.06876	0.4647	0.02900	34.31	22.32	0.00343	1.00937	0.7563
44	2.9297	424.5	0.06854	0.4696	0.02703	35.58	23.14	0.00381	1.00885	0.7462
46	2.7803	448.0	0.06830	0.4736	0.02535	36.87	23.95	0.00420	1.00840	0.7377
48	2.6468	471.6	0.06801	0.4769	0.02389	38.17	24.74	0.00461	1.00800	0.7304
50	2.5266	495.3	0.06767	0.4797	0.02261	39.48	25.53	0.00502	1.00763	0.7241
52	2.4176	519.2	0.06728	0.4822	0.02148	40.80	26.30	0.00545	1.00730	0.7189
54	2.3184	543.4	0.06682	0.4843	0.02046	42.12	27.07	0.00588	1.00700	0.7146
56	2.2275	568.1	0.06629	0.4861	0.01955	43.46	27.82	0.00632	1.00673	0.7110
58	2.1438	593.2	0.06570	0.4876	0.01872	44.79	28.56	0.00677	1.00647	0.7084
60	2.0666	619.0	0.06503	0.4890	0.01797	46.20	29.30	0.00723	1.00624	0.7055
65	1.8970	686.7	0.06306	0.4918	0.01635	49.96	31.10	0.00844	1.00573	0.6988
70	1.7541	759.9	0.06069	0.4939	0.01502	53.73	32.84	0.00967	1.00529	0.6974
75	1.6319	839.3	0.05805	0.4954	0.01389	57.44	34.54	0.01086	1.00493	0.7013
80	1.5261	925.5	0.05529	0.4967	0.01294	61.28	36.19	0.01207	1.00461	0.7073
85	1.4335	1018.3	0.05252	0.4976	0.01211	66.19	37.81	0.01348	1.00433	0.7044
90	1.3518	1117.3	0.04986	0.4984	0.01139	71.11	39.39	0.01488	1.00408	0.7048
95	1.2790	1221.7	0.04738	0.4990	0.01075	76.03	40.93	0.01629	1.00386	0.7070
100	1.2138	1330.7	0.04513	0.4995	0.01018	89.72	45.75	0.01964	1.00366	0.6909
120	1.0090	1788.8	0.03854	0.5007	0.00842	117.09	54.04	0.02775	1.00304	0.6948
140	0.8639	2230.1	0.03524	0.5012	0.00718	136.09	59.17	0.03541	1.00261	0.6964
160	0.7555	2614.9	0.03406	0.5015	0.00627	148.28	63.00	0.04311	1.00228	0.6964
180	0.6714	2939.7	0.03409	0.5016	0.00556	155.68	66.19	0.05105	1.00202	0.6953
200	0.6042	3219.9	0.03475	0.5016	0.00500	160.27	69.05	0.05931	1.00182	0.6937
220	0.5493	3473.9	0.03566	0.5016	0.00454	163.51	71.70	0.06790	1.00166	0.6921
240	0.5035	3717.5	0.03659	0.5016	0.00416	166.35	74.24	0.07686	1.00152	0.6906
260	0.4648	3961.1	0.03742	0.5016	0.00384	169.29	76.69	0.08617	1.00140	0.6893
280	0.4317	4209.5	0.03810	0.5015	0.00357	172.53	79.10	0.09585	1.00130	0.6883
300	0.4030	4464.5	0.03862	0.5015	0.00333	176.10	81.48	0.10589	1.00121	0.6874
350	0.3455	5133.6	0.03938	0.5013	0.00285	186.35	87.32	0.13263	1.00104	0.6860
400	0.3024	5832.7	0.03967	0.5012	0.00250	197.69	93.07	0.16171	1.00091	0.6853
450	0.2688	6551.7	0.03974	0.5011	0.00222	209.61	98.76	0.19314	1.00081	0.6848
500	0.2420	7276.0	0.03974	0.5010	0.00200	221.59	104.39	0.22691	1.00073	0.6844
550	0.2200	8007.3	0.03970	0.5009	0.00182	233.63	109.96	0.26298	1.00066	0.6842
600	0.2017	8743.9	0.03963	0.5009	0.00166	245.67	115.47	0.30131	1.00061	0.6840
700	0.1729	10238.4	0.03941	0.5007	0.00143	269.82	126.30	0.38454	1.00052	0.6837
800	0.1513	11772.4	0.03906	0.5007	0.00125	294.26	136.87	0.47623	1.00046	0.6837
900	0.1345	13356.6	0.03859	0.5006	0.00111	319.14	147.20	0.57601	1.00041	0.6837
1000	0.1211	14998.5	0.03802	0.5005	0.00100	440.56	157.28	0.87403	1.00036	0.5349
1200	0.1009	18456.9	0.03672	0.5004	0.00083	511.14	176.81	1.18624	1.00030	0.5316
1400	0.0865	22144.5	0.03533	0.5004	0.00071	583.48	195.56	1.53583	1.00026	0.5297
1600	0.0757	26007.4	0.03406	0.5003	0.00062	656.97	213.64	1.92282	1.00023	0.5282
1800	0.0673	30140.9	0.03275	0.5003	0.00056	736.42	231.13	2.35342	1.00020	0.5253
2000	0.0606	34998.9	0.03097	0.5001	0.00050	839.11	248.14	2.85075	1.00018	0.5174
2500	0.0482	58129.2	0.02218	0.4974	0.00040	1493.02	289.44	4.77084	1.00015	0.4531
3000	0.0390	125516.5	0.01185	0.4830	0.00035	3780.36	333.43	8.05486	1.00012	0.3820

* TWO-PHASE BOUNDARY

Table 4.1 (continued)
THERMODYNAMIC PROPERTIES OF PARAHYDROGEN (ISOBARS, SI UNITS)

0.60 MPA ISOBAR

TEMPERATURE DEG. K	VOLUME CU M/KG	ISOTHERM DERIVATIVE CU M-MPA/KG	ISOCHORE DERIVATIVE MPA/K	INTERNAL ENERGY KJ/KG-K	ENTHALPY KJ/KG	ENTROPY KJ/KG-K	CV KJ / KG-K	CP KJ / KG-K	VELOCITY OF SOUND M/S
* 14.001	0.01293	1.2078	0.9302	-308.7	-300.9	4.980	4.70	6.38	1280
15	0.01306	1.0920	0.9281	-302.2	-294.3	5.437	4.90	6.92	1242
16	0.01321	1.0321	0.9095	-295.2	-287.2	5.894	5.10	7.33	1219
17	0.01337	0.9660	0.8938	-287.7	-279.7	6.354	5.28	7.79	1194
18	0.01354	0.9027	0.8790	-279.7	-271.6	6.814	5.44	8.27	1171
19	0.01373	0.8398	0.8638	-271.3	-263.1	7.275	5.59	8.77	1148
20	0.01393	0.7755	0.8468	-262.4	-254.0	7.739	5.72	9.31	1123
21	0.01416	0.7107	0.8280	-252.9	-244.4	8.207	5.84	9.90	1098
22	0.01440	0.6428	0.8069	-242.8	-234.2	8.683	5.93	10.56	1069
23	0.01468	0.5760	0.7829	-232.1	-223.3	9.168	6.01	11.29	1040
24	0.01499	0.5082	0.7545	-220.6	-211.6	9.666	6.08	12.12	1007
25	0.01535	0.4376	0.7211	-208.2	-199.0	10.180	6.14	13.14	968
26	0.01577	0.3665	0.6813	-194.7	-185.3	10.719	6.19	14.38	923
27	0.01629	0.2933	0.6346	-179.9	-170.1	11.291	6.25	16.08	869
28	0.01694	0.2216	0.5810	-163.1	-152.9	11.916	6.32	18.56	807
* 28.123	0.01704	0.2189	0.5735	-160.8	-150.6	11.999	6.34	18.60	802
* 28.123	0.13376	0.0472	0.0369	115.5	195.8	24.322	7.13	21.64	378
29	0.14530	0.0570	0.0332	126.3	213.5	24.941	6.95	18.80	393
30	0.15699	0.0666	0.0302	137.1	231.3	25.545	6.81	16.93	407
31	0.16776	0.0752	0.0279	146.9	247.6	26.079	6.71	15.73	420
32	0.17792	0.0831	0.0260	156.1	262.8	26.564	6.64	14.88	431
33	0.18761	0.0905	0.0244	164.9	277.4	27.013	6.59	14.25	442
34	0.19695	0.0974	0.0231	173.2	291.4	27.430	6.53	13.74	453
35	0.20602	0.1041	0.0219	181.3	305.0	27.823	6.48	13.35	463
36	0.21486	0.1104	0.0209	189.2	318.1	28.194	6.45	13.03	472
37	0.22352	0.1166	0.0200	196.9	331.0	28.548	6.42	12.77	482
38	0.23202	0.1226	0.0192	204.5	343.7	28.885	6.40	12.55	490
39	0.24038	0.1285	0.0185	211.9	356.1	29.209	6.38	12.36	499
40	0.24864	0.1342	0.0178	219.2	368.4	29.520	6.37	12.20	507
42	0.26485	0.1453	0.0166	233.7	392.6	30.109	6.35	11.95	523
44	0.28074	0.1560	0.0156	247.8	416.3	30.660	6.34	11.75	538
46	0.29637	0.1665	0.0147	261.8	439.6	31.178	6.34	11.59	552
48	0.31179	0.1767	0.0139	275.6	462.7	31.669	6.35	11.47	565
50	0.32704	0.1867	0.0132	289.3	485.5	32.135	6.36	11.38	578
52	0.34214	0.1965	0.0126	302.9	508.2	32.580	6.38	11.31	590
54	0.35712	0.2062	0.0121	316.5	530.8	33.006	6.41	11.27	602
56	0.37198	0.2158	0.0115	330.1	553.3	33.416	6.45	11.24	613
58	0.38675	0.2252	0.0111	343.7	575.7	33.810	6.49	11.23	624
60	0.40143	0.2346	0.0107	357.3	598.2	34.190	6.55	11.23	634
65	0.43784	0.2576	0.0097	391.8	654.5	35.092	6.73	11.31	658
70	0.47391	0.2802	0.0090	427.1	711.5	35.936	6.97	11.48	679
75	0.50972	0.3025	0.0083	463.6	769.5	36.736	7.27	11.72	698
80	0.54533	0.3246	0.0078	501.6	828.8	37.502	7.62	12.03	716
85	0.58077	0.3465	0.0073	541.4	889.8	38.242	8.01	12.38	732
90	0.61608	0.3682	0.0068	583.0	952.7	38.960	8.42	12.76	747
95	0.65128	0.3899	0.0065	626.7	1017.5	39.661	8.85	13.17	761
100	0.68639	0.4113	0.0061	672.5	1084.4	40.347	9.29	13.58	776
120	0.82612	0.4965	0.0051	875.9	1371.6	42.959	10.86	15.07	831
140	0.96515	0.5808	0.0043	1104.6	1683.7	45.362	11.83	16.03	887
160	1.10375	0.6646	0.0038	1346.6	2008.9	47.533	12.22	16.40	945
180	1.24207	0.7481	0.0034	1591.8	2337.0	49.466	12.19	16.36	1002
200	1.38020	0.8314	0.0030	1833.8	2661.9	51.177	11.95	16.11	1059
220	1.51820	0.9145	0.0027	2070.0	2980.9	52.698	11.64	15.79	1114
240	1.65609	0.9975	0.0025	2299.9	3293.5	54.058	11.33	15.48	1167
260	1.79391	1.0804	0.0023	2524.1	3600.4	55.286	11.08	15.22	1218
280	1.93167	1.1632	0.0021	2743.7	3902.7	56.407	10.88	15.01	1267
300	2.06928	1.2459	0.0020	2959.9	4201.4	57.438	10.72	14.86	1314
350	2.41338	1.4526	0.0017	3490.2	4938.2	59.708	10.51	14.64	1423
400	2.75734	1.6592	0.0015	4013.5	5667.9	61.658	10.43	14.56	1522
450	3.10122	1.8656	0.0013	4533.9	6394.6	63.371	10.41	14.53	1614
500	3.44506	2.0720	0.0012	5054.4	7121.4	64.901	10.40	14.53	1701
550	3.78885	2.2783	0.0011	5574.4	7847.8	66.286	10.41	14.54	1784
600	4.13262	2.4847	0.0010	6095.5	8575.1	67.550	10.43	14.55	1862
700	4.82011	2.8972	0.0009	7140.5	10032.6	69.798	10.48	14.61	2009
800	5.50756	3.3097	0.0008	8193.1	11497.6	71.753	10.57	14.70	2145
900	6.19499	3.7222	0.0007	9256.5	12973.5	73.490	10.70	14.82	2271
1000	6.88239	4.1347	0.0006	10333.7	14463.2	75.062	10.86	14.98	2389
1200	8.25718	4.9596	0.0005	12544.0	17498.3	77.816	11.24	15.37	2604
1400	9.63195	5.7845	0.0004	14836.8	20615.9	80.223	11.68	15.81	2798
1600	11.00681	6.6094	0.0004	17216.9	23821.0	82.339	12.12	16.24	2977
1800	12.38234	7.4342	0.0003	19692.0	27121.4	84.245	12.59	16.72	3142
2000	13.76076	8.2591	0.0003	22290.0	30546.5	86.115	13.28	17.45	3293
2500	17.28176	10.3212	0.0002	29969.2	40338.2	92.811	18.27	22.88	3595
3000	21.29856	12.3833	0.0002	42830.8	55610.0	101.702	34.18	41.14	3857

* TWO-PHASE BOUNDARY

Table 4.1 (continued)
THERMODYNAMIC PROPERTIES OF PARAHYDROGEN (ISOBARS, SI UNITS)

0.60 MPA ISOBAR

TEMPERATURE DEG. K	DENSITY KG/CU M	V(DH/DV)$_P$ KJ/KG	V(DP/DU)$_V$ MPA-CU M/KJ X 10^2	-V(DP/DV)$_T$ MPA	-(DV/DT)$_P$/V 1/DEG. K	THERMAL CONDUCTIVITY W/K-M X 10^3	VISCOSITY KG/M-S X 10^7	THERMAL DIFFUSIVITY SQ M/HR	DIELECTRIC CONSTANT	PRANDTL NUMBER
* 14.001	77.3529	640.6	0.25573	93.4243	0.00996	75.07	258.50	0.00055	1.25276	2.1963
15	76.5504	623.6	0.24723	83.5919	0.01110	83.73	227.26	0.00057	1.24994	1.8790
16	75.7022	630.0	0.23579	78.1349	0.01164	89.81	202.51	0.00058	1.24696	1.6534
17	74.7902	629.7	0.22649	72.2485	0.01237	93.84	182.26	0.00058	1.24376	1.5130
18	73.8345	627.0	0.21873	66.6472	0.01319	96.71	165.61	0.00057	1.24041	1.4160
19	72.8306	621.3	0.21212	61.1633	0.01412	98.41	151.65	0.00055	1.23691	1.3522
20	71.7697	612.1	0.20621	55.6607	0.01521	99.92	139.71	0.00054	1.23321	1.3020
21	70.6402	600.0	0.20087	50.2035	0.01649	101.66	129.29	0.00052	1.22928	1.2585
22	69.4268	583.9	0.19588	44.6299	0.01808	102.62	120.02	0.00050	1.22506	1.2345
23	68.1216	565.8	0.19108	39.2410	0.01995	102.89	111.66	0.00048	1.22054	1.2249
24	66.7023	544.7	0.18601	33.8976	0.02226	102.51	103.99	0.00046	1.21563	1.2298
25	65.1405	519.3	0.18040	28.5044	0.02530	101.48	96.80	0.00043	1.21025	1.2531
26	63.3921	490.4	0.17372	23.2325	0.02933	99.75	89.93	0.00039	1.20424	1.2967
27	61.3956	456.4	0.16546	18.0098	0.03524	97.28	83.19	0.00035	1.19741	1.3751
28	59.0308	417.9	0.15564	13.0801	0.04442	93.92	76.36	0.00031	1.18935	1.5092
* 28.123	58.6975	416.7	0.15422	12.8515	0.04462	93.42	75.48	0.00031	1.18821	1.5024
* 28.123	7.4761	207.0	0.06917	0.3528	0.10454	30.27	17.01	0.00067	1.02271	1.2160
29	6.8825	222.0	0.06951	0.3925	0.08467	29.46	17.28	0.00082	1.02089	1.1027
30	6.3700	238.0	0.06964	0.4245	0.07115	29.18	17.64	0.00097	1.01933	1.0232
31	5.9608	253.0	0.06965	0.4484	0.06216	29.32	18.02	0.00113	1.01808	0.9662
32	5.6207	267.4	0.06962	0.4671	0.05565	29.62	18.41	0.00128	1.01704	0.9249
33	5.3301	281.3	0.06951	0.4822	0.05064	30.00	18.81	0.00142	1.01615	0.8935
34	5.0773	294.5	0.06964	0.4946	0.04667	30.43	19.22	0.00157	1.01538	0.8682
35	4.8539	307.4	0.06970	0.5051	0.04342	30.91	19.63	0.00172	1.01470	0.8479
36	4.6542	320.2	0.06970	0.5140	0.04069	31.43	20.05	0.00187	1.01409	0.8312
37	4.4739	332.9	0.06967	0.5217	0.03836	31.97	20.46	0.00201	1.01355	0.8172
38	4.3100	345.4	0.06962	0.5285	0.03633	32.53	20.87	0.00216	1.01305	0.8053
39	4.1600	357.8	0.06954	0.5344	0.03455	33.10	21.28	0.00232	1.01259	0.7949
40	4.0220	370.1	0.06946	0.5397	0.03297	33.69	21.69	0.00247	1.01217	0.7857
42	3.7758	394.5	0.06926	0.5486	0.03028	34.89	22.51	0.00279	1.01142	0.7705
44	3.5621	418.7	0.06902	0.5558	0.02806	36.12	23.31	0.00311	1.01077	0.7581
46	3.3742	442.7	0.06874	0.5617	0.02619	37.38	24.11	0.00344	1.01020	0.7478
48	3.2072	466.7	0.06843	0.5667	0.02458	38.65	24.90	0.00378	1.00970	0.7391
50	3.0577	490.7	0.06807	0.5708	0.02319	39.93	25.67	0.00413	1.00924	0.7317
52	2.9228	515.0	0.06766	0.5744	0.02197	41.23	26.44	0.00449	1.00883	0.7256
54	2.8002	539.5	0.06718	0.5774	0.02088	42.53	27.20	0.00485	1.00846	0.7204
56	2.6883	564.4	0.06663	0.5800	0.01991	43.85	27.95	0.00523	1.00812	0.7162
58	2.5857	589.8	0.06602	0.5823	0.01904	45.17	28.69	0.00560	1.00781	0.7130
60	2.4911	615.7	0.06534	0.5843	0.01824	46.56	29.42	0.00599	1.00753	0.7096
65	2.2839	683.8	0.06333	0.5883	0.01655	50.29	31.21	0.00701	1.00690	0.7020
70	2.1101	757.3	0.06094	0.5913	0.01516	54.04	32.94	0.00803	1.00637	0.6999
75	1.9618	837.0	0.05828	0.5935	0.01401	57.71	34.63	0.00903	1.00592	0.7036
80	1.8338	923.3	0.05549	0.5953	0.01303	61.54	36.28	0.01005	1.00554	0.7091
85	1.7219	1016.3	0.05270	0.5966	0.01218	66.44	37.89	0.01122	1.00520	0.7059
90	1.6232	1115.4	0.05002	0.5977	0.01144	71.35	39.46	0.01240	1.00490	0.7059
95	1.5354	1220.0	0.04753	0.5986	0.01079	76.26	41.00	0.01358	1.00463	0.7080
100	1.4569	1329.1	0.04526	0.5993	0.01022	89.77	45.76	0.01633	1.00440	0.6922
120	1.2105	1787.9	0.03862	0.6010	0.00843	117.09	54.03	0.02310	1.00365	0.6957
140	1.0361	2229.8	0.03530	0.6018	0.00719	136.08	59.17	0.02949	1.00313	0.6971
160	0.9060	2615.2	0.03411	0.6022	0.00627	148.28	63.00	0.03592	1.00273	0.6968
180	0.8051	2940.6	0.03414	0.6023	0.00556	155.69	66.20	0.04255	1.00243	0.6956
200	0.7245	3221.1	0.03479	0.6024	0.00500	160.29	69.05	0.04944	1.00218	0.6940
220	0.6587	3475.5	0.03570	0.6024	0.00454	163.54	71.71	0.05661	1.00199	0.6923
240	0.6038	3719.3	0.03663	0.6023	0.00416	166.38	74.25	0.06408	1.00182	0.6908
260	0.5574	3963.1	0.03746	0.6022	0.00384	169.33	76.71	0.07185	1.00168	0.6895
280	0.5177	4211.6	0.03813	0.6022	0.00357	172.57	79.12	0.07993	1.00156	0.6884
300	0.4833	4466.8	0.03865	0.6021	0.00333	176.14	81.49	0.08830	1.00146	0.6875
350	0.4144	5136.1	0.03940	0.6019	0.00285	186.40	87.34	0.11060	1.00125	0.6861
400	0.3627	5835.3	0.03970	0.6017	0.00249	197.75	93.10	0.13485	1.00109	0.6853
450	0.3225	6554.5	0.03976	0.6016	0.00222	209.68	98.79	0.16106	1.00097	0.6848
500	0.2903	7278.8	0.03976	0.6014	0.00200	221.66	104.43	0.18922	1.00087	0.6844
550	0.2639	8010.2	0.03972	0.6013	0.00181	233.72	110.00	0.21931	1.00080	0.6842
600	0.2420	8746.8	0.03965	0.6012	0.00166	245.76	115.51	0.25127	1.00073	0.6840
700	0.2075	10241.4	0.03942	0.6011	0.00143	269.93	126.35	0.32067	1.00063	0.6837
800	0.1816	11775.4	0.03907	0.6009	0.00125	294.38	136.93	0.39712	1.00055	0.6837
900	0.1614	13359.7	0.03860	0.6008	0.00111	319.27	147.26	0.48032	1.00049	0.6837
1000	0.1453	15001.7	0.03803	0.6008	0.00100	440.56	157.35	0.72851	1.00044	0.5351
1200	0.1211	18460.1	0.03672	0.6006	0.00083	511.13	176.89	0.98870	1.00036	0.5318
1400	0.1038	22147.5	0.03534	0.6006	0.00071	583.46	195.65	1.28004	1.00031	0.5300
1600	0.0909	26006.8	0.03407	0.6005	0.00062	656.81	213.74	1.60242	1.00027	0.5285
1800	0.0808	30120.9	0.03279	0.6004	0.00056	735.50	231.25	1.96050	1.00024	0.5258
2000	0.0727	34905.7	0.03108	0.6002	0.00050	835.37	248.26	2.37186	1.00022	0.5185
2500	0.0579	56937.1	0.02271	0.5972	0.00040	1447.69	289.52	3.93639	1.00017	0.4576
3000	0.0470	119609.2	0.01246	0.5814	0.00034	3558.42	333.10	6.63130	1.00014	0.3851

* TWO-PHASE BOUNDARY

Table 4.1 (continued)
THERMODYNAMIC PROPERTIES OF PARAHYDROGEN (ISOBARS, SI UNITS)

0.70 MPA ISOBAR

TEMPERATURE DEG. K	VOLUME CU M/KG	ISOTHERM DERIVATIVE CU M-MPA/KG	ISOCHORE DERIVATIVE MPA/K	INTERNAL ENERGY KJ/KG-K	ENTHALPY KJ/KG	ENTROPY KJ/KG-K	CV KJ / KG-K	CP KJ / KG-K	VELOCITY OF SOUND M/S
* 14.034	0.01292	1.2145	0.9313	-308.7	-299.6	4.983	4.71	6.38	1283
15	0.01305	1.1212	0.9307	-302.4	-293.2	5.423	4.90	6.88	1254
16	0.01319	1.0426	0.9125	-295.4	-286.2	5.879	5.09	7.32	1224
17	0.01335	0.9761	0.8968	-287.9	-278.6	6.338	5.27	7.77	1199
18	0.01352	0.9125	0.8820	-280.0	-270.6	6.796	5.44	8.24	1176
19	0.01371	0.8500	0.8668	-271.7	-262.1	7.256	5.59	8.74	1153
20	0.01391	0.7857	0.8500	-262.8	-253.1	7.718	5.72	9.28	1129
21	0.01413	0.7213	0.8313	-253.4	-243.5	8.184	5.83	9.85	1104
22	0.01437	0.6554	0.8104	-243.4	-233.4	8.657	5.93	10.48	1076
23	0.01464	0.5871	0.7866	-232.7	-222.5	9.139	6.01	11.21	1046
24	0.01495	0.5200	0.7588	-221.4	-210.9	9.633	6.08	12.02	1014
25	0.01530	0.4498	0.7262	-209.1	-198.4	10.142	6.13	13.00	976
26	0.01571	0.3816	0.6878	-195.9	-184.9	10.673	6.18	14.14	934
27	0.01620	0.3102	0.6429	-181.4	-170.0	11.234	6.24	15.68	883
28	0.01682	0.2364	0.5913	-165.0	-153.2	11.844	6.31	18.03	822
29	0.01765	0.1660	0.5302	-145.9	-133.6	12.533	6.41	21.71	750
* 29.019	0.01767	0.1659	0.5291	-145.6	-133.2	12.546	6.41	21.70	749
* 29.019	0.11211	0.0412	0.0451	110.8	189.3	23.666	7.28	25.29	378
30	0.12448	0.0537	0.0395	124.6	211.7	24.425	7.04	20.53	396
31	0.13519	0.0641	0.0356	136.3	230.9	25.055	6.89	18.10	411
32	0.14493	0.0733	0.0327	146.7	248.2	25.604	6.78	16.59	424
33	0.15402	0.0817	0.0304	156.5	264.3	26.099	6.70	15.56	435
34	0.16265	0.0894	0.0285	165.6	279.4	26.552	6.61	14.79	447
35	0.17092	0.0967	0.0269	174.3	293.9	26.972	6.55	14.21	458
36	0.17893	0.1036	0.0255	182.6	307.9	27.365	6.50	13.75	468
37	0.18672	0.1102	0.0243	190.8	321.5	27.737	6.47	13.39	478
38	0.19432	0.1166	0.0232	198.7	334.7	28.090	6.44	13.09	487
39	0.20178	0.1229	0.0223	206.4	347.7	28.427	6.42	12.84	496
40	0.20911	0.1289	0.0214	214.0	360.4	28.749	6.40	12.63	504
42	0.22344	0.1406	0.0199	228.9	385.3	29.357	6.38	12.29	521
44	0.23742	0.1518	0.0186	243.4	409.6	29.922	6.36	12.04	536
46	0.25114	0.1626	0.0175	257.7	433.5	30.453	6.36	11.84	550
48	0.26463	0.1732	0.0166	271.8	457.0	30.953	6.36	11.69	564
50	0.27793	0.1835	0.0157	285.7	480.3	31.428	6.38	11.57	577
52	0.29109	0.1936	0.0149	299.5	503.3	31.880	6.40	11.48	590
54	0.30411	0.2036	0.0143	313.3	526.2	32.312	6.42	11.42	601
56	0.31701	0.2133	0.0136	327.1	549.0	32.726	6.46	11.37	613
58	0.32982	0.2230	0.0131	340.8	571.7	33.125	6.50	11.35	624
60	0.34255	0.2325	0.0126	354.6	594.4	33.510	6.56	11.34	634
65	0.37406	0.2560	0.0115	389.4	651.2	34.419	6.74	11.40	658
70	0.40522	0.2789	0.0105	424.9	708.6	35.269	6.98	11.55	680
75	0.43612	0.3015	0.0098	461.6	766.9	36.074	7.28	11.78	699
80	0.46681	0.3238	0.0091	499.8	826.5	36.844	7.62	12.08	716
85	0.49734	0.3459	0.0085	539.6	887.9	37.586	8.01	12.42	732
90	0.52774	0.3678	0.0080	581.4	950.8	38.307	8.43	12.80	748
95	0.55802	0.3896	0.0076	625.2	1015.9	39.010	8.86	13.20	762
100	0.58822	0.4112	0.0072	671.1	1082.9	39.697	9.29	13.61	776
120	0.70831	0.4968	0.0059	874.8	1370.6	42.314	10.84	15.10	832
140	0.82769	0.5814	0.0051	1103.7	1683.0	44.720	11.83	16.05	888
160	0.94664	0.6655	0.0044	1345.8	2008.5	46.893	12.22	16.41	945
180	1.06532	0.7491	0.0039	1591.1	2336.9	48.827	12.19	16.37	1003
200	1.18381	0.8325	0.0035	1833.2	2661.9	50.539	11.95	16.12	1059
220	1.30216	0.9157	0.0032	2069.5	2981.0	52.060	11.64	15.79	1115
240	1.42041	0.9988	0.0029	2299.4	3293.7	53.420	11.33	15.48	1168
260	1.53859	1.0817	0.0027	2523.7	3600.7	54.649	11.08	15.22	1219
280	1.65671	1.1646	0.0025	2743.3	3903.0	55.770	10.88	15.02	1268
300	1.77470	1.2474	0.0023	2959.6	4201.8	56.801	10.72	14.86	1315
350	2.06970	1.4541	0.0020	3489.9	4938.7	59.072	10.51	14.64	1423
400	2.36456	1.6607	0.0018	4013.3	5668.5	61.022	10.43	14.56	1523
450	2.65935	1.8672	0.0016	4533.7	6395.3	62.735	10.41	14.54	1615
500	2.95408	2.0737	0.0014	5054.2	7122.1	64.265	10.40	14.53	1702
550	3.24878	2.2800	0.0013	5574.3	7848.5	65.650	10.41	14.54	1784
600	3.54346	2.4863	0.0012	6095.4	8575.9	66.914	10.43	14.55	1863
700	4.13275	2.8989	0.0010	7140.5	10033.4	69.162	10.48	14.61	2010
800	4.72201	3.3115	0.0009	8193.0	11498.4	71.117	10.57	14.70	2146
900	5.31123	3.7240	0.0008	9256.4	12974.3	72.854	10.70	14.82	2271
1000	5.90045	4.1365	0.0007	10333.7	14464.0	74.426	10.86	14.98	2389
1200	7.07884	4.9614	0.0006	12544.0	17499.2	77.180	11.24	15.37	2604
1400	8.25722	5.7863	0.0005	14836.7	20616.8	79.587	11.68	15.81	2798
1600	9.43567	6.6111	0.0004	17216.7	23821.7	81.704	12.11	16.24	2977
1800	10.61465	7.4360	0.0004	19690.8	27121.1	83.609	12.58	16.71	3143
2000	11.79592	8.2608	0.0004	22284.7	30541.9	85.476	13.25	17.41	3295
2500	14.80859	10.3230	0.0003	29886.3	40252.3	92.138	17.92	22.50	3599
3000	18.21479	12.3851	0.0002	42312.9	55063.2	100.864	32.73	39.46	3860

* TWO-PHASE BOUNDARY

Table 4.1 (continued)
THERMODYNAMIC PROPERTIES OF PARAHYDROGEN (ISOBARS, SI UNITS)

0.70 MPA ISOBAR

TEMPERATURE DEG. K	DENSITY KG/CU M	V(DH/DV)$_P$ KJ/KG	V(DP/DU)$_V$ MPA-CU M/KJ x 10^2	-V(DP/DV)$_T$ MPA	-(DV/DT)$_P$/V 1/DEG. K	THERMAL CONDUCTIVITY W/K-M x 10^3	VISCOSITY KG/M-S x 10^7	THERMAL DIFFUSIVITY SQ M/HR	DIELECTRIC CONSTANT	PRANDTL NUMBER
* 14.034	77.4091	644.0	0.25560	94.0105	0.00991	75.48	259.12	0.00055	1.25296	2.1901
15	76.6375	634.7	0.24774	85.9246	0.01083	83.85	228.80	0.00057	1.25024	1.8761
16	75.7962	633.6	0.23643	79.0247	0.01155	89.96	203.88	0.00058	1.24729	1.6580
17	74.8918	633.3	0.22711	73.0991	0.01227	94.03	183.49	0.00058	1.24412	1.5163
18	73.9440	630.7	0.21933	67.4756	0.01307	96.93	166.73	0.00057	1.24080	1.4182
19	72.9482	625.4	0.21269	62.0050	0.01398	98.66	152.68	0.00056	1.23732	1.3530
20	71.8967	616.4	0.20678	56.4899	0.01505	100.21	140.66	0.00054	1.23365	1.3020
21	70.7792	604.8	0.20142	51.0545	0.01628	101.99	130.19	0.00053	1.22976	1.2571
22	69.5856	589.9	0.19642	45.6088	0.01777	103.00	120.92	0.00051	1.22561	1.2306
23	68.2935	571.3	0.19162	40.0958	0.01962	103.31	112.52	0.00049	1.22113	1.2208
24	66.8966	550.9	0.18662	34.7861	0.02181	102.98	104.85	0.00046	1.21630	1.2235
25	65.3638	526.1	0.18114	29.4004	0.02470	102.02	97.68	0.00043	1.21102	1.2442
26	63.6599	499.3	0.17471	24.2906	0.02832	100.38	90.86	0.00040	1.20516	1.2796
27	61.7337	467.0	0.16688	19.1479	0.03357	98.07	84.23	0.00036	1.19857	1.3467
28	59.4604	428.5	0.15755	14.0551	0.04207	94.89	77.51	0.00032	1.19081	1.4724
29	56.6553	385.1	0.14604	9.4055	0.05638	90.57	70.44	0.00027	1.18129	1.6885
* 29.019	56.5997	385.0	0.14582	9.3871	0.05636	90.49	70.31	0.00027	1.18110	1.6862
* 29.019	8.9195	206.1	0.06948	0.3678	0.12267	33.25	17.94	0.00053	1.02714	1.3638
30	8.0336	224.4	0.06982	0.4315	0.09151	31.56	18.14	0.00069	1.02442	1.1802
31	7.3969	241.0	0.06993	0.4744	0.07508	31.10	18.44	0.00084	1.02247	1.0728
32	6.8999	256.6	0.06995	0.5059	0.06468	31.07	18.78	0.00098	1.02095	1.0031
33	6.4925	271.4	0.06985	0.5302	0.05733	31.23	19.15	0.00111	1.01970	0.9539
34	6.1483	285.2	0.07008	0.5496	0.05186	31.51	19.53	0.00125	1.01865	0.9163
35	5.8506	298.6	0.07020	0.5655	0.04757	31.88	19.91	0.00138	1.01774	0.8873
36	5.5888	311.9	0.07024	0.5789	0.04409	32.31	20.31	0.00151	1.01694	0.8643
37	5.3557	325.0	0.07022	0.5904	0.04119	32.78	20.71	0.00165	1.01623	0.8455
38	5.1461	337.9	0.07017	0.6002	0.03873	33.29	21.10	0.00178	1.01559	0.8297
39	4.9559	350.7	0.07010	0.6088	0.03661	33.82	21.50	0.00191	1.01501	0.8163
40	4.7823	363.3	0.07000	0.6164	0.03476	34.37	21.90	0.00205	1.01448	0.8047
42	4.4755	388.3	0.06977	0.6291	0.03165	35.51	22.70	0.00232	1.01355	0.7858
44	4.2119	413.0	0.06951	0.6393	0.02914	36.69	23.49	0.00261	1.01275	0.7708
46	3.9819	437.5	0.06920	0.6476	0.02706	37.90	24.28	0.00289	1.01205	0.7584
48	3.7789	461.9	0.06886	0.6545	0.02530	39.14	25.06	0.00319	1.01143	0.7482
50	3.5980	486.3	0.06848	0.6603	0.02379	40.40	25.82	0.00349	1.01088	0.7396
52	3.4354	510.8	0.06804	0.6652	0.02247	41.67	26.58	0.00380	1.01039	0.7324
54	3.2883	535.6	0.06754	0.6694	0.02131	42.95	27.34	0.00412	1.00994	0.7265
56	3.1544	560.8	0.06698	0.6730	0.02028	44.25	28.08	0.00444	1.00954	0.7216
58	3.0319	586.4	0.06635	0.6761	0.01935	45.55	28.81	0.00477	1.00916	0.7178
60	2.9193	612.5	0.06566	0.6788	0.01852	46.93	29.54	0.00510	1.00882	0.7139
65	2.6734	681.0	0.06361	0.6843	0.01674	50.63	31.32	0.00598	1.00808	0.7053
70	2.4678	754.7	0.06119	0.6883	0.01531	54.36	33.05	0.00686	1.00746	0.7024
75	2.2929	834.7	0.05850	0.6914	0.01412	57.99	34.73	0.00773	1.00693	0.7058
80	2.1422	921.2	0.05569	0.6937	0.01311	61.80	36.37	0.00860	1.00647	0.7109
85	2.0107	1014.3	0.05288	0.6956	0.01225	66.69	37.97	0.00961	1.00607	0.7074
90	1.8949	1113.6	0.05018	0.6970	0.01150	71.59	39.54	0.01067	1.00572	0.7071
95	1.7920	1218.3	0.04767	0.6982	0.01084	76.49	41.08	0.01164	1.00541	0.7090
100	1.7000	1327.6	0.04539	0.6991	0.01025	89.83	45.76	0.01398	1.00513	0.6934
120	1.4118	1787.0	0.03871	0.7014	0.00845	117.09	54.03	0.01978	1.00426	0.6965
140	1.2082	2229.6	0.03537	0.7025	0.00720	136.08	59.17	0.02527	1.00364	0.6977
160	1.0564	2615.6	0.03417	0.7030	0.00627	148.28	63.00	0.03079	1.00319	0.6973
180	0.9387	2941.5	0.03419	0.7032	0.00556	155.70	66.20	0.03648	1.00283	0.6960
200	0.8447	3222.4	0.03484	0.7032	0.00500	160.31	69.06	0.04239	1.00255	0.6943
220	0.7680	3477.1	0.03574	0.7032	0.00454	163.56	71.72	0.04855	1.00232	0.6926
240	0.7040	3721.1	0.03667	0.7031	0.00416	166.42	74.26	0.05496	1.00212	0.6910
260	0.6499	3965.1	0.03749	0.7030	0.00384	169.37	76.72	0.06162	1.00196	0.6896
280	0.6036	4213.8	0.03816	0.7029	0.00356	172.61	79.14	0.06855	1.00182	0.6885
300	0.5635	4469.2	0.03868	0.7029	0.00333	176.19	81.51	0.07574	1.00170	0.6876
350	0.4832	5138.6	0.03943	0.7026	0.00285	186.45	87.36	0.09485	1.00146	0.6861
400	0.4229	5838.0	0.03972	0.7023	0.00249	197.82	93.12	0.11567	1.00127	0.6853
450	0.3760	6557.2	0.03977	0.7021	0.00222	209.75	98.82	0.13815	1.00113	0.6848
500	0.3385	7281.6	0.03978	0.7020	0.00200	221.74	104.46	0.16231	1.00102	0.6844
550	0.3078	8013.0	0.03974	0.7018	0.00181	233.80	110.04	0.18811	1.00093	0.6842
600	0.2822	8749.7	0.03966	0.7017	0.00166	245.85	115.55	0.21552	1.00085	0.6840
700	0.2420	10244.4	0.03944	0.7015	0.00143	270.04	126.40	0.27504	1.00073	0.6837
800	0.2118	11778.5	0.03908	0.7013	0.00125	294.50	136.98	0.34061	1.00064	0.6836
900	0.1883	13362.8	0.03861	0.7012	0.00111	319.41	147.32	0.41197	1.00057	0.6837
1000	0.1695	15004.8	0.03804	0.7010	0.00100	440.56	157.42	0.62457	1.00051	0.5354
1200	0.1413	18463.4	0.03673	0.7009	0.00083	511.13	176.97	0.94761	1.00043	0.5321
1400	0.1211	22150.6	0.03535	0.7008	0.00071	583.45	195.74	1.09733	1.00036	0.5302
1600	0.1060	26007.2	0.03408	0.7007	0.00062	656.69	213.84	1.37359	1.00032	0.5288
1800	0.0942	30106.1	0.03281	0.7005	0.00056	734.78	231.36	1.68002	1.00028	0.5262
2000	0.0848	34834.1	0.03116	0.7003	0.00050	832.46	248.38	2.03055	1.00026	0.5194
2500	0.0675	56010.7	0.02314	0.6971	0.00040	1412.42	289.62	3.34689	1.00020	0.4613
3000	0.0549	114990.9	0.01299	0.6799	0.00034	3384.84	332.87	5.62467	1.00017	0.3881

* TWO-PHASE BOUNDARY

Table 4.1 (continued)
THERMODYNAMIC PROPERTIES OF PARAHYDROGEN (ISOBARS, SI UNITS)

0.80 MPA ISOBAR

TEMPERATURE DEG. K	VOLUME CU M/KG	ISOTHERM DERIVATIVE CU M-MPA/KG	ISOCHORE DERIVATIVE MPA/K	INTERNAL ENERGY KJ/KG-K	ENTHALPY KJ/KG	ENTROPY KJ/KG-K	CV KJ / KG-K	CP KJ / KG-K	VELOCITY OF SOUND M/S
* 14.067	0.01291	1.2211	0.9325	-308.6	-298.3	4.985	4.71	6.38	1286
15	0.01303	1.1342	0.9319	-302.6	-292.1	5.410	4.90	6.85	1259
16	0.01318	1.0530	0.9155	-295.6	-285.1	5.865	5.09	7.30	1229
17	0.01333	0.9860	0.8998	-288.2	-277.5	6.322	5.27	7.75	1204
18	0.01350	0.9223	0.8850	-280.3	-269.5	6.779	5.43	8.22	1181
19	0.01369	0.8601	0.8698	-272.0	-261.1	7.237	5.58	8.71	1159
20	0.01388	0.7958	0.8531	-263.2	-252.1	7.697	5.71	9.24	1134
21	0.01410	0.7318	0.8346	-253.9	-242.6	8.161	5.83	9.80	1109
22	0.01434	0.6662	0.8139	-243.9	-232.5	8.631	5.93	10.42	1082
23	0.01461	0.5980	0.7903	-233.4	-221.7	9.111	6.01	11.13	1053
24	0.01491	0.5315	0.7631	-222.1	-210.2	9.601	6.08	11.92	1021
25	0.01525	0.4638	0.7313	-210.0	-197.8	10.105	6.13	12.83	985
26	0.01565	0.3950	0.6939	-197.0	-184.4	10.630	6.18	13.94	944
27	0.01612	0.3242	0.6505	-182.7	-169.8	11.181	6.24	15.39	895
28	0.01670	0.2545	0.6010	-166.9	-153.5	11.775	6.30	17.39	838
29	0.01747	0.1831	0.5432	-148.6	-134.6	12.437	6.39	20.66	769
* 29.828	0.01837	0.1248	0.4856	-130.4	-115.7	13.080	6.49	25.51	700
* 29.828	0.09519	0.0349	0.0545	104.8	180.9	23.030	7.46	30.49	378
30	0.09783	0.0379	0.0526	108.2	186.5	23.214	7.38	28.36	382
31	0.10956	0.0516	0.0455	123.5	211.2	24.025	7.12	22.06	400
32	0.11945	0.0627	0.0408	136.1	231.6	24.674	6.95	19.10	415
33	0.12833	0.0723	0.0374	147.2	249.8	25.235	6.84	17.33	428
34	0.13656	0.0810	0.0347	157.3	266.5	25.733	6.71	16.12	441
35	0.14434	0.0890	0.0325	166.7	282.2	26.187	6.63	15.25	453
36	0.15178	0.0966	0.0306	175.7	297.1	26.608	6.56	14.61	464
37	0.15896	0.1037	0.0290	184.3	311.4	27.001	6.52	14.10	474
38	0.16594	0.1106	0.0276	192.6	325.3	27.371	6.48	13.70	484
39	0.17273	0.1172	0.0264	200.7	338.9	27.723	6.45	13.37	493
40	0.17939	0.1236	0.0253	208.6	352.1	28.058	6.43	13.10	502
42	0.19234	0.1358	0.0234	224.0	377.8	28.686	6.40	12.67	518
44	0.20491	0.1475	0.0218	238.9	402.9	29.267	6.39	12.35	534
46	0.21720	0.1588	0.0205	253.5	427.3	29.811	6.38	12.10	549
48	0.22925	0.1697	0.0193	267.9	451.3	30.322	6.38	11.91	563
50	0.24110	0.1804	0.0183	282.1	475.0	30.805	6.39	11.77	576
52	0.25280	0.1908	0.0173	296.1	498.4	31.264	6.41	11.65	589
54	0.26436	0.2010	0.0165	310.1	521.6	31.702	6.44	11.57	601
56	0.27580	0.2110	0.0158	324.0	544.7	32.122	6.47	11.51	613
58	0.28715	0.2208	0.0151	337.9	567.7	32.525	6.51	11.47	624
60	0.29840	0.2305	0.0145	351.9	590.6	32.914	6.57	11.45	634
65	0.32623	0.2544	0.0132	386.9	647.9	33.831	6.75	11.49	658
70	0.35372	0.2777	0.0121	422.7	705.7	34.688	6.98	11.63	680
75	0.38093	0.3005	0.0112	459.6	764.3	35.497	7.28	11.85	699
80	0.40794	0.3231	0.0105	497.9	824.3	36.270	7.63	12.13	717
85	0.43478	0.3454	0.0098	537.9	885.7	37.016	8.02	12.47	733
90	0.46149	0.3675	0.0092	579.8	949.0	37.739	8.43	12.84	748
95	0.48810	0.3894	0.0087	623.7	1014.2	38.444	8.86	13.24	763
100	0.51461	0.4112	0.0082	669.7	1081.4	39.133	9.29	13.64	777
120	0.61995	0.4972	0.0068	873.7	1369.6	41.755	10.84	15.12	832
140	0.72459	0.5821	0.0058	1102.7	1682.4	44.163	11.83	16.06	889
160	0.82881	0.6663	0.0050	1345.1	2008.1	46.337	12.22	16.42	946
180	0.93276	0.7501	0.0045	1590.4	2336.6	48.272	12.19	16.38	1004
200	1.03651	0.8336	0.0040	1832.7	2661.9	49.985	11.95	16.12	1060
220	1.14013	0.9169	0.0037	2069.0	2981.1	51.507	11.64	15.80	1116
240	1.24366	1.0000	0.0033	2299.0	3293.9	52.868	11.33	15.49	1169
260	1.34710	1.0831	0.0031	2523.3	3601.0	54.097	11.08	15.23	1220
280	1.45049	1.1660	0.0029	2743.0	3903.4	55.218	10.88	15.02	1269
300	1.55377	1.2488	0.0027	2959.3	4202.3	56.249	10.72	14.86	1316
350	1.81194	1.4557	0.0023	3489.7	4939.3	58.520	10.51	14.65	1424
400	2.06998	1.6623	0.0020	4013.1	5669.1	60.471	10.43	14.56	1523
450	2.32794	1.8689	0.0018	4533.6	6395.9	62.184	10.41	14.54	1616
500	2.58586	2.0753	0.0016	5054.1	7122.8	63.714	10.40	14.53	1703
550	2.84373	2.2817	0.0015	5574.2	7849.2	65.099	10.41	14.54	1785
600	3.10158	2.4880	0.0013	6095.3	8576.6	66.363	10.43	14.55	1864
700	3.61723	2.9007	0.0011	7140.4	10034.2	68.612	10.48	14.61	2011
800	4.13284	3.3132	0.0010	8193.0	11499.2	70.567	10.57	14.70	2146
900	4.64842	3.7257	0.0009	9256.4	12975.1	72.303	10.70	14.82	2272
1000	5.16399	4.1382	0.0008	10333.6	14464.8	73.876	10.86	14.98	2390
1200	6.19509	4.9631	0.0007	12543.9	17500.0	76.629	11.24	15.37	2605
1400	7.22618	5.7880	0.0006	14836.7	20617.6	79.036	11.68	15.81	2799
1600	8.25732	6.6129	0.0005	17216.5	23822.4	81.153	12.11	16.24	2977
1800	9.28889	7.4378	0.0004	19689.8	27121.0	83.057	12.57	16.70	3144
2000	10.32234	8.2626	0.0004	22280.3	30538.2	84.923	13.22	17.38	3296
2500	12.95466	10.3247	0.0003	29819.5	40183.3	91.556	17.64	22.19	3603
3000	15.90915	12.3869	0.0003	41895.2	54622.5	100.150	31.55	38.10	3863

* TWO-PHASE BOUNDARY

Table 4.1 (continued)
THERMODYNAMIC PROPERTIES OF PARAHYDROGEN (ISOBARS, SI UNITS)

0.80 MPA ISOBAR

TEMPERATURE DEG. K	DENSITY KG/CU M	$V(DH/DV)_P$ KJ/KG	$V(DP/DU)_V$ MPA-CU M/KJ X 10^2	$-V(DP/DV)_T$ MPA	$-(DV/DT)_P/V$ 1/DEG. K	THERMAL CONDUCTIVITY W/K-M X 10^3	VISCOSITY KG/M-S X 10^7	THERMAL DIFFUSIVITY SQ M/HR	DIELECTRIC CONSTANT	PRANDTL NUMBER
* 14.067	77.4651	647.3	0.25548	94.5954	0.00386	75.88	259.74	0.00055	1.25315	2.1840
15	76.7248	639.7	0.24789	87.0229	0.01071	83.96	230.36	0.00058	1.25055	1.8798
16	75.8894	637.2	0.23706	79.9090	0.01146	90.11	205.24	0.00059	1.24761	1.6627
17	74.9923	637.0	0.22772	73.9430	0.01217	94.21	184.71	0.00058	1.24447	1.5196
18	74.0521	634.5	0.21992	68.2963	0.01296	97.15	167.84	0.00057	1.24118	1.4204
19	73.0644	629.5	0.21325	62.8392	0.01384	98.91	153.70	0.00056	1.23772	1.3540
20	72.0222	620.7	0.20734	57.3120	0.01489	100.49	141.61	0.00054	1.23409	1.3021
21	70.9163	609.5	0.20196	51.8962	0.01608	102.31	131.10	0.00053	1.23024	1.2559
22	69.7362	595.0	0.19695	46.4549	0.01752	103.37	121.79	0.00051	1.22613	1.2292
23	68.4619	576.7	0.19215	40.9370	0.01931	103.72	113.38	0.00049	1.22172	1.2171
24	67.0865	556.9	0.18721	35.6589	0.02140	103.44	105.70	0.00047	1.21696	1.2177
25	65.5817	533.8	0.18185	30.4154	0.02404	102.55	98.55	0.00044	1.21177	1.2332
26	63.9171	507.2	0.17564	25.2504	0.02748	101.00	91.76	0.00041	1.20605	1.2664
27	62.0450	475.9	0.16816	20.1166	0.03234	98.90	85.20	0.00037	1.19963	1.3271
28	59.8722	440.8	0.15929	15.2350	0.03945	95.82	78.64	0.00033	1.19221	1.4271
29	57.2336	398.5	0.14856	10.4781	0.05184	91.83	71.81	0.00029	1.18324	1.6155
* 29.828	54.4335	356.9	0.13744	6.7925	0.07148	87.32	65.59	0.00023	1.17378	1.9164
* 29.828	10.5055	205.0	0.06963	0.3666	0.14877	36.90	18.93	0.00041	1.03202	1.5642
30	10.2218	208.9	0.06974	0.3877	0.13577	36.00	18.91	0.00045	1.03114	1.4895
31	9.1271	228.2	0.07006	0.4710	0.09664	33.73	19.00	0.00060	1.02777	1.2426
32	8.3720	245.4	0.07018	0.5246	0.07784	32.98	19.24	0.00074	1.02545	1.1146
33	7.7923	261.3	0.07011	0.5634	0.06630	32.75	19.54	0.00087	1.02368	1.0339
34	7.3227	275.8	0.07049	0.5930	0.05844	32.79	19.88	0.00100	1.02224	0.9770
35	6.9280	289.9	0.07069	0.6167	0.05262	32.99	20.23	0.00112	1.02103	0.9353
36	6.5883	303.6	0.07078	0.6361	0.04810	33.30	20.60	0.00125	1.01999	0.9034
37	6.2907	317.2	0.07079	0.6525	0.04447	33.69	20.98	0.00137	1.01908	0.8781
38	6.0264	330.5	0.07074	0.6664	0.04146	34.12	21.36	0.00149	1.01828	0.8575
39	5.7893	343.6	0.07067	0.6784	0.03891	34.60	21.74	0.00161	1.01755	0.8404
40	5.5746	356.6	0.07056	0.6889	0.03672	35.10	22.13	0.00173	1.01690	0.8258
42	5.1992	382.2	0.07031	0.7062	0.03314	36.16	22.91	0.00198	1.01575	0.8024
44	4.8801	407.5	0.07001	0.7200	0.03030	37.28	23.68	0.00223	1.01473	0.7843
46	4.6041	432.4	0.06968	0.7312	0.02799	38.45	24.45	0.00248	1.01394	0.7697
48	4.3621	457.2	0.06931	0.7405	0.02605	39.65	25.22	0.00275	1.01320	0.7577
50	4.1476	482.0	0.06890	0.7482	0.02441	40.88	25.98	0.00302	1.01255	0.7478
52	3.9557	506.8	0.06843	0.7547	0.02299	42.12	26.73	0.00329	1.01197	0.7395
54	3.7827	531.9	0.06791	0.7602	0.02175	43.38	27.48	0.00357	1.01144	0.7327
56	3.6258	557.3	0.06732	0.7649	0.02065	44.66	28.21	0.00385	1.01097	0.7270
58	3.4826	583.1	0.06668	0.7690	0.01967	45.94	28.94	0.00414	1.01053	0.7226
60	3.3512	609.4	0.06597	0.7726	0.01880	47.31	29.66	0.00444	1.01013	0.7182
65	3.0653	678.3	0.06389	0.7797	0.01694	50.98	31.43	0.00521	1.00927	0.7086
70	2.8271	752.3	0.06144	0.7850	0.01546	54.67	33.15	0.00599	1.00854	0.7050
75	2.6251	832.5	0.05873	0.7889	0.01423	58.26	34.82	0.00674	1.00793	0.7090
80	2.4513	919.2	0.05589	0.7920	0.01320	62.06	36.46	0.00751	1.00741	0.7127
85	2.3000	1012.4	0.05306	0.7944	0.01232	66.95	38.06	0.00840	1.00695	0.7088
90	2.1669	1111.9	0.05035	0.7962	0.01155	71.83	39.62	0.00929	1.00654	0.7093
95	2.0488	1216.7	0.04782	0.7978	0.01088	76.73	41.15	0.01018	1.00619	0.7100
100	1.9432	1326.1	0.04553	0.7990	0.01029	89.38	45.77	0.01221	1.00587	0.6947
120	1.6130	1786.2	0.03880	0.8019	0.00846	117.09	54.02	0.01729	1.00487	0.6974
140	1.3801	2229.4	0.03544	0.8033	0.00720	136.07	59.17	0.02210	1.00416	0.6983
160	1.2065	2616.0	0.03423	0.8039	0.00628	148.28	63.01	0.02694	1.00364	0.6978
180	1.0721	2942.4	0.03424	0.8042	0.00557	155.71	66.21	0.03193	1.00323	0.6964
200	0.9648	3223.7	0.03488	0.8042	0.00500	160.32	69.07	0.03711	1.00291	0.6946
220	0.8771	3478.7	0.03578	0.8042	0.00454	163.59	71.73	0.04250	1.00265	0.6928
240	0.8041	3723.0	0.03671	0.8041	0.00416	166.45	74.28	0.04811	1.00242	0.6911
260	0.7423	3967.1	0.03752	0.8040	0.00384	169.41	76.74	0.05395	1.00224	0.6898
280	0.6894	4215.9	0.03819	0.8039	0.00356	172.66	79.15	0.06002	1.00208	0.6886
300	0.6436	4471.5	0.03871	0.8037	0.00332	176.23	81.53	0.06632	1.00194	0.6877
350	0.5519	5141.1	0.03945	0.8034	0.00285	186.51	87.33	0.08306	1.00165	0.6862
400	0.4831	5840.6	0.03974	0.8031	0.00249	197.38	93.15	0.10128	1.00146	0.6853
450	0.4296	6559.9	0.03979	0.8028	0.00222	209.32	98.85	0.12097	1.00129	0.6848
500	0.3867	7284.5	0.03980	0.8026	0.00199	221.82	104.49	0.14212	1.00117	0.6844
550	0.3517	8015.9	0.03975	0.8024	0.00181	233.88	110.07	0.16471	1.00106	0.6842
600	0.3224	8752.6	0.03968	0.8022	0.00166	245.95	115.59	0.18871	1.00097	0.6840
700	0.2765	10247.4	0.03945	0.8019	0.00143	270.14	126.44	0.24082	1.00083	0.6837
800	0.2420	11781.5	0.03910	0.8017	0.00125	294.62	137.04	0.29823	1.00073	0.6836
900	0.2151	13365.9	0.03862	0.8015	0.00111	319.55	147.38	0.36071	1.00065	0.6837
1000	0.1936	15008.0	0.03805	0.8014	0.00100	440.56	157.48	0.54661	1.00058	0.5356
1200	0.1614	18466.6	0.03674	0.8011	0.00083	511.13	177.05	0.74179	1.00049	0.5323
1400	0.1384	22153.7	0.03535	0.8010	0.00071	583.44	195.83	0.96030	1.00042	0.5305
1600	0.1211	26008.2	0.03409	0.8009	0.00062	656.59	213.94	1.20199	1.00036	0.5291
1800	0.1077	30094.9	0.03284	0.8007	0.00056	734.20	231.47	1.46976	1.00032	0.5266
2000	0.0969	34777.1	0.03123	0.8005	0.00050	830.12	248.50	1.77502	1.00029	0.5202
2500	0.0772	55264.2	0.02350	0.7970	0.00040	1383.98	289.72	2.90881	1.00023	0.4645
3000	0.0629	111251.3	0.01345	0.7786	0.00034	3244.24	332.71	4.87643	1.00019	0.3908

* TWO-PHASE BOUNDARY

Table 4.1 (continued)
THERMODYNAMIC PROPERTIES OF PARAHYDROGEN (ISOBARS, SI UNITS)

0.90 MPA ISOBAR

TEMPERATURE DEG. K	VOLUME CU M/KG	ISOTHERM DERIVATIVE CU M-MPA/KG	ISOCHORE DERIVATIVE MPA/K	INTERNAL ENERGY KJ/KG-K	ENTHALPY KJ/KG	ENTROPY KJ/KG-K	CV KJ / KG-K	CP KJ / KG-K	VELOCITY OF SOUND M/S
* 14.100	0.01290	1.2278	0.9336	-308.6	-297.0	4.987	4.72	6.38	1289
15	0.01302	1.1472	0.9331	-302.7	-291.0	5.396	4.90	6.83	1265
16	0.01316	1.0633	0.9185	-295.8	-284.0	5.850	5.09	7.28	1234
17	0.01332	0.9959	0.9027	-288.5	-276.5	6.306	5.27	7.73	1209
18	0.01348	0.9319	0.8879	-280.6	-268.5	6.761	5.43	8.20	1186
19	0.01367	0.8700	0.8728	-272.4	-260.1	7.218	5.58	8.68	1164
20	0.01386	0.8057	0.8562	-263.6	-251.1	7.677	5.71	9.20	1140
21	0.01407	0.7421	0.8377	-254.3	-241.6	8.139	5.82	9.76	1115
22	0.01431	0.6767	0.8172	-244.5	-231.6	8.607	5.92	10.37	1088
23	0.01457	0.6109	0.7941	-234.0	-220.9	9.082	6.00	11.04	1060
24	0.01486	0.5428	0.7672	-222.8	-209.4	9.569	6.07	11.82	1028
25	0.01520	0.4762	0.7360	-210.9	-197.2	10.069	6.13	12.70	993
26	0.01558	0.4081	0.6998	-198.0	-184.0	10.588	6.18	13.76	953
27	0.01604	0.3377	0.6579	-184.0	-169.6	11.131	6.23	15.13	906
28	0.01660	0.2691	0.6099	-168.5	-153.6	11.711	6.29	16.96	851
29	0.01731	0.2011	0.5551	-151.0	-135.4	12.350	6.37	19.69	788
30	0.01833	0.1312	0.4891	-129.9	-113.4	13.096	6.48	24.86	709
* 30.567	0.01918	0.0889	0.4424	-114.9	-97.7	13.615	6.59	31.34	650
* 30.567	0.08139	0.0281	0.0656	97.2	170.5	22.391	7.67	38.60	376
31	0.08768	0.0365	0.0595	106.9	185.8	22.890	7.46	30.60	387
32	0.09864	0.0507	0.0511	123.3	212.1	23.726	7.18	23.21	405
33	0.10775	0.0622	0.0457	136.6	233.6	24.386	7.01	19.86	420
34	0.11588	0.0721	0.0418	148.1	252.4	24.948	6.83	17.87	434
35	0.12339	0.0811	0.0387	158.5	269.6	25.446	6.71	16.57	447
36	0.13047	0.0893	0.0362	168.2	285.6	25.899	6.63	15.64	459
37	0.13723	0.0971	0.0342	177.4	300.9	26.317	6.57	14.94	470
38	0.14374	0.1044	0.0324	186.2	315.6	26.708	6.52	14.40	480
39	0.15006	0.1114	0.0308	194.7	329.8	27.077	6.49	13.97	490
40	0.15620	0.1182	0.0294	203.0	343.6	27.426	6.46	13.62	499
42	0.16811	0.1311	0.0271	218.9	370.2	28.076	6.43	13.07	516
44	0.17961	0.1433	0.0252	234.3	395.9	28.675	6.41	12.68	533
46	0.19079	0.1550	0.0235	249.3	421.0	29.232	6.40	12.38	548
48	0.20173	0.1663	0.0221	264.0	445.5	29.753	6.40	12.15	562
50	0.21246	0.1773	0.0209	278.4	469.6	30.245	6.41	11.97	576
52	0.22303	0.1880	0.0198	292.7	493.4	30.712	6.42	11.83	588
54	0.23345	0.1984	0.0189	306.9	517.0	31.157	6.45	11.73	601
56	0.24376	0.2086	0.0180	321.0	540.4	31.582	6.48	11.65	612
58	0.25396	0.2187	0.0172	335.0	563.6	31.990	6.53	11.60	623
60	0.26408	0.2286	0.0165	349.1	586.8	32.382	6.58	11.57	634
65	0.28905	0.2528	0.0150	384.5	644.0	33.308	6.75	11.58	658
70	0.31367	0.2764	0.0138	420.5	702.8	34.171	6.99	11.70	680
75	0.33802	0.2996	0.0127	457.6	761.8	34.985	7.29	11.91	700
80	0.36216	0.3224	0.0118	496.0	822.0	35.762	7.63	12.18	717
85	0.38614	0.3448	0.0111	536.2	883.7	36.510	8.02	12.51	734
90	0.40998	0.3671	0.0104	578.2	947.2	37.235	8.43	12.88	749
95	0.43371	0.3892	0.0098	622.2	1012.6	37.942	8.86	13.27	763
100	0.45736	0.4111	0.0093	668.3	1079.9	38.633	9.29	13.67	778
120	0.55124	0.4975	0.0077	872.6	1368.7	41.260	10.85	15.14	833
140	0.64442	0.5827	0.0065	1101.8	1681.8	43.671	11.83	16.07	890
160	0.73717	0.6671	0.0057	1344.3	2007.8	45.847	12.22	16.43	947
180	0.82966	0.7511	0.0050	1589.8	2336.5	47.783	12.20	16.39	1005
200	0.92195	0.8347	0.0045	1832.1	2661.8	49.496	11.95	16.13	1061
220	1.01411	0.9181	0.0041	2068.5	2981.2	51.019	11.64	15.80	1117
240	1.10618	1.0013	0.0038	2298.5	3294.1	52.380	11.34	15.49	1170
260	1.19817	1.0844	0.0035	2522.9	3601.2	53.610	11.08	15.23	1221
280	1.29010	1.1674	0.0032	2742.6	3903.7	54.731	10.88	15.02	1270
300	1.38193	1.2503	0.0030	2958.9	4202.7	55.762	10.72	14.87	1316
350	1.61146	1.4572	0.0026	3489.5	4939.8	58.034	10.51	14.65	1425
400	1.84086	1.6639	0.0023	4012.9	5669.7	59.985	10.43	14.56	1524
450	2.07019	1.8705	0.0020	4533.4	6396.6	61.698	10.41	14.54	1616
500	2.29945	2.0770	0.0018	5054.0	7123.5	63.228	10.40	14.53	1703
550	2.52869	2.2834	0.0016	5574.1	7849.9	64.613	10.41	14.54	1786
600	2.75790	2.4897	0.0015	6095.2	8577.3	65.878	10.43	14.55	1864
700	3.21627	2.9024	0.0013	7140.3	10035.0	68.126	10.48	14.61	2011
800	3.67460	3.3149	0.0011	8192.9	11500.1	70.081	10.57	14.70	2147
900	4.13290	3.7275	0.0010	9256.3	12975.9	71.818	10.70	14.82	2273
1000	4.59118	4.1400	0.0009	10333.6	14465.7	73.390	10.86	14.98	2390
1200	5.50773	4.9649	0.0008	12543.9	17500.9	76.143	11.24	15.37	2605
1400	6.42425	5.7898	0.0006	14836.7	20618.5	78.550	11.68	15.81	2799
1600	7.34082	6.6147	0.0006	17216.4	23823.1	80.667	12.11	16.24	2978
1800	8.25775	7.4396	0.0005	19689.0	27121.0	82.571	12.57	16.70	3144
2000	9.17625	8.2644	0.0005	22276.7	30535.4	84.435	13.20	17.35	3296
2500	11.51332	10.3265	0.0004	29764.2	40126.2	91.045	17.41	21.93	3606
3000	14.12044	12.3886	0.0003	41549.2	54257.6	99.529	30.58	36.98	3867

* TWO-PHASE BOUNDARY

Table 4.1 (continued)

THERMODYNAMIC PROPERTIES OF PARAHYDROGEN (ISOBARS, SI UNITS)

0.90 MPA ISOBAR

TEMPERATURE DEG. K	DENSITY KG/CU M	$V(DH/DV)_P$ KJ/KG	$V(DP/DU)_V$ MPA-CU M/KJ X 10[illegible]	$-V(DP/DV)_T$ MPA	$-(DV/DT)_P/V$ 1/DEG. K	THERMAL CONDUCTIVITY W/K-M $\times 10^3$	VISCOSITY KG/M-S $\times 10^7$	THERMAL DIFFUSIVITY SQ M/HR	DIELECTRIC CONSTANT	PRANDTL NUMBER
* 14.100	77.5208	650.6	0.25535	95.1791	0.00981	76.29	260.36	0.00056	1.25335	2.1781
15	76.8115	644.7	0.24804	88.1146	0.01059	84.07	231.93	0.00058	1.25085	1.8836
16	75.9818	640.8	0.23767	80.7877	0.01137	90.26	206.61	0.00059	1.24794	1.6675
17	75.0917	640.5	0.22831	74.7803	0.01207	94.39	185.94	0.00059	1.24482	1.5231
18	74.1591	638.2	0.22044	69.1097	0.01285	97.36	168.95	0.00058	1.24155	1.4227
19	73.1793	633.5	0.21380	63.6661	0.01371	99.15	154.72	0.00056	1.23812	1.3551
20	72.1462	624.9	0.20788	58.1271	0.01473	100.77	142.56	0.00055	1.23452	1.3023
21	71.0514	614.1	0.20249	52.7287	0.01589	102.62	132.00	0.00053	1.23071	1.2550
22	69.8844	599.9	0.19747	47.2907	0.01728	103.72	122.66	0.00052	1.22665	1.2260
23	68.6311	583.2	0.19269	41.9288	0.01894	104.13	114.25	0.00049	1.22230	1.2118
24	67.2723	562.8	0.18778	36.5169	0.02101	103.90	106.55	0.00047	1.21760	1.2125
25	65.7941	540.6	0.18251	31.3292	0.02349	103.07	99.40	0.00044	1.21250	1.2249
26	64.1659	514.8	0.17651	26.1854	0.02673	101.60	92.65	0.00041	1.20690	1.2546
27	62.3439	484.3	0.16936	21.0554	0.03124	99.51	86.15	0.00038	1.20065	1.3101
28	60.2507	450.8	0.16085	16.2152	0.03761	96.69	79.70	0.00034	1.19350	1.3975
29	57.7615	412.0	0.15082	11.6174	0.04778	92.98	73.10	0.00029	1.18504	1.5477
30	54.5650	363.7	0.13827	7.1562	0.06834	87.88	65.89	0.00023	1.17422	1.8639
* 30.567	52.1365	328.3	0.12885	4.6342	0.09546	85.67	61.14	0.00019	1.16605	2.2367
* 30.567	12.2868	203.6	0.06362	0.3458	0.18963	41.99	20.06	0.00032	1.03752	1.8447
31	11.4051	213.9	0.06992	0.4158	0.14310	38.62	19.89	0.00040	1.03479	1.5758
32	10.1382	233.6	0.07027	0.5144	0.09938	35.76	19.87	0.00055	1.03088	1.2896
33	9.2804	251.0	0.07027	0.5773	0.07913	34.73	20.04	0.00068	1.02824	1.1459
34	8.6298	266.3	0.07085	0.6224	0.06711	34.35	20.30	0.00080	1.02625	1.0563
35	8.1045	281.1	0.07117	0.6569	0.05894	34.30	20.60	0.00092	1.02463	0.9951
36	7.6647	295.4	0.07132	0.6846	0.05293	34.44	20.93	0.00103	1.02329	0.9505
37	7.2871	309.4	0.07137	0.7073	0.04829	34.70	21.28	0.00115	1.02213	0.9164
38	6.9569	323.2	0.07134	0.7264	0.04457	35.04	21.64	0.00126	1.02112	0.8894
39	6.6642	336.7	0.07126	0.7427	0.04149	35.44	22.00	0.00137	1.02022	0.8674
40	6.4019	350.1	0.07115	0.7568	0.03890	35.88	22.37	0.00148	1.01942	0.8491
42	5.9485	376.3	0.07087	0.7799	0.03474	36.85	23.12	0.00171	1.01804	0.8204
44	5.5677	402.0	0.07054	0.7980	0.03153	37.91	23.88	0.00193	1.01688	0.7987
46	5.2414	427.5	0.07017	0.8126	0.02896	39.02	24.64	0.00217	1.01588	0.7815
48	4.9572	452.7	0.06977	0.8245	0.02684	40.18	25.39	0.00240	1.01502	0.7677
50	4.7068	477.8	0.06932	0.8344	0.02505	41.37	26.14	0.00264	1.01425	0.7563
52	4.4838	503.0	0.06883	0.8428	0.02352	42.59	26.88	0.00289	1.01358	0.7469
54	4.2835	528.3	0.06829	0.8499	0.02220	43.82	27.62	0.00314	1.01297	0.7391
56	4.1025	554.0	0.06768	0.8559	0.02103	45.08	28.35	0.00340	1.01242	0.7326
58	3.9376	579.9	0.06702	0.8611	0.02000	46.34	29.07	0.00365	1.01191	0.7275
60	3.7868	606.5	0.06629	0.8656	0.01908	47.69	29.79	0.00392	1.01146	0.7226
65	3.4596	675.6	0.06418	0.8746	0.01714	51.[illegible]2	31.54	0.00461	1.01046	0.7119
70	3.1881	750.0	0.06169	0.8813	0.01560	55.00	33.25	0.00531	1.00964	0.7076
75	2.9584	830.3	0.05895	0.8863	0.01434	58.54	34.92	0.00598	1.00894	0.7103
80	2.7612	917.2	0.05609	0.8901	0.01328	62.32	36.55	0.00667	1.00834	0.7145
85	2.5898	1010.6	0.05324	0.8931	0.01239	67.20	38.14	0.00747	1.00782	0.7103
90	2.4391	1110.2	0.05051	0.8954	0.01160	72.08	39.70	0.00826	1.00737	0.7095
95	2.3057	1215.2	0.04797	0.8973	0.01092	76.96	41.23	0.00905	1.00696	0.7110
100	2.1865	1324.7	0.04566	0.8988	0.01032	89.94	45.78	0.01083	1.00660	0.6960
120	1.8141	1785.5	0.03889	0.9025	0.00848	117.10	54.02	0.01535	1.00548	0.6982
140	1.5518	2229.3	0.03551	0.9042	0.00721	136.07	59.17	0.01964	1.00468	0.6990
160	1.3565	2616.5	0.03428	0.9050	0.00628	148.28	63.01	0.02395	1.00409	0.6983
180	1.2053	2943.4	0.03429	0.9053	0.00557	155.72	66.22	0.02839	1.00364	0.6968
200	1.0847	3225.0	0.03492	0.9054	0.00500	160.35	69.08	0.03300	1.00327	0.6949
220	0.9861	3480.3	0.03582	0.9053	0.00454	163.62	71.75	0.03780	1.00297	0.6930
240	0.9040	3724.8	0.03674	0.9052	0.00416	166.48	74.29	0.04279	1.00273	0.6913
260	0.8346	3969.2	0.03756	0.9051	0.00384	169.44	76.75	0.04799	1.00252	0.6899
280	0.7751	4218.1	0.03823	0.9049	0.00356	172.70	79.17	0.05339	1.00234	0.6887
300	0.7236	4473.8	0.03874	0.9047	0.00332	176.28	81.55	0.05899	1.00218	0.6878
350	0.6206	5143.7	0.03948	0.9043	0.00285	186.56	87.40	0.07389	1.00187	0.6862
400	0.5432	5843.3	0.03976	0.9039	0.00249	197.94	93.17	0.09009	1.00164	0.6854
450	0.4830	6562.7	0.03981	0.9035	0.00222	209.89	98.88	0.10760	1.00146	0.6848
500	0.4349	7287.3	0.03982	0.9032	0.00199	221.89	104.52	0.12641	1.00131	0.6845
550	0.3955	8018.8	0.03977	0.9030	0.00181	233.97	110.11	0.14651	1.00119	0.6842
600	0.3626	8755.6	0.03969	0.9028	0.00166	246.04	115.63	0.16786	1.00109	0.6840
700	0.3109	10250.3	0.03946	0.9024	0.00143	270.25	126.49	0.21421	1.00094	0.6837
800	0.2721	11784.6	0.03911	0.9021	0.00125	294.74	137.09	0.26527	1.00082	0.6836
900	0.2420	13389.0	0.03863	0.9019	0.00111	319.68	147.44	0.32084	1.00073	0.6837
1000	0.2178	15011.1	0.03806	0.9017	0.00100	440.56	157.55	0.48597	1.00066	0.5358
1200	0.1816	18469.9	0.03674	0.9014	0.00083	511.13	177.13	0.65948	1.00055	0.5325
1400	0.1557	22156.9	0.03536	0.9012	0.00071	583.43	195.92	0.85373	1.00047	0.5307
1600	0.1362	26009.5	0.03410	0.9011	0.00062	656.51	214.04	1.06853	1.00041	0.5294
1800	0.1211	30086.2	0.03286	0.9009	0.00055	733.73	231.58	1.30629	1.00036	0.5270
2000	0.1090	34730.5	0.03129	0.9006	0.00050	828.18	248.62	1.57655	1.00033	0.5210
2500	0.0869	54646.3	0.02381	0.8969	0.00040	1360.41	289.83	2.57075	1.00026	0.4673
3000	0.0708	108142.4	0.01385	0.8774	0.00034	3127.32	332.61	4.29912	1.00021	0.3933

* TWO-PHASE BOUNDARY

Table 4.1 (continued)
THERMODYNAMIC PROPERTIES OF PARAHYDROGEN (ISOBARS, SI UNITS)

1.00 MPA ISOBAR

TEMPERATURE	VOLUME	ISOTHERM DERIVATIVE	ISOCHORE DERIVATIVE	INTERNAL ENERGY	ENTHALPY	ENTROPY	CV	CP	VELOCITY OF SOUND
DEG. K	CU M/KG	CU M-MPA/KG	MPA/K	KJ/KG-K	KJ/KG	KJ/KG-K	KJ /	KG-K	M/S
* 14.133	0.01289	1.2344	0.9347	-308.5	-295.7	4.990	4.72	6.38	1292
15	0.01300	1.1600	0.9343	-302.9	-289.9	5.382	4.90	6.80	1270
16	0.01315	1.0735	0.9214	-296.1	-282.9	5.835	5.08	7.27	1239
17	0.01330	1.0056	0.9056	-288.7	-275.4	6.290	5.26	7.71	1214
18	0.01347	0.9414	0.8908	-280.9	-267.5	6.744	5.43	8.18	1191
19	0.01364	0.8798	0.8757	-272.7	-259.0	7.199	5.57	8.66	1169
20	0.01384	0.8155	0.8593	-264.0	-250.1	7.656	5.70	9.17	1145
21	0.01405	0.7523	0.8409	-254.7	-240.7	8.117	5.82	9.71	1121
22	0.01428	0.6871	0.8205	-245.0	-230.7	8.582	5.92	10.31	1094
23	0.01454	0.6219	0.7976	-234.6	-220.1	9.055	6.00	10.97	1066
24	0.01482	0.5539	0.7713	-223.5	-208.7	9.538	6.07	11.74	1035
25	0.01515	0.4883	0.7407	-211.7	-196.6	10.034	6.13	12.58	1001
26	0.01553	0.4207	0.7056	-199.0	-183.5	10.547	6.18	13.59	962
27	0.01597	0.3531	0.6648	-185.3	-169.3	11.082	6.23	14.84	917
28	0.01650	0.2850	0.6185	-170.2	-153.7	11.650	6.29	16.52	865
29	0.01717	0.2166	0.5658	-153.1	-136.0	12.271	6.36	19.00	804
30	0.01809	0.1485	0.5041	-133.2	-115.1	12.979	6.45	23.26	731
31	0.01957	0.0784	0.4243	-107.2	-87.6	13.878	6.63	33.89	633
* 31.248	0.02016	0.0594	0.3982	-98.7	-78.5	14.171	6.71	40.60	600
* 31.248	0.06965	0.0213	0.0788	87.6	157.2	21.719	7.93	52.16	374
32	0.08043	0.0367	0.0654	107.0	187.4	22.675	7.50	31.66	393
33	0.09052	0.0511	0.0561	124.2	214.7	23.515	7.22	23.86	411
34	0.09887	0.0627	0.0502	137.8	236.6	24.170	6.97	20.33	427
35	0.10632	0.0727	0.0459	149.5	255.9	24.727	6.81	18.27	442
36	0.11320	0.0818	0.0426	160.2	273.4	25.221	6.70	16.91	454
37	0.11969	0.0903	0.0398	170.1	289.8	25.671	6.62	15.94	466
38	0.12587	0.0982	0.0375	179.5	305.4	26.086	6.57	15.22	477
39	0.13183	0.1057	0.0356	188.5	320.3	26.474	6.53	14.65	487
40	0.13759	0.1128	0.0339	197.1	334.7	26.839	6.49	14.20	497
42	0.14869	0.1264	0.0310	213.7	362.4	27.514	6.45	13.52	515
44	0.15934	0.1391	0.0287	229.6	388.9	28.131	6.43	13.03	531
46	0.16965	0.1513	0.0267	245.0	414.6	28.702	6.42	12.67	547
48	0.17971	0.1629	0.0251	260.0	439.7	29.235	6.41	12.39	561
50	0.18954	0.1742	0.0236	274.7	464.2	29.736	6.42	12.18	575
52	0.19921	0.1852	0.0224	289.2	488.4	30.211	6.44	12.02	588
54	0.20873	0.1959	0.0213	303.6	512.3	30.662	6.46	11.89	600
56	0.21813	0.2063	0.0203	317.9	536.0	31.093	6.49	11.79	612
58	0.22742	0.2166	0.0194	332.1	559.5	31.506	6.54	11.73	623
60	0.23663	0.2267	0.0185	346.3	582.9	31.902	6.59	11.69	634
65	0.25931	0.2513	0.0168	382.0	641.3	32.836	6.76	11.68	659
70	0.28164	0.2752	0.0154	418.3	699.9	33.705	7.00	11.78	681
75	0.30370	0.2986	0.0142	455.5	759.2	34.524	7.29	11.97	700
80	0.32554	0.3217	0.0132	494.2	819.7	35.304	7.64	12.24	718
85	0.34723	0.3444	0.0123	534.5	881.7	36.056	8.02	12.56	734
90	0.36878	0.3668	0.0116	576.6	945.4	36.783	8.44	12.92	749
95	0.39022	0.3890	0.0109	620.7	1010.9	37.492	8.86	13.31	764
100	0.41157	0.4111	0.0103	666.9	1078.5	38.185	9.30	13.70	778
120	0.49627	0.4979	0.0085	871.5	1367.7	40.816	10.85	15.16	834
140	0.58028	0.5833	0.0073	1100.9	1681.2	43.230	11.84	16.09	891
160	0.66386	0.6680	0.0063	1343.5	2007.4	45.407	12.22	16.44	948
180	0.74718	0.7521	0.0056	1589.1	2336.3	47.345	12.20	16.39	1005
200	0.83031	0.8358	0.0050	1831.5	2661.8	49.059	11.95	16.14	1062
220	0.91330	0.9193	0.0046	2068.0	2981.3	50.582	11.64	15.81	1117
240	0.99620	1.0026	0.0042	2298.1	3294.3	51.944	11.34	15.50	1171
260	1.07902	1.0858	0.0039	2522.5	3601.5	53.173	11.08	15.23	1222
280	1.16179	1.1688	0.0036	2742.3	3904.0	54.295	10.88	15.03	1271
300	1.24446	1.2517	0.0033	2958.6	4203.1	55.327	10.73	14.87	1317
350	1.45108	1.4588	0.0029	3489.2	4940.3	57.599	10.51	14.65	1426
400	1.65757	1.6655	0.0025	4012.7	5670.3	59.550	10.43	14.56	1525
450	1.86398	1.8721	0.0022	4533.3	6397.2	61.263	10.41	14.54	1617
500	2.07033	2.0786	0.0020	5053.8	7124.2	62.794	10.40	14.53	1704
550	2.27666	2.2851	0.0018	5574.0	7850.7	64.178	10.41	14.54	1786
600	2.48295	2.4914	0.0017	6095.1	8578.1	65.443	10.43	14.55	1865
700	2.89550	2.9041	0.0014	7140.2	10035.7	67.691	10.48	14.61	2012
800	3.30800	3.3167	0.0013	8192.9	11500.9	69.646	10.57	14.70	2147
900	3.72048	3.7292	0.0011	9256.3	12976.8	71.383	10.70	14.82	2273
1000	4.13294	4.1417	0.0010	10333.6	14466.5	72.955	10.86	14.98	2391
1200	4.95784	4.9667	0.0008	12543.9	17501.7	75.709	11.24	15.37	2605
1400	5.78271	5.7916	0.0007	14836.6	20619.3	78.116	11.68	15.80	2799
1600	6.60763	6.6165	0.0006	17216.3	23823.9	80.232	12.11	16.24	2978
1800	7.43284	7.4413	0.0006	19688.4	27121.2	82.136	12.56	16.69	3145
2000	8.25941	8.2662	0.0005	22273.7	30533.1	83.999	13.18	17.33	3297
2500	10.36066	10.3283	0.0004	29717.4	40078.1	90.589	17.22	21.72	3609
3000	12.69252	12.3904	0.0003	41256.5	53949.0	98.980	29.75	36.03	3869

* TWO-PHASE BOUNDARY

Table 4.1 (continued)
THERMODYNAMIC PROPERTIES OF PARAHYDROGEN (ISOBARS, SI UNITS)

1.00 MPA ISOBAR

TEMPERATURE DEG. K	DENSITY KG/CU M	$V(DH/DV)_P$ KJ/KG	$V(DP/DU)_V$ MPA-CU M/KJ X 10^2	$-V(DP/DV)_T$ MPA	$-(DV/DT)_P/V$ 1/DEG. K	THERMAL CONDUCTIVITY W/K-M X 10^3	VISCOSITY KG/M-S X 10^7	THERMAL DIFFUSIVITY SQ M/HR	DIELECTRIC CONSTANT	PRANDTL NUMBER
* 14.133	77.5763	654.0	0.25522	95.7614	0.00976	76.69	260.99	0.00056	1.25355	2.1723
15	76.8975	649.7	0.24818	89.1999	0.01047	84.18	233.50	0.00058	1.25116	1.8875
16	76.0734	644.3	0.23827	81.6612	0.01128	90.41	207.98	0.00059	1.24826	1.6724
17	75.1902	644.1	0.22890	75.6113	0.01199	94.58	187.17	0.00059	1.24516	1.5266
18	74.2648	641.8	0.22106	69.9161	0.01274	97.58	170.06	0.00058	1.24192	1.4251
19	73.2929	637.5	0.21433	64.4858	0.01358	99.40	155.74	0.00056	1.23852	1.3564
20	72.2689	629.1	0.20842	58.9352	0.01458	101.04	143.51	0.00055	1.23495	1.3027
21	71.1846	618.6	0.20301	53.5525	0.01570	102.94	132.90	0.00054	1.23117	1.2541
22	70.0303	604.8	0.19798	48.1167	0.01705	104.07	123.52	0.00052	1.22716	1.2241
23	68.7931	588.6	0.19320	42.7839	0.01864	104.53	115.09	0.00050	1.22286	1.2083
24	67.4541	568.5	0.18834	37.3608	0.02064	104.35	107.38	0.00047	1.21823	1.2077
25	66.0015	547.2	0.18315	32.2261	0.02298	103.58	100.25	0.00045	1.21322	1.2173
26	64.4069	522.0	0.17735	27.0974	0.02604	102.18	93.53	0.00042	1.20773	1.2442
27	62.6325	493.8	0.17046	22.1181	0.03006	100.19	87.08	0.00039	1.20164	1.2900
28	60.6157	461.3	0.16232	17.2762	0.03580	97.53	80.74	0.00035	1.19475	1.3671
29	58.2362	423.5	0.15281	12.6128	0.04486	94.04	74.28	0.00031	1.18665	1.5007
30	55.2770	378.6	0.14128	8.2063	0.06142	89.37	67.41	0.00025	1.17662	1.7544
31	51.0906	320.0	0.12529	4.0067	0.10589	85.69	59.29	0.00018	1.16254	2.3450
* 31.248	49.6094	300.4	0.11970	2.9465	0.13513	86.22	56.77	0.00015	1.15759	2.6729
* 31.248	14.3567	202.2	0.06923	0.3054	0.25798	49.56	21.45	0.00024	1.04393	2.2578
32	12.4337	220.6	0.07010	0.4557	0.14350	40.71	20.84	0.00037	1.03797	1.6202
33	11.0478	240.2	0.07027	0.5646	0.09932	37.54	20.72	0.00051	1.03369	1.3169
34	10.1147	256.8	0.07115	0.6340	0.07916	36.34	20.83	0.00064	1.03081	1.1650
35	9.4056	272.3	0.07164	0.6842	0.06708	35.87	21.05	0.00075	1.02863	1.0719
36	8.8336	287.3	0.07188	0.7230	0.05886	35.75	21.32	0.00086	1.02687	1.0083
37	8.3550	301.8	0.07197	0.7541	0.05282	35.84	21.62	0.00097	1.02540	0.9618
38	7.9444	316.0	0.07196	0.7798	0.04815	36.05	21.95	0.00107	1.02414	0.9263
39	7.5856	329.9	0.07188	0.8014	0.04440	36.36	22.29	0.00118	1.02304	0.8982
40	7.2677	343.6	0.07177	0.8199	0.04132	36.72	22.64	0.00128	1.02207	0.8752
42	6.7254	370.5	0.07145	0.8499	0.03648	37.58	23.36	0.00149	1.02041	0.8401
44	6.2759	396.8	0.07108	0.8731	0.03284	38.56	24.09	0.00170	1.01904	0.8142
46	5.8943	422.6	0.07068	0.8917	0.02998	39.62	24.83	0.00191	1.01787	0.7941
48	5.5647	448.2	0.07024	0.9067	0.02765	40.73	25.57	0.00213	1.01687	0.7781
50	5.2758	473.7	0.06976	0.9192	0.02571	41.88	26.31	0.00235	1.01599	0.7651
52	5.0198	499.2	0.06924	0.9296	0.02407	43.06	27.04	0.00257	1.01521	0.7545
54	4.7908	524.8	0.06868	0.9384	0.02265	44.27	27.77	0.00280	1.01451	0.7457
56	4.5844	550.7	0.06804	0.9459	0.02141	45.50	28.49	0.00303	1.01388	0.7383
58	4.3971	576.9	0.06736	0.9524	0.02033	46.75	29.21	0.00326	1.01331	0.7326
60	4.2261	603.6	0.06662	0.9580	0.01936	48.08	29.91	0.00350	1.01279	0.7270
65	3.8564	673.1	0.06447	0.9691	0.01734	51.68	31.66	0.00413	1.01167	0.7152
70	3.5506	747.7	0.06195	0.9772	0.01575	55.32	33.36	0.00476	1.01074	0.7102
75	3.2928	828.3	0.05918	0.9834	0.01445	58.83	35.02	0.00537	1.00996	0.7126
80	3.0718	915.3	0.05630	0.9881	0.01337	62.59	36.64	0.00599	1.00929	0.7163
85	2.8800	1008.9	0.05342	0.9917	0.01245	67.45	38.23	0.00671	1.00870	0.7117
90	2.7117	1108.6	0.05067	0.9946	0.01166	72.32	39.78	0.00743	1.00819	0.7107
95	2.5627	1213.7	0.04811	0.9969	0.01096	77.19	41.30	0.00815	1.00774	0.7120
100	2.4297	1323.3	0.04579	0.9988	0.01035	90.00	45.80	0.00973	1.00734	0.6972
120	2.0150	1784.8	0.03898	1.0032	0.00849	117.10	54.01	0.01380	1.00608	0.6991
140	1.7233	2229.2	0.03557	1.0053	0.00722	136.07	59.17	0.01767	1.00520	0.6996
160	1.5063	2617.0	0.03434	1.0062	0.00628	148.29	63.01	0.02155	1.00455	0.6988
180	1.3384	2944.3	0.03433	1.0066	0.00557	155.73	66.22	0.02555	1.00404	0.6971
200	1.2044	3226.4	0.03497	1.0067	0.00500	160.37	69.09	0.02971	1.00363	0.6952
220	1.0949	3481.9	0.03586	1.0066	0.00454	163.64	71.76	0.03403	1.00330	0.6932
240	1.0038	3726.7	0.03678	1.0064	0.00416	166.52	74.30	0.03853	1.00303	0.6915
260	0.9268	3971.2	0.03759	1.0062	0.00384	169.48	76.77	0.04321	1.00280	0.6900
280	0.8607	4220.3	0.03826	1.0060	0.00356	172.74	79.18	0.04808	1.00260	0.6889
300	0.8036	4476.1	0.03877	1.0059	0.00332	176.33	81.56	0.05312	1.00242	0.6879
350	0.6891	5146.2	0.03950	1.0053	0.00285	186.62	87.42	0.06654	1.00208	0.6863
400	0.6033	5845.9	0.03978	1.0048	0.00249	198.00	93.20	0.08114	1.00182	0.6854
450	0.5365	6565.4	0.03983	1.0044	0.00221	209.96	98.91	0.09691	1.00162	0.6849
500	0.4830	7290.1	0.03983	1.0040	0.00199	221.97	104.56	0.11385	1.00146	0.6845
550	0.4392	8021.6	0.03978	1.0037	0.00181	234.05	110.15	0.13195	1.00132	0.6842
600	0.4027	8758.5	0.03971	1.0034	0.00166	246.13	115.67	0.15117	1.00121	0.6840
700	0.3454	10253.3	0.03947	1.0030	0.00142	270.36	126.54	0.19292	1.00104	0.6837
800	0.3023	11787.6	0.03912	1.0026	0.00125	294.86	137.14	0.23890	1.00091	0.6836
900	0.2688	13372.1	0.03864	1.0023	0.00111	319.82	147.50	0.28894	1.00081	0.6837
1000	0.2420	15014.3	0.03806	1.0021	0.00100	440.56	157.62	0.43747	1.00073	0.5361
1200	0.2017	18473.1	0.03675	1.0018	0.00083	511.13	177.21	0.59363	1.00061	0.5328
1400	0.1729	22160.0	0.03536	1.0015	0.00071	583.42	196.01	0.76847	1.00052	0.5310
1600	0.1513	26011.3	0.03410	1.0013	0.00062	656.44	214.15	0.96177	1.00046	0.5297
1800	0.1345	30079.4	0.03288	1.0011	0.00055	733.32	231.70	1.17555	1.00041	0.5274
2000	0.1211	34691.6	0.03134	1.0008	0.00050	826.53	248.74	1.41797	1.00036	0.5216
2500	0.0965	54123.9	0.02407	0.9969	0.00040	1340.45	289.95	2.30214	1.00029	0.4698
3000	0.0788	105504.7	0.01422	0.9762	0.00034	3028.08	332.54	3.84065	1.00024	0.3956

* TWO-PHASE BOUNDARY

Table 4.1 (continued)
THERMODYNAMIC PROPERTIES OF PARAHYDROGEN (ISOBARS, SI UNITS)

1.292800 MPA ISOBAR

TEMPERATURE	VOLUME	ISOTHERM DERIVATIVE	ISOCHORE DERIVATIVE	INTERNAL ENERGY	ENTHALPY	ENTROPY	CV	CP	VELOCITY OF SOUND
DEG. K	CU M/KG	CU. M-MPA/KG	MPA/K	KJ/KG-K	KJ/KG	KJ/KG-K	KJ /	KG-K	M/S
* 14.223	0.01286	1.2537	0.9367	-308.4	-291.8	4.996	4.73	6.38	1300
15	0.01296	1.1970	0.9378	-303.5	-286.7	5.343	4.89	6.74	1285
16	0.01310	1.1028	0.9299	-296.7	-279.7	5.793	5.08	7.23	1253
17	0.01325	1.0336	0.9139	-289.4	-272.3	6.245	5.25	7.66	1228
18	0.01341	0.9688	0.8991	-281.7	-264.4	6.695	5.42	8.12	1205
19	0.01358	0.9102	0.8842	-273.6	-256.1	7.145	5.56	8.57	1184
20	0.01377	0.8467	0.8680	-265.0	-247.2	7.599	5.69	9.07	1161
21	0.01397	0.7813	0.8499	-256.0	-237.9	8.054	5.81	9.60	1136
22	0.01420	0.7166	0.8300	-246.4	-228.0	8.513	5.91	10.17	1111
23	0.01444	0.6530	0.8077	-236.3	-217.6	8.978	5.99	10.79	1084
24	0.01471	0.5882	0.7826	-225.5	-206.5	9.451	6.06	11.47	1055
25	0.01502	0.5232	0.7537	-214.0	-194.6	9.936	6.12	12.25	1023
26	0.01537	0.4570	0.7209	-201.8	-181.9	10.433	6.17	13.16	987
27	0.01577	0.3916	0.6835	-188.6	-168.3	10.948	6.22	14.23	946
28	0.01624	0.3269	0.6411	-174.4	-153.4	11.489	6.27	15.56	901
29	0.01682	0.2603	0.5941	-158.7	-136.9	12.066	6.33	17.45	847
30	0.01755	0.1976	0.5409	-141.0	-118.3	12.697	6.40	20.08	787
31	0.01856	0.1349	0.4787	-120.2	-96.2	13.422	6.50	24.64	715
32	0.02020	0.0718	0.3996	-93.1	-67.0	14.348	6.68	35.71	620
* 32.975	0.03128	-0.0001	0.1927	-5.6	34.9	17.461	9.88	-14658.08	348
* 32.975	0.03241	-0.0001	0.1868	-0.1	41.8	17.671	9.80	-13261.84	351
33	0.03934	0.0025	0.1548	30.3	81.1	18.865	9.44	501.14	363
34	0.06012	0.0298	0.0915	93.8	171.6	21.578	7.72	42.28	404
35	0.06941	0.0456	0.0760	116.0	205.7	22.568	7.23	28.61	425
36	0.07672	0.0584	0.0669	132.1	231.3	23.239	6.98	23.23	441
37	0.08307	0.0694	0.0607	145.5	252.9	23.883	6.82	20.36	455
38	0.08886	0.0793	0.0559	157.4	272.3	24.400	6.72	18.54	468
39	0.09425	0.0884	0.0521	168.4	290.2	24.865	6.64	17.28	479
40	0.09935	0.0969	0.0489	178.5	307.0	25.290	6.59	16.35	490
42	0.10894	0.1126	0.0439	197.5	338.3	26.054	6.53	15.07	510
44	0.11795	0.1270	0.0401	215.1	367.5	26.734	6.49	14.22	528
46	0.12654	0.1405	0.0369	231.8	395.4	27.353	6.47	13.63	544
48	0.13482	0.1533	0.0344	247.9	422.2	27.923	6.46	13.18	559
50	0.14286	0.1655	0.0322	263.5	448.2	28.454	6.46	12.85	574
52	0.15070	0.1773	0.0303	278.8	473.6	28.953	6.48	12.59	587
54	0.15839	0.1888	0.0287	293.8	498.6	29.424	6.50	12.39	600
56	0.16594	0.1999	0.0272	308.7	523.2	29.872	6.53	12.23	612
58	0.17338	0.2107	0.0259	323.4	547.6	30.299	6.57	12.12	624
60	0.18072	0.2214	0.0247	338.1	571.7	30.709	6.62	12.04	635
65	0.19875	0.2471	0.0223	374.7	631.6	31.668	6.78	11.95	660
70	0.21640	0.2720	0.0203	411.7	691.5	32.555	7.02	12.00	682
75	0.23379	0.2961	0.0187	449.6	751.8	33.387	7.31	12.15	702
80	0.25096	0.3198	0.0173	488.7	813.1	34.179	7.65	12.39	720
85	0.26796	0.3431	0.0162	529.4	875.8	34.939	8.03	12.69	736
90	0.28483	0.3660	0.0152	571.9	940.1	35.674	8.45	13.04	752
95	0.30159	0.3886	0.0143	616.3	1006.2	36.388	8.87	13.41	766
100	0.31826	0.4111	0.0135	662.8	1074.2	37.086	9.30	13.79	781
120	0.38425	0.4990	0.0111	868.2	1365.0	39.731	10.85	15.22	836
140	0.44956	0.5853	0.0094	1098.3	1679.4	42.152	11.84	16.13	893
160	0.51445	0.6705	0.0082	1341.3	2006.4	44.335	12.23	16.48	950
180	0.57907	0.7551	0.0073	1587.2	2335.8	46.275	12.20	16.42	1008
200	0.64351	0.8392	0.0065	1829.8	2661.9	47.992	11.96	16.16	1065
220	0.70781	0.9229	0.0059	2066.5	2981.6	49.516	11.64	15.83	1120
240	0.77202	1.0064	0.0054	2296.8	3294.8	50.879	11.34	15.51	1173
260	0.83616	1.0897	0.0050	2521.3	3602.3	52.110	11.08	15.25	1224
280	0.90024	1.1729	0.0046	2741.2	3905.0	53.232	10.88	15.04	1273
300	0.96425	1.2560	0.0043	2957.8	4204.3	54.265	10.73	14.88	1320
350	1.12416	1.4633	0.0037	3488.5	4941.9	56.538	10.51	14.66	1428
400	1.28395	1.6702	0.0032	4012.1	5672.0	58.489	10.43	14.57	1527
450	1.44365	1.8769	0.0029	4532.8	6399.2	60.203	10.41	14.54	1619
500	1.60330	2.0835	0.0026	5053.5	7126.2	61.734	10.40	14.53	1706
550	1.76292	2.2900	0.0024	5573.7	7852.8	63.119	10.41	14.54	1788
600	1.92252	2.4964	0.0022	6094.8	8580.3	64.383	10.43	14.55	1867
700	2.24165	2.9091	0.0018	7140.0	10038.0	66.632	10.48	14.61	2014
800	2.56075	3.3218	0.0016	8192.7	11503.2	68.587	10.57	14.70	2149
900	2.87982	3.7343	0.0014	9256.2	12979.2	70.324	10.70	14.83	2275
1000	3.19888	4.1469	0.0013	10333.5	14463.0	71.896	10.86	14.98	2392
1200	3.83696	4.9718	0.0011	12543.8	17504.2	74.650	11.24	15.37	2607
1400	4.47502	5.7968	0.0009	14836.5	20621.9	77.056	11.68	15.80	2801
1600	5.11310	6.6217	0.0008	17216.0	23926.2	79.173	12.11	16.23	2980
1800	5.75139	7.4465	0.0007	19686.8	27122.2	81.076	12.55	16.68	3146
2000	6.39059	8.2714	0.0006	22266.9	30528.7	82.936	13.13	17.28	3299
2500	8.01231	10.3335	0.0005	29613.1	39971.5	89.483	16.78	21.24	3616
3000	9.79099	12.3957	0.0004	40604.3	53262.1	97.666	27.91	33.90	3877

* TWO-PHASE BOUNDARY

Table 4.1 (continued)

THERMODYNAMIC PROPERTIES OF PARAHYDROGEN (ISOBARS, SI UNITS)

1.292800 MPA ISOBAR

TEMPERATURE DEG. K	DENSITY KG/CU M	$V(DH/DV)_P$ KJ/KG	$V(DP/DU)_V$ MPA-CU M/KJ X 10^2	$-V(DP/DV)_T$ MPA	$-(DV/DT)_P/V$ 1/DEG. K	THERMAL CONDUCTIVITY W/K-M X 10^3	VISCOSITY KG/M-S X 10^7	THERMAL DIFFUSIVITY SQ M/HR	DIELECTRIC CONSTANT	PRANDTL NUMBER
* 14.229	77.7376	664.1	0.25449	97.4607	0.00961	77.85	262.81	0.00056	1.25411	2.1545
15	77.1455	663.8	0.24860	92.3420	0.01016	84.50	238.11	0.00058	1.25203	1.8998
16	76.3375	654.5	0.23996	84.1885	0.01105	90.83	212.01	0.00059	1.24919	1.6873
17	75.4732	654.2	0.23054	78.0093	0.01172	95.10	190.76	0.00059	1.24615	1.5374
18	74.5678	652.2	0.22264	72.2386	0.01245	98.19	173.31	0.00058	1.24298	1.4327
19	73.6240	649.8	0.21587	67.0094	0.01319	100.11	158.77	0.00057	1.23968	1.3597
20	72.6196	642.4	0.20992	61.4904	0.01412	101.84	146.28	0.00056	1.23617	1.3025
21	71.5645	631.5	0.20447	55.9166	0.01520	103.83	135.51	0.00054	1.23249	1.2526
22	70.4452	618.6	0.19942	50.4817	0.01644	105.08	126.02	0.00053	1.22860	1.2197
23	69.2507	603.8	0.19462	45.2201	0.01786	105.66	117.53	0.00051	1.22445	1.1997
24	67.9678	586.1	0.18987	39.9775	0.01958	105.62	109.81	0.00049	1.22001	1.1929
25	66.5794	566.0	0.18490	34.8312	0.02164	105.01	102.67	0.00046	1.21521	1.1974
26	65.0753	542.6	0.17949	29.7379	0.02424	103.82	96.02	0.00044	1.21003	1.2167
27	63.4185	517.0	0.17329	24.8318	0.02753	102.08	89.69	0.00041	1.20434	1.2502
28	61.5718	488.4	0.16604	20.1288	0.03185	99.77	83.56	0.00037	1.19801	1.3029
29	59.4606	454.6	0.15784	15.4766	0.03838	96.80	77.47	0.00034	1.19081	1.3965
30	56.9770	418.0	0.14833	11.2605	0.04803	93.03	71.26	0.00029	1.18237	1.5379
31	53.8752	374.2	0.13669	7.2696	0.06585	87.74	64.60	0.00024	1.17190	1.8140
32	49.5003	317.8	0.12082	3.5564	0.11236	85.51	56.75	0.00017	1.15723	2.3699
* 32.976	31.9741	198.6	0.06097	-0.0026	-73.80653	-5850.18	35.91	0.00004	1.09971	8.9969
* 32.975	30.8504	199.5	0.06180	-0.0028	-66.46463	-5289.75	34.93	0.00005	1.09609	8.7585
33	25.4217	204.8	0.06448	0.0633	2.44661	241.60	30.45	0.00007	1.07873	6.3152
34	16.6336	228.8	0.07129	0.4952	0.18481	50.08	24.13	0.00026	1.05103	2.0376
35	14.4079	247.2	0.07296	0.6567	0.11577	43.73	23.27	0.00038	1.04409	1.5227
36	13.0352	264.3	0.07355	0.7615	0.08789	41.38	23.04	0.00049	1.03983	1.2929
37	12.0376	280.5	0.07387	0.8358	0.07258	40.31	23.04	0.00059	1.03675	1.1633
38	11.2542	296.0	0.07395	0.8927	0.06263	39.82	23.16	0.00069	1.03432	1.0785
39	10.6104	311.1	0.07389	0.9380	0.05554	39.64	23.36	0.00079	1.03234	1.0183
40	10.0653	325.9	0.07374	0.9753	0.05017	39.65	23.60	0.00087	1.03066	0.9731
42	9.1792	354.5	0.07332	1.0334	0.04250	40.03	24.17	0.00104	1.02793	0.9096
44	8.4784	382.3	0.07281	1.0767	0.03720	40.69	24.80	0.00121	1.02578	0.8668
46	7.9028	409.5	0.07227	1.1102	0.03328	41.51	25.46	0.00139	1.02402	0.8356
48	7.4174	436.2	0.07170	1.1370	0.03023	42.45	26.14	0.00156	1.02253	0.8118
50	7.0001	462.7	0.07111	1.1588	0.02777	43.47	26.83	0.00174	1.02125	0.7931
52	6.6357	489.0	0.07050	1.1768	0.02575	44.54	27.53	0.00192	1.02014	0.7782
54	6.3136	515.4	0.06984	1.1919	0.02404	45.66	28.23	0.00210	1.01915	0.7660
56	6.0264	541.9	0.06912	1.2046	0.02258	46.81	28.92	0.00229	1.01828	0.7559
58	5.7678	568.6	0.06839	1.2154	0.02131	47.98	29.61	0.00247	1.01749	0.7479
60	5.5334	595.8	0.06759	1.2248	0.02020	49.26	30.30	0.00266	1.01677	0.7405
65	5.0315	666.3	0.06533	1.2433	0.01793	52.74	32.01	0.00316	1.01524	0.7252
70	4.6210	741.6	0.06271	1.2567	0.01618	56.29	33.68	0.00365	1.01399	0.7178
75	4.2774	822.8	0.05986	1.2667	0.01477	59.67	35.31	0.00413	1.01295	0.7192
80	3.9848	910.3	0.05690	1.2743	0.01361	63.38	36.91	0.00462	1.01206	0.7216
85	3.7319	1004.2	0.05396	1.2802	0.01264	68.21	38.48	0.00518	1.01129	0.7160
90	3.5109	1104.2	0.05115	1.2849	0.01180	73.05	40.02	0.00575	1.01062	0.7141
95	3.3158	1209.7	0.04855	1.2886	0.01108	77.89	41.53	0.00631	1.01003	0.7148
100	3.1421	1319.7	0.04618	1.2916	0.01045	90.19	45.84	0.00749	1.00950	0.7009
120	2.6025	1783.0	0.03923	1.2987	0.00853	117.13	54.01	0.01065	1.00786	0.7016
140	2.2244	2229.2	0.03577	1.3020	0.00724	136.07	59.17	0.01365	1.00672	0.7015
160	1.9438	2618.5	0.03450	1.3034	0.00629	148.31	63.03	0.01667	1.00587	0.7002
180	1.7269	2947.3	0.03448	1.3039	0.00557	155.78	66.25	0.01978	1.00521	0.6982
200	1.5540	3230.4	0.03510	1.3040	0.00500	160.43	69.12	0.02301	1.00469	0.6960
220	1.4128	3486.8	0.03598	1.3039	0.00454	163.73	71.79	0.02636	1.00426	0.6939
240	1.2953	3732.1	0.03689	1.3036	0.00416	166.62	74.34	0.02986	1.00391	0.6920
260	1.1959	3977.2	0.03769	1.3033	0.00383	169.60	76.81	0.03349	1.00361	0.6905
280	1.1108	4226.7	0.03835	1.3029	0.00356	172.87	79.23	0.03726	1.00335	0.6892
300	1.0371	4483.0	0.03886	1.3026	0.00332	176.46	81.62	0.04117	1.00313	0.6881
350	0.8896	5153.6	0.03958	1.3017	0.00284	186.78	87.49	0.05158	1.00268	0.6864
400	0.7788	5853.7	0.03985	1.3008	0.00249	198.18	93.27	0.06289	1.00235	0.6855
450	0.6927	6573.5	0.03989	1.3001	0.00221	210.16	98.99	0.07511	1.00209	0.6849
500	0.6237	7298.3	0.03988	1.2995	0.00199	222.20	104.65	0.08825	1.00188	0.6845
550	0.5672	8030.0	0.03983	1.2990	0.00181	234.30	110.25	0.10227	1.00171	0.6842
600	0.5202	8767.0	0.03975	1.2985	0.00166	246.40	115.79	0.11717	1.00157	0.6840
700	0.4461	10262.1	0.03951	1.2978	0.00142	270.67	126.68	0.14952	1.00134	0.6837
800	0.3905	11796.5	0.03915	1.2972	0.00125	295.22	137.30	0.18515	1.00118	0.6836
900	0.3472	13381.2	0.03867	1.2967	0.00111	320.22	147.68	0.22393	1.00105	0.6837
1000	0.3126	15023.5	0.03809	1.2963	0.00100	440.56	157.81	0.33859	1.00094	0.5368
1200	0.2606	18482.6	0.03677	1.2958	0.00083	511.13	177.44	0.45942	1.00079	0.5335
1400	0.2235	22169.5	0.03538	1.2954	0.00071	583.41	196.28	0.59468	1.00067	0.5317
1600	0.1956	26017.5	0.03412	1.2950	0.00062	656.28	214.45	0.74417	1.00059	0.5304
1800	0.1739	30066.8	0.03292	1.2947	0.00055	732.42	232.03	0.90921	1.00052	0.5284
2000	0.1565	34607.6	0.03145	1.2943	0.00050	822.87	249.10	1.09531	1.00047	0.5232
2500	0.1248	52962.2	0.02469	1.2897	0.00040	1295.97	290.31	1.76030	1.00038	0.4757
3000	0.1021	99603.9	0.01512	1.2660	0.00034	2805.93	332.51	2.91717	1.00031	0.4018

* TWO-PHASE BOUNDARY

Table 4.1 (continued)
THERMODYNAMIC PROPERTIES OF PARAHYDROGEN (ISOBARS, SI UNITS)

1.50 MPA ISOBAR

TEMPERATURE DEG. K	VOLUME CU M/KG	ISOTHERM DERIVATIVE CU M-MPA/KG	ISOCHORE DERIVATIVE MPA/K	INTERNAL ENERGY KJ/KG-K	ENTHALPY KJ/KG	ENTROPY KJ/KG-K	CV KJ /	CP KG-K	VELOCITY OF SOUND M/S
* 14.296	0.01285	1.2672	0.9378	-308.3	-289.1	5.001	4.74	6.38	1306
15	0.01293	1.2227	0.9402	-303.8	-284.4	5.316	4.89	6.70	1295
16	0.01307	1.1232	0.9358	-297.1	-277.5	5.764	5.07	7.20	1263
17	0.01321	1.0597	0.9197	-289.9	-270.1	6.212	5.25	7.62	1240
18	0.01337	0.9938	0.9049	-282.3	-262.3	6.660	5.41	8.06	1217
19	0.01354	0.9302	0.8898	-274.3	-254.0	7.109	5.56	8.52	1194
20	0.01372	0.8675	0.8738	-265.8	-245.2	7.559	5.69	9.00	1172
21	0.01392	0.8044	0.8561	-256.8	-235.9	8.010	5.80	9.51	1148
22	0.01414	0.7407	0.8365	-247.4	-226.2	8.465	5.90	10.06	1123
23	0.01437	0.6765	0.8147	-237.4	-215.5	8.925	5.99	10.65	1097
24	0.01464	0.6103	0.7902	-226.8	-204.8	9.393	6.06	11.32	1068
25	0.01493	0.5466	0.7624	-215.5	-193.1	9.870	6.12	12.05	1037
26	0.01526	0.4806	0.7310	-203.6	-180.7	10.358	6.17	12.91	1003
27	0.01564	0.4190	0.6953	-190.8	-167.3	10.862	6.22	13.84	966
28	0.01608	0.3544	0.6555	-177.0	-152.9	11.386	6.26	15.04	923
29	0.01661	0.2909	0.6113	-162.1	-137.2	11.939	6.32	16.59	874
30	0.01726	0.2295	0.5622	-145.5	-119.6	12.535	6.38	18.68	820
31	0.01810	0.1686	0.5067	-126.6	-99.4	13.196	6.45	21.92	757
32	0.01930	0.1106	0.4417	-103.9	-75.0	13.971	6.57	27.61	682
33	0.02140	0.0534	0.3575	-73.3	-41.2	15.009	6.80	42.98	581
34	0.03016	0.0078	0.2048	-1.9	43.4	17.524	8.76	174.96	395
35	0.04884	0.0245	0.1167	77.1	150.4	20.638	7.70	54.07	415
36	0.05798	0.0408	0.0942	104.8	191.8	21.806	7.27	33.56	434
37	0.06494	0.0542	0.0817	123.7	221.2	22.612	7.02	26.25	450
38	0.07082	0.0656	0.0737	138.9	245.1	23.250	6.84	22.59	465
39	0.07614	0.0761	0.0668	152.0	266.2	23.798	6.74	20.01	475
40	0.08105	0.0857	0.0620	163.8	285.4	24.284	6.67	18.44	487
42	0.09006	0.1030	0.0546	185.0	320.1	25.131	6.58	16.45	508
44	0.09836	0.1186	0.0492	204.1	351.7	25.866	6.53	15.23	526
46	0.10618	0.1331	0.0450	222.0	381.2	26.524	6.50	14.40	543
48	0.11365	0.1467	0.0416	238.9	409.4	27.123	6.49	13.81	559
50	0.12085	0.1597	0.0388	255.3	436.6	27.678	6.49	13.37	573
52	0.12785	0.1721	0.0363	271.2	463.0	28.195	6.51	13.03	587
54	0.13467	0.1840	0.0343	286.7	488.7	28.682	6.52	12.77	600
56	0.14135	0.1956	0.0324	302.1	514.1	29.143	6.55	12.56	612
58	0.14792	0.2068	0.0308	317.2	539.1	29.581	6.59	12.41	624
60	0.15439	0.2178	0.0294	332.2	563.8	30.000	6.64	12.29	635
65	0.17022	0.2443	0.0263	369.5	624.8	30.977	6.80	12.15	661
70	0.18567	0.2698	0.0239	407.0	685.5	31.877	7.03	12.16	683
75	0.20084	0.2945	0.0220	445.3	746.6	32.720	7.32	12.28	703
80	0.21580	0.3186	0.0203	484.8	808.5	33.519	7.66	12.50	721
85	0.23060	0.3423	0.0190	525.8	871.7	34.285	8.04	12.78	738
90	0.24526	0.3655	0.0177	568.6	936.5	35.025	8.45	13.12	753
95	0.25981	0.3885	0.0167	613.2	1002.9	35.744	8.88	13.48	768
100	0.27427	0.4112	0.0158	659.9	1071.3	36.445	9.31	13.85	782
120	0.33142	0.4999	0.0129	865.9	1363.1	39.099	10.86	15.26	838
140	0.38790	0.5868	0.0110	1096.4	1678.2	41.526	11.84	16.16	895
160	0.44397	0.6724	0.0095	1339.7	2005.7	43.712	12.23	16.50	952
180	0.49977	0.7572	0.0084	1585.8	2335.5	45.655	12.20	16.44	1010
200	0.55538	0.8415	0.0076	1828.7	2661.7	47.373	11.96	16.17	1067
220	0.61087	0.9255	0.0069	2065.5	2981.8	48.898	11.64	15.84	1122
240	0.66627	1.0091	0.0063	2295.9	3295.3	50.262	11.34	15.52	1175
260	0.72159	1.0926	0.0058	2520.5	3602.9	51.494	11.08	15.25	1226
280	0.77685	1.1759	0.0054	2740.5	3905.8	52.616	10.88	15.04	1275
300	0.83206	1.2591	0.0050	2957.1	4205.2	53.650	10.73	14.88	1322
350	0.96994	1.4665	0.0043	3488.1	4943.0	55.923	10.51	14.66	1430
400	1.10769	1.6735	0.0038	4011.8	5673.3	57.875	10.43	14.57	1529
450	1.24536	1.8803	0.0033	4532.5	6400.5	59.589	10.41	14.54	1621
500	1.38297	2.0869	0.0030	5053.2	7127.7	61.120	10.40	14.53	1708
550	1.52056	2.2934	0.0027	5573.5	7854.3	62.505	10.41	14.54	1790
600	1.65812	2.4999	0.0025	6094.6	8581.8	63.770	10.43	14.56	1868
700	1.93319	2.9127	0.0021	7139.9	10039.7	66.018	10.48	14.61	2015
800	2.20822	3.3253	0.0019	8192.6	11504.9	67.974	10.57	14.70	2150
900	2.48322	3.7379	0.0017	9256.1	12980.9	69.710	10.70	14.83	2276
1000	2.75821	4.1505	0.0015	10333.4	14470.7	71.283	10.86	14.98	2393
1200	3.30816	4.9755	0.0013	12543.8	17506.0	74.036	11.24	15.37	2608
1400	3.85809	5.8004	0.0011	14836.5	20623.6	76.443	11.68	15.80	2802
1600	4.40804	6.6253	0.0009	17215.9	23827.9	78.560	12.11	16.23	2980
1800	4.95814	7.4502	0.0008	19686.0	27123.3	80.462	12.54	16.67	3147
2000	5.50838	8.2751	0.0008	22263.3	30526.8	82.321	13.11	17.26	3301
2500	6.90505	10.3372	0.0006	29558.6	39916.2	88.845	16.55	20.98	3620
3000	8.42675	12.3993	0.0005	40263.2	52903.4	96.920	26.94	32.79	3881

* TWO-PHASE BOUNDARY

Table 4.1 (continued)
THERMODYNAMIC PROPERTIES OF PARAHYDROGEN (ISOBARS, SI UNITS)

3.00 MPA ISOBAR

TEMPERATURE DEG. K	VOLUME CU M/KG	ISOTHERM DERIVATIVE CU M-MPA/KG	ISOCHORE DERIVATIVE MPA/K	INTERNAL ENERGY KJ/KG-K	ENTHALPY KJ/KG	ENTROPY KJ/KG-K	CV KJ / KG-K	CP KJ / KG-K	VELOCITY OF SOUND M/S
* 14.772	0.01272	1.3623	0.9470	-307.5	-269.4	5.036	4.81	6.38	1345
15	0.01274	1.3585	0.9492	-306.1	-267.9	5.135	4.86	6.47	1345
16	0.01286	1.2794	0.9544	-299.8	-261.2	5.568	5.05	6.93	1326
17	0.01299	1.2075	0.9510	-293.0	-254.1	6.000	5.22	7.37	1306
18	0.01313	1.1408	0.9404	-285.9	-246.5	6.433	5.37	7.78	1285
19	0.01327	1.0731	0.9266	-278.3	-238.5	6.865	5.52	8.20	1262
20	0.01343	1.0093	0.9118	-270.4	-230.1	7.296	5.65	8.62	1241
21	0.01360	0.9488	0.8960	-262.1	-221.3	7.726	5.76	9.05	1221
22	0.01378	0.8875	0.8787	-253.3	-212.0	8.157	5.86	9.50	1199
23	0.01397	0.8271	0.8595	-244.2	-202.3	8.590	5.95	9.97	1177
24	0.01419	0.7655	0.8386	-234.6	-192.1	9.024	6.03	10.47	1153
25	0.01442	0.7053	0.8155	-224.6	-181.4	9.462	6.10	11.00	1128
26	0.01467	0.6461	0.7904	-214.1	-170.1	9.904	6.16	11.57	1102
27	0.01494	0.5884	0.7629	-203.0	-158.2	10.352	6.21	12.17	1074
28	0.01525	0.5315	0.7331	-191.4	-145.7	10.806	6.25	12.84	1044
29	0.01559	0.4762	0.7011	-179.3	-132.5	11.270	6.29	13.57	1013
30	0.01597	0.4222	0.6670	-166.4	-118.5	11.744	6.33	14.39	980
31	0.01640	0.3688	0.6309	-152.9	-103.7	12.230	6.36	15.36	943
32	0.01689	0.3184	0.5926	-138.4	-87.8	12.735	6.40	16.47	905
33	0.01747	0.2698	0.5524	-123.1	-70.7	13.261	6.44	17.82	864
34	0.01815	0.2258	0.5101	-106.5	-52.1	13.815	6.48	19.38	822
35	0.01896	0.1858	0.4661	-88.7	-31.8	14.403	6.53	21.25	778
36	0.01997	0.1498	0.4207	-69.4	-9.4	15.034	6.59	23.55	732
37	0.02123	0.1196	0.3747	-48.3	15.4	15.714	6.66	26.24	686
38	0.02281	0.0967	0.3293	-25.5	43.0	16.450	6.75	28.93	644
39	0.02479	0.0809	0.2866	-1.2	73.1	17.233	6.84	31.20	607
40	0.02717	0.0718	0.2490	23.6	105.1	18.042	6.91	32.43	580
42	0.03273	0.0711	0.1916	69.8	168.0	19.579	6.93	30.16	556
44	0.03841	0.0830	0.1546	108.5	223.[illegible]	20.876	6.86	25.57	556
46	0.04369	0.0984	0.1306	140.0	271.1	21.928	6.79	22.00	565
48	0.04856	0.1144	0.1138	166.8	312.5	22.810	6.73	19.56	576
50	0.05311	0.1299	0.1015	190.5	349.8	23.572	6.70	17.89	589
52	0.05740	0.1449	0.0920	212.2	384.4	24.250	6.69	16.70	601
54	0.06150	0.1593	0.0844	232.4	416.9	24.863	6.70	15.83	614
56	0.06543	0.1732	0.0782	251.8	448.1	25.431	6.72	15.18	626
58	0.06925	0.1867	0.0731	270.3	478.0	25.956	6.74	14.70	638
60	0.07295	0.1998	0.0686	288.2	507.0	26.448	6.77	14.31	649
65	0.08185	0.2307	0.0598	331.2	576.7	27.564	6.91	13.65	675
70	0.09035	0.2599	0.0532	373.0	644.1	28.562	7.12	13.33	698
75	0.09858	0.2877	0.0480	414.7	710.4	29.478	7.39	13.23	718
80	0.10659	0.3145	0.0439	456.9	776.7	30.333	7.72	13.29	736
85	0.11445	0.3405	0.0405	500.2	843.5	31.143	8.09	13.45	752
90	0.12217	0.3658	0.0376	544.8	911.3	31.918	8.50	13.69	768
95	0.12980	0.3905	0.0351	591.1	980.5	32.666	8.92	13.97	782
100	0.13733	0.4149	0.0330	639.1	1051.1	33.391	9.34	14.29	797
120	0.16683	0.5085	0.0266	849.8	1350.3	36.113	10.90	15.55	852
140	0.19571	0.5987	0.0224	1083.1	1670.2	38.577	11.87	16.36	908
160	0.22421	0.6868	0.0194	1328.5	2001.1	40.786	12.25	16.65	966
180	0.25246	0.7736	0.0171	1576.2	2333.6	42.745	12.22	16.55	1024
200	0.28054	0.8595	0.0153	1820.3	2661.9	44.474	11.98	16.26	1080
220	0.30851	0.9446	0.0139	2058.1	2983.7	46.007	11.66	15.91	1135
240	0.33638	1.0293	0.0127	2289.4	3298.5	47.377	11.35	15.58	1189
260	0.36419	1.1135	0.0117	2514.7	3607.3	48.613	11.10	15.31	1239
280	0.39195	1.1975	0.0108	2735.3	3911.1	49.739	10.89	15.09	1288
300	0.41969	1.2816	0.0101	2952.7	4211.8	50.776	10.74	14.92	1334
350	0.48881	1.4900	0.0086	3484.6	4951.1	53.055	10.52	14.68	1442
400	0.55781	1.6977	0.0075	4009.0	5682.4	55.009	10.43	14.59	1540
450	0.62674	1.9049	0.0067	4530.3	6410.5	56.725	10.41	14.56	1632
500	0.69561	2.1119	0.0060	5051.3	7138.2	58.258	10.41	14.54	1718
550	0.76446	2.3187	0.0055	5571.9	7865.3	59.644	10.41	14.55	1800
600	0.83328	2.5254	0.0050	6093.3	8593.1	60.909	10.43	14.56	1878
700	0.97087	2.9385	0.0043	7138.9	10051.5	63.158	10.48	14.61	2024
800	1.10842	3.3513	0.0038	8191.8	11517.1	65.114	10.57	14.70	2159
900	1.24595	3.7640	0.0033	9255.4	12993.3	66.851	10.70	14.83	2284
1000	1.38347	4.1767	0.0030	10332.9	14483.3	68.423	10.86	14.99	2401
1200	1.65847	5.0018	0.0025	12543.4	17518.8	71.177	11.24	15.37	2615
1400	1.93345	5.8268	0.0021	14836.2	20636.6	73.584	11.68	15.80	2808
1600	2.20843	6.6518	0.0019	17215.2	23840.5	75.700	12.10	16.23	2986
1800	2.48346	7.4767	0.0017	19682.9	27133.3	77.602	12.52	16.65	3153
2000	2.75875	8.3016	0.0015	22249.8	30526.1	79.455	13.02	17.16	3308
2500	3.45369	10.3638	0.0012	29351.7	39712.7	85.891	15.69	20.03	3637
3000	4.19314	12.4259	0.0010	38968.3	51547.7	93.555	23.26	28.58	3904

* TWO-PHASE BOUNDARY

Table 4.1 (continued)
THERMODYNAMIC PROPERTIES OF PARAHYDROGEN (ISOBARS, SI UNITS)

2.00 MPA ISOBAR

TEMPERATURE DEG. K	VOLUME CU M/KG	ISOTHERM DERIVATIVE CU M-MPA/KG	ISOCHORE DERIVATIVE MPA/K	INTERNAL ENERGY KJ/KG-K	ENTHALPY KJ/KG	ENTROPY KJ/KG-K	CV KJ /	CP KG-K	VELOCITY OF SOUND M/S
* 14.457	0.01280	1.2994	0.9407	-308.1	-282.5	5.013	4.77	6.38	1319
15	0.01287	1.2650	0.9449	-304.7	-278.9	5.254	4.88	6.63	1311
16	0.01300	1.1850	0.9435	-298.1	-272.1	5.696	5.06	7.09	1288
17	0.01313	1.1096	0.9322	-291.0	-264.8	6.138	5.23	7.53	1264
18	0.01328	1.0414	0.9177	-283.6	-257.0	6.581	5.39	7.96	1240
19	0.01345	0.9770	0.9029	-275.7	-248.8	7.024	5.54	8.41	1218
20	0.01362	0.9157	0.8873	-267.4	-240.2	7.467	5.67	8.86	1196
21	0.01381	0.8527	0.8704	-258.7	-231.1	7.910	5.79	9.34	1173
22	0.01401	0.7908	0.8514	-249.5	-221.5	8.356	5.89	9.85	1150
23	0.01423	0.7273	0.8307	-239.9	-211.4	8.805	5.97	10.39	1125
24	0.01447	0.6647	0.8075	-229.7	-200.7	9.260	6.05	10.98	1098
25	0.01474	0.6021	0.7817	-218.9	-189.4	9.721	6.11	11.63	1070
26	0.01504	0.5392	0.7529	-207.5	-177.5	10.191	6.16	12.35	1039
27	0.01537	0.4780	0.7209	-195.5	-164.7	10.672	6.21	13.15	1006
28	0.01575	0.4174	0.6856	-182.6	-151.1	11.166	6.26	14.08	969
29	0.01619	0.3585	0.6467	-168.9	-136.5	11.678	6.30	15.17	929
30	0.01671	0.2994	0.6046	-154.1	-120.7	12.216	6.34	16.57	884
31	0.01733	0.2429	0.5588	-137.9	-103.2	12.787	6.40	18.37	835
32	0.01812	0.1881	0.5085	-119.9	-83.7	13.408	6.46	20.89	780
33	0.01915	0.1373	0.4528	-99.4	-61.1	14.102	6.54	24.60	719
34	0.02062	0.0915	0.3897	-75.0	-33.8	14.918	6.67	30.65	649
35	0.02303	0.0526	0.3166	-43.9	2.1	15.958	6.89	42.21	568
36	0.02752	0.0300	0.2395	-2.1	53.0	17.389	7.30	59.51	494
37	0.03460	0.0290	0.1764	43.8	113.0	19.036	7.32	54.86	466
38	0.04137	0.0388	0.1415	77.1	159.8	20.286	7.23	40.85	468
39	0.04711	0.0505	0.1203	101.5	195.8	21.219	7.06	31.86	477
40	0.05206	0.0618	0.1061	120.7	224.8	21.956	6.92	26.66	488
42	0.06054	0.0824	0.0877	150.8	271.9	23.105	6.73	21.10	508
44	0.06776	0.1002	0.0762	175.1	310.6	24.005	6.62	18.34	527
46	0.07446	0.1169	0.0680	196.5	345.4	24.780	6.58	16.66	544
48	0.08071	0.1323	0.0617	216.2	377.6	25.465	6.57	15.56	560
50	0.08664	0.1468	0.0567	234.6	407.9	26.083	6.56	14.77	575
52	0.09232	0.1606	0.0526	252.2	436.8	26.651	6.57	14.20	589
54	0.09781	0.1737	0.0491	269.1	464.8	27.178	6.59	13.76	602
56	0.10314	0.1863	0.0461	285.7	492.0	27.674	6.61	13.41	615
58	0.10835	0.1984	0.0435	301.9	518.6	28.139	6.64	13.14	627
60	0.11345	0.2102	0.0413	317.7	544.6	28.581	6.68	12.94	638
65	0.12584	0.2385	0.0367	356.8	608.5	29.604	6.84	12.64	664
70	0.13784	0.2654	0.0331	395.7	671.4	30.536	7.06	12.55	687
75	0.14956	0.2913	0.0302	435.1	734.2	31.403	7.34	12.60	707
80	0.16106	0.3164	0.0278	475.5	797.6	32.221	7.68	12.77	725
85	0.17240	0.3409	0.0258	517.2	862.0	33.002	8.06	13.01	742
90	0.18360	0.3650	0.0241	560.6	927.8	33.754	8.47	13.31	757
95	0.19470	0.3886	0.0227	605.8	995.2	34.483	8.89	13.65	772
100	0.20571	0.4119	0.0214	652.9	1064.3	35.192	9.32	14.00	787
120	0.24906	0.5024	0.0174	860.5	1358.6	37.869	10.87	15.36	842
140	0.29176	0.5905	0.0147	1091.9	1675.4	40.309	11.85	16.23	899
160	0.33405	0.6770	0.0128	1335.9	2004.0	42.503	12.24	16.55	957
180	0.37609	0.7625	0.0113	1582.6	2334.8	44.451	12.21	16.48	1014
200	0.41794	0.8474	0.0101	1825.8	2661.7	46.172	11.97	16.20	1071
220	0.45968	0.9317	0.0092	2063.0	2982.4	47.701	11.65	15.86	1126
240	0.50131	1.0157	0.0084	2293.7	3296.3	49.067	11.34	15.54	1180
260	0.54288	1.0995	0.0077	2518.6	3604.3	50.300	11.09	15.27	1231
280	0.58439	1.1830	0.0072	2738.7	3907.5	51.424	10.88	15.06	1279
300	0.62587	1.2665	0.0067	2955.6	4207.4	52.458	10.73	14.90	1326
350	0.72937	1.4743	0.0057	3486.9	4945.6	54.733	10.52	14.67	1434
400	0.83275	1.6815	0.0050	4010.8	5676.3	56.686	10.43	14.57	1533
450	0.93605	1.8885	0.0045	4531.7	6403.8	58.401	10.41	14.55	1624
500	1.03929	2.0952	0.0040	5052.6	7131.2	59.932	10.40	14.54	1711
550	1.14251	2.3019	0.0036	5572.9	7857.9	61.318	10.41	14.54	1793
600	1.24570	2.5084	0.0033	6094.2	8585.6	62.583	10.43	14.56	1871
700	1.45203	2.9213	0.0029	7139.5	10043.6	64.831	10.48	14.61	2018
800	1.65832	3.3340	0.0025	8192.3	11509.0	66.787	10.57	14.70	2153
900	1.86459	3.7466	0.0022	9255.8	12985.0	68.524	10.70	14.83	2278
1000	2.07084	4.1592	0.0020	10333.2	14474.9	70.096	10.86	14.98	2396
1200	2.48332	4.9843	0.0017	12543.6	17510.3	72.850	11.24	15.37	2610
1400	2.89578	5.8092	0.0014	14836.4	20627.9	75.257	11.68	15.80	2804
1600	3.30824	6.6342	0.0013	17215.6	23832.1	77.373	12.10	16.23	2982
1800	3.72080	7.4591	0.0011	19684.6	27126.2	79.275	12.53	16.66	3149
2000	4.13384	8.2840	0.0010	22257.2	30524.9	81.131	13.07	17.22	3303
2500	5.17877	10.3461	0.0008	29464.0	39821.5	87.615	16.16	20.55	3627
3000	6.30535	12.4082	0.0007	39671.0	52281.7	95.502	25.26	30.87	3890

* TWO-PHASE BOUNDARY

Table 4.1 (continued)
THERMODYNAMIC PROPERTIES OF PARAHYDROGEN (ISOBARS, SI UNITS)

2.00 MPA ISOBAR

TEMPERATURE DEG. K	DENSITY KG/CU M	V(DH/DV)$_P$ KJ/KG	V(DP/DU)$_V$ MPA-CU M/KJ X 10^2	-V(DP/DV)$_T$ MPA	-(DV/DT)$_P$/V 1/DEG. K	THERMAL CONDUCTIVITY W/K-M X 10^3	VISCOSITY KG/M-S X 10^7	THERMAL DIFFUSIVITY SQ M/HR	DIELECTRIC CONSTANT	PRANDTL NUMBER
* 14.457	78.1196	688.5	0.25262	101.5121	0.00927	80.58	267.20	0.00058	1.25546	2.1155
15	77.7130	689.7	0.24935	98.3063	0.00961	85.23	249.22	0.00060	1.25403	1.9386
16	76.9464	685.5	0.24215	91.1801	0.01035	91.81	221.79	0.00061	1.25133	1.7135
17	76.1334	682.5	0.23391	84.4778	0.01104	96.32	199.58	0.00060	1.24847	1.5606
18	75.2732	680.3	0.22598	78.3875	0.01171	99.64	181.27	0.00060	1.24545	1.4489
19	74.3695	676.5	0.21912	72.6572	0.01243	101.74	165.95	0.00059	1.24229	1.3714
20	73.4224	671.4	0.21311	67.2294	0.01320	103.67	152.96	0.00057	1.23897	1.3073
21	72.4296	662.9	0.20771	61.7632	0.01409	105.90	141.77	0.00056	1.23551	1.2507
22	71.3823	652.7	0.20261	56.4471	0.01508	107.38	131.98	0.00055	1.23186	1.2100
23	70.2758	639.5	0.19784	51.1094	0.01625	108.22	123.28	0.00053	1.22801	1.1840
24	69.0976	624.6	0.19317	45.9266	0.01758	108.47	115.43	0.00051	1.22392	1.1686
25	67.8400	607.5	0.18853	40.8465	0.01914	108.19	108.27	0.00049	1.21956	1.1634
26	66.4945	588.0	0.18367	35.8550	0.02100	107.37	101.65	0.00047	1.21492	1.1689
27	65.0436	567.2	0.17842	31.0906	0.02319	106.09	95.46	0.00045	1.20992	1.1832
28	63.4743	544.2	0.17267	26.4928	0.02588	104.38	89.61	0.00042	1.20453	1.2090
29	61.7543	519.4	0.16625	22.1397	0.02921	102.20	83.98	0.00039	1.19864	1.2466
30	59.8446	491.1	0.15926	17.9155	0.03375	99.54	78.48	0.00036	1.19212	1.3066
31	57.6930	460.7	0.15142	14.0151	0.03987	95.97	73.03	0.00033	1.18480	1.3977
32	55.2006	426.7	0.14257	10.3835	0.04897	91.79	67.50	0.00029	1.17637	1.5366
33	52.2285	389.7	0.13249	7.1714	0.06313	87.82	61.76	0.00025	1.16636	1.7303
34	48.4961	349.2	0.12049	4.4390	0.08778	84.92	55.59	0.00021	1.15388	2.0066
35	43.4287	304.9	0.10580	2.2862	0.13846	82.44	48.63	0.00016	1.13707	2.4902
36	36.3370	270.5	0.09025	1.0885	0.22002	80.21	40.96	0.00013	1.11384	3.0388
37	28.9056	260.6	0.08339	0.8378	0.21056	70.15	34.76	0.00016	1.08985	2.7180
38	24.1693	270.4	0.08095	0.9366	0.15110	60.34	31.40	0.00022	1.07475	2.1255
39	21.2266	284.0	0.08025	1.0722	0.11220	54.70	29.55	0.00029	1.06544	1.7216
40	19.2072	298.3	0.07980	1.1871	0.08935	51.56	28.51	0.00036	1.05909	1.4742
42	16.5171	327.5	0.07896	1.3614	0.06444	48.56	27.62	0.00050	1.05066	1.2002
44	14.7578	355.7	0.07799	1.4784	0.05155	47.51	27.48	0.00063	1.04518	1.0604
46	13.4303	384.8	0.07688	1.5698	0.04329	47.26	27.66	0.00076	1.04106	0.9748
48	12.3903	413.5	0.07582	1.6397	0.03762	47.48	28.01	0.00089	1.03784	0.9179
50	11.5426	441.8	0.07482	1.6949	0.03344	47.96	28.47	0.00101	1.03522	0.8771
52	10.8320	469.7	0.07386	1.7393	0.03022	48.62	29.00	0.00114	1.03302	0.8466
54	10.2239	497.6	0.07292	1.7759	0.02765	49.42	29.56	0.00126	1.03115	0.8230
56	9.6956	525.2	0.07193	1.8060	0.02553	50.30	30.15	0.00139	1.02952	0.8037
58	9.2297	553.0	0.07102	1.8315	0.02377	51.25	30.76	0.00152	1.02809	0.7888
60	8.8146	581.0	0.07008	1.8533	0.02228	52.34	31.37	0.00165	1.02681	0.7758
65	7.9467	653.4	0.06750	1.8955	0.01935	55.46	32.93	0.00199	1.02415	0.7506
70	7.2548	730.1	0.06462	1.9256	0.01719	58.75	34.50	0.00232	1.02203	0.7369
75	6.6864	812.3	0.06155	1.9479	0.01551	61.79	36.06	0.00264	1.02029	0.7354
80	6.2088	900.7	0.05839	1.9647	0.01417	65.35	37.60	0.00297	1.01883	0.7344
85	5.8005	995.4	0.05529	1.9776	0.01307	70.09	39.12	0.00334	1.01759	0.7261
90	5.4465	1096.1	0.05234	1.9877	0.01214	74.84	40.61	0.00372	1.01651	0.7223
95	5.1361	1202.3	0.04961	1.9958	0.01135	79.60	42.08	0.00409	1.01556	0.7214
100	4.8612	1312.9	0.04714	2.0021	0.01066	90.72	45.98	0.00480	1.01472	0.7096
120	4.0150	1780.4	0.03986	2.0171	0.00863	117.28	54.04	0.00685	1.01215	0.7076
140	3.4275	2230.4	0.03625	2.0238	0.00728	136.14	59.21	0.00881	1.01037	0.7059
160	2.9935	2623.4	0.03489	2.0266	0.00631	148.39	63.09	0.01078	1.00905	0.7036
180	2.6590	2955.2	0.03482	2.0275	0.00558	155.90	66.32	0.01281	1.00803	0.7008
200	2.3927	3240.7	0.03540	2.0275	0.00500	160.60	69.20	0.01492	1.00723	0.6981
220	2.1754	3498.9	0.03626	2.0270	0.00453	163.94	71.88	0.01710	1.00657	0.6955
240	1.9948	3745.8	0.03715	2.0262	0.00415	166.87	74.44	0.01938	1.00602	0.6933
260	1.8420	3991.9	0.03793	2.0253	0.00383	169.88	76.92	0.02174	1.00556	0.6915
280	1.7112	4242.3	0.03858	2.0243	0.00355	173.18	79.35	0.02419	1.00517	0.6900
300	1.5978	4499.8	0.03907	2.0236	0.00331	176.79	81.75	0.02674	1.00482	0.6887
350	1.3710	5171.6	0.03976	2.0213	0.00284	187.16	87.64	0.03351	1.00414	0.6868
400	1.2008	5872.5	0.04000	2.0192	0.00248	198.62	93.45	0.04086	1.00362	0.6857
450	1.0683	6593.0	0.04003	2.0175	0.00221	210.65	99.20	0.04879	1.00322	0.6850
500	0.9622	7318.3	0.04001	2.0160	0.00199	222.74	104.88	0.05732	1.00290	0.6846
550	0.8753	8050.4	0.03994	2.0147	0.00181	234.89	110.51	0.06643	1.00264	0.6842
600	0.8028	8787.6	0.03985	2.0136	0.00166	247.05	116.07	0.07610	1.00242	0.6840
700	0.6887	10283.2	0.03960	2.0119	0.00142	271.42	127.01	0.09711	1.00208	0.6837
800	0.6030	11818.0	0.03922	2.0105	0.00124	296.08	137.68	0.12024	1.00182	0.6836
900	0.5363	13403.1	0.03873	2.0094	0.00111	321.18	148.11	0.14541	1.00162	0.6837
1000	0.4829	15045.8	0.03815	2.0085	0.00100	440.56	158.29	0.21918	1.00146	0.5384
1200	0.4027	18505.6	0.03682	2.0071	0.00083	511.13	178.00	0.29733	1.00121	0.5352
1400	0.3453	22192.8	0.03542	2.0061	0.00071	583.38	196.92	0.38480	1.00104	0.5335
1600	0.3023	26036.4	0.03416	2.0053	0.00062	656.06	215.17	0.48142	1.00091	0.5323
1800	0.2688	30059.0	0.03299	2.0047	0.00055	731.13	232.83	0.58781	1.00081	0.5306
2000	0.2419	34497.9	0.03163	2.0039	0.00050	817.64	249.97	0.70681	1.00073	0.5263
2500	0.1931	51309.9	0.02564	1.9978	0.00040	1232.26	291.26	1.11812	1.00058	0.4856
3000	0.1586	91109.1	0.01664	1.9679	0.00034	2485.57	332.94	1.82796	1.00048	0.4134

* TWO-PHASE BOUNDARY

Table 4.1 (continued)
THERMODYNAMIC PROPERTIES OF PARAHYDROGEN (ISOBARS, SI UNITS)

2.50 MPA ISOBAR

TEMPERATURE	VOLUME	ISOTHERM DERIVATIVE	ISOCHORE DERIVATIVE	INTERNAL ENERGY	ENTHALPY	ENTROPY	CV	CP	VELOCITY OF SOUND
DEG. K	CU M/KG	CU M-MPA/KG	MPA/K	KJ/KG-K	KJ/KG	KJ/KG-K	KJ /	KG-K	M/S
* 14.615	0.01276	1.3311	0.9438	-307.8	-275.9	5.024	4.79	6.38	1332
15	0.01280	1.3125	0.9471	-305.4	-273.4	5.193	4.87	6.55	1329
16	0.01293	1.2382	0.9495	-299.0	-266.6	5.630	5.05	7.00	1310
17	0.01306	1.1575	0.9442	-292.1	-259.4	6.068	5.23	7.46	1285
18	0.01320	1.0930	0.9296	-284.8	-251.8	6.505	5.38	7.86	1264
19	0.01336	1.0271	0.9152	-277.1	-243.7	6.942	5.53	8.29	1241
20	0.01352	0.9641	0.8999	-269.0	-235.2	7.379	5.66	8.73	1220
21	0.01370	0.9029	0.8836	-260.5	-226.2	7.816	5.77	9.18	1198
22	0.01389	0.8411	0.8655	-251.5	-216.8	8.254	5.87	9.65	1176
23	0.01410	0.7795	0.8456	-242.1	-206.9	8.694	5.96	10.16	1152
24	0.01432	0.7170	0.8236	-232.3	-196.5	9.138	6.04	10.70	1127
25	0.01457	0.6554	0.7993	-221.9	-185.5	9.586	6.11	11.28	1100
26	0.01484	0.5935	0.7725	-211.0	-173.9	10.041	6.16	11.92	1072
27	0.01514	0.5348	0.7431	-199.5	-161.6	10.503	6.21	12.60	1042
28	0.01548	0.4768	0.7109	-187.4	-148.6	10.976	6.25	13.37	1009
29	0.01586	0.4193	0.6761	-174.5	-134.8	11.460	6.29	14.25	974
30	0.01630	0.3619	0.6386	-160.8	-120.1	11.960	6.33	15.31	935
31	0.01680	0.3088	0.5984	-146.2	-104.2	12.480	6.37	16.52	895
32	0.01740	0.2566	0.5555	-130.4	-86.9	13.029	6.42	18.07	850
33	0.01813	0.2071	0.5095	-113.2	-67.9	13.614	6.47	20.07	802
34	0.01904	0.1628	0.4605	-94.2	-46.6	14.249	6.53	22.59	750
35	0.02022	0.1226	0.4081	-72.9	-22.4	14.953	6.61	26.06	695
36	0.02184	0.0888	0.3529	-48.6	6.0	15.751	6.73	30.80	638
37	0.02410	0.0645	0.2985	-20.7	39.6	16.672	6.88	36.58	586
38	0.02724	0.0509	0.2453	10.1	78.2	17.701	7.05	40.39	540
39	0.03114	0.0477	0.2036	41.0	118.8	18.756	7.14	39.99	517
40	0.03530	0.0523	0.1728	68.0	156.2	19.705	7.06	35.52	513
42	0.04305	0.0690	0.1338	110.7	218.3	21.221	6.91	27.10	520
44	0.04977	0.0872	0.1111	142.7	267.2	22.358	6.77	22.20	534
46	0.05576	0.1048	0.0962	169.0	308.4	23.276	6.69	19.31	550
48	0.06125	0.1213	0.0855	192.0	345.1	24.057	6.64	17.49	565
50	0.06631	0.1365	0.0774	212.8	378.6	24.739	6.63	16.28	579
52	0.07120	0.1512	0.0713	232.3	410.3	25.363	6.63	15.49	594
54	0.07589	0.1653	0.0660	250.9	440.6	25.934	6.65	14.83	607
56	0.08040	0.1787	0.0614	268.9	469.9	26.467	6.67	14.31	619
58	0.08478	0.1916	0.0576	286.2	498.1	26.962	6.69	13.92	631
60	0.08905	0.2042	0.0544	303.0	525.7	27.429	6.73	13.62	643
65	0.09936	0.2339	0.0478	344.0	592.4	28.498	6.87	13.15	669
70	0.10928	0.2621	0.0428	384.4	657.6	29.464	7.09	12.94	692
75	0.11891	0.2890	0.0389	424.9	722.2	30.355	7.37	12.92	712
80	0.12832	0.3151	0.0357	466.2	787.0	31.192	7.70	13.03	730
85	0.13758	0.3403	0.0330	508.7	852.6	31.987	8.08	13.23	747
90	0.14670	0.3650	0.0308	552.7	919.4	32.751	8.48	13.50	762
95	0.15572	0.3893	0.0288	598.4	987.7	33.489	8.90	13.81	777
100	0.16465	0.4131	0.0271	646.0	1057.6	34.206	9.33	14.15	792
120	0.19970	0.5052	0.0220	855.1	1354.3	36.906	10.88	15.45	847
140	0.23411	0.5944	0.0185	1087.5	1672.7	39.358	11.86	16.30	904
160	0.26813	0.6818	0.0161	1332.2	2002.5	41.560	12.25	16.60	961
180	0.30190	0.7680	0.0142	1579.4	2334.1	43.513	12.22	16.51	1019
200	0.33550	0.8533	0.0127	1823.1	2661.8	45.239	11.97	16.23	1076
220	0.36897	0.9381	0.0115	2060.6	2983.0	46.770	11.65	15.89	1131
240	0.40235	1.0225	0.0105	2291.5	3297.4	48.138	11.35	15.56	1184
260	0.43566	1.1065	0.0097	2516.6	3605.8	49.372	11.09	15.29	1235
280	0.46892	1.1902	0.0090	2737.0	3909.3	50.497	10.89	15.07	1284
300	0.50216	1.2740	0.0084	2954.2	4209.6	51.533	10.74	14.91	1330
350	0.58503	1.4821	0.0072	3485.8	4948.3	53.810	10.52	14.68	1438
400	0.66779	1.6896	0.0063	4009.9	5679.4	55.764	10.43	14.58	1537
450	0.75046	1.8967	0.0056	4531.0	6407.1	57.479	10.41	14.55	1628
500	0.83309	2.1036	0.0050	5052.0	7134.7	59.011	10.40	14.54	1715
550	0.91568	2.3103	0.0046	5572.4	7861.6	60.397	10.41	14.55	1796
600	0.99825	2.5169	0.0042	6093.7	8589.3	61.662	10.43	14.56	1875
700	1.16334	2.9299	0.0036	7139.2	10047.5	63.911	10.48	14.61	2021
800	1.32839	3.3427	0.0031	8192.0	11513.0	65.866	10.57	14.70	2156
900	1.49341	3.7553	0.0028	9255.6	12989.2	67.603	10.70	14.83	2281
1000	1.65842	4.1680	0.0025	10333.0	14479.1	69.176	10.86	14.99	2398
1200	1.98841	4.9931	0.0021	12543.5	17514.5	71.929	11.24	15.37	2612
1400	2.31838	5.8180	0.0018	14836.3	20632.2	74.336	11.68	15.80	2806
1600	2.64836	6.6430	0.0016	17215.4	23836.3	76.453	12.10	16.23	2984
1800	2.97840	7.4679	0.0014	19683.7	27129.7	78.354	12.52	16.65	3151
2000	3.30878	8.2928	0.0013	22253.0	30524.9	80.209	13.04	17.19	3306
2500	4.14356	10.3550	0.0010	29399.4	39758.3	86.665	15.89	20.25	3632
3000	5.03682	12.4171	0.0008	39266.7	51858.8	94.424	24.11	29.55	3898

* TWO-PHASE BOUNDARY

Table 4.1 (continued)
THERMODYNAMIC PROPERTIES OF PARAHYDROGEN (ISOBARS, SI UNITS)

2.50 MPA ISOBAR

TEMPERATURE DEG. K	DENSITY KG/CU M	V(DH/DV)$_P$ KJ/KG	V(DP/DU)$_V$ MPA-CU M/KJ X 10^2	-V(DP/DV)$_T$ MPA	-(DV/DT)$_P$/V 1/DEG. K	THERMAL CONDUCTIVITY W/K-M X 10^3	VISCOSITY KG/M-S X 10^7	THERMAL DIFFUSIVITY SQ M/HR	DIELECTRIC CONSTANT	PRANDTL NUMBER
* 14.615	78.3836	705.3	0.25148	104.3378	0.00905	82.46	270.30	0.00059	1.25639	2.0913
15	78.1041	708.6	0.24920	102.5141	0.00924	85.73	257.35	0.00060	1.25541	1.9652
16	77.3614	706.3	0.24281	95.7904	0.00991	92.48	228.86	0.00061	1.25279	1.7326
17	76.5720	700.2	0.23596	88.6329	0.01065	97.14	205.81	0.00061	1.25001	1.5803
18	75.7402	700.3	0.22800	82.7863	0.01123	100.60	186.86	0.00061	1.24709	1.4608
19	74.8684	696.8	0.22111	76.9010	0.01190	102.83	171.05	0.00060	1.24403	1.3793
20	73.9536	691.6	0.21508	71.2964	0.01262	104.90	157.65	0.00058	1.24083	1.3118
21	72.9991	684.8	0.20968	65.9100	0.01341	107.28	146.15	0.00058	1.23750	1.2507
22	71.9949	675.5	0.20462	60.5553	0.01429	108.91	136.11	0.00056	1.23399	1.2066
23	70.9403	664.1	0.19987	55.2969	0.01529	109.91	127.25	0.00055	1.23032	1.1758
24	69.8221	650.3	0.19526	50.0652	0.01645	110.33	119.28	0.00053	1.22643	1.1565
25	68.6372	634.8	0.19073	44.9873	0.01777	110.24	112.05	0.00051	1.22232	1.1464
26	67.3769	617.1	0.18610	39.9903	0.01932	109.64	105.41	0.00049	1.21796	1.1460
27	66.0317	599.0	0.18121	35.3142	0.02104	108.61	99.24	0.00047	1.21332	1.1516
28	64.5913	579.1	0.17599	30.7942	0.02309	107.17	93.46	0.00045	1.20836	1.1658
29	63.0401	557.1	0.17040	26.4313	0.02558	105.35	87.98	0.00042	1.20304	1.1900
30	61.3589	532.4	0.16439	22.2056	0.02876	103.15	82.72	0.00040	1.19728	1.2278
31	59.5118	507.4	0.15779	18.3786	0.03256	100.15	77.61	0.00037	1.19098	1.2803
32	57.4635	479.8	0.15060	14.7479	0.03767	96.76	72.58	0.00034	1.18402	1.3555
33	55.1554	449.9	0.14279	11.4246	0.04460	92.91	67.56	0.00030	1.17621	1.4591
34	52.5200	419.4	0.13424	8.5493	0.05386	89.17	62.51	0.00027	1.16734	1.5836
35	49.4439	387.1	0.12479	6.0624	0.06732	86.23	57.37	0.00024	1.15704	1.7338
36	45.7965	355.0	0.11450	4.0682	0.08675	83.05	52.11	0.00021	1.14490	1.9321
37	41.4871	327.9	0.10455	2.6760	0.11154	79.63	46.84	0.00019	1.13068	2.1516
38	36.7123	307.7	0.09474	1.8688	0.13127	75.30	42.00	0.00018	1.11506	2.2526
39	32.1093	301.1	0.08882	1.5331	0.13282	70.09	38.12	0.00020	1.10014	2.1749
40	28.3274	304.5	0.08639	1.4810	0.11665	64.70	35.39	0.00023	1.08800	1.9429
42	23.2312	324.6	0.08336	1.6025	0.08349	57.52	32.25	0.00033	1.07177	1.5196
44	20.0928	350.1	0.08161	1.7516	0.06342	54.07	30.78	0.00044	1.06187	1.2640
46	17.9350	377.3	0.08014	1.8787	0.05118	52.45	30.17	0.00055	1.05510	1.1107
48	16.3274	405.2	0.07877	1.9804	0.04315	51.80	30.01	0.00065	1.05007	1.0132
50	15.0882	432.8	0.07748	2.0578	0.03762	51.72	30.14	0.00076	1.04618	0.9490
52	14.0443	461.5	0.07656	2.1241	0.03356	51.98	30.42	0.00086	1.04296	0.9062
54	13.1774	489.8	0.07533	2.1785	0.03028	52.45	30.80	0.00097	1.04027	0.8710
56	12.4383	517.8	0.07408	2.2229	0.02764	53.07	31.25	0.00107	1.03799	0.8427
58	11.7954	545.9	0.07303	2.2606	0.02550	53.81	31.75	0.00118	1.03600	0.8214
60	11.2294	574.3	0.07196	2.2926	0.02372	54.72	32.28	0.00129	1.03425	0.8035
65	10.0642	647.3	0.06911	2.3545	0.02031	57.53	33.69	0.00157	1.03066	0.7697
70	9.1511	724.6	0.06603	2.3985	0.01786	60.58	35.15	0.00184	1.02785	0.7508
75	8.4099	807.3	0.06278	2.4308	0.01600	63.37	36.63	0.00210	1.02557	0.7469
80	7.7927	896.1	0.05948	2.4551	0.01454	66.80	38.12	0.00237	1.02368	0.7434
85	7.2686	991.2	0.05625	2.4738	0.01335	71.46	39.59	0.00267	1.02207	0.7331
90	6.8166	1092.3	0.05320	2.4883	0.01236	76.14	41.05	0.00298	1.02069	0.7279
95	6.4219	1198.8	0.05038	2.4998	0.01152	80.84	42.49	0.00328	1.01948	0.7260
100	6.0735	1309.9	0.04783	2.5089	0.01080	91.16	46.12	0.00382	1.01842	0.7157
120	5.0075	1779.9	0.04030	2.5300	0.00868	117.44	54.10	0.00546	1.01517	0.7118
140	4.2715	2232.3	0.03659	2.5391	0.00730	136.23	59.26	0.00704	1.01293	0.7090
160	3.7295	2627.6	0.03517	2.5428	0.00632	148.48	63.14	0.00863	1.01128	0.7059
180	3.3124	2961.4	0.03506	2.5438	0.00558	156.01	66.38	0.01027	1.01002	0.7026
200	2.9807	3248.5	0.03562	2.5435	0.00500	160.74	69.27	0.01196	1.00901	0.6995
220	2.7103	3507.9	0.03646	2.5425	0.00453	164.11	71.96	0.01372	1.00819	0.6966
240	2.4854	3755.7	0.03733	2.5412	0.00414	167.06	74.52	0.01555	1.00751	0.6942
260	2.2954	4002.6	0.03810	2.5397	0.00382	170.09	77.00	0.01745	1.00693	0.6922
280	2.1326	4253.6	0.03874	2.5382	0.00354	173.40	79.44	0.01942	1.00644	0.6906
300	1.9914	4511.9	0.03921	2.5371	0.00330	177.03	81.84	0.02147	1.00601	0.6892
350	1.7093	5184.5	0.03988	2.5333	0.00283	187.44	87.75	0.02690	1.00516	0.6870
400	1.4975	5886.0	0.04011	2.5301	0.00248	198.93	93.58	0.03280	1.00452	0.6859
450	1.3325	6606.8	0.04013	2.5274	0.00220	211.00	99.34	0.03917	1.00402	0.6851
500	1.2004	7332.5	0.04009	2.5250	0.00198	223.12	105.05	0.04602	1.00362	0.6846
550	1.0921	8064.8	0.04002	2.5230	0.00180	235.31	110.69	0.05333	1.00329	0.6843
600	1.0018	8802.3	0.03992	2.5213	0.00165	247.50	116.27	0.06109	1.00302	0.6840
700	0.8596	10298.1	0.03966	2.5185	0.00142	271.95	127.24	0.07794	1.00259	0.6837
800	0.7528	11833.2	0.03928	2.5163	0.00124	296.68	137.95	0.09651	1.00227	0.6836
900	0.6696	13418.5	0.03878	2.5146	0.00110	321.86	148.41	0.11671	1.00202	0.6837
1000	0.6030	15061.5	0.03819	2.5132	0.00099	440.56	158.63	0.17552	1.00182	0.5396
1200	0.5029	18521.9	0.03685	2.5111	0.00083	511.13	178.40	0.23807	1.00152	0.5364
1400	0.4313	22209.3	0.03545	2.5095	0.00071	583.37	197.38	0.30807	1.00130	0.5347
1600	0.3776	26051.3	0.03419	2.5083	0.00062	655.96	215.68	0.38537	1.00114	0.5336
1800	0.3358	30062.4	0.03303	2.5074	0.00055	730.57	233.39	0.47039	1.00101	0.5320
2000	0.3022	34457.4	0.03171	2.5063	0.00050	815.37	250.59	0.56514	1.00091	0.5282
2500	0.2413	50601.4	0.02608	2.4990	0.00040	1204.64	291.96	0.88745	1.00073	0.4907
3000	0.1985	87415.3	0.01741	2.4653	0.00034	2345.90	333.45	1.43955	1.00060	0.4200

* TWO-PHASE BOUNDARY

THERMODYNAMIC PROPERTIES OF PARAHYDROGEN (ISOBARS, SI UNITS)

3.00 MPA ISOBAR

TEMPERATURE DEG. K	VOLUME CU M/KG	ISOTHERM DERIVATIVE CU M-MPA/KG	ISOCHORE DERIVATIVE MPA/K	INTERNAL ENERGY KJ/KG-K	ENTHALPY KJ/KG	ENTROPY KJ/KG-K	CV KJ /	CP KG-K	VELOCITY OF SOUND M/S
* 14.772	0.01272	1.3623	0.9470	-307.5	-269.4	5.036	4.81	6.38	1345
15	0.01274	1.3585	0.9492	-306.1	-267.9	5.135	4.86	6.47	1345
16	0.01286	1.2794	0.9544	-299.8	-261.2	5.568	5.05	6.93	1326
17	0.01299	1.2075	0.9510	-293.0	-254.1	6.000	5.22	7.37	1306
18	0.01313	1.1408	0.9404	-285.9	-246.5	6.433	5.37	7.78	1285
19	0.01327	1.0731	0.9266	-278.3	-238.5	6.865	5.52	8.20	1262
20	0.01343	1.0093	0.9118	-270.4	-230.1	7.296	5.65	8.62	1241
21	0.01360	0.9488	0.8960	-262.1	-221.3	7.726	5.76	9.05	1221
22	0.01378	0.8875	0.8787	-253.3	-212.0	8.157	5.86	9.50	1199
23	0.01397	0.8271	0.8595	-244.2	-202.3	8.590	5.95	9.97	1177
24	0.01419	0.7655	0.8386	-234.6	-192.1	9.024	6.03	10.47	1153
25	0.01442	0.7053	0.8155	-224.6	-181.4	9.462	6.10	11.00	1128
26	0.01467	0.6461	0.7904	-214.1	-170.1	9.904	6.16	11.57	1102
27	0.01494	0.5884	0.7629	-203.0	-158.2	10.352	6.21	12.17	1074
28	0.01525	0.5315	0.7331	-191.4	-145.7	10.806	6.25	12.84	1044
29	0.01559	0.4762	0.7011	-179.3	-132.5	11.270	6.29	13.57	1013
30	0.01597	0.4222	0.6670	-166.4	-118.5	11.744	6.33	14.39	980
31	0.01640	0.3688	0.6309	-152.9	-103.7	12.230	6.36	15.36	943
32	0.01689	0.3184	0.5926	-138.4	-87.8	12.735	6.40	16.47	905
33	0.01747	0.2698	0.5524	-123.1	-70.7	13.261	6.44	17.82	864
34	0.01815	0.2258	0.5101	-106.5	-52.1	13.815	6.48	19.38	822
35	0.01896	0.1858	0.4661	-88.7	-31.8	14.403	6.53	21.25	778
36	0.01997	0.1498	0.4207	-69.4	-9.4	15.034	6.59	23.55	732
37	0.02123	0.1196	0.3747	-48.3	15.4	15.714	6.66	26.24	686
38	0.02281	0.0967	0.3293	-25.5	43.0	16.450	6.75	28.93	644
39	0.02479	0.0809	0.2866	-1.2	73.1	17.233	6.84	31.20	607
40	0.02717	0.0718	0.2490	23.6	105.1	18.042	6.91	32.43	580
42	0.03273	0.0711	0.1916	69.8	168.0	19.579	6.93	30.16	556
44	0.03841	0.0830	0.1546	108.5	223.9	20.876	6.86	25.57	556
46	0.04369	0.0984	0.1306	140.0	271.1	21.928	6.79	22.00	565
48	0.04856	0.1144	0.1138	166.8	312.5	22.810	6.73	19.56	576
50	0.05311	0.1299	0.1015	190.5	349.8	23.572	6.70	17.89	589
52	0.05740	0.1449	0.0920	212.2	384.4	24.250	6.69	16.70	601
54	0.06150	0.1593	0.0844	232.4	416.9	24.863	6.70	15.83	614
56	0.06543	0.1732	0.0782	251.8	448.1	25.431	6.72	15.18	626
58	0.06925	0.1867	0.0731	270.3	478.0	25.956	6.74	14.70	638
60	0.07295	0.1998	0.0686	288.2	507.0	26.448	6.77	14.31	649
65	0.08185	0.2307	0.0598	331.2	576.7	27.564	6.91	13.65	675
70	0.09035	0.2599	0.0532	373.0	644.1	28.562	7.12	13.33	698
75	0.09858	0.2877	0.0480	414.7	710.4	29.478	7.39	13.23	718
80	0.10659	0.3145	0.0439	456.9	776.7	30.333	7.72	13.29	736
85	0.11445	0.3405	0.0405	500.2	843.5	31.143	8.09	13.45	752
90	0.12217	0.3658	0.0376	544.8	911.3	31.918	8.50	13.69	768
95	0.12980	0.3905	0.0351	591.1	980.5	32.666	8.92	13.97	782
100	0.13733	0.4149	0.0330	639.1	1051.1	33.391	9.34	14.29	797
120	0.16683	0.5085	0.0266	849.8	1350.3	36.113	10.90	15.55	852
140	0.19571	0.5987	0.0224	1083.1	1670.2	38.577	11.87	16.36	908
160	0.22421	0.6868	0.0194	1328.5	2001.1	40.786	12.25	16.65	966
180	0.25246	0.7736	0.0171	1576.2	2333.6	42.745	12.22	16.55	1024
200	0.28054	0.8595	0.0153	1820.3	2661.9	44.474	11.98	16.26	1080
220	0.30851	0.9446	0.0139	2058.1	2983.7	46.007	11.66	15.91	1135
240	0.33638	1.0293	0.0127	2289.4	3298.5	47.377	11.35	15.58	1189
260	0.36419	1.1135	0.0117	2514.7	3607.3	48.613	11.10	15.31	1239
280	0.39195	1.1975	0.0108	2735.3	3911.1	49.739	10.89	15.09	1288
300	0.41969	1.2816	0.0101	2952.7	4211.8	50.776	10.74	14.92	1334
350	0.48881	1.4900	0.0086	3484.6	4951.1	53.055	10.52	14.68	1442
400	0.55781	1.6977	0.0075	4009.0	5682.4	55.009	10.43	14.59	1540
450	0.62674	1.9049	0.0067	4530.3	6410.5	56.725	10.41	14.56	1632
500	0.69561	2.1119	0.0060	5051.3	7138.2	58.258	10.41	14.54	1718
550	0.76446	2.3187	0.0055	5571.9	7865.3	59.644	10.41	14.55	1800
600	0.83328	2.5254	0.0050	6093.3	8593.1	60.909	10.43	14.56	1878
700	0.97087	2.9385	0.0043	7138.9	10051.5	63.158	10.48	14.61	2024
800	1.10842	3.3513	0.0038	8191.8	11517.1	65.114	10.57	14.70	2159
900	1.24595	3.7640	0.0033	9255.4	12993.3	66.851	10.70	14.83	2284
1000	1.38347	4.1767	0.0030	10332.9	14483.3	68.423	10.86	14.99	2401
1200	1.65847	5.0018	0.0025	12543.4	17518.8	71.177	11.24	15.37	2615
1400	1.93345	5.8268	0.0021	14836.2	20636.6	73.584	11.68	15.80	2808
1600	2.20843	6.6518	0.0019	17215.2	23840.5	75.700	12.10	16.23	2986
1800	2.48346	7.4767	0.0017	19682.9	27133.3	77.602	12.52	16.65	3153
2000	2.75875	8.3016	0.0015	22249.8	30526.1	79.455	13.02	17.16	3308
2500	3.45369	10.3638	0.0012	29351.7	39712.7	85.891	15.69	20.03	3637
3000	4.19314	12.4259	0.0010	38968.3	51547.7	93.555	23.26	28.58	3904

* TWO-PHASE BOUNDARY

Table 4.1 (continued)
THERMODYNAMIC PROPERTIES OF PARAHYDROGEN (ISOBARS, SI UNITS)

3.00 MPA ISOBAR

TEMPERATURE DEG. K	DENSITY KG/CU M	V(DH/DV)$_P$ KJ/KG	V(DP/DU)$_V$ MPA-CU M/KJ X 10^2	-V(DP/DV)$_T$ MPA	-(DV/DT)$_P$/V 1/DEG. K	THERMAL CONDUCTIVITY W/K-M X 10^3	VISCOSITY KG/M-S X 10^7	THERMAL DIFFUSIVITY SQ M/HR	DIELECTRIC CONSTANT	PRANDTL NUMBER
* 14.772	78.6428	721.9	0.25043	107.1338	0.00884	84.29	273.40	0.00060	1.25731	2.0698
15	78.4822	726.8	0.24911	106.6179	0.00890	86.21	265.60	0.00061	1.25674	1.9933
16	77.7515	722.4	0.24323	99.4716	0.00959	93.11	235.84	0.00062	1.25416	1.7554
17	76.9926	720.1	0.23670	92.9694	0.01023	97.93	212.07	0.00062	1.25149	1.5952
18	76.1868	718.9	0.22967	86.9158	0.01087	101.52	192.47	0.00062	1.24866	1.4746
19	75.3438	715.1	0.22288	80.8495	0.01146	103.89	176.15	0.00061	1.24570	1.3896
20	74.4611	710.3	0.21687	75.1511	0.01213	106.09	162.34	0.00060	1.24261	1.3188
21	73.5386	704.5	0.21146	69.7716	0.01284	108.60	150.50	0.00059	1.23938	1.2539
22	72.5731	696.3	0.20646	64.4117	0.01364	110.37	140.21	0.00058	1.23601	1.2066
23	71.5604	686.3	0.20172	59.1907	0.01452	111.51	131.14	0.00056	1.23248	1.1719
24	70.4955	673.7	0.19717	53.9663	0.01554	112.09	123.04	0.00055	1.22877	1.1492
25	69.3718	659.9	0.19271	48.9308	0.01667	112.17	115.72	0.00053	1.22487	1.1347
26	68.1839	644.6	0.18824	44.0539	0.01794	111.75	109.03	0.00051	1.22075	1.1284
27	66.9234	628.3	0.18359	39.3788	0.01937	110.92	102.85	0.00049	1.21640	1.1286
28	65.5849	610.4	0.17873	34.8590	0.02103	109.72	97.09	0.00047	1.21178	1.1361
29	64.1571	591.2	0.17362	30.5492	0.02295	108.16	91.68	0.00045	1.20687	1.1499
30	62.6303	570.4	0.16832	26.4435	0.02523	106.27	86.54	0.00042	1.20163	1.1717
31	60.9904	547.6	0.16254	22.4936	0.02805	103.67	81.63	0.00040	1.19603	1.2093
32	59.2057	523.9	0.15635	18.8488	0.03144	100.76	76.87	0.00037	1.18994	1.2565
33	57.2569	498.5	0.14980	15.4471	0.03576	97.54	72.21	0.00034	1.18332	1.3195
34	55.1099	472.8	0.14279	12.4434	0.04099	94.03	67.64	0.00032	1.17606	1.3942
35	52.7287	446.7	0.13536	9.7964	0.04758	90.55	63.12	0.00029	1.16804	1.4814
36	50.0718	420.0	0.12752	7.5024	0.05608	87.76	58.66	0.00027	1.15913	1.5742
37	47.1064	394.5	0.11939	5.6338	0.06651	84.75	54.27	0.00025	1.14925	1.6802
38	43.8311	372.3	0.11133	4.2386	0.07769	81.47	50.05	0.00023	1.13840	1.7772
39	40.3347	354.9	0.10392	3.2613	0.08789	77.98	46.15	0.00022	1.12689	1.8463
40	36.8045	344.0	0.09785	2.6409	0.09429	74.36	42.76	0.00022	1.11536	1.8651
42	30.5512	341.8	0.09054	2.1711	0.08823	67.12	37.90	0.00026	1.09513	1.7028
44	26.0341	357.2	0.08655	2.1601	0.07158	61.74	35.12	0.00033	1.08067	1.4545
46	22.8887	379.4	0.08406	2.2517	0.05799	58.56	33.56	0.00042	1.07069	1.2609
48	20.5912	404.6	0.08213	2.3549	0.04834	56.82	32.72	0.00051	1.06344	1.1262
50	18.8289	431.0	0.08045	2.4449	0.04152	55.98	32.34	0.00060	1.05790	1.0337
52	17.4203	458.3	0.07892	2.5248	0.03644	55.67	32.25	0.00069	1.05349	0.9676
54	16.2609	486.0	0.07748	2.5911	0.03258	55.74	32.37	0.00078	1.04986	0.9194
56	15.2835	514.0	0.07614	2.6476	0.02953	56.05	32.62	0.00087	1.04682	0.8831
58	14.4407	542.2	0.07508	2.6965	0.02711	56.56	32.95	0.00096	1.04420	0.8564
60	13.7071	570.7	0.07392	2.7380	0.02507	57.27	33.36	0.00105	1.04192	0.8333
65	12.2174	643.9	0.07078	2.8188	0.02120	59.70	34.55	0.00129	1.03730	0.7899
70	11.0676	721.4	0.06747	2.8764	0.01848	62.50	35.87	0.00152	1.03375	0.7653
75	10.1443	804.3	0.06404	2.9187	0.01645	65.00	37.26	0.00174	1.03090	0.7586
80	9.3815	893.3	0.06059	2.9505	0.01487	68.30	38.67	0.00197	1.02856	0.7524
85	8.7377	988.6	0.05723	2.9749	0.01360	72.86	40.09	0.00223	1.02658	0.7401
90	8.1851	1090.0	0.05406	2.9939	0.01256	77.47	41.51	0.00249	1.02488	0.7334
95	7.7044	1196.8	0.05115	3.0089	0.01168	82.10	42.92	0.00275	1.02341	0.7305
100	7.2815	1308.1	0.04853	3.0208	0.01092	91.66	46.29	0.00317	1.02211	0.7217
120	5.9941	1780.6	0.04075	3.0477	0.00873	117.65	54.18	0.00455	1.01818	0.7159
140	5.1097	2235.1	0.03692	3.0591	0.00732	136.36	59.33	0.00587	1.01548	0.7120
160	4.4602	2632.5	0.03545	3.0635	0.00632	148.59	63.21	0.00720	1.01350	0.7082
180	3.9610	2968.1	0.03530	3.0643	0.00558	156.14	66.45	0.00857	1.01199	0.7044
200	3.5645	3256.7	0.03584	3.0635	0.00499	160.90	69.35	0.00999	1.01078	0.7009
220	3.2414	3517.2	0.03666	3.0618	0.00452	164.28	72.03	0.01147	1.00980	0.6977
240	2.9728	3765.8	0.03751	3.0598	0.00414	167.25	74.60	0.01300	1.00899	0.6950
260	2.7458	4013.4	0.03827	3.0575	0.00381	170.30	77.09	0.01459	1.00830	0.6929
280	2.5514	4265.0	0.03889	3.0552	0.00354	173.63	79.53	0.01624	1.00771	0.6911
300	2.3827	4524.1	0.03936	3.0536	0.00330	177.28	81.94	0.01795	1.00720	0.6896
350	2.0458	5197.5	0.04001	3.0481	0.00283	187.71	87.86	0.02250	1.00618	0.6873
400	1.7927	5899.5	0.04022	3.0434	0.00247	199.24	93.70	0.02743	1.00541	0.6860
450	1.5956	6620.7	0.04022	3.0394	0.00220	211.34	99.49	0.03276	1.00482	0.6852
500	1.4376	7346.7	0.04018	3.0360	0.00198	223.50	105.21	0.03848	1.00434	0.6847
550	1.3081	8079.3	0.04010	3.0331	0.00180	235.73	110.87	0.04459	1.00395	0.6843
600	1.2001	8816.9	0.03999	3.0307	0.00165	247.96	116.47	0.05108	1.00362	0.6840
700	1.0300	10313.1	0.03972	3.0266	0.00142	272.48	127.48	0.06517	1.00311	0.6837
800	0.9022	11848.5	0.03933	3.0235	0.00124	297.28	138.22	0.08069	1.00272	0.6836
900	0.8026	13434.0	0.03883	3.0210	0.00110	322.53	148.71	0.09757	1.00242	0.6837
1000	0.7228	15077.2	0.03823	3.0190	0.00099	440.56	158.96	0.14642	1.00218	0.5407
1200	0.6030	18538.1	0.03688	3.0159	0.00083	511.13	178.80	0.19856	1.00182	0.5376
1400	0.5172	22225.9	0.03548	3.0137	0.00071	583.37	197.84	0.25692	1.00156	0.5360
1600	0.4528	26066.8	0.03422	3.0120	0.00062	655.89	216.20	0.32134	1.00136	0.5349
1800	0.4027	30069.6	0.03307	3.0106	0.00055	730.16	233.96	0.39214	1.00121	0.5334
2000	0.3625	34432.3	0.03178	3.0092	0.00050	813.70	251.21	0.47083	1.00109	0.5299
2500	0.2895	50083.3	0.02642	3.0008	0.00040	1184.25	292.67	0.73517	1.00087	0.4950
3000	0.2385	84685.3	0.01803	2.9634	0.00034	2242.45	334.06	1.18453	1.00072	0.4257

* TWO-PHASE BOUNDARY

Table 4.1 (continued)

THERMODYNAMIC PROPERTIES OF PARAHYDROGEN (ISOBARS, SI UNITS)

3.50 MPA ISOBAR

TEMPERATURE DEG. K	VOLUME CU M/KG	ISOTHERM DERIVATIVE CU M-MPA/KG	ISOCHORE DERIVATIVE MPA/K	INTERNAL ENERGY KJ/KG-K	ENTHALPY KJ/KG	ENTROPY KJ/KG-K	CV KJ / KG-K	CP KJ / KG-K	VELOCITY OF SOUND M/S
* 14.927	0.01267	1.3929	0.9496	-307.2	-262.9	5.047	4.83	6.38	1357
15	0.01268	1.3951	0.9505	-306.8	-262.4	5.079	4.84	6.41	1358
16	0.01280	1.3237	0.9576	-300.5	-255.7	5.507	5.04	6.85	1342
17	0.01292	1.2538	0.9572	-293.9	-248.7	5.935	5.21	7.28	1324
18	0.01305	1.1850	0.9506	-286.9	-241.2	6.364	5.37	7.71	1304
19	0.01319	1.1224	0.9370	-279.5	-233.3	6.791	5.51	8.10	1284
20	0.01334	1.0577	0.9228	-271.7	-225.0	7.217	5.64	8.50	1263
21	0.01350	0.9958	0.9075	-263.5	-216.3	7.641	5.75	8.92	1243
22	0.01368	0.9357	0.8908	-255.0	-207.2	8.066	5.86	9.35	1222
23	0.01386	0.8759	0.8726	-246.1	-197.6	8.491	5.95	9.79	1201
24	0.01406	0.8161	0.8525	-236.8	-187.6	8.917	6.03	10.25	1178
25	0.01427	0.7561	0.8307	-227.0	-177.1	9.346	6.10	10.74	1154
26	0.01451	0.6973	0.8068	-216.9	-166.1	9.777	6.16	11.26	1130
27	0.01476	0.6399	0.7809	-206.2	-154.5	10.213	6.21	11.82	1104
28	0.01504	0.5828	0.7530	-195.1	-142.4	10.653	6.26	12.42	1076
29	0.01535	0.5285	0.7232	-183.4	-129.7	11.100	6.30	13.06	1047
30	0.01569	0.4773	0.6917	-171.2	-116.3	11.554	6.33	13.73	1017
31	0.01606	0.4258	0.6585	-158.4	-102.2	12.016	6.36	14.51	985
32	0.01649	0.3763	0.6236	-145.0	-87.2	12.491	6.40	15.39	951
33	0.01697	0.3294	0.5872	-130.8	-71.4	12.980	6.43	16.37	916
34	0.01752	0.2847	0.5494	-115.8	-54.4	13.485	6.46	17.53	879
35	0.01816	0.2442	0.5103	-99.8	-36.3	14.011	6.50	18.81	841
36	0.01891	0.2077	0.4703	-83.0	-16.8	14.560	6.54	20.25	802
37	0.01979	0.1756	0.4300	-65.0	4.3	15.137	6.58	21.83	763
38	0.02084	0.1481	0.3899	-46.0	27.0	15.743	6.63	23.57	726
39	0.02209	0.1257	0.3508	-25.9	51.4	16.377	6.68	25.32	690
40	0.02356	0.1099	0.3140	-5.1	77.4	17.035	6.73	26.65	660
42	0.02717	0.0926	0.2509	37.3	132.4	18.376	6.82	27.90	615
44	0.03140	0.0917	0.2035	77.2	187.1	19.649	6.84	26.43	595
46	0.03578	0.1007	0.1699	111.5	236.8	20.753	6.83	23.72	591
48	0.04000	0.1135	0.1462	141.6	281.6	21.707	6.79	21.24	596
50	0.04401	0.1278	0.1288	168.0	322.1	22.534	6.76	19.33	605
52	0.04782	0.1421	0.1156	191.9	359.2	23.263	6.75	17.92	614
54	0.05145	0.1564	0.1052	213.9	394.0	23.919	6.75	16.86	625
56	0.05495	0.1703	0.0967	234.8	427.1	24.521	6.76	16.05	636
58	0.05832	0.1839	0.0897	254.4	458.6	25.073	6.78	15.42	647
60	0.06159	0.1970	0.0838	273.3	488.9	25.587	6.81	14.93	657
65	0.06947	0.2289	0.0725	318.4	561.5	26.751	6.94	14.14	683
70	0.07695	0.2588	0.0640	361.7	631.0	27.781	7.15	13.71	705
75	0.08415	0.2874	0.0576	404.6	699.1	28.720	7.42	13.54	724
80	0.09115	0.3148	0.0524	447.7	766.7	29.593	7.74	13.54	742
85	0.09799	0.3413	0.0482	491.7	834.7	30.417	8.11	13.66	758
90	0.10472	0.3672	0.0447	537.0	903.5	31.204	8.51	13.87	774
95	0.11134	0.3924	0.0416	583.8	973.5	31.961	8.93	14.13	788
100	0.11788	0.4172	0.0390	632.3	1044.9	32.693	9.35	14.43	802
120	0.14339	0.5120	0.0313	844.5	1346.4	35.437	10.91	15.64	857
140	0.16830	0.6032	0.0263	1078.8	1667.9	37.913	11.88	16.43	913
160	0.19285	0.6921	0.0227	1324.9	1999.9	40.129	12.26	16.70	971
180	0.21716	0.7794	0.0200	1573.1	2333.2	42.093	12.23	16.59	1028
200	0.24130	0.8657	0.0179	1817.6	2662.2	43.825	11.98	16.29	1085
220	0.26533	0.9512	0.0162	2055.8	2984.4	45.361	11.66	15.94	1140
240	0.28927	1.0361	0.0148	2287.3	3299.7	46.733	11.36	15.60	1193
260	0.31314	1.1206	0.0136	2512.9	3608.9	47.970	11.10	15.32	1244
280	0.33697	1.2048	0.0126	2733.6	3913.0	49.098	10.90	15.10	1292
300	0.36079	1.2892	0.0118	2951.2	4214.0	50.136	10.74	14.93	1339
350	0.42009	1.4979	0.0101	3483.5	4953.8	52.416	10.52	14.69	1446
400	0.47926	1.7058	0.0088	4008.1	5685.5	54.371	10.44	14.59	1544
450	0.53837	1.9132	0.0078	4529.5	6413.8	56.088	10.41	14.56	1636
500	0.59742	2.1203	0.0070	5050.7	7141.7	57.621	10.41	14.55	1722
550	0.65644	2.3272	0.0064	5571.4	7868.9	59.007	10.41	14.55	1803
600	0.71544	2.5339	0.0058	6092.9	8596.9	60.272	10.43	14.56	1881
700	0.83339	2.9470	0.0050	7138.5	10055.4	62.522	10.48	14.61	2027
800	0.95131	3.3600	0.0044	8191.5	11521.1	64.478	10.57	14.70	2161
900	1.06920	3.7727	0.0039	9255.2	12997.4	66.215	10.70	14.83	2286
1000	1.18707	4.1854	0.0035	10332.7	14487.5	67.787	10.86	14.99	2403
1200	1.42280	5.0106	0.0029	12543.3	17523.1	70.542	11.24	15.37	2617
1400	1.65850	5.8356	0.0025	14836.1	20640.9	72.948	11.68	15.80	2810
1600	1.89420	6.6606	0.0022	17215.1	23844.8	75.065	12.10	16.23	2988
1800	2.12994	7.4855	0.0019	19682.4	27137.2	76.966	12.51	16.64	3155
2000	2.36588	8.3104	0.0018	22247.4	30528.0	78.818	13.01	17.15	3310
2500	2.96107	10.3726	0.0014	29314.6	39678.3	85.239	15.53	19.86	3641
3000	3.59160	12.4347	0.0012	38736.3	51306.9	92.829	22.59	27.82	3910

* TWO-PHASE BOUNDARY

Table 4.1 (continued)
THERMODYNAMIC PROPERTIES OF PARAHYDROGEN (ISOBARS, SI UNITS)

3.50 MPA ISOBAR

TEMPERATURE DEG. K	DENSITY KG/CU M	$V(DH/DV)_P$ KJ/KG	$V(DP/DU)_V$ MPA-CU M/KJ X 10^2	$-V(DP/DV)_T$ MPA	$-(DV/DT)_P/V$ 1/DEG. K	THERMAL CONDUCTIVITY W/K-M X 10^3	VISCOSITY KG/M-S X 10^7	THERMAL DIFFUSIVITY SQ M/HR	DIELECTRIC CONSTANT	PRANDTL NUMBER
* 14.927	78.8975	738.4	0.24927	109.8930	0.00864	86.07	276.50	0.00062	1.25820	2.0499
15	78.8440	741.3	0.24890	109.9949	0.00864	86.68	273.88	0.00062	1.25802	2.0242
16	78.1394	740.2	0.24329	103.4356	0.00926	93.74	243.10	0.00063	1.25553	1.7771
17	77.3997	738.6	0.23732	97.0466	0.00986	98.70	218.42	0.00063	1.25292	1.6121
18	76.6149	736.0	0.23117	90.7901	0.01047	102.42	198.10	0.00062	1.25016	1.4905
19	75.7943	735.1	0.22439	85.0704	0.01101	104.90	181.21	0.00062	1.24728	1.3987
20	74.9428	730.5	0.21842	79.2684	0.01164	107.22	167.01	0.00061	1.24429	1.3246
21	74.0524	724.8	0.21303	73.7426	0.01231	109.86	154.84	0.00060	1.24118	1.2571
22	73.1206	717.8	0.20806	68.4215	0.01302	111.77	144.27	0.00059	1.23792	1.2063
23	72.1477	708.8	0.20339	63.1958	0.01381	113.04	134.99	0.00058	1.23452	1.1687
24	71.1269	697.9	0.19890	58.0441	0.01469	113.76	126.73	0.00056	1.23097	1.1420
25	70.0562	685.1	0.19453	52.9693	0.01568	113.99	119.30	0.00055	1.22725	1.1244
26	68.9286	671.0	0.19016	48.0622	0.01679	113.73	112.54	0.00053	1.22333	1.1146
27	67.7393	655.9	0.18568	43.3471	0.01802	113.08	106.33	0.00051	1.21922	1.1111
28	66.4831	639.0	0.18105	38.7445	0.01944	112.07	100.56	0.00049	1.21488	1.1145
29	65.1537	621.7	0.17629	34.4343	0.02100	110.73	95.18	0.00047	1.21030	1.1223
30	63.7444	604.0	0.17143	30.4265	0.02273	109.09	90.10	0.00045	1.20545	1.1340
31	62.2492	584.1	0.16625	26.5052	0.02485	106.77	85.30	0.00043	1.20033	1.1593
32	60.6468	563.1	0.16074	22.8203	0.02733	104.19	80.69	0.00040	1.19485	1.1917
33	58.9266	541.4	0.15500	19.4131	0.03025	101.38	76.24	0.00038	1.18899	1.2314
34	57.0738	518.4	0.14897	16.2475	0.03382	98.37	71.93	0.00035	1.18270	1.2818
35	55.0661	495.5	0.14263	13.4449	0.03795	95.12	67.73	0.00033	1.17591	1.3393
36	52.8898	472.8	0.13606	10.9834	0.04282	91.89	63.65	0.00031	1.16858	1.4023
37	50.5307	450.6	0.12932	8.8738	0.04845	89.25	59.69	0.00029	1.16067	1.4608
38	47.9855	429.5	0.12258	7.1055	0.05487	86.47	55.87	0.00028	1.15217	1.5228
39	45.2729	410.5	0.11603	5.6889	0.06167	83.58	52.26	0.00026	1.14317	1.5829
40	42.4430	396.0	0.10991	4.6658	0.06730	80.55	48.93	0.00026	1.13382	1.6192
42	36.8054	379.0	0.09996	3.4086	0.07362	74.55	43.44	0.00026	1.11536	1.6258
44	31.8437	379.4	0.09348	2.9211	0.06968	69.12	39.67	0.00030	1.09929	1.5170
46	27.9523	392.7	0.08907	2.8134	0.06040	64.98	37.31	0.00035	1.08680	1.3616
48	25.0017	412.5	0.08614	2.8385	0.05150	62.26	35.85	0.00042	1.07739	1.2230
50	22.7238	435.8	0.08386	2.9033	0.04437	60.62	34.97	0.00050	1.07017	1.1152
52	20.9139	460.8	0.08192	2.9720	0.03889	59.71	34.48	0.00057	1.06445	1.0348
54	19.4351	487.3	0.08018	3.0397	0.03460	59.32	34.27	0.00065	1.05980	0.9739
56	18.1994	514.3	0.07859	3.0997	0.03120	59.27	34.26	0.00073	1.05593	0.9276
58	17.1469	541.7	0.07719	3.1526	0.02846	59.48	34.40	0.00081	1.05263	0.8917
60	16.2361	569.7	0.07582	3.1993	0.02621	59.96	34.64	0.00089	1.04979	0.8627
65	14.3943	643.0	0.07248	3.2958	0.02199	61.98	35.54	0.00110	1.04485	0.8107
70	12.9961	720.4	0.06895	3.3640	0.01904	64.48	36.68	0.00130	1.03971	0.7802
75	11.8836	803.2	0.06533	3.4150	0.01686	66.68	37.95	0.00149	1.03627	0.7705
80	10.9709	892.3	0.06171	3.4537	0.01517	69.83	39.27	0.00169	1.03345	0.7615
85	10.2046	987.6	0.05822	3.4833	0.01383	74.38	40.63	0.00192	1.03109	0.7471
90	9.5497	1089.1	0.05494	3.5064	0.01273	78.82	41.99	0.00214	1.02907	0.7389
95	8.9818	1196.1	0.05193	3.5245	0.01182	83.38	43.36	0.00236	1.02733	0.7349
100	8.4835	1307.7	0.04923	3.5398	0.01103	92.22	46.50	0.00271	1.02580	0.7275
120	6.9740	1782.2	0.04119	3.5718	0.00877	117.91	54.29	0.00389	1.02117	0.7208
140	5.9417	2238.7	0.03726	3.5842	0.00734	136.52	59.42	0.00503	1.01802	0.7150
160	5.1854	2638.1	0.03573	3.5888	0.00633	148.73	63.30	0.00618	1.01571	0.7185
180	4.6050	2975.3	0.03554	3.5892	0.00558	156.29	66.53	0.00737	1.01394	0.7062
200	4.1442	3265.2	0.03605	3.5876	0.00499	161.06	69.43	0.00859	1.01254	0.7023
220	3.7689	3526.7	0.03685	3.5850	0.00452	164.47	72.12	0.00986	1.01140	0.6988
240	3.4570	3776.2	0.03769	3.5819	0.00413	167.45	74.69	0.01118	1.01045	0.6959
260	3.1934	4024.4	0.03844	3.5787	0.00381	170.52	77.18	0.01255	1.00965	0.6935
280	2.9676	4276.5	0.03905	3.5754	0.00353	173.86	79.63	0.01396	1.00897	0.6917
300	2.7717	4536.5	0.03951	3.5733	0.00329	177.52	82.03	0.01544	1.00838	0.6900
350	2.3805	5210.5	0.04013	3.5657	0.00282	187.99	87.97	0.01935	1.00719	0.6875
400	2.0865	5913.1	0.04033	3.5592	0.00247	199.55	93.83	0.02359	1.00630	0.6861
450	1.8575	6634.7	0.04032	3.5537	0.00219	211.69	99.63	0.02818	1.00561	0.6853
500	1.6739	7360.9	0.04027	3.5490	0.00198	223.88	105.37	0.03310	1.00505	0.6847
550	1.5234	8093.7	0.04018	3.5451	0.00180	236.15	111.05	0.03835	1.00460	0.6843
600	1.3977	8831.6	0.04007	3.5417	0.00165	248.41	116.67	0.04393	1.00422	0.6840
700	1.1999	10328.0	0.03978	3.5362	0.00142	273.01	127.71	0.05605	1.00362	0.6837
800	1.0512	11863.7	0.03938	3.5319	0.00124	297.89	138.49	0.06938	1.00317	0.6836
900	0.9353	13449.4	0.03887	3.5285	0.00110	323.21	149.02	0.08390	1.00282	0.6837
1000	0.8424	15092.9	0.03827	3.5258	0.00099	440.56	159.30	0.12563	1.00254	0.5419
1200	0.7028	18554.3	0.03691	3.5216	0.00083	511.13	179.20	0.17034	1.00212	0.5388
1400	0.6030	22242.5	0.03550	3.5186	0.00071	583.36	198.30	0.22038	1.00182	0.5372
1600	0.5279	26082.6	0.03424	3.5163	0.00062	655.84	216.71	0.27561	1.00159	0.5362
1800	0.4695	30079.1	0.03310	3.5144	0.00055	729.84	234.53	0.33626	1.00142	0.5348
2000	0.4227	34416.8	0.03183	3.5126	0.00050	812.40	251.83	0.40353	1.00127	0.5315
2500	0.3377	49684.8	0.02669	3.5030	0.00040	1168.39	293.40	0.62723	1.00102	0.4986
3000	0.2784	82563.1	0.01855	3.4622	0.00034	2161.83	334.72	1.00469	1.00084	0.4308

* TWO-PHASE BOUNDARY

Table 4.1 (continued)
THERMODYNAMIC PROPERTIES OF PARAHYDROGEN (ISOBARS, SI UNITS)

4.00 MPA ISOBAR

TEMPERATURE DEG. K	VOLUME CU M/KG	ISOTHERM DERIVATIVE CU M-MPA/KG	ISOCHORE DERIVATIVE MPA/K	INTERNAL ENERGY KJ/KG-K	ENTHALPY KJ/KG	ENTROPY KJ/KG-K	CV KJ / KG-K	CP KJ / KG-K	VELOCITY OF SOUND M/S
* 15.080	0.01263	1.4112	0.9524	-306.9	-256.4	5.059	4.85	6.39	1364
16	0.01274	1.3667	0.9608	-301.3	-250.3	5.448	5.03	6.78	1358
17	0.01286	1.2950	0.9624	-294.7	-243.3	5.873	5.20	7.21	1340
18	0.01298	1.2281	0.9572	-287.8	-235.9	6.297	5.36	7.62	1322
19	0.01312	1.1684	0.9466	-280.5	-228.1	6.720	5.50	8.01	1304
20	0.01326	1.1021	0.9333	-272.9	-219.9	7.140	5.63	8.41	1283
21	0.01341	1.0393	0.9185	-264.9	-211.3	7.560	5.74	8.81	1263
22	0.01358	0.9817	0.9023	-256.6	-202.2	7.979	5.85	9.21	1244
23	0.01375	0.9227	0.8848	-247.8	-192.8	8.398	5.94	9.63	1223
24	0.01394	0.8638	0.8656	-238.7	-183.0	8.817	6.02	10.07	1202
25	0.01414	0.8051	0.8448	-229.3	-172.7	9.237	6.09	10.52	1179
26	0.01436	0.7467	0.8221	-219.4	-161.9	9.660	6.15	11.01	1156
27	0.01460	0.6877	0.7975	-209.1	-150.7	10.085	6.21	11.53	1130
28	0.01486	0.6334	0.7712	-198.3	-138.9	10.513	6.26	12.06	1105
29	0.01514	0.5804	0.7431	-187.1	-126.5	10.946	6.30	12.62	1078
30	0.01545	0.5290	0.7135	-175.4	-113.6	11.384	6.33	13.22	1051
31	0.01578	0.4784	0.6827	-163.2	-100.1	11.828	6.37	13.89	1022
32	0.01616	0.4311	0.6504	-150.4	-85.8	12.281	6.40	14.60	992
33	0.01657	0.3833	0.6169	-137.1	-70.8	12.742	6.43	15.43	959
34	0.01704	0.3410	0.5822	-123.2	-55.0	13.215	6.46	16.27	927
35	0.01757	0.2997	0.5466	-108.5	-38.2	13.700	6.48	17.26	893
36	0.01817	0.2623	0.5102	-93.1	-20.5	14.201	6.51	18.31	859
37	0.01886	0.2287	0.4736	-77.0	-1.6	14.718	6.55	19.45	824
38	0.01965	0.1993	0.4371	-60.1	18.5	15.253	6.58	20.64	791
39	0.02056	0.1742	0.4012	-42.5	39.7	15.805	6.61	21.84	759
40	0.02161	0.1532	0.3666	-24.3	62.1	16.372	6.64	23.02	729
42	0.02413	0.1259	0.3036	13.4	110.0	17.539	6.71	24.62	680
44	0.02719	0.1135	0.2519	51.0	159.8	18.698	6.76	24.95	647
46	0.03060	0.1131	0.2117	85.8	208.3	19.774	6.81	23.87	630
48	0.03412	0.1204	0.1813	117.7	254.2	20.753	6.81	22.06	625
50	0.03759	0.1314	0.1586	146.3	296.6	21.619	6.79	20.33	627
52	0.04095	0.1439	0.1413	172.0	335.8	22.387	6.79	18.88	633
54	0.04418	0.1571	0.1277	195.6	372.4	23.077	6.79	17.73	641
56	0.04730	0.1702	0.1168	217.9	407.1	23.710	6.80	16.83	649
58	0.05032	0.1836	0.1078	238.8	440.1	24.288	6.81	16.11	659
60	0.05325	0.1967	0.1003	258.7	471.7	24.824	6.84	15.54	668
65	0.06027	0.2283	0.0859	305.7	546.7	26.026	6.97	14.59	691
70	0.06699	0.2590	0.0754	350.6	618.5	27.090	7.17	14.07	713
75	0.07341	0.2880	0.0675	394.6	688.2	28.051	7.44	13.83	732
80	0.07964	0.3159	0.0612	438.6	757.2	28.942	7.76	13.78	749
85	0.08572	0.3430	0.0561	483.4	826.3	29.779	8.13	13.87	765
90	0.09168	0.3692	0.0519	529.3	896.0	30.577	8.52	14.04	780
95	0.09754	0.3949	0.0483	576.7	966.8	31.342	8.94	14.28	794
100	0.10333	0.4200	0.0452	625.6	1038.9	32.082	9.36	14.56	808
120	0.12584	0.5160	0.0361	839.3	1342.7	34.847	10.92	15.73	862
140	0.14777	0.6080	0.0303	1074.5	1665.6	37.334	11.89	16.49	918
160	0.16935	0.6975	0.0261	1321.3	1998.7	39.557	12.27	16.74	976
180	0.19069	0.7854	0.0230	1570.0	2332.8	41.526	12.23	16.63	1033
200	0.21188	0.8721	0.0205	1814.9	2662.4	43.261	11.99	16.32	1090
220	0.23295	0.9579	0.0186	2053.4	2985.2	44.800	11.67	15.96	1145
240	0.25394	1.0431	0.0169	2285.2	3300.9	46.174	11.36	15.62	1198
260	0.27486	1.1278	0.0156	2511.0	3610.4	47.413	11.10	15.34	1248
280	0.29574	1.2122	0.0144	2732.0	3914.9	48.541	10.90	15.12	1297
300	0.31662	1.2969	0.0135	2949.8	4216.3	49.581	10.74	14.94	1343
350	0.36854	1.5058	0.0115	3482.4	4956.6	51.862	10.53	14.70	1450
400	0.42035	1.7139	0.0100	4007.2	5688.6	53.819	10.44	14.60	1548
450	0.47209	1.9215	0.0089	4528.8	6417.2	55.536	10.42	14.56	1639
500	0.52378	2.1287	0.0080	5050.1	7145.2	57.069	10.41	14.55	1725
550	0.57543	2.3356	0.0073	5570.9	7872.6	58.455	10.41	14.55	1807
600	0.62706	2.5424	0.0067	6092.4	8600.7	59.721	10.43	14.57	1884
700	0.73029	2.9556	0.0057	7138.2	10059.4	61.971	10.48	14.62	2030
800	0.83347	3.3686	0.0050	8191.3	11525.2	63.927	10.57	14.70	2164
900	0.93663	3.7814	0.0044	9255.0	13001.6	65.664	10.70	14.83	2289
1000	1.03977	4.1941	0.0040	10332.6	14491.6	67.237	10.86	14.99	2406
1200	1.24604	5.0193	0.0033	12543.2	17527.3	69.991	11.24	15.37	2619
1400	1.45228	5.8443	0.0029	14836.1	20645.2	72.398	11.68	15.80	2812
1600	1.65852	6.6693	0.0025	17215.0	23849.1	74.514	12.10	16.23	2990
1800	1.86479	7.4942	0.0022	19681.9	27141.1	76.415	12.51	16.64	3157
2000	2.07123	8.3192	0.0020	22245.5	30530.4	78.266	12.99	17.13	3312
2500	2.59169	10.3813	0.0016	29284.7	39651.5	84.674	15.41	19.72	3645
3000	3.14109	12.4435	0.0013	38549.3	51113.7	92.205	22.06	27.21	3915

* TWO-PHASE BOUNDARY

Table 4.1 (continued)
THERMODYNAMIC PROPERTIES OF PARAHYDROGEN (ISOBARS, SI UNITS)

4.00 MPA ISOBAR

TEMPERATURE DEG. K	DENSITY KG/CU M	$V(DH/DV)_P$ KJ/KG	$V(DP/DU)_V$ MPA-CU M/KJ X 10^2	$-V(DP/DV)_T$ MPA	$-(DV/DT)_P/V$ 1/DEG. K	THERMAL CONDUCTIVITY W/K-M X 10^3	VISCOSITY KG/M-S X 10^7	THERMAL DIFFUSIVITY SQ M/HR	DIELECTRIC CONSTANT	PRANDTL NUMBER
* 15.080	79.1477	750.0	0.24825	111.6967	0.00853	87.81	279.60	0.00062	1.25909	2.0363
16	78.5151	757.3	0.24339	107.3080	0.00895	94.35	250.48	0.00064	1.25686	1.8001
17	77.7893	755.0	0.23774	100.7400	0.00955	99.43	224.78	0.00064	1.25430	1.6305
18	77.0297	753.5	0.23179	94.5996	0.01012	103.29	203.80	0.00063	1.25162	1.5044
19	76.2330	753.7	0.22567	89.0699	0.01063	105.88	186.36	0.00062	1.24882	1.4097
20	75.4072	748.8	0.21984	83.1052	0.01123	108.33	171.72	0.00061	1.24592	1.3330
21	74.5439	743.3	0.21449	77.4730	0.01186	111.09	159.18	0.00061	1.24290	1.2627
22	73.6412	738.1	0.20952	72.2903	0.01248	113.11	148.31	0.00060	1.23974	1.2080
23	72.7031	730.2	0.20492	67.0801	0.01319	114.51	138.81	0.00059	1.23646	1.1676
24	71.7214	720.6	0.20047	61.9556	0.01397	115.35	130.37	0.00058	1.23304	1.1378
25	70.6965	709.1	0.19619	56.9159	0.01484	115.72	122.82	0.00056	1.22947	1.1170
26	69.6206	696.1	0.19190	51.9873	0.01581	115.60	115.96	0.00054	1.22573	1.1041
27	68.4918	681.1	0.18756	47.1047	0.01693	115.10	109.69	0.00052	1.22182	1.0989
28	67.3048	666.8	0.18310	42.6332	0.01809	114.26	103.90	0.00051	1.21771	1.0968
29	66.0553	651.3	0.17857	38.3405	0.01938	113.10	98.51	0.00049	1.21340	1.0995
30	64.7413	634.6	0.17400	34.2491	0.02083	111.67	93.47	0.00047	1.20888	1.1067
31	63.3553	616.7	0.16924	30.3107	0.02252	109.56	88.72	0.00045	1.20412	1.1247
32	61.8867	598.8	0.16425	26.6805	0.02438	107.24	84.19	0.00043	1.19909	1.1459
33	60.3319	578.4	0.15908	23.1272	0.02667	104.73	79.88	0.00041	1.19378	1.1766
34	58.6740	559.2	0.15369	20.0072	0.02910	102.05	75.72	0.00038	1.18813	1.2073
35	56.9093	538.5	0.14811	17.0574	0.03204	99.20	71.72	0.00036	1.18215	1.2476
36	55.0254	518.1	0.14234	14.4332	0.03535	96.17	67.85	0.00034	1.17577	1.2920
37	53.0180	498.1	0.13646	12.1255	0.03906	93.18	64.12	0.00033	1.16901	1.3387
38	50.8854	479.0	0.13059	10.1427	0.04309	90.72	60.55	0.00031	1.16186	1.3777
39	48.6348	461.3	0.12484	8.4737	0.04735	88.15	57.15	0.00030	1.15434	1.4159
40	46.2851	445.2	0.11926	7.0906	0.05170	85.54	53.96	0.00029	1.14652	1.4520
42	41.4430	423.0	0.10917	5.2174	0.05819	80.22	48.39	0.00028	1.13053	1.4850
44	36.7771	413.5	0.10127	4.1750	0.06034	75.26	44.10	0.00030	1.11527	1.4621
46	32.6747	416.9	0.09513	3.6962	0.05727	70.99	41.07	0.00033	1.10197	1.3812
48	29.3046	429.5	0.09090	3.5296	0.05137	67.69	39.07	0.00038	1.09113	1.2733
50	26.6002	447.8	0.08777	3.4940	0.04540	65.41	37.76	0.00044	1.08248	1.1734
52	24.4207	469.6	0.08527	3.5143	0.04020	63.94	36.91	0.00050	1.07554	1.0900
54	22.6331	493.6	0.08313	3.5550	0.03592	63.07	36.40	0.00057	1.06988	1.0233
56	21.1407	518.8	0.08124	3.5988	0.03244	62.65	36.13	0.00063	1.06517	0.9707
58	19.8723	545.2	0.07962	3.6482	0.02955	62.55	36.05	0.00070	1.06118	0.9284
60	18.7786	572.4	0.07805	3.6932	0.02715	62.78	36.10	0.00077	1.05774	0.8938
65	16.5932	643.9	0.07423	3.7886	0.02266	64.35	36.67	0.00096	1.05090	0.8315
70	14.9280	721.3	0.07042	3.8661	0.01950	66.52	37.59	0.00114	1.04571	0.7949
75	13.6221	804.0	0.06663	3.9234	0.01721	68.42	38.70	0.00131	1.04165	0.7826
80	12.5567	892.8	0.06286	3.9672	0.01544	71.41	39.92	0.00149	1.03835	0.7706
85	11.6663	988.1	0.05923	4.0010	0.01403	75.77	41.20	0.00169	1.03560	0.7541
90	10.9079	1089.6	0.05583	4.0274	0.01289	80.20	42.51	0.00188	1.03326	0.7444
95	10.2521	1196.7	0.05272	4.0481	0.01194	84.69	43.83	0.00208	1.03123	0.7392
100	9.6782	1308.5	0.04993	4.0646	0.01113	92.82	46.74	0.00237	1.02947	0.7332
120	7.9469	1784.9	0.04163	4.1004	0.00881	118.22	54.43	0.00341	1.02415	0.7240
140	6.7673	2243.1	0.03760	4.1147	0.00735	136.72	59.53	0.00441	1.02054	0.7180
160	5.9050	2644.2	0.03600	4.1190	0.00633	148.90	63.39	0.00542	1.01791	0.7128
180	5.2440	2983.0	0.03578	4.1185	0.00557	156.46	66.63	0.00646	1.01589	0.7080
200	4.7197	3274.1	0.03626	4.1158	0.00498	161.24	69.52	0.00754	1.01429	0.7036
220	4.2927	3536.6	0.03705	4.1120	0.00451	164.66	72.21	0.00865	1.01299	0.6999
240	3.9379	3786.8	0.03788	4.1077	0.00413	167.66	74.78	0.00981	1.01192	0.6967
260	3.6382	4035.6	0.03861	4.1033	0.00380	170.74	77.27	0.01101	1.01100	0.6942
280	3.3814	4288.2	0.03921	4.0989	0.00353	174.10	79.72	0.01226	1.01023	0.6922
300	3.1584	4548.9	0.03965	4.0961	0.00329	177.77	82.13	0.01356	1.00955	0.6905
350	2.7134	5223.7	0.04026	4.0859	0.00281	188.27	88.09	0.01699	1.00820	0.6878
400	2.3790	5926.7	0.04044	4.0774	0.00246	199.86	93.96	0.02072	1.00719	0.6863
450	2.1183	6648.7	0.04041	4.0702	0.00219	212.03	99.78	0.02474	1.00640	0.6854
500	1.9092	7375.2	0.04035	4.0641	0.00197	224.26	105.54	0.02906	1.00576	0.6848
550	1.7378	8108.2	0.04025	4.0589	0.00179	236.57	111.24	0.03367	1.00525	0.6843
600	1.5947	8846.2	0.04014	4.0544	0.00165	248.87	116.87	0.03857	1.00481	0.6840
700	1.3693	10343.0	0.03984	4.0472	0.00141	273.54	127.95	0.04920	1.00413	0.6837
800	1.1998	11878.9	0.03943	4.0417	0.00124	298.49	138.76	0.06091	1.00362	0.6836
900	1.0677	13464.9	0.03892	4.0372	0.00110	323.89	149.32	0.07365	1.00322	0.6837
1000	0.9617	15108.6	0.03831	4.0336	0.00099	440.56	159.63	0.11004	1.00290	0.5430
1200	0.8025	18570.5	0.03695	4.0282	0.00083	511.13	179.59	0.14918	1.00242	0.5400
1400	0.6886	22259.1	0.03553	4.0242	0.00071	583.36	198.76	0.19297	1.00208	0.5385
1600	0.6029	26098.7	0.03427	4.0212	0.00062	655.79	217.23	0.24131	1.00182	0.5375
1800	0.5363	30090.2	0.03312	4.0188	0.00055	729.58	235.10	0.29436	1.00162	0.5362
2000	0.4828	34407.7	0.03188	4.0165	0.00050	811.35	252.45	0.35310	1.00146	0.5331
2500	0.3858	49367.3	0.02691	4.0056	0.00040	1155.60	294.12	0.54677	1.00116	0.5019
3000	0.3184	80853.3	0.01899	3.9615	0.00034	2096.71	335.42	0.87127	1.00096	0.4353

* TWO-PHASE BOUNDARY

Table 4.1 (continued)
THERMODYNAMIC PROPERTIES OF PARAHYDROGEN (ISOBARS, SI UNITS)

5.00 MPA ISOBAR

TEMPERATURE DEG. K	VOLUME CU M/KG	ISOTHERM DERIVATIVE CU M-MPA/KG	ISOCHORE DERIVATIVE MPA/K	INTERNAL ENERGY KJ/KG-K	ENTHALPY KJ/KG	ENTROPY KJ/KG-K	CV KJ /	CP KG-K	VELOCITY OF SOUND M/S
* 15.382	0.01256	1.4746	0.9589	-306.2	-243.4	5.081	4.88	6.40	1390
16	0.01262	1.4419	0.9656	-302.5	-239.4	5.338	5.01	6.66	1384
17	0.01273	1.3771	0.9705	-296.2	-232.5	5.754	5.19	7.07	1370
18	0.01285	1.3108	0.9690	-289.5	-225.2	6.170	5.35	7.48	1354
19	0.01298	1.2480	0.9623	-282.5	-217.6	6.585	5.49	7.87	1337
20	0.01311	1.1937	0.9514	-275.1	-209.5	6.998	5.62	8.22	1322
21	0.01325	1.1291	0.9383	-267.4	-201.1	7.408	5.73	8.61	1302
22	0.01340	1.0703	0.9233	-259.3	-192.3	7.818	5.84	8.98	1284
23	0.01356	1.0116	0.9071	-250.9	-183.1	8.225	5.93	9.37	1264
24	0.01373	0.9542	0.8895	-242.2	-173.6	8.632	6.01	9.76	1245
25	0.01391	0.8976	0.8703	-233.2	-163.6	9.039	6.08	10.17	1225
26	0.01411	0.8409	0.8497	-223.8	-153.2	9.446	6.15	10.59	1204
27	0.01432	0.7852	0.8274	-214.0	-142.4	9.854	6.21	11.03	1181
28	0.01454	0.7307	0.8035	-203.9	-131.2	10.264	6.26	11.49	1158
29	0.01478	0.6782	0.7782	-193.3	-119.4	10.675	6.31	11.97	1134
30	0.01504	0.6262	0.7516	-182.4	-107.2	11.089	6.34	12.47	1109
31	0.01533	0.5776	0.7238	-171.1	-94.5	11.506	6.38	12.98	1084
32	0.01563	0.5319	0.6951	-159.4	-81.3	11.927	6.41	13.51	1059
33	0.01597	0.4873	0.6657	-147.3	-67.4	12.352	6.44	14.09	1033
34	0.01633	0.4434	0.6355	-134.7	-53.1	12.781	6.46	14.72	1005
35	0.01673	0.4034	0.6045	-121.7	-38.0	13.217	6.49	15.36	977
36	0.01717	0.3645	0.5731	-108.2	-22.3	13.659	6.51	16.07	949
37	0.01766	0.3303	0.5413	-94.2	-5.9	14.109	6.53	16.77	921
38	0.01820	0.2981	0.5095	-79.7	11.2	14.566	6.55	17.51	893
39	0.01879	0.2690	0.4779	-64.8	29.1	15.030	6.56	18.26	865
40	0.01945	0.2443	0.4470	-49.5	47.7	15.502	6.58	18.96	839
42	0.02098	0.2041	0.3883	-17.9	87.0	16.461	6.62	20.28	791
44	0.02281	0.1771	0.3356	14.6	128.6	17.428	6.66	21.22	751
46	0.02491	0.1615	0.2903	46.5	171.0	18.370	6.73	21.63	720
48	0.02724	0.1555	0.2525	77.9	214.1	19.287	6.78	21.39	701
50	0.02970	0.1565	0.2217	107.7	256.2	20.147	6.80	20.66	689
52	0.03222	0.1620	0.1967	135.5	296.6	20.939	6.82	19.71	684
54	0.03473	0.1706	0.1766	161.4	335.1	21.664	6.83	18.74	684
56	0.03721	0.1808	0.1604	185.9	371.9	22.335	6.85	17.88	687
58	0.03964	0.1920	0.1470	208.7	406.9	22.948	6.86	17.12	692
60	0.04202	0.2035	0.1358	230.4	440.5	23.518	6.89	16.50	698
65	0.04776	0.2333	0.1147	281.1	519.9	24.789	7.01	15.37	715
70	0.05325	0.2631	0.0997	328.7	594.9	25.902	7.21	14.72	733
75	0.05853	0.2918	0.0886	374.9	667.5	26.904	7.47	14.38	749
80	0.06365	0.3204	0.0799	420.8	739.0	27.827	7.79	14.25	765
85	0.06867	0.3483	0.0728	467.1	810.5	28.693	8.16	14.26	780
90	0.07355	0.3752	0.0670	514.2	882.0	29.511	8.55	14.37	794
95	0.07834	0.4014	0.0621	562.6	954.3	30.293	8.96	14.57	808
100	0.08306	0.4271	0.0580	612.5	1027.8	31.046	9.38	14.81	821
120	0.10133	0.5249	0.0459	829.2	1335.8	33.851	10.94	15.90	873
140	0.11907	0.6184	0.0383	1066.2	1661.6	36.360	11.91	16.61	929
160	0.13648	0.7090	0.0329	1314.3	1996.7	38.597	12.28	16.83	986
180	0.15367	0.7977	0.0289	1564.0	2332.4	40.574	12.25	16.70	1043
200	0.17071	0.8851	0.0258	1809.7	2663.2	42.317	12.00	16.38	1099
220	0.18764	0.9716	0.0233	2048.8	2987.0	43.860	11.68	16.01	1154
240	0.20449	1.0573	0.0213	2281.1	3303.5	45.237	11.37	15.66	1207
260	0.22128	1.1424	0.0196	2507.3	3613.7	46.479	11.11	15.37	1257
280	0.23802	1.2271	0.0181	2728.7	3918.8	47.610	10.91	15.14	1305
300	0.25478	1.3124	0.0169	2947.0	4220.9	48.651	10.75	14.97	1352
350	0.29639	1.5219	0.0144	3480.2	4962.1	50.936	10.53	14.72	1458
400	0.33788	1.7303	0.0126	4005.5	5694.9	52.894	10.44	14.61	1556
450	0.37930	1.9381	0.0112	4527.4	6423.9	54.612	10.42	14.57	1646
500	0.42067	2.1455	0.0100	5049.0	7152.3	56.146	10.41	14.56	1732
550	0.46201	2.3525	0.0091	5569.9	7880.0	57.533	10.42	14.56	1813
600	0.50333	2.5594	0.0083	6091.6	8608.3	58.799	10.43	14.57	1891
700	0.58593	2.9728	0.0072	7137.6	10067.3	61.049	10.49	14.62	2036
800	0.66849	3.3858	0.0063	8190.8	11533.3	63.006	10.58	14.71	2170
900	0.75103	3.7987	0.0056	9254.7	13009.8	64.743	10.70	14.83	2294
1000	0.83355	4.2114	0.0050	10332.3	14500.0	66.316	10.86	14.99	2411
1200	0.99857	5.0367	0.0042	12543.0	17535.9	69.070	11.24	15.37	2624
1400	1.16357	5.8618	0.0036	14835.9	20653.8	71.477	11.68	15.81	2816
1600	1.32856	6.6867	0.0031	17214.8	23857.6	73.593	12.10	16.23	2994
1800	1.49358	7.5117	0.0028	19681.2	27149.1	75.494	12.51	16.63	3161
2000	1.65871	8.3366	0.0025	22242.5	30536.1	77.344	12.97	17.11	3316
2500	2.07467	10.3988	0.0020	29239.1	39612.4	83.733	15.22	19.51	3651
3000	2.51137	12.4609	0.0017	38263.4	50820.3	91.173	21.24	26.28	3924

* TWO-PHASE BOUNDARY

Table 4.1 (continued)
THERMODYNAMIC PROPERTIES OF PARAHYDROGEN (ISOBARS, SI UNITS)

5.00 MPA ISOBAR

TEMPERATURE DEG. K	DENSITY KG/CU M	$V(DH/DV)_P$ KJ/KG	$V(DP/DU)_V$ MPA-CU M/KJ X 10^2	$-V(DP/DV)_T$ MPA	$-(DV/DT)_P/V$ 1/DEG. K	THERMAL CONDUCTIVITY W/K-M X 10^3	VISCOSITY KG/M-S X 10^7	THERMAL DIFFUSIVITY SQ M/HR	DIELECTRIC CONSTANT	PRANDTL NUMBER
* 15.382	79.6358	783.3	0.24654	117.4329	0.00817	91.17	285.79	0.00064	1.26081	2.0050
16	79.2264	787.3	0.24343	114.2398	0.00845	95.52	265.42	0.00065	1.25937	1.8493
17	78.5488	788.2	0.23818	108.1565	0.00897	100.86	237.88	0.00065	1.25695	1.6680
18	77.8155	787.1	0.23283	101.9981	0.00950	104.95	215.35	0.00065	1.25439	1.5345
19	77.0622	786.1	0.22740	96.1733	0.01001	107.77	196.75	0.00064	1.25174	1.4359
20	76.2761	787.1	0.22203	91.0477	0.01045	110.42	181.10	0.00063	1.24897	1.3490
21	75.4646	781.7	0.21690	85.2096	0.01101	113.42	167.84	0.00063	1.24612	1.2738
22	74.6146	777.0	0.21204	79.8570	0.01156	115.66	156.35	0.00062	1.24314	1.2145
23	73.7364	770.4	0.20750	74.5884	0.01216	117.28	146.37	0.00061	1.24007	1.1694
24	72.8222	762.8	0.20319	69.4868	0.01280	118.35	137.55	0.00060	1.23688	1.1348
25	71.8715	753.7	0.19904	64.5099	0.01349	118.95	129.69	0.00059	1.23356	1.1087
26	70.8805	743.0	0.19495	59.6058	0.01425	119.08	122.61	0.00057	1.23011	1.0906
27	69.8496	731.3	0.19081	54.8474	0.01509	118.84	116.19	0.00056	1.22653	1.0786
28	68.7720	718.7	0.18661	50.2511	0.01599	118.27	110.29	0.00054	1.22279	1.0717
29	67.6476	705.4	0.18239	45.8796	0.01696	117.41	104.84	0.00052	1.21890	1.0684
30	66.4752	690.6	0.17822	41.6295	0.01805	116.30	99.78	0.00051	1.21485	1.0697
31	65.2583	676.0	0.17391	37.6892	0.01920	114.51	95.05	0.00049	1.21063	1.0776
32	63.9697	661.5	0.16952	34.0273	0.02043	112.55	90.59	0.00047	1.20623	1.0877
33	62.6318	645.9	0.16512	30.5213	0.02181	110.43	86.38	0.00045	1.20164	1.1019
34	61.2349	629.0	0.16060	27.1527	0.02341	108.22	82.39	0.00043	1.19686	1.1208
35	59.7690	612.7	0.15593	24.1108	0.02507	105.87	78.58	0.00042	1.19186	1.1401
36	58.2355	595.4	0.15119	21.2278	0.02700	103.41	74.94	0.00040	1.18664	1.1648
37	56.6288	579.3	0.14635	18.7024	0.02894	100.85	71.46	0.00038	1.18120	1.1879
38	54.9544	562.9	0.14154	16.3795	0.03110	98.22	68.14	0.00037	1.17554	1.2145
39	53.2122	546.9	0.13682	14.3153	0.03339	95.62	64.97	0.00035	1.16967	1.2408
40	51.4071	532.7	0.13215	12.5594	0.03559	93.45	61.98	0.00035	1.16360	1.2573
42	47.6583	507.9	0.12317	9.7251	0.03993	89.06	56.55	0.00033	1.15109	1.2877
44	43.8420	490.8	0.11495	7.7629	0.04323	84.77	51.94	0.00033	1.13844	1.3002
46	40.1370	483.0	0.10743	6.4812	0.04479	80.84	48.24	0.00034	1.12625	1.2908
48	36.7078	483.4	0.10151	5.7089	0.04424	77.38	45.39	0.00035	1.11504	1.2545
50	33.6647	490.9	0.09677	5.2680	0.04208	74.53	43.31	0.00039	1.10517	1.2004
52	31.0368	503.9	0.09293	5.0278	0.03912	72.32	41.84	0.00043	1.09669	1.1404
54	28.7934	521.3	0.08977	4.9125	0.03596	70.71	40.81	0.00047	1.08949	1.0819
56	26.8763	541.6	0.08712	4.8583	0.03301	69.61	40.11	0.00052	1.08336	1.0301
58	25.2271	564.1	0.08487	4.8432	0.03034	68.90	39.64	0.00057	1.07811	0.9848
60	23.7978	588.2	0.08283	4.8435	0.02804	68.64	39.37	0.00063	1.07357	0.9461
65	20.9367	654.7	0.07812	4.8841	0.02348	69.25	39.24	0.00077	1.06453	0.8711
70	18.7801	728.8	0.07368	4.9402	0.02019	70.78	39.67	0.00092	1.05775	0.8248
75	17.0856	809.5	0.06937	4.9860	0.01776	72.04	40.43	0.00106	1.05244	0.8069
80	15.7118	897.8	0.06524	5.0342	0.01587	74.67	41.39	0.00120	1.04815	0.7897
85	14.5616	993.2	0.06130	5.0718	0.01435	78.78	42.47	0.00137	1.04457	0.7685
90	13.5961	1094.5	0.05763	5.1010	0.01313	83.01	43.63	0.00153	1.04157	0.7554
95	12.7650	1201.5	0.05432	5.1242	0.01213	87.34	44.83	0.00169	1.03900	0.7480
100	12.0401	1313.4	0.05136	5.1426	0.01128	94.20	47.31	0.00190	1.03675	0.7440
120	9.8688	1792.9	0.04252	5.1798	0.00887	118.99	54.78	0.00273	1.03085	0.7318
140	8.3984	2254.0	0.03826	5.1932	0.00737	137.23	59.80	0.00354	1.02554	0.7238
160	7.3272	2658.1	0.03655	5.1949	0.00633	149.31	63.63	0.00436	1.02225	0.7173
180	6.5075	2999.5	0.03625	5.1912	0.00557	156.85	66.84	0.00520	1.01975	0.7115
200	5.8579	3292.8	0.03669	5.1850	0.00497	161.64	69.72	0.00607	1.01776	0.7063
220	5.3293	3557.0	0.03744	5.1778	0.00450	165.08	72.40	0.00697	1.01615	0.7020
240	4.8902	3808.5	0.03824	5.1703	0.00411	168.10	74.97	0.00790	1.01481	0.6984
260	4.5191	4058.5	0.03895	5.1628	0.00379	171.20	77.47	0.00887	1.01368	0.6956
280	4.2013	4312.0	0.03952	5.1556	0.00351	174.58	79.92	0.00989	1.01272	0.6933
300	3.9249	4574.3	0.03994	5.1512	0.00327	178.27	82.34	0.01093	1.01188	0.6913
350	3.3740	5250.3	0.04051	5.1348	0.00280	188.83	88.32	0.01369	1.01020	0.6883
400	2.9596	5954.2	0.04066	5.1212	0.00245	200.48	94.22	0.01669	1.00895	0.6866
450	2.6365	6676.8	0.04060	5.1097	0.00218	212.72	100.07	0.01993	1.00797	0.6856
500	2.3772	7403.8	0.04052	5.1001	0.00197	225.02	105.86	0.02341	1.00718	0.6849
550	2.1644	8137.3	0.04041	5.0919	0.00179	237.40	111.60	0.02712	1.00654	0.6844
600	1.9868	8875.6	0.04028	5.0849	0.00164	249.77	117.27	0.03106	1.00600	0.6840
700	1.7067	10372.9	0.03996	5.0736	0.00141	274.59	128.42	0.03962	1.00515	0.6837
800	1.4959	11909.3	0.03953	5.0649	0.00123	299.69	139.30	0.04904	1.00451	0.6835
900	1.3315	13495.8	0.03901	5.0580	0.00110	325.24	149.93	0.05929	1.00402	0.6836
1000	1.1997	15140.0	0.03839	5.0524	0.00099	440.56	160.30	0.08821	1.00362	0.5454
1200	1.0014	18602.9	0.03701	5.0439	0.00083	511.13	180.39	0.11955	1.00302	0.5425
1400	0.8594	22292.3	0.03558	5.0377	0.00071	583.35	199.68	0.15461	1.00259	0.5410
1600	0.7527	26131.4	0.03431	5.0331	0.00062	655.72	218.26	0.19329	1.00227	0.5401
1800	0.6695	30115.3	0.03318	5.0293	0.00055	729.19	236.25	0.23572	1.00202	0.5389
2000	0.6029	34402.4	0.03196	5.0259	0.00050	809.75	253.70	0.28256	1.00182	0.5362
2500	0.4820	48890.6	0.02727	5.0123	0.00040	1136.04	295.60	0.43494	1.00145	0.5076
3000	0.3982	78243.2	0.01970	4.9618	0.00034	1996.91	336.92	0.68695	1.00120	0.4434

* TWO-PHASE BOUNDARY

Table 4.1 (continued)
THERMODYNAMIC PROPERTIES OF PARAHYDROGEN (ISOBARS, SI UNITS)

6.00 MPA ISOBAR

TEMPERATURE DEG. K	VOLUME CU M/KG	ISOTHERM DERIVATIVE CU M-MPA/KG	ISOCHORE DERIVATIVE MPA/K	INTERNAL ENERGY KJ/KG-K	ENTHALPY KJ/KG	ENTROPY KJ/KG-K	CV KJ /	CP KG-K	VELOCITY OF SOUND M/S
* 15.678	0.01248	1.5352	0.9665	-305.5	-230.6	5.103	4.92	6.40	1414
16	0.01252	1.5184	0.9702	-303.6	-228.5	5.235	4.98	6.54	1411
17	0.01262	1.4521	0.9773	-297.4	-221.7	5.644	5.17	6.95	1397
18	0.01273	1.3891	0.9784	-291.0	-214.6	6.052	5.33	7.34	1383
19	0.01285	1.3275	0.9746	-284.1	-207.1	6.459	5.48	7.72	1368
20	0.01297	1.2678	0.9665	-277.0	-199.1	6.865	5.61	8.09	1352
21	0.01310	1.2171	0.9551	-269.5	-190.9	7.268	5.72	8.43	1339
22	0.01324	1.1548	0.9421	-261.7	-182.3	7.668	5.83	8.79	1320
23	0.01339	1.0975	0.9270	-253.6	-173.3	8.068	5.92	9.15	1302
24	0.01355	1.0404	0.9107	-245.2	-164.0	8.464	6.00	9.52	1284
25	0.01371	0.9847	0.8930	-236.5	-154.3	8.860	6.08	9.89	1265
26	0.01389	0.9301	0.8740	-227.5	-144.2	9.256	6.15	10.26	1246
27	0.01407	0.8757	0.8536	-218.2	-133.7	9.650	6.21	10.66	1226
28	0.01427	0.8223	0.8318	-208.5	-122.9	10.045	6.26	11.06	1205
29	0.01449	0.7702	0.8086	-198.5	-111.6	10.440	6.31	11.48	1183
30	0.01471	0.7202	0.7842	-188.2	-99.9	10.837	6.35	11.90	1161
31	0.01496	0.6722	0.7588	-177.6	-87.8	11.234	6.39	12.33	1139
32	0.01522	0.6260	0.7325	-166.6	-75.2	11.633	6.42	12.78	1116
33	0.01550	0.5802	0.7055	-155.2	-62.2	12.033	6.45	13.25	1092
34	0.01580	0.5386	0.6782	-143.6	-48.7	12.436	6.48	13.73	1068
35	0.01613	0.4990	0.6504	-131.6	-34.8	12.841	6.50	14.22	1045
36	0.01648	0.4609	0.6222	-119.2	-20.3	13.249	6.52	14.74	1020
37	0.01687	0.4244	0.5936	-106.5	-5.3	13.660	6.54	15.28	996
38	0.01728	0.3915	0.5651	-93.4	10.3	14.074	6.56	15.81	972
39	0.01773	0.3604	0.5364	-80.0	26.3	14.492	6.57	16.36	947
40	0.01822	0.3333	0.5082	-66.3	42.9	14.912	6.58	16.86	924
42	0.01931	0.2862	0.4536	-38.2	77.7	15.760	6.60	17.86	880
44	0.02058	0.2504	0.4027	-9.2	114.3	16.611	6.63	18.70	841
46	0.02204	0.2252	0.3568	19.6	151.9	17.446	6.70	19.33	806
48	0.02366	0.2085	0.3165	48.9	190.9	18.276	6.75	19.65	779
50	0.02542	0.2003	0.2818	77.7	230.2	19.078	6.79	19.60	760
52	0.02728	0.1977	0.2523	105.4	269.1	19.842	6.82	19.29	747
54	0.02921	0.1995	0.2273	132.0	307.2	20.560	6.85	18.78	739
56	0.03116	0.2049	0.2063	157.5	344.4	21.237	6.88	18.17	736
58	0.03311	0.2122	0.1887	181.4	380.1	21.864	6.90	17.56	735
60	0.03505	0.2211	0.1738	204.4	414.7	22.450	6.93	17.00	737
65	0.03981	0.2460	0.1456	257.9	496.8	23.764	7.05	15.92	745
70	0.04442	0.2737	0.1256	307.9	574.4	24.916	7.24	15.21	758
75	0.04887	0.3015	0.1109	356.1	649.3	25.949	7.50	14.81	771
80	0.05321	0.3291	0.0994	403.6	722.8	26.898	7.82	14.62	785
85	0.05743	0.3565	0.0903	451.2	795.8	27.783	8.18	14.60	798
90	0.06156	0.3834	0.0829	499.5	868.9	28.619	8.57	14.68	810
95	0.06562	0.4098	0.0766	549.0	942.7	29.417	8.98	14.85	823
100	0.06965	0.4362	0.0713	599.8	1017.7	30.186	9.40	15.05	836
120	0.08506	0.5350	0.0560	819.3	1329.7	33.027	10.97	16.05	885
140	0.09999	0.6296	0.0464	1058.1	1658.1	35.556	11.93	16.72	939
160	0.11460	0.7212	0.0398	1307.5	1995.1	37.806	12.30	16.92	996
180	0.12901	0.8107	0.0349	1558.1	2332.2	39.793	12.26	16.76	1053
200	0.14328	0.8987	0.0311	1804.6	2664.3	41.541	12.01	16.43	1109
220	0.15745	0.9856	0.0281	2044.3	2989.0	43.089	11.69	16.05	1163
240	0.17154	1.0718	0.0256	2277.1	3306.3	44.470	11.38	15.70	1216
260	0.18557	1.1573	0.0235	2503.8	3617.2	45.714	11.12	15.40	1266
280	0.19955	1.2423	0.0218	2725.5	3922.8	46.847	10.91	15.17	1314
300	0.21357	1.3282	0.0203	2944.2	4225.7	47.891	10.76	14.99	1360
350	0.24829	1.5381	0.0173	3478.1	4967.8	50.178	10.54	14.73	1466
400	0.28290	1.7469	0.0151	4003.8	5701.2	52.138	10.45	14.62	1564
450	0.31744	1.9548	0.0134	4526.0	6430.7	53.857	10.42	14.58	1654
500	0.35194	2.1623	0.0120	5047.8	7159.5	55.392	10.41	14.56	1739
550	0.38640	2.3695	0.0109	5569.0	7887.4	56.780	10.42	14.56	1820
600	0.42085	2.5765	0.0100	6090.8	8615.9	58.046	10.43	14.57	1897
700	0.48969	2.9900	0.0086	7137.0	10075.2	60.297	10.49	14.62	2042
800	0.55850	3.4031	0.0075	8190.4	11541.4	62.253	10.58	14.71	2175
900	0.62729	3.8160	0.0067	9254.4	13018.1	63.991	10.70	14.83	2300
1000	0.69606	4.2287	0.0060	10332.0	14508.4	65.564	10.86	14.99	2416
1200	0.83359	5.0540	0.0050	12542.9	17544.4	68.318	11.24	15.37	2628
1400	0.97109	5.8791	0.0043	14835.8	20662.4	70.725	11.68	15.81	2821
1600	1.10859	6.7041	0.0038	17214.7	23866.2	72.841	12.10	16.22	2998
1800	1.24610	7.5290	0.0033	19680.8	27157.4	74.741	12.50	16.63	3165
2000	1.38370	8.3539	0.0030	22240.3	30542.5	76.591	12.96	17.10	3320
2500	1.73009	10.4161	0.0024	29205.4	39585.9	82.966	15.08	19.35	3656
3000	2.09227	12.4783	0.0020	38052.4	50606.0	90.339	20.64	25.59	3932

* TWO-PHASE BOUNDARY

Table 4.1 (continued)
THERMODYNAMIC PROPERTIES OF PARAHYDROGEN (ISOBARS, SI UNITS)

6.00 MPA ISOBAR

TEMPERATURE DEG. K	DENSITY KG/CU M	$V(DH/DV)_P$ KJ/KG	$V(DP/DU)_V$ MPA-CU M/KJ X 10^2	$-V(DP/DV)_T$ MPA	$-(DV/DT)_P/V$ 1/DEG. K	THERMAL CONDUCTIVITY W/K-M X 10^3	VISCOSITY KG/M-S X 10^7	THERMAL DIFFUSIVITY SQ M/HR	DIELECTRIC CONSTANT	PRANDTL NUMBER
* 15.678	80.1084	815.0	0.24531	122.9816	0.00786	94.38	291.98	0.00066	1.26248	1.9813
16	79.9026	817.4	0.24368	121.3254	0.00800	96.62	280.92	0.00067	1.26176	1.9004
17	79.2468	818.3	0.23857	115.0744	0.00849	102.22	251.31	0.00067	1.25944	1.7087
18	78.5592	819.2	0.23347	109.1289	0.00897	106.54	227.24	0.00066	1.25701	1.5666
19	77.8398	819.0	0.22846	103.3346	0.00943	109.56	207.34	0.00066	1.25447	1.4619
20	77.0922	818.0	0.22351	97.7382	0.00989	112.41	190.68	0.00065	1.25184	1.3721
21	76.3144	819.4	0.21868	92.8849	0.01028	115.60	176.53	0.00065	1.24911	1.2867
22	75.5170	813.9	0.21407	87.2047	0.01080	118.06	164.45	0.00064	1.24631	1.2247
23	74.6827	809.0	0.20964	81.9628	0.01131	119.87	153.88	0.00063	1.24338	1.1746
24	73.8252	802.5	0.20543	76.8050	0.01186	121.15	144.65	0.00062	1.24038	1.1361
25	72.9346	795.0	0.20141	71.8215	0.01243	121.95	136.45	0.00061	1.23727	1.1061
26	72.0101	786.6	0.19746	66.9741	0.01305	122.28	129.09	0.00060	1.23404	1.0836
27	71.0541	776.9	0.19351	62.2237	0.01372	122.26	122.45	0.00058	1.23072	1.0675
28	70.0607	766.3	0.18952	57.6127	0.01444	121.91	116.40	0.00057	1.22726	1.0564
29	69.0310	754.8	0.18553	53.1672	0.01521	121.28	110.84	0.00055	1.22369	1.0492
30	67.9619	742.7	0.18162	48.9441	0.01602	120.41	105.70	0.00054	1.21999	1.0447
31	66.8519	730.4	0.17759	44.9390	0.01688	118.86	100.92	0.00052	1.21615	1.0470
32	65.7014	717.5	0.17353	41.1262	0.01781	117.16	96.46	0.00050	1.21218	1.0521
33	64.5097	703.3	0.16947	37.4312	0.01885	115.32	92.27	0.00049	1.20808	1.0605
34	63.2732	690.0	0.16545	34.0776	0.01990	113.39	88.31	0.00047	1.20384	1.0694
35	61.9922	676.5	0.16138	30.9365	0.02103	111.37	84.57	0.00045	1.19945	1.0801
36	60.6643	662.4	0.15723	27.9606	0.02225	109.25	81.02	0.00044	1.19491	1.0931
37	59.2908	647.8	0.15304	25.1656	0.02359	107.06	77.65	0.00043	1.19023	1.1083
38	57.8698	634.0	0.14889	22.6546	0.02494	104.80	74.43	0.00041	1.18540	1.1231
39	56.4047	619.8	0.14480	20.3267	0.02639	102.51	71.38	0.00040	1.18044	1.1390
40	54.8979	607.1	0.14075	18.2964	0.02778	100.19	68.49	0.00039	1.17534	1.1527
42	51.7818	583.5	0.13275	14.8206	0.03061	95.96	63.18	0.00037	1.16486	1.1757
44	48.5797	565.0	0.12507	12.1657	0.03310	92.17	58.53	0.00037	1.15415	1.1877
46	45.3752	553.4	0.11743	10.2169	0.03492	88.57	54.57	0.00036	1.14351	1.1908
48	42.2687	547.3	0.11097	8.8143	0.03591	85.29	51.32	0.00037	1.13325	1.1825
50	39.3388	548.0	0.10549	7.8782	0.03577	82.37	48.71	0.00038	1.12363	1.1591
52	36.6521	553.9	0.10085	7.2446	0.03482	79.91	46.70	0.00041	1.11486	1.1272
54	34.2378	564.4	0.09686	6.8315	0.03327	77.92	45.19	0.00044	1.10702	1.0890
56	32.0953	579.2	0.09346	6.5765	0.03137	76.36	44.08	0.00047	1.10010	1.0486
58	30.2835	596.6	0.09057	6.4089	0.02944	75.22	43.27	0.00051	1.09401	1.0105
60	28.5293	616.9	0.08797	6.3073	0.02756	74.52	42.71	0.00055	1.08864	0.9744
65	25.1194	675.9	0.08223	6.1803	0.02355	74.27	42.01	0.00067	1.07776	0.9003
70	22.5144	745.7	0.07708	6.1615	0.02039	75.12	41.98	0.00079	1.06951	0.8497
75	20.4609	823.7	0.07224	6.1681	0.01797	75.78	42.37	0.00090	1.06303	0.8279
80	18.7948	909.6	0.06768	6.1862	0.01608	78.02	43.05	0.00102	1.05779	0.8068
85	17.4128	1003.1	0.06342	6.2078	0.01455	81.89	43.91	0.00116	1.05346	0.7827
90	16.2435	1103.5	0.05951	6.2279	0.01331	85.91	44.89	0.00130	1.04981	0.7673
95	15.2387	1209.8	0.05599	6.2454	0.01227	90.05	45.96	0.00143	1.04668	0.7578
100	14.3578	1322.2	0.05282	6.2622	0.01138	95.79	48.01	0.00160	1.04394	0.7543
120	11.7562	1804.3	0.04339	6.2901	0.00890	119.95	55.24	0.00229	1.03588	0.7392
140	10.0012	2267.5	0.03893	6.2970	0.00737	137.88	60.15	0.00297	1.03046	0.7295
160	8.7257	2674.1	0.03709	6.2927	0.00633	149.83	63.92	0.00365	1.02654	0.7217
180	7.7511	3017.7	0.03673	6.2834	0.00555	157.31	67.09	0.00436	1.02355	0.7148
200	6.9792	3312.8	0.03711	6.2720	0.00496	162.09	69.95	0.00509	1.02119	0.7090
220	6.3512	3578.4	0.03782	6.2600	0.00449	165.54	72.62	0.00585	1.01927	0.7041
240	5.8296	3831.0	0.03859	6.2480	0.00410	168.57	75.18	0.00663	1.01768	0.7001
260	5.3888	4081.9	0.03928	6.2364	0.00377	171.68	77.68	0.00745	1.01633	0.6969
280	5.0112	4336.2	0.03983	6.2255	0.00350	175.08	80.13	0.00829	1.01518	0.6944
300	4.6823	4600.2	0.04023	6.2192	0.00326	178.79	82.56	0.00917	1.01418	0.6921
350	4.0275	5277.2	0.04075	6.1949	0.00279	189.39	88.55	0.01149	1.01219	0.6888
400	3.5348	5981.9	0.04087	6.1748	0.00244	201.11	94.48	0.01401	1.01069	0.6869
450	3.1502	6705.2	0.04079	6.1581	0.00217	213.41	100.36	0.01673	1.00952	0.6857
500	2.8414	7432.6	0.04069	6.1441	0.00196	225.78	106.19	0.01964	1.00859	0.6850
550	2.5880	8166.5	0.04056	6.1323	0.00178	238.23	111.96	0.02275	1.00782	0.6844
600	2.3762	8905.1	0.04041	6.1222	0.00164	250.67	117.67	0.02606	1.00718	0.6841
700	2.0421	10402.9	0.04007	6.1058	0.00141	275.64	128.88	0.03323	1.00617	0.6836
800	1.7905	11939.7	0.03963	6.0932	0.00123	300.89	139.84	0.04113	1.00541	0.6835
900	1.5942	13526.6	0.03909	6.0833	0.00110	326.59	150.53	0.04973	1.00481	0.6836
1000	1.4367	15171.3	0.03847	6.0752	0.00099	440.56	160.98	0.07365	1.00434	0.5477
1200	1.1996	18635.1	0.03707	6.0630	0.00082	511.13	181.19	0.09979	1.00362	0.5449
1400	1.0298	22325.5	0.03563	6.0541	0.00071	583.34	200.60	0.12903	1.00311	0.5435
1600	0.9020	26164.3	0.03436	6.0474	0.00062	655.67	219.30	0.16128	1.00272	0.5427
1800	0.8025	30142.9	0.03322	6.0421	0.00055	728.90	237.40	0.19663	1.00242	0.5416
2000	0.7227	34407.7	0.03203	6.0374	0.00050	808.56	254.96	0.23558	1.00218	0.5391
2500	0.5780	48548.7	0.02754	6.0206	0.00040	1121.60	297.09	0.36097	1.00174	0.5126
3000	0.4780	76323.3	0.02028	5.9640	0.00034	1923.06	338.48	0.56595	1.00144	0.4505

* TWO-PHASE BOUNDARY

Table 4.1 (continued)
THERMODYNAMIC PROPERTIES OF PARAHYDROGEN (ISOBARS, SI UNITS)

7.00 MPA ISOBAR

TEMPERATURE DEG. K	VOLUME CU M/KG	ISOTHERM DERIVATIVE CU M-MPA/KG	ISOCHORE DERIVATIVE MPA/K	INTERNAL ENERGY KJ/KG-K	ENTHALPY KJ/KG	ENTROPY KJ/KG-K	CV KJ / KG-K	CP KJ / KG-K	VELOCITY OF SOUND M/S
* 15.968	0.01241	1.5924	0.9744	-304.7	-217.8	5.125	4.95	6.42	1437
16	0.01242	1.5910	0.9748	-304.5	-217.6	5.137	4.96	6.43	1437
17	0.01251	1.5282	0.9838	-298.5	-210.9	5.540	5.15	6.84	1424
18	0.01262	1.4642	0.9869	-292.2	-203.9	5.942	5.32	7.22	1410
19	0.01273	1.4027	0.9850	-285.6	-196.5	6.342	5.47	7.60	1396
20	0.01284	1.3450	0.9792	-278.6	-188.7	6.740	5.60	7.95	1382
21	0.01297	1.2874	0.9699	-271.4	-180.6	7.137	5.72	8.30	1367
22	0.01310	1.2395	0.9580	-263.8	-172.1	7.531	5.82	8.62	1354
23	0.01323	1.1794	0.9449	-256.0	-163.3	7.921	5.91	8.96	1337
24	0.01338	1.1240	0.9297	-247.8	-154.2	8.310	6.00	9.30	1320
25	0.01353	1.0684	0.9134	-239.4	-144.7	8.697	6.08	9.65	1303
26	0.01369	1.0143	0.8957	-230.7	-134.9	9.082	6.15	10.00	1285
27	0.01386	0.9618	0.8768	-221.8	-124.7	9.466	6.21	10.36	1267
28	0.01404	0.9094	0.8568	-212.5	-114.2	9.850	6.27	10.72	1247
29	0.01423	0.8584	0.8355	-202.9	-103.3	10.232	6.32	11.10	1228
30	0.01444	0.8083	0.8129	-193.1	-92.0	10.615	6.36	11.47	1207
31	0.01465	0.7604	0.7894	-182.9	-80.3	10.997	6.40	11.86	1187
32	0.01488	0.7147	0.7650	-172.5	-68.3	11.379	6.44	12.24	1166
33	0.01513	0.6708	0.7400	-161.8	-55.9	11.762	6.47	12.63	1144
34	0.01539	0.6287	0.7144	-150.7	-43.0	12.145	6.50	13.03	1123
35	0.01567	0.5885	0.6887	-139.5	-29.8	12.529	6.52	13.44	1101
36	0.01596	0.5506	0.6628	-127.9	-16.1	12.914	6.54	13.86	1080
37	0.01628	0.5134	0.6367	-116.0	-2.1	13.299	6.56	14.31	1058
38	0.01662	0.4800	0.6104	-103.9	12.4	13.686	6.58	14.72	1037
39	0.01698	0.4485	0.5841	-91.5	27.4	14.074	6.59	15.14	1015
40	0.01737	0.4194	0.5579	-78.9	42.7	14.463	6.59	15.55	995
42	0.01823	0.3681	0.5067	-53.0	74.6	15.241	6.61	16.34	954
44	0.01920	0.3261	0.4581	-26.4	108.1	16.019	6.63	17.07	916
46	0.02030	0.2935	0.4130	0.2	142.3	16.779	6.69	17.71	881
48	0.02152	0.2706	0.3721	27.5	178.2	17.542	6.74	18.11	853
50	0.02285	0.2544	0.3357	54.8	214.7	18.287	6.79	18.34	829
52	0.02427	0.2442	0.3036	81.5	251.4	19.008	6.83	18.39	811
54	0.02576	0.2400	0.2758	107.7	288.0	19.699	6.87	18.22	798
56	0.02731	0.2398	0.2516	133.3	324.5	20.362	6.90	17.92	789
58	0.02889	0.2428	0.2307	157.7	360.0	20.984	6.92	17.53	784
60	0.03048	0.2480	0.2127	181.3	394.6	21.572	6.96	17.12	781
65	0.03445	0.2667	0.1776	236.6	477.8	22.904	7.08	16.20	781
70	0.03836	0.2901	0.1526	288.5	557.0	24.078	7.27	15.54	788
75	0.04218	0.3157	0.1341	338.3	633.6	25.135	7.53	15.13	797
80	0.04592	0.3420	0.1198	387.2	708.6	26.104	7.84	14.92	807
85	0.04957	0.3686	0.1085	436.0	783.0	27.006	8.20	14.87	817
90	0.05315	0.3949	0.0992	485.4	857.5	27.857	8.59	14.93	828
95	0.05667	0.4211	0.0915	535.8	932.5	28.668	9.00	15.08	840
100	0.06013	0.4469	0.0850	587.4	1008.3	29.446	9.42	15.27	851
120	0.07350	0.5464	0.0662	809.8	1324.3	32.324	11.00	16.20	897
140	0.08640	0.6418	0.0547	1050.3	1655.1	34.872	11.95	16.83	951
160	0.09901	0.7340	0.0468	1300.8	1993.9	37.134	12.31	17.00	1006
180	0.11143	0.8241	0.0410	1552.4	2332.4	39.128	12.27	16.82	1063
200	0.12371	0.9126	0.0365	1799.6	2665.5	40.882	12.02	16.48	1119
220	0.13590	1.0000	0.0329	2039.9	2991.2	42.435	11.70	16.09	1173
240	0.14801	1.0865	0.0300	2273.2	3309.2	43.819	11.39	15.73	1225
260	0.16007	1.1724	0.0275	2500.3	3620.7	45.066	11.13	15.43	1275
280	0.17208	1.2577	0.0255	2722.3	3926.9	46.200	10.92	15.20	1323
300	0.18414	1.3443	0.0237	2941.5	4230.5	47.247	10.77	15.01	1369
350	0.21394	1.5545	0.0202	3476.0	4973.6	49.537	10.54	14.75	1475
400	0.24363	1.7635	0.0176	4002.2	5707.6	51.498	10.45	14.63	1571
450	0.27326	1.9716	0.0156	4524.7	6437.5	53.219	10.42	14.59	1661
500	0.30284	2.1793	0.0141	5046.8	7166.6	54.754	10.41	14.57	1746
550	0.33239	2.3865	0.0128	5568.0	7894.8	56.142	10.42	14.57	1827
600	0.36192	2.5936	0.0117	6090.0	8623.5	57.409	10.43	14.58	1903
700	0.42095	3.0071	0.0100	7136.5	10083.1	59.660	10.49	14.62	2048
800	0.47993	3.4203	0.0088	8190.0	11549.5	61.617	10.58	14.71	2181
900	0.53890	3.8332	0.0078	9254.0	13026.4	63.355	10.70	14.83	2305
1000	0.59785	4.2460	0.0070	10331.8	14516.8	64.928	10.86	14.99	2421
1200	0.71574	5.0713	0.0058	12542.7	17552.9	67.682	11.24	15.37	2633
1400	0.83360	5.8964	0.0050	14835.8	20671.0	70.089	11.68	15.81	2825
1600	0.95146	6.7213	0.0044	17214.6	23874.8	72.205	12.10	16.22	3002
1800	1.06933	7.5463	0.0039	19680.4	27165.7	74.105	12.50	16.63	3168
2000	1.18726	8.3712	0.0035	22238.7	30549.5	75.954	12.95	17.09	3323
2500	1.48400	10.4334	0.0028	29179.3	39567.3	82.318	14.97	19.23	3661
3000	1.79328	12.4955	0.0023	37888.4	50441.4	89.639	20.17	25.06	3938

* TWO-PHASE BOUNDARY

Table 4.1 (continued)

THERMODYNAMIC PROPERTIES OF PARAHYDROGEN (ISOBARS, SI UNITS)

7.00 MPA ISOBAR

TEMPERATURE DEG. K	DENSITY KG/CU M	$V(DH/DV)_P$ KJ/KG	$V(DP/DU)_V$ MPA-CU M/KJ X 10^2	$-V(DP/DV)_T$ MPA	$-(DV/DT)_P/V$ 1/DEG. K	THERMAL CONDUCTIVITY W/K-M X 10^3	VISCOSITY KG/M-S X 10^7	THERMAL DIFFUSIVITY SQ M/HR	DIELECTRIC CONSTANT	PRANDTL NUMBER
* 15.968	80.5666	844.9	0.24430	128.2933	0.00760	97.46	298.16	0.00068	1.26411	1.9633
16	80.5475	845.3	0.24413	128.1504	0.00761	97.68	297.03	0.00068	1.26404	1.9553
17	79.9143	848.5	0.23909	122.1251	0.00806	103.51	265.12	0.00068	1.26180	1.7507
18	79.2588	849.6	0.23410	116.0489	0.00850	108.05	239.40	0.00068	1.25948	1.6008
19	78.5724	850.1	0.22924	110.2150	0.00894	111.26	218.17	0.00067	1.25706	1.4898
20	77.8601	850.4	0.22459	104.7199	0.00935	114.31	200.45	0.00066	1.25455	1.3945
21	77.1150	849.3	0.22001	99.2809	0.00977	117.69	185.39	0.00066	1.25193	1.3069
22	76.3500	851.1	0.21557	94.6381	0.01012	120.32	172.53	0.00066	1.24923	1.2354
23	75.5655	845.5	0.21141	89.1232	0.01060	122.33	161.46	0.00065	1.24648	1.1831
24	74.7481	840.7	0.20730	84.0149	0.01107	123.78	151.72	0.00064	1.24361	1.1404
25	73.9086	834.2	0.20338	78.9602	0.01157	124.76	143.14	0.00063	1.24067	1.1072
26	73.0392	827.1	0.19956	74.0861	0.01209	125.27	135.47	0.00062	1.23764	1.0814
27	72.1426	819.5	0.19575	69.3888	0.01264	125.42	128.58	0.00060	1.23451	1.0617
28	71.2161	810.7	0.19194	64.7641	0.01323	125.26	122.32	0.00059	1.23128	1.0473
29	70.2596	801.1	0.18816	60.3127	0.01385	124.82	116.61	0.00058	1.22795	1.0366
30	69.2709	790.3	0.18445	55.9937	0.01452	124.14	111.36	0.00056	1.22452	1.0292
31	68.2493	779.5	0.18064	51.8953	0.01521	122.79	106.49	0.00055	1.22098	1.0283
32	67.1955	768.5	0.17681	48.0249	0.01593	121.28	101.97	0.00053	1.21734	1.0293
33	66.1095	757.1	0.17302	44.3451	0.01669	119.65	97.75	0.00052	1.21359	1.0321
34	64.9888	745.3	0.16922	40.8590	0.01749	117.94	93.78	0.00050	1.20973	1.0363
35	63.8360	733.3	0.16545	37.5682	0.01833	116.14	90.04	0.00049	1.20577	1.0423
36	62.6474	721.4	0.16171	34.4949	0.01921	114.25	86.51	0.00047	1.20169	1.0496
37	61.4259	708.6	0.15796	31.5377	0.02019	112.31	83.17	0.00046	1.19751	1.0594
38	60.1709	696.7	0.15423	28.8802	0.02113	110.31	80.00	0.00045	1.19323	1.0679
39	58.8836	684.7	0.15060	26.4106	0.02212	108.27	77.00	0.00044	1.18885	1.0768
40	57.5659	673.0	0.14699	24.1448	0.02311	106.21	74.14	0.00043	1.18437	1.0856
42	54.8551	651.4	0.13975	20.1923	0.02509	102.06	68.90	0.00041	1.17520	1.1033
44	52.0711	633.0	0.13262	16.9828	0.02697	98.31	64.24	0.00040	1.16583	1.1157
46	49.2584	620.1	0.12523	14.4590	0.02856	95.01	60.18	0.00039	1.15642	1.1217
48	46.4718	612.2	0.11872	12.5758	0.02958	91.89	56.70	0.00039	1.14714	1.1176
50	43.7729	608.7	0.11296	11.1370	0.03014	89.06	53.79	0.00040	1.13821	1.1080
52	41.2891	609.5	0.10789	10.0635	0.03017	86.56	51.41	0.00041	1.12976	1.0924
54	38.8158	615.6	0.10347	9.3166	0.02960	84.40	49.51	0.00043	1.12192	1.0689
56	36.6163	625.5	0.09960	8.7809	0.02865	82.62	48.02	0.00045	1.11475	1.0418
58	34.6166	638.8	0.09625	8.4036	0.02745	81.18	46.88	0.00048	1.10825	1.0126
60	32.8099	655.1	0.09319	8.1382	0.02613	80.18	46.02	0.00051	1.10241	0.9826
65	29.0283	706.2	0.08645	7.7421	0.02294	79.23	44.77	0.00061	1.09024	0.9156
70	26.0692	770.1	0.08055	7.5635	0.02018	79.48	44.34	0.00071	1.08079	0.8671
75	23.7860	844.3	0.07516	7.4838	0.01792	79.59	44.42	0.00080	1.07328	0.8445
80	21.7779	927.7	0.07016	7.4491	0.01609	81.44	44.82	0.00090	1.06718	0.8214
85	20.1741	1019.4	0.06555	7.4365	0.01459	85.04	45.46	0.00102	1.06212	0.7948
90	18.8139	1118.3	0.06137	7.4304	0.01336	88.84	46.26	0.00114	1.05785	0.7776
95	17.6464	1223.8	0.05761	7.4312	0.01232	92.80	47.17	0.00126	1.05419	0.7664
100	16.6298	1334.6	0.05429	7.4325	0.01144	97.56	48.83	0.00138	1.05102	0.7642
120	13.6060	1818.5	0.04426	7.4343	0.00891	121.08	55.79	0.00198	1.04160	0.7464
140	11.5745	2283.4	0.03958	7.4281	0.00737	138.67	60.57	0.00256	1.03531	0.7350
160	10.1003	2691.9	0.03763	7.4137	0.00631	150.44	64.26	0.00315	1.03077	0.7259
180	8.9746	3037.3	0.03719	7.3960	0.00554	157.85	67.38	0.00376	1.02730	0.7182
200	8.0834	3333.9	0.03753	7.3773	0.00494	162.59	70.21	0.00439	1.02457	0.7116
220	7.3585	3600.6	0.03821	7.3588	0.00447	166.03	72.86	0.00505	1.02235	0.7061
240	6.7563	3854.2	0.03895	7.3410	0.00408	169.07	75.41	0.00573	1.02051	0.7017
260	6.2474	4105.8	0.03961	7.3243	0.00376	172.19	77.90	0.00643	1.01895	0.6981
280	5.8112	4360.8	0.04013	7.3087	0.00348	175.59	80.35	0.00716	1.01762	0.6954
300	5.4307	4626.5	0.04051	7.3002	0.00324	179.32	82.78	0.00792	1.01646	0.6929
350	4.6743	5304.5	0.04100	7.2663	0.00278	189.96	88.79	0.00992	1.01415	0.6892
400	4.1046	6009.9	0.04108	7.2384	0.00243	201.73	94.75	0.01209	1.01242	0.6871
450	3.6596	6733.7	0.04098	7.2154	0.00217	214.10	100.66	0.01444	1.01107	0.6859
500	3.3021	7461.6	0.04086	7.1961	0.00195	226.53	106.52	0.01695	1.00998	0.6851
550	3.0085	8195.7	0.04071	7.1799	0.00178	239.05	112.32	0.01964	1.00909	0.6845
600	2.7630	8934.7	0.04055	7.1661	0.00163	251.57	118.06	0.02249	1.00835	0.6841
700	2.3756	10432.9	0.04019	7.1437	0.00140	276.69	129.35	0.02867	1.00718	0.6836
800	2.0836	11970.1	0.03974	7.1266	0.00123	302.09	140.37	0.03548	1.00629	0.6835
900	1.8556	13557.4	0.03918	7.1130	0.00109	327.94	151.14	0.04289	1.00560	0.6836
1000	1.6726	15202.5	0.03854	7.1020	0.00099	440.56	161.65	0.06326	1.00505	0.5500
1200	1.3972	18667.3	0.03714	7.0854	0.00082	511.13	181.99	0.08568	1.00422	0.5473
1400	1.1996	22358.6	0.03569	7.0733	0.00071	583.34	201.52	0.11076	1.00362	0.5460
1600	1.0510	26197.5	0.03440	7.0643	0.00062	655.63	220.34	0.13841	1.00317	0.5453
1800	0.9352	30172.1	0.03327	7.0570	0.00055	728.67	238.55	0.16871	1.00282	0.5443
2000	0.8423	34419.8	0.03209	7.0508	0.00050	807.64	256.23	0.20205	1.00254	0.5420
2500	0.6739	48291.4	0.02776	7.0306	0.00040	1110.37	298.60	0.30846	1.00203	0.5172
3000	0.5576	74837.9	0.02075	6.9679	0.00033	1865.56	340.11	0.48061	1.00168	0.4568

* TWO-PHASE BOUNDARY

Table 4.1 (continued)
THERMODYNAMIC PROPERTIES OF PARAHYDROGEN (ISOBARS, SI UNITS)

8.00 MPA ISOBAR

TEMPERATURE DEG. K	VOLUME CU M/KG	ISOTHERM DERIVATIVE CU M-MPA/KG	ISOCHORE DERIVATIVE MPA/K	INTERNAL ENERGY KJ/KG-K	ENTHALPY KJ/KG	ENTROPY KJ/KG-K	CV KJ / KG-K	CP KJ / KG-K	VELOCITY OF SOUND M/S
* 16.253	0.01234	1.6420	1.010	-303.9	-205.1	5.142	4.99	6.53	1465
17	0.01241	1.6009	0.9901	-299.4	-200.1	5.442	5.13	6.73	1450
18	0.01251	1.5400	0.9949	-293.3	-193.2	5.838	5.30	7.11	1437
19	0.01262	1.4780	0.9947	-286.8	-185.9	6.232	5.46	7.48	1424
20	0.01273	1.4176	0.9904	-280.1	-178.3	6.624	5.59	7.83	1409
21	0.01284	1.3626	0.9829	-273.0	-170.3	7.014	5.71	8.16	1396
22	0.01296	1.3074	0.9726	-265.7	-161.9	7.402	5.82	8.49	1382
23	0.01309	1.2612	0.9603	-258.0	-153.3	7.786	5.91	8.79	1370
24	0.01323	1.2048	0.9469	-250.1	-144.3	8.168	6.00	9.12	1354
25	0.01337	1.1497	0.9317	-242.0	-135.0	8.547	6.07	9.45	1337
26	0.01352	1.0957	0.9153	-233.6	-125.4	8.924	6.14	9.78	1320
27	0.01367	1.0434	0.8978	-224.9	-115.5	9.299	6.21	10.11	1303
28	0.01384	0.9928	0.8791	-215.9	-105.2	9.672	6.27	10.44	1286
29	0.01401	0.9423	0.8593	-206.7	-94.6	10.045	6.33	10.79	1268
30	0.01420	0.8935	0.8385	-197.2	-83.7	10.416	6.37	11.13	1249
31	0.01439	0.8453	0.8166	-187.5	-72.4	10.786	6.41	11.48	1230
32	0.01460	0.7993	0.7939	-177.5	-60.7	11.156	6.45	11.83	1210
33	0.01481	0.7555	0.7704	-167.2	-48.7	11.525	6.48	12.17	1191
34	0.01504	0.7134	0.7465	-156.7	-36.4	11.893	6.51	12.52	1171
35	0.01529	0.6734	0.7222	-146.0	-23.7	12.261	6.54	12.87	1151
36	0.01554	0.6345	0.6978	-135.0	-10.7	12.628	6.56	13.24	1131
37	0.01582	0.5986	0.6734	-123.8	2.8	12.996	6.58	13.59	1112
38	0.01611	0.5644	0.6489	-112.3	16.5	13.363	6.60	13.95	1092
39	0.01641	0.5326	0.6244	-100.6	30.7	13.730	6.61	14.30	1073
40	0.01674	0.5024	0.5998	-88.8	45.1	14.097	6.62	14.64	1054
42	0.01745	0.4474	0.5516	-64.5	75.1	14.828	6.63	15.33	1017
44	0.01824	0.4026	0.5050	-39.6	106.4	15.555	6.65	15.93	982
46	0.01912	0.3647	0.4611	-14.7	138.3	16.263	6.71	16.51	947
48	0.02009	0.3360	0.4206	11.1	171.7	16.976	6.76	16.95	918
50	0.02114	0.3135	0.3835	36.9	206.0	17.675	6.80	17.28	893
52	0.02227	0.2983	0.3503	62.6	240.7	18.356	6.84	17.45	872
54	0.02346	0.2882	0.3206	88.0	275.7	19.015	6.88	17.49	856
56	0.02471	0.2824	0.2943	113.2	310.9	19.655	6.92	17.41	843
58	0.02600	0.2806	0.2712	137.5	345.5	20.264	6.95	17.23	834
60	0.02732	0.2819	0.2508	161.2	379.8	20.844	6.98	16.98	828
65	0.03068	0.2941	0.2100	217.5	462.9	22.176	7.11	16.28	821
70	0.03403	0.3128	0.1802	270.6	542.8	23.360	7.30	15.71	821
75	0.03734	0.3348	0.1580	321.7	620.3	24.430	7.55	15.35	825
80	0.04059	0.3590	0.1408	371.7	696.5	25.413	7.87	15.15	831
85	0.04380	0.3840	0.1271	421.6	772.0	26.329	8.22	15.09	839
90	0.04694	0.4096	0.1160	472.0	847.5	27.191	8.62	15.14	848
95	0.05004	0.4349	0.1068	523.1	923.5	28.013	9.02	15.27	858
100	0.05309	0.4605	0.0991	575.5	1000.2	28.800	9.44	15.44	868
120	0.06494	0.5604	0.0771	800.5	1320.1	31.714	11.02	16.39	913
140	0.07623	0.6547	0.0632	1042.7	1652.6	34.275	11.97	16.92	962
160	0.08733	0.7475	0.0539	1294.4	1993.0	36.548	12.33	17.07	1017
180	0.09825	0.8380	0.0471	1546.9	2332.9	38.550	12.29	16.88	1073
200	0.10904	0.9270	0.0419	1794.7	2667.1	40.310	12.03	16.53	1128
220	0.11974	1.0148	0.0377	2035.6	2993.5	41.866	11.71	16.13	1182
240	0.13037	1.1016	0.0343	2269.4	3312.3	43.253	11.40	15.76	1234
260	0.14095	1.1877	0.0315	2496.9	3624.4	44.503	11.13	15.46	1284
280	0.15148	1.2733	0.0292	2719.3	3931.1	45.639	10.93	15.22	1332
300	0.16207	1.3605	0.0271	2938.9	4235.5	46.688	10.77	15.03	1378
350	0.18818	1.5711	0.0231	3474.0	4979.4	48.980	10.55	14.76	1483
400	0.21418	1.7803	0.0202	4000.5	5714.0	50.944	10.45	14.64	1579
450	0.24012	1.9885	0.0179	4523.4	6444.4	52.665	10.43	14.60	1668
500	0.26602	2.1963	0.0161	5045.7	7173.8	54.201	10.42	14.58	1753
550	0.29188	2.4036	0.0146	5567.2	7902.2	55.590	10.42	14.57	1833
600	0.31773	2.6107	0.0134	6089.3	8631.2	56.857	10.44	14.58	1910
700	0.36939	3.0243	0.0114	7136.0	10091.0	59.108	10.49	14.63	2054
800	0.42101	3.4375	0.0100	8189.6	11557.7	61.065	10.58	14.71	2186
900	0.47261	3.8504	0.0089	9253.8	13034.6	62.803	10.70	14.83	2310
1000	0.52419	4.2632	0.0080	10331.6	14525.1	64.377	10.86	14.99	2426
1200	0.62735	5.0885	0.0067	12542.6	17561.4	67.131	11.25	15.37	2637
1400	0.73048	5.9135	0.0057	14835.7	20679.6	69.538	11.68	15.81	2829
1600	0.83361	6.7385	0.0050	17214.6	23883.4	71.655	12.10	16.22	3006
1800	0.93674	7.5634	0.0044	19680.1	27174.1	73.554	12.50	16.62	3172
2000	1.03993	8.3883	0.0040	22237.3	30556.8	75.402	12.94	17.08	3327
2500	1.29946	10.4505	0.0032	29158.2	39553.9	81.757	14.88	19.13	3666
3000	1.56927	12.5126	0.0027	37756.2	50310.4	89.037	19.79	24.63	3944

* TWO-PHASE BOUNDARY

Table 4.1 (continued)
THERMODYNAMIC PROPERTIES OF PARAHYDROGEN (ISOBARS, SI UNITS)

8.00 MPA ISOBAR

TEMPERATURE DEG. K	DENSITY KG/CU M	$V(DH/DV)_P$ KJ/KG	$V(DP/DU)_V$ MPA-CU M/KJ X 10^2	$-V(DP/DV)_T$ MPA	$-(DV/DT)_P/V$ 1/DEG. K	THERMAL CONDUCTIVITY W/K-M X 10^3	VISCOSITY KG/M-S X 10^7	THERMAL DIFFUSIVITY SQ M/HR	DIELECTRIC CONSTANT	PRANDTL NUMBER
* 16.253	81.0262	860.5	0.24956	133.0474	0.00759	100.44	304.73	0.00068	1.26573	1.9811
17	80.5578	876.8	0.23972	128.9649	0.00768	104.76	279.55	0.00070	1.26407	1.7963
18	79.9223	879.9	0.23480	123.0814	0.00808	109.49	251.89	0.00069	1.26183	1.6363
19	79.2669	880.9	0.23004	117.1587	0.00849	112.90	229.28	0.00069	1.25951	1.5190
20	78.5812	880.8	0.22547	111.3942	0.00889	116.11	210.37	0.00068	1.25709	1.4188
21	77.8729	881.4	0.22110	106.1125	0.00926	119.69	194.42	0.00068	1.25459	1.3262
22	77.1386	880.4	0.21681	100.8483	0.00964	122.49	180.78	0.00067	1.25201	1.2531
23	76.3826	882.0	0.21271	96.3300	0.00997	124.65	169.02	0.00067	1.24935	1.1922
24	75.6049	877.4	0.20886	91.0901	0.01040	126.26	158.77	0.00066	1.24662	1.1470
25	74.8090	872.1	0.20506	86.0107	0.01083	127.40	149.79	0.00065	1.24383	1.1107
26	73.9871	865.9	0.20134	81.0675	0.01129	128.07	141.79	0.00064	1.24095	1.0823
27	73.1404	859.3	0.19765	76.3177	0.01176	128.38	134.61	0.00063	1.23799	1.0599
28	72.2675	852.4	0.19397	71.7474	0.01225	128.38	128.13	0.00061	1.23494	1.0424
29	71.3703	844.2	0.19035	67.2527	0.01278	128.11	122.23	0.00060	1.23182	1.0292
30	70.4457	835.3	0.18684	62.9424	0.01332	127.59	116.83	0.00059	1.22860	1.0189
31	69.4946	825.6	0.18323	58.7436	0.01390	126.39	111.85	0.00057	1.22530	1.0156
32	68.5164	815.8	0.17958	54.7641	0.01450	125.04	107.23	0.00056	1.22191	1.0142
33	67.5114	805.9	0.17598	51.0066	0.01510	123.57	102.94	0.00054	1.21843	1.0140
34	66.4798	795.6	0.17238	47.4290	0.01574	122.03	98.93	0.00053	1.21487	1.0152
35	65.4218	785.3	0.16880	44.0567	0.01639	120.40	95.16	0.00051	1.21122	1.0175
36	64.3373	774.4	0.16524	40.8199	0.01709	118.70	91.62	0.00050	1.20749	1.0217
37	63.2256	764.1	0.16176	37.8475	0.01779	116.93	88.27	0.00049	1.20367	1.0262
38	62.0887	753.6	0.15835	35.0456	0.01852	115.12	85.10	0.00048	1.19978	1.0315
39	60.9269	743.1	0.15505	32.4475	0.01924	113.27	82.10	0.00047	1.19581	1.0365
40	59.7429	732.7	0.15174	30.0121	0.01999	111.40	79.26	0.00046	1.19177	1.0418
42	57.3140	712.5	0.14512	25.6429	0.02151	107.61	74.03	0.00044	1.18352	1.0543
44	54.8247	696.1	0.13848	22.0737	0.02288	103.88	69.35	0.00043	1.17510	1.0632
46	52.3052	683.1	0.13138	19.0754	0.02418	100.54	65.22	0.00042	1.16662	1.0714
48	49.7865	674.2	0.12502	16.7280	0.02514	97.60	61.61	0.00042	1.15818	1.0701
50	47.3102	668.3	0.11920	14.8314	0.02586	94.88	58.51	0.00042	1.14993	1.0658
52	44.9111	667.3	0.11397	13.3977	0.02614	92.38	55.88	0.00042	1.14197	1.0552
54	42.6222	669.9	0.10928	12.2818	0.02610	90.19	53.70	0.00044	1.13441	1.0410
56	40.4657	676.0	0.10512	11.4269	0.02576	88.30	51.91	0.00045	1.12732	1.0237
58	38.4552	685.6	0.10147	10.7920	0.02513	86.70	50.47	0.00047	1.12074	1.0029
60	36.5979	698.5	0.09812	10.3163	0.02431	85.51	49.34	0.00050	1.11469	0.9797
65	32.5995	743.2	0.09062	9.5875	0.02190	84.01	47.52	0.00057	1.10173	0.9207
70	29.3897	801.5	0.08403	9.1934	0.01960	83.75	46.69	0.00065	1.09140	0.8759
75	26.7843	871.1	0.07810	8.9679	0.01762	83.41	46.48	0.00073	1.08307	0.8550
80	24.6339	951.3	0.07266	8.8430	0.01592	84.88	46.64	0.00082	1.07622	0.8322
85	22.8327	1040.6	0.06770	8.7688	0.01450	88.22	47.08	0.00092	1.07051	0.8051
90	21.3028	1138.0	0.06323	8.7248	0.01330	91.80	47.70	0.00102	1.06568	0.7865
95	19.9830	1242.0	0.05924	8.6903	0.01229	95.56	48.46	0.00113	1.06152	0.7742
100	18.8344	1352.2	0.05572	8.6726	0.01142	99.49	49.75	0.00123	1.05792	0.7723
120	15.3987	1834.1	0.04545	8.6301	0.00893	122.37	56.44	0.00175	1.04718	0.7558
140	13.1174	2301.3	0.04023	8.5880	0.00735	139.58	61.06	0.00226	1.04009	0.7403
160	11.4504	2711.4	0.03817	8.5589	0.00630	151.15	64.65	0.00278	1.03493	0.7301
180	10.1779	3058.2	0.03766	8.5295	0.00552	158.45	67.71	0.00332	1.03101	0.7214
200	9.1706	3355.9	0.03795	8.5012	0.00492	163.14	70.49	0.00387	1.02791	0.7141
220	8.3512	3623.6	0.03859	8.4745	0.00445	166.56	73.12	0.00445	1.02539	0.7081
240	7.6703	3878.0	0.03930	8.4496	0.00407	169.59	75.65	0.00505	1.02330	0.7032
260	7.0948	4130.3	0.03993	8.4266	0.00374	172.71	78.13	0.00567	1.02154	0.6994
280	6.6015	4385.8	0.04044	8.4053	0.00347	176.12	80.58	0.00631	1.02003	0.6964
300	6.1701	4653.2	0.04080	8.3944	0.00323	179.86	83.01	0.00698	1.01872	0.6938
350	5.3141	5332.0	0.04124	8.3490	0.00277	190.54	89.03	0.00875	1.01610	0.6897
400	4.6690	6038.0	0.04129	8.3119	0.00242	202.36	95.02	0.01066	1.01414	0.6874
450	4.1646	6762.3	0.04116	8.2815	0.00216	214.78	100.95	0.01272	1.01260	0.6861
500	3.7592	7490.6	0.04102	8.2561	0.00195	227.28	106.85	0.01493	1.01137	0.6852
550	3.4260	8225.1	0.04086	8.2348	0.00177	239.87	112.68	0.01730	1.01036	0.6846
600	3.1473	8964.3	0.04069	8.2166	0.00163	252.47	118.46	0.01981	1.00951	0.6841
700	2.7072	10462.9	0.04030	8.1873	0.00140	277.73	129.82	0.02525	1.00818	0.6836
800	2.3753	12000.5	0.03983	8.1649	0.00123	303.29	140.91	0.03125	1.00717	0.6835
900	2.1159	13588.2	0.03927	8.1471	0.00109	329.29	151.74	0.03777	1.00639	0.6836
1000	1.9077	15233.8	0.03862	8.1328	0.00098	440.56	162.32	0.05546	1.00576	0.5523
1200	1.5940	18699.5	0.03720	8.1111	0.00082	511.13	182.80	0.07510	1.00481	0.5497
1400	1.3090	22391.6	0.03574	8.0954	0.00071	583.34	202.45	0.09705	1.00413	0.5486
1600	1.1996	26230.7	0.03445	8.0836	0.00062	655.60	221.39	0.12127	1.00362	0.5479
1800	1.0675	30202.3	0.03331	8.0742	0.00055	728.49	239.71	0.14778	1.00322	0.5470
2000	0.9616	34436.3	0.03214	8.0662	0.00050	806.90	257.50	0.17691	1.00290	0.5449
2500	0.7696	48091.1	0.02795	8.0422	0.00040	1101.32	300.12	0.26926	1.00232	0.5214
3000	0.6372	73647.0	0.02115	7.9735	0.00033	1819.14	341.77	0.41728	1.00192	0.4627

* TWO-PHASE BOUNDARY

Table 4.1 (continued)
THERMODYNAMIC PROPERTIES OF PARAHYDROGEN (ISOBARS, SI UNITS)

10.00 MPA ISOBAR

TEMPERATURE DEG. K	VOLUME CU M/KG	ISOTHERM DERIVATIVE CU M-MPA/KG	ISOCHORE DERIVATIVE MPA/K	INTERNAL ENERGY KJ/KG-K	ENTHALPY KJ/KG	ENTROPY KJ/KG-K	CV KJ / KG-K	CP KJ / KG-K	VELOCITY OF SOUND M/S
* 16.808	0.01221	1.7491	1.024	-302.1	-179.9	5.181	5.07	6.58	1506
17	0.01223	1.7383	1.024	-301.0	-178.6	5.256	5.11	6.64	1503
18	0.01232	1.6821	1.021	-295.1	-171.8	5.645	5.27	6.97	1491
19	0.01241	1.6243	1.012	-288.9	-164.7	6.029	5.42	7.27	1476
20	0.01251	1.5672	1.011	-282.4	-157.3	6.411	5.57	7.61	1464
21	0.01262	1.5088	1.006	-275.7	-149.5	6.790	5.69	7.93	1450
22	0.01272	1.4503	0.9983	-268.7	-141.4	7.166	5.80	8.25	1436
23	0.01284	1.3999	0.9884	-261.4	-133.1	7.539	5.90	8.55	1424
24	0.01296	1.3481	0.9764	-253.9	-124.4	7.909	5.99	8.84	1410
25	0.01308	1.3034	0.9632	-246.2	-115.4	8.276	6.07	9.12	1399
26	0.01321	1.2515	0.9495	-238.2	-106.1	8.640	6.14	9.41	1385
27	0.01335	1.2002	0.9342	-230.0	-96.5	9.000	6.21	9.71	1370
28	0.01349	1.1494	0.9179	-221.6	-86.7	9.359	6.28	10.01	1354
29	0.01364	1.1003	0.9005	-212.9	-76.5	9.715	6.34	10.31	1338
30	0.01379	1.0532	0.8822	-204.0	-66.1	10.070	6.38	10.60	1322
31	0.01396	1.0079	0.8630	-194.9	-55.3	10.422	6.43	10.89	1307
32	0.01413	0.9630	0.8431	-185.5	-44.3	10.773	6.47	11.19	1290
33	0.01431	0.9196	0.8225	-176.0	-32.9	11.121	6.51	11.48	1273
34	0.01449	0.8772	0.8012	-166.3	-21.3	11.468	6.54	11.77	1256
35	0.01469	0.8367	0.7794	-156.3	-9.4	11.814	6.57	12.06	1239
36	0.01490	0.7978	0.7574	-146.2	2.8	12.157	6.60	12.34	1221
37	0.01511	0.7606	0.7352	-135.8	15.3	12.499	6.62	12.63	1204
38	0.01534	0.7254	0.7129	-125.3	28.0	12.840	6.64	12.90	1187
39	0.01557	0.6901	0.6906	-114.7	41.1	13.179	6.66	13.19	1170
40	0.01582	0.6588	0.6686	-103.8	54.4	13.516	6.66	13.46	1153
42	0.01635	0.6014	0.6249	-81.7	81.8	14.185	6.68	13.98	1121
44	0.01693	0.5501	0.5821	-59.1	110.3	14.847	6.70	14.47	1090
46	0.01757	0.5076	0.5408	-36.5	139.2	15.489	6.76	14.94	1059
48	0.01825	0.4709	0.5016	-13.0	169.5	16.133	6.81	15.35	1031
50	0.01899	0.4405	0.4649	10.7	200.5	16.767	6.85	15.69	1005
52	0.01977	0.4159	0.4308	34.5	232.2	17.388	6.89	15.96	982
54	0.02061	0.3968	0.3996	58.3	264.3	17.994	6.93	16.15	962
56	0.02148	0.3825	0.3711	82.2	297.0	18.589	6.97	16.27	945
58	0.02239	0.3723	0.3452	105.7	329.6	19.161	7.00	16.31	931
60	0.02334	0.3655	0.3218	128.9	362.3	19.714	7.04	16.30	920
65	0.02580	0.3615	0.2729	185.3	443.3	21.012	7.17	16.09	901
70	0.02835	0.3702	0.2352	239.6	523.0	22.193	7.36	15.76	891
75	0.03090	0.3855	0.2062	292.2	601.2	23.272	7.61	15.50	886
80	0.03345	0.4039	0.1835	343.8	678.3	24.267	7.91	15.38	886
85	0.03598	0.4247	0.1653	395.3	755.0	25.198	8.27	15.35	888
90	0.03848	0.4468	0.1505	447.1	831.9	26.077	8.66	15.41	892
95	0.04096	0.4702	0.1382	499.6	909.3	26.913	9.06	15.54	898
100	0.04342	0.4938	0.1278	553.2	987.4	27.714	9.48	15.71	905
120	0.05297	0.5907	0.0987	782.4	1312.1	30.672	11.05	16.61	942
140	0.06221	0.6861	0.0807	1028.0	1650.1	33.276	12.00	17.15	990
160	0.07104	0.7761	0.0682	1282.0	1992.4	35.562	12.36	17.21	1039
180	0.07985	0.8673	0.0595	1536.2	2334.7	37.578	12.31	16.99	1094
200	0.08854	0.9568	0.0528	1785.4	2670.8	39.348	12.05	16.62	1148
220	0.09715	1.0451	0.0475	2027.3	2998.8	40.912	11.73	16.21	1202
240	0.10570	1.1324	0.0432	2262.0	3319.0	42.305	11.41	15.83	1253
260	0.11419	1.2190	0.0396	2490.3	3632.2	43.559	11.15	15.51	1302
280	0.12265	1.3049	0.0366	2713.3	3939.9	44.699	10.94	15.27	1349
300	0.13119	1.3936	0.0340	2933.8	4245.7	45.752	10.79	15.07	1395
350	0.15212	1.6046	0.0290	3470.0	4991.2	48.050	10.56	14.79	1499
400	0.17295	1.8141	0.0252	3997.5	5727.0	50.016	10.46	14.66	1594
450	0.19373	2.0226	0.0224	4520.9	6458.2	51.740	10.43	14.61	1683
500	0.21446	2.2304	0.0201	5043.7	7188.3	53.277	10.42	14.59	1767
550	0.23517	2.4378	0.0182	5565.5	7917.2	54.666	10.43	14.58	1846
600	0.25586	2.6449	0.0167	6087.9	8646.5	55.934	10.44	14.59	1922
700	0.29720	3.0586	0.0143	7135.0	10106.9	58.187	10.49	14.63	2065
800	0.33850	3.4718	0.0125	8188.9	11573.9	60.144	10.58	14.71	2197
900	0.37979	3.8847	0.0111	9253.3	13051.2	61.882	10.71	14.84	2320
1000	0.42106	4.2974	0.0100	10331.2	14541.9	63.456	10.86	14.99	2435
1200	0.50359	5.1227	0.0083	12542.5	17578.4	66.211	11.25	15.37	2646
1400	0.58610	5.9477	0.0071	14835.7	20696.7	68.618	11.68	15.81	2837
1600	0.66860	6.7726	0.0062	17214.6	23900.6	70.734	12.10	16.22	3014
1800	0.75111	7.5975	0.0056	19679.8	27190.9	72.634	12.49	16.62	3179
2000	0.83366	8.4223	0.0050	22235.5	30572.0	74.481	12.93	17.06	3334
2500	1.04113	10.4844	0.0040	29126.2	39537.5	80.822	14.74	18.98	3674
3000	1.25599	12.5464	0.0033	37554.2	50114.1	88.037	19.21	23.97	3955

* TWO-PHASE BOUNDARY

Table 4.1 (continued)
THERMODYNAMIC PROPERTIES OF PARAHYDROGEN (ISOBARS, SI UNITS)

10.00 MPA ISOBAR

TEMPERATURE DEG. K	DENSITY KG/CU M	$V(DH/DV)_P$ KJ/KG	$V(DP/DU)_V$ MPA-CU M/KJ $\times 10^2$	$-V(DP/DV)_T$ MPA	$-(DV/DT)_P/V$ 1/DEG. K	THERMAL CONDUCTIVITY W/K-M $\times 10^3$	VISCOSITY KG/M-S $\times 10^7$	THERMAL DIFFUSIVITY SQ M/HR	DIELECTRIC CONSTANT	PRANDTL NUMBER
* 16.808	81.8729	919.9	0.24645	143.2003	0.00715	106.03	316.91	0.00071	1.26874	1.9661
17	81.7600	922.0	0.24510	142.1202	0.00720	107.12	309.94	0.00071	1.26834	1.9216
18	81.1621	931.5	0.23860	136.5241	0.00748	112.22	278.08	0.00071	1.26621	1.7262
19	80.5558	939.9	0.23167	130.8463	0.00774	115.98	252.38	0.00071	1.26407	1.5825
20	79.9218	942.6	0.22727	125.2542	0.00807	119.53	230.97	0.00071	1.26182	1.4701
21	79.2664	943.0	0.22304	119.5946	0.00841	123.45	212.99	0.00071	1.25951	1.3687
22	78.5887	941.9	0.21893	113.9783	0.00876	126.57	197.68	0.00070	1.25711	1.2884
23	77.8936	942.9	0.21502	109.0437	0.00906	129.04	184.56	0.00070	1.25466	1.2223
24	77.1767	941.9	0.21121	104.0384	0.00939	130.93	173.15	0.00069	1.25214	1.1691
25	76.4428	942.9	0.20758	99.6355	0.00967	132.34	163.18	0.00068	1.24956	1.1240
26	75.6908	939.1	0.20418	94.7233	0.01002	133.27	154.38	0.00067	1.24692	1.0904
27	74.9239	934.7	0.20070	89.9250	0.01039	133.85	146.57	0.00066	1.24423	1.0633
28	74.1366	929.4	0.19723	85.2159	0.01077	134.11	139.55	0.00065	1.24147	1.0417
29	73.3292	923.7	0.19384	80.6838	0.01116	134.10	133.20	0.00064	1.23865	1.0241
30	72.5015	917.6	0.19059	76.3574	0.01155	133.85	127.43	0.00063	1.23576	1.0093
31	71.6523	911.6	0.18727	72.2213	0.01195	132.88	122.13	0.00061	1.23280	1.0011
32	70.7876	904.6	0.18397	68.1673	0.01237	131.78	117.27	0.00060	1.22979	0.9956
33	69.9029	897.2	0.18070	64.2854	0.01279	130.55	112.77	0.00059	1.22671	0.9915
34	68.9988	889.2	0.17743	60.5267	0.01324	129.26	108.59	0.00057	1.22358	0.9888
35	68.0764	881.1	0.17416	56.9585	0.01369	127.89	104.70	0.00056	1.22038	0.9871
36	67.1359	872.9	0.17091	53.5628	0.01414	126.44	101.05	0.00055	1.21713	0.9864
37	66.1783	864.5	0.16770	50.3323	0.01461	124.93	97.62	0.00054	1.21383	0.9867
38	65.2037	856.2	0.16457	47.2967	0.01507	123.38	94.40	0.00053	1.21047	0.9874
39	64.2133	846.5	0.16161	44.3160	0.01559	121.79	91.36	0.00052	1.20706	0.9896
40	63.2056	838.2	0.15871	41.6392	0.01606	120.18	88.48	0.00051	1.20360	0.9908
42	61.1498	822.5	0.15292	36.7751	0.01699	116.88	83.17	0.00049	1.19657	0.9945
44	59.0530	807.9	0.14703	32.4879	0.01792	113.60	78.42	0.00048	1.18942	0.9991
46	56.9271	798.2	0.14050	28.8948	0.01872	110.41	74.16	0.00047	1.18221	1.0034
48	54.7942	789.5	0.13451	25.8049	0.01944	107.36	70.37	0.00046	1.17499	1.0058
50	52.6712	783.0	0.12889	23.2021	0.02004	104.64	67.02	0.00046	1.16785	1.0048
52	50.5770	779.2	0.12366	21.0339	0.02043	102.27	64.09	0.00046	1.16082	1.0001
54	48.5318	778.5	0.11881	19.2554	0.02075	100.11	61.53	0.00046	1.15399	0.9930
56	46.5526	780.7	0.11439	17.8052	0.02084	98.15	59.34	0.00047	1.14741	0.9836
58	44.6539	785.5	0.11041	16.6230	0.02077	96.42	57.47	0.00048	1.14112	0.9723
60	42.8456	793.3	0.10669	15.6610	0.02055	95.05	55.91	0.00049	1.13515	0.9588
65	38.7532	825.8	0.09822	14.0104	0.01948	92.92	53.06	0.00054	1.12172	0.9186
70	35.2794	875.2	0.09063	13.0614	0.01801	91.90	51.41	0.00059	1.11040	0.8818
75	32.3603	938.1	0.08376	12.4759	0.01653	90.89	50.57	0.00065	1.10095	0.8627
80	29.8978	1011.8	0.07755	12.0748	0.01520	91.69	50.28	0.00072	1.09303	0.8431
85	27.7951	1096.0	0.07193	11.8046	0.01401	94.55	50.34	0.00080	1.08629	0.8172
90	25.9853	1189.0	0.06691	11.6098	0.01296	97.71	50.65	0.00088	1.08052	0.7991
95	24.4121	1290.5	0.06246	11.4780	0.01204	101.10	51.15	0.00096	1.07552	0.7861
100	23.0330	1398.1	0.05857	11.3739	0.01124	103.72	51.83	0.00103	1.07115	0.7852
120	18.8773	1876.1	0.04731	11.1500	0.00885	125.38	57.97	0.00144	1.05805	0.7680
140	16.0754	2343.2	0.04183	11.0295	0.00732	141.76	62.22	0.00185	1.04929	0.7527
160	14.0768	2754.6	0.03922	10.9245	0.00625	152.83	65.56	0.00227	1.04306	0.7382
180	12.5242	3103.3	0.03857	10.8616	0.00548	159.85	68.46	0.00270	1.03825	0.7278
200	11.2944	3402.7	0.03877	10.8066	0.00489	164.39	71.14	0.00315	1.03445	0.7191
220	10.2934	3671.7	0.03934	10.7578	0.00441	167.73	73.69	0.00362	1.03136	0.7120
240	9.4610	3927.2	0.03999	10.7141	0.00403	170.71	76.18	0.00410	1.02880	0.7063
260	8.7570	4180.5	0.04058	10.6747	0.00371	173.82	78.63	0.00461	1.02664	0.7019
280	8.1531	4436.9	0.04104	10.6392	0.00344	177.23	81.07	0.00513	1.02478	0.6984
300	7.6226	4707.8	0.04136	10.6229	0.00320	180.97	83.49	0.00567	1.02316	0.6954
350	6.5738	5387.9	0.04172	10.5486	0.00274	191.71	89.53	0.00710	1.01995	0.6906
400	5.7819	6095.0	0.04171	10.4889	0.00241	203.61	95.56	0.00865	1.01753	0.6880
450	5.1619	6820.1	0.04153	10.4403	0.00214	216.15	101.55	0.01032	1.01564	0.6864
500	4.6628	7549.0	0.04135	10.4000	0.00193	228.78	107.50	0.01211	1.01412	0.6854
550	4.2523	8284.0	0.04115	10.3662	0.00176	241.51	113.40	0.01402	1.01287	0.6847
600	3.9084	9023.6	0.04095	10.3375	0.00162	254.25	119.25	0.01605	1.01183	0.6842
700	3.3648	10523.1	0.04053	10.2915	0.00139	279.81	130.75	0.02046	1.01017	0.6836
800	2.9542	12061.3	0.04003	10.2563	0.00122	305.67	141.98	0.02532	1.00893	0.6834
900	2.6330	13649.8	0.03944	10.2286	0.00109	331.98	152.95	0.03059	1.00796	0.6835
1000	2.3749	15296.1	0.03877	10.2062	0.00099	440.56	163.66	0.04454	1.00717	0.5569
1200	1.9857	18763.6	0.03732	10.1723	0.00082	511.13	184.40	0.06028	1.00600	0.5546
1400	1.7062	22457.5	0.03584	10.1479	0.00070	583.33	204.31	0.07787	1.00515	0.5536
1600	1.4957	26297.3	0.03453	10.1295	0.00062	655.56	223.49	0.09726	1.00451	0.5531
1800	1.3314	30264.6	0.03339	10.1150	0.00055	728.21	242.05	0.11847	1.00402	0.5524
2000	1.1995	34478.0	0.03224	10.1029	0.00049	805.77	260.06	0.14174	1.00362	0.5506
2500	0.9605	47802.4	0.02824	10.0701	0.00040	1087.47	303.19	0.21469	1.00290	0.5293
3000	0.7962	71842.6	0.02179	9.9893	0.00033	1748.06	345.18	0.32975	1.00240	0.4733

* TWO-PHASE BOUNDARY

Table 4.1 (continued)
THERMODYNAMIC PROPERTIES OF PARAHYDROGEN (ISOBARS, SI UNITS)

12.00 MPA ISOBAR

TEMPERATURE DEG. K	VOLUME CU M/KG	ISOTHERM DERIVATIVE CU M-MPA/KG	ISOCHORE DERIVATIVE MPA/K	INTERNAL ENERGY KJ/KG-K	ENTHALPY KJ/KG	ENTROPY KJ/KG-K	CV KJ / KG-K	CP KJ / KG-K	VELOCITY OF SOUND M/S
* 17.345	0.01210	1.8521	1.038	-300.1	-154.9	5.220	5.15	6.63	1544
18	0.01215	1.8168	1.037	-296.3	-150.5	5.469	5.25	6.83	1537
19	0.01224	1.7629	1.035	-290.4	-143.6	5.847	5.41	7.13	1525
20	0.01233	1.7091	1.031	-284.2	-136.3	6.220	5.54	7.43	1514
21	0.01242	1.6510	1.027	-277.7	-128.7	6.588	5.67	7.74	1501
22	0.01252	1.5973	1.021	-271.0	-120.9	6.955	5.79	8.04	1489
23	0.01262	1.5426	1.013	-264.1	-112.7	7.319	5.89	8.33	1477
24	0.01272	1.4862	1.003	-256.9	-104.2	7.679	5.98	8.61	1463
25	0.01284	1.4391	0.9916	-249.5	-95.5	8.036	6.07	8.88	1451
26	0.01295	1.3900	0.9786	-241.9	-86.4	8.390	6.14	9.15	1439
27	0.01307	1.3454	0.9647	-234.0	-77.2	8.740	6.21	9.41	1427
28	0.01320	1.2977	0.9507	-226.0	-67.6	9.087	6.28	9.68	1414
29	0.01333	1.2500	0.9353	-217.7	-57.8	9.431	6.34	9.95	1400
30	0.01346	1.2023	0.9191	-209.3	-47.7	9.773	6.40	10.22	1386
31	0.01360	1.1560	0.9019	-200.6	-37.4	10.112	6.45	10.48	1371
32	0.01375	1.1136	0.8840	-191.8	-26.8	10.449	6.49	10.74	1357
33	0.01391	1.0708	0.8655	-182.8	-15.9	10.784	6.53	11.00	1343
34	0.01406	1.0288	0.8465	-173.5	-4.8	11.116	6.57	11.25	1327
35	0.01423	0.9889	0.8269	-164.2	6.6	11.446	6.60	11.50	1313
36	0.01440	0.9495	0.8069	-154.6	18.2	11.773	6.63	11.75	1297
37	0.01458	0.9134	0.7866	-144.9	30.1	12.098	6.66	11.99	1282
38	0.01477	0.8776	0.7660	-135.0	42.2	12.421	6.68	12.22	1267
39	0.01496	0.8431	0.7455	-125.0	54.6	12.742	6.70	12.45	1252
40	0.01517	0.8101	0.7249	-114.9	67.1	13.060	6.71	12.68	1237
42	0.01559	0.7470	0.6841	-94.2	92.9	13.689	6.73	13.13	1207
44	0.01606	0.6919	0.6441	-73.1	119.6	14.309	6.76	13.56	1178
46	0.01655	0.6457	0.6052	-52.1	146.6	14.909	6.82	13.96	1150
48	0.01708	0.6042	0.5677	-30.1	174.9	15.511	6.86	14.33	1124
50	0.01765	0.5690	0.5319	-7.9	203.9	16.103	6.90	14.65	1099
52	0.01825	0.5385	0.4982	14.5	233.5	16.683	6.94	14.93	1076
54	0.01888	0.5129	0.4667	37.0	263.6	17.251	6.98	15.16	1055
56	0.01955	0.4916	0.4373	59.8	294.3	17.811	7.02	15.35	1036
58	0.02024	0.4753	0.4101	82.3	325.2	18.352	7.06	15.46	1021
60	0.02096	0.4623	0.3851	104.7	356.2	18.877	7.10	15.55	1007
65	0.02285	0.4435	0.3313	159.9	434.2	20.127	7.23	15.63	979
70	0.02485	0.4390	0.2881	214.1	512.3	21.284	7.42	15.59	961
75	0.02689	0.4458	0.2537	267.3	589.9	22.355	7.67	15.49	949
80	0.02895	0.4593	0.2260	319.8	667.2	23.353	7.97	15.42	943
85	0.03101	0.4755	0.2037	372.2	744.3	24.288	8.32	15.45	940
90	0.03307	0.4942	0.1853	425.0	821.8	25.173	8.70	15.54	939
95	0.03511	0.5142	0.1700	478.5	899.8	26.017	9.10	15.69	941
100	0.03714	0.5351	0.1571	533.0	978.7	26.826	9.51	15.88	945
120	0.04511	0.6260	0.1208	765.5	1306.8	29.815	11.09	16.78	973
140	0.05285	0.7197	0.0984	1013.8	1648.0	32.443	12.04	17.30	1017
160	0.06040	0.8125	0.0832	1270.3	1995.0	34.760	12.40	17.36	1067
180	0.06761	0.8980	0.0723	1526.1	2337.4	36.778	12.34	17.13	1116
200	0.07490	0.9879	0.0638	1776.5	2675.3	38.558	12.08	16.70	1169
220	0.08211	1.0765	0.0573	2019.5	3004.8	40.128	11.75	16.27	1221
240	0.08926	1.1642	0.0521	2255.0	3326.1	41.527	11.43	15.89	1272
260	0.09637	1.2510	0.0477	2484.0	3640.4	42.785	11.17	15.57	1321
280	0.10344	1.3372	0.0441	2707.7	3949.0	43.928	10.96	15.31	1367
300	0.11061	1.4275	0.0409	2928.9	4256.2	44.985	10.80	15.11	1413
350	0.12809	1.6388	0.0348	3466.3	5003.3	47.288	10.57	14.81	1516
400	0.14547	1.8484	0.0303	3994.5	5740.2	49.257	10.47	14.68	1610
450	0.16280	2.0569	0.0269	4518.6	6472.2	50.983	10.44	14.62	1697
500	0.18009	2.2648	0.0241	5041.8	7202.9	52.521	10.43	14.60	1780
550	0.19736	2.4722	0.0219	5564.0	7932.3	53.911	10.43	14.59	1859
600	0.21461	2.6793	0.0201	6086.7	8661.9	55.180	10.44	14.59	1935
700	0.24906	3.0929	0.0172	7134.1	10122.9	57.433	10.49	14.63	2077
800	0.28349	3.5060	0.0150	8188.3	11590.2	59.391	10.58	14.72	2208
900	0.31790	3.9188	0.0133	9252.9	13067.7	61.130	10.71	14.84	2330
1000	0.35230	4.3315	0.0120	10331.0	14558.6	62.703	10.87	14.99	2445
1200	0.42107	5.1566	0.0100	12542.5	17595.4	65.458	11.25	15.37	2655
1400	0.48983	5.9815	0.0086	14835.8	20713.8	67.866	11.68	15.81	2845
1600	0.55859	6.8063	0.0075	17214.8	23917.8	69.982	12.10	16.22	3021
1800	0.62734	7.6311	0.0067	19679.7	27207.8	71.881	12.49	16.62	3186
2000	0.69613	8.4559	0.0060	22234.2	30587.7	73.728	12.92	17.05	3341
2500	0.86894	10.5178	0.0048	29102.7	39530.0	80.059	14.64	18.87	3682
3000	1.04737	12.5798	0.0040	37405.3	49973.8	87.227	18.78	23.48	3965

* TWO-PHASE BOUNDARY

Table 4.1 (continued)
THERMODYNAMIC PROPERTIES OF PARAHYDROGEN (ISOBARS, SI UNITS)

12.00 MPA ISOBAR

TEMPERATURE DEG. K	DENSITY KG/CU M	$V(DH/DV)_P$ KJ/KG	$V(DP/DU)_V$ MPA-CU M/KJ X 10^2	$-V(DP/DV)_T$ MPA	$-(DV/DT)_P/V$ 1/DEG. K	THERMAL CONDUCTIVITY W/K-M X 10^3	VISCOSITY KG/M-S X 10^7	THERMAL DIFFUSIVITY SQ M/HR	DIELECTRIC CONSTANT	PRANDTL NUMBER
* 17.346	82.6756	977.1	0.24393	153.1232	0.00678	111.38	329.05	0.00073	1.27159	1.9572
18	82.3055	984.2	0.23990	149.5335	0.00694	114.78	306.11	0.00074	1.27027	1.8211
19	81.7264	993.1	0.23428	144.0750	0.00718	118.93	276.63	0.00073	1.26822	1.6609
20	81.1312	999.7	0.22926	138.6583	0.00744	122.68	252.39	0.00073	1.26611	1.5293
21	80.5315	1001.9	0.22488	132.9611	0.00772	126.94	232.38	0.00073	1.26398	1.4161
22	79.9020	1004.2	0.22092	127.6311	0.00800	130.37	215.25	0.00073	1.26175	1.3267
23	79.2534	1004.6	0.21711	122.2533	0.00829	133.10	200.58	0.00073	1.25946	1.2547
24	78.5874	1002.8	0.21339	116.7973	0.00859	135.25	187.90	0.00072	1.25711	1.1966
25	77.9078	1004.0	0.20984	112.1188	0.00884	136.90	176.88	0.00071	1.25471	1.1472
26	77.2094	1003.1	0.20637	107.3243	0.00912	138.06	167.16	0.00070	1.25225	1.1075
27	76.4976	1003.4	0.20295	102.9186	0.00937	138.85	158.57	0.00069	1.24975	1.0742
28	75.7724	1001.0	0.19973	98.3287	0.00967	139.32	150.91	0.00068	1.24720	1.0483
29	75.0316	997.5	0.19654	93.7885	0.00997	139.52	144.03	0.00067	1.24461	1.0269
30	74.2761	992.6	0.19348	89.3044	0.01029	139.48	137.80	0.00066	1.24196	1.0093
31	73.5041	987.7	0.19035	84.9711	0.01061	138.71	132.13	0.00065	1.23926	0.9987
32	72.7171	983.8	0.18725	80.9802	0.01092	137.78	126.94	0.00064	1.23651	0.9894
33	71.9154	978.4	0.18422	77.0050	0.01124	136.73	122.16	0.00062	1.23371	0.9825
34	71.1002	972.5	0.18122	73.1453	0.01157	135.62	117.75	0.00061	1.23088	0.9771
35	70.2703	966.7	0.17822	69.4894	0.01190	134.42	113.66	0.00060	1.22799	0.9727
36	69.4279	960.2	0.17523	65.9188	0.01224	133.15	109.84	0.00059	1.22507	0.9696
37	68.5720	954.7	0.17226	62.6369	0.01256	131.82	106.28	0.00058	1.22210	0.9665
38	67.7045	948.1	0.16936	59.4184	0.01289	130.44	102.93	0.00057	1.21910	0.9645
39	66.8261	941.1	0.16662	56.3441	0.01323	129.03	99.79	0.00056	1.21606	0.9629
40	65.9356	934.0	0.16391	53.4161	0.01357	127.60	96.83	0.00055	1.21299	0.9619
42	64.1275	919.3	0.15852	47.9026	0.01428	124.63	91.39	0.00053	1.20677	0.9627
44	62.2808	907.0	0.15309	43.0902	0.01495	121.65	86.50	0.00052	1.20044	0.9640
46	60.4123	900.1	0.14699	39.0098	0.01551	118.72	82.11	0.00051	1.19405	0.9658
48	58.5352	893.0	0.14137	35.3675	0.01605	115.90	78.17	0.00050	1.18766	0.9668
50	56.6572	887.8	0.13601	32.2400	0.01650	113.21	74.64	0.00049	1.18129	0.9658
52	54.7950	884.0	0.13096	29.5076	0.01688	110.69	71.49	0.00049	1.17500	0.9641
54	52.9589	882.3	0.12618	27.1606	0.01718	108.43	68.71	0.00049	1.16881	0.9606
56	51.1592	882.6	0.12172	25.1500	0.01739	106.49	66.24	0.00049	1.16277	0.9546
58	49.4060	885.5	0.11764	23.4846	0.01746	104.72	64.09	0.00049	1.15691	0.9464
60	47.7096	890.7	0.11377	22.0559	0.01746	103.29	62.21	0.00050	1.15126	0.9369
65	43.7550	915.6	0.10473	19.4053	0.01707	100.88	58.59	0.00053	1.13815	0.9078
70	40.2475	956.0	0.09651	17.6700	0.01631	99.42	56.20	0.00057	1.12661	0.8813
75	37.1910	1012.6	0.08898	16.5807	0.01530	98.00	54.74	0.00061	1.11662	0.8652
80	34.5429	1082.7	0.08212	15.8651	0.01425	98.23	53.93	0.00066	1.10801	0.8469
85	32.2500	1163.1	0.07593	15.3363	0.01328	100.70	53.61	0.00073	1.10060	0.8222
90	30.2433	1253.2	0.07043	14.9467	0.01240	103.49	53.61	0.00079	1.09414	0.8048
95	28.4826	1351.2	0.06558	14.6446	0.01161	106.56	53.85	0.00086	1.08849	0.7928
100	26.9263	1455.9	0.06135	14.4084	0.01091	108.27	54.15	0.00091	1.08352	0.7941
120	22.1672	1927.9	0.04913	13.8769	0.00870	128.94	59.77	0.00125	1.06841	0.7783
140	18.9214	2394.3	0.04319	13.6180	0.00722	144.34	63.60	0.00159	1.05819	0.7621
160	16.5567	2808.8	0.04053	13.4531	0.00618	154.82	66.65	0.00194	1.05079	0.7475
180	14.7907	3147.2	0.03960	13.2823	0.00544	161.48	69.34	0.00229	1.04528	0.7354
200	13.3517	3452.3	0.03957	13.1900	0.00484	165.82	71.88	0.00268	1.04081	0.7239
220	12.1790	3722.1	0.04007	13.1110	0.00437	169.03	74.35	0.00307	1.03718	0.7158
240	11.2029	3978.2	0.04067	13.0422	0.00399	171.94	76.77	0.00348	1.03417	0.7093
260	10.3766	4232.1	0.04121	12.9814	0.00368	175.00	79.19	0.00390	1.03162	0.7043
280	9.6672	4489.2	0.04163	12.9272	0.00341	178.39	81.60	0.00434	1.02943	0.7004
300	9.0407	4763.9	0.04191	12.9053	0.00317	182.13	84.01	0.00480	1.02751	0.6969
350	7.8071	5444.9	0.04219	12.7940	0.00272	192.90	90.05	0.00600	1.02372	0.6915
400	6.8741	6152.6	0.04212	12.7058	0.00239	204.88	96.11	0.00731	1.02087	0.6885
450	6.1424	6878.4	0.04189	12.6343	0.00213	217.52	102.15	0.00872	1.01863	0.6868
500	5.5527	7607.9	0.04167	12.5755	0.00192	230.27	108.16	0.01023	1.01683	0.6856
550	5.0669	8343.3	0.04144	12.5264	0.00175	243.14	114.12	0.01184	1.01535	0.6848
600	4.6597	9083.2	0.04122	12.4847	0.00161	256.01	120.03	0.01355	1.01411	0.6842
700	4.0150	10583.3	0.04075	12.4180	0.00138	281.87	131.67	0.01727	1.01215	0.6836
800	3.5275	12122.1	0.04022	12.3672	0.00121	308.03	143.04	0.02136	1.01067	0.6834
900	3.1456	13711.2	0.03961	12.3272	0.00108	334.66	154.15	0.02581	1.00951	0.6835
1000	2.8385	15358.3	0.03892	12.2950	0.00098	440.56	165.01	0.03727	1.00858	0.5616
1200	2.3749	18827.4	0.03744	12.2463	0.00082	511.13	186.02	0.05040	1.00717	0.5595
1400	2.0415	22523.1	0.03593	12.2113	0.00070	583.33	206.18	0.06507	1.00616	0.5587
1600	1.7902	26363.8	0.03462	12.1849	0.00062	655.52	225.61	0.08125	1.00540	0.5584
1800	1.5940	30328.2	0.03347	12.1642	0.00055	728.00	244.41	0.09894	1.00481	0.5579
2000	1.4365	34526.8	0.03232	12.1471	0.00049	804.93	262.64	0.11831	1.00433	0.5563
2500	1.1508	47608.5	0.02848	12.1042	0.00040	1077.25	306.32	0.17854	1.00347	0.5367
3000	0.9548	70529.5	0.02231	12.0108	0.00033	1695.49	348.70	0.27223	1.00288	0.4830

* TWO-PHASE BOUNDARY

Table 4.1 (continued)
THERMODYNAMIC PROPERTIES OF PARAHYDROGEN (ISOBARS, SI UNITS)

14.00 MPA ISOBAR

TEMPERATURE	VOLUME	ISOTHERM DERIVATIVE	ISOCHORE DERIVATIVE	INTERNAL ENERGY	ENTHALPY	ENTROPY	CV	CP	VELOCITY OF SOUND
DEG. K	CU M/KG	CU M-MPA/KG	MPA/K	KJ/KG-K	KJ/KG	KJ/KG-K	KJ / KG-K		M/S
* 17.863	0.01198	1.9517	1.052	-298.0	-130.2	5.258	5.21	6.67	1580
18	0.01199	1.9449	1.052	-297.2	-129.3	5.307	5.24	6.71	1579
19	0.01207	1.8929	1.051	-291.5	-122.5	5.678	5.39	7.00	1569
20	0.01216	1.8408	1.049	-285.5	-115.3	6.045	5.53	7.29	1559
21	0.01224	1.7889	1.045	-279.3	-107.9	6.407	5.65	7.57	1548
22	0.01233	1.7372	1.039	-272.8	-100.2	6.766	5.77	7.85	1537
23	0.01242	1.6804	1.036	-266.1	-92.2	7.119	5.87	8.14	1526
24	0.01252	1.6297	1.027	-259.2	-83.9	7.471	5.97	8.41	1515
25	0.01262	1.5784	1.017	-252.1	-75.4	7.820	6.06	8.67	1503
26	0.01273	1.5244	1.006	-244.8	-66.6	8.165	6.14	8.93	1489
27	0.01283	1.4799	0.9933	-237.2	-57.6	8.506	6.21	9.18	1479
28	0.01295	1.4333	0.9796	-229.5	-48.3	8.845	6.28	9.43	1466
29	0.01306	1.3879	0.9652	-221.6	-38.7	9.180	6.35	9.67	1454
30	0.01318	1.3439	0.9509	-213.5	-28.9	9.512	6.40	9.91	1442
31	0.01331	1.2995	0.9355	-205.2	-18.9	9.841	6.46	10.16	1430
32	0.01344	1.2548	0.9194	-196.8	-8.6	10.167	6.51	10.40	1416
33	0.01357	1.2134	0.9025	-188.1	1.9	10.491	6.55	10.63	1403
34	0.01371	1.1721	0.8851	-179.3	12.7	10.812	6.59	10.86	1390
35	0.01386	1.1320	0.8672	-170.4	23.6	11.130	6.63	11.09	1377
36	0.01401	1.0925	0.8490	-161.3	34.9	11.446	6.66	11.32	1363
37	0.01416	1.0568	0.8303	-152.0	46.3	11.759	6.69	11.53	1350
38	0.01432	1.0212	0.8114	-142.6	57.9	12.069	6.71	11.74	1336
39	0.01449	0.9864	0.7923	-133.1	69.8	12.377	6.73	11.94	1323
40	0.01466	0.9529	0.7731	-123.5	81.8	12.682	6.74	12.14	1310
42	0.01502	0.8909	0.7347	-103.9	106.4	13.283	6.77	12.51	1283
44	0.01541	0.8333	0.6969	-83.8	131.8	13.874	6.80	12.89	1257
46	0.01582	0.7799	0.6598	-63.9	157.5	14.443	6.86	13.29	1229
48	0.01625	0.7351	0.6238	-43.1	184.4	15.016	6.91	13.62	1204
50	0.01671	0.6957	0.5891	-22.0	212.0	15.579	6.96	13.92	1180
52	0.01720	0.6606	0.5560	-0.7	240.1	16.130	7.00	14.20	1158
54	0.01771	0.6313	0.5246	20.8	268.7	16.671	7.04	14.42	1137
56	0.01824	0.6055	0.4951	42.7	298.1	17.204	7.08	14.62	1118
58	0.01880	0.5834	0.4674	64.3	327.5	17.720	7.11	14.79	1101
60	0.01937	0.5657	0.4416	85.9	357.2	18.223	7.15	14.91	1086
65	0.02089	0.5344	0.3847	139.9	432.4	19.427	7.29	15.14	1054
70	0.02250	0.5198	0.3378	193.3	508.4	20.554	7.47	15.26	1030
75	0.02418	0.5164	0.2992	246.4	584.9	21.609	7.72	15.32	1012
80	0.02589	0.5221	0.2676	299.2	661.6	22.600	8.02	15.37	1000
85	0.02761	0.5335	0.2415	352.2	738.7	23.534	8.37	15.45	992
90	0.02933	0.5479	0.2199	405.6	816.2	24.421	8.75	15.58	988
95	0.03106	0.5648	0.2018	459.7	894.5	25.267	9.15	15.75	986
100	0.03278	0.5833	0.1865	514.9	973.8	26.081	9.55	15.96	987
120	0.03958	0.6666	0.1431	750.0	1304.1	29.089	11.12	16.90	1006
140	0.04623	0.7567	0.1163	1000.5	1647.7	31.736	12.07	17.42	1045
160	0.05272	0.8476	0.0980	1258.9	1997.0	34.068	12.43	17.47	1092
180	0.05909	0.9381	0.0849	1516.5	2343.8	36.112	12.37	17.20	1142
200	0.06537	1.0274	0.0749	1768.3	2683.4	37.900	12.10	16.77	1193
220	0.07138	1.1089	0.0672	2012.0	3011.3	39.463	11.77	16.34	1241
240	0.07753	1.1967	0.0610	2248.3	3333.8	40.866	11.45	15.94	1291
260	0.08365	1.2837	0.0559	2478.0	3649.1	42.128	11.18	15.61	1339
280	0.08973	1.3701	0.0516	2702.3	3958.5	43.275	10.97	15.35	1385
300	0.09592	1.4620	0.0479	2924.2	4267.1	44.336	10.82	15.14	1431
350	0.11093	1.6734	0.0407	3462.7	5015.7	46.643	10.58	14.84	1532
400	0.12585	1.8830	0.0354	3991.7	5753.6	48.615	10.48	14.70	1625
450	0.14071	2.0915	0.0314	4516.4	6486.4	50.342	10.45	14.64	1712
500	0.15554	2.2993	0.0282	5040.0	7217.6	51.882	10.43	14.61	1794
550	0.17035	2.5067	0.0256	5562.5	7947.4	53.273	10.44	14.60	1872
600	0.18514	2.7137	0.0234	6085.5	8677.4	54.542	10.45	14.60	1947
700	0.21468	3.1271	0.0200	7133.4	10138.9	56.796	10.50	14.64	2088
800	0.24419	3.5401	0.0175	8187.9	11606.5	58.754	10.59	14.72	2219
900	0.27369	3.9528	0.0156	9252.6	13084.3	60.493	10.71	14.84	2340
1000	0.30317	4.3654	0.0140	10330.9	14575.3	62.067	10.87	15.00	2454
1200	0.36213	5.1903	0.0117	12542.5	17612.3	64.822	11.25	15.37	2663
1400	0.42106	6.0150	0.0100	14836.0	20730.9	67.230	11.68	15.81	2853
1600	0.47999	6.8397	0.0087	17215.0	23934.9	69.346	12.10	16.22	3028
1800	0.53893	7.6644	0.0078	19679.8	27224.8	71.245	12.49	16.62	3193
2000	0.59788	8.4891	0.0070	22233.4	30603.8	73.091	12.91	17.04	3347
2500	0.74595	10.5509	0.0056	29084.7	39527.9	79.415	14.57	18.79	3689
3000	0.89849	12.6127	0.0047	37289.7	49868.5	86.546	18.44	23.11	3973

* TWO-PHASE BOUNDARY

Table 4.1 (continued)
THERMODYNAMIC PROPERTIES OF PARAHYDROGEN (ISOBARS, SI UNITS)

14.00 MPA ISOBAR

TEMPERATURE DEG. K	DENSITY KG/CU M	$V(\partial H/\partial V)_P$ KJ/KG	$V(\partial P/\partial U)_V$ MPA-CU M/KJ X 10^2	$-V(\partial P/\partial V)_T$ MPA	$-(\partial V/\partial T)_P/V$ 1/DEG. K	THERMAL CONDUCTIVITY W/K-M X 10^3	VISCOSITY KG/M-S X 10^7	THERMAL DIFFUSIVITY SQ M/HR	DIELECTRIC CONSTANT	PRANDTL NUMBER
* 17.869	83.4399	1032.3	0.24187	162.8525	0.00646	116.52	341.17	0.00075	1.27431	1.9531
18	83.3689	1033.9	0.24110	162.1439	0.00649	117.20	336.21	0.00075	1.27405	1.9250
19	82.8207	1044.6	0.23561	156.7682	0.00671	121.54	302.68	0.00075	1.27210	1.7445
20	82.2581	1052.9	0.23071	151.4246	0.00693	125.67	275.26	0.00075	1.27010	1.5972
21	81.6813	1059.2	0.22626	146.1222	0.00715	130.19	252.51	0.00076	1.26806	1.4687
22	81.0903	1063.9	0.22214	140.8714	0.00739	133.89	233.37	0.00076	1.26596	1.3678
23	80.4940	1063.1	0.21902	135.2625	0.00766	136.91	217.21	0.00075	1.26385	1.2913
24	79.8708	1065.2	0.21544	130.1661	0.00789	139.29	203.13	0.00075	1.26164	1.2261
25	79.2334	1065.7	0.21197	125.0644	0.00814	141.16	190.92	0.00074	1.25939	1.1725
26	78.5812	1063.8	0.20856	119.7910	0.00840	142.53	180.21	0.00073	1.25709	1.1295
27	77.9182	1065.5	0.20518	115.3118	0.00861	143.51	170.79	0.00072	1.25475	1.0932
28	77.2394	1065.2	0.20185	110.7064	0.00885	144.16	162.39	0.00071	1.25236	1.0617
29	76.5489	1064.3	0.19865	106.2418	0.00909	144.53	154.88	0.00070	1.24993	1.0362
30	75.8487	1062.5	0.19578	101.9367	0.00933	144.66	148.14	0.00069	1.24747	1.0150
31	75.1334	1059.9	0.19284	97.6386	0.00959	144.03	142.01	0.00068	1.24496	1.0013
32	74.4075	1056.2	0.18990	93.3670	0.00985	143.25	136.44	0.00067	1.24242	0.9906
33	73.6684	1053.1	0.18703	89.3899	0.01010	142.34	131.33	0.00065	1.23983	0.9810
34	72.9202	1049.1	0.18418	85.4730	0.01036	141.37	126.64	0.00064	1.23722	0.9732
35	72.1585	1044.8	0.18137	81.6823	0.01062	140.31	122.29	0.00063	1.23456	0.9668
36	71.3871	1039.8	0.17859	77.9869	0.01089	139.17	118.26	0.00062	1.23187	0.9619
37	70.6058	1036.2	0.17583	74.6181	0.01113	137.97	114.51	0.00061	1.22916	0.9570
38	69.8145	1031.4	0.17314	71.2911	0.01138	136.72	111.01	0.00060	1.22641	0.9532
39	69.0158	1025.9	0.17060	68.0748	0.01164	135.43	107.73	0.00059	1.22364	0.9498
40	68.2087	1020.3	0.16808	64.9973	0.01189	134.12	104.65	0.00058	1.22084	0.9470
42	66.5716	1010.1	0.16301	59.3071	0.01239	131.39	99.00	0.00057	1.21518	0.9429
44	64.9100	1000.2	0.15787	54.0908	0.01288	128.64	93.96	0.00055	1.20946	0.9412
46	63.2267	993.1	0.15202	49.3116	0.01338	125.91	89.42	0.00054	1.20368	0.9436
48	61.5326	987.9	0.14665	45.2351	0.01379	123.25	85.33	0.00053	1.19788	0.9432
50	59.8340	983.9	0.14153	41.6279	0.01415	120.69	81.64	0.00052	1.19208	0.9419
52	58.1427	980.7	0.13664	38.4099	0.01448	118.26	78.32	0.00052	1.18633	0.9402
54	56.4686	980.0	0.13197	35.6485	0.01472	115.98	75.35	0.00051	1.18065	0.9368
56	54.8188	980.3	0.12761	33.1922	0.01492	113.87	72.69	0.00051	1.17508	0.9333
58	53.1998	981.8	0.12355	31.0358	0.01506	111.97	70.31	0.00051	1.16962	0.9285
60	51.6176	986.1	0.11963	29.1990	0.01512	110.52	68.21	0.00052	1.16431	0.9204
65	47.8611	1006.8	0.11034	25.5765	0.01504	107.99	63.98	0.00054	1.15176	0.8973
70	44.4354	1043.3	0.10169	23.0981	0.01462	106.25	60.99	0.00056	1.14040	0.8757
75	41.3625	1093.8	0.09368	21.3608	0.01401	104.65	58.97	0.00059	1.13027	0.8634
80	38.6321	1158.9	0.08634	20.1711	0.01327	104.45	57.67	0.00063	1.12132	0.8489
85	36.2217	1236.3	0.07968	19.3230	0.01250	106.59	56.93	0.00069	1.11346	0.8253
90	34.0980	1323.5	0.07375	18.6790	0.01177	109.07	56.60	0.00074	1.10655	0.8085
95	32.1975	1419.6	0.06852	18.1847	0.01110	111.86	56.57	0.00079	1.10043	0.7967
100	30.5058	1522.9	0.06398	17.7952	0.01048	112.96	56.59	0.00084	1.09498	0.7994
120	25.2655	1988.9	0.05092	16.8430	0.00850	132.61	61.75	0.00112	1.07823	0.7867
140	21.6318	2452.0	0.04452	16.3678	0.00710	147.23	65.14	0.00141	1.06672	0.7706
160	18.9674	2865.2	0.04159	16.0775	0.00610	157.08	67.87	0.00171	1.05833	0.7549
180	16.9222	3217.1	0.04054	15.8753	0.00535	163.33	70.34	0.00202	1.05193	0.7407
200	15.2983	3519.1	0.04045	15.7172	0.00476	167.40	72.72	0.00235	1.04686	0.7284
220	14.0096	3774.3	0.04079	15.5346	0.00433	170.45	75.07	0.00268	1.04286	0.7195
240	12.8974	4030.7	0.04134	15.4341	0.00395	173.26	77.42	0.00303	1.03941	0.7123
260	11.9548	4284.9	0.04183	15.3466	0.00364	176.25	79.78	0.00340	1.03649	0.7067
280	11.1448	4542.4	0.04221	15.2694	0.00338	179.60	82.16	0.00378	1.03399	0.7023
300	10.4251	4821.1	0.04246	15.2420	0.00314	183.32	84.55	0.00418	1.03177	0.6984
350	9.0149	5502.6	0.04265	15.0854	0.00270	194.10	90.58	0.00522	1.02743	0.6924
400	7.9461	6210.9	0.04252	14.9625	0.00237	206.15	96.66	0.00636	1.02415	0.6891
450	7.1067	6937.2	0.04224	14.8637	0.00211	218.89	102.75	0.00758	1.02158	0.6871
500	6.4291	7667.1	0.04198	14.7826	0.00191	231.76	108.81	0.00888	1.01951	0.6858
550	5.8704	8402.8	0.04172	14.7150	0.00174	244.75	114.84	0.01028	1.01780	0.6849
600	5.4014	9143.0	0.04147	14.6579	0.00160	257.77	120.82	0.01177	1.01637	0.6843
700	4.6581	10643.5	0.04097	14.5667	0.00138	283.92	132.59	0.01499	1.01411	0.6836
800	4.0951	12182.9	0.04041	14.4973	0.00121	310.39	144.11	0.01854	1.01239	0.6834
900	3.6538	13772.6	0.03977	14.4428	0.00109	337.33	155.36	0.02240	1.01105	0.6835
1000	3.2984	15420.3	0.03906	14.3990	0.00097	440.56	166.35	0.03207	1.00997	0.5662
1200	2.7615	18891.0	0.03755	14.3328	0.00081	511.13	187.63	0.04334	1.00834	0.5644
1400	2.3749	22588.4	0.03603	14.2853	0.00070	583.33	208.06	0.05594	1.00717	0.5638
1600	2.0834	26430.1	0.03470	14.2496	0.00061	655.49	227.74	0.06982	1.00629	0.5637
1800	1.8555	30392.7	0.03355	14.2216	0.00055	727.84	246.78	0.08499	1.00560	0.5634
2000	1.6726	34580.1	0.03240	14.1986	0.00049	804.28	265.26	0.10158	1.00505	0.5621
2500	1.3406	47473.8	0.02867	14.1443	0.00040	1069.30	309.48	0.15283	1.00404	0.5438
3000	1.1130	69525.3	0.02273	14.0377	0.00033	1654.59	352.30	0.23163	1.00336	0.4920

* TWO-PHASE BOUNDARY

Table 4.1 (continued)
THERMODYNAMIC PROPERTIES OF PARAHYDROGEN (ISOBARS, SI UNITS)

16.00 MPA ISOBAR

TEMPERATURE DEG. K	VOLUME CU M/KG	ISOTHERM DERIVATIVE CU M-MPA/KG	ISOCHORE DERIVATIVE MPA/K	INTERNAL ENERGY KJ/KG-K	ENTHALPY KJ/KG	ENTROPY KJ/KG-K	CV KJ /	CP KG-K	VELOCITY OF SOUND. M/S
* 18.378	0.01188	2.0484	1.066	-295.7	-105.7	5.295	5.28	6.71	1615
19	0.01193	2.0171	1.066	-292.3	-101.4	5.522	5.37	6.89	1609
20	0.01200	1.9667	1.065	-286.4	-94.4	5.883	5.51	7.17	1600
21	0.01208	1.9163	1.062	-280.4	-87.1	6.239	5.64	7.44	1591
22	0.01217	1.8661	1.058	-274.2	-79.5	6.591	5.75	7.70	1581
23	0.01225	1.8161	1.052	-267.7	-71.7	6.940	5.86	7.96	1571
24	0.01234	1.7664	1.045	-261.0	-63.6	7.284	5.96	8.22	1561
25	0.01243	1.7119	1.041	-254.1	-55.3	7.624	6.05	8.49	1550
26	0.01253	1.6642	1.031	-247.1	-46.7	7.961	6.13	8.74	1540
27	0.01262	1.6158	1.020	-239.8	-37.8	8.296	6.21	8.98	1528
28	0.01273	1.5644	1.007	-232.3	-28.7	8.627	6.28	9.22	1515
29	0.01283	1.5218	0.9937	-224.7	-19.4	8.954	6.35	9.45	1505
30	0.01294	1.4775	0.9796	-216.9	-9.8	9.278	6.41	9.67	1493
31	0.01306	1.4313	0.9650	-208.9	-0.0	9.599	6.46	9.90	1481
32	0.01317	1.3906	0.9504	-200.8	10.0	9.917	6.52	10.12	1470
33	0.01329	1.3492	0.9351	-192.4	20.2	10.232	6.56	10.34	1458
34	0.01342	1.3072	0.9192	-184.0	30.7	10.544	6.61	10.56	1446
35	0.01355	1.2689	0.9026	-175.4	41.3	10.853	6.65	10.77	1434
36	0.01368	1.2299	0.8857	-166.6	52.2	11.159	6.68	10.98	1421
37	0.01381	1.1921	0.8684	-157.7	63.3	11.463	6.71	11.18	1409
38	0.01396	1.1571	0.8508	-148.7	74.6	11.764	6.74	11.37	1397
39	0.01410	1.1222	0.8330	-139.6	86.0	12.061	6.76	11.55	1385
40	0.01425	1.0890	0.8151	-130.3	97.7	12.356	6.78	11.73	1373
42	0.01456	1.0265	0.7789	-111.5	121.5	12.937	6.81	12.07	1349
44	0.01490	0.9676	0.7429	-92.4	146.0	13.506	6.84	12.41	1325
46	0.01525	0.9144	0.7075	-73.3	170.6	14.054	6.91	12.76	1300
48	0.01562	0.8633	0.6729	-53.4	196.5	14.604	6.96	13.10	1275
50	0.01601	0.8176	0.6391	-33.1	223.0	15.144	7.01	13.40	1251
52	0.01641	0.7824	0.6068	-12.6	250.0	15.675	7.05	13.64	1230
54	0.01684	0.7466	0.5758	8.1	277.5	16.194	7.09	13.89	1209
56	0.01729	0.7195	0.5463	29.2	305.7	16.708	7.13	14.07	1192
58	0.01775	0.6931	0.5184	50.1	334.1	17.204	7.16	14.25	1174
60	0.01823	0.6719	0.4922	71.0	362.7	17.690	7.20	14.39	1159
65	0.01949	0.6301	0.4336	123.6	435.5	18.855	7.34	14.70	1124
70	0.02083	0.6056	0.3840	176.2	509.6	19.953	7.53	14.93	1096
75	0.02224	0.5943	0.3425	228.8	584.6	20.988	7.78	15.09	1074
80	0.02368	0.5913	0.3076	281.6	660.5	21.967	8.08	15.25	1057
85	0.02515	0.5968	0.2785	334.8	737.1	22.896	8.42	15.40	1045
90	0.02663	0.6071	0.2539	388.5	814.5	23.781	8.79	15.57	1037
95	0.02811	0.6208	0.2332	443.1	892.9	24.629	9.19	15.76	1032
100	0.02959	0.6361	0.2155	498.7	972.2	25.443	9.59	15.99	1030
120	0.03550	0.7105	0.1654	735.7	1303.7	28.461	11.16	16.98	1040
140	0.04131	0.7966	0.1342	988.1	1649.0	31.122	12.10	17.51	1073
160	0.04700	0.8856	0.1130	1248.1	2000.2	33.466	12.46	17.56	1117
180	0.05259	0.9744	0.0977	1507.3	2348.8	35.520	12.41	17.28	1165
200	0.05810	1.0633	0.0861	1760.3	2689.9	37.317	12.14	16.84	1215
220	0.06353	1.1504	0.0770	2005.3	3021.7	38.899	11.79	16.37	1264
240	0.06875	1.2299	0.0700	2242.0	3341.9	40.292	11.47	15.98	1309
260	0.07411	1.3170	0.0641	2472.2	3658.0	41.558	11.20	15.66	1357
280	0.07945	1.4035	0.0592	2697.1	3968.2	42.707	10.98	15.39	1402
300	0.08491	1.4973	0.0548	2919.8	4278.4	43.772	10.83	15.18	1448
350	0.09806	1.7085	0.0466	3459.3	5028.3	46.083	10.59	14.86	1548
400	0.11113	1.9180	0.0405	3989.1	5767.2	48.058	10.49	14.71	1640
450	0.12415	2.1264	0.0359	4514.3	6500.6	49.787	10.45	14.65	1726
500	0.13713	2.3341	0.0322	5038.4	7232.4	51.328	10.44	14.62	1808
550	0.15009	2.5413	0.0292	5561.2	7962.6	52.720	10.44	14.60	1885
600	0.16303	2.7482	0.0268	6084.5	8693.0	53.989	10.45	14.61	1960
700	0.18889	3.1614	0.0229	7132.7	10154.9	56.244	10.50	14.64	2100
800	0.21471	3.5742	0.0200	8187.5	11622.9	58.203	10.59	14.72	2229
900	0.24052	3.9867	0.0178	9252.4	13100.8	59.942	10.71	14.84	2350
1000	0.26632	4.3991	0.0160	10330.8	14592.0	61.516	10.87	15.00	2464
1200	0.31791	5.2237	0.0133	12542.7	17629.2	64.271	11.25	15.38	2672
1400	0.36948	6.0483	0.0114	14836.3	20748.0	66.679	11.68	15.81	2860
1600	0.42104	6.8728	0.0100	17215.3	23952.0	68.795	12.10	16.22	3036
1800	0.47261	7.6974	0.0089	19680.0	27241.8	70.695	12.49	16.61	3200
2000	0.52419	8.5219	0.0080	22232.9	30620.0	72.540	12.91	17.04	3354
2500	0.65371	10.5835	0.0064	29070.3	39529.6	78.858	14.50	18.72	3696
3000	0.78689	12.6451	0.0053	37196.8	49787.0	85.959	18.18	22.80	3981

* TWO-PHASE BOUNDARY

Table 4.1 (continued)
THERMODYNAMIC PROPERTIES OF PARAHYDROGEN (ISOBARS, SI UNITS)

16.00 MPA ISOBAR

TEMPERATURE DEG. K	DENSITY KG/CU M	$V(DH/DV)_P$ KJ/KG	$V(DP/DH)_V$ MPA-CU M/KJ X 10	$-V(DP/DV)_T$ MPA	$-(DV/DT)_P/V$ 1/DEG. K	THERMAL CONDUCTIVITY W/K-M X 10^3	VISCOSITY KG/M-S X 10^7	THERMAL DIFFUSIVITY SQ M/HR	DIELECTRIC CONSTANT	PRANDTL NUMBER
* 18.378	84.1703	1085.6	0.24014	172.4170	0.00618	121.39	353.30	0.00077	1.27691	1.9543
19	83.8438	1093.2	0.23686	169.1204	0.00630	124.11	330.57	0.00077	1.27574	1.8357
20	83.3088	1103.2	0.23204	163.8434	0.00650	128.51	299.59	0.00077	1.27384	1.6714
21	82.7610	1111.2	0.22767	158.5978	0.00670	133.30	273.99	0.00078	1.27189	1.5293
22	82.2006	1117.5	0.22362	153.3933	0.00689	137.24	252.55	0.00078	1.26990	1.4178
23	81.6276	1122.3	0.21989	148.2402	0.00710	140.47	234.38	0.00078	1.26787	1.3287
24	81.0422	1125.8	0.21637	143.1502	0.00730	143.08	218.82	0.00077	1.26579	1.2567
25	80.4489	1123.8	0.21368	137.7179	0.00756	145.17	205.41	0.00076	1.26369	1.2016
26	79.8365	1126.0	0.21059	132.8603	0.00776	146.73	193.63	0.00076	1.26152	1.1529
27	79.2100	1127.0	0.20728	127.9866	0.00797	147.88	183.25	0.00075	1.25931	1.1125
28	78.5728	1125.7	0.20401	122.9176	0.00819	148.69	174.06	0.00074	1.25706	1.0796
29	77.9254	1127.5	0.20094	118.5882	0.00838	149.22	165.88	0.00073	1.25478	1.0503
30	77.2670	1127.1	0.19786	114.1641	0.00858	149.50	158.54	0.00072	1.25246	1.0257
31	76.5983	1125.0	0.19488	109.6379	0.00880	148.99	151.91	0.00071	1.25011	1.0096
32	75.9211	1124.5	0.19209	105.5768	0.00900	148.33	145.91	0.00069	1.24773	0.9958
33	75.2310	1122.6	0.18937	101.4996	0.00921	147.53	140.41	0.00068	1.24530	0.9844
34	74.5334	1119.6	0.18667	97.4322	0.00943	146.67	135.39	0.00067	1.24286	0.9750
35	73.8266	1117.6	0.18398	93.6801	0.00964	145.72	130.76	0.00066	1.24039	0.9663
36	73.1104	1114.4	0.18132	89.9185	0.00985	144.69	126.48	0.00065	1.23788	0.9596
37	72.3861	1110.9	0.17870	86.2894	0.01006	143.59	122.51	0.00064	1.23536	0.9539
38	71.6542	1108.0	0.17618	82.9097	0.01026	142.43	118.81	0.00063	1.23280	0.9485
39	70.9152	1103.9	0.17379	79.5832	0.01047	141.24	115.36	0.00062	1.23023	0.9437
40	70.1694	1099.9	0.17144	76.4173	0.01067	140.01	112.12	0.00061	1.22764	0.9394
42	68.6604	1092.3	0.16668	70.4824	0.01105	137.46	106.22	0.00060	1.22240	0.9329
44	67.1339	1084.9	0.16182	64.9599	0.01144	134.87	100.98	0.00058	1.21712	0.9290
46	65.5916	1081.6	0.15618	59.9771	0.01180	132.29	96.27	0.00057	1.21180	0.9285
48	64.0397	1076.0	0.15100	55.2845	0.01217	129.77	92.03	0.00056	1.20647	0.9289
50	62.4792	1071.4	0.14602	51.0822	0.01251	127.31	88.19	0.00055	1.20112	0.9286
52	60.9229	1071.7	0.14129	47.6678	0.01273	124.96	84.72	0.00054	1.19580	0.9249
54	59.3772	1069.6	0.13674	44.3230	0.01299	122.74	81.59	0.00054	1.19053	0.9235
56	57.8466	1072.1	0.13244	41.6200	0.01313	120.65	78.75	0.00053	1.18532	0.9185
58	56.3397	1073.4	0.12844	39.0511	0.01328	118.72	76.21	0.00053	1.18022	0.9147
60	54.8590	1077.9	0.12455	36.8625	0.01335	117.10	73.92	0.00053	1.17521	0.9085
65	51.3040	1096.4	0.11516	32.3268	0.01341	114.40	69.21	0.00055	1.16326	0.8896
70	47.9986	1129.9	0.10628	29.0695	0.01321	112.50	65.71	0.00057	1.15222	0.8718
75	44.9729	1177.9	0.09793	26.7252	0.01281	110.83	63.20	0.00059	1.14217	0.8607
80	42.2333	1238.2	0.09021	24.9716	0.01232	110.30	61.47	0.00062	1.13313	0.8500
85	39.7686	1312.7	0.08318	23.7322	0.01173	112.20	60.33	0.00066	1.12504	0.8282
90	37.5586	1398.2	0.07687	22.8006	0.01114	114.43	59.66	0.00070	1.11782	0.8118
95	35.5736	1493.2	0.07132	22.0338	0.01056	116.97	59.34	0.00075	1.11136	0.7998
100	33.7906	1594.7	0.06648	21.4951	0.01003	117.66	59.08	0.00078	1.10558	0.8029
120	28.1711	2054.6	0.05263	20.0154	0.00826	136.57	63.85	0.00103	1.08750	0.7938
140	24.2091	2515.4	0.04581	19.2857	0.00696	150.35	66.82	0.00128	1.07487	0.7781
160	21.2755	2927.2	0.04263	18.8405	0.00600	159.55	69.21	0.00154	1.06559	0.7618
180	19.0141	3277.5	0.04142	18.5275	0.00527	165.35	71.43	0.00181	1.05848	0.7466
200	17.2120	3580.3	0.04121	18.3010	0.00470	169.12	73.63	0.00210	1.05283	0.7332
220	15.7408	3849.9	0.04148	18.1087	0.00425	171.97	75.85	0.00240	1.04824	0.7221
240	14.5460	4086.3	0.04195	17.8896	0.00391	174.66	78.12	0.00270	1.04452	0.7148
260	13.4930	4338.7	0.04244	17.7700	0.00361	177.57	80.41	0.00303	1.04125	0.7090
280	12.5871	4596.3	0.04278	17.6655	0.00335	180.87	82.75	0.00336	1.03845	0.7042
300	11.7769	4879.4	0.04300	17.6334	0.00311	184.56	85.12	0.00372	1.03594	0.7000
350	10.1977	5561.1	0.04311	17.4229	0.00267	195.32	91.13	0.00464	1.03107	0.6933
400	8.9985	6269.8	0.04291	17.2590	0.00235	207.42	97.23	0.00564	1.02738	0.6896
450	8.0551	6996.5	0.04259	17.1280	0.00209	220.25	103.35	0.00672	1.02448	0.6874
500	7.2924	7726.6	0.04229	17.0203	0.00189	233.23	109.47	0.00785	1.02215	0.6860
550	6.6628	8462.5	0.04200	16.9319	0.00173	246.35	115.55	0.00911	1.02022	0.6850
600	6.1338	9203.0	0.04173	16.8568	0.00159	259.51	121.60	0.01043	1.01860	0.6844
700	5.2942	10703.9	0.04118	16.7372	0.00137	285.96	133.51	0.01328	1.01604	0.6836
800	4.6574	12243.7	0.04059	16.6464	0.00120	312.74	145.16	0.01642	1.01410	0.6834
900	4.1576	13833.9	0.03993	16.5752	0.00107	339.99	156.56	0.01984	1.01258	0.6834
1000	3.7548	15482.3	0.03920	16.5179	0.00097	440.56	167.69	0.02817	1.01136	0.5708
1200	3.1456	18954.4	0.03766	16.4316	0.00081	511.13	189.25	0.03805	1.00951	0.5693
1400	2.7065	22653.4	0.03612	16.3698	0.00070	583.32	209.95	0.04908	1.00818	0.5690
1600	2.3751	26496.2	0.03478	16.3233	0.00061	655.47	229.89	0.06124	1.00717	0.5690
1800	2.1159	30457.5	0.03362	16.2870	0.00055	727.71	249.18	0.07452	1.00639	0.5689
2000	1.9077	34636.1	0.03248	16.2573	0.00049	803.76	267.90	0.08904	1.00576	0.5678
2500	1.5297	47379.1	0.02883	16.1900	0.00040	1062.89	312.69	0.13362	1.00462	0.5507
3000	1.2708	68730.0	0.02308	16.0698	0.00033	1621.58	355.97	0.20147	1.00383	0.5005

* TWO-PHASE BOUNDARY

Table 4.1 (continued)
THERMODYNAMIC PROPERTIES OF PARAHYDROGEN (ISOBARS, SI UNITS)

18.00 MPA ISOBAR

TEMPERATURE DEG. K	VOLUME CU M/KG	ISOTHERM DERIVATIVE CU M-MPA/KG	ISOCHORE DERIVATIVE MPA/K	INTERNAL ENERGY KJ/KG-K	ENTHALPY KJ/KG	ENTROPY KJ/KG-K	CV KJ / KG-K	CP KJ / KG-K	VELOCITY OF SOUND M/S
* 18.874	0.01178	2.1426	1.080	-293.4	-81.3	5.332	5.33	6.76	1648
19	0.01179	2.1364	1.080	-292.7	-80.5	5.377	5.35	6.79	1647
20	0.01186	2.0875	1.080	-287.1	-73.5	5.732	5.49	7.06	1639
21	0.01194	2.0386	1.078	-281.2	-66.3	6.083	5.62	7.32	1630
22	0.01201	1.9897	1.075	-275.1	-58.9	6.430	5.74	7.58	1621
23	0.01209	1.9409	1.070	-268.9	-51.2	6.772	5.85	7.83	1612
24	0.01217	1.8923	1.064	-262.4	-43.2	7.111	5.95	8.07	1603
25	0.01226	1.8441	1.057	-255.7	-35.0	7.445	6.04	8.31	1593
26	0.01235	1.7964	1.049	-248.9	-26.6	7.776	6.12	8.55	1584
27	0.01244	1.7451	1.043	-241.8	-17.9	8.103	6.20	8.81	1574
28	0.01253	1.7001	1.032	-234.6	-9.0	8.427	6.28	9.04	1564
29	0.01263	1.6542	1.020	-227.2	0.1	8.749	6.35	9.26	1553
30	0.01273	1.6057	1.007	-219.6	9.5	9.066	6.41	9.48	1541
31	0.01283	1.5650	0.9933	-211.9	19.1	9.380	6.47	9.69	1531
32	0.01294	1.5226	0.9789	-204.0	28.9	9.691	6.52	9.90	1520
33	0.01305	1.4807	0.9642	-196.0	38.9	9.999	6.57	10.10	1508
34	0.01316	1.4380	0.9496	-187.8	49.1	10.303	6.62	10.31	1497
35	0.01328	1.3992	0.9343	-179.5	59.5	10.605	6.66	10.51	1486
36	0.01339	1.3617	0.9186	-171.0	70.1	10.904	6.70	10.70	1475
37	0.01352	1.3241	0.9024	-162.4	80.9	11.200	6.73	10.89	1463
38	0.01364	1.2873	0.8859	-153.7	91.9	11.493	6.76	11.08	1452
39	0.01377	1.2532	0.8692	-144.9	103.0	11.783	6.78	11.24	1441
40	0.01391	1.2192	0.8523	-136.0	114.4	12.070	6.80	11.41	1430
42	0.01419	1.1548	0.8181	-117.8	137.5	12.634	6.84	11.74	1408
44	0.01448	1.0969	0.7838	-99.3	161.3	13.187	6.87	12.04	1386
46	0.01478	1.0427	0.7499	-81.0	185.2	13.718	6.94	12.37	1363
48	0.01511	0.9914	0.7165	-61.7	210.2	14.250	7.00	12.67	1340
50	0.01544	0.9454	0.6840	-42.2	235.8	14.773	7.05	12.95	1318
52	0.01580	0.9019	0.6523	-22.4	262.0	15.286	7.09	13.22	1296
54	0.01617	0.8648	0.6217	-2.3	288.6	15.789	7.14	13.45	1276
56	0.01655	0.8327	0.5924	18.1	316.0	16.287	7.18	13.64	1258
58	0.01694	0.8083	0.5646	38.5	343.5	16.769	7.21	13.82	1241
60	0.01735	0.7794	0.5381	58.9	371.3	17.240	7.25	13.97	1225
65	0.01843	0.7289	0.4784	110.2	442.0	18.373	7.39	14.32	1189
70	0.01958	0.6962	0.4270	161.9	514.3	19.444	7.58	14.61	1158
75	0.02078	0.6762	0.3832	214.0	588.0	20.460	7.82	14.86	1133
80	0.02202	0.6677	0.3459	266.5	662.8	21.426	8.12	15.07	1113
85	0.02329	0.6663	0.3142	319.6	738.7	22.347	8.47	15.30	1097
90	0.02457	0.6711	0.2872	373.5	815.8	23.228	8.84	15.52	1085
95	0.02587	0.6806	0.2641	428.3	894.0	24.073	9.24	15.75	1077
100	0.02717	0.6927	0.2442	484.3	973.3	24.887	9.64	15.99	1072
120	0.03237	0.7587	0.1876	722.7	1305.3	27.910	11.19	17.02	1074
140	0.03751	0.8391	0.1523	976.6	1651.8	30.580	12.14	17.58	1102
160	0.04258	0.9252	0.1281	1238.1	2004.5	32.934	12.49	17.64	1143
180	0.04756	1.0128	0.1106	1498.5	2354.6	34.997	12.44	17.35	1189
200	0.05246	1.1000	0.0974	1752.8	2697.1	36.801	12.17	16.91	1236
220	0.05730	1.1874	0.0870	1998.8	3030.2	38.389	11.83	16.43	1284
240	0.06208	1.2728	0.0786	2236.6	3354.1	39.798	11.50	15.99	1331
260	0.06670	1.3507	0.0720	2467.0	3667.6	41.054	11.22	15.66	1373
280	0.07145	1.4373	0.0667	2692.1	3978.3	42.205	11.00	15.43	1420
300	0.07635	1.5332	0.0618	2915.5	4289.9	43.274	10.85	15.21	1466
350	0.08806	1.7441	0.0524	3456.1	5041.1	45.589	10.60	14.88	1565
400	0.09968	1.9533	0.0456	3986.6	5780.9	47.566	10.50	14.73	1655
450	0.11126	2.1615	0.0404	4512.4	6515.0	49.297	10.46	14.66	1740
500	0.12281	2.3690	0.0362	5036.8	7247.3	50.839	10.45	14.62	1821
550	0.13433	2.5760	0.0329	5560.0	7977.9	52.231	10.45	14.61	1898
600	0.14584	2.7827	0.0301	6083.5	8708.6	53.501	10.46	14.61	1972
700	0.16882	3.1956	0.0258	7132.2	10170.9	55.757	10.50	14.65	2111
800	0.19178	3.6082	0.0225	8187.2	11639.2	57.716	10.59	14.72	2240
900	0.21472	4.0205	0.0200	9252.3	13117.4	59.456	10.71	14.84	2360
1000	0.23766	4.4327	0.0180	10330.9	14608.7	61.030	10.87	15.00	2473
1200	0.28351	5.2570	0.0150	12543.0	17646.1	63.785	11.25	15.38	2680
1400	0.32935	6.0812	0.0128	14836.7	20765.0	66.193	11.69	15.81	2868
1600	0.37518	6.9056	0.0112	17215.8	23969.1	68.309	12.10	16.22	3043
1800	0.42102	7.7300	0.0100	19680.4	27258.7	70.209	12.49	16.61	3207
2000	0.46687	8.5544	0.0090	22232.7	30636.3	72.054	12.90	17.03	3360
2500	0.58196	10.6157	0.0072	29058.5	39533.8	78.366	14.45	18.66	3782
3000	0.70014	12.6772	0.0060	37119.9	49722.4	85.443	17.95	22.55	3989

* TWO-PHASE BOUNDARY

Table 4.1 (continued)
THERMODYNAMIC PROPERTIES OF PARAHYDROGEN (ISOBARS, SI UNITS)

18.00 MPA ISOBAR

TEMPERATURE DEG. K	DENSITY KG/CU M	V(DH/DV)$_P$ KJ/KG	V(DP/DU)$_V$ MPA-CU M/KJ X 10^2	-V(DP/DV)$_T$ MPA	-(DV/DT)$_P$/V 1/DEG. K	THERMAL CONDUCTIVITY W/K-M X 10^3	VISCOSITY KG/M-S X 10^7	THERMAL DIFFUSIVITY SQ M/HR	DIELECTRIC CONSTANT	PRANDTL NUMBER
* 18.874	84.8704	1137.7	0.23871	181.8403	0.00594	126.03	365.44	0.00079	1.27940	1.9595
19	84.8070	1139.4	0.23807	181.1834	0.00596	126.57	360.52	0.00079	1.27918	1.9347
20	84.2955	1151.0	0.23330	175.9684	0.00614	131.23	325.55	0.00079	1.27735	1.7520
21	83.7725	1160.5	0.22897	170.7764	0.00631	136.27	296.78	0.00080	1.27549	1.5953
22	83.2381	1168.4	0.22497	165.6158	0.00649	140.45	272.79	0.00080	1.27359	1.4725
23	82.6924	1174.6	0.22129	160.4964	0.00667	143.88	252.54	0.00080	1.27165	1.3744
24	82.1357	1179.6	0.21783	155.4285	0.00685	146.69	235.26	0.00080	1.26967	1.2951
25	81.5680	1183.3	0.21458	150.4231	0.00703	148.95	220.37	0.00079	1.26765	1.2300
26	80.9895	1186.2	0.21145	145.4917	0.00721	150.69	207.42	0.00078	1.26560	1.1769
27	80.4001	1184.7	0.20915	140.3040	0.00744	152.01	196.07	0.00077	1.26352	1.1362
28	79.7998	1187.5	0.20598	135.6647	0.00761	152.98	186.04	0.00076	1.26139	1.0988
29	79.1848	1188.9	0.20288	130.9867	0.00779	153.64	177.07	0.00075	1.25922	1.0670
30	78.5637	1187.5	0.19997	126.1467	0.00798	154.05	169.09	0.00074	1.25703	1.0405
31	77.9345	1189.4	0.19701	121.9677	0.00814	153.66	161.92	0.00073	1.25481	1.0207
32	77.2941	1189.6	0.19413	117.6904	0.00832	153.09	155.41	0.00072	1.25255	1.0045
33	76.6471	1188.8	0.19138	113.4883	0.00850	152.40	149.51	0.00071	1.25028	0.9909
34	75.9926	1186.6	0.18878	109.2804	0.00869	151.63	144.12	0.00070	1.24798	0.9800
35	75.3265	1185.5	0.18621	105.3932	0.00886	150.76	139.15	0.00069	1.24564	0.9700
36	74.6560	1184.3	0.18368	101.6578	0.00904	149.82	134.59	0.00068	1.24329	0.9614
37	73.9771	1182.2	0.18116	97.9564	0.00921	148.80	130.37	0.00066	1.24091	0.9542
38	73.2918	1179.5	0.17874	94.3467	0.00939	147.72	126.45	0.00066	1.23852	0.9480
39	72.6014	1177.1	0.17646	90.9873	0.00955	146.61	122.80	0.00065	1.23611	0.9418
40	71.9051	1172.9	0.17423	87.6682	0.00972	145.45	119.39	0.00064	1.23368	0.9367
42	70.4961	1167.8	0.16972	81.4092	0.01005	143.03	113.19	0.00062	1.22877	0.9287
44	69.0729	1163.7	0.16510	75.7670	0.01035	140.56	107.70	0.00061	1.22383	0.9225
46	67.6380	1163.0	0.15967	70.5258	0.01063	138.09	102.80	0.00059	1.21887	0.9205
48	66.1964	1160.5	0.15468	65.6252	0.01092	135.66	98.39	0.00058	1.21389	0.9190
50	64.7504	1158.9	0.14988	61.2150	0.01117	133.30	94.41	0.00057	1.20891	0.9171
52	63.3028	1156.8	0.14524	57.0946	0.01142	131.01	90.79	0.00056	1.20394	0.9158
54	61.8615	1157.0	0.14078	53.5005	0.01162	128.84	87.51	0.00056	1.19900	0.9132
56	60.4295	1158.8	0.13655	50.3192	0.01177	126.78	84.53	0.00055	1.19411	0.9096
58	59.0159	1160.6	0.13261	47.4072	0.01191	124.85	81.83	0.00055	1.18930	0.9059
60	57.6214	1165.7	0.12873	44.9082	0.01198	123.23	79.38	0.00055	1.18456	0.8998
65	54.2514	1183.9	0.11935	39.5450	0.01210	120.30	74.26	0.00056	1.17316	0.8840
70	51.0762	1216.2	0.11033	35.5587	0.01201	118.25	70.34	0.00057	1.16250	0.8689
75	48.1267	1261.7	0.10175	32.5426	0.01177	116.58	67.41	0.00059	1.15264	0.8591
80	45.4163	1321.5	0.09375	30.3250	0.01141	115.80	65.28	0.00061	1.14364	0.8497
85	42.9418	1392.9	0.08643	28.6114	0.01098	117.52	63.78	0.00064	1.13547	0.8302
90	40.6944	1475.8	0.07982	27.3084	0.01052	119.55	62.78	0.00068	1.12807	0.8151
95	38.6540	1569.0	0.07397	26.3060	0.01004	121.88	62.18	0.00072	1.12139	0.8035
100	36.8044	1669.4	0.06887	25.4928	0.00958	122.30	61.57	0.00075	1.11536	0.8052
120	30.8955	2126.8	0.05427	23.4407	0.00800	140.63	66.03	0.00096	1.09624	0.7993
140	26.6567	2582.5	0.04707	22.3681	0.00681	153.64	68.60	0.00118	1.08266	0.7849
160	23.4862	2992.0	0.04366	21.7295	0.00590	162.19	70.64	0.00141	1.07258	0.7682
180	21.0273	3341.9	0.04229	21.2970	0.00519	167.52	72.60	0.00165	1.06481	0.7521
200	19.0621	3642.5	0.04197	20.9689	0.00464	170.97	74.60	0.00191	1.05863	0.7380
220	17.4520	3914.2	0.04214	20.7219	0.00420	173.60	76.69	0.00218	1.05359	0.7259
240	16.1070	4169.2	0.04246	20.5002	0.00384	176.14	78.86	0.00246	1.04938	0.7161
260	14.9925	4404.1	0.04280	20.2510	0.00355	178.95	81.09	0.00274	1.04591	0.7094
280	13.9953	4650.8	0.04334	20.1147	0.00332	182.18	83.38	0.00304	1.04281	0.7060
300	13.0969	4938.7	0.04353	20.0796	0.00308	185.83	85.71	0.00336	1.04002	0.7014
350	11.3563	5620.3	0.04356	19.8062	0.00265	196.56	91.69	0.00419	1.03464	0.6942
400	10.0318	6329.1	0.04330	19.5952	0.00233	208.70	97.80	0.00509	1.03056	0.6902
450	8.9880	7056.1	0.04293	19.4272	0.00208	221.61	103.96	0.00605	1.02735	0.6878
500	8.1430	7786.4	0.04260	19.2904	0.00188	234.70	110.12	0.00709	1.02475	0.6862
550	7.4445	8522.5	0.04227	19.1769	0.00171	247.95	116.27	0.00821	1.02261	0.6851
600	6.8570	9263.1	0.04197	19.0812	0.00158	261.24	122.37	0.00939	1.02081	0.6844
700	5.9234	10764.3	0.04139	18.9292	0.00136	287.98	134.42	0.01195	1.01796	0.6836
800	5.2143	12304.5	0.04077	18.8141	0.00120	315.07	146.22	0.01477	1.01580	0.6833
900	4.6572	13895.1	0.04009	18.7239	0.00107	342.65	157.76	0.01784	1.01410	0.6834
1000	4.2077	15544.1	0.03934	18.6515	0.00096	440.56	169.04	0.02513	1.01274	0.5754
1200	3.5272	19017.6	0.03777	18.5425	0.00081	511.13	190.88	0.03393	1.01067	0.5742
1400	3.0363	22718.2	0.03621	18.4645	0.00070	583.32	211.85	0.04375	1.00918	0.5741
1600	2.6654	26562.1	0.03485	18.4059	0.00061	655.45	232.06	0.05457	1.00805	0.5744
1800	2.3752	30522.5	0.03369	18.3602	0.00054	727.61	251.61	0.06638	1.00717	0.5745
2000	2.1419	34694.0	0.03255	18.3229	0.00049	803.32	270.57	0.07928	1.00647	0.5736
2500	1.7183	47312.7	0.02897	18.2413	0.00039	1057.58	315.94	0.11872	1.00519	0.5575
3000	1.4283	68083.8	0.02339	18.1067	0.00033	1594.22	359.72	0.17820	1.00431	0.5088

* TWO-PHASE BOUNDARY

Table 4.1 (continued)
THERMODYNAMIC PROPERTIES OF PARAHYDROGEN (ISOBARS, SI UNITS)

20.00 MPA ISOBAR

TEMPERATURE DEG. K	VOLUME CU M/KG	ISOTHERM DERIVATIVE CU M-MPA/KG	ISOCHORE DERIVATIVE MPA/K	INTERNAL ENERGY KJ/KG-K	ENTHALPY KJ/KG	ENTROPY KJ/KG-K	CV KJ / KG-K	CP KJ / KG-K	VELOCITY OF SOUND M/S
* 19.360	0.01169	2.2344	1.094	-291.0	-57.2	5.367	5.38	6.80	1680
20	0.01173	2.2040	1.094	-287.4	-52.8	5.591	5.47	6.97	1675
21	0.01180	2.1564	1.093	-281.7	-45.7	5.937	5.60	7.22	1667
22	0.01187	2.1087	1.090	-275.8	-38.3	6.279	5.72	7.47	1659
23	0.01195	2.0611	1.087	-269.7	-30.7	6.616	5.83	7.71	1651
24	0.01203	2.0136	1.082	-263.4	-22.9	6.950	5.93	7.95	1642
25	0.01210	1.9664	1.075	-256.9	-14.8	7.279	6.03	8.18	1634
26	0.01219	1.9195	1.068	-250.2	-6.5	7.604	6.11	8.41	1625
27	0.01227	1.8731	1.060	-243.4	2.0	7.926	6.20	8.63	1615
28	0.01236	1.8274	1.051	-236.4	10.7	8.244	6.28	8.86	1606
29	0.01245	1.7797	1.044	-229.2	19.7	8.560	6.35	9.10	1597
30	0.01254	1.7369	1.032	-221.8	28.9	8.872	6.41	9.30	1588
31	0.01263	1.6937	1.019	-214.3	38.3	9.180	6.47	9.51	1577
32	0.01273	1.6480	1.006	-206.7	47.9	9.485	6.53	9.71	1566
33	0.01283	1.6088	0.9921	-198.8	57.8	9.787	6.58	9.90	1556
34	0.01293	1.5685	0.9777	-190.9	67.7	10.085	6.63	10.09	1546
35	0.01304	1.5277	0.9630	-182.8	77.9	10.381	6.67	10.28	1535
36	0.01315	1.4861	0.9483	-174.6	88.3	10.673	6.71	10.48	1523
37	0.01326	1.4494	0.9331	-166.3	98.9	10.963	6.75	10.66	1513
38	0.01337	1.4146	0.9177	-157.8	109.6	11.249	6.78	10.83	1503
39	0.01349	1.3793	0.9019	-149.3	120.5	11.533	6.81	10.99	1493
40	0.01361	1.3444	0.8859	-140.6	131.6	11.813	6.83	11.15	1482
42	0.01386	1.2803	0.8533	-123.0	154.2	12.364	6.86	11.45	1462
44	0.01412	1.2210	0.8206	-105.1	177.4	12.904	6.90	11.74	1441
46	0.01440	1.1649	0.7880	-87.3	200.7	13.421	6.98	12.06	1419
48	0.01468	1.1145	0.7558	-68.6	225.1	13.940	7.03	12.34	1398
50	0.01498	1.0671	0.7242	-49.6	250.0	14.449	7.09	12.60	1378
52	0.01529	1.0229	0.6935	-30.4	275.5	14.948	7.13	12.85	1358
54	0.01562	0.9819	0.6636	-10.9	301.4	15.438	7.18	13.09	1338
56	0.01595	0.9463	0.6346	9.0	328.1	15.923	7.22	13.29	1319
58	0.01630	0.9139	0.6067	28.8	354.8	16.392	7.26	13.47	1302
60	0.01666	0.8863	0.5803	48.8	381.9	16.851	7.30	13.63	1286
65	0.01760	0.8298	0.5197	99.0	451.0	17.957	7.44	13.99	1249
70	0.01860	0.7893	0.4671	149.9	521.8	19.006	7.62	14.32	1217
75	0.01964	0.7626	0.4216	201.3	594.1	20.004	7.87	14.62	1190
80	0.02073	0.7461	0.3823	253.4	667.9	20.956	8.17	14.90	1167
85	0.02184	0.7398	0.3486	306.3	743.1	21.867	8.51	15.17	1148
90	0.02297	0.7396	0.3195	360.2	819.6	22.742	8.89	15.44	1134
95	0.02412	0.7444	0.2943	415.1	897.5	23.584	9.28	15.71	1123
100	0.02527	0.7532	0.2725	471.3	976.7	24.397	9.68	15.97	1115
120	0.02990	0.8101	0.2096	710.7	1308.7	27.421	11.22	17.04	1109
140	0.03451	0.8842	0.1702	965.9	1656.0	30.096	12.17	17.63	1132
160	0.03906	0.9669	0.1432	1228.6	2009.8	32.458	12.52	17.70	1169
180	0.04355	1.0521	0.1235	1490.2	2361.1	34.528	12.47	17.42	1212
200	0.04796	1.1390	0.1087	1745.6	2704.9	36.339	12.20	16.97	1259
220	0.05233	1.2245	0.0970	1992.7	3039.2	37.932	11.86	16.49	1305
240	0.05664	1.3105	0.0876	2231.4	3364.3	39.347	11.53	16.05	1350
260	0.06091	1.3950	0.0800	2462.8	3681.0	40.615	11.25	15.67	1394
280	0.06514	1.4793	0.0735	2688.2	3991.0	41.764	11.02	15.36	1436
300	0.06951	1.5696	0.0688	2911.4	4301.6	42.828	10.86	15.24	1484
350	0.08005	1.7800	0.0583	3453.0	5054.1	45.147	10.61	14.90	1581
400	0.09053	1.9889	0.0507	3984.2	5794.8	47.126	10.51	14.74	1671
450	0.10095	2.1968	0.0449	4510.5	6529.5	48.858	10.47	14.67	1755
500	0.11135	2.4040	0.0403	5035.4	7262.3	50.401	10.45	14.63	1835
550	0.12172	2.6108	0.0365	5558.9	7993.3	51.795	10.45	14.62	1911
600	0.13208	2.8173	0.0334	6082.7	8724.2	53.065	10.46	14.62	1984
700	0.15277	3.2299	0.0286	7131.7	10187.0	55.321	10.51	14.65	2122
800	0.17343	3.6420	0.0250	8187.0	11655.6	57.281	10.59	14.73	2250
900	0.19408	4.0541	0.0222	9252.3	13133.9	59.021	10.72	14.85	2370
1000	0.21472	4.4660	0.0200	10331.0	14625.4	60.595	10.87	15.00	2482
1200	0.25599	5.2899	0.0167	12543.3	17663.0	63.351	11.25	15.38	2688
1400	0.29724	6.1139	0.0143	14837.2	20781.9	65.758	11.69	15.81	2876
1600	0.33849	6.9380	0.0125	17216.3	23986.1	67.875	12.10	16.22	3050
1800	0.37974	7.7622	0.0111	19680.9	27275.7	69.774	12.49	16.61	3213
2000	0.42100	8.5865	0.0100	22232.7	30652.7	71.619	12.90	17.02	3367
2500	0.52456	10.6475	0.0080	29048.8	39540.0	77.927	14.41	18.61	3709
3000	0.63076	12.7087	0.0067	37055.1	49670.3	84.983	17.77	22.33	3996

* TWO-PHASE BOUNDARY

Table 4.1 (continued)
THERMODYNAMIC PROPERTIES OF PARAHYDROGEN (ISOBARS, SI UNITS)

20.00 MPA ISOBAR

TEMPERATURE DEG. K	DENSITY KG/CU M	V(DH/DV)$_P$ KJ/KG	V(DP/DU)$_V$ MPA-CU M/KJ X 10^2	-V(DP/DV)$_T$ MPA	-(DV/DT)$_P$/V 1/DEG. K	THERMAL CONDUCTIVITY W/K-M X 10^3	VISCOSITY KG/M-S X 10^7	THERMAL DIFFUSIVITY SQ M/HR	DIELECTRIC CONSTANT	PRANDTL NUMBER
* 19.360	85.5433	1188.4	0.23748	191.1415	0.00572	130.46	377.62	0.00081	1.28181	1.9680
20	85.2277	1196.5	0.23452	187.8410	0.00582	133.83	353.32	0.00081	1.28068	1.8393
21	84.7262	1207.5	0.23021	182.7004	0.00598	139.12	321.04	0.00082	1.27889	1.6668
22	84.2142	1216.9	0.22624	177.5827	0.00614	143.52	294.22	0.00082	1.27706	1.5318
23	83.6921	1224.6	0.22258	172.4970	0.00630	147.16	271.67	0.00082	1.27520	1.4241
24	83.1599	1231.0	0.21916	167.4527	0.00646	150.15	252.50	0.00082	1.27331	1.3370
25	82.6179	1236.0	0.21594	162.4598	0.00662	152.58	236.03	0.00081	1.27138	1.2657
26	82.0662	1240.2	0.21286	157.5289	0.00678	154.47	221.76	0.00081	1.26942	1.2072
27	81.5050	1243.8	0.20984	152.6707	0.00694	155.94	209.29	0.00080	1.26743	1.1589
28	80.9346	1246.8	0.20687	147.8964	0.00710	157.06	198.32	0.00079	1.26541	1.1185
29	80.3501	1246.2	0.20469	142.9994	0.00730	157.84	188.54	0.00078	1.26334	1.0868
30	79.7603	1248.6	0.20186	138.5345	0.00745	158.37	179.85	0.00077	1.26125	1.0564
31	79.1617	1250.3	0.19899	134.0749	0.00760	158.07	172.05	0.00076	1.25914	1.0348
32	78.5555	1249.9	0.19617	129.4586	0.00777	157.59	165.02	0.00074	1.25700	1.0172
33	77.9419	1251.7	0.19344	125.3936	0.00791	156.98	158.65	0.00073	1.25483	1.0009
34	77.3220	1252.1	0.19078	121.2815	0.00806	156.30	152.86	0.00072	1.25265	0.9871
35	76.6946	1251.2	0.18820	117.1683	0.00822	155.51	147.55	0.00071	1.25044	0.9757
36	76.0619	1249.0	0.18574	113.0376	0.00839	154.64	142.68	0.00070	1.24822	0.9667
37	75.4195	1248.5	0.18331	109.3153	0.00854	153.69	138.17	0.00069	1.24597	0.9581
38	74.7739	1247.9	0.18099	105.7719	0.00868	152.68	134.00	0.00068	1.24370	0.9503
39	74.1217	1245.9	0.17881	102.2326	0.00882	151.62	130.13	0.00067	1.24142	0.9434
40	73.4653	1243.3	0.17665	98.7645	0.00897	150.53	126.52	0.00066	1.23912	0.9374
42	72.1398	1239.8	0.17232	92.3572	0.00924	148.21	119.99	0.00065	1.23450	0.9274
44	70.8011	1237.3	0.16788	86.4488	0.00949	145.83	114.23	0.00063	1.22984	0.9200
46	69.4518	1238.2	0.16263	80.9077	0.00974	143.45	109.10	0.00062	1.22515	0.9172
48	68.0970	1239.0	0.15780	75.8970	0.00996	141.09	104.50	0.00060	1.22045	0.9139
50	66.7399	1239.3	0.15316	71.2198	0.01017	138.79	100.35	0.00059	1.21576	0.9113
52	65.3829	1239.6	0.14868	66.8821	0.01037	136.56	96.59	0.00059	1.21109	0.9091
54	64.0296	1239.9	0.14433	62.8680	0.01056	134.42	93.17	0.00058	1.20643	0.9072
56	62.6829	1242.1	0.14015	59.3154	0.01070	132.39	90.06	0.00057	1.20181	0.9040
58	61.3466	1244.4	0.13624	56.0628	0.01082	130.47	87.22	0.00057	1.19724	0.9003
60	60.0282	1249.4	0.13241	53.2057	0.01091	128.85	84.63	0.00057	1.19274	0.8950
65	56.8213	1269.0	0.12301	47.1485	0.01102	125.90	79.15	0.00057	1.18185	0.8795
70	53.7734	1300.9	0.11393	42.4454	0.01100	123.59	74.87	0.00058	1.17155	0.8673
75	50.9103	1345.9	0.10521	38.8224	0.01086	121.93	71.57	0.00059	1.16194	0.8580
80	48.2479	1403.2	0.09699	36.0001	0.01062	120.97	69.08	0.00061	1.15305	0.8510
85	45.7912	1474.3	0.08943	33.8776	0.01029	122.55	67.25	0.00064	1.14488	0.8324
90	43.5352	1556.0	0.08259	32.2001	0.00992	124.43	65.95	0.00067	1.13742	0.8182
95	41.4668	1647.3	0.07650	30.8672	0.00953	126.60	65.07	0.00070	1.13061	0.8074
100	39.5742	1747.1	0.07116	29.8080	0.00914	126.84	64.04	0.00072	1.12440	0.8064
120	33.4477	2202.9	0.05584	27.0964	0.00774	144.72	68.24	0.00091	1.10447	0.8036
140	28.9793	2654.0	0.04827	25.6237	0.00664	157.04	70.44	0.00111	1.09008	0.7907
160	25.6015	3060.1	0.04465	24.7531	0.00578	164.96	72.14	0.00131	1.07930	0.7741
180	22.9647	3407.2	0.04314	24.1615	0.00511	169.80	73.83	0.00153	1.07093	0.7574
200	20.8486	3708.7	0.04272	23.7458	0.00458	172.91	75.64	0.00176	1.06425	0.7424
220	19.1108	3978.1	0.04279	23.4010	0.00415	175.31	77.58	0.00200	1.05878	0.7297
240	17.6549	4235.6	0.04305	23.1363	0.00379	177.68	79.64	0.00226	1.05422	0.7192
260	16.4175	4489.0	0.04328	22.9019	0.00349	180.38	81.79	0.00252	1.05035	0.7106
280	15.3509	4746.1	0.04344	22.7079	0.00324	183.53	84.03	0.00280	1.04703	0.7035
300	14.3862	4998.7	0.04405	22.5803	0.00305	187.13	86.33	0.00307	1.04403	0.7029
350	12.4914	5680.0	0.04400	22.2353	0.00262	197.82	92.27	0.00383	1.03815	0.6951
400	11.0465	6388.9	0.04368	21.9708	0.00231	209.98	98.38	0.00464	1.03368	0.6907
450	9.9058	7116.0	0.04326	21.7611	0.00206	222.96	104.57	0.00552	1.03017	0.6881
500	8.9811	7846.4	0.04289	21.5907	0.00186	236.16	110.78	0.00647	1.02732	0.6864
550	8.2157	8582.7	0.04254	21.4496	0.00170	249.53	116.98	0.00748	1.02497	0.6853
600	7.5713	9323.4	0.04222	21.3309	0.00157	262.96	123.15	0.00855	1.02300	0.6845
700	6.5460	10824.8	0.04160	21.1426	0.00135	289.99	135.33	0.01089	1.01986	0.6836
800	5.7660	12365.2	0.04094	21.0002	0.00119	317.40	147.27	0.01346	1.01748	0.6833
900	5.1525	13956.3	0.04024	20.8888	0.00106	345.30	158.95	0.01625	1.01561	0.6834
1000	4.6572	15605.7	0.03948	20.7994	0.00096	440.56	170.38	0.02270	1.01410	0.5801
1200	3.9065	19080.6	0.03788	20.6651	0.00081	511.13	192.51	0.03063	1.01182	0.5791
1400	3.3643	22782.6	0.03630	20.5690	0.00069	583.32	213.76	0.03948	1.01017	0.5793
1600	2.9543	26627.7	0.03493	20.4971	0.00061	655.44	234.24	0.04923	1.00893	0.5798
1800	2.6334	30587.5	0.03375	20.4409	0.00054	727.51	254.05	0.05987	1.00796	0.5801
2000	2.3753	34753.2	0.03261	20.3953	0.00049	802.95	273.27	0.07148	1.00717	0.5794
2500	1.9054	47267.3	0.02910	20.2979	0.00039	1053.09	319.24	0.10683	1.00576	0.5643
3000	1.5854	67548.4	0.02365	20.1484	0.00033	1571.05	363.54	0.15972	1.00478	0.5168

* TWO-PHASE BOUNDARY

Table 4.1 (continued)
THERMODYNAMIC PROPERTIES OF PARAHYDROGEN (ISOBARS, SI UNITS)

22.00 MPA ISOBAR

TEMPERATURE DEG. K	VOLUME CU M/KG	ISOTHERM DERIVATIVE CU M-MPA/KG	ISOCHORE DERIVATIVE MPA/K	INTERNAL ENERGY KJ/KG-K	ENTHALPY KJ/KG	ENTROPY KJ/KG-K	CV KJ /	CP KG-K	VELOCITY OF SOUND M/S
* 19.835	0.01160	2.3243	1.107	-288.5	-33.2	5.401	5.43	6.84	1711
20	0.01161	2.3167	1.107	-287.6	-32.1	5.458	5.45	6.88	1710
21	0.01168	2.2703	1.107	-282.0	-25.1	5.800	5.59	7.13	1702
22	0.01175	2.2238	1.105	-276.2	-17.8	6.138	5.71	7.37	1695
23	0.01182	2.1773	1.102	-270.3	-10.3	6.471	5.82	7.61	1687
24	0.01189	2.1308	1.098	-264.1	-2.6	6.799	5.92	7.84	1680
25	0.01196	2.0846	1.093	-257.8	5.4	7.124	6.02	8.07	1672
26	0.01204	2.0385	1.086	-251.3	13.5	7.445	6.11	8.29	1663
27	0.01212	1.9928	1.079	-244.6	21.9	7.762	6.19	8.50	1655
28	0.01220	1.9476	1.070	-237.8	30.5	8.075	6.27	8.72	1646
29	0.01228	1.9030	1.061	-230.8	39.4	8.384	6.34	8.93	1637
30	0.01236	1.8591	1.051	-223.6	48.4	8.691	6.41	9.14	1628
31	0.01245	1.8157	1.043	-216.3	57.7	8.995	6.47	9.35	1620
32	0.01254	1.7748	1.031	-208.8	67.1	9.295	6.53	9.55	1611
33	0.01264	1.7338	1.018	-201.2	76.8	9.592	6.58	9.73	1601
34	0.01273	1.6912	1.004	-193.5	86.6	9.885	6.63	9.92	1590
35	0.01283	1.6534	0.9903	-185.6	96.6	10.175	6.68	10.10	1581
36	0.01293	1.6150	0.9760	-177.6	106.8	10.462	6.72	10.27	1571
37	0.01303	1.5756	0.9615	-169.5	117.2	10.746	6.76	10.45	1560
38	0.01314	1.5350	0.9468	-161.3	127.7	11.027	6.80	10.62	1549
39	0.01324	1.5001	0.9318	-152.9	138.4	11.305	6.82	10.78	1540
40	0.01335	1.4676	0.9167	-144.5	149.2	11.580	6.85	10.93	1531
42	0.01358	1.4030	0.8856	-127.4	171.4	12.120	6.89	11.22	1512
44	0.01382	1.3406	0.8542	-109.9	194.1	12.649	6.93	11.50	1492
46	0.01407	1.2848	0.8227	-92.6	216.9	13.154	7.01	11.80	1471
48	0.01433	1.2332	0.7915	-74.4	240.8	13.663	7.07	12.07	1451
50	0.01459	1.1839	0.7609	-55.9	265.2	14.161	7.12	12.33	1432
52	0.01487	1.1394	0.7309	-37.2	290.1	14.649	7.17	12.56	1413
54	0.01516	1.0982	0.7017	-18.2	315.4	15.127	7.22	12.78	1395
56	0.01546	1.0601	0.6734	1.3	341.5	15.601	7.26	12.99	1377
58	0.01577	1.0243	0.6458	20.7	367.6	16.060	7.30	13.17	1359
60	0.01609	0.9927	0.6193	40.2	394.1	16.509	7.34	13.34	1343
65	0.01692	0.9299	0.5583	89.5	461.8	17.592	7.48	13.72	1306
70	0.01780	0.8831	0.5046	139.5	531.2	18.621	7.67	14.07	1273
75	0.01873	0.8498	0.4577	190.3	602.4	19.603	7.92	14.40	1243
80	0.01969	0.8291	0.4170	242.0	675.2	20.542	8.21	14.72	1219
85	0.02068	0.8156	0.3814	294.7	749.5	21.444	8.55	15.04	1197
90	0.02168	0.8110	0.3506	348.4	825.5	22.312	8.93	15.34	1181
95	0.02271	0.8116	0.3237	403.4	902.9	23.149	9.32	15.64	1167
100	0.02374	0.8163	0.3002	459.6	981.9	23.959	9.72	15.94	1157
120	0.02790	0.8643	0.2314	699.9	1313.7	26.982	11.26	17.05	1144
140	0.03207	0.9309	0.1880	955.9	1661.4	29.660	12.20	17.66	1161
160	0.03620	1.0099	0.1582	1219.7	2016.0	32.028	12.55	17.75	1195
180	0.04028	1.0936	0.1365	1482.4	2368.5	34.104	12.50	17.47	1236
200	0.04430	1.1784	0.1200	1738.8	2713.3	35.920	12.23	17.03	1281
220	0.04827	1.2638	0.1071	1986.9	3048.7	37.519	11.89	16.54	1326
240	0.05219	1.3483	0.0966	2226.5	3374.8	38.938	11.56	16.09	1370
260	0.05608	1.4326	0.0881	2458.8	3692.5	40.209	11.28	15.72	1412
280	0.05993	1.5155	0.0810	2684.9	4003.3	41.362	11.06	15.41	1453
300	0.06392	1.6065	0.0758	2907.5	4313.7	42.423	10.88	15.26	1581
350	0.07351	1.8164	0.0642	3450.1	5067.3	44.746	10.63	14.92	1597
400	0.08303	2.0248	0.0558	3982.0	5808.8	46.728	10.52	14.76	1686
450	0.09252	2.2323	0.0494	4508.8	6544.2	48.461	10.48	14.68	1769
500	0.10197	2.4392	0.0443	5034.1	7277.4	50.005	10.46	14.64	1848
550	0.11140	2.6457	0.0402	5557.9	8008.7	51.399	10.46	14.62	1923
600	0.12082	2.8520	0.0368	6081.9	8740.0	52.670	10.47	14.62	1996
700	0.13963	3.2641	0.0315	7131.3	10203.1	54.927	10.51	14.65	2133
800	0.15841	3.6759	0.0275	8186.9	11672.0	56.887	10.60	14.73	2260
900	0.17718	4.0876	0.0244	9252.4	13150.5	58.627	10.72	14.85	2379
1000	0.19595	4.4993	0.0220	10331.3	14642.1	60.201	10.88	15.00	2491
1200	0.23346	5.3227	0.0183	12543.7	17679.9	62.957	11.26	15.38	2697
1400	0.27096	6.1463	0.0157	14837.7	20798.9	65.365	11.69	15.81	2883
1600	0.30846	6.9702	0.0137	17217.0	24003.1	67.482	12.10	16.22	3057
1800	0.34596	7.7941	0.0122	19681.5	27292.6	69.381	12.49	16.61	3220
2000	0.38347	8.6182	0.0110	22232.9	30669.2	71.226	12.89	17.02	3373
2500	0.47759	10.6788	0.0088	29040.7	39547.7	77.530	14.37	18.57	3715
3000	0.57401	12.7398	0.0073	36999.6	49627.8	84.567	17.60	22.15	4083

* TWO-PHASE BOUNDARY

Table 4.1 (continued)
THERMODYNAMIC PROPERTIES OF PARAHYDROGEN (ISOBARS, SI UNITS)

22.00 MPA ISOBAR

TEMPERATURE DEG. K	DENSITY KG/CU M	$V(DH/DV)_P$ KJ/KG	$V(DP/DU)_V$ MPA-CU M/KJ X 10^2	$-V(DP/DV)_T$ MPA	$-(DV/DT)_P/V$ 1/DEG. K	THERMAL CONDUCTIVITY W/K-M X 10^3	VISCOSITY KG/M-S X 10^7	THERMAL DIFFUSIVITY SQ M/HR	DIELECTRIC CONSTANT	PRANDTL NUMBER
* 19.835	86.1915	1237.7	0.23646	200.3367	0.00553	135.33	389.82	0.00083	1.28412	1.9701
20	86.1126	1240.0	0.23571	199.4935	0.00555	136.34	383.11	0.00083	1.28384	1.9336
21	85.6298	1252.5	0.23141	194.4029	0.00569	141.36	346.91	0.00084	1.28211	1.7439
22	85.1375	1263.3	0.22745	189.3277	0.00584	146.47	316.97	0.00084	1.28036	1.5959
23	84.6359	1272.4	0.22380	184.2761	0.00598	150.31	291.88	0.00084	1.27857	1.4779
24	84.1251	1280.1	0.22039	179.2570	0.00613	153.47	270.62	0.00084	1.27675	1.3826
25	83.6054	1286.5	0.21719	174.2797	0.00627	156.07	252.43	0.00083	1.27489	1.3045
26	83.0769	1291.9	0.21414	169.3539	0.00641	158.10	236.71	0.00083	1.27301	1.2405
27	82.5398	1296.7	0.21115	164.4893	0.00656	159.71	223.02	0.00082	1.27110	1.1875
28	81.9943	1300.9	0.20822	159.6962	0.00670	160.95	211.00	0.00081	1.26917	1.1432
29	81.4407	1304.3	0.20543	154.9848	0.00685	161.87	200.38	0.00080	1.26720	1.1057
30	80.8791	1306.6	0.20282	150.3656	0.00699	162.51	190.94	0.00079	1.26521	1.0734
31	80.3017	1307.3	0.20073	145.8000	0.00716	162.27	182.41	0.00078	1.26317	1.0515
32	79.7230	1310.3	0.19798	141.4923	0.00729	161.88	174.79	0.00077	1.26112	1.0308
33	79.1391	1312.1	0.19533	137.2131	0.00742	161.34	167.92	0.00075	1.25906	1.0131
34	78.5491	1312.2	0.19275	132.8393	0.00756	160.73	161.68	0.00074	1.25697	0.9981
35	77.9518	1314.1	0.19018	128.8886	0.00768	160.01	155.98	0.00073	1.25487	0.9843
36	77.3507	1314.7	0.18769	124.9214	0.00781	159.20	150.76	0.00072	1.25275	0.9728
37	76.7426	1314.0	0.18527	120.9139	0.00795	158.31	145.96	0.00071	1.25061	0.9633
38	76.1310	1311.4	0.18300	116.8606	0.00810	157.37	141.53	0.00070	1.24846	0.9556
39	75.5108	1310.6	0.18088	113.2708	0.00823	156.36	137.41	0.00069	1.24629	0.9475
40	74.8894	1310.3	0.17882	109.9089	0.00834	155.31	133.59	0.00068	1.24411	0.9400
42	73.6316	1308.6	0.17463	103.3082	0.00857	153.08	126.68	0.00067	1.23970	0.9284
44	72.3632	1306.5	0.17031	97.0078	0.00881	150.78	120.52	0.00065	1.23528	0.9203
46	71.0858	1310.2	0.16518	91.3284	0.00901	148.46	115.24	0.00064	1.23083	0.9162
48	69.8029	1312.7	0.16049	86.0786	0.00920	146.15	110.43	0.00062	1.22637	0.9120
50	68.5179	1314.3	0.15598	81.1197	0.00938	143.89	106.10	0.00061	1.22191	0.9090
52	67.2339	1316.7	0.15163	76.6029	0.00954	141.70	102.18	0.00060	1.21747	0.9060
54	65.9530	1319.6	0.14741	72.4282	0.00969	139.60	98.62	0.00060	1.21305	0.9032
56	64.6783	1322.5	0.14336	68.5652	0.00982	137.58	95.38	0.00059	1.20866	0.9004
58	63.4126	1324.9	0.13950	64.9510	0.00994	135.68	92.41	0.00058	1.20431	0.8973
60	62.1590	1329.5	0.13569	61.7071	0.01004	134.07	89.70	0.00058	1.20002	0.8928
65	59.0971	1350.3	0.12630	54.9559	0.01016	131.10	83.90	0.00058	1.18957	0.8779
70	56.1677	1382.7	0.11713	49.6016	0.01017	128.72	79.30	0.00059	1.17964	0.8665
75	53.3927	1427.6	0.10831	45.3741	0.01009	126.94	75.67	0.00059	1.17027	0.8585
80	50.7897	1486.2	0.09996	42.1081	0.00990	125.83	72.86	0.00061	1.16154	0.8522
85	48.3638	1555.0	0.09219	39.4447	0.00967	127.32	70.72	0.00063	1.15343	0.8353
90	46.1163	1636.5	0.08516	37.3989	0.00937	129.07	69.13	0.00066	1.14596	0.8218
95	44.0399	1727.2	0.07888	35.7421	0.00906	131.11	68.00	0.00069	1.13909	0.8114
100	42.1248	1825.7	0.07335	34.3878	0.00873	131.27	66.46	0.00070	1.13277	0.8071
120	35.8380	2281.6	0.05737	30.9735	0.00747	148.80	70.45	0.00088	1.11222	0.8072
140	31.1839	2727.2	0.04943	29.0286	0.00648	160.50	72.33	0.00105	1.09716	0.7960
160	27.6265	3130.0	0.04562	27.8988	0.00567	167.82	73.70	0.00123	1.08575	0.7796
180	24.8290	3476.2	0.04397	27.1517	0.00503	172.19	75.12	0.00143	1.07684	0.7623
200	22.5752	3775.1	0.04345	26.6016	0.00451	174.95	76.73	0.00164	1.06970	0.7467
220	20.7183	4045.6	0.04345	26.1833	0.00409	177.09	78.51	0.00186	1.06384	0.7333
240	19.1592	4301.9	0.04362	25.8330	0.00374	179.29	80.46	0.00209	1.05894	0.7222
260	17.8323	4556.1	0.04379	25.5464	0.00345	181.86	82.53	0.00234	1.05478	0.7132
280	16.6866	4811.8	0.04389	25.2890	0.00320	184.93	84.71	0.00259	1.05119	0.7059
300	15.6457	5059.5	0.04456	25.1355	0.00302	188.46	86.97	0.00284	1.04795	0.7043
350	13.6037	5740.3	0.04444	24.7099	0.00260	199.09	92.86	0.00353	1.04160	0.6960
400	12.0431	6449.1	0.04406	24.3855	0.00229	211.27	98.97	0.00428	1.03676	0.6913
450	10.8089	7176.2	0.04359	24.1293	0.00205	224.32	105.18	0.00509	1.03295	0.6885
500	9.8070	7906.7	0.04319	23.9216	0.00185	237.61	111.43	0.00596	1.02986	0.6866
550	8.9766	8643.0	0.04281	23.7498	0.00169	251.10	117.69	0.00689	1.02731	0.6854
600	8.2769	9383.8	0.04246	23.6056	0.00156	264.66	123.92	0.00787	1.02516	0.6846
700	7.1619	10885.3	0.04180	23.3771	0.00135	291.99	136.23	0.01002	1.02175	0.6836
800	6.3126	12426.0	0.04112	23.2044	0.00119	319.71	148.32	0.01238	1.01915	0.6833
900	5.6438	14017.4	0.04039	23.0696	0.00106	347.94	160.15	0.01495	1.01711	0.6834
1000	5.1034	15667.3	0.03961	22.9615	0.00096	440.56	171.72	0.02072	1.01546	0.5847
1200	4.2834	19143.4	0.03799	22.7992	0.00080	511.13	194.15	0.02794	1.01297	0.5841
1400	3.6905	22846.9	0.03639	22.6834	0.00069	583.32	215.68	0.03599	1.01116	0.5845
1600	3.2419	26693.0	0.03500	22.5966	0.00061	655.42	236.44	0.04486	1.00980	0.5853
1800	2.8905	30652.5	0.03382	22.5290	0.00054	727.44	256.52	0.05454	1.00874	0.5857
2000	2.6078	34813.3	0.03268	22.4743	0.00049	802.64	276.00	0.06510	1.00788	0.5853
2500	2.0938	47237.8	0.02921	22.3597	0.00039	1049.23	322.58	0.09713	1.00632	0.5710
3000	1.7421	67097.7	0.02389	22.1945	0.00033	1551.11	367.42	0.14470	1.00526	0.5247

* TWO-PHASE BOUNDARY

Table 4.1 (continued)
THERMODYNAMIC PROPERTIES OF PARAHYDROGEN (ISOBARS, SI UNITS)

24.00 MPA ISOBAR

TEMPERATURE DEG. K	VOLUME CU M/KG	ISOTHERM DERIVATIVE CU M-MPA/KG	ISOCHORE DERIVATIVE MPA/K	INTERNAL ENERGY KJ/KG-K	ENTHALPY KJ/KG	ENTROPY KJ/KG-K	CV KJ / KG-K	CP KJ / KG-K	VELOCITY OF SOUND M/S
* 20.300	0.01152	2.4124	1.120	-285.9	-9.4	5.435	5.48	6.88	1741
21	0.01156	2.3807	1.120	-282.0	-4.5	5.671	5.57	7.05	1736
22	0.01163	2.3354	1.119	-276.4	2.6	6.004	5.69	7.29	1729
23	0.01169	2.2899	1.117	-270.6	10.0	6.333	5.81	7.52	1722
24	0.01176	2.2445	1.113	-264.6	17.7	6.658	5.91	7.74	1715
25	0.01183	2.1991	1.109	-258.4	25.5	6.979	6.01	7.96	1707
26	0.01190	2.1539	1.103	-252.0	33.6	7.295	6.10	8.18	1700
27	0.01197	2.1089	1.096	-245.5	41.9	7.608	6.18	8.39	1692
28	0.01205	2.0643	1.089	-238.8	50.4	7.917	6.26	8.60	1683
29	0.01213	2.0202	1.080	-232.0	59.1	8.222	6.34	8.80	1675
30	0.01221	1.9767	1.071	-225.0	68.0	8.524	6.41	9.00	1666
31	0.01229	1.9338	1.061	-217.9	77.1	8.822	6.47	9.20	1658
32	0.01237	1.8918	1.051	-210.6	86.4	9.117	6.53	9.39	1649
33	0.01246	1.8526	1.042	-203.1	95.9	9.411	6.59	9.59	1642
34	0.01255	1.8136	1.029	-195.6	105.6	9.700	6.64	9.76	1633
35	0.01264	1.7746	1.016	-187.9	115.4	9.986	6.69	9.94	1624
36	0.01273	1.7350	1.002	-180.1	125.5	10.268	6.73	10.11	1614
37	0.01283	1.6987	0.9882	-172.2	135.7	10.547	6.77	10.27	1605
38	0.01292	1.6620	0.9740	-164.2	146.0	10.823	6.81	10.43	1596
39	0.01302	1.6240	0.9596	-156.0	156.5	11.096	6.84	10.59	1586
40	0.01312	1.5846	0.9451	-147.8	167.2	11.366	6.86	10.74	1575
42	0.01333	1.5207	0.9154	-131.1	188.9	11.897	6.91	11.02	1558
44	0.01355	1.4602	0.8853	-114.0	211.2	12.416	6.95	11.29	1540
46	0.01378	1.4025	0.8548	-97.1	233.6	12.912	7.03	11.58	1520
40	0.01401	1.3475	0.8245	-79.3	257.0	13.411	7.09	11.85	1500
50	0.01426	1.2986	0.7946	-61.2	281.0	13.900	7.15	12.09	1482
52	0.01451	1.2533	0.7653	-42.9	305.4	14.379	7.20	12.32	1464
54	0.01477	1.2109	0.7367	-24.3	330.3	14.848	7.25	12.54	1447
56	0.01504	1.1701	0.7089	-5.2	355.8	15.312	7.30	12.74	1429
58	0.01532	1.1345	0.6820	13.7	381.5	15.763	7.34	12.92	1413
60	0.01561	1.1015	0.6558	32.9	407.5	16.203	7.38	13.09	1398
65	0.01636	1.0314	0.5943	81.3	473.9	17.267	7.52	13.48	1359
70	0.01715	0.9787	0.5398	130.6	542.2	18.279	7.71	13.84	1325
75	0.01798	0.9401	0.4919	180.8	612.3	19.245	7.96	14.20	1295
80	0.01884	0.9126	0.4499	232.1	684.1	20.173	8.26	14.55	1268
85	0.01972	0.8956	0.4130	284.4	757.7	21.065	8.60	14.89	1246
90	0.02063	0.8860	0.3806	338.0	833.0	21.926	8.97	15.23	1227
95	0.02155	0.8823	0.3522	392.8	910.0	22.759	9.36	15.56	1211
100	0.02248	0.8836	0.3273	449.1	988.7	23.565	9.76	15.88	1199
120	0.02626	0.9203	0.2531	689.9	1320.1	26.584	11.29	17.05	1179
140	0.03005	0.9814	0.2057	946.7	1667.9	29.263	12.23	17.68	1191
160	0.03382	1.0549	0.1732	1211.4	2023.2	31.635	12.58	17.79	1221
180	0.03756	1.1359	0.1494	1475.0	2376.5	33.716	12.53	17.52	1260
200	0.04125	1.2194	0.1313	1732.4	2722.3	35.538	12.26	17.07	1303
220	0.04489	1.3032	0.1171	1981.3	3058.6	37.141	11.92	16.59	1346
240	0.04849	1.3873	0.1057	2221.8	3385.6	38.564	11.60	16.14	1389
260	0.05206	1.4706	0.0963	2454.9	3704.2	39.839	11.32	15.76	1431
280	0.05559	1.5537	0.0885	2681.8	4015.9	40.994	11.09	15.45	1471
300	0.05925	1.6440	0.0828	2903.8	4325.9	42.054	10.89	15.29	1519
350	0.06806	1.8531	0.0701	3447.3	5080.6	44.380	10.64	14.94	1613
400	0.07679	2.0610	0.0609	3979.9	5822.9	46.364	10.53	14.77	1701
450	0.08549	2.2681	0.0539	4507.2	6558.9	48.099	10.49	14.69	1783
500	0.09415	2.4746	0.0483	5032.9	7292.6	49.644	10.47	14.65	1861
550	0.10280	2.6808	0.0438	5557.0	8024.2	51.039	10.46	14.63	1936
600	0.11143	2.8867	0.0401	6081.3	8755.7	52.310	10.47	14.63	2008
700	0.12868	3.2983	0.0343	7131.1	10219.3	54.567	10.52	14.66	2144
800	0.14589	3.7096	0.0300	8186.9	11688.4	56.528	10.60	14.73	2270
900	0.16310	4.1209	0.0266	9252.6	13167.0	58.268	10.72	14.85	2389
1000	0.18030	4.5323	0.0240	10331.6	14658.8	59.842	10.88	15.00	2500
1200	0.21468	5.3553	0.0200	12544.3	17696.7	62.598	11.26	15.38	2705
1400	0.24906	6.1785	0.0171	14838.4	20815.8	65.006	11.69	15.81	2890
1600	0.28343	7.0020	0.0150	17217.7	24020.0	67.123	12.10	16.22	3863
1800	0.31780	7.8257	0.0133	19682.2	27309.5	69.022	12.49	16.61	3226
2000	0.35219	8.6496	0.0120	22233.2	30685.7	70.866	12.89	17.02	3379
2500	0.43845	10.7098	0.0096	29033.7	39556.5	77.167	14.34	18.54	3721
3000	0.52673	12.7705	0.0080	36951.3	49592.8	84.189	17.46	21.99	4009

* TWO-PHASE BOUNDARY

Table 4.1 (continued)
THERMODYNAMIC PROPERTIES OF PARAHYDROGEN (ISOBARS, SI UNITS)

24.00 MPA ISOBAR

TEMPERATURE DEG. K	DENSITY KG/CU M	$V(DH/DV)_P$ KJ/KG	$V(DP/DU)_V$ MPA-CU M/KJ X 10^2	$-V(DP/DV)_T$ MPA	$-(DV/DT)_P/V$ 1/DEG. K	THERMAL CONDUCTIVITY W/K-M X 10^3	VISCOSITY KG/M-S X 10^7	THERMAL DIFFUSIVITY SQ M/HR	DIELECTRIC CONSTANT	PRANDTL NUMBER
* 20.300	86.8172	1285.9	0.23558	209.4394	0.00535	140.60	402.08	0.00085	1.28636	1.9669
21	86.4899	1295.6	0.23260	205.9105	0.00544	144.52	374.57	0.00085	1.28519	1.8271
22	86.0150	1307.8	0.22862	200.8776	0.00557	149.33	341.16	0.00086	1.28349	1.6649
23	85.5314	1318.2	0.22497	195.8609	0.00570	153.35	313.28	0.00086	1.28176	1.5359
24	85.0394	1327.3	0.22156	190.8687	0.00583	156.68	289.73	0.00086	1.28001	1.4318
25	84.5393	1334.9	0.21837	185.9097	0.00596	159.43	269.64	0.00085	1.27822	1.3465
26	84.0311	1341.6	0.21532	180.9929	0.00609	161.61	252.34	0.00085	1.27641	1.2766
27	83.5151	1347.6	0.21235	176.1274	0.00622	163.34	237.31	0.00084	1.27457	1.2187
28	82.9915	1352.9	0.20944	171.3226	0.00635	164.70	224.16	0.00083	1.27271	1.1702
29	82.4604	1357.4	0.20668	166.5879	0.00649	165.72	212.57	0.00082	1.27082	1.1291
30	81.9220	1360.7	0.20411	161.9329	0.00662	166.47	202.29	0.00081	1.26891	1.0938
31	81.3766	1363.9	0.20153	157.3672	0.00674	166.34	193.11	0.00080	1.26698	1.0679
32	80.8245	1366.7	0.19900	152.9007	0.00687	166.02	184.88	0.00079	1.26502	1.0460
33	80.2547	1368.7	0.19700	148.6770	0.00701	165.51	177.35	0.00077	1.26300	1.0275
34	79.6883	1371.6	0.19447	144.5215	0.00712	164.95	170.61	0.00076	1.26100	1.0100
35	79.1188	1373.7	0.19199	140.4044	0.00723	164.29	164.49	0.00075	1.25899	0.9950
36	78.5450	1374.6	0.18957	136.2770	0.00735	163.54	158.91	0.00074	1.25696	0.9824
37	77.9637	1376.6	0.18717	132.4359	0.00746	162.70	153.77	0.00073	1.25491	0.9708
38	77.3810	1377.2	0.18490	128.6050	0.00757	161.81	149.05	0.00072	1.25286	0.9608
39	76.7910	1375.8	0.18280	124.7123	0.00769	160.85	144.67	0.00071	1.25078	0.9521
40	76.2003	1372.7	0.18078	120.7474	0.00783	159.85	140.62	0.00070	1.24871	0.9452
42	75.0009	1373.2	0.17671	114.0513	0.00803	157.69	133.31	0.00069	1.24450	0.9318
44	73.7929	1374.3	0.17251	107.7547	0.00822	155.45	126.92	0.00067	1.24027	0.9218
46	72.5767	1379.2	0.16747	101.7880	0.00840	153.18	121.26	0.00066	1.23602	0.9169
48	71.3536	1382.0	0.16287	96.1475	0.00858	150.92	116.22	0.00064	1.23176	0.9127
50	70.1302	1386.1	0.15845	91.0690	0.00873	148.69	111.70	0.00063	1.22750	0.9085
52	68.9070	1390.3	0.15420	86.3606	0.00886	146.53	107.61	0.00062	1.22326	0.9049
54	67.6870	1394.6	0.15008	81.9636	0.00899	144.44	103.90	0.00061	1.21904	0.9018
56	66.4728	1398.0	0.14612	77.7819	0.00911	142.44	100.52	0.00061	1.21484	0.8992
58	65.2660	1402.7	0.14240	74.0437	0.00921	140.54	97.43	0.00060	1.21068	0.8957
60	64.0694	1408.5	0.13866	70.5749	0.00929	138.93	94.60	0.00060	1.20657	0.8912
65	61.1381	1429.8	0.12926	63.0573	0.00943	135.96	88.52	0.00059	1.19653	0.8774
70	58.3171	1463.3	0.12004	57.0761	0.00946	133.52	83.62	0.00060	1.18692	0.8667
75	55.6308	1509.2	0.11113	52.3013	0.00941	131.66	79.71	0.00060	1.17782	0.8594
80	53.0883	1566.9	0.10266	48.4474	0.00929	130.42	76.60	0.00061	1.16925	0.8547
85	50.7035	1637.5	0.09477	45.4103	0.00910	131.84	74.19	0.00063	1.16125	0.8380
90	48.4772	1718.6	0.08756	42.9526	0.00886	133.51	72.33	0.00065	1.15381	0.8252
95	46.4056	1808.9	0.08111	40.9421	0.00860	135.43	70.95	0.00068	1.14692	0.8153
100	44.4809	1907.3	0.07541	39.3019	0.00833	135.57	68.84	0.00069	1.14055	0.8064
120	38.0809	2361.2	0.05885	35.0463	0.00722	152.83	72.66	0.00085	1.11952	0.8106
140	33.2795	2806.4	0.05056	32.6598	0.00630	163.99	74.23	0.00100	1.10392	0.8003
160	29.5650	3203.1	0.04656	31.1885	0.00555	170.74	75.30	0.00117	1.09196	0.7845
180	26.6240	3546.3	0.04479	30.2420	0.00494	174.65	76.46	0.00135	1.08255	0.7670
200	24.2438	3844.2	0.04416	29.5624	0.00444	177.05	77.86	0.00154	1.07498	0.7507
220	22.2769	4112.8	0.04408	29.0318	0.00403	178.94	79.48	0.00174	1.06876	0.7368
240	20.6220	4369.3	0.04419	28.6092	0.00369	180.95	81.31	0.00196	1.06353	0.7252
260	19.2104	4623.5	0.04429	28.2515	0.00341	183.38	83.30	0.00218	1.05910	0.7158
280	17.9895	4881.2	0.04433	27.9499	0.00316	186.35	85.42	0.00241	1.05527	0.7080
300	16.8763	5120.8	0.04506	27.7448	0.00299	189.82	87.63	0.00265	1.05179	0.7057
350	14.6938	5800.9	0.04487	27.2296	0.00258	200.38	93.46	0.00329	1.04498	0.6968
400	13.0222	6509.5	0.04443	26.8391	0.00227	212.56	99.56	0.00398	1.03979	0.6918
450	11.6977	7236.7	0.04392	26.5315	0.00203	225.67	105.80	0.00473	1.03569	0.6888
500	10.6210	7967.2	0.04347	26.2827	0.00184	239.06	112.09	0.00553	1.03237	0.6869
550	9.7276	8703.5	0.04306	26.0774	0.00168	252.66	118.40	0.00639	1.02962	0.6856
600	8.9739	9444.3	0.04269	25.9051	0.00155	266.35	124.68	0.00731	1.02730	0.6847
700	7.7715	10945.9	0.04199	25.6324	0.00134	293.98	137.14	0.00929	1.02361	0.6837
800	6.8542	12486.7	0.04128	25.4266	0.00118	322.01	149.37	0.01148	1.02081	0.6833
900	6.1311	14078.5	0.04053	25.2661	0.00105	350.57	161.34	0.01386	1.01860	0.6834
1000	5.5463	15728.8	0.03974	25.1375	0.00095	440.56	173.07	0.01906	1.01681	0.5893
1200	4.6580	19206.1	0.03809	24.9447	0.00080	511.13	195.79	0.02569	1.01411	0.5890
1400	4.0151	22910.9	0.03647	24.8072	0.00069	583.32	217.61	0.03308	1.01215	0.5898
1600	3.5282	26758.1	0.03507	24.7043	0.00061	655.41	238.65	0.04122	1.01067	0.5908
1800	3.1466	30717.4	0.03388	24.6243	0.00054	727.37	259.01	0.05010	1.00951	0.5915
2000	2.8394	34874.1	0.03274	24.5597	0.00049	802.36	278.76	0.05978	1.00858	0.5912
2500	2.2808	47220.8	0.02931	24.4266	0.00039	1045.86	325.98	0.08905	1.00689	0.5778
3000	1.8985	66713.6	0.02411	24.2450	0.00033	1533.71	371.37	0.13225	1.00573	0.5325

* TWO-PHASE BOUNDARY

Table 4.1 (continued)

THERMODYNAMIC PROPERTIES OF PARAHYDROGEN (ISOBARS, SI UNITS)

26.00 MPA ISOBAR

TEMPERATURE DEG. K	VOLUME CU M/KG	ISOTHERM DERIVATIVE CU M-MFA/KG	ISOCHORE DERIVATIVE MPA/K	INTERNAL ENERGY KJ/KG-K	ENTHALPY KJ/KG	ENTROPY KJ/KG-K	CV KJ /	CP KG-K	VELOCITY OF SOUND M/S
* 20.756	0.01144	2.4989	1.133	-283.2	14.2	5.467	5.52	6.92	1769
21	0.01145	2.4882	1.133	-281.9	15.9	5.548	5.55	6.97	1768
22	0.01151	2.4439	1.133	-276.4	23.0	5.878	5.68	7.21	1762
23	0.01158	2.3994	1.131	-270.7	30.3	6.204	5.79	7.43	1755
24	0.01164	2.3549	1.128	-264.8	37.9	6.525	5.90	7.65	1748
25	0.01171	2.3104	1.124	-258.7	45.6	6.841	5.99	7.87	1741
26	0.01177	2.2660	1.119	-252.5	53.6	7.154	6.09	8.08	1734
27	0.01184	2.2218	1.113	-246.1	61.8	7.463	6.17	8.28	1727
28	0.01191	2.1778	1.106	-239.6	70.2	7.768	6.26	8.49	1719
29	0.01199	2.1342	1.098	-232.9	78.7	8.069	6.33	8.69	1711
30	0.01206	2.0911	1.090	-226.1	87.5	8.367	6.40	8.88	1703
31	0.01214	2.0485	1.080	-219.1	96.5	8.662	6.47	9.07	1695
32	0.01222	2.0066	1.071	-212.0	105.7	8.953	6.53	9.26	1687
33	0.01230	1.9655	1.060	-204.7	115.0	9.240	6.59	9.44	1678
34	0.01238	1.9252	1.049	-197.3	124.6	9.525	6.64	9.62	1670
35	0.01247	1.8904	1.039	-189.8	134.4	9.809	6.69	9.80	1664
36	0.01255	1.8533	1.026	-182.2	144.2	10.087	6.74	9.96	1655
37	0.01264	1.8160	1.013	-174.4	154.3	10.363	6.78	10.12	1647
38	0.01273	1.7794	0.9998	-166.5	164.5	10.634	6.82	10.28	1638
39	0.01282	1.7444	0.9858	-158.6	174.8	10.904	6.85	10.42	1629
40	0.01292	1.7093	0.9718	-150.5	185.3	11.169	6.87	10.56	1621
42	0.01311	1.6348	0.9433	-134.2	206.7	11.691	6.92	10.85	1601
44	0.01331	1.5740	0.9141	-117.5	228.7	12.202	6.97	11.11	1584
46	0.01352	1.5171	0.8848	-100.9	250.7	12.690	7.06	11.40	1565
48	0.01374	1.4625	0.8553	-83.5	273.7	13.180	7.12	11.65	1547
50	0.01396	1.4113	0.8261	-65.8	297.3	13.661	7.18	11.89	1529
52	0.01420	1.3625	0.7972	-47.8	321.3	14.132	7.23	12.12	1511
54	0.01444	1.3197	0.7692	-29.6	345.8	14.594	7.28	12.33	1495
56	0.01468	1.2797	0.7419	-10.9	370.9	15.051	7.33	12.52	1478
58	0.01494	1.2422	0.7153	7.8	396.1	15.493	7.37	12.70	1463
60	0.01520	1.2077	0.6895	26.6	421.7	15.927	7.42	12.87	1448
65	0.01588	1.1337	0.6285	74.3	487.0	16.973	7.56	13.27	1411
70	0.01659	1.0741	0.5733	122.9	554.3	17.970	7.75	13.65	1375
75	0.01734	1.0313	0.5243	172.5	623.4	18.924	8.00	14.01	1344
80	0.01812	0.9976	0.4812	223.3	694.5	19.840	8.29	14.39	1316
85	0.01892	0.9759	0.4432	275.3	767.3	20.724	8.63	14.76	1292
90	0.01975	0.9621	0.4096	328.6	842.0	21.578	9.01	15.12	1271
95	0.02058	0.9548	0.3798	383.4	918.5	22.405	9.40	15.48	1254
100	0.02143	0.9522	0.3535	439.6	996.8	23.208	9.79	15.82	1240
120	0.02488	0.9781	0.2744	680.8	1327.7	26.221	11.32	17.04	1213
140	0.02836	1.0328	0.2233	938.3	1675.6	28.901	12.26	17.69	1221
160	0.03183	1.1016	0.1881	1203.6	2031.1	31.274	12.61	17.81	1247
180	0.03527	1.1794	0.1623	1468.1	2385.0	33.360	12.56	17.56	1284
200	0.03867	1.2608	0.1426	1726.2	2731.8	35.186	12.29	17.12	1325
220	0.04204	1.3439	0.1271	1976.0	3069.0	36.793	11.95	16.63	1367
240	0.04536	1.4264	0.1147	2217.3	3396.7	38.219	11.63	16.18	1409
260	0.04866	1.5096	0.1045	2451.1	3716.2	39.498	11.35	15.80	1450
280	0.05191	1.5918	0.0959	2678.8	4028.6	40.656	11.12	15.49	1489
300	0.05531	1.6819	0.0898	2900.3	4338.4	41.714	10.91	15.31	1537
350	0.06344	1.8902	0.0760	3444.6	5094.1	44.043	10.65	14.96	1629
400	0.07151	2.0975	0.0660	3977.9	5837.1	46.029	10.54	14.78	1716
450	0.07954	2.3040	0.0584	4505.7	6573.7	47.765	10.49	14.70	1797
500	0.08754	2.5101	0.0523	5031.8	7307.8	49.311	10.47	14.66	1874
550	0.09552	2.7159	0.0475	5556.2	8039.8	50.707	10.47	14.64	1948
600	0.10349	2.9215	0.0435	6080.7	8771.5	51.979	10.48	14.63	2020
700	0.11941	3.3324	0.0372	7130.9	10235.4	54.236	10.52	14.66	2155
800	0.13530	3.7433	0.0325	8187.0	11704.8	56.197	10.60	14.73	2280
900	0.15118	4.1542	0.0289	9252.9	13183.6	57.937	10.73	14.85	2398
1000	0.16706	4.5652	0.0260	10332.0	14675.5	59.512	10.88	15.00	2509
1200	0.19879	5.3876	0.0216	12544.9	17713.5	62.268	11.26	15.38	2713
1400	0.23052	6.2104	0.0185	14839.1	20832.7	64.676	11.69	15.81	2898
1600	0.26225	7.0335	0.0162	17218.5	24037.0	66.792	12.11	16.22	3070
1800	0.29397	7.8569	0.0144	19683.0	27326.3	68.692	12.49	16.61	3232
2000	0.32571	8.6806	0.0130	22233.7	30702.2	70.536	12.89	17.01	3385
2500	0.40532	10.7403	0.0104	29027.8	39566.2	76.834	14.31	18.51	3726
3000	0.48673	12.8008	0.0087	36909.0	49563.9	83.842	17.34	21.85	4015

* TWO-PHASE BOUNDARY

Table 4.1 (continued)

THERMODYNAMIC PROPERTIES OF PARAHYDROGEN (ISOBARS, SI UNITS)

26.00 MPA ISOBAR

TEMPERATURE DEG. K	DENSITY KG/CU M	$V(\partial H/\partial V)_P$ KJ/KG	$V(\partial P/\partial U)_V$ MPA-CU M/KJ $\times 10^2$	$-V(\partial P/\partial V)_T$ MPA	$-(\partial V/\partial T)_P/V$ 1/DEG. K	THERMAL CONDUCTIVITY W/K-M $\times 10^3$	VISCOSITY KG/M-S $\times 10^7$	THERMAL DIFFUSIVITY SQ M/HR	DIELECTRIC CONSTANT	PRANDTL NUMBER
* 20.756	87.4223	1333.3	0.23481	218.4609	0.00519	145.73	414.39	0.00087	1.28853	1.9666
21	87.3115	1336.9	0.23377	217.2446	0.00522	147.09	404.17	0.00087	1.28813	1.9164
22	86.8520	1350.5	0.22978	212.2542	0.00534	152.10	366.94	0.00087	1.28649	1.7390
23	86.3845	1362.3	0.22612	207.2732	0.00546	156.30	335.97	0.00089	1.28481	1.5982
24	85.9092	1372.7	0.22269	202.3095	0.00559	159.79	309.91	0.00087	1.28311	1.4846
25	85.4264	1381.6	0.21949	197.3713	0.00570	162.69	287.75	0.00087	1.28139	1.3917
26	84.9362	1389.5	0.21644	192.4670	0.00581	164.99	268.72	0.00087	1.27964	1.3156
27	84.4389	1396.6	0.21347	187.6052	0.00593	166.85	252.24	0.00086	1.27786	1.2524
28	83.9345	1403.1	0.21057	182.7946	0.00605	168.31	237.85	0.00085	1.27607	1.1996
29	83.4234	1408.7	0.20782	178.0440	0.00617	169.44	225.21	0.00084	1.27425	1.1548
30	82.9055	1413.0	0.20527	173.3623	0.00629	170.28	214.02	0.00083	1.27240	1.1162
31	82.3812	1417.1	0.20272	168.7583	0.00640	170.22	204.06	0.00082	1.27054	1.0876
32	81.8507	1420.7	0.20023	164.2410	0.00652	169.97	195.14	0.00081	1.26866	1.0633
33	81.3142	1423.7	0.19786	159.8192	0.00663	169.56	187.11	0.00079	1.26675	1.0422
34	80.7719	1426.2	0.19555	155.5019	0.00675	169.07	179.85	0.00078	1.26483	1.0238
35	80.2107	1430.0	0.19358	151.6293	0.00685	168.41	173.15	0.00077	1.26285	1.0077
36	79.6580	1433.1	0.19121	147.6268	0.00695	167.70	167.14	0.00076	1.26089	0.9930
37	79.1022	1435.3	0.18889	143.6535	0.00705	166.91	161.65	0.00075	1.25893	0.9804
38	78.5436	1436.8	0.18670	139.7629	0.00715	166.06	156.61	0.00074	1.25696	0.9693
39	77.9779	1437.9	0.18462	136.0262	0.00725	165.13	151.94	0.00073	1.25496	0.9588
40	77.4132	1438.1	0.18262	132.3252	0.00734	164.18	147.64	0.00072	1.25297	0.9498
42	76.2690	1434.7	0.17862	124.6851	0.00757	162.09	139.91	0.00070	1.24895	0.9369
44	75.1118	1437.4	0.17451	118.2239	0.00773	159.90	133.16	0.00069	1.24489	0.9256
46	73.9473	1445.0	0.16957	112.1823	0.00789	157.67	127.21	0.00067	1.24081	0.9195
48	72.7792	1450.3	0.16503	106.4422	0.00804	155.44	121.92	0.00066	1.23673	0.9140
50	71.6093	1455.1	0.16068	101.0649	0.00817	153.25	117.19	0.00065	1.23265	0.9096
52	70.4365	1459.3	0.15648	95.9713	0.00831	151.09	112.92	0.00064	1.22857	0.9060
54	69.2688	1465.4	0.15244	91.4121	0.00841	149.02	109.05	0.00063	1.22451	0.9024
56	68.1060	1471.4	0.14857	87.1538	0.00851	147.02	105.53	0.00062	1.22049	0.8990
58	66.9501	1476.9	0.14492	83.1667	0.00860	145.12	102.32	0.00061	1.21649	0.8955
60	65.8029	1483.6	0.14126	79.4700	0.00868	143.51	99.37	0.00061	1.21253	0.8912
65	62.9868	1507.4	0.13200	71.4067	0.00880	140.54	93.00	0.00061	1.20286	0.8780
70	60.2661	1541.0	0.12273	64.7324	0.00886	138.06	87.84	0.00060	1.19355	0.8684
75	57.6618	1589.0	0.11371	59.4679	0.00882	136.11	83.67	0.00061	1.18470	0.8612
80	55.1834	1646.5	0.10514	55.0509	0.00874	134.78	80.31	0.00061	1.17631	0.8576
85	52.8431	1717.6	0.09713	51.5670	0.00859	136.15	77.63	0.00063	1.16842	0.8417
90	50.6437	1799.2	0.08980	48.7231	0.00841	137.74	75.53	0.00065	1.16105	0.8294
95	48.5853	1890.2	0.08320	46.3870	0.00819	139.57	73.92	0.00067	1.15417	0.8197
100	46.6621	1988.3	0.07737	44.4301	0.00796	139.74	71.16	0.00068	1.14777	0.8057
120	40.1885	2441.5	0.06030	39.3070	0.00698	156.80	74.85	0.00082	1.12641	0.8136
140	35.2629	2885.5	0.05166	36.4196	0.00613	167.47	76.15	0.00097	1.11035	0.8043
160	31.4218	3278.5	0.04747	34.6155	0.00543	173.71	76.92	0.00112	1.09793	0.7889
180	28.3535	3617.8	0.04558	33.4409	0.00485	177.16	77.83	0.00128	1.08808	0.7713
200	25.8571	3913.0	0.04487	32.6018	0.00437	179.21	79.02	0.00146	1.08011	0.7547
220	23.7883	4181.4	0.04471	31.9687	0.00398	180.84	80.49	0.00165	1.07354	0.7402
240	22.0442	4436.4	0.04475	31.4449	0.00365	182.65	82.20	0.00184	1.06802	0.7282
260	20.5527	4691.8	0.04479	31.0273	0.00337	184.35	84.09	0.00205	1.06332	0.7183
280	19.2624	4949.6	0.04477	30.6617	0.00313	187.81	86.14	0.00227	1.05926	0.7102
300	18.0791	5182.7	0.04556	30.4077	0.00295	191.21	88.30	0.00249	1.05555	0.7071
350	15.7624	5862.0	0.04529	29.7942	0.00255	201.67	94.07	0.00308	1.04831	0.6977
400	13.9841	6570.2	0.04479	29.3312	0.00225	213.86	100.16	0.00372	1.04278	0.6924
450	12.5726	7297.3	0.04424	28.9676	0.00201	227.02	106.41	0.00442	1.03840	0.6892
500	11.4235	8027.8	0.04376	28.6740	0.00183	240.50	112.75	0.00517	1.03485	0.6871
550	10.4688	8764.1	0.04332	28.4319	0.00167	254.21	119.10	0.00597	1.03190	0.6857
600	9.6626	9504.9	0.04292	28.2290	0.00154	268.04	125.45	0.00683	1.02942	0.6847
700	8.3747	11006.5	0.04218	27.9082	0.00133	295.95	138.03	0.00868	1.02546	0.6837
800	7.3910	12547.5	0.04145	27.6665	0.00117	324.30	150.41	0.01072	1.02245	0.6833
900	6.6145	14139.5	0.04068	27.4781	0.00105	353.20	162.54	0.01295	1.02007	0.6833
1000	5.9860	15790.2	0.03986	27.3273	0.00095	440.56	174.41	0.01766	1.01815	0.5939
1200	5.0303	19268.5	0.03819	27.1013	0.00080	511.13	197.43	0.02379	1.01524	0.5940
1400	4.3380	22974.7	0.03655	26.9404	0.00069	583.32	219.56	0.03062	1.01313	0.5951
1600	3.8132	26823.0	0.03514	26.8201	0.00060	655.40	240.89	0.03814	1.01154	0.5963
1800	3.4017	30782.2	0.03394	26.7266	0.00054	727.31	261.52	0.04634	1.01029	0.5972
2000	3.0702	34935.3	0.03280	26.6512	0.00049	802.12	281.55	0.05528	1.00928	0.5972
2500	2.4672	47213.7	0.02941	26.4983	0.00039	1042.89	329.42	0.08223	1.00745	0.5845
3000	2.0545	66383.1	0.02430	26.2996	0.00033	1518.35	375.39	0.12176	1.00620	0.5402

* TWO-PHASE BOUNDARY

Table 4.1 (continued)
THERMODYNAMIC PROPERTIES OF PARAHYDROGEN (ISOBARS, SI UNITS)

28.00 MPA ISOBAR

TEMPERATURE DEG. K	VOLUME CU M/KG	ISOTHERM DERIVATIVE CU M-MPA/KG	ISOCHORE DERIVATIVE MPA/K	INTERNAL ENERGY KJ/KG-K	ENTHALPY KJ/KG	ENTROPY KJ/KG-K	CV KJ / KG-K	CP KJ / KG-K	VELOCITY OF SOUND M/S
* 21.205	0.01136	2.5840	1.146	-280.5	37.7	5.499	5.56	6.95	1797
22	0.01141	2.5495	1.146	-276.2	43.3	5.758	5.66	7.14	1793
23	0.01147	2.5061	1.145	-270.6	50.5	6.080	5.78	7.36	1787
24	0.01153	2.4625	1.142	-264.8	58.0	6.398	5.88	7.57	1780
25	0.01159	2.4189	1.139	-258.9	65.7	6.712	5.98	7.78	1774
26	0.01166	2.3753	1.134	-252.8	73.6	7.021	6.08	7.99	1767
27	0.01172	2.3318	1.129	-246.5	81.6	7.326	6.17	8.19	1760
28	0.01179	2.2884	1.122	-240.1	89.9	7.628	6.25	8.39	1753
29	0.01186	2.2454	1.115	-233.6	98.4	7.926	6.33	8.59	1745
30	0.01193	2.2027	1.107	-226.9	107.1	8.220	6.40	8.77	1738
31	0.01200	2.1604	1.098	-220.0	116.0	8.511	6.47	8.96	1730
32	0.01207	2.1188	1.089	-213.1	125.0	8.798	6.53	9.14	1722
33	0.01215	2.0777	1.079	-205.9	134.3	9.082	6.59	9.32	1714
34	0.01223	2.0374	1.069	-198.7	143.7	9.363	6.65	9.50	1706
35	0.01231	1.9979	1.058	-191.3	153.3	9.641	6.70	9.67	1698
36	0.01239	1.9594	1.047	-183.9	163.0	9.916	6.74	9.84	1690
37	0.01247	1.9290	1.036	-176.2	173.0	10.190	6.79	9.99	1685
38	0.01256	1.8935	1.023	-168.5	183.1	10.459	6.83	10.14	1677
39	0.01264	1.8580	1.010	-160.7	193.3	10.724	6.86	10.28	1669
40	0.01273	1.8243	0.9969	-152.8	203.6	10.986	6.89	10.42	1661
42	0.01291	1.7570	0.9693	-136.8	224.7	11.500	6.94	10.68	1645
44	0.01310	1.6857	0.9413	-120.4	246.4	12.003	6.99	10.96	1626
46	0.01329	1.6273	0.9127	-104.2	268.0	12.484	7.08	11.24	1608
48	0.01350	1.5737	0.8841	-87.1	290.8	12.968	7.14	11.49	1591
50	0.01370	1.5221	0.8556	-69.7	314.0	13.442	7.20	11.72	1574
52	0.01392	1.4731	0.8273	-52.1	337.6	13.906	7.26	11.94	1556
54	0.01414	1.4282	0.7996	-34.2	361.7	14.360	7.31	12.15	1540
56	0.01437	1.3862	0.7726	-15.8	386.5	14.811	7.36	12.34	1524
58	0.01460	1.3465	0.7463	2.6	411.4	15.247	7.40	12.52	1509
60	0.01484	1.3108	0.7209	21.1	436.6	15.674	7.45	12.69	1494
65	0.01546	1.2344	0.6606	68.1	501.0	16.706	7.59	13.09	1458
70	0.01612	1.1720	0.6050	116.1	567.4	17.690	7.79	13.47	1424
75	0.01680	1.1230	0.5553	165.2	635.7	18.632	8.03	13.85	1391
80	0.01751	1.0848	0.5112	215.6	705.9	19.538	8.33	14.24	1362
85	0.01824	1.0576	0.4721	267.2	778.0	20.413	8.67	14.63	1336
90	0.01900	1.0395	0.4374	320.3	852.2	21.260	9.04	15.02	1314
95	0.01976	1.0284	0.4065	374.9	928.2	22.082	9.43	15.39	1295
100	0.02054	1.0229	0.3790	431.0	1006.1	22.881	9.83	15.75	1280
120	0.02371	1.0378	0.2955	672.3	1336.3	25.888	11.36	17.03	1248
140	0.02692	1.0866	0.2407	930.4	1684.1	28.567	12.28	17.69	1251
160	0.03012	1.1497	0.2029	1196.3	2039.8	30.942	12.64	17.83	1274
180	0.03331	1.2234	0.1751	1461.5	2394.3	33.030	12.58	17.59	1308
200	0.03647	1.3033	0.1539	1720.4	2741.6	34.859	12.32	17.15	1347
220	0.03960	1.3851	0.1372	1970.9	3079.6	36.471	11.98	16.67	1388
240	0.04269	1.4673	0.1237	2213.0	3408.2	37.901	11.66	16.22	1429
260	0.04574	1.5489	0.1127	2447.6	3728.4	39.182	11.38	15.84	1468
280	0.04877	1.6308	0.1034	2676.0	4041.6	40.343	11.15	15.52	1506
300	0.05193	1.7203	0.0968	2896.9	4351.1	41.399	10.92	15.33	1554
350	0.05949	1.9276	0.0819	3442.1	5107.7	43.731	10.66	14.97	1645
400	0.06698	2.1341	0.0711	3976.0	5851.5	45.719	10.55	14.80	1730
450	0.07444	2.3401	0.0629	4504.3	6588.6	47.457	10.50	14.71	1811
500	0.08187	2.5457	0.0564	5030.8	7323.1	49.003	10.48	14.66	1887
550	0.08928	2.7511	0.0511	5555.5	8055.4	50.399	10.48	14.64	1961
600	0.09668	2.9563	0.0468	6080.3	8787.4	51.672	10.48	14.63	2032
700	0.11146	3.3666	0.0400	7130.8	10251.6	53.930	10.53	14.66	2165
800	0.12622	3.7769	0.0350	8187.1	11721.2	55.891	10.61	14.73	2290
900	0.14096	4.1873	0.0311	9253.2	13200.2	57.631	10.73	14.85	2407
1000	0.15570	4.5980	0.0279	10332.5	14692.2	59.206	10.88	15.00	2518
1200	0.18517	5.4197	0.0233	12545.6	17730.3	61.962	11.26	15.38	2720
1400	0.21463	6.2420	0.0200	14839.9	20849.6	64.370	11.69	15.81	2905
1600	0.24409	7.0648	0.0175	17219.4	24053.9	66.487	12.11	16.22	3077
1800	0.27354	7.8879	0.0155	19683.9	27343.1	68.386	12.49	16.61	3238
2000	0.30301	8.7112	0.0140	22234.4	30718.7	70.230	12.89	17.01	3391
2500	0.37692	10.7705	0.0112	29022.8	39576.7	76.525	14.29	18.48	3732
3000	0.45245	12.8306	0.0093	36871.4	49540.0	83.522	17.23	21.72	4021

* TWO-PHASE BOUNDARY

Table 4.1 (continued)
THERMODYNAMIC PROPERTIES OF PARAHYDROGEN (ISOBARS, SI UNITS)

28.00 MPA ISOBAR

TEMPERATURE DEG. K	DENSITY KG/CU M	$V(DH/DV)_P$ KJ/KG	$V(DP/DU)_V$ MPA-CU M/KJ X 10^2	$-V(DP/DV)_T$ MPA	$-(DV/DT)_P/V$ 1/DEG. K	THERMAL CONDUCTIVITY W/K-M X 10^3	VISCOSITY KG/M-S X 10^7	THERMAL DIFFUSIVITY SQ M/HR	DIELECTRIC CONSTANT	PRANDTL NUMBER
* 21.205	88.0083	1379.8	0.23412	227.4110	0.00504	150.73	426.76	0.00089	1.29063	1.9688
22	87.6531	1391.6	0.23094	223.4755	0.00513	154.79	394.45	0.00089	1.28936	1.8186
23	87.1999	1404.8	0.22725	218.5311	0.00524	159.16	360.09	0.00089	1.28773	1.6649
24	86.7396	1416.4	0.22379	213.5974	0.00535	162.81	331.27	0.00089	1.28608	1.5413
25	86.2722	1426.6	0.22058	208.6821	0.00546	165.85	306.83	0.00089	1.28441	1.4402
26	85.7981	1435.7	0.21752	203.7934	0.00556	168.28	285.92	0.00088	1.28272	1.3574
27	85.3174	1444.0	0.21454	198.9393	0.00567	170.25	267.85	0.00088	1.28100	1.2887
28	84.8302	1451.7	0.21164	194.1280	0.00578	171.82	252.13	0.00087	1.27926	1.2313
29	84.3368	1458.3	0.20889	189.3677	0.00589	173.05	238.34	0.00086	1.27750	1.1825
30	83.8372	1463.6	0.20634	184.6668	0.00599	173.97	226.17	0.00085	1.27572	1.1406
31	83.3317	1468.7	0.20381	180.0335	0.00610	173.98	215.36	0.00084	1.27392	1.1091
32	82.8205	1473.2	0.20134	175.4760	0.00621	173.79	205.70	0.00083	1.27210	1.0823
33	82.3037	1476.9	0.19899	171.0027	0.00631	173.44	197.03	0.00081	1.27027	1.0591
34	81.7815	1480.2	0.19671	166.6217	0.00642	172.99	189.21	0.00080	1.26841	1.0388
35	81.2543	1483.2	0.19449	162.3412	0.00652	172.43	182.11	0.00079	1.26654	1.0212
36	80.7222	1485.8	0.19232	158.1692	0.00662	171.77	175.65	0.00078	1.26466	1.0059
37	80.1708	1491.4	0.19040	154.6531	0.00670	170.95	169.63	0.00077	1.26270	0.9916
38	79.6309	1494.2	0.18825	150.7823	0.00679	170.13	164.23	0.00076	1.26080	0.9790
39	79.0888	1495.7	0.18626	146.9473	0.00687	169.24	159.27	0.00075	1.25888	0.9676
40	78.5454	1497.4	0.18434	143.2904	0.00696	168.32	154.69	0.00074	1.25696	0.9574
42	77.4471	1499.6	0.18039	136.0715	0.00712	166.28	146.49	0.00072	1.25309	0.9411
44	76.3388	1498.3	0.17637	128.6814	0.00731	164.15	139.37	0.00071	1.24919	0.9306
46	75.2198	1507.2	0.17147	122.4022	0.00746	161.95	133.10	0.00069	1.24527	0.9236
48	74.0969	1514.9	0.16702	116.6041	0.00758	159.75	127.55	0.00068	1.24133	0.9171
50	72.9730	1521.6	0.16273	111.0704	0.00770	157.57	122.59	0.00066	1.23740	0.9119
52	71.8491	1527.7	0.15859	105.8386	0.00782	155.44	118.14	0.00065	1.23348	0.9076
54	70.7270	1534.5	0.15457	101.0133	0.00792	153.37	114.11	0.00064	1.22958	0.9037
56	69.6092	1541.2	0.15074	96.4948	0.00801	151.38	110.44	0.00063	1.22569	0.9003
58	68.4956	1547.0	0.14715	92.2294	0.00809	149.47	107.09	0.00063	1.22183	0.8969
60	67.3919	1554.9	0.14357	88.3401	0.00816	147.86	104.02	0.00062	1.21802	0.8926
65	64.6769	1581.7	0.13449	79.8351	0.00827	144.87	97.38	0.00062	1.20866	0.8797
70	62.0472	1618.6	0.12520	72.7208	0.00832	142.35	91.97	0.00061	1.19964	0.8701
75	59.5203	1666.8	0.11613	66.8417	0.00831	140.34	87.56	0.00061	1.19101	0.8640
80	57.1053	1725.9	0.10743	61.9491	0.00825	138.92	83.96	0.00061	1.18281	0.8607
85	54.8122	1796.9	0.09932	57.9683	0.00814	140.26	81.05	0.00063	1.17506	0.8457
90	52.6444	1879.1	0.09188	54.7214	0.00799	141.79	78.73	0.00065	1.16776	0.8340
95	50.6046	1970.7	0.08516	52.0429	0.00781	143.55	76.89	0.00066	1.16092	0.8246
100	48.6892	2070.0	0.07920	49.8064	0.00761	143.80	73.44	0.00067	1.15452	0.8044
120	42.1725	2522.8	0.06170	43.7660	0.00675	160.69	77.00	0.00081	1.13293	0.8163
140	37.1500	2967.2	0.05275	40.3688	0.00596	170.94	78.06	0.00094	1.11648	0.8079
160	33.1965	3355.3	0.04836	38.1655	0.00532	176.69	78.56	0.00107	1.10365	0.7930
180	30.0178	3689.0	0.04636	36.7232	0.00477	179.71	79.22	0.00123	1.09341	0.7755
200	27.4181	3983.0	0.04556	35.7338	0.00431	181.42	80.21	0.00139	1.08509	0.7584
220	25.2544	4250.6	0.04533	34.9799	0.00392	182.78	81.52	0.00156	1.07819	0.7434
240	23.4260	4506.1	0.04531	34.3718	0.00360	184.39	83.10	0.00175	1.07239	0.7310
260	21.8607	4759.9	0.04529	33.8600	0.00333	186.54	84.91	0.00194	1.06744	0.7208
280	20.5035	5018.8	0.04521	33.4365	0.00309	189.30	86.88	0.00214	1.06316	0.7124
300	19.2549	5245.1	0.04604	33.1238	0.00292	192.62	89.00	0.00235	1.05924	0.7085
350	16.8102	5923.3	0.04570	32.4031	0.00253	202.98	94.69	0.00290	1.05158	0.6986
400	14.9294	6631.2	0.04515	31.8614	0.00223	215.16	100.76	0.00351	1.04572	0.6929
450	13.4340	7358.2	0.04455	31.4371	0.00200	228.37	107.03	0.00416	1.04107	0.6896
500	12.2147	8088.6	0.04404	31.0949	0.00181	241.93	113.40	0.00486	1.03729	0.6873
550	11.2005	8824.9	0.04357	30.8132	0.00166	255.75	119.81	0.00561	1.03416	0.6858
600	10.3431	9565.6	0.04315	30.5772	0.00153	269.71	126.21	0.00641	1.03152	0.6848
700	8.9719	11067.2	0.04237	30.2045	0.00132	297.91	138.93	0.00815	1.02730	0.6837
800	7.9229	12608.2	0.04161	29.9238	0.00117	326.58	151.45	0.01007	1.02408	0.6833
900	7.0941	14200.4	0.04082	29.7053	0.00105	355.82	163.72	0.01216	1.02154	0.6833
1000	6.4225	15851.5	0.03998	29.5305	0.00095	440.56	175.75	0.01646	1.01949	0.5985
1200	5.4005	19330.9	0.03828	29.2688	0.00080	511.13	199.08	0.02216	1.01637	0.5990
1400	4.6592	23038.3	0.03663	29.0827	0.00069	583.32	221.51	0.02851	1.01411	0.6004
1600	4.0969	26887.6	0.03521	28.9436	0.00060	655.39	243.14	0.03550	1.01240	0.6019
1800	3.6557	30846.8	0.03400	28.8357	0.00054	727.26	264.06	0.04312	1.01106	0.6030
2000	3.3002	34996.8	0.03285	28.7488	0.00049	801.90	284.37	0.05142	1.00998	0.6032
2500	2.6531	47214.6	0.02950	28.5748	0.00039	1040.24	332.90	0.07640	1.00802	0.5913
3000	2.2102	66096.2	0.02448	28.3582	0.00033	1504.66	379.48	0.11282	1.00668	0.5479

* TWO-PHASE BOUNDARY

Table 4.1 (continued)
THERMODYNAMIC PROPERTIES OF PARAHYDROGEN (ISOBARS, SI UNITS)

30.00 MPA ISOBAR

TEMPERATURE DEG. K	VOLUME CU M/KG	ISOTHERM DERIVATIVE CU M-MPA/KG	ISOCHORE DERIVATIVE MPA/K	INTERNAL ENERGY KJ/KG-K	ENTHALPY KJ/KG	ENTROPY KJ/KG-K	CV KJ /	CP KG-K	VELOCITY OF SOUND M/S
* 21.645	0.01129	2.6677	1.159	-277.7	61.0	5.529	5.60	6.99	1824
22	0.01131	2.6527	1.159	-275.8	63.5	5.644	5.65	7.07	1823
23	0.01137	2.6102	1.158	-270.3	70.7	5.963	5.76	7.29	1817
24	0.01142	2.5675	1.156	-264.7	78.0	6.278	5.87	7.50	1811
25	0.01148	2.5247	1.153	-258.9	85.7	6.588	5.97	7.71	1805
26	0.01154	2.4819	1.149	-252.9	93.5	6.895	6.07	7.91	1799
27	0.01161	2.4391	1.144	-246.7	101.5	7.197	6.16	8.11	1792
28	0.01167	2.3964	1.138	-240.4	109.7	7.495	6.24	8.30	1785
29	0.01174	2.3539	1.131	-234.0	118.1	7.790	6.32	8.49	1778
30	0.01180	2.3117	1.123	-227.4	126.7	8.081	6.39	8.68	1771
31	0.01187	2.2699	1.115	-220.7	135.4	8.368	6.46	8.86	1764
32	0.01194	2.2285	1.107	-213.9	144.4	8.652	6.53	9.04	1756
33	0.01201	2.1876	1.097	-206.9	153.5	8.933	6.59	9.21	1749
34	0.01209	2.1473	1.088	-199.8	162.8	9.211	6.65	9.38	1741
35	0.01216	2.1078	1.077	-192.6	172.3	9.485	6.70	9.55	1733
36	0.01224	2.0691	1.067	-185.2	181.9	9.757	6.75	9.71	1726
37	0.01232	2.0313	1.056	-177.8	191.7	10.025	6.79	9.87	1718
38	0.01240	1.9944	1.044	-170.2	201.6	10.290	6.83	10.03	1710
39	0.01248	1.9684	1.033	-162.5	211.8	10.555	6.87	10.16	1706
40	0.01256	1.9328	1.020	-154.8	222.1	10.814	6.90	10.29	1699
42	0.01273	1.8695	0.9938	-139.0	242.9	11.322	6.95	10.55	1684
44	0.01291	1.8048	0.9667	-122.9	264.2	11.819	7.01	10.80	1668
46	0.01309	1.7369	0.9392	-107.0	285.6	12.293	7.09	11.10	1648
48	0.01328	1.6806	0.9112	-90.2	308.0	12.770	7.16	11.34	1631
50	0.01347	1.6300	0.8834	-73.1	331.0	13.238	7.23	11.57	1615
52	0.01367	1.5811	0.8557	-55.8	354.3	13.696	7.29	11.79	1599
54	0.01387	1.5340	0.8284	-38.1	378.1	14.145	7.34	11.99	1583
56	0.01408	1.4913	0.8017	-20.0	402.5	14.589	7.39	12.18	1567
58	0.01430	1.4507	0.7756	-2.0	427.1	15.020	7.43	12.35	1553
60	0.01452	1.4139	0.7504	16.3	451.9	15.441	7.48	12.52	1538
65	0.01510	1.3324	0.6905	62.7	515.6	16.460	7.63	12.93	1503
70	0.01570	1.2694	0.6352	110.1	581.2	17.433	7.82	13.31	1470
75	0.01633	1.2140	0.5848	158.8	648.7	18.364	8.07	13.71	1436
80	0.01698	1.1744	0.5399	208.7	718.2	19.261	8.37	14.10	1406
85	0.01766	1.1411	0.4998	260.0	789.7	20.128	8.71	14.51	1379
90	0.01835	1.1182	0.4642	312.8	863.3	20.969	9.08	14.92	1355
95	0.01905	1.1033	0.4323	367.2	938.8	21.786	9.47	15.31	1336
100	0.01977	1.0945	0.4038	423.2	1016.4	22.581	9.86	15.68	1319
120	0.02270	1.0997	0.3162	664.6	1345.7	25.580	11.39	17.01	1282
140	0.02568	1.1425	0.2580	923.0	1693.4	28.258	12.31	17.69	1281
160	0.02865	1.2008	0.2176	1189.5	2049.2	30.633	12.66	17.84	1301
180	0.03162	1.2699	0.1879	1455.3	2403.9	32.723	12.61	17.62	1332
200	0.03457	1.3471	0.1652	1714.9	2752.0	34.556	12.35	17.19	1369
220	0.03749	1.4270	0.1472	1966.1	3090.7	36.171	12.01	16.71	1409
240	0.04037	1.5078	0.1327	2208.9	3420.0	37.604	11.69	16.26	1448
260	0.04322	1.5886	0.1208	2444.2	3740.9	38.888	11.41	15.87	1487
280	0.04605	1.6701	0.1109	2673.3	4054.8	40.052	11.19	15.56	1524
300	0.04901	1.7590	0.1038	2893.7	4363.9	41.106	10.94	15.36	1571
350	0.05606	1.9653	0.0878	3439.7	5121.5	43.441	10.67	14.99	1661
400	0.06306	2.1710	0.0762	3974.2	5865.9	45.430	10.56	14.81	1745
450	0.07002	2.3764	0.0673	4503.0	6603.5	47.169	10.51	14.72	1824
500	0.07695	2.5814	0.0604	5029.8	7338.4	48.716	10.49	14.67	1900
550	0.08387	2.7863	0.0548	5554.9	8071.0	50.113	10.48	14.65	1973
600	0.09078	2.9911	0.0501	6079.9	8803.3	51.386	10.49	14.64	2043
700	0.10457	3.4007	0.0429	7130.7	10267.9	53.645	10.53	14.66	2176
800	0.11834	3.8104	0.0375	8187.3	11737.6	55.606	10.61	14.74	2300
900	0.13210	4.2204	0.0333	9253.6	13216.8	57.347	10.73	14.85	2417
1000	0.14586	4.6306	0.0299	10333.1	14708.8	58.921	10.89	15.00	2526
1200	0.17336	5.4516	0.0249	12546.3	17747.1	61.678	11.26	15.38	2728
1400	0.20085	6.2734	0.0214	14840.8	20866.4	64.086	11.70	15.81	2912
1600	0.22834	7.0957	0.0187	17220.4	24070.7	66.202	12.11	16.22	3083
1800	0.25583	7.9185	0.0166	19684.9	27359.9	68.101	12.49	16.61	3245
2000	0.28334	8.7415	0.0150	22235.1	30735.2	69.946	12.89	17.01	3396
2500	0.35231	10.8003	0.0120	29018.4	39587.7	76.238	14.27	18.45	3737
3000	0.42274	12.8600	0.0100	36837.9	49520.2	83.224	17.13	21.61	4027

* TWO-PHASE BOUNDARY

Table 4.1 (continued)
THERMODYNAMIC PROPERTIES OF PARAHYDROGEN (ISOBARS, SI UNITS)

30.00 MPA ISOBAR

TEMPERATURE DEG. K	DENSITY KG/CU M	$V(DH/DV)_P$ KJ/KG	$V(DP/DU)_V$ MPA-CU M/KJ X 10^2	$-V(DP/DV)_T$ MPA	$-(DV/DT)_P/V$ 1/DEG. K	THERMAL CONDUCTIVITY W/K-M X 10^3	VISCOSITY KG/M-S X 10^7	THERMAL DIFFUSIVITY SQ M/HR	DIELECTRIC CONSTANT	PRANDTL NUMBER
* 21.645	88.5767	1425.6	0.23351	236.2981	0.00490	155.60	439.19	0.00090	1.29267	1.9730
22	88.4220	1431.2	0.23210	234.5564	0.00494	157.41	423.85	0.00091	1.29212	1.9038
23	87.9818	1445.7	0.22837	229.6496	0.00504	161.95	385.75	0.00091	1.29054	1.7363
24	87.5348	1458.6	0.22489	224.7472	0.00514	165.75	353.91	0.00091	1.28893	1.6018
25	87.0814	1470.0	0.22165	219.8570	0.00524	168.92	326.99	0.00091	1.28731	1.4920
26	86.6217	1480.3	0.21857	214.9865	0.00534	171.48	304.01	0.00090	1.28566	1.4022
27	86.1559	1489.8	0.21557	210.1434	0.00544	173.55	284.22	0.00089	1.28399	1.3276
28	85.6841	1498.6	0.21266	205.3357	0.00554	175.23	267.04	0.00089	1.28231	1.2653
29	85.2066	1506.3	0.20990	200.5710	0.00564	176.54	252.01	0.00088	1.28060	1.2123
30	84.7234	1512.6	0.20736	195.8572	0.00574	177.55	238.78	0.00087	1.27888	1.1668
31	84.2347	1518.7	0.20482	191.2021	0.00583	177.62	227.05	0.00086	1.27714	1.1324
32	83.7407	1524.1	0.20236	186.6133	0.00593	177.49	216.60	0.00084	1.27538	1.1030
33	83.2416	1528.7	0.20003	182.0987	0.00603	177.19	207.24	0.00083	1.27360	1.0775
34	82.7375	1532.8	0.19777	177.6659	0.00612	176.80	198.81	0.00082	1.27181	1.0551
35	82.2286	1536.4	0.19557	173.3223	0.00622	176.28	191.18	0.00081	1.27000	1.0358
36	81.7153	1539.7	0.19342	169.0756	0.00631	175.65	184.24	0.00080	1.26818	1.0189
37	81.1976	1542.6	0.19134	164.9329	0.00640	174.93	177.92	0.00079	1.26634	1.0042
38	80.6760	1545.0	0.18937	160.9016	0.00649	174.13	172.12	0.00077	1.26449	0.9911
39	80.1347	1551.3	0.18770	157.7348	0.00655	173.21	166.68	0.00077	1.26258	0.9775
40	79.6102	1552.7	0.18581	153.8691	0.00663	172.31	161.81	0.00076	1.26072	0.9666
42	78.5503	1558.5	0.18203	146.8531	0.00677	170.32	153.10	0.00074	1.25698	0.9481
44	77.4833	1562.5	0.17806	139.8430	0.00691	168.22	145.57	0.00072	1.25322	0.9346
46	76.4083	1567.9	0.17326	132.7170	0.00708	166.06	138.97	0.00071	1.24944	0.9286
48	75.3261	1576.0	0.16884	126.5935	0.00720	163.88	133.13	0.00069	1.24564	0.9216
50	74.2419	1585.1	0.16462	121.0169	0.00730	161.71	127.93	0.00068	1.24184	0.9154
52	73.1587	1593.2	0.16052	115.6721	0.00740	159.59	123.28	0.00067	1.23805	0.9104
54	72.0777	1600.5	0.15655	110.5687	0.00749	157.52	119.07	0.00066	1.23428	0.9064
56	70.9997	1608.6	0.15275	105.8852	0.00757	155.53	115.25	0.00065	1.23053	0.9025
58	69.9274	1615.7	0.14919	101.4425	0.00765	153.67	111.77	0.00064	1.22680	0.8988
60	68.8615	1624.7	0.14564	97.3636	0.00771	152.00	108.57	0.00063	1.22310	0.8944
65	66.2364	1652.6	0.13667	88.2558	0.00782	149.00	101.67	0.00063	1.21403	0.8822
70	63.6849	1693.8	0.12750	80.8416	0.00786	146.43	96.01	0.00062	1.20525	0.8726
75	61.2321	1742.2	0.11834	74.3388	0.00787	144.37	91.38	0.00062	1.19685	0.8675
80	58.8765	1805.3	0.10957	69.1426	0.00781	142.87	87.56	0.00062	1.18882	0.8639
85	56.6333	1876.1	0.10135	64.6270	0.00773	144.19	84.44	0.00063	1.18121	0.8497
90	54.5005	1958.5	0.09381	60.9434	0.00762	145.67	81.91	0.00065	1.17400	0.8387
95	52.4837	2050.6	0.08700	57.9045	0.00747	147.37	79.87	0.00066	1.16722	0.8297
100	50.5805	2150.6	0.08094	55.3620	0.00729	147.75	75.66	0.00067	1.16084	0.8032
120	44.0441	2605.9	0.06304	48.4344	0.00653	164.50	79.13	0.00079	1.13910	0.8183
140	38.9443	3050.7	0.05381	44.4929	0.00580	174.37	79.96	0.00091	1.12234	0.8112
160	34.8991	3436.6	0.04923	41.9073	0.00519	179.68	80.21	0.00104	1.10917	0.7965
180	31.6242	3764.4	0.04712	40.1595	0.00468	182.28	80.63	0.00118	1.09858	0.7792
200	28.9273	4054.6	0.04625	38.9666	0.00424	183.66	81.42	0.00133	1.08992	0.7620
220	26.6772	4319.7	0.04595	38.0675	0.00387	184.75	82.57	0.00149	1.08272	0.7467
240	24.7712	4574.1	0.04586	37.3495	0.00355	186.17	84.03	0.00166	1.07666	0.7338
260	23.1356	4828.2	0.04578	36.7534	0.00329	188.17	85.74	0.00184	1.07147	0.7232
280	21.7153	5088.1	0.04565	36.2665	0.00306	190.81	87.65	0.00203	1.06698	0.7145
300	20.4046	5307.8	0.04652	35.8926	0.00289	194.05	89.70	0.00223	1.06285	0.7098
350	17.8377	5984.9	0.04611	35.0560	0.00250	204.30	95.33	0.00275	1.05479	0.6994
400	15.8585	6692.4	0.04551	34.4295	0.00221	216.46	101.38	0.00332	1.04861	0.6935
450	14.2821	7419.2	0.04486	33.9397	0.00198	229.71	107.66	0.00393	1.04370	0.6899
500	12.9948	8149.5	0.04431	33.5454	0.00180	243.35	114.06	0.00460	1.03971	0.6876
550	11.9229	8885.7	0.04381	33.2209	0.00165	257.29	120.51	0.00530	1.03639	0.6860
600	11.0157	9626.4	0.04337	32.9494	0.00152	271.36	126.97	0.00606	1.03359	0.6849
700	9.5629	11127.9	0.04256	32.5208	0.00132	299.86	139.82	0.00770	1.02911	0.6837
800	8.4501	12669.0	0.04177	32.1984	0.00116	328.85	152.48	0.00951	1.02569	0.6833
900	7.5698	14261.3	0.04095	31.9475	0.00104	358.44	164.91	0.01148	1.02300	0.6833
1000	6.8559	15912.7	0.04010	31.7470	0.00094	440.56	177.09	0.01542	1.02081	0.6031
1200	5.7684	19393.0	0.03838	31.4470	0.00079	511.13	200.74	0.02074	1.01749	0.6040
1400	4.9788	23101.6	0.03671	31.2338	0.00068	583.31	223.48	0.02668	1.01508	0.6057
1600	4.3794	26952.0	0.03527	31.0748	0.00060	655.38	245.41	0.03321	1.01326	0.6075
1800	3.9088	30911.2	0.03406	30.9514	0.00054	727.21	266.63	0.04033	1.01183	0.6089
2000	3.5294	35058.6	0.03290	30.8523	0.00049	801.71	287.22	0.04808	1.01067	0.6094
2500	2.8384	47222.1	0.02958	30.6557	0.00039	1037.86	336.44	0.07134	1.00858	0.5981
3000	2.3655	65845.4	0.02464	30.4206	0.00033	1492.37	383.64	0.10510	1.00715	0.5555

* TWO-PHASE BOUNDARY

Table 4.1 (continued)
THERMODYNAMIC PROPERTIES OF PARAHYDROGEN (ISOBARS, SI UNITS)

35.00 MPA ISOBAR

TEMPERATURE DEG. K	VOLUME CU M/KG	ISOTHERM DERIVATIVE CU M-MPA/KG	ISOCHORE DERIVATIVE MPA/K	INTERNAL ENERGY KJ/KG-K	ENTHALPY KJ/KG	ENTROPY KJ/KG-K	CV KJ /	CP KG-K	VELOCITY OF SOUND M/S
* 22.715	0.01112	2.8721	1.190	-270.6	118.6	5.603	5.70	7.08	1889
23	0.01113	2.8606	1.190	-269.1	120.6	5.691	5.73	7.14	1888
24	0.01119	2.8201	1.189	-263.6	127.9	6.000	5.84	7.35	1883
25	0.01124	2.7794	1.186	-258.1	135.3	6.304	5.94	7.54	1878
26	0.01129	2.7385	1.183	-252.3	143.0	6.603	6.04	7.74	1873
27	0.01135	2.6975	1.179	-246.5	150.8	6.899	6.14	7.93	1867
28	0.01141	2.6564	1.174	-240.4	158.8	7.191	6.22	8.11	1861
29	0.01147	2.6154	1.168	-234.3	167.0	7.478	6.31	8.30	1855
30	0.01152	2.5745	1.161	-227.9	175.4	7.763	6.38	8.47	1848
31	0.01159	2.5338	1.154	-221.5	184.0	8.043	6.46	8.64	1842
32	0.01165	2.4933	1.147	-214.9	192.7	8.320	6.52	8.81	1835
33	0.01171	2.4531	1.138	-208.2	201.6	8.594	6.59	8.98	1828
34	0.01177	2.4133	1.130	-201.4	210.7	8.865	6.65	9.14	1822
35	0.01184	2.3740	1.121	-194.5	219.9	9.132	6.70	9.30	1815
36	0.01191	2.3352	1.111	-187.5	229.3	9.396	6.76	9.45	1808
37	0.01197	2.2970	1.101	-180.3	238.8	9.657	6.81	9.61	1801
38	0.01204	2.2595	1.091	-173.1	248.5	9.915	6.85	9.75	1794
39	0.01211	2.2228	1.080	-165.7	258.3	10.170	6.88	9.89	1787
40	0.01219	2.1869	1.069	-158.3	268.2	10.423	6.92	10.02	1780
42	0.01233	2.1179	1.047	-143.2	288.6	10.918	6.98	10.28	1767
44	0.01249	2.0695	1.023	-127.7	309.5	11.405	7.04	10.51	1758
46	0.01265	2.0085	0.9981	-112.4	330.2	11.866	7.13	10.78	1742
48	0.01281	1.9548	0.9725	-96.2	352.0	12.330	7.21	11.02	1729
50	0.01297	1.8968	0.9466	-79.7	374.3	12.784	7.28	11.25	1713
52	0.01314	1.8424	0.9204	-63.0	397.0	13.229	7.34	11.47	1697
54	0.01332	1.7924	0.8941	-46.0	420.2	13.666	7.40	11.67	1681
56	0.01350	1.7490	0.8682	-28.4	444.0	14.099	7.46	11.85	1667
58	0.01368	1.7066	0.8428	-11.0	467.8	14.518	7.50	12.02	1654
60	0.01387	1.6653	0.8179	6.7	492.0	14.928	7.55	12.19	1639
65	0.01435	1.5755	0.7585	51.7	554.0	15.919	7.70	12.59	1605
70	0.01486	1.5009	0.7037	98.0	618.0	16.868	7.90	13.00	1571
75	0.01538	1.4414	0.6533	145.5	684.0	17.779	8.15	13.41	1540
80	0.01593	1.3958	0.6068	194.5	752.0	18.657	8.45	13.81	1510
85	0.01649	1.3528	0.5648	245.1	822.1	19.507	8.79	14.24	1480
90	0.01706	1.3226	0.5269	297.2	894.4	20.333	9.16	14.66	1455
95	0.01765	1.2957	0.4930	351.1	968.8	21.137	9.55	15.10	1431
100	0.01825	1.2775	0.4624	406.7	1045.4	21.922	9.94	15.52	1412
120	0.02071	1.2580	0.3662	647.7	1372.5	24.901	11.46	16.95	1364
140	0.02322	1.2843	0.3004	906.9	1719.5	27.574	12.38	17.68	1354
160	0.02574	1.3330	0.2538	1174.4	2075.3	29.949	12.73	17.85	1367
180	0.02826	1.3923	0.2195	1441.3	2430.4	32.041	12.67	17.65	1392
200	0.03077	1.4599	0.1932	1702.3	2779.4	33.879	12.41	17.25	1425
220	0.03327	1.5347	0.1722	1955.0	3119.5	35.501	12.08	16.78	1460
240	0.03574	1.6127	0.1553	2199.4	3450.4	36.940	11.75	16.34	1497
260	0.03819	1.6921	0.1413	2436.3	3773.0	38.231	11.48	15.95	1533
280	0.04061	1.7709	0.1296	2667.0	4088.5	39.401	11.26	15.64	1568
300	0.04316	1.8575	0.1213	2886.3	4396.8	40.450	10.98	15.40	1614
350	0.04921	2.0607	0.1025	3434.2	5156.5	42.791	10.71	15.03	1701
400	0.05521	2.2642	0.0889	3970.2	5902.5	44.785	10.58	14.83	1782
450	0.06117	2.4677	0.0785	4500.1	6641.2	46.527	10.53	14.74	1858
500	0.06712	2.6712	0.0704	5027.9	7377.1	48.076	10.51	14.69	1932
550	0.07305	2.8748	0.0638	5553.7	8110.4	49.474	10.50	14.66	2003
600	0.07897	3.0784	0.0584	6079.2	8843.1	50.747	10.50	14.65	2072
700	0.09078	3.4860	0.0499	7131.0	10308.5	53.008	10.54	14.67	2202
800	0.10258	3.8940	0.0436	8188.3	11778.7	54.970	10.62	14.74	2325
980	0.11438	4.3025	0.0388	9255.0	13258.2	56.711	10.74	14.85	2439
1000	0.12616	4.7115	0.0349	10334.8	14750.5	58.285	10.90	15.01	2547
1200	0.14972	5.5305	0.0291	12548.6	17789.0	61.042	11.27	15.38	2747
1400	0.17328	6.3507	0.0249	14843.5	20908.4	63.450	11.70	15.81	2929
1600	0.19684	7.1718	0.0218	17223.3	24112.7	65.567	12.11	16.22	3099
1800	0.22040	7.9936	0.0194	19687.8	27401.8	67.466	12.49	16.61	3259
2000	0.24397	8.8159	0.0174	22237.6	30776.4	69.309	12.89	17.00	3410
2500	0.30306	10.8731	0.0140	29010.1	39617.4	75.597	14.22	18.40	3750
3000	0.36333	12.9317	0.0116	36768.0	49484.4	82.559	16.92	21.37	4041

* TWO-PHASE BOUNDARY

Table 4.1 (continued)
THERMODYNAMIC PROPERTIES OF PARAHYDROGEN (ISOBARS, SI UNITS)

35.00 MPA ISOBAR

TEMPERATURE DEG. K	DENSITY KG/CU M	V(DH/DV)$_P$ KJ/KG	V(DP/DU)$_V$ MPA-CU M/KJ X 10^2	-V(DP/DV)$_T$ MPA	-(DV/DT)$_P$/V 1/DEG. K	THERMAL CONDUCTIVITY W/K-M X 10^3	VISCOSITY KG/M-S X 10^7	THERMAL DIFFUSIVITY SQ M/HR	DIELECTRIC CONSTANT	PRANDTL NUMBER
* 22.715	89.9284	1537.2	0.23225	258.2851	0.00461	167.31	470.59	0.00095	1.29753	1.9915
23	89.8102	1541.8	0.23123	256.9143	0.00463	168.63	457.54	0.00095	1.29711	1.9371
24	89.3915	1558.0	0.22764	252.0961	0.00471	172.79	416.80	0.00095	1.29560	1.7717
25	88.9674	1572.5	0.22430	247.2759	0.00480	176.29	382.62	0.00095	1.29407	1.6373
26	88.5380	1585.7	0.22114	242.4608	0.00488	179.13	353.65	0.00094	1.29253	1.5275
27	88.1035	1598.2	0.21807	237.6579	0.00496	181.46	328.86	0.00094	1.29097	1.4366
28	87.6641	1609.9	0.21510	232.8743	0.00504	183.37	307.48	0.00093	1.28940	1.3606
29	87.2198	1620.3	0.21229	228.1169	0.00512	184.90	288.89	0.00092	1.28780	1.2962
30	86.7710	1629.2	0.20972	223.3927	0.00520	186.10	272.62	0.00091	1.28620	1.2408
31	86.3176	1637.7	0.20716	218.7086	0.00528	186.31	258.27	0.00090	1.28457	1.1982
32	85.8598	1645.5	0.20468	214.0714	0.00536	186.31	245.56	0.00089	1.28294	1.1617
33	85.3978	1652.2	0.20235	209.4879	0.00543	186.12	234.22	0.00087	1.28129	1.1300
34	84.9317	1658.3	0.20010	204.9647	0.00551	185.84	224.06	0.00086	1.27962	1.1021
35	84.4616	1663.9	0.19791	200.5084	0.00559	185.41	214.92	0.00085	1.27794	1.0779
36	83.9876	1668.9	0.19580	196.1255	0.00567	184.88	206.64	0.00084	1.27626	1.0568
37	83.5100	1673.4	0.19376	191.8221	0.00574	184.23	199.13	0.00083	1.27455	1.0383
38	83.0289	1677.1	0.19183	187.6046	0.00581	183.51	192.28	0.00082	1.27284	1.0218
39	82.5445	1679.7	0.19009	183.4789	0.00589	182.71	186.01	0.00081	1.27112	1.0068
40	82.0570	1682.1	0.18839	179.4508	0.00596	181.84	180.25	0.00080	1.26939	0.9935
42	81.0737	1687.1	0.18499	171.7099	0.00610	179.91	170.04	0.00078	1.26590	0.9719
44	80.0660	1702.1	0.18157	165.6944	0.00618	177.79	161.17	0.00076	1.26233	0.9529
46	79.0744	1715.5	0.17698	158.8177	0.00628	175.68	153.61	0.00074	1.25883	0.9427
48	78.0791	1729.1	0.17280	152.6322	0.00637	173.54	146.99	0.00073	1.25532	0.9331
50	77.0835	1738.0	0.16875	146.2088	0.00647	171.40	141.14	0.00071	1.25181	0.9265
52	76.0855	1747.0	0.16479	140.1801	0.00657	169.29	135.92	0.00070	1.24830	0.9210
54	75.0875	1757.0	0.16090	134.5895	0.00664	167.21	131.23	0.00069	1.24480	0.9160
56	74.0914	1768.9	0.15718	129.5832	0.00670	165.20	126.99	0.00068	1.24131	0.9110
58	73.1007	1779.2	0.15368	124.7533	0.00676	163.26	123.14	0.00067	1.23785	0.9065
60	72.1159	1789.5	0.15016	120.0965	0.00681	161.63	119.63	0.00066	1.23441	0.9019
65	69.6848	1822.7	0.14128	109.7904	0.00691	158.59	112.04	0.00065	1.22596	0.8897
70	67.3054	1866.5	0.13228	101.0215	0.00697	155.92	105.81	0.00064	1.21772	0.8824
75	65.0001	1923.2	0.12326	93.6896	0.00697	153.70	100.64	0.00063	1.20977	0.8781
80	62.7793	1993.9	0.11435	87.6305	0.00692	152.05	96.33	0.00063	1.20214	0.8748
85	60.6511	2068.8	0.10591	82.0505	0.00688	153.33	92.74	0.00064	1.19487	0.8614
90	58.6090	2156.9	0.09813	77.5149	0.00680	154.73	89.75	0.00065	1.18791	0.8505
95	56.6637	2248.7	0.09111	73.4173	0.00672	156.32	87.27	0.00066	1.18131	0.8430
100	54.8072	2349.4	0.08484	70.0140	0.00660	157.18	81.02	0.00067	1.17504	0.7998
120	48.2939	2811.9	0.06614	60.7526	0.00603	173.69	84.29	0.00076	1.15320	0.8226
140	43.0694	3256.5	0.05633	55.3153	0.00543	182.77	84.65	0.00086	1.13589	0.8190
160	38.8496	3642.1	0.05133	51.7872	0.00490	187.09	84.33	0.00097	1.12203	0.8045
180	35.3863	3960.6	0.04896	49.2697	0.00446	188.77	84.21	0.00109	1.11075	0.7872
200	32.4953	4236.2	0.04791	47.4392	0.00407	189.34	84.52	0.00122	1.10139	0.7700
220	30.0560	4494.6	0.04746	46.1257	0.00373	189.80	85.28	0.00135	1.09354	0.7542
240	27.9774	4747.6	0.04722	45.1204	0.00344	190.71	86.43	0.00150	1.08688	0.7405
260	26.1847	5003.3	0.04699	44.3060	0.00319	192.33	87.90	0.00166	1.08115	0.7290
280	24.6224	5262.5	0.04674	43.6041	0.00297	194.68	89.62	0.00182	1.07619	0.7198
300	23.1706	5466.0	0.04768	43.0404	0.00282	197.69	91.53	0.00199	1.07158	0.7131
350	20.3223	6139.9	0.04710	41.8772	0.00245	207.63	96.94	0.00245	1.06259	0.7015
400	18.1136	6846.0	0.04636	41.0123	0.00217	219.72	102.92	0.00294	1.05566	0.6949
450	16.3468	7572.4	0.04561	40.3390	0.00195	233.05	109.22	0.00348	1.05013	0.6909
500	14.8989	8302.3	0.04497	39.7984	0.00177	246.88	115.69	0.00406	1.04562	0.6882
550	13.6895	9038.2	0.04441	39.3544	0.00162	261.08	122.25	0.00468	1.04186	0.6864
600	12.6634	9778.7	0.04390	38.9832	0.00150	275.46	128.85	0.00535	1.03868	0.6852
700	11.0151	11279.8	0.04300	38.3982	0.00130	304.68	142.02	0.00679	1.03359	0.6838
800	9.7480	12820.8	0.04215	37.9589	0.00115	334.48	155.05	0.00838	1.02968	0.6833
900	8.7431	14413.5	0.04128	37.6175	0.00103	364.94	167.87	0.01012	1.02659	0.6833
1000	7.9264	16065.5	0.04039	37.3450	0.00093	440.56	180.45	0.01333	1.02409	0.6146
1200	6.6790	13547.9	0.03860	36.9381	0.00079	511.13	204.90	0.01791	1.02027	0.6166
1400	5.7709	23259.3	0.03689	36.6497	0.00068	583.31	228.44	0.02302	1.01750	0.6192
1600	5.0803	27111.9	0.03543	36.4349	0.00060	655.37	251.16	0.02863	1.01539	0.6217
1800	4.5372	31071.5	0.03419	36.2687	0.00053	727.11	273.15	0.03474	1.01374	0.6238
2000	4.0989	35213.4	0.03303	36.1356	0.00048	801.30	294.50	0.04139	1.01240	0.6249
2500	3.2996	47263.1	0.02976	35.8771	0.00039	1032.83	345.52	0.06125	1.00998	0.6154
3000	2.7523	65341.6	0.02500	35.5924	0.00033	1466.36	394.37	0.08974	1.00832	0.5748

* TWO-PHASE BOUNDARY

Table 4.1 (continued)
THERMODYNAMIC PROPERTIES OF PARAHYDROGEN (ISOBARS, SI UNITS)

40.00 MPA ISOBAR

TEMPERATURE DEG. K	VOLUME CU M/KG	ISOTHERM DERIVATIVE CU M-MPA/KG	ISOCHORE DERIVATIVE MPA/K	INTERNAL ENERGY KJ/KG-K	ENTHALPY KJ/KG	ENTROPY KJ/KG-K	CV KJ /	CP KG-K	VELOCITY OF SOUND M/S
* 23.746	0.01097	3.0705	1.220	-263.2	175.4	5.672	5.78	7.16	1951
24	0.01098	3.0607	1.220	-261.9	177.3	5.748	5.81	7.22	1950
25	0.01103	3.0220	1.218	-256.5	184.6	6.047	5.92	7.41	1945
26	0.01108	2.9829	1.215	-251.0	192.1	6.341	6.02	7.60	1941
27	0.01113	2.9437	1.212	-245.3	199.8	6.631	6.11	7.78	1936
28	0.01118	2.9043	1.208	-239.5	207.6	6.917	6.21	7.96	1930
29	0.01123	2.8648	1.202	-233.5	215.7	7.200	6.29	8.14	1925
30	0.01128	2.8253	1.197	-227.4	223.9	7.479	6.37	8.31	1919
31	0.01134	2.7858	1.190	-221.2	232.3	7.754	6.45	8.47	1913
32	0.01139	2.7464	1.183	-214.9	240.8	8.025	6.52	8.63	1907
33	0.01145	2.7071	1.176	-208.4	249.6	8.293	6.58	8.79	1901
34	0.01151	2.6681	1.168	-201.9	258.4	8.558	6.65	8.95	1895
35	0.01157	2.6293	1.159	-195.2	267.5	8.820	6.71	9.10	1889
36	0.01163	2.5909	1.151	-188.4	276.6	9.078	6.76	9.25	1882
37	0.01169	2.5529	1.142	-181.5	285.9	9.334	6.81	9.39	1876
38	0.01175	2.5154	1.132	-174.5	295.4	9.586	6.86	9.53	1870
39	0.01181	2.4784	1.122	-167.4	305.0	9.835	6.90	9.66	1863
40	0.01187	2.4420	1.112	-160.2	314.7	10.081	6.93	9.79	1857
42	0.01200	2.3712	1.091	-145.6	334.6	10.565	7.00	10.04	1844
44	0.01214	2.3036	1.070	-130.7	354.9	11.038	7.07	10.29	1831
46	0.01228	2.2396	1.047	-115.9	375.2	11.489	7.16	10.56	1817
48	0.01242	2.2010	1.025	-100.2	396.7	11.947	7.24	10.78	1810
50	0.01257	2.1414	1.001	-84.3	418.5	12.391	7.32	11.01	1795
52	0.01272	2.0954	0.9765	-68.0	440.7	12.827	7.39	11.21	1784
54	0.01287	2.0472	0.9519	-51.5	463.3	13.254	7.45	11.41	1771
56	0.01303	1.9978	0.9274	-34.5	486.6	13.677	7.51	11.60	1757
58	0.01319	1.9532	0.9026	-17.5	510.0	14.087	7.56	11.77	1744
60	0.01335	1.9092	0.8783	-0.3	533.7	14.489	7.61	11.93	1730
65	0.01377	1.8163	0.8195	43.6	594.3	15.460	7.77	12.33	1697
70	0.01420	1.7308	0.7646	88.9	657.0	16.388	7.98	12.75	1663
75	0.01466	1.6626	0.7139	135.5	721.8	17.282	8.23	13.17	1631
80	0.01513	1.6074	0.6672	183.7	788.7	18.146	8.53	13.60	1601
85	0.01561	1.5682	0.6242	233.5	857.8	18.983	8.87	14.01	1574
90	0.01610	1.5299	0.5848	285.1	929.0	19.797	9.24	14.45	1547
95	0.01660	1.4914	0.5491	338.5	1002.4	20.591	9.63	14.92	1520
100	0.01711	1.4676	0.5167	393.7	1078.0	21.366	10.82	15.34	1499
120	0.01922	1.4216	0.4137	633.9	1402.7	24.322	11.54	16.87	1442
140	0.02139	1.4312	0.3413	893.3	1748.8	26.988	12.45	17.66	1425
160	0.02357	1.4693	0.2892	1161.6	2104.5	29.363	12.79	17.85	1432
180	0.02576	1.5206	0.2505	1429.3	2459.7	31.455	12.73	17.66	1452
200	0.02794	1.5804	0.2207	1691.3	2809.0	33.295	12.47	17.28	1480
220	0.03012	1.6472	0.1971	1945.2	3150.0	34.920	12.13	16.84	1512
240	0.03228	1.7212	0.1778	2190.9	3482.2	36.365	11.82	16.41	1546
260	0.03442	1.7982	0.1617	2429.2	3806.1	37.662	11.54	16.02	1580
280	0.03654	1.8760	0.1482	2661.4	4123.2	38.837	11.33	15.71	1613
300	0.03877	1.9580	0.1386	2879.8	4430.6	39.881	11.02	15.44	1657
350	0.04407	2.1575	0.1171	3429.4	5192.1	42.228	10.74	15.06	1739
400	0.04932	2.3584	0.1015	3966.8	5939.5	44.226	10.61	14.86	1817
450	0.05454	2.5599	0.0897	4497.8	6679.3	45.970	10.56	14.76	1892
500	0.05974	2.7616	0.0804	5026.4	7416.0	47.521	10.53	14.70	1964
550	0.06492	2.9637	0.0729	5552.9	8149.9	48.920	10.52	14.67	2033
600	0.07010	3.1659	0.0667	6079.1	8883.2	50.195	10.52	14.66	2180
700	0.08044	3.5711	0.0570	7131.8	10349.2	52.456	10.56	14.68	2228
800	0.09076	3.9772	0.0498	8189.6	11819.9	54.418	10.63	14.74	2348
900	0.10107	4.3840	0.0442	9256.9	13299.6	56.160	10.75	14.86	2461
1000	0.11137	4.7915	0.0398	10337.1	14792.1	57.735	10.90	15.01	2568
1200	0.13198	5.6082	0.0332	12551.4	17830.7	60.492	11.28	15.38	2765
1400	0.15259	6.4267	0.0284	14846.6	20950.2	62.900	11.71	15.81	2946
1600	0.17320	7.2463	0.0249	17226.7	24154.5	65.016	12.12	16.22	3114
1800	0.19380	8.0669	0.0221	19691.2	27443.4	66.915	12.50	16.60	3274
2000	0.21442	8.8881	0.0199	22240.7	30817.5	68.759	12.89	17.00	3424
2500	0.26611	10.9435	0.0159	29004.7	39649.1	75.043	14.19	18.35	3763
3000	0.31876	13.0008	0.0133	36712.9	49463.4	81.986	16.75	21.18	4053

* TWO-PHASE BOUNDARY

Table 4.1 (continued)
THERMODYNAMIC PROPERTIES OF PARAHYDROGEN (ISOBARS, SI UNITS)

40.00 MPA ISOBAR

TEMPERATURE DEG. K	DENSITY KG/CU M	$V(DH/DV)_P$ KJ/KG	$V(DP/DU)_V$ MPA-CU M/KJ X 10^2	$-V(DP/DV)_T$ MPA	$-(DV/DT)_P/V$ 1/DEG. K	THERMAL CONDUCTIVITY W/K-M X 10^3	VISCOSITY KG/M-S X 10^7	THERMAL DIFFUSIVITY SQ M/HR	DIELECTRIC CONSTANT	PRANDTL NUMBER
* 23.746	91.1935	1644.6	0.23141	280.0107	0.00436	178.41	502.49	0.00098	1.30209	2.0180
24	91.0924	1649.4	0.23046	278.8086	0.00437	179.48	490.21	0.00098	1.30173	1.9708
25	90.6916	1667.0	0.22700	274.0675	0.00444	183.27	446.98	0.00098	1.30028	1.8070
26	90.2863	1683.3	0.22373	269.3192	0.00451	186.38	410.60	0.00098	1.29882	1.6736
27	89.8767	1698.7	0.22056	264.5703	0.00458	188.96	379.68	0.00097	1.29735	1.5635
28	89.4630	1713.2	0.21750	259.8275	0.00465	191.07	353.17	0.00097	1.29586	1.4716
29	89.0452	1726.4	0.21462	255.0973	0.00471	192.80	330.25	0.00096	1.29435	1.3939
30	88.6236	1737.9	0.21198	250.3862	0.00478	194.17	310.30	0.00095	1.29284	1.3272
31	88.1982	1748.9	0.20936	245.7003	0.00484	194.51	292.80	0.00094	1.29131	1.2753
32	87.7692	1759.2	0.20684	241.0461	0.00491	194.62	277.37	0.00092	1.28977	1.2306
33	87.3366	1768.2	0.20447	236.4296	0.00497	194.53	263.68	0.00091	1.28822	1.1918
34	86.9007	1776.5	0.20219	231.8569	0.00504	194.33	251.47	0.00090	1.28666	1.1578
35	86.4613	1784.1	0.19999	227.3338	0.00510	193.99	240.52	0.00089	1.28509	1.1281
36	86.0188	1791.0	0.19787	222.8662	0.00516	193.53	230.66	0.00088	1.28350	1.1021
37	85.5732	1797.3	0.19582	218.4597	0.00523	192.95	221.75	0.00086	1.28191	1.0793
38	85.1246	1802.6	0.19390	214.1198	0.00529	192.28	213.65	0.00085	1.28031	1.0590
39	84.6731	1806.6	0.19217	209.8520	0.00535	191.53	206.27	0.00084	1.27870	1.0405
40	84.2189	1810.2	0.19049	205.6614	0.00541	190.71	199.52	0.00083	1.27708	1.0242
42	83.3031	1817.2	0.18715	197.5321	0.00552	188.85	187.62	0.00081	1.27382	0.9974
44	82.3785	1825.1	0.18372	189.7700	0.00564	186.84	177.47	0.00079	1.27053	0.9771
46	81.4466	1839.6	0.17941	182.4096	0.00574	184.72	168.71	0.00077	1.26722	0.9643
48	80.4895	1863.4	0.17577	177.1544	0.00579	182.48	160.96	0.00076	1.26383	0.9509
50	79.5574	1874.7	0.17189	170.3629	0.00587	180.33	154.32	0.00074	1.26054	0.9423
52	78.6261	1891.9	0.16816	164.7520	0.00593	178.21	148.44	0.00073	1.25725	0.9340
54	77.6931	1906.3	0.16446	159.0507	0.00598	176.11	143.19	0.00072	1.25396	0.9276
56	76.7649	1918.1	0.16091	153.3596	0.00605	174.08	138.48	0.00070	1.25069	0.9228
58	75.8377	1930.8	0.15746	148.1271	0.00609	172.10	134.22	0.00069	1.24743	0.9175
60	74.9161	1943.2	0.15400	143.0324	0.00614	170.44	130.34	0.00069	1.24420	0.9125
65	72.6362	1984.5	0.14518	131.9255	0.00621	167.34	122.02	0.00067	1.23623	0.8989
70	70.4040	2031.5	0.13615	121.8556	0.00627	164.59	115.23	0.00066	1.22845	0.8924
75	68.2267	2092.6	0.12713	113.4349	0.00629	162.24	109.56	0.00065	1.22090	0.8894
80	66.1148	2166.3	0.11828	106.2738	0.00628	160.42	104.80	0.00064	1.21361	0.8885
85	64.0771	2256.1	0.10980	100.4830	0.00621	161.66	100.78	0.00065	1.20660	0.8737
90	62.1255	2349.0	0.10187	95.0470	0.00615	163.02	97.40	0.00065	1.19991	0.8635
95	60.2513	2441.3	0.09466	89.8598	0.00611	164.54	94.54	0.00066	1.19350	0.8572
100	58.4545	2547.7	0.08821	85.7859	0.00602	166.04	86.10	0.00067	1.18739	0.7957
120	52.0300	3017.0	0.06891	73.9680	0.00559	182.40	89.25	0.00075	1.16569	0.8256
140	46.7548	3462.8	0.05864	66.9139	0.00510	190.84	89.21	0.00083	1.14808	0.8257
160	42.4196	3847.2	0.05330	62.3257	0.00464	194.35	88.40	0.00092	1.13374	0.8119
180	38.8206	4161.3	0.05069	59.0296	0.00424	195.22	87.79	0.00103	1.12194	0.7941
200	35.7865	4427.5	0.04948	56.5563	0.00390	195.07	87.67	0.00114	1.11205	0.7766
220	33.2007	4673.3	0.04891	54.6870	0.00360	194.93	88.07	0.00126	1.10367	0.7608
240	30.9777	4921.5	0.04856	53.3182	0.00333	195.36	88.91	0.00138	1.09650	0.7468
260	29.0508	5176.9	0.04821	52.2382	0.00310	196.58	90.13	0.00152	1.09031	0.7347
280	27.3648	5439.8	0.04782	51.3371	0.00289	198.63	91.66	0.00166	1.08492	0.7247
300	25.7922	5625.3	0.04880	50.5022	0.00275	201.42	93.43	0.00182	1.07990	0.7163
350	22.6935	6295.7	0.04806	48.9622	0.00239	211.01	98.61	0.00222	1.07007	0.7037
400	20.2773	7000.3	0.04719	47.8225	0.00212	222.99	104.50	0.00266	1.06245	0.6963
450	18.3361	7726.0	0.04633	46.9380	0.00191	236.38	110.79	0.00314	1.05635	0.6918
500	16.7397	8455.5	0.04561	46.2291	0.00174	250.37	117.32	0.00366	1.05136	0.6889
550	15.4024	9191.1	0.04498	45.6475	0.00160	264.81	123.99	0.00422	1.04719	0.6869
600	14.2650	9931.3	0.04442	45.1617	0.00148	279.49	130.72	0.00481	1.04365	0.6855
700	12.4322	11431.9	0.04343	44.3967	0.00128	309.42	144.21	0.00611	1.03797	0.6839
800	11.0185	12972.7	0.04251	43.8227	0.00114	340.05	157.60	0.00754	1.03360	0.6833
900	9.8944	14565.5	0.04159	43.3771	0.00102	371.40	170.82	0.00910	1.03013	0.6833
1000	8.9787	16218.0	0.04066	43.0218	0.00093	440.56	183.81	0.01177	1.02732	0.6261
1200	7.5767	19702.1	0.03882	42.4922	0.00078	511.13	209.10	0.01579	1.02302	0.6292
1400	6.5536	23415.8	0.03706	42.1174	0.00068	583.31	233.48	0.02027	1.01989	0.6329
1600	5.7738	27270.6	0.03557	41.8388	0.00059	655.35	257.04	0.02519	1.01751	0.6363
1800	5.1599	31230.6	0.03431	41.6238	0.00053	727.02	279.85	0.03055	1.01563	0.6391
2000	4.6637	35368.3	0.03315	41.4520	0.00048	800.96	301.99	0.03637	1.01412	0.6409
2500	3.7578	47326.9	0.02992	41.1238	0.00039	1028.77	354.94	0.05370	1.01137	0.6332
3000	3.1371	64367.4	0.02530	40.7851	0.00033	1445.38	405.57	0.07832	1.00948	0.5943

* TWO-PHASE BOUNDARY

Table 4.1 (continued)
THERMODYNAMIC PROPERTIES OF PARAHYDROGEN (ISOBARS, SI UNITS)

45.00 MPA ISOBAR

TEMPERATURE DEG. K	VOLUME CU M/KG	ISOTHERM DERIVATIVE CU M-MPA/KG	ISOCHORE DERIVATIVE MPA/K	INTERNAL ENERGY KJ/KG-K	ENTHALPY KJ/KG	ENTROPY KJ/KG-K	CV KJ / KG-K	CP KJ / KG-K	VELOCITY OF SOUND M/S
* 24.742	0.01082	3.2640	1.249	-255.6	231.5	5.737	5.86	7.25	2009
25	0.01084	3.2545	1.249	-254.3	233.3	5.812	5.89	7.30	2008
26	0.01088	3.2173	1.247	-248.9	240.7	6.102	5.99	7.48	2004
27	0.01093	3.1798	1.244	-243.4	248.3	6.387	6.09	7.66	2000
28	0.01097	3.1420	1.240	-237.8	256.1	6.669	6.19	7.84	1995
29	0.01102	3.1041	1.235	-232.1	264.0	6.947	6.28	8.01	1990
30	0.01107	3.0660	1.230	-226.2	272.1	7.221	6.36	8.17	1985
31	0.01112	3.0278	1.224	-220.1	280.3	7.492	6.43	8.33	1980
32	0.01117	2.9896	1.217	-214.0	288.7	7.759	6.51	8.49	1975
33	0.01122	2.9514	1.210	-207.7	297.3	8.022	6.58	8.64	1969
34	0.01127	2.9133	1.203	-201.4	306.0	8.283	6.64	8.79	1963
35	0.01133	2.8753	1.195	-194.9	314.9	8.540	6.70	8.93	1958
36	0.01138	2.8375	1.187	-188.3	323.9	8.793	6.76	9.08	1952
37	0.01144	2.8000	1.178	-181.6	333.0	9.044	6.82	9.22	1946
38	0.01149	2.7628	1.170	-174.8	342.3	9.291	6.86	9.35	1940
39	0.01155	2.7260	1.160	-167.9	351.7	9.536	6.91	9.48	1934
40	0.01161	2.6896	1.151	-161.0	361.2	9.777	6.94	9.60	1928
42	0.01172	2.6182	1.131	-146.8	380.7	10.251	7.02	9.84	1916
44	0.01184	2.5492	1.111	-132.3	400.6	10.715	7.09	10.08	1904
46	0.01197	2.4829	1.090	-118.0	420.5	11.156	7.19	10.34	1890
48	0.01209	2.4198	1.068	-102.8	441.4	11.602	7.28	10.59	1876
50	0.01222	2.3601	1.046	-87.3	462.8	12.039	7.35	10.82	1863
52	0.01236	2.3042	1.023	-71.5	484.7	12.467	7.43	11.03	1850
54	0.01250	2.2794	1.002	-55.3	507.1	12.890	7.49	11.21	1846
56	0.01264	2.2290	0.9786	-38.7	530.0	13.306	7.55	11.40	1834
58	0.01278	2.1913	0.9554	-22.1	552.9	13.709	7.61	11.55	1824
60	0.01292	2.1432	0.9323	-5.3	576.2	14.104	7.67	11.73	1811
65	0.01329	2.0456	0.8749	37.7	635.9	15.058	7.83	12.13	1780
70	0.01368	1.9626	0.8202	82.0	697.5	15.972	8.04	12.53	1749
75	0.01407	1.8870	0.7692	127.8	761.2	16.851	8.30	12.9[illegible]	1716
80	0.01448	1.8198	0.7219	175.3	827.1	17.701	8.60	13.41	1684
85	0.01491	1.7672	0.6782	224.5	895.3	18.528	8.95	13.86	1655
90	0.01534	1.7229	0.6379	275.6	965.8	19.333	9.32	14.32	1627
95	0.01578	1.6938	0.6010	328.5	1038.4	20.119	9.70	14.74	1604
100	0.01622	1.6639	0.5671	383.3	1113.3	20.887	10.09	15.18	1582
120	0.01806	1.5871	0.4586	622.5	1435.4	23.819	11.61	16.80	1515
140	0.01997	1.5823	0.3806	881.9	1780.5	26.477	12.52	17.63	1493
160	0.02190	1.6082	0.3236	1150.6	2135.9	28.850	12.85	17.85	1494
180	0.02382	1.6518	0.2807	1419.1	2491.2	30.942	12.79	17.66	1510
200	0.02575	1.7050	0.2477	1681.8	2840.6	32.783	12.52	17.29	1535
220	0.02768	1.7654	0.2215	1936.6	3182.0	34.410	12.19	16.87	1563
240	0.02959	1.8336	0.2000	2183.3	3515.0	35.859	11.87	16.46	1594
260	0.03150	1.9071	0.1821	2422.8	3840.2	37.161	11.60	16.09	1626
280	0.03338	1.9838	0.1669	2656.3	4158.5	38.340	11.39	15.77	1657
300	0.03536	2.0602	0.1559	2874.0	4465.2	39.380	11.05	15.48	1698
350	0.04006	2.2557	0.1316	3425.3	5228.2	41.732	10.77	15.09	1778
400	0.04473	2.4537	0.1141	3964.0	5976.9	43.733	10.64	14.88	1853
450	0.04937	2.6528	0.1008	4496.0	6717.7	45.479	10.58	14.78	1925
500	0.05399	2.8526	0.0903	5025.5	7455.1	47.032	10.55	14.71	1995
550	0.05860	3.0529	0.0818	5552.6	8189.7	48.432	10.54	14.68	2062
600	0.06320	3.2536	0.0749	6079.4	8923.4	49.707	10.54	14.67	2128
700	0.07238	3.6562	0.0640	7132.9	10390.1	51.970	10.57	14.68	2253
800	0.08155	4.0600	0.0559	8191.5	11861.1	53.933	10.65	14.75	2371
900	0.09071	4.4649	0.0497	9259.2	13341.1	55.674	10.76	14.86	2483
1000	0.09986	4.8708	0.0447	10339.8	14833.6	57.249	10.91	15.01	2588
1200	0.11817	5.6849	0.0373	12554.7	17872.4	60.006	11.29	15.38	2783
1400	0.13648	6.5012	0.0320	14850.3	20991.9	62.415	11.72	15.81	2962
1600	0.15479	7.3191	0.0280	17230.5	24196.1	64.531	12.13	16.22	3129
1800	0.17310	8.1383	0.0249	19695.2	27484.8	66.430	12.50	16.60	3287
2000	0.19142	8.9584	0.0224	22244.5	30858.5	68.273	12.89	16.99	3437
2500	0.23735	11.0116	0.0179	29001.4	39682.2	74.554	14.16	18.32	3775
3000	0.28410	13.0674	0.0149	36668.5	49452.7	81.481	16.61	21.02	4065

* TWO-PHASE BOUNDARY

Table 4.1 (continued)
THERMODYNAMIC PROPERTIES OF PARAHYDROGEN (ISOBARS, SI UNITS)

45.00 MPA ISOBAR

TEMPERATURE DEG. K	DENSITY KG/CU M	V(DH/DV)$_P$ KJ/KG	V(DP/DU)$_V$ MPA-CU M/KJ x 10^2	-V(DP/DV)$_T$ MPA	-(DV/DT)$_P$/V 1/DEG. K	THERMAL CONDUCTIVITY W/K-M x 10^3	VISCOSITY KG/M-S x 10^7	THERMAL DIFFUSIVITY SQ M/HR	DIELECTRIC CONSTANT	PRANDTL NUMBER
* 24.742	92.3840	1749.4	0.23071	301.5396	0.00414	188.97	534.95	0.00102	1.30639	2.0519
25	92.2852	1754.5	0.22980	300.3381	0.00416	189.95	521.85	0.00102	1.30603	2.0045
26	91.8995	1773.8	0.22639	295.6654	0.00422	193.31	476.31	0.00101	1.30464	1.8431
27	91.5102	1792.2	0.22310	290.9811	0.00427	196.11	437.86	0.00101	1.30323	1.7103
28	91.1173	1809.6	0.21993	286.2917	0.00433	198.43	405.09	0.00100	1.30182	1.5998
29	90.7211	1825.6	0.21695	281.6032	0.00439	200.33	376.93	0.00099	1.30039	1.5065
30	90.3216	1839.6	0.21423	276.9220	0.00444	201.86	352.54	0.00098	1.29895	1.4267
31	89.9189	1853.2	0.21154	272.2539	0.00450	202.31	331.26	0.00097	1.29750	1.3640
32	89.5132	1866.0	0.20895	267.6047	0.00455	202.52	312.59	0.00096	1.29604	1.3102
33	89.1045	1877.4	0.20652	262.9803	0.00460	202.52	296.09	0.00095	1.29457	1.2633
34	88.6930	1887.9	0.20419	258.3862	0.00465	202.40	281.45	0.00093	1.29309	1.2222
35	88.2787	1897.6	0.20195	253.8280	0.00471	202.13	268.38	0.00092	1.29160	1.1863
36	87.8617	1906.6	0.19979	249.3110	0.00476	201.72	256.65	0.00091	1.29010	1.1549
37	87.4422	1914.8	0.19771	244.8403	0.00481	201.20	246.09	0.00090	1.28860	1.1272
38	87.0201	1921.9	0.19578	240.4212	0.00486	200.58	236.54	0.00089	1.28709	1.1026
39	86.5955	1927.6	0.19403	236.0586	0.00492	199.87	227.86	0.00088	1.28557	1.0803
40	86.1687	1932.6	0.19235	231.7573	0.00497	199.09	219.96	0.00087	1.28404	1.0604
42	85.3086	1942.3	0.18901	223.3571	0.00507	197.28	206.09	0.00085	1.28097	1.0278
44	84.4405	1952.4	0.18560	215.2561	0.00516	195.30	194.34	0.00083	1.27787	1.0027
46	83.5657	1969.2	0.18132	207.4866	0.00525	193.20	184.27	0.00080	1.27475	0.9865
48	82.6853	1983.3	0.17749	200.0778	0.00534	191.04	175.55	0.00079	1.27162	0.9728
50	81.8007	1997.0	0.17381	193.0548	0.00542	188.86	167.94	0.00077	1.26848	0.9619
52	80.9135	2010.7	0.17026	186.4382	0.00549	186.67	161.22	0.00075	1.26533	0.9529
54	80.0086	2040.1	0.16712	182.3746	0.00549	184.44	155.17	0.00074	1.26213	0.9429
56	79.1301	2054.2	0.16370	176.3816	0.00555	182.36	149.90	0.00073	1.25903	0.9369
58	78.2527	2073.7	0.16046	171.4717	0.00557	180.33	145.15	0.00072	1.25593	0.9300
60	77.3824	2086.5	0.15717	165.8425	0.00562	178.63	140.87	0.00071	1.25286	0.9250
65	75.2269	2133.5	0.14849	153.8824	0.00569	175.46	131.74	0.00069	1.24529	0.9108
70	73.1119	2192.1	0.13950	143.4871	0.00572	172.60	124.33	0.00068	1.23789	0.9027
75	71.0481	2258.8	0.13042	134.0696	0.00574	170.14	118.20	0.00067	1.23069	0.9002
80	69.0373	2334.1	0.12151	125.6357	0.00575	168.20	113.03	0.00065	1.22371	0.9012
85	67.0851	2423.4	0.11299	118.5543	0.00572	169.41	108.62	0.00066	1.21696	0.8888
90	65.1997	2521.2	0.10501	112.3300	0.00568	170.71	104.86	0.00066	1.21046	0.8795
95	63.3895	2634.2	0.09770	107.3672	0.00560	172.17	101.66	0.00066	1.20424	0.8707
100	61.6472	2745.7	0.09113	102.5731	0.00553	174.40	90.94	0.00067	1.19827	0.7917
120	55.3571	3217.9	0.07137	87.8588	0.00522	190.68	94.01	0.00074	1.17690	0.8281
140	50.0760	3669.5	0.06073	79.2331	0.00480	198.59	93.63	0.00081	1.15915	0.8312
160	45.6714	4050.6	0.05513	73.4478	0.00441	201.41	92.40	0.00089	1.14449	0.8187
180	41.9732	4361.9	0.05231	69.3320	0.00405	201.57	91.35	0.00098	1.13228	0.8005
200	38.8312	4622.3	0.05095	66.2089	0.00374	200.78	90.84	0.00108	1.12197	0.7824
220	36.1318	4859.2	0.05029	63.7886	0.00347	200.08	90.88	0.00118	1.11317	0.7663
240	33.7920	5098.1	0.04986	61.9605	0.00323	200.05	91.44	0.00129	1.10558	0.7523
260	31.7500	5349.7	0.04943	60.5518	0.00301	200.90	92.42	0.00142	1.09899	0.7401
280	29.9569	5615.0	0.04892	59.4287	0.00281	202.65	93.76	0.00154	1.09322	0.7297
300	28.2816	5785.2	0.04986	58.2664	0.00268	205.20	95.37	0.00169	1.08785	0.7193
350	24.9599	6451.9	0.04898	56.3024	0.00234	214.42	100.31	0.00205	1.07726	0.7057
400	22.3557	7155.0	0.04798	54.8534	0.00208	226.27	106.09	0.00245	1.06900	0.6978
450	20.2547	7880.0	0.04703	53.7312	0.00188	239.68	112.37	0.00288	1.06238	0.6928
500	18.5211	8609.0	0.04622	52.8327	0.00171	253.82	118.95	0.00335	1.05693	0.6896
550	17.0646	9344.3	0.04552	52.0962	0.00157	268.48	125.71	0.00386	1.05237	0.6873
600	15.8228	10084.1	0.04491	51.4811	0.00145	283.46	132.56	0.00440	1.04850	0.6858
700	13.8159	11584.2	0.04383	50.5130	0.00127	314.10	146.37	0.00558	1.04225	0.6841
800	12.2628	13124.6	0.04285	49.7869	0.00112	345.56	160.13	0.00688	1.03744	0.6834
900	11.0245	14717.4	0.04188	49.2236	0.00101	377.81	173.75	0.00830	1.03362	0.6833
1000	10.0136	16370.2	0.04092	48.7748	0.00092	440.56	187.17	0.01055	1.03050	0.6376
1200	8.4622	19855.7	0.03901	48.1065	0.00077	511.13	213.33	0.01414	1.02573	0.6420
1400	7.3271	23571.5	0.03722	47.6344	0.00067	583.31	238.60	0.01813	1.02225	0.6467
1600	6.4604	27428.1	0.03570	47.2842	0.00059	655.34	263.04	0.02251	1.01960	0.6511
1800	5.7769	31388.6	0.03443	47.0144	0.00053	726.96	286.72	0.02728	1.01751	0.6549
2000	5.2241	35522.9	0.03325	46.7994	0.00048	800.69	309.72	0.03247	1.01583	0.6574
2500	4.2132	47406.5	0.03005	46.3936	0.00039	1025.41	364.72	0.04784	1.01275	0.6515
3000	3.5199	64685.0	0.02556	45.9966	0.00032	1427.99	417.28	0.06949	1.01065	0.6142

* TWO-PHASE BOUNDARY

Table 4.1 (continued)
THERMODYNAMIC PROPERTIES OF PARAHYDROGEN (ISOBARS, SI UNITS)

50.00 MPA ISOBAR

TEMPERATURE	VOLUME	ISOTHERM DERIVATIVE	ISOCHORE DERIVATIVE	INTERNAL ENERGY	ENTHALPY	ENTROPY	CV	CP	VELOCITY OF SOUND
DEG. K	CU M/KG	CU M-MPA/KG	MPA/K	KJ/KG-K	KJ/KG	KJ/KG-K	KJ /	KG-K	M/S
* 25.708	0.01069	3.4533	1.278	-247.9	286.8	5.798	5.94	7.33	2064
26	0.01071	3.4430	1.278	-246.4	289.0	5.881	5.97	7.38	2063
27	0.01075	3.4072	1.275	-241.0	296.4	6.163	6.07	7.56	2060
28	0.01079	3.3711	1.271	-235.6	304.1	6.441	6.17	7.73	2056
29	0.01084	3.3346	1.267	-230.0	311.9	6.715	6.26	7.90	2051
30	0.01088	3.2980	1.262	-224.2	319.9	6.986	6.34	8.06	2047
31	0.01093	3.2612	1.256	-218.4	328.0	7.252	6.42	8.21	2042
32	0.01097	3.2242	1.250	-212.4	336.3	7.515	6.50	8.37	2037
33	0.01102	3.1872	1.243	-206.3	344.7	7.775	6.57	8.51	2032
34	0.01107	3.1501	1.236	-200.1	353.3	8.032	6.64	8.66	2027
35	0.01112	3.1131	1.229	-193.8	362.0	8.285	6.70	8.80	2022
36	0.01117	3.0761	1.221	-187.4	370.9	8.534	6.76	8.94	2016
37	0.01122	3.0393	1.213	-180.9	379.9	8.781	6.82	9.07	2011
38	0.01127	3.0027	1.205	-174.3	389.1	9.025	6.87	9.20	2005
39	0.01132	2.9663	1.196	-167.6	398.3	9.265	6.91	9.32	2000
40	0.01137	2.9302	1.187	-160.8	407.7	9.503	6.95	9.44	1994
42	0.01148	2.8590	1.168	-147.0	426.8	9.969	7.03	9.67	1983
44	0.01159	2.7895	1.149	-132.9	446.4	10.424	7.11	9.90	1971
46	0.01170	2.7220	1.129	-119.0	465.9	10.858	7.22	10.16	1958
48	0.01181	2.6570	1.108	-104.2	486.5	11.295	7.31	10.40	1945
50	0.01193	2.5948	1.087	-89.0	507.5	11.725	7.39	10.63	1932
52	0.01205	2.5356	1.065	-73.6	529.0	12.146	7.46	10.84	1919
54	0.01217	2.4797	1.043	-57.9	550.9	12.559	7.53	11.05	1907
56	0.01230	2.4274	1.021	-41.6	573.4	12.969	7.60	11.24	1895
58	0.01243	2.4097	1.002	-25.3	596.3	13.371	7.65	11.39	1893
60	0.01256	2.3577	0.9796	-8.9	619.3	13.760	7.71	11.57	1880
65	0.01290	2.2638	0.9248	33.3	678.1	14.701	7.89	11.97	1854
70	0.01324	2.1798	0.8710	76.9	738.9	15.602	8.10	12.37	1824
75	0.01360	2.1048	0.8199	122.0	801.8	16.470	8.36	12.79	1794
80	0.01396	2.0360	0.7723	168.8	866.9	17.310	8.67	13.24	1763
85	0.01433	1.9734	0.7280	217.5	934.2	18.126	9.02	13.71	1732
90	0.01472	1.9221	0.6869	268.0	1003.9	18.923	9.39	14.18	1703
95	0.01511	1.8810	0.6489	320.6	1076.0	19.702	9.78	14.63	1678
100	0.01550	1.8546	0.6139	375.0	1150.2	20.464	10.17	15.05	1657
120	0.01714	1.7591	0.5009	613.2	1470.1	23.376	11.68	16.70	1586
140	0.01883	1.7337	0.4184	872.3	1814.0	26.024	12.58	17.60	1557
160	0.02055	1.7491	0.3569	1141.3	2169.0	28.394	12.91	17.84	1554
180	0.02228	1.7838	0.3103	1410.3	2524.2	30.487	12.85	17.67	1566
200	0.02400	1.8323	0.2740	1673.6	2873.8	32.328	12.58	17.30	1588
220	0.02573	1.8865	0.2453	1929.1	3215.4	33.956	12.24	16.89	1613
240	0.02744	1.9485	0.2218	2176.6	3548.8	35.406	11.92	16.49	1641
260	0.02916	2.0170	0.2022	2417.1	3874.8	36.712	11.66	16.14	1671
280	0.03085	2.0909	0.1856	2651.7	4196.4	37.896	11.45	15.83	1701
300	0.03263	2.1638	0.1730	2869.1	4500.4	38.932	11.09	15.51	1739
350	0.03686	2.3550	0.1461	3421.8	5264.8	41.288	10.80	15.11	1815
400	0.04106	2.5498	0.1266	3961.6	6014.6	43.292	10.66	14.90	1888
450	0.04523	2.7463	0.1118	4494.7	6756.4	45.040	10.60	14.79	1957
500	0.04939	2.9440	0.1002	5024.9	7494.5	46.594	10.57	14.73	2025
550	0.05354	3.1424	0.0908	5552.7	8229.6	47.996	10.56	14.69	2091
600	0.05767	3.3414	0.0830	6080.1	8963.7	49.272	10.55	14.67	2155
700	0.06593	3.7411	0.0710	7134.5	10431.0	51.535	10.58	14.68	2278
800	0.07417	4.1425	0.0620	8193.7	11902.4	53.498	10.66	14.75	2394
900	0.08241	4.5453	0.0551	9262.0	13382.5	55.240	10.77	14.86	2504
1000	0.09065	4.9494	0.0496	10342.9	14875.2	56.815	10.92	15.01	2688
1200	0.10711	5.7604	0.0413	12558.4	17914.0	59.572	11.30	15.38	2801
1400	0.12358	6.5743	0.0354	14854.4	21033.4	61.981	11.72	15.81	2978
1600	0.14005	7.3904	0.0310	17234.9	24237.5	64.097	12.13	16.22	3143
1800	0.15653	8.2080	0.0276	19699.7	27526.1	65.996	12.51	16.60	3301
2000	0.17301	9.0268	0.0248	22248.8	30899.3	67.839	12.89	16.99	3449
2500	0.21433	11.0775	0.0199	28999.8	39716.2	74.116	14.13	18.29	3786
3000	0.25635	13.1315	0.0166	36632.2	49449.7	81.030	16.50	20.88	4076

* TWO-PHASE BOUNDARY

Table 4.1 (continued)
THERMODYNAMIC PROPERTIES OF PARAHYDROGEN (ISOBARS, SI UNITS)

50.00 MPA ISOBAR

TEMPERATURE DEG. K	DENSITY KG/CU M	$V(DH/DV)_P$ KJ/KG	$V(DP/DU)_V$ MPA-CU M/KJ X 10^2	$-V(DP/DV)_T$ MPA	$-(DV/DT)_P/V$ 1/DEG. K	THERMAL CONDUCTIVITY W/K-M X 10^3	VISCOSITY KG/M-S X 10^7	THERMAL DIFFUSIVITY SQ M/HR	DIELECTRIC CONSTANT	PRANDTL NUMBER
* 25.708	93.5095	1851.7	0.23018	322.9196	0.00396	198.99	567.99	0.00105	1.31046	2.0921
26	93.4012	1858.0	0.22917	321.5761	0.00397	199.98	552.43	0.00104	1.31007	2.0392
27	93.0286	1879.3	0.22574	316.9646	0.00402	202.99	504.76	0.00104	1.30872	1.8795
28	92.6530	1899.6	0.22244	312.3381	0.00407	205.49	464.37	0.00103	1.30736	1.7471
29	92.2745	1918.3	0.21935	307.7025	0.00412	207.56	429.87	0.00103	1.30600	1.6357
30	91.8933	1935.1	0.21652	303.0638	0.00416	209.24	400.14	0.00102	1.30462	1.5407
31	91.5094	1951.3	0.21374	298.4278	0.00421	209.79	374.34	0.00100	1.30323	1.4655
32	91.1229	1966.5	0.21107	293.8000	0.00425	210.09	351.80	0.00099	1.30184	1.4010
33	90.7340	1980.3	0.20857	289.1859	0.00430	210.16	331.98	0.00098	1.30043	1.3450
34	90.3427	1993.2	0.20617	284.5909	0.00434	210.11	314.46	0.00097	1.29902	1.2959
35	89.9491	2005.2	0.20387	280.0202	0.00439	209.90	298.88	0.00095	1.29761	1.2530
36	89.5533	2016.2	0.20166	275.4789	0.00443	209.55	284.97	0.00094	1.29618	1.2154
37	89.1553	2026.5	0.19954	270.9719	0.00448	209.07	272.48	0.00093	1.29475	1.1822
38	88.7551	2035.5	0.19757	266.5040	0.00452	208.49	261.23	0.00092	1.29331	1.1527
39	88.3530	2042.9	0.19579	262.0800	0.00456	207.81	251.05	0.00091	1.29187	1.1260
40	87.9489	2049.5	0.19409	257.7042	0.00461	207.06	241.80	0.00090	1.29042	1.1023
42	87.1352	2062.2	0.19072	249.1153	0.00469	205.29	225.65	0.00088	1.28750	1.0630
44	86.3146	2075.1	0.18728	240.7706	0.00477	203.33	212.06	0.00086	1.28456	1.0326
46	85.4880	2094.8	0.18300	232.7010	0.00485	201.25	200.49	0.00083	1.28161	1.0123
48	84.6562	2111.2	0.17917	224.9347	0.00493	199.09	190.53	0.00081	1.27864	0.9953
50	83.8203	2126.7	0.17552	217.4966	0.00500	196.89	181.87	0.00080	1.27566	0.9816
52	82.9812	2141.7	0.17200	210.4081	0.00506	194.68	174.28	0.00078	1.27267	0.9707
54	82.1403	2156.9	0.16857	203.6869	0.00512	192.48	167.58	0.00076	1.26969	0.9618
56	81.2989	2172.1	0.16527	197.3461	0.00517	190.32	161.62	0.00075	1.26670	0.9543
58	80.4310	2203.2	0.16274	193.8173	0.00517	188.12	156.13	0.00074	1.26363	0.9452
60	79.6003	2216.3	0.15952	187.6705	0.00522	186.36	151.37	0.00073	1.26069	0.9396
65	77.5440	2272.4	0.15120	175.5411	0.00527	183.08	141.30	0.00071	1.25343	0.9239
70	75.5271	2338.7	0.14234	164.6374	0.00529	180.12	133.23	0.00069	1.24634	0.9152
75	73.5548	2415.4	0.13327	154.8164	0.00530	177.53	126.58	0.00068	1.23944	0.9121
80	71.6309	2500.3	0.12432	145.8433	0.00530	175.47	121.00	0.00067	1.23272	0.9129
85	69.7620	2592.3	0.11572	137.6651	0.00529	176.68	116.25	0.00067	1.22622	0.9020
90	67.9453	2695.2	0.10766	130.5979	0.00526	177.93	112.16	0.00067	1.21993	0.8936
95	66.1884	2807.4	0.10027	124.5019	0.00521	179.32	108.64	0.00067	1.21386	0.8865
100	64.4956	2933.1	0.09360	119.6114	0.00513	182.33	95.57	0.00068	1.20803	0.7891
120	58.3468	3422.5	0.07353	102.6367	0.00488	198.60	98.60	0.00073	1.18702	0.8293
140	53.0937	3871.4	0.06263	92.0502	0.00455	206.05	97.93	0.00079	1.16927	0.8363
160	48.6518	4252.8	0.05681	85.0971	0.00419	208.25	96.31	0.00086	1.15439	0.8249
180	44.8845	4559.2	0.05381	80.0658	0.00388	207.78	94.87	0.00094	1.14188	0.8067
200	41.6582	4818.3	0.05231	76.3289	0.00359	206.42	93.99	0.00103	1.13124	0.7877
220	38.8704	5047.7	0.05155	73.3272	0.00335	205.21	93.71	0.00113	1.12210	0.7711
240	36.4373	5277.2	0.05106	70.9973	0.00312	204.74	93.99	0.00123	1.11416	0.7570
260	34.2992	5520.7	0.05058	69.1826	0.00292	205.23	94.75	0.00133	1.10722	0.7450
280	32.4108	5783.2	0.05002	67.7689	0.00274	206.69	95.89	0.00145	1.10112	0.7346
300	30.6495	5945.2	0.05089	66.3210	0.00261	209.00	97.35	0.00158	1.09544	0.7223
350	27.1292	6608.2	0.04986	63.8894	0.00229	217.84	102.04	0.00191	1.08417	0.7078
400	24.3547	7309.7	0.04875	62.0985	0.00204	229.54	107.70	0.00228	1.07534	0.6992
450	22.1071	8034.1	0.04769	60.7131	0.00184	242.96	113.95	0.00267	1.06822	0.6939
500	20.2464	8762.7	0.04681	59.6048	0.00168	257.23	120.57	0.00311	1.06235	0.6903
550	18.6789	9497.6	0.04605	58.6964	0.00155	272.10	127.41	0.00357	1.05743	0.6878
600	17.3392	10237.1	0.04538	57.9379	0.00143	287.36	134.38	0.00407	1.05323	0.6862
700	15.1678	11736.6	0.04422	56.7441	0.00125	318.70	148.50	0.00515	1.04646	0.6842
800	13.4820	13276.7	0.04317	55.8489	0.00111	351.00	162.64	0.00635	1.04122	0.6834
900	12.1343	14869.3	0.04216	55.1545	0.00100	384.19	176.67	0.00767	1.03704	0.6833
1000	11.0320	16522.3	0.04115	54.6014	0.00091	440.56	190.52	0.00958	1.03364	0.6491
1200	9.3359	20008.9	0.03920	53.7787	0.00077	511.13	217.60	0.01281	1.02842	0.6548
1400	8.0919	23726.5	0.03737	53.1985	0.00067	583.31	243.81	0.01641	1.02460	0.6608
1600	7.1402	27584.6	0.03582	52.7687	0.00059	655.33	269.18	0.02037	1.02163	0.6663
1800	6.3887	31545.5	0.03453	52.4382	0.00053	726.90	293.79	0.02467	1.01936	0.6710
2000	5.7801	35676.9	0.03335	52.1754	0.00048	800.46	317.70	0.02934	1.01752	0.6744
2500	4.6657	47497.4	0.03017	51.6845	0.00038	1022.57	374.91	0.04315	1.01413	0.6704
3000	3.9009	64470.2	0.02578	51.2248	0.00032	1413.28	429.54	0.06246	1.01180	0.6347

* TWO-PHASE BOUNDARY

THERMODYNAMIC PROPERTIES OF PARAHYDROGEN (ISOBARS, SI UNITS)

55.00 MPA ISOBAR

TEMPERATURE DEG. K	VOLUME CU M/KG	ISOTHERM DERIVATIVE CU M-MPA/KG	ISOCHORE DERIVATIVE MPA/K	INTERNAL ENERGY KJ/KG-K	ENTHALPY KJ/KG	ENTROPY KJ/KG-K	CV KJ /	CP KG-K	VELOCITY OF SOUND M/S
* 26.646	0.01057	3.6392	1.307	-240.0	341.5	5.856	6.01	7.41	2118
27	0.01059	3.6271	1.305	-238.2	344.1	5.954	6.05	7.47	2117
28	0.01063	3.5926	1.302	-232.9	351.7	6.229	6.15	7.64	2113
29	0.01067	3.5577	1.298	-227.4	359.4	6.500	6.24	7.81	2109
30	0.01071	3.5225	1.293	-221.8	367.3	6.767	6.33	7.96	2105
31	0.01075	3.4871	1.288	-216.1	375.3	7.031	6.41	8.11	2101
32	0.01080	3.4514	1.282	-210.3	383.5	7.291	6.49	8.26	2097
33	0.01084	3.4156	1.275	-204.4	391.9	7.547	6.56	8.41	2092
34	0.01088	3.3796	1.268	-198.3	400.3	7.800	6.63	8.55	2087
35	0.01093	3.3436	1.261	-192.2	409.0	8.050	6.70	8.69	2082
36	0.01097	3.3076	1.254	-185.9	417.7	8.297	6.76	8.82	2077
37	0.01102	3.2716	1.246	-179.5	426.6	8.540	6.82	8.95	2072
38	0.01107	3.2357	1.238	-173.1	435.6	8.781	6.87	9.07	2067
39	0.01111	3.1999	1.229	-166.5	444.7	9.018	6.92	9.19	2062
40	0.01116	3.1643	1.221	-159.9	454.0	9.252	6.96	9.31	2057
42	0.01126	3.0937	1.203	-146.4	472.8	9.712	7.04	9.53	2046
44	0.01136	3.0244	1.184	-132.6	492.1	10.160	7.12	9.75	2035
46	0.01146	2.9566	1.165	-119.0	511.3	10.587	7.23	10.01	2022
48	0.01157	2.8906	1.145	-104.6	531.6	11.018	7.33	10.24	2010
50	0.01167	2.8269	1.125	-89.7	552.3	11.440	7.41	10.46	1997
52	0.01178	2.7656	1.104	-74.6	573.4	11.855	7.50	10.68	1985
54	0.01189	2.7071	1.083	-59.2	595.0	12.262	7.57	10.88	1972
56	0.01201	2.6515	1.061	-43.3	617.2	12.666	7.64	11.07	1960
58	0.01213	2.5990	1.040	-27.4	639.5	13.058	7.70	11.25	1948
60	0.01224	2.5499	1.018	-11.3	662.2	13.442	7.76	11.42	1937
65	0.01255	2.4754	0.9691	30.2	720.6	14.377	7.94	11.83	1920
70	0.01287	2.3895	0.9170	73.1	780.8	15.269	8.16	12.24	1893
75	0.01319	2.3173	0.8665	117.6	843.1	16.129	8.43	12.65	1865
80	0.01352	2.2438	0.8187	163.9	907.5	16.959	8.74	13.10	1835
85	0.01386	2.1803	0.7743	212.0	974.1	17.767	9.08	13.57	1805
90	0.01420	2.1219	0.7327	262.1	1043.2	18.556	9.46	14.05	1775
95	0.01455	2.0737	0.6939	314.3	1114.6	19.329	9.85	14.52	1748
100	0.01491	2.0336	0.6579	368.4	1188.4	20.085	10.24	14.97	1724
120	0.01638	1.9310	0.5405	605.6	1506.4	22.980	11.74	16.61	1653
140	0.01790	1.8884	0.4545	864.3	1849.0	25.618	12.65	17.55	1619
160	0.01945	1.8907	0.3891	1133.4	2203.4	27.984	12.98	17.82	1612
180	0.02102	1.9181	0.3390	1402.8	2558.6	30.077	12.90	17.67	1620
200	0.02257	1.9596	0.2998	1666.6	2908.2	31.918	12.63	17.30	1638
220	0.02413	2.0095	0.2685	1922.6	3250.0	33.547	12.29	16.89	1662
240	0.02569	2.0648	0.2431	2170.8	3583.6	34.998	11.97	16.51	1687
260	0.02724	2.1280	0.2220	2412.0	3910.1	36.305	11.71	16.17	1715
280	0.02878	2.1975	0.2039	2647.4	4230.4	37.493	11.50	15.89	1743
300	0.03039	2.2687	0.1899	2864.7	4536.1	38.526	11.13	15.53	1780
350	0.03424	2.4553	0.1605	3418.7	5301.8	40.886	10.83	15.13	1852
400	0.03805	2.6466	0.1390	3959.7	6052.6	42.893	10.69	14.92	1922
450	0.04185	2.8404	0.1228	4493.7	6795.2	44.643	10.63	14.81	1989
500	0.04562	3.0358	0.1100	5024.8	7534.0	46.199	10.59	14.74	2055
550	0.04939	3.2322	0.0997	5553.2	8269.6	47.601	10.58	14.70	2120
600	0.05315	3.4294	0.0912	6081.1	9004.1	48.878	10.57	14.68	2182
700	0.06065	3.8259	0.0779	7136.4	10471.9	51.142	10.60	14.69	2302
800	0.06813	4.2247	0.0681	8196.3	11943.6	53.106	10.67	14.75	2416
900	0.07562	4.6252	0.0605	9265.1	13423.9	54.848	10.79	14.86	2524
1000	0.08310	5.0272	0.0545	10346.4	14916.7	56.423	10.94	15.01	2627
1200	0.09805	5.8349	0.0454	12562.5	17955.5	59.180	11.31	15.38	2817
1400	0.11302	6.6462	0.0389	14858.9	21074.8	61.588	11.73	15.81	2993
1600	0.12798	7.4601	0.0341	17239.8	24278.8	63.705	12.14	16.22	3157
1800	0.14295	8.2760	0.0303	19704.7	27567.1	65.603	12.51	16.60	3313
2000	0.15793	9.0933	0.0273	22253.7	30939.9	67.446	12.90	16.99	3461
2500	0.19548	11.1412	0.0219	28999.7	39751.0	73.721	14.11	18.26	3796
3000	0.23364	13.1933	0.0182	36602.2	49452.5	80.624	16.40	20.77	4087

* TWO-PHASE BOUNDARY

Table 4.1 (continued)
THERMODYNAMIC PROPERTIES OF PARAHYDROGEN (ISOBARS, SI UNITS)

55.00 MPA ISOBAR

TEMPERATURE DEG. K	DENSITY KG/CU M	$V(DH/DV)_P$ KJ/KG	$V(DP/DU)_V$ MPA-CU M/KJ X 10^2	$-V(DP/DV)_T$ MPA	$-(DV/DT)_P/V$ 1/DEG. K	THERMAL CONDUCTIVITY W/K-M X 10^3	VISCOSITY KG/M-S X 10^7	THERMAL DIFFUSIVITY SQ M/HR	DIELECTRIC CONSTANT	PRANDTL NUMBER
* 26.646	94.5778	1952.1	0.22975	344.1862	0.00380	208.55	601.68	0.00107	1.31434	2.1378
27	94.4505	1960.6	0.22849	342.5778	0.00381	209.63	581.97	0.00107	1.31388	2.0742
28	94.0893	1983.8	0.22505	338.0217	0.00385	212.31	532.32	0.00106	1.31257	1.9159
29	93.7257	2005.4	0.22183	333.4476	0.00389	214.53	490.14	0.00106	1.31125	1.7832
30	93.3597	2024.8	0.21889	328.8612	0.00393	216.35	454.00	0.00105	1.30992	1.6705
31	92.9915	2043.6	0.21600	324.2682	0.00397	217.00	422.79	0.00104	1.30859	1.5810
32	92.6212	2061.4	0.21324	319.6741	0.00401	217.38	395.65	0.00102	1.30725	1.5042
33	92.2489	2077.6	0.21065	315.0841	0.00405	217.52	371.90	0.00101	1.30590	1.4376
34	91.8746	2092.8	0.20818	310.5035	0.00409	217.54	350.98	0.00100	1.30455	1.3794
35	91.4983	2107.1	0.20581	305.9372	0.00412	217.38	332.47	0.00098	1.30319	1.3285
36	91.1202	2120.3	0.20354	301.3902	0.00416	217.07	315.99	0.00097	1.30183	1.2839
37	90.7403	2132.7	0.20137	296.8672	0.00420	216.63	301.26	0.00096	1.30046	1.2447
38	90.3586	2143.7	0.19934	292.3729	0.00423	216.07	288.03	0.00095	1.29908	1.2097
39	89.9752	2152.8	0.19752	287.9117	0.00427	215.42	276.09	0.00094	1.29770	1.1781
40	89.5903	2161.2	0.19578	283.4879	0.00431	214.69	265.29	0.00093	1.29631	1.1499
42	88.8157	2177.1	0.19235	274.7696	0.00438	212.95	246.52	0.00091	1.29353	1.1033
44	88.0353	2193.1	0.18887	266.2499	0.00445	211.01	230.82	0.00088	1.29073	1.0669
46	87.2498	2216.1	0.18455	257.9587	0.00452	208.92	217.52	0.00086	1.28791	1.0419
48	86.4597	2235.2	0.18071	249.9233	0.00458	206.76	206.14	0.00084	1.28508	1.0210
50	85.6657	2253.1	0.17705	242.1683	0.00464	204.54	196.31	0.00082	1.28224	1.0042
52	84.8687	2270.1	0.17353	234.7155	0.00470	202.30	187.73	0.00080	1.27940	0.9907
54	84.0695	2286.8	0.17011	227.5835	0.00476	200.08	180.19	0.00079	1.27655	0.9798
56	83.2691	2303.1	0.16683	220.7875	0.00481	197.88	173.51	0.00077	1.27370	0.9708
58	82.4687	2318.6	0.16374	214.3387	0.00485	195.73	167.55	0.00076	1.27085	0.9629
60	81.6692	2335.7	0.16060	208.2446	0.00489	193.90	162.21	0.00075	1.26801	0.9553
65	79.6560	2406.1	0.15324	197.1767	0.00491	190.35	150.86	0.00073	1.26088	0.9372
70	77.7140	2478.4	0.14461	185.6980	0.00494	187.23	142.00	0.00071	1.25403	0.9281
75	75.8132	2565.5	0.13565	175.6818	0.00493	184.50	134.76	0.00069	1.24735	0.9242
80	73.9677	2656.4	0.12670	165.9722	0.00493	182.32	128.76	0.00068	1.24088	0.9254
85	72.1689	2758.0	0.11810	157.3462	0.00492	183.52	123.66	0.00067	1.23460	0.9145
90	70.4197	2865.4	0.11000	149.4223	0.00490	184.76	119.28	0.00067	1.22851	0.9072
95	68.7183	2982.1	0.10252	142.5027	0.00487	186.10	115.48	0.00067	1.22261	0.9010
100	67.0733	3104.5	0.09576	136.4020	0.00482	189.87	99.99	0.00068	1.21691	0.7885
120	61.0551	3624.0	0.07538	117.8980	0.00458	206.19	103.03	0.00073	1.19625	0.8302
140	55.8554	4074.1	0.06434	105.4773	0.00431	213.23	102.11	0.00078	1.17858	0.8406
160	51.4027	4452.2	0.05833	97.1866	0.00400	214.89	100.14	0.00084	1.16359	0.8306
180	47.5849	4757.1	0.05520	91.2726	0.00371	213.86	98.34	0.00092	1.15084	0.8123
200	44.2982	5011.1	0.05357	86.8068	0.00345	211.96	97.12	0.00100	1.13994	0.7929
220	41.4370	5238.2	0.05271	83.2688	0.00322	210.29	96.54	0.00108	1.13051	0.7754
240	38.9298	5457.9	0.05215	80.3804	0.00302	209.42	96.56	0.00117	1.12229	0.7611
260	36.7130	5692.8	0.05164	78.1244	0.00284	209.57	97.09	0.00127	1.11506	0.7493
280	34.7442	5947.8	0.05105	76.3489	0.00267	210.74	98.05	0.00137	1.10866	0.7392
300	32.9061	6104.8	0.05188	74.6540	0.00254	212.82	99.36	0.00150	1.10272	0.7252
350	29.2084	6764.3	0.05071	71.7146	0.00224	221.27	103.79	0.00180	1.09082	0.7099
400	26.2794	7464.3	0.04949	69.5512	0.00200	232.80	109.32	0.00214	1.08146	0.7007
450	23.8972	8188.1	0.04834	67.8786	0.00181	246.21	115.54	0.00250	1.07388	0.6949
500	21.9189	8916.3	0.04738	66.5409	0.00165	260.59	122.19	0.00290	1.06762	0.6910
550	20.2477	9651.0	0.04655	65.4444	0.00152	275.67	129.11	0.00333	1.06236	0.6884
600	18.8162	10390.2	0.04583	64.5288	0.00141	291.20	136.19	0.00380	1.05786	0.6865
700	16.4894	11889.1	0.04459	63.0873	0.00124	323.24	150.61	0.00480	1.05058	0.6844
800	14.6771	13428.8	0.04348	62.0060	0.00110	356.38	165.13	0.00593	1.04493	0.6835
900	13.2248	15021.2	0.04242	61.1673	0.00099	390.53	179.58	0.00715	1.04042	0.6834
1000	12.0343	16674.3	0.04138	60.4995	0.00090	440.56	193.88	0.00878	1.03674	0.6605
1200	10.1984	20161.8	0.03937	59.5066	0.00076	511.13	221.91	0.01173	1.03107	0.6677
1400	8.8483	23880.8	0.03751	58.8073	0.00066	583.31	249.10	0.01501	1.02692	0.6751
1600	7.8136	27740.2	0.03593	58.2902	0.00058	655.32	275.46	0.01861	1.02374	0.6818
1800	6.9953	31701.4	0.03463	57.8931	0.00052	726.85	301.05	0.02253	1.02124	0.6876
2000	6.3319	35830.4	0.03343	57.5781	0.00047	800.26	325.94	0.02678	1.01921	0.6919
2500	5.1157	47596.5	0.03028	56.9945	0.00038	1020.12	385.52	0.03932	1.01550	0.6900
3000	4.2801	64306.6	0.02598	56.4680	0.00032	1400.61	442.39	0.05673	1.01296	0.6559

* TWO-PHASE BOUNDARY

Table 4.1 (continued)
THERMODYNAMIC PROPERTIES OF PARAHYDROGEN (ISOBARS, SI UNITS)

60.00 MPA ISOBAR

TEMPERATURE DEG. K	VOLUME CU M/KG	ISOTHERM DERIVATIVE CU M-MPA/KG	ISOCHORE DERIVATIVE MPA/K	INTERNAL ENERGY KJ/KG-K	ENTHALPY KJ/KG	ENTROPY KJ/KG-K	CV KJ /	CP KG-K	VELOCITY OF SOUND M/S
* 27.553	0.01046	3.8220	1.334	-232.1	395.6	5.911	6.09	7.49	2169
28	0.01048	3.8074	1.333	-229.8	398.9	6.031	6.13	7.56	2167
29	0.01052	3.7741	1.329	-224.4	406.5	6.299	6.23	7.73	2164
30	0.01056	3.7404	1.324	-219.0	414.3	6.564	6.31	7.88	2161
31	0.01060	3.7063	1.319	-213.4	422.3	6.824	6.40	8.03	2157
32	0.01064	3.6720	1.313	-207.7	430.4	7.082	6.48	8.18	2153
33	0.01068	3.6374	1.306	-201.9	438.6	7.336	6.55	8.32	2149
34	0.01072	3.6026	1.300	-196.0	447.0	7.586	6.63	8.46	2144
35	0.01076	3.5677	1.293	-190.0	455.6	7.833	6.69	8.59	2140
36	0.01080	3.5327	1.285	-183.9	464.2	8.077	6.76	8.72	2135
37	0.01084	3.4976	1.278	-177.6	473.0	8.318	6.82	8.85	2131
38	0.01089	3.4625	1.270	-171.3	481.9	8.555	6.87	8.97	2126
39	0.01093	3.4274	1.261	-164.9	490.9	8.790	6.92	9.08	2121
40	0.01098	3.3924	1.253	-158.4	500.1	9.021	6.96	9.19	2116
42	0.01107	3.3228	1.236	-145.2	518.7	9.475	7.05	9.41	2106
44	0.01116	3.2541	1.217	-131.7	537.7	9.918	7.18	9.63	2096
46	0.01125	3.1864	1.199	-118.4	556.7	10.339	7.25	9.88	2083
48	0.01135	3.1202	1.180	-104.2	576.7	10.764	7.35	10.10	2071
50	0.01145	3.0557	1.160	-89.7	597.1	11.181	7.44	10.32	2059
52	0.01155	2.9932	1.140	-74.8	618.0	11.590	7.52	10.53	2047
54	0.01165	2.9329	1.119	-59.7	639.2	11.991	7.60	10.73	2035
56	0.01175	2.8751	1.099	-44.1	661.1	12.390	7.68	10.92	2023
58	0.01186	2.8199	1.078	-28.4	683.2	12.776	7.74	11.10	2011
60	0.01197	2.7676	1.057	-12.6	705.5	13.156	7.81	11.27	1999
65	0.01225	2.6498	1.004	28.1	763.0	14.075	7.99	11.70	1970
70	0.01254	2.5870	0.9585	70.4	823.0	14.964	8.21	12.12	1954
75	0.01284	2.5083	0.9089	114.4	884.8	15.816	8.48	12.56	1927
80	0.01314	2.4474	0.8616	160.2	948.7	16.641	8.80	12.99	1901
85	0.01345	2.3784	0.8168	207.9	1014.7	17.442	9.15	13.46	1871
90	0.01376	2.3204	0.7751	257.6	1083.3	18.225	9.52	13.94	1843
95	0.01408	2.2690	0.7361	309.3	1154.1	18.991	9.92	14.41	1816
100	0.01440	2.2209	0.6994	363.1	1227.3	19.743	10.31	14.88	1790
120	0.01574	2.1021	0.5771	599.4	1543.8	22.623	11.81	16.52	1715
140	0.01712	2.0411	0.4886	857.6	1885.0	25.250	12.71	17.51	1677
160	0.01854	2.0338	0.4201	1126.7	2238.9	27.613	13.04	17.81	1667
180	0.01996	2.0523	0.3668	1396.4	2593.8	29.704	12.96	17.66	1672
200	0.02138	2.0865	0.3249	1660.7	2943.5	31.545	12.69	17.31	1687
220	0.02280	2.1322	0.2912	1917.2	3285.4	33.175	12.35	16.90	1708
240	0.02422	2.1830	0.2638	2165.9	3619.2	34.627	12.03	16.52	1731
260	0.02564	2.2398	0.2411	2407.7	3946.0	35.936	11.76	16.19	1756
280	0.02705	2.3043	0.2218	2643.8	4266.9	37.125	11.55	15.92	1783
300	0.02852	2.3746	0.2067	2861.0	4572.3	38.156	11.16	15.55	1819
350	0.03205	2.5564	0.1747	3416.2	5339.0	40.519	10.86	15.16	1888
400	0.03554	2.7441	0.1514	3958.3	6090.9	42.529	10.72	14.94	1956
450	0.03902	2.9351	0.1337	4493.1	6834.2	44.281	10.65	14.82	2021
500	0.04248	3.1280	0.1197	5025.0	7573.7	45.838	10.61	14.75	2085
550	0.04593	3.3222	0.1085	5554.1	8309.7	47.241	10.60	14.71	2147
600	0.04937	3.5175	0.0992	6082.5	9044.6	48.519	10.59	14.69	2208
700	0.05624	3.9107	0.0848	7138.7	10512.9	50.784	10.62	14.69	2326
800	0.06309	4.3066	0.0741	8199.3	11984.9	52.748	10.69	14.75	2438
900	0.06995	4.7046	0.0659	9268.6	13465.4	54.490	10.80	14.86	2544
1000	0.07680	5.1045	0.0593	10350.4	14958.2	56.065	10.95	15.01	2645
1200	0.09030	5.9085	0.0494	12567.1	17997.0	58.822	11.32	15.38	2834
1400	0.10420	6.7169	0.0424	14863.9	21116.1	61.230	11.74	15.81	3007
1600	0.11791	7.5284	0.0371	17245.1	24319.9	63.346	12.15	16.22	3171
1800	0.13163	8.3423	0.0330	19710.2	27608.0	65.245	12.52	16.60	3325
2000	0.14535	9.1580	0.0297	22259.2	30980.3	67.087	12.90	16.98	3472
2500	0.17976	11.2028	0.0238	29000.8	39786.2	73.360	14.10	18.23	3807
3000	0.21471	13.2526	0.0199	36577.3	49459.7	80.253	16.31	20.66	4096

* TWO-PHASE BOUNDARY

Table 4.1 (continued)
THERMODYNAMIC PROPERTIES OF PARAHYDROGEN (ISOBARS, SI UNITS)

60.00 MPA ISOBAR

TEMPERATURE DEG. K	DENSITY KG/CU M	$V(DH/DV)_P$ KJ/KG	$V(DP/DU)_V$ MPA-CU M/KJ X 10^{-2}	$-V(DP/DV)_T$ MPA	$-(DV/DT)_P/V$ 1/DEG. K	THERMAL CONDUCTIVITY W/K-M X 10^3	VISCOSITY KG/M-S X 10^7	THERMAL DIFFUSIVITY SQ M/HR	DIELECTRIC CONSTANT	PRANDTL NUMBER
* 27.559	95.5951	2051.5	0.22930	365.3665	0.00365	217.73	636.02	0.00109	1.31803	2.1885
28	95.4408	2062.7	0.22776	363.3461	0.00367	218.92	610.44	0.00109	1.31747	2.1094
29	95.0898	2087.0	0.22441	358.8802	0.00370	221.30	558.99	0.00109	1.31620	1.9517
30	94.7368	2109.1	0.22134	354.3538	0.00374	223.25	515.14	0.00108	1.31492	1.8182
31	94.3819	2130.6	0.21834	349.8125	0.00377	223.98	477.48	0.00106	1.31363	1.7119
32	94.0253	2150.9	0.21547	345.2615	0.00380	224.43	444.87	0.00105	1.31233	1.6210
33	93.6670	2169.6	0.21279	340.7362	0.00383	224.64	416.47	0.00104	1.31104	1.5423
34	93.3070	2187.2	0.21023	336.1516	0.00387	224.71	391.56	0.00103	1.30973	1.4736
35	92.9455	2203.8	0.20779	331.6027	0.00390	224.60	369.60	0.00101	1.30842	1.4137
36	92.5824	2219.3	0.20545	327.0642	0.00393	224.33	350.12	0.00100	1.30711	1.3612
37	92.2179	2233.8	0.20321	322.5406	0.00396	223.92	332.77	0.00099	1.30579	1.3149
38	91.8519	2246.8	0.20113	318.0367	0.00399	223.39	317.25	0.00098	1.30447	1.2738
39	91.4845	2257.8	0.19926	313.5566	0.00402	222.76	303.29	0.00097	1.30314	1.2367
40	91.1158	2268.0	0.19747	309.1046	0.00405	222.04	290.69	0.00095	1.30181	1.2036
42	90.3746	2287.4	0.19396	300.3011	0.00411	220.32	268.91	0.00093	1.29914	1.1487
44	89.6286	2306.6	0.19042	291.6575	0.00417	218.38	250.78	0.00091	1.29645	1.1057
46	88.8782	2333.1	0.18605	283.2026	0.00423	216.29	235.52	0.00089	1.29375	1.0754
48	88.1239	2355.3	0.18217	274.9634	0.00429	214.10	222.53	0.00087	1.29105	1.0501
50	87.3663	2375.9	0.17848	266.9642	0.00434	211.86	211.35	0.00085	1.28833	1.0297
52	86.6059	2395.3	0.17494	259.2269	0.00440	209.60	201.66	0.00083	1.28560	1.0132
54	85.8434	2414.1	0.17152	251.7707	0.00445	207.34	193.17	0.00081	1.28288	1.0000
56	85.0795	2432.1	0.16824	244.6121	0.00449	205.10	185.69	0.00079	1.28015	0.9890
58	84.3151	2448.9	0.16515	237.7642	0.00453	202.91	179.04	0.00078	1.27742	0.9796
60	83.5509	2467.1	0.16202	231.2368	0.00457	201.04	173.11	0.00077	1.27470	0.9708
65	81.6480	2521.4	0.15386	216.3540	0.00464	197.50	160.71	0.00074	1.26794	0.9520
70	79.7274	2609.0	0.14637	206.2574	0.00465	194.06	150.75	0.00072	1.26114	0.9418
75	77.8890	2699.0	0.13753	195.3693	0.00465	191.18	142.87	0.00070	1.25465	0.9383
80	76.0998	2807.6	0.12868	186.2454	0.00463	188.85	136.37	0.00069	1.24835	0.9379
85	74.3631	2914.4	0.12007	176.8644	0.00462	190.03	130.91	0.00068	1.24225	0.9272
90	72.6669	3032.0	0.11199	168.6176	0.00460	191.23	126.21	0.00068	1.23634	0.9199
95	71.0221	3155.6	0.10451	161.1474	0.00457	192.55	122.16	0.00068	1.23060	0.9144
100	69.4259	3280.8	0.09769	154.1877	0.00454	197.07	104.23	0.00069	1.22506	0.7872
120	63.5332	3824.0	0.07689	133.5510	0.00432	213.48	107.32	0.00073	1.20473	0.8307
140	58.4025	4272.0	0.06583	119.2041	0.00410	220.17	106.17	0.00078	1.18721	0.8444
160	53.9490	4650.9	0.05973	109.7236	0.00383	221.33	103.88	0.00083	1.17215	0.8358
180	50.1062	4952.2	0.05647	102.8327	0.00357	219.79	101.75	0.00089	1.15925	0.8177
200	46.7707	5200.3	0.05475	97.5868	0.00333	217.41	100.21	0.00097	1.14813	0.7980
220	43.8538	5425.4	0.05378	93.5038	0.00311	215.30	99.35	0.00105	1.13847	0.7798
240	41.2843	5641.9	0.05314	90.1241	0.00293	214.06	99.12	0.00113	1.13001	0.7648
260	39.0033	5867.3	0.05258	87.3592	0.00276	213.98	99.44	0.00122	1.12253	0.7529
280	36.9667	6114.0	0.05197	85.1829	0.00260	214.78	100.22	0.00131	1.11589	0.7430
300	35.0602	6263.9	0.05283	83.2541	0.00248	216.63	101.38	0.00143	1.10969	0.7280
350	31.2041	6920.1	0.05153	79.7700	0.00219	224.70	105.55	0.00171	1.09723	0.7119
400	28.1346	7618.7	0.05020	77.2052	0.00196	236.05	110.95	0.00202	1.08738	0.7022
450	25.6289	8342.0	0.04896	75.2225	0.00178	249.44	117.13	0.00236	1.07938	0.6960
500	23.5414	9069.9	0.04793	73.6366	0.00163	263.92	123.80	0.00274	1.07276	0.6918
550	21.7735	9804.4	0.04703	72.3365	0.00150	279.19	130.78	0.00314	1.06717	0.6889
600	20.2558	10543.3	0.04626	71.2505	0.00139	294.98	137.98	0.00357	1.06238	0.6869
700	17.7820	12041.8	0.04494	69.5398	0.00122	327.72	152.70	0.00452	1.05462	0.6846
800	15.8493	13581.0	0.04377	68.2560	0.00109	361.71	167.59	0.00557	1.04858	0.6836
900	14.2966	15173.2	0.04267	67.2599	0.00098	396.82	182.47	0.00672	1.04375	0.6834
1000	13.0213	16826.3	0.04159	66.4668	0.00087	440.56	197.24	0.00812	1.03979	0.6720
1200	11.0499	20314.4	0.03953	65.2883	0.00076	511.13	226.26	0.01083	1.03369	0.6808
1400	9.5966	24034.7	0.03763	64.4590	0.00066	583.31	254.47	0.01384	1.02922	0.6897
1600	8.4807	27895.1	0.03604	63.8466	0.00058	655.31	281.88	0.01715	1.02579	0.6977
1800	7.5971	31856.4	0.03472	63.3771	0.00052	726.81	308.52	0.02075	1.02308	0.7046
2000	6.8798	35983.1	0.03351	63.0054	0.00047	800.08	334.45	0.02465	1.02088	0.7100
2500	5.5631	47701.8	0.03038	62.3218	0.00038	1017.99	396.58	0.03613	1.01686	0.7104
3000	4.6575	64183.0	0.02615	61.7244	0.00032	1389.57	455.88	0.05198	1.01410	0.6780

* TWO-PHASE BOUNDARY

Table 4.1 (continued)
THERMODYNAMIC PROPERTIES OF PARAHYDROGEN (ISOBARS, SI UNITS)

65.00 MPA ISOBAR

TEMPERATURE DEG. K	VOLUME CU M/KG	ISOTHERM DERIVATIVE CU M-MPA/KG	ISOCHORE DERIVATIVE MPA/K	INTERNAL ENERGY KJ/KG-K	ENTHALPY KJ/KG	ENTROPY KJ/KG-K	CV KJ / KG-K	CP KJ / KG-K	VELOCITY OF SOUND M/S
* 28.449	0.01036	4.0022	1.362	-224.0	449.1	5.964	6.16	7.57	2218
29	0.01038	3.9846	1.359	-221.1	453.3	6.110	6.21	7.66	2217
30	0.01041	3.9524	1.355	-215.8	461.0	6.372	6.30	7.81	2213
31	0.01045	3.9197	1.349	-210.4	468.9	6.631	6.39	7.96	2210
32	0.01049	3.8867	1.343	-204.8	476.9	6.886	6.47	8.10	2206
33	0.01053	3.8534	1.337	-199.1	485.1	7.137	6.55	8.24	2203
34	0.01056	3.8198	1.330	-193.3	493.4	7.385	6.62	8.38	2199
35	0.01060	3.7860	1.323	-187.4	501.9	7.630	6.69	8.51	2195
36	0.01064	3.7520	1.316	-181.4	510.4	7.872	6.75	8.64	2190
37	0.01068	3.7179	1.309	-175.3	519.1	8.110	6.82	8.76	2186
38	0.01072	3.6836	1.301	-169.1	528.0	8.345	6.87	8.88	2182
39	0.01076	3.6494	1.293	-162.8	536.9	8.577	6.92	8.99	2177
40	0.01081	3.6151	1.284	-156.4	545.9	8.806	6.97	9.10	2173
42	0.01089	3.5468	1.267	-143.5	564.3	9.255	7.06	9.31	2163
44	0.01098	3.4788	1.249	-130.2	583.2	9.693	7.14	9.52	2153
46	0.01106	3.4117	1.231	-117.2	601.9	10.110	7.26	9.76	2142
48	0.01115	3.3456	1.212	-103.2	621.7	10.530	7.36	9.99	2130
50	0.01124	3.2809	1.193	-88.9	641.9	10.942	7.46	10.20	2118
52	0.01134	3.2177	1.174	-74.4	662.5	11.346	7.55	10.41	2106
54	0.01143	3.1564	1.154	-59.5	683.5	11.743	7.63	10.61	2095
56	0.01153	3.0971	1.134	-44.1	705.2	12.137	7.71	10.79	2083
58	0.01162	3.0401	1.113	-28.7	726.9	12.519	7.78	10.97	2071
60	0.01172	2.9855	1.093	-13.1	749.0	12.894	7.85	11.15	2059
65	0.01198	2.8605	1.041	27.1	805.8	13.803	8.04	11.58	2029
70	0.01225	2.7527	0.9896	68.8	864.8	14.676	8.27	12.00	1999
75	0.01253	2.6984	0.9479	112.2	926.7	15.530	8.54	12.46	1984
80	0.01281	2.6250	0.9006	157.5	990.1	16.348	8.86	12.91	1956
85	0.01309	2.5726	0.8561	204.8	1055.8	17.145	9.21	13.36	1932
90	0.01338	2.5105	0.8141	254.1	1123.8	17.923	9.59	13.84	1904
95	0.01367	2.4550	0.7749	305.5	1194.2	18.684	9.98	14.33	1877
100	0.01397	2.4087	0.7382	359.0	1267.1	19.431	10.38	14.79	1853
120	0.01519	2.2696	0.6121	594.6	1582.1	22.298	11.89	16.46	1773
140	0.01646	2.1952	0.5204	852.2	1921.9	24.915	12.78	17.45	1732
160	0.01775	2.1754	0.4500	1121.1	2275.1	27.273	13.10	17.79	1719
180	0.01906	2.1875	0.3937	1391.0	2630.1	29.364	13.02	17.66	1722
200	0.02037	2.2138	0.3493	1655.7	2979.6	31.204	12.75	17.32	1734
220	0.02168	2.2546	0.3134	1912.7	3321.6	32.834	12.40	16.91	1753
240	0.02298	2.3018	0.2841	2161.8	3655.6	34.288	12.08	16.52	1774
260	0.02428	2.3531	0.2597	2404.1	3982.5	35.596	11.81	16.20	1797
280	0.02558	2.4115	0.2392	2640.8	4303.7	36.787	11.60	15.94	1821
300	0.02694	2.4814	0.2234	2857.8	4608.9	37.816	11.20	15.57	1858
350	0.03019	2.6583	0.1888	3414.1	5376.6	40.182	10.89	15.17	1924
400	0.03342	2.8423	0.1636	3957.2	6129.3	42.194	10.75	14.95	1989
450	0.03662	3.0302	0.1445	4492.9	6873.4	43.948	10.68	14.84	2052
500	0.03981	3.2205	0.1294	5025.5	7613.4	45.506	10.64	14.76	2114
550	0.04300	3.4125	0.1173	5555.3	8350.0	46.911	10.62	14.72	2175
600	0.04617	3.6058	0.1073	6084.2	9085.2	48.189	10.61	14.69	2234
700	0.05250	3.9954	0.0917	7141.4	10554.0	50.454	10.63	14.70	2350
800	0.05883	4.3882	0.0801	8202.6	12026.2	52.419	10.70	14.76	2460
900	0.06514	4.7836	0.0712	9272.4	13506.8	54.161	10.81	14.86	2564
1000	0.07146	5.1811	0.0641	10354.7	14999.6	55.737	10.96	15.01	2664
1200	0.08410	5.9812	0.0534	12572.0	18038.3	58.493	11.33	15.38	2850
1400	0.09674	6.7864	0.0458	14869.3	21157.3	60.901	11.75	15.81	3022
1600	0.10939	7.5953	0.0402	17250.8	24360.9	63.017	12.16	16.22	3183
1800	0.12204	8.4071	0.0357	19716.2	27648.8	64.916	12.53	16.60	3337
2000	0.13470	9.2210	0.0322	22265.1	31020.6	66.758	12.91	16.98	3483
2500	0.16644	11.2623	0.0258	29003.0	39821.8	73.029	14.09	18.21	3816
3000	0.19867	13.3096	0.0215	36556.7	49470.5	79.913	16.24	20.57	4186

* TWO-PHASE BOUNDARY

Table 4.1 (continued)
THERMODYNAMIC PROPERTIES OF PARAHYDROGEN (ISOBARS, SI UNITS)

65.00 MPA ISOBAR

TEMPERATURE DEG. K	DENSITY KG/CU M	$V(DH/DV)_P$ KJ/KG	$V(DP/DU)_V$ MPA-CU M/KJ X 10^2	$-V(DP/DV)_T$ MPA	$-(DV/DT)_P/V$ 1/DEG. K	THERMAL CONDUCTIVITY W/K-M X 10^3	VISCOSITY KG/M-S X 10^7	THERMAL DIFFUSIVITY SQ M/HR	DIELECTRIC CONSTANT	PRANDTL NUMBER
* 28.449	96.5666	2149.1	0.22897	386.4808	0.00352	226.54	671.06	0.00112	1.32157	2.2429
29	96.3788	2163.7	0.22708	384.0341	0.00354	227.87	637.85	0.00111	1.32089	2.1438
30	96.0369	2188.5	0.22388	379.5738	0.00357	229.95	584.76	0.00110	1.31964	1.9860
31	95.6934	2212.6	0.22076	375.0909	0.00360	230.77	539.38	0.00109	1.31839	1.8602
32	95.3485	2235.6	0.21778	370.5907	0.00362	231.29	500.29	0.00108	1.31714	1.7529
33	95.0022	2256.7	0.21500	366.0785	0.00365	231.55	466.38	0.00106	1.31588	1.6602
34	94.6545	2276.7	0.21235	361.5592	0.00369	231.67	436.78	0.00105	1.31462	1.5795
35	94.3056	2295.6	0.20982	357.0376	0.00371	231.60	410.78	0.00104	1.31335	1.5092
36	93.9554	2313.4	0.20740	352.5185	0.00373	231.36	387.81	0.00103	1.31208	1.4477
37	93.6040	2330.0	0.20509	348.0064	0.00376	230.98	367.42	0.00101	1.31081	1.3936
38	93.2515	2345.1	0.20294	343.5058	0.00379	230.47	349.23	0.00100	1.30953	1.3455
39	92.8979	2358.1	0.20101	339.0208	0.00381	229.85	332.94	0.00099	1.30825	1.3023
40	92.5432	2370.1	0.19918	334.5557	0.00384	229.15	318.27	0.00098	1.30697	1.2637
42	91.8306	2393.1	0.19557	325.7010	0.00389	227.43	293.03	0.00096	1.30439	1.1995
44	91.1142	2415.6	0.19196	316.9722	0.00394	225.49	272.14	0.00094	1.30180	1.1491
46	90.3942	2445.9	0.18751	308.3978	0.00399	223.39	254.64	0.00091	1.29921	1.1130
48	89.6710	2471.3	0.18358	300.0041	0.00404	221.18	239.82	0.00089	1.29660	1.0828
50	88.9450	2494.9	0.17985	291.8155	0.00409	218.91	227.13	0.00087	1.29399	1.0583
52	88.2165	2517.0	0.17629	283.8540	0.00413	216.62	216.17	0.00085	1.29138	1.0385
54	87.4862	2538.2	0.17285	276.1393	0.00418	214.32	206.62	0.00083	1.28876	1.0224
56	86.7546	2558.4	0.16955	268.6887	0.00422	212.04	198.24	0.00082	1.28614	1.0092
58	86.0223	2577.0	0.16646	261.5165	0.00426	209.80	190.83	0.00080	1.28352	0.9979
60	85.2899	2596.8	0.16333	254.6345	0.00429	207.89	184.23	0.00079	1.28090	0.9877
65	83.4637	2654.4	0.15517	238.7487	0.00436	204.26	170.54	0.00076	1.27439	0.9665
70	81.6539	2726.7	0.14655	224.7719	0.00440	200.86	159.83	0.00074	1.26796	0.9552
75	79.8094	2831.2	0.13906	215.3549	0.00440	197.60	150.94	0.00072	1.26143	0.9519
80	78.0755	2938.8	0.13021	204.9503	0.00439	195.13	143.92	0.00070	1.25531	0.9525
85	76.3829	3067.0	0.12167	196.5034	0.00436	196.26	138.02	0.00069	1.24935	0.9397
90	74.7374	3190.4	0.11359	187.6264	0.00434	197.42	133.01	0.00069	1.24358	0.9326
95	73.1402	3319.5	0.10614	179.5626	0.00432	198.71	128.69	0.00068	1.23799	0.9278
100	71.5815	3455.3	0.09936	172.4154	0.00428	203.94	108.30	0.00069	1.23255	0.7856
120	65.8245	4017.1	0.07823	149.3972	0.00410	220.50	111.47	0.00073	1.21261	0.8321
140	60.7642	4473.6	0.06704	133.3902	0.00390	226.89	110.13	0.00077	1.19525	0.8472
160	56.3265	4845.1	0.06099	122.5338	0.00367	227.60	107.55	0.00082	1.18017	0.8407
180	52.4598	5146.9	0.05763	114.7553	0.00343	225.58	105.10	0.00088	1.16714	0.8227
200	49.0968	5388.9	0.05582	108.6924	0.00321	222.76	103.27	0.00094	1.15588	0.8028
220	46.1349	5610.8	0.05477	104.0147	0.00301	220.25	102.14	0.00102	1.14602	0.7840
240	43.5148	5825.6	0.05405	100.1616	0.00284	218.64	101.67	0.00109	1.13736	0.7683
260	41.1807	6045.4	0.05341	96.9032	0.00269	218.16	101.79	0.00118	1.12967	0.7560
280	39.0881	6282.9	0.05277	94.2613	0.00254	218.79	102.40	0.00126	1.12281	0.7462
300	37.1199	6422.2	0.05375	92.1109	0.00243	220.43	103.42	0.00137	1.11639	0.7307
350	33.1220	7075.3	0.05233	88.0478	0.00214	228.12	107.33	0.00163	1.10341	0.7139
400	29.9249	7772.7	0.05089	85.0545	0.00192	239.28	112.59	0.00192	1.09312	0.7037
450	27.3054	8495.7	0.04956	82.7398	0.00175	252.64	118.71	0.00225	1.08473	0.6971
500	25.1167	9223.4	0.04845	80.8880	0.00160	267.21	125.40	0.00259	1.07776	0.6926
550	23.2585	9957.7	0.04750	79.3692	0.00148	282.66	132.45	0.00297	1.07186	0.6895
600	21.6597	10696.6	0.04668	78.1000	0.00137	298.71	139.75	0.00338	1.06681	0.6874
700	19.0469	12194.6	0.04528	76.0993	0.00121	332.12	154.77	0.00427	1.05858	0.6848
800	16.9995	13733.4	0.04405	74.5967	0.00107	366.98	170.04	0.00527	1.05217	0.6837
900	15.3506	15325.3	0.04290	73.4304	0.00097	403.08	185.35	0.00636	1.04703	0.6834
1000	13.9936	16978.4	0.04179	72.5016	0.00088	440.56	200.60	0.00755	1.04281	0.6834
1200	11.8910	20466.9	0.03968	71.1217	0.00075	511.13	230.64	0.01006	1.03629	0.6940
1400	10.3371	24188.1	0.03775	70.1517	0.00065	583.31	259.95	0.01285	1.03150	0.7045
1600	9.1420	28049.4	0.03613	69.4362	0.00058	655.31	288.47	0.01591	1.02782	0.7140
1800	8.1941	32010.6	0.03479	68.8884	0.00052	726.77	316.23	0.01924	1.02491	0.7222
2000	7.4239	36135.2	0.03358	68.4555	0.00047	799.93	343.27	0.02284	1.02255	0.7287
2500	6.0081	47811.9	0.03046	67.6646	0.00038	1016.11	408.13	0.03343	1.01822	0.7316
3000	5.0333	64091.2	0.02631	66.9921	0.00032	1379.81	470.07	0.04797	1.01525	0.7009

* TWO-PHASE BOUNDARY

Table 4.1 (continued)
THERMODYNAMIC PROPERTIES OF PARAHYDROGEN (ISOBARS, SI UNITS)

70.00 MPA ISOBAR

TEMPERATURE DEG. K	VOLUME CU M/KG	ISOTHERM DERIVATIVE CU M-MPA/KG	ISOCHORE DERIVATIVE MPA/K	INTERNAL ENERGY KJ/KG-K	ENTHALPY KJ/KG	ENTROPY KJ/KG-K	CV KJ /	CP KG-K	VELOCITY OF SOUND M/S
* 29.318	0.01026	4.1801	1.389	-215.9	502.1	6.014	6.22	7.65	2266
30	0.01028	4.1590	1.385	-212.3	507.3	6.191	6.29	7.75	2264
31	0.01032	4.1278	1.380	-207.0	515.1	6.448	6.38	7.90	2261
32	0.01035	4.0961	1.374	-201.5	523.1	6.701	6.46	8.04	2258
33	0.01039	4.0640	1.368	-195.9	531.2	6.950	6.54	8.18	2254
34	0.01042	4.0316	1.361	-190.2	539.5	7.197	6.61	8.31	2251
35	0.01046	3.9989	1.354	-184.5	547.8	7.439	6.68	8.44	2247
36	0.01050	3.9660	1.346	-178.6	556.3	7.679	6.75	8.56	2243
37	0.01054	3.9329	1.339	-172.6	565.0	7.915	6.81	8.69	2239
38	0.01057	3.8996	1.331	-166.5	573.7	8.148	6.87	8.80	2235
39	0.01061	3.8663	1.323	-160.3	582.6	8.378	6.92	8.91	2231
40	0.01065	3.8328	1.315	-154.0	591.5	8.605	6.97	9.02	2227
42	0.01073	3.7658	1.298	-141.3	609.8	9.050	7.06	9.22	2218
44	0.01081	3.6990	1.280	-128.3	628.4	9.484	7.15	9.43	2208
46	0.01089	3.6326	1.262	-115.4	647.0	9.896	7.27	9.67	2197
48	0.01097	3.5669	1.244	-101.7	666.5	10.312	7.38	9.89	2186
50	0.01106	3.5023	1.225	-87.6	686.5	10.720	7.48	10.10	2175
52	0.01115	3.4389	1.206	-73.3	706.9	11.120	7.57	10.30	2163
54	0.01123	3.3770	1.186	-58.6	727.7	11.513	7.65	10.49	2152
56	0.01132	3.3169	1.167	-43.5	749.2	11.903	7.73	10.68	2140
58	0.01141	3.2586	1.147	-28.3	770.7	12.281	7.81	10.86	2129
60	0.01151	3.2024	1.127	-12.9	792.6	12.651	7.88	11.03	2117
65	0.01174	3.0721	1.076	26.7	848.8	13.552	8.08	11.46	2087
70	0.01199	2.9572	1.026	68.0	907.2	14.417	8.32	11.90	2056
75	0.01224	2.8583	0.9754	110.9	967.8	15.253	8.60	12.34	2025
80	0.01251	2.8104	0.9375	155.8	1031.7	16.077	8.92	12.83	2011
85	0.01278	2.7389	0.8922	202.7	1097.0	16.869	9.27	13.31	1982
90	0.01304	2.6879	0.8500	251.6	1164.7	17.643	9.65	13.77	1958
95	0.01331	2.6388	0.8108	302.7	1234.7	18.400	10.05	14.24	1934
100	0.01359	2.5856	0.7738	355.9	1307.2	19.143	10.44	14.72	1909
120	0.01472	2.4419	0.6464	590.8	1621.0	21.999	11.96	16.41	1838
140	0.01588	2.3428	0.5496	847.9	1959.5	24.606	12.84	17.40	1781
160	0.01708	2.3086	0.4783	1116.6	2312.0	26.959	13.16	17.78	1766
180	0.01829	2.3219	0.4199	1386.6	2666.8	29.049	13.08	17.65	1770
200	0.01950	2.3430	0.3730	1651.6	3016.3	30.890	12.80	17.32	1788
220	0.02071	2.3767	0.3350	1909.0	3358.4	32.520	12.46	16.91	1796
240	0.02191	2.4198	0.3039	2158.6	3692.5	33.974	12.13	16.53	1816
260	0.02312	2.4681	0.2779	2401.3	4019.6	35.284	11.86	16.21	1837
280	0.02432	2.5196	0.2561	2638.4	4340.9	36.474	11.65	15.96	1858
300	0.02558	2.5891	0.2399	2855.1	4645.7	37.502	11.23	15.59	1896
350	0.02860	2.7609	0.2029	3412.5	5414.3	39.871	10.93	15.19	1959
400	0.03159	2.9409	0.1758	3956.4	6167.8	41.885	10.77	14.97	2021
450	0.03457	3.1257	0.1552	4493.0	6912.7	43.640	10.70	14.85	2082
500	0.03753	3.3133	0.1391	5026.3	7653.2	45.200	10.66	14.77	2143
550	0.04048	3.5030	0.1260	5556.7	8390.2	46.605	10.64	14.72	2202
600	0.04342	3.6941	0.1153	6086.3	9125.8	47.883	10.63	14.70	2268
700	0.04930	4.0800	0.0985	7144.3	10595.1	50.150	10.65	14.70	2373
800	0.05516	4.4695	0.0861	8206.2	12067.6	52.114	10.72	14.76	2481
900	0.06102	4.8621	0.0765	9276.6	13548.2	53.857	10.83	14.86	2583
1000	0.06688	5.2571	0.0689	10359.3	15041.1	55.432	10.97	15.01	2681
1200	0.07860	6.0529	0.0574	12577.3	18079.7	58.189	11.34	15.38	2865
1400	0.09033	6.8547	0.0493	14875.1	21198.4	60.597	11.76	15.81	3035
1600	0.10207	7.6609	0.0432	17257.0	24401.8	62.713	12.16	16.22	3196
1800	0.11381	8.4703	0.0384	19722.6	27689.3	64.611	12.54	16.60	3349
2000	0.12556	9.2822	0.0346	22271.6	31060.8	66.453	12.91	16.98	3494
2500	0.15502	11.3197	0.0277	29006.2	39857.6	72.722	14.08	18.19	3825
3000	0.18492	13.3644	0.0231	36539.5	49484.2	79.598	16.17	20.49	4114

* TWO-PHASE BOUNDARY

Table 4.1 (continued)

THERMODYNAMIC PROPERTIES OF PARAHYDROGEN (ISOBARS, SI UNITS)

70.00 MPA ISOBAR

TEMPERATURE DEG. K	DENSITY KG/CU M	$V(DH/DV)_P$ KJ/KG	$V(DP/DU)_V$ MPA-CU M/KJ X 10^2	$-V(DP/DV)_T$ MPA	$-(DV/DT)_P/V$ 1/DEG. K	THERMAL CONDUCTIVITY W/K-M X 10^3	VISCOSITY KG/M-S X 10^7	THERMAL DIFFUSIVITY SQ M/HR	DIELECTRIC CONSTANT	PRANDTL NUMBER
* 29.318	97.4967	2244.6	0.22878	407.5451	0.00341	235.03	706.81	0.00113	1.32496	2.2999
30	97.2698	2263.3	0.22652	404.5479	0.00342	236.48	664.21	0.00113	1.32413	2.1766
31	96.9362	2290.1	0.22326	400.1290	0.00345	237.38	609.63	0.00112	1.32292	2.0281
32	96.6013	2315.6	0.22016	395.6857	0.00347	237.96	562.84	0.00110	1.32170	1.9017
33	96.2654	2339.2	0.21728	391.2231	0.00350	238.28	522.44	0.00109	1.32047	1.7928
34	95.9283	2361.7	0.21452	386.7464	0.00352	238.44	487.31	0.00108	1.31925	1.6983
35	95.5903	2382.9	0.21190	382.2602	0.00354	238.41	456.58	0.00106	1.31802	1.6162
36	95.2513	2402.9	0.20940	377.7693	0.00356	238.20	429.54	0.00105	1.31678	1.5445
37	94.9113	2421.8	0.20701	373.2781	0.00359	237.84	405.62	0.00104	1.31555	1.4814
38	94.5704	2439.0	0.20480	368.7910	0.00361	237.35	384.35	0.00103	1.31431	1.4255
39	94.2287	2454.0	0.20280	364.3122	0.00363	236.74	365.36	0.00101	1.31307	1.3753
40	93.8860	2467.9	0.20091	359.8459	0.00365	236.04	348.32	0.00100	1.31183	1.3306
42	93.1984	2494.5	0.19720	350.9664	0.00370	234.33	319.11	0.00098	1.30934	1.2561
44	92.5077	2520.5	0.19350	342.1825	0.00374	232.37	295.08	0.00096	1.30684	1.1974
46	91.8141	2554.6	0.18898	333.5225	0.00378	230.25	275.04	0.00093	1.30433	1.1548
48	91.1180	2583.4	0.18498	325.0125	0.00383	228.02	258.15	0.00091	1.30182	1.1192
50	90.4196	2610.1	0.18120	316.6769	0.00387	225.72	243.76	0.00089	1.29930	1.0902
52	89.7192	2635.1	0.17760	308.5377	0.00391	223.38	231.37	0.00087	1.29678	1.0666
54	89.0173	2659.1	0.17413	300.6150	0.00395	221.04	220.63	0.00085	1.29425	1.0474
56	88.3142	2681.6	0.17081	292.9268	0.00398	218.72	211.24	0.00083	1.29173	1.0315
58	87.6105	2702.3	0.16771	285.4884	0.00402	216.43	202.97	0.00082	1.28920	1.0180
60	86.9866	2724.0	0.16456	278.3130	0.00405	214.49	195.64	0.00081	1.28668	1.0060
65	85.1501	2786.0	0.15639	261.5869	0.00411	210.78	180.52	0.00078	1.28040	0.9818
70	83.4061	2862.2	0.14776	246.6504	0.00416	207.27	168.78	0.00075	1.27418	0.9692
75	81.6851	2954.4	0.13883	233.4795	0.00418	204.10	159.41	0.00073	1.26807	0.9640
80	79.9139	3074.7	0.13156	224.5879	0.00417	201.19	151.41	0.00071	1.26180	0.9659
85	78.2693	3197.1	0.12292	214.3757	0.00416	202.29	145.08	0.00070	1.25599	0.9544
90	76.6635	3338.2	0.11486	206.0676	0.00413	203.40	139.71	0.00069	1.25033	0.9458
95	75.1048	3481.5	0.10745	198.1874	0.00409	204.63	135.10	0.00069	1.24486	0.9403
100	73.5871	3619.7	0.10068	190.2657	0.00407	210.52	112.22	0.00070	1.23955	0.7847
120	67.9484	4211.3	0.07954	165.9205	0.00390	227.28	115.50	0.00073	1.21994	0.8337
140	62.9718	4669.6	0.06796	147.5332	0.00373	233.41	114.00	0.00077	1.20280	0.8497
160	58.5562	5027.6	0.06203	135.1813	0.00354	233.70	111.15	0.00081	1.18773	0.8456
180	54.6784	5338.3	0.05869	126.9584	0.00331	231.24	108.41	0.00086	1.17460	0.8276
200	51.2917	5579.8	0.05679	120.1748	0.00310	228.00	106.29	0.00092	1.16322	0.8072
220	48.2948	5794.6	0.05568	114.7806	0.00292	225.11	104.90	0.00099	1.15320	0.7882
240	45.6339	6007.1	0.05488	110.4232	0.00275	223.18	104.21	0.00107	1.14436	0.7719
260	43.2550	6226.3	0.05418	106.7565	0.00260	222.40	104.13	0.00114	1.13650	0.7589
280	41.1164	6455.3	0.05348	103.5957	0.00247	222.79	104.58	0.00122	1.12946	0.7490
300	39.0924	6579.6	0.05464	101.2148	0.00237	224.21	105.45	0.00132	1.12283	0.7334
350	34.9676	7230.0	0.05310	96.5405	0.00210	231.53	109.11	0.00157	1.10939	0.7159
400	31.6542	7926.3	0.05155	93.0932	0.00189	242.50	114.23	0.00184	1.09868	0.7052
450	28.9380	8649.0	0.05014	90.4261	0.00172	255.82	120.30	0.00214	1.08993	0.6983
500	26.6473	9376.7	0.04896	88.2911	0.00158	270.46	126.99	0.00247	1.08263	0.6935
550	24.7046	10111.0	0.04795	86.5392	0.00146	286.08	134.10	0.00283	1.07645	0.6902
600	23.0296	10849.8	0.04708	85.0743	0.00135	302.38	141.50	0.00322	1.07114	0.6878
700	20.2852	12347.5	0.04560	82.7633	0.00119	336.47	156.81	0.00406	1.06247	0.6850
800	18.1285	13886.0	0.04431	81.0261	0.00106	372.19	172.46	0.00501	1.05570	0.6838
900	16.3874	15477.6	0.04312	79.6768	0.00096	409.31	188.22	0.00605	1.05026	0.6835
1000	14.9516	17130.6	0.04197	78.6020	0.00088	440.56	203.95	0.00707	1.04578	0.6948
1200	12.7219	20619.3	0.03982	77.0053	0.00075	511.13	235.07	0.00941	1.03886	0.7073
1400	11.0702	24341.3	0.03785	75.8835	0.00065	583.31	265.52	0.01200	1.03376	0.7195
1600	9.7974	28203.1	0.03621	75.0570	0.00058	655.30	295.22	0.01485	1.02983	0.7306
1800	8.7866	32164.1	0.03486	74.4252	0.00052	726.74	324.17	0.01794	1.02673	0.7403
2000	7.9643	36286.7	0.03364	73.9267	0.00047	799.79	352.41	0.02129	1.02420	0.7481
2500	6.4508	47925.6	0.03053	73.0211	0.00038	1014.43	420.22	0.03111	1.01957	0.7537
3000	5.4076	64025.0	0.02645	72.2696	0.00032	1371.13	485.01	0.04454	1.01639	0.7249

* TWO-PHASE BOUNDARY

Table 4.1 (continued)
THERMODYNAMIC PROPERTIES OF PARAHYDROGEN (ISOBARS, SI UNITS)

80.00 MPA ISOBAR

TEMPERATURE DEG. K	VOLUME CU M/KG	ISOTHERM DERIVATIVE CU M-MPA/KG	ISOCHORE DERIVATIVE MPA/K	INTERNAL ENERGY KJ/KG-K	ENTHALPY KJ/KG	ENTROPY KJ/KG-K	CV KJ / KG-K	CP KJ / KG-K	VELOCITY OF SOUND M/S
* 30.999	0.01008	4.5298	1.441	-199.5	606.6	6.109	6.35	7.80	2358
31	0.01008	4.5298	1.441	-199.5	606.6	6.109	6.35	7.80	2358
32	0.01011	4.5007	1.435	-194.2	614.4	6.359	6.44	7.94	2355
33	0.01014	4.4712	1.428	-188.8	622.5	6.605	6.52	8.07	2352
34	0.01017	4.4412	1.421	-183.3	630.6	6.848	6.60	8.20	2349
35	0.01021	4.4109	1.414	-177.7	638.9	7.088	6.67	8.33	2346
36	0.01024	4.3802	1.406	-172.0	647.2	7.324	6.74	8.45	2343
37	0.01027	4.3491	1.399	-166.2	655.8	7.557	6.81	8.57	2339
38	0.01031	4.3178	1.391	-160.3	664.4	7.787	6.87	8.68	2335
39	0.01034	4.2863	1.382	-154.3	673.1	8.014	6.92	8.78	2332
40	0.01038	4.2545	1.374	-148.3	681.9	8.238	6.97	8.89	2328
42	0.01045	4.1906	1.357	-135.9	699.9	8.676	7.07	9.08	2320
44	0.01052	4.1263	1.339	-123.2	718.3	9.103	7.17	9.28	2312
46	0.01059	4.0620	1.322	-110.8	736.5	9.509	7.29	9.51	2302
48	0.01067	3.9979	1.303	-97.4	755.8	9.918	7.40	9.72	2291
50	0.01074	3.9342	1.285	-83.8	775.4	10.319	7.51	9.93	2281
52	0.01082	3.8712	1.266	-69.8	795.5	10.712	7.60	10.12	2270
54	0.01089	3.8092	1.247	-55.5	815.9	11.098	7.69	10.31	2259
56	0.01097	3.7484	1.228	-40.7	837.0	11.481	7.78	10.49	2248
58	0.01105	3.6888	1.209	-25.9	858.1	11.852	7.86	10.66	2238
60	0.01113	3.6308	1.190	-10.9	879.6	12.217	7.94	10.84	2226
65	0.01134	3.4936	1.141	27.9	934.9	13.101	8.16	11.27	2197
70	0.01155	3.3690	1.092	68.3	992.4	13.953	8.41	11.72	2167
75	0.01177	3.2580	1.043	110.5	1052.1	14.777	8.71	12.18	2135
80	0.01199	3.1609	0.9949	154.6	1114.2	15.578	9.04	12.64	2103
85	0.01222	3.0766	0.9473	200.7	1178.6	16.359	9.40	13.11	2071
90	0.01247	3.0327	0.9169	249.0	1246.9	17.140	9.78	13.66	2058
95	0.01271	2.9769	0.8751	299.6	1316.3	17.890	10.18	14.13	2033
100	0.01295	2.9212	0.8370	352.3	1388.2	18.628	10.58	14.60	2008
120	0.01393	2.7645	0.7095	585.7	1699.9	21.465	12.10	16.33	1932
140	0.01493	2.6714	0.6055	842.3	2036.9	24.060	12.99	17.27	1885
160	0.01596	2.5674	0.5274	1110.3	2387.0	26.397	13.29	17.71	1849
180	0.01701	2.5527	0.4682	1380.1	2741.1	28.483	13.20	17.68	1849
200	0.01807	2.5993	0.4183	1645.6	3091.1	30.326	12.92	17.32	1866
220	0.01912	2.6204	0.3764	1903.7	3433.3	31.957	12.57	16.92	1878
240	0.02017	2.6518	0.3419	2154.1	3767.6	33.412	12.24	16.55	1893
260	0.02122	2.6955	0.3131	2397.7	4095.1	34.723	11.97	16.22	1912
280	0.02226	2.7419	0.2886	2635.8	4416.7	35.915	11.75	15.97	1930
300	0.02336	2.8067	0.2723	2851.0	4727.1	36.935	11.30	15.62	1970
350	0.02600	2.9678	0.2305	3410.2	5490.3	39.309	10.99	15.22	2028
400	0.02862	3.1398	0.1999	3956.0	6245.3	41.327	10.83	15.00	2085
450	0.03122	3.3179	0.1765	4494.1	6991.4	43.086	10.75	14.87	2142
500	0.03380	3.4998	0.1581	5028.9	7733.0	44.648	10.71	14.79	2199
550	0.03638	3.6845	0.1433	5560.5	8470.9	46.054	10.68	14.74	2255
600	0.03895	3.8711	0.1310	6091.1	9207.2	47.334	10.67	14.71	2310
700	0.04408	4.2490	0.1120	7151.0	10677.3	49.602	10.69	14.71	2418
800	0.04920	4.6316	0.0979	8214.4	12150.2	51.567	10.75	14.76	2521
900	0.05431	5.0180	0.0870	9286.0	13631.0	53.310	10.86	14.86	2621
1000	0.05943	5.4076	0.0783	10369.6	15123.9	54.885	11.00	15.01	2716
1200	0.06966	6.1941	0.0653	12589.1	18162.1	57.642	11.36	15.38	2895
1400	0.07990	6.9883	0.0561	14888.0	21280.4	60.049	11.78	15.80	3062
1600	0.09016	7.7882	0.0491	17270.7	24483.1	62.164	12.18	16.21	3219
1800	0.10041	8.5924	0.0437	19736.8	27770.0	64.062	12.55	16.59	3370
2000	0.11068	9.3998	0.0394	22286.0	31140.6	65.904	12.92	16.97	3513
2500	0.13643	11.4287	0.0316	29015.2	39929.7	72.169	14.06	18.16	3842
3000	0.16255	13.4672	0.0264	36513.9	49518.2	79.032	16.06	20.36	4130

* TWO-PHASE BOUNDARY

Table 4.1 (continued)
THERMODYNAMIC PROPERTIES OF PARAHYDROGEN (ISOBARS, SI UNITS)

80.00 MPA ISOBAR

TEMPERATURE DEG. K	DENSITY KG/CU M	$V(DH/DV)_P$ KJ/KG	$V(DP/DU)_V$ MPA-CU M/KJ X 10^2	$-V(DP/DV)_T$ MPA	$-(DV/DT)_P/V$ 1/DEG. K	THERMAL CONDUCTIVITY W/K-M X 10^3	VISCOSITY KG/M-S X 10^7	THERMAL DIFFUSIVITY SQ M/HR	DIELECTRIC CONSTANT	PRANDTL NUMBER
* 30.999	99.2474	2432.3	0.22852	449.5687	0.00321	250.16	780.58	0.00116	1.33135	2.4329
31	99.2472	2432.4	0.22852	449.5664	0.00321	250.16	780.54	0.00116	1.33135	2.4328
32	98.9287	2463.0	0.22516	445.2508	0.00322	250.86	713.79	0.00115	1.33019	2.2585
33	98.6096	2491.7	0.22204	440.9031	0.00324	251.28	656.67	0.00114	1.32902	2.1094
34	98.2898	2518.9	0.21907	436.5283	0.00326	251.52	607.43	0.00112	1.32785	1.9807
35	97.9695	2544.9	0.21625	432.1310	0.00327	251.55	564.71	0.00111	1.32668	1.8694
36	97.6487	2569.5	0.21356	427.7161	0.00329	251.39	527.40	0.00110	1.32551	1.7726
37	97.3273	2592.9	0.21101	423.2880	0.00330	251.06	494.63	0.00108	1.32434	1.6879
38	97.0055	2614.3	0.20863	418.8511	0.00332	250.59	465.69	0.00107	1.32317	1.6131
39	96.6832	2633.3	0.20650	414.4095	0.00334	250.00	440.01	0.00106	1.32199	1.5462
40	96.3604	2651.1	0.20447	409.9674	0.00335	249.31	417.12	0.00105	1.32082	1.4868
42	95.7136	2685.2	0.20053	401.0970	0.00338	247.57	378.19	0.00102	1.31846	1.3878
44	95.0651	2718.5	0.19662	392.2698	0.00341	245.58	346.50	0.00100	1.31611	1.3098
46	94.4152	2760.4	0.19192	383.5138	0.00345	243.40	320.34	0.00098	1.31375	1.2519
48	93.7638	2796.3	0.18777	374.8552	0.00348	241.09	298.49	0.00095	1.31139	1.2037
50	93.1112	2829.6	0.18387	366.3183	0.00351	238.71	280.02	0.00093	1.30902	1.1643
52	92.4575	2860.9	0.18017	357.9256	0.00354	236.29	264.26	0.00091	1.30666	1.1319
54	91.8029	2890.6	0.17661	349.6978	0.00357	233.86	250.70	0.00089	1.30429	1.1053
56	91.1477	2918.5	0.17323	341.6538	0.00360	231.44	238.92	0.00087	1.30193	1.0832
58	90.4922	2944.0	0.17007	333.8105	0.00362	229.05	228.63	0.00085	1.29956	1.0645
60	89.8367	2970.3	0.16689	326.1830	0.00365	227.02	219.57	0.00084	1.29720	1.0480
65	88.2001	3043.0	0.15866	308.1377	0.00370	223.14	201.10	0.00081	1.29132	1.0158
70	86.5721	3129.1	0.15001	291.6594	0.00375	219.43	186.99	0.00078	1.28548	0.9987
75	84.9602	3231.1	0.14104	276.8026	0.00377	216.05	175.89	0.00075	1.27972	0.9916
80	83.3722	3349.3	0.13200	263.5304	0.00378	213.19	166.95	0.00073	1.27406	0.9902
85	81.8151	3482.3	0.12316	251.7159	0.00376	214.22	159.60	0.00072	1.26853	0.9764
90	80.1666	3622.4	0.11696	243.1203	0.00377	214.82	152.93	0.00071	1.26269	0.9725
95	78.6848	3781.3	0.10926	234.2384	0.00374	215.98	147.72	0.00070	1.25745	0.9663
100	77.2243	3935.0	0.10246	225.5857	0.00371	222.91	119.62	0.00071	1.25231	0.7835
120	71.7961	4569.5	0.08170	198.4820	0.00357	240.19	123.22	0.00074	1.23330	0.8380
140	66.9708	5104.0	0.06960	178.9029	0.00338	245.91	121.47	0.00077	1.21656	0.8532
160	62.6633	5401.5	0.06332	160.8803	0.00328	245.46	118.14	0.00080	1.20175	0.8523
180	58.7805	5665.4	0.06032	150.0471	0.00312	242.20	114.85	0.00084	1.18850	0.8383
200	55.3429	5954.7	0.05851	143.8526	0.00291	238.20	112.22	0.00089	1.17685	0.8158
220	52.3020	6161.1	0.05724	137.0497	0.00275	234.62	110.35	0.00095	1.16661	0.7960
240	49.5830	6363.3	0.05632	131.4833	0.00260	232.07	109.24	0.00102	1.15750	0.7790
260	47.1312	6583.5	0.05550	127.0411	0.00246	230.75	108.79	0.00109	1.14933	0.7649
280	44.9199	6813.4	0.05468	123.1659	0.00234	230.67	108.91	0.00116	1.14200	0.7538
300	42.8013	6891.6	0.05633	120.1304	0.00227	231.69	109.52	0.00125	1.13500	0.7385
350	38.4605	7537.1	0.05457	114.1447	0.00202	238.28	112.68	0.00147	1.12076	0.7199
400	34.9447	8231.9	0.05283	109.7182	0.00182	248.86	117.51	0.00171	1.10932	0.7082
450	32.0350	8954.7	0.05124	106.2877	0.00166	262.08	123.45	0.00198	1.09991	0.7006
500	29.5836	9682.6	0.04993	103.5380	0.00153	276.84	130.14	0.00228	1.09202	0.6952
550	27.4879	10417.3	0.04880	101.2786	0.00141	292.78	137.36	0.00260	1.08531	0.6915
600	25.6738	11156.2	0.04783	99.3867	0.00132	309.55	144.94	0.00295	1.07953	0.6888
700	22.6868	12653.8	0.04620	96.3963	0.00116	344.98	160.82	0.00372	1.07005	0.6855
800	20.3262	14191.8	0.04480	94.1434	0.00104	382.45	177.25	0.00459	1.06260	0.6841
900	18.4117	15783.0	0.04352	92.3908	0.00094	421.65	193.92	0.00555	1.05659	0.6836
1000	16.8270	17435.7	0.04231	90.9932	0.00086	440.56	210.68	0.00628	1.05163	0.7177
1200	14.3550	20924.5	0.04006	88.9164	0.00073	511.13	244.07	0.00834	1.04393	0.7342
1400	12.5150	24647.4	0.03803	87.4585	0.00064	583.31	276.98	0.01062	1.03822	0.7504
1600	11.0919	28509.6	0.03636	86.3863	0.00057	655.29	309.25	0.01312	1.03382	0.7652
1800	9.9587	32469.6	0.03498	85.5686	0.00051	726.68	340.83	0.01583	1.03033	0.7782
2000	9.0348	36587.9	0.03374	84.9255	0.00046	799.56	371.72	0.01877	1.02749	0.7891
2500	7.3297	48161.0	0.03065	83.7689	0.00038	1011.56	446.17	0.02736	1.02226	0.8011
3000	6.1518	63952.0	0.02668	82.8475	0.00032	1356.25	517.44	0.03899	1.01866	0.7766

* TWO-PHASE BOUNDARY

Table 4.1 (continued)
THERMODYNAMIC PROPERTIES OF PARAHYDROGEN (ISOBARS, SI UNITS)

90.00 MPA ISOBAR

TEMPERATURE DEG. K	VOLUME CU M/KG	ISOTHERM DERIVATIVE CU M-MPA/KG	ISOCHORE DERIVATIVE MPA/K	INTERNAL ENERGY KJ/KG-K	ENTHALPY KJ/KG	ENTROPY KJ/KG-K	CV KJ / KG-K	CP KJ / KG-K	VELOCITY OF SOUND M/S
* 32.614	0.00991	4.8726	1.492	-182.9	709.3	6.197	6.48	7.94	2444
33	0.00993	4.8621	1.489	-180.9	712.4	6.291	6.51	7.99	2443
34	0.00996	4.8345	1.482	-175.6	720.4	6.531	6.59	8.12	2441
35	0.00999	4.8065	1.474	-170.1	728.6	6.769	6.67	8.24	2438
36	0.01002	4.7780	1.466	-164.6	736.9	7.003	6.74	8.36	2435
37	0.01005	4.7491	1.458	-158.9	745.3	7.233	6.81	8.48	2432
38	0.01008	4.7198	1.450	-153.2	753.9	7.461	6.87	8.59	2429
39	0.01011	4.6902	1.441	-147.4	762.5	7.685	6.93	8.69	2426
40	0.01014	4.6602	1.432	-141.5	771.3	7.907	6.98	8.79	2423
42	0.01021	4.5996	1.415	-129.4	789.0	8.340	7.08	8.98	2416
44	0.01027	4.5382	1.397	-117.1	807.2	8.762	7.18	9.17	2408
46	0.01033	4.4764	1.379	-104.9	825.2	9.163	7.31	9.40	2399
48	0.01040	4.4143	1.360	-91.9	844.2	9.567	7.42	9.60	2389
50	0.01047	4.3522	1.342	-78.5	863.6	9.963	7.53	9.80	2379
52	0.01054	4.2904	1.323	-64.9	883.4	10.351	7.63	9.99	2370
54	0.01061	4.2290	1.304	-50.9	903.6	10.731	7.73	10.17	2359
56	0.01068	4.1684	1.286	-36.4	924.3	11.109	7.82	10.35	2349
58	0.01075	4.1087	1.267	-21.9	945.2	11.475	7.90	10.51	2339
60	0.01082	4.0500	1.248	-7.2	966.4	11.835	7.98	10.68	2328
65	0.01100	3.9093	1.201	30.8	1020.9	12.707	8.21	11.11	2300
70	0.01119	3.7786	1.153	70.6	1077.6	13.547	8.48	11.57	2270
75	0.01138	3.6597	1.105	112.2	1136.6	14.361	8.79	12.04	2238
80	0.01158	3.5532	1.058	155.8	1198.0	15.153	9.14	12.52	2206
85	0.01178	3.4592	1.011	201.5	1261.8	15.927	9.52	13.00	2174
90	0.01199	3.3767	0.9648	249.3	1328.0	16.684	9.91	13.48	2143
95	0.01219	3.3041	0.9199	299.3	1396.6	17.425	10.32	13.93	2112
100	0.01243	3.2490	0.8965	351.4	1469.9	18.176	10.71	14.53	2100
120	0.01329	3.0689	0.7615	583.5	1779.7	20.996	12.23	16.24	2018
140	0.01418	2.9813	0.6595	839.6	2116.1	23.587	13.13	17.24	1978
160	0.01508	2.8959	0.5722	1107.7	2464.6	25.913	13.44	17.55	1945
180	0.01599	2.7697	0.5084	1377.0	2816.1	27.983	13.34	17.63	1913
200	0.01693	2.7672	0.4587	1642.3	3165.9	29.826	13.04	17.40	1921
220	0.01787	2.8492	0.4157	1900.9	3509.1	31.462	12.69	16.95	1951
240	0.01880	2.8914	0.3780	2152.1	3844.1	32.920	12.36	16.55	1968
260	0.01972	2.9172	0.3463	2396.6	4171.6	34.231	12.08	16.24	1980
280	0.02065	2.9541	0.3196	2635.4	4493.5	35.424	11.86	15.99	1995
300	0.02163	3.0270	0.3042	2848.3	4795.1	36.437	11.36	15.65	2042
350	0.02397	3.1770	0.2578	3409.2	5566.6	38.815	11.04	15.25	2095
400	0.02629	3.3404	0.2236	3956.6	6323.0	40.836	10.88	15.02	2147
450	0.02860	3.5114	0.1975	4496.3	7070.3	42.598	10.81	14.89	2200
500	0.03090	3.6874	0.1769	5032.4	7813.0	44.162	10.75	14.81	2253
550	0.03318	3.8667	0.1603	5565.3	8551.6	45.570	10.73	14.75	2306
600	0.03546	4.0486	0.1466	6097.0	9288.5	46.851	10.71	14.72	2358
700	0.04001	4.4178	0.1253	7158.7	10759.5	49.120	10.73	14.71	2461
800	0.04455	4.7929	0.1096	8223.7	12232.8	51.086	10.79	14.76	2561
900	0.04908	5.1726	0.0974	9296.5	13713.8	52.828	10.89	14.86	2657
1000	0.05362	5.5561	0.0877	10381.2	15206.6	54.404	11.03	15.01	2749
1200	0.06269	6.3323	0.0731	12602.3	18244.4	57.160	11.39	15.37	2924
1400	0.07177	7.1179	0.0628	14902.4	21362.0	59.567	11.80	15.80	3087
1600	0.08087	7.9107	0.0550	17286.0	24564.0	61.682	12.20	16.21	3241
1800	0.08997	8.7088	0.0490	19752.8	27850.2	63.579	12.57	16.59	3390
2000	0.09909	9.5111	0.0442	22302.2	31219.9	65.420	12.94	16.97	3531
2500	0.12194	11.5300	0.0354	29027.3	40002.2	71.683	14.05	18.13	3857
3000	0.14512	13.5613	0.0296	36497.5	49558.7	78.534	15.98	20.24	4144

* TWO-PHASE BOUNDARY

Table 4.1 (continued)
THERMODYNAMIC PROPERTIES OF PARAHYDROGEN (ISOBARS, SI UNITS)

90.00 MPA ISOBAR

TEMPERATURE DEG. K	DENSITY KG/CU M	$V(DH/DV)_P$ KJ/KG	$V(DP/DU)_V$ MPA-CU M/KJ X 10^2	$-V(DP/DV)_T$ MPA	$-(DV/DT)_P/V$ 1/DEG. K	THERMAL CONDUCTIVITY W/K-M X 10^3	VISCOSITY KG/M-S X 10^7	THERMAL DIFFUSIVITY SQ M/HR	DIELECTRIC CONSTANT	PRANDTL NUMBER
* 32.614	100.8714	2616.4	0.22833	491.5039	0.00304	263.61	857.52	0.00118	1.33730	2.5839
33	100.7531	2629.0	0.22707	489.8700	0.00304	263.77	827.99	0.00118	1.33687	2.5091
34	100.4467	2661.2	0.22386	485.6133	0.00305	264.08	759.30	0.00117	1.33575	2.3349
35	100.1401	2691.8	0.22082	481.3228	0.00306	264.16	700.18	0.00115	1.33462	2.1852
36	99.8333	2721.1	0.21794	477.0030	0.00307	264.04	648.95	0.00114	1.33350	2.0556
37	99.5264	2748.9	0.21519	472.6586	0.00308	263.73	604.29	0.00113	1.33238	1.9428
38	99.2193	2774.7	0.21265	468.2939	0.00310	263.28	565.12	0.00111	1.33125	1.8436
39	98.9121	2797.8	0.21036	463.9132	0.00311	262.69	530.59	0.00110	1.33013	1.7554
40	98.6048	2819.6	0.20819	459.5205	0.00312	261.99	499.99	0.00109	1.32900	1.6774
42	97.9899	2861.3	0.20398	450.7152	0.00314	260.22	448.40	0.00106	1.32676	1.5477
44	97.3745	2902.1	0.19984	441.9079	0.00316	258.16	406.84	0.00104	1.32451	1.4456
46	96.7587	2951.9	0.19492	433.1268	0.00318	255.91	372.86	0.00101	1.32227	1.3690
48	96.1426	2995.2	0.19059	424.3983	0.00321	253.52	344.72	0.00099	1.32003	1.3053
50	95.5262	3035.4	0.18654	415.7470	0.00323	251.04	321.14	0.00097	1.31778	1.2531
52	94.9095	3073.2	0.18270	407.1958	0.00325	248.52	301.17	0.00094	1.31554	1.2101
54	94.2927	3109.1	0.17904	398.7661	0.00327	245.98	284.10	0.00092	1.31330	1.1746
56	93.6758	3142.8	0.17557	390.4774	0.00329	243.45	269.39	0.00090	1.31107	1.1450
58	93.0590	3173.6	0.17234	382.3478	0.00331	240.96	256.61	0.00089	1.30883	1.1198
60	92.4426	3205.0	0.16910	374.3937	0.00333	238.83	245.43	0.00087	1.30660	1.0978
65	90.9041	3289.9	0.16078	355.3682	0.00338	234.76	222.87	0.00084	1.30105	1.0552
70	89.3730	3387.9	0.15207	337.7071	0.00341	230.85	205.89	0.00080	1.29553	1.0316
75	87.8545	3501.5	0.14308	321.5186	0.00344	227.26	192.71	0.00077	1.29008	1.0209
80	86.3545	3631.4	0.13402	306.8348	0.00345	224.18	182.23	0.00075	1.28471	1.0178
85	84.8790	3776.9	0.12513	293.6117	0.00344	225.18	173.71	0.00073	1.27943	1.0032
90	83.4336	3935.9	0.11663	281.7316	0.00342	226.16	166.66	0.00072	1.27428	0.9933
95	82.0230	4105.4	0.10869	271.0114	0.00339	227.24	160.73	0.00072	1.26927	0.9856
100	80.4651	4237.4	0.10403	261.4332	0.00343	234.40	126.52	0.00072	1.26375	0.7843
120	75.2353	4924.3	0.08272	230.8886	0.00330	252.31	130.51	0.00074	1.24532	0.8400
140	70.5036	5494.3	0.07124	210.1927	0.00314	257.78	128.61	0.00076	1.22880	0.8601
160	66.3292	5892.5	0.06418	192.0795	0.00298	256.71	124.87	0.00079	1.21435	0.8538
180	62.5400	6008.2	0.06094	173.2180	0.00293	252.74	121.11	0.00083	1.20132	0.8450
200	59.0710	6200.8	0.05954	163.4613	0.00281	248.05	118.00	0.00087	1.18948	0.8278
220	55.9659	6502.8	0.05852	159.4606	0.00261	243.85	115.69	0.00093	1.17895	0.8042
240	53.1923	6734.3	0.05750	153.7984	0.00246	240.73	114.19	0.00098	1.16960	0.7851
260	50.7026	6935.2	0.05654	147.9087	0.00234	238.91	113.39	0.00104	1.16124	0.7707
280	48.4373	7157.2	0.05564	143.0900	0.00223	238.39	113.22	0.00111	1.15368	0.7592
300	46.2315	7199.7	0.05793	139.9443	0.00217	239.04	113.57	0.00119	1.14634	0.7435
350	41.7167	7841.3	0.05595	132.5354	0.00194	244.93	116.24	0.00139	1.13143	0.7238
400	38.0322	8535.3	0.05402	127.0413	0.00176	255.13	120.78	0.00161	1.11936	0.7112
450	34.9644	9258.7	0.05228	122.7753	0.00161	268.21	126.58	0.00185	1.10938	0.7029
500	32.3671	9987.5	0.05083	119.3503	0.00148	283.08	133.26	0.00213	1.10098	0.6970
550	30.1370	10723.0	0.04959	116.5315	0.00138	299.30	140.56	0.00242	1.09380	0.6928
600	28.1996	11462.5	0.04853	114.1675	0.00128	316.52	148.32	0.00274	1.08759	0.6898
700	24.9947	12960.7	0.04675	110.4230	0.00114	353.26	164.75	0.00346	1.07737	0.6861
800	22.4485	14498.6	0.04524	107.5944	0.00102	392.50	181.95	0.00426	1.06930	0.6843
900	20.3744	16089.6	0.04388	105.3896	0.00092	433.85	199.57	0.00516	1.06276	0.6837
1000	18.6513	17742.0	0.04260	103.6291	0.00085	440.56	217.40	0.00567	1.05734	0.7405
1200	15.9516	21230.6	0.04026	101.0104	0.00072	511.13	253.25	0.00750	1.04890	0.7617
1400	13.9327	24953.6	0.03818	99.1723	0.00063	583.30	288.89	0.00954	1.04262	0.7825
1600	12.3659	28815.3	0.03646	97.8224	0.00056	655.28	324.06	0.01177	1.03776	0.8017
1800	11.1146	32773.5	0.03506	96.7950	0.00051	726.63	358.65	0.01419	1.03389	0.8187
2000	10.0923	36887.5	0.03381	95.9893	0.00046	799.36	392.61	0.01680	1.03074	0.8334
2500	8.2006	48404.0	0.03074	94.5528	0.00037	1009.19	474.83	0.02443	1.02493	0.8532
3000	6.8906	63938.3	0.02687	93.4458	0.00032	1343.93	553.83	0.03469	1.02092	0.8341

* TWO-PHASE BOUNDARY

Table 4.1 (continued)
THERMODYNAMIC PROPERTIES OF PARAHYDROGEN (ISOBARS, SI UNITS)

100 MPA ISOBAR

TEMPERATURE DEG. K	VOLUME CU M/KG	ISOTHERM DERIVATIVE CU M-MPA/KG	ISOCHORE DERIVATIVE MPA/K	INTERNAL ENERGY KJ/KG-K	ENTHALPY KJ/KG	ENTROPY KJ/KG-K	CV KJ / KG-K	CP KJ / KG-K	VELOCITY OF SOUND M/S
* 34.169	0.00977	5.2095	1.542	-166.3	810.4	6.279	6.60	8.08	2527
35	0.00979	5.1881	1.535	-161.9	817.2	6.474	6.66	8.18	2525
36	0.00982	5.1618	1.526	-156.5	825.4	6.706	6.73	8.30	2522
37	0.00985	5.1350	1.517	-151.0	833.8	6.935	6.81	8.41	2520
38	0.00988	5.1078	1.508	-145.4	842.2	7.161	6.87	8.52	2517
39	0.00990	5.0801	1.499	-139.7	850.8	7.384	6.93	8.62	2514
40	0.00993	5.0520	1.490	-133.9	859.5	7.603	6.98	8.72	2511
42	0.00999	4.9948	1.472	-122.1	877.1	8.033	7.09	8.90	2505
44	0.01005	4.9365	1.453	-110.1	895.1	8.452	7.19	9.09	2498
46	0.01011	4.8773	1.434	-98.2	913.0	8.848	7.32	9.31	2490
48	0.01017	4.8175	1.415	-85.4	931.8	9.249	7.44	9.51	2481
50	0.01023	4.7573	1.396	-72.3	951.0	9.641	7.55	9.70	2472
52	0.01029	4.6971	1.377	-58.9	970.6	10.025	7.66	9.88	2462
54	0.01036	4.6369	1.359	-45.2	990.5	10.401	7.76	10.06	2453
56	0.01042	4.5771	1.340	-31.0	1011.1	10.775	7.85	10.23	2443
58	0.01048	4.5178	1.321	-16.7	1031.7	11.137	7.93	10.40	2433
60	0.01055	4.4591	1.302	-2.2	1052.7	11.493	8.02	10.56	2423
65	0.01071	4.3169	1.255	35.2	1106.5	12.355	8.26	10.99	2396
70	0.01088	4.1826	1.209	74.3	1162.6	13.185	8.54	11.44	2367
75	0.01106	4.0581	1.162	115.4	1221.0	13.991	8.87	11.92	2336
80	0.01123	3.9448	1.116	158.5	1281.8	14.776	9.23	12.41	2304
85	0.01141	3.8431	1.070	203.9	1345.1	15.543	9.62	12.91	2272
90	0.01160	3.7530	1.024	251.4	1411.0	16.296	10.03	13.41	2240
95	0.01178	3.6737	0.9796	301.2	1479.2	17.034	10.44	13.89	2210
100	0.01197	3.6035	0.9357	353.3	1549.8	17.758	10.86	14.34	2181
120	0.01276	3.3346	0.8066	584.0	1860.0	20.580	12.38	16.19	2088
140	0.01355	3.1987	0.7030	839.3	2194.3	23.155	13.28	17.25	2039
160	0.01437	3.1972	0.6180	1107.5	2544.5	25.493	13.59	17.53	2031
180	0.01517	3.1143	0.5439	1377.3	2893.9	27.551	13.49	17.42	2005
200	0.01597	2.9502	0.4914	1642.2	3239.6	29.372	13.18	17.36	1971
220	0.01684	2.9419	0.4488	1900.6	3584.7	31.017	12.82	17.09	1981
240	0.01768	3.0865	0.4119	2152.3	3920.5	32.478	12.48	16.60	2026
260	0.01852	3.1323	0.3779	2397.5	4249.2	33.794	12.20	16.26	2044
280	0.01934	3.1686	0.3488	2637.2	4571.2	34.988	11.97	15.99	2057
300	0.02024	3.2500	0.3354	2847.0	4870.5	35.991	11.42	15.67	2112
350	0.02234	3.3883	0.2846	3409.3	5643.3	38.373	11.10	15.28	2159
400	0.02443	3.5426	0.2470	3958.2	6400.9	40.398	10.94	15.05	2208
450	0.02650	3.7063	0.2182	4499.3	7149.3	42.162	10.86	14.92	2257
500	0.02856	3.8759	0.1955	5036.7	7892.9	43.728	10.80	14.82	2306
550	0.03061	4.0497	0.1771	5570.9	8632.3	45.137	10.77	14.77	2356
600	0.03266	4.2264	0.1620	6103.7	9369.9	46.419	10.76	14.73	2406
700	0.03674	4.5866	0.1385	7167.4	10841.7	48.690	10.77	14.72	2504
800	0.04082	4.9536	0.1210	8233.9	12315.4	50.656	10.82	14.76	2599
900	0.04488	5.3261	0.1076	9308.1	13796.5	52.399	10.93	14.86	2692
1000	0.04895	5.7030	0.0968	10393.9	15289.2	53.974	11.06	15.01	2781
1200	0.05710	6.4678	0.0808	12616.8	18326.5	56.730	11.42	15.37	2951
1400	0.06525	7.2439	0.0694	14918.2	21443.4	59.136	11.83	15.80	3110
1600	0.07342	8,0286	0.0609	17302.8	24644.6	61.251	12.23	16.21	3262
1800	0.08160	8.8200	0.0542	19770.4	27930.0	63.148	12.59	16.58	3408
2000	0.08978	9.6165	0.0489	22320.2	31298.6	64.988	12.96	16.96	3548
2500	0.11033	11.6239	0.0392	29042.2	40074.7	71.248	14.05	18.11	3871
3000	0.13115	13.6469	0.0328	36488.4	49603.8	78.089	15.91	20.14	4156

* TWO-PHASE BOUNDARY

Table 4.1 (continued)
THERMODYNAMIC PROPERTIES OF PARAHYDROGEN (ISOBARS, SI UNITS)

100 MPA ISOBAR

TEMPERATURE DEG. K	DENSITY KG/CU M	$V(DH/DV)_P$ KJ/KG	$V(DP/DU)_V$ MPA-CU M/KJ X 10²	$-V(DP/DV)_T$ MPA	$-(DV/DT)_P/V$ 1/DEG. K	THERMAL CONDUCTIVITY W/K-M X 10³	VISCOSITY KG/M-S X 10⁷	THERMAL DIFFUSIVITY SQ M/HR	DIELECTRIC CONSTANT	PRANDTL NUMBER
* 34.169	102.3875	2796.3	0.22830	533.3893	0.00289	276.26	937.81	0.00120	1.34287	2.7436
35	102.1417	2825.5	0.22560	529.9203	0.00290	276.34	870.98	0.00119	1.34197	2.5791
36	101.8460	2859.4	0.22250	525.7090	0.00290	276.25	800.91	0.00118	1.34088	2.4065
37	101.5504	2891.7	0.21955	521.4625	0.00291	275.96	740.27	0.00116	1.33980	2.2570
38	101.2550	2921.8	0.21682	517.1855	0.00292	275.51	687.46	0.00115	1.33871	2.1263
39	100.9598	2949.0	0.21436	512.8820	0.00292	274.92	641.22	0.00114	1.33763	2.0108
40	100.6647	2974.8	0.21203	508.5564	0.00293	274.20	600.50	0.00112	1.33655	1.9090
42	100.0751	3024.3	0.20754	499.8546	0.00294	272.38	532.43	0.00110	1.33439	1.7406
44	99.4861	3072.7	0.20314	491.1104	0.00296	270.25	478.19	0.00108	1.33223	1.6086
46	98.8978	3130.6	0.19798	482.3525	0.00297	267.90	434.26	0.00105	1.33008	1.5088
48	98.3099	3181.3	0.19345	473.6076	0.00299	265.41	398.20	0.00102	1.32793	1.4263
50	97.7226	3228.6	0.18922	464.9006	0.00300	262.83	368.22	0.00100	1.32578	1.3586
52	97.1359	3273.1	0.18524	456.2550	0.00302	260.20	343.02	0.00098	1.32364	1.3028
54	96.5496	3315.5	0.18144	447.6927	0.00303	257.54	321.63	0.00095	1.32151	1.2565
56	95.9639	3355.2	0.17786	439.2340	0.00305	254.90	303.31	0.00093	1.31937	1.2178
58	95.3788	3391.6	0.17454	430.8979	0.00307	252.29	287.50	0.00092	1.31725	1.1849
60	94.7943	3428.5	0.17122	422.7017	0.00308	250.06	273.74	0.00090	1.31512	1.1563
65	93.3370	3526.6	0.16276	402.9244	0.00312	245.78	246.24	0.00086	1.30984	1.1009
70	91.8870	3637.2	0.15398	384.3249	0.00315	241.67	225.80	0.00083	1.30459	1.0690
75	90.4480	3763.4	0.14496	367.0497	0.00317	237.86	210.14	0.00079	1.29940	1.0529
80	89.0241	3905.9	0.13589	351.1813	0.00318	234.58	197.82	0.00076	1.29428	1.0468
85	87.6203	4064.2	0.12700	336.7366	0.00318	235.54	187.92	0.00075	1.28924	1.0304
90	86.2414	4236.7	0.11848	323.6679	0.00317	236.46	179.81	0.00074	1.28430	1.0197
95	84.8920	4421.2	0.11049	311.8664	0.00314	237.48	173.06	0.00073	1.27948	1.0120
100	83.5760	4614.4	0.10311	301.1690	0.00311	245.13	132.99	0.00074	1.27479	0.7778
120	78.3682	5246.8	0.08312	261.3250	0.00309	263.75	137.43	0.00075	1.25634	0.8438
140	73.7994	5791.3	0.07176	236.0605	0.00298	269.09	135.46	0.00076	1.24029	0.8682
160	69.5909	6312.1	0.06536	222.4929	0.00278	267.50	131.38	0.00079	1.22563	0.8611
180	65.9363	6578.4	0.06114	205.3481	0.00265	262.90	127.19	0.00082	1.21299	0.8429
200	62.6024	6524.4	0.05955	184.6887	0.00266	257.59	123.64	0.00085	1.20154	0.8333
220	59.3817	6650.9	0.05898	174.6977	0.00257	252.80	120.93	0.00090	1.19054	0.8174
240	56.5559	7036.2	0.05836	174.5602	0.00236	249.16	119.06	0.00096	1.18095	0.7933
260	54.0061	7278.9	0.05737	169.1637	0.00223	246.88	117.94	0.00101	1.17234	0.7768
280	51.7071	7513.4	0.05633	163.8406	0.00213	245.96	117.48	0.00107	1.16461	0.7639
300	49.4193	7504.2	0.05944	160.6133	0.00209	246.25	117.58	0.00114	1.15695	0.7483
350	44.7643	8142.5	0.05726	151.6740	0.00188	251.48	119.78	0.00132	1.14148	0.7276
400	40.9388	8836.4	0.05515	145.0297	0.00170	261.29	124.03	0.00153	1.12888	0.7143
450	37.7361	9561.1	0.05326	139.8609	0.00156	274.23	129.68	0.00175	1.11840	0.7053
500	35.0120	10291.3	0.05168	135.7041	0.00144	289.17	136.33	0.00201	1.10953	0.6989
550	32.6639	11028.1	0.05033	132.2775	0.00134	305.65	143.70	0.00228	1.10193	0.6943
600	30.6169	11768.7	0.04918	129.3991	0.00125	323.29	151.62	0.00258	1.09534	0.6909
700	27.2161	13268.3	0.04725	124.8291	0.00111	361.32	168.59	0.00325	1.08444	0.6867
800	24.5007	14806.7	0.04564	121.3674	0.00100	402.34	186.57	0.00400	1.07580	0.6847
900	22.2795	16397.7	0.04419	118.6628	0.00091	445.92	205.17	0.00485	1.06876	0.6839
1000	20.4277	18050.0	0.04285	116.4995	0.00083	440.56	224.14	0.00517	1.06292	0.7634
1200	17.5141	21538.1	0.04043	113.2774	0.00071	511.13	262.64	0.00684	1.05378	0.7898
1400	15.3253	25260.6	0.03829	111.0148	0.00063	583.30	301.29	0.00867	1.04695	0.8159
1600	13.6206	29121.1	0.03654	109.3547	0.00056	655.28	339.74	0.01069	1.04165	0.8403
1800	12.2556	33076.7	0.03512	108.0935	0.00050	726.59	377.78	0.01287	1.03742	0.8622
2000	11.1379	37185.9	0.03385	107.1068	0.00046	799.20	415.31	0.01523	1.03397	0.8814
2500	9.0641	48652.2	0.03080	105.3609	0.00037	1007.18	506.73	0.02209	1.02758	0.9111
3000	7.6246	63967.6	0.02702	104.0528	0.00031	1333.50	595.02	0.03126	1.02316	0.8988

* TWO-PHASE BOUNDARY

Table 4.2
THERMODYNAMIC PROPERTIES OF COEXISTING GASEOUS AND LIQUID PARAHYDROGEN (SI UNITS)

Temperature, K	Density, kg/m³	V(DH/DV)P, kJ/kg	V(DP/DU)V, MPA-m³/kJ × 10²	−V(DP/DV)T, MPA	(DV/DT)P/V, 1/K	Thermal conductivity, W/K-m × 10³	Viscosity, kg/m-sec × 10⁷	Thermal diffusivity, m²/hr	Dielectric constant	Prandtl number
13.800[a]	77.0235	622.6	0.25654	90.3401	0.01022	72.59	255.05	0.00053	1.25160	2.2359
	0.1254	139.8	0.06684	0.0069	0.07523	12.43	7.36	0.03395	1.00038	0.6230
14	76.8554	619.2	0.25480	88.3211	0.01046	74.62	248.01	0.00054	1.25101	2.1517
	0.1388	141.6	0.06684	0.0077	0.07441	12.54	7.48	0.03084	1.00042	0.6291
15	75.9963	621.9	0.24237	81.5880	0.01111	83.02	217.84	0.00057	1.24799	1.8128
	0.2228	150.5	0.06683	0.0131	0.07089	13.08	8.08	0.01981	1.00067	0.6591
16	75.1030	623.0	0.23130	75.3083	0.01181	88.86	194.15	0.00058	1.24486	1.6074
	0.3385	159.1	0.06680	0.0208	0.06820	13.66	8.68	0.01339	1.00102	0.6891
17	74.1710	613.1	0.22244	68.0612	0.01286	92.71	175.07	0.00057	1.24159	1.4884
	0.4922	167.1	0.06674	0.0313	0.06624	14.27	9.28	0.00943	1.00148	0.7197
18	73.1950	604.3	0.21509	61.7874	0.01394	95.43	159.34	0.00056	0.23818	1.4063
	0.6902	174.6	0.06669	0.0452	0.06495	14.97	9.88	0.00689	1.00208	0.7484
19	72.1693	601.0	0.20874	56.9619	0.01486	97.00	146.10	0.00054	1.23460	1.3451
	0.9395	181.4	0.06665	0.0626	0.06427	15.72	10.49	0.00517	1.00283	0.7783
20	71.0870	596.1	0.20314	52.3516	0.01585	98.40	134.76	0.00053	1.23083	1.2942
	1.2474	187.6	0.06666	0.0840	0.06420	16.64	11.11	0.00399	1.00376	0.8042
20.268[a]	70.7864	587.9	0.20176	50.2079	0.01644	98.92	131.98	0.00052	1.22978	1.2893
	1.3378	189.1	0.06667	0.0903	0.06424	16.94	11.28	0.00375	1.00404	0.8091
21	69.9403	577.8	0.19814	46.2935	0.01754	100.05	124.87	0.00051	1.22684	1.2648
	1.6189	193.0	0.06672	0.1093	0.06471	17.79	11.74	0.00317	1.00489	0.8241
22	68.7200	563.1	0.19347	41.1835	0.01921	100.95	116.11	0.00049	1.22261	1.2444
	2.0711	197.5	0.06686	0.1388	0.06597	19.02	12.38	0.00254	1.00625	0.8481
23	67.4149	540.5	0.18882	35.4854	0.02162	101.20	108.22	0.00046	1.21810	1.2499
	2.6119	201.2	0.06708	0.1719	0.06804	20.36	13.04	0.00205	1.00789	0.8769

[a] Two-phase boundary.

Table 4.2 (continued)
THERMODYNAMIC PROPERTIES OF COEXISTING GASEOUS AND LIQUID PARAHYDROGEN (SI UNITS)

Temperature, K	Density, kg/m³	V(DH/DV)p, kJ/kg	$V(DP/DU)_V$, MPA-m³/kJ × 10²	$-V(DP/DT)_T$, MPA	(DV/DT)p/V, 1/K	Thermal conductivity, W/K-m × 10³	Viscosity, kg/m-sec × 10⁷	Thermal diffusivity, m²/hr	Dielectric constant	Prandtl number
24	66.0112	523.0	0.18378	30.8574	0.02394	100.84	101.01	0.00044	1.21325	1.2543
	3.2548	203.9	0.06739	0.2081	0.07108	21.81	13.73	0.00166	1.00984	0.9123
25	64.4917	504.2	0.17820	26.4945	0.02665	99.93	94.32	0.00042	1.20802	1.2682
	4.0171	205.8	0.06777	0.2460	0.07540	23.43	14.44	0.00135	1.01216	0.9567
26	62.8337	475.1	0.17151	21.4244	0.03116	98.43	88.03	0.00038	1.20233	1.3240
	4.9215	206.9	0.06822	0.2842	0.08144	25.24	15.19	0.00110	1.01491	1.0145
27	61.0065	451.1	0.16381	17.4346	0.03585	96.38	82.02	0.00035	1.19608	1.3764
	5.9999	207.3	0.06868	0.3201	0.09002	27.36	16.00	0.00088	1.01820	1.0916
28	58.9665	418.4	0.15535	13.1188	0.04417	93.78	76.19	0.00031	1.18913	1.5012
	7.2979	207.1	0.06912	0.3498	0.10259	29.91	16.89	0.00069	1.02216	1.1997
29	56.6460	382.6	0.14600	9.1967	0.05763	90.55	70.42	0.00026	1.18125	1.7146
	8.8866	206.2	0.06947	0.3676	0.12220	33.18	17.91	0.00053	1.02703	1.3601
30	53.9303	350.6	0.13555	6.2743	0.07585	86.57	64.57	0.00022	1.17208	1.9834
	10.8872	204.7	0.06965	0.3638	0.15631	37.87	19.17	0.00039	1.03319	1.6195
31	50.5892	310.8	0.12320	3.5282	0.11759	85.85	58.40	0.00017	1.16087	2.4889
	13.5411	202.6	0.06952	0.3205	0.22979	46.48	20.89	0.00027	1.04140	2.0930
32	45.9927	265.8	0.10744	1.3886	0.24599	91.46	51.31	0.00011	1.14555	3.6672
	17.4983	199.1	0.06952	0.2258	0.43710	66.77	23.79	0.00016	1.05373	3.1004
32.976[a]	31.4285	199.0					35.43		1.09795	

[a] Two-phase boundary.

Table 4.2 (continued)
THERMODYNAMIC PROPERTIES OF COEXISTING GASEOUS AND LIQUID PARAHYDROGEN

Temperature, K	Pressure, MPA	Isotherm derivative, m^3/MPA/kg	Isochore derivative, MPA/K	Internal energy, kJ/kg-K	Enthalpy, kJ/kg	Entropy, kJ/kg-K	CV, kJ/kg-K	CP, kJ/kg-K	Velocity of sound, m/sec	Surface tension, N/m × 10^3
13.800[a]	0.0070	1.1729	0.9234	−309.0	−308.9	4.964	4.67	6.36	1264	2.9910
	0.0070	0.0552	0.0005	84.2	140.3	37.520	6.21	10.52	306	
14	0.0079	1.1492	0.9234	−307.6	−307.5	5.057	4.72	6.47	1256	2.9578
	0.0079	0.0558	0.0006	85.3	142.1	37.135	6.21	10.54	308	
15	0.0134	1.0736	0.9064	−300.9	−300.7	5.525	4.92	6.91	1228	2.7921
	0.0134	0.0588	0.0009	90.7	151.1	35.648	6.24	10.67	317	
16	0.0216	1.0027	0.8893	−293.7	−293.4	5.989	5.12	7.36	1200	2.6270
	0.0216	0.0615	0.0014	95.9	159.6	34.303	6.27	10.85	326	
17	0.0329	0.9176	0.8750	−286.0	−285.6	6.453	5.30	7.88	1168	2.4625
	0.0329	0.0637	0.0021	100.7	167.6	33.111	6.32	11.07	334	
18	0.0482	0.8441	0.8612	−277.9	−277.3	6.917	5.47	8.42	1140	2.2987
	0.0482	0.0654	0.0029	105.1	175.0	32.042	6.37	11.34	341	
19	0.0682	0.7893	0.8465	−269.3	−268.4	7.382	5.62	8.93	1120	2.1356
	0.0682	0.0667	0.0040	109.1	181.7	31.073	6.43	11.66	348	
20	0.0935	0.7364	0.8300	−260.2	−258.9	7.851	5.75	9.45	1100	1.9732
	0.0935	0.0674	0.0054	112.7	187.6	30.184	6.49	12.04	354	
20.268[a]	0.1013	0.7093	0.8253	−257.7	−256.2	7.977	5.78	9.66	1089	1.9299
	0.1013	0.0675	0.0058	113.6	189.3	29.967	6.50	12.15	355	
21	0.1250	0.6619	0.8119	−250.5	−248.8	8.323	5.86	10.13	1070	1.8117
	0.1250	0.0675	0.0071	115.8	193.0	29.369	6.55	12.49	359	
22	0.1634	0.5993	0.7913	−240.3	−237.9	8.802	5.95	10.82	1044	1.6510
	0.1634	0.0670	0.0092	118.3	197.2	28.592	6.61	13.03	363	
23	0.2096	0.5264	0.7673	−229.4	−226.3	9.287	6.03	11.69	1010	1.4913
	0.2096	0.0658	0.0117	120.2	200.5	27.855	6.68	13.69	367	

[a] Two-phase boundary.

Table 4.2 (continued)
THERMODYNAMIC PROPERTIES OF COEXISTING GASEOUS AND LIQUID PARAHYDROGEN

Temperature, K	Pressure, MPA	Isotherm derivative, m^3/MPA/kg	Isochore derivative, MPA/K	Internal energy, kJ/kg-K	Enthalpy, kJ/kg	Entropy, kJ/kg-K	CV, kJ/kg-K	CP, kJ/kg-K	Velocity of sound, m/sec	Surface tension, N/m × 10
24	0.2645	0.4675	0.7389	−217.9	−213.8	9.783	6.09	12.52	980	1.3327
	0.2645	0.0639	0.0148	121.4	202.6	27.148	6.74	14.49	371	
25	0.3288	0.4108	0.7060	−205.5	−200.4	10.291	6.14	13.44	948	1.1751
	0.3288	0.0613	0.0186	121.7	203.6	26.461	6.81	15.52	373	
26	0.4035	0.3410	0.6676	−192.3	−185.9	10.813	6.20	14.80	903	1.0189
	0.4035	0.0578	0.0231	121.1	203.1	25.784	6.89	16.85	376	
27	0.4892	0.2858	0.6251	−178.2	−170.1	11.356	6.26	16.17	860	0.8641
	0.4892	0.0533	0.0288	119.3	200.8	25.104	6.99	18.66	377	
28	0.5871	0.2225	0.5794	−162.8	−152.8	11.926	6.33	18.48	806	0.7110
	0.5871	0.0479	0.0359	116.1	196.5	24.409	7.11	21.24	378	
29	0.6978	0.1624	0.5300	−145.9	−133.6	12.535	6.41	22.05	747	0.5599
	0.6978	0.0414	0.0449	110.9	189.5	23.680	7.28	25.19	378	
30	0.8225	0.1163	0.4759	−126.9	−111.7	13.200	6.51	26.59	689	0.4112
	0.8225	0.0334	0.0569	103.2	178.8	22.887	7.50	31.99	378	
31	0.9627	0.0697	0.4149	−104.9	−85.8	13.960	6.66	36.55	619	0.2659
	0.9627	0.0237	0.0736	91.5	162.6	21.976	7.82	46.56	375	
32	1.1198	0.0302	0.3416	−76.7	−52.4	14.917	6.91	65.37	534	0.1254
	1.1198	0.0129	0.0987	72.1	136.1	20.811	8.11	87.02	372	
32.976[a]	1.2928		0.1898	−2.8	38.3	17.565				0.0000

[a] Two-phase boundary.

Table 4.3
SECOND AND THIRD VIRIAL COEFFICIENTS FOR PARAHYDROGEN (VIRIAL COEFFICIENTS)

Temperature, K	B (cm^3/g mol)	C (cm^3/g mol)^2
14	-2.372×10^2	-6.713×10^4
16	-1.992×10^2	-2.265×10^4
18	-1.698×10^2	-7.121×10^3
20	-1.467×10^2	-1.503×10^3
22	-1.281×10^2	5.656×10^2
24	-1.129×10^2	1.323×10^3
26	-1.003×10^2	1.580×10^3
28	-8.969×10^1	1.640×10^3
30	-8.066×10^1	1.615×10^3
32	-7.290×10^1	1.550×10^3
34	-6.615×10^1	1.467×10^3
36	-6.025×10^1	1.376×10^3
38	-5.504×10^1	1.290×10^3
40	-5.042×10^1	1.214×10^3
45	-4.086×10^1	1.066×10^3
50	-3.343×10^1	9.638×10^2
60	-2.264×10^1	8.351×10^2
70	-1.522×10^1	7.510×10^2
80	-9.824×10	6.878×10^2
90	-5.725×10	6.405×10^2
100	-2.510×10	6.085×10^2
120	2.144×10	5.491×10^2
140	5.342×10	5.063×10^2
160	7.654×10	4.726×10^2
180	9.388×10	4.449×10^2
200	1.073×10^1	4.217×10^2
250	1.300×10^1	3.768×10^2
300	1.438×10^1	3.438×10^2
400	1.589×10^1	2.976×10^2
500	1.660×10^1	2.661×10^2

Table 4.4
FIXED POINT PROPERTIES OF PARAHYDROGEN

Properties	Triple point: Solid	Triple point: Liquid	Triple point: Vapor	Normal boiling point: Liquid	Normal boiling point: Vapor	Critical point[a]	Standard conditions: STP(0°C)	Standard conditions: NTP(20°C)	References and notes[b]
Temperature (K)	13.803	13.803	13.803	20.268		32.976	273.15	293.15	1
Pressure (mm Hg)	52.82	52.82	52.85	760		9696.8	760	760	1
Density (mol/cm³) × 10³	42.91[277]	38.207	0.0623	35.11	0.6636	15.59	0.5459	0.04155	1, 278
Specific volume (cm³/mol) × 10^{-3}	0.02330[277]	0.026173	16.057	0.028482	1.5069	0.064144	22.425	24.069	1, 278
Compressibility factor, $Z = \frac{PV}{RT}$		0.001606	0.9850	0.01712	0.9061	0.3025	1.0005	1.0006	f
Heats of fusion and vaporization (J/mol)	117.5[114]	905.5		898.3		0	–	–	g
Specific heat (J/mol-K)									
C_s, at saturation	5.73[279]	13.85	–46.94	18.91	–33.28	(Very large)	–	–	106, 279
C_p, at constant pressure	–	13.13	21.20	19.53	24.50	(Very large)	30.35	30.02	1, 278
C_v, at constant volume	–	9.50	12.52	11.57	13.11	19.7	21.87	21.70	1, 278
Specific heat ratio, $\gamma = C_p/C_v$	–	1.382	1.693	1.688	1.869	(Large)	1.388	1.383	h
Enthalpy (J/mol)	–740.2[114]	–622.7	282.8	–516.6	381.7	77.6	7656.6	8260.6	1, 278[i]
Internal energy (J/mol)	–740.4[h]	–622.9	169.8	–519.5	229.0	5.7	5384.5	5822.0	1.278[i]
Entropy (J/mol-K)	1.49[114]	10.00	75.63	16.08	60.41	35.4	127.77	129.90	1,278[i]
Velocity of sound (m/sec)		1273	305	1093	35.5	350	1246	1294	1, 2
Viscosity, μ	–								
N-sec/m² × 10³	–	0.026	0.00073	0.0133	0.0011	0.0035	0.00834	0.00875	127, 280, 281
Centipoise[c]	–	0.026	0.00073	0.0133	0.0011	0.0035	0.00834	0.00875	127, 280, 281
Thermal conductivity (mW/cm-K), k	9.0[282]	0.73	0.104	1.00	0.163	e	1.826	1.897	123, 127, 280
Prandtl number, $N_{pr} = \mu C_p/k$	–	2.33	0.738	1.288	0.8202		0.6873	0.6867	h
Dielectric constant, ϵ	1.286	1.252	1.00038	1.230	1.0040	1.098	1.00027	1.00026	283
Index of refraction, $n = \sqrt{\epsilon}$[d]	1.134	1.119	1.00019	1.109	1.0020	1.048	1.00013	1.00012	110[j]
Surface tension (N/m) × 10³	–	2.99	–	1.93	–	0	–	–	121
Equiv. vol/vol liquid at NBT	0.8181	0.9190	563.8	1	52.91	2.252	787.4	845.1	k

Note: Gas constant[1] – R = 62, 362.7 cm³-mm Hg/mol-K; Molecular weight[107] = 2.01594; Mole = gram mole.

[a] Values in parentheses are estimates.
[b] Apply to all items on the line except for specific values separately referenced in parentheses.
[c] Units for poise are g/cm-sec.
[d] Long wavelengths.
[e] Anomalously large.
[f] Compressibility factor calculated from the tabulated data using the gas constant, R, from Reference 1.
[g] Heats of vaporization calculated from enthalpy differences given in References 1 and 278.
[h] Calculated form property values given in this table.
[i] Base point (zero values) for enthalpy, internal energy, and entropy are 0 K for the ideal gas at 1 atm pressure.
[j] Index of refraction calculated from the dielectric constant data given in Reference 283, in good agreement with the values reported in Reference 110.
[k] Ratio of specific volumes reported in References 1 and 278 or 5 and 284

Table 4.5
FIXED POINT PROPERTIES OF NORMAL HYDROGEN

Properties	Conditions								References and notes[b]
	Triple point			Normal boiling point		Critical point[a]	Standard conditions		
	Solid	Liquid	Vapor	Liquid	Vapor		STP(0°C)	NTP(20°C)	
Temperature (K)	13.957[c]	13.957[c]	13.957[c]	20.390[c]		33.19[c]	273.15	293.15	106
Pressure (mm Hg)	54.04	54.04	54.04	760		9865	760	760	106
Density (mol/cm³) × 10³	43.01[277]	38.30	0.0644	35.20	0.6604	14.94	0.04460	0.04155	106, 284
Specific volume (cm³/mol) × 10⁻³	0.02325[277]	0.026108	15.519	0.028409	1.5143	0.066949	22.423	24.066	106, 284
Compressibility factor, $Z = \frac{PV}{RT}$	–	0.001621	0.9635	0.01698	0.9051	0.3191	1.00042	1.00049	d
Heats of fusion and vaporization (J/mol)	117.1	911.3		899.1		0	–	–	114[e]
Specific heat (J/mol-K)									
C_s, at saturation	5.73	13.85	-46.94	18.91	-33.28	(Very large)	–	–	106, 279
C_p, at constant pressure	–	13.23	21.22	19.70	24.60	(Very large)	28.59	28.89	106, 279, 284
C_v, at constant volume	–	9.53	12.52	11.60	13.2	(19.7)	20.30	20.40	106, 279, 284
Specific heat ratio, $\gamma = C_p/C_v$	–	1.388	1.695	1.698	1.863	(Large)	1.408	1.416	f
Enthalpy (J/mol)	321.6	438.7	1350.0	548.3	1447.4	1164	7749.2	8324.1	106, 284[g]
Internal energy (J/mol)	317.9[f]	435.0	1234.8	545.7	1294.0	–	5477.1	5885.4	106, 284[g]
Entropy (J/mol-K)	20.3	28.7	93.6	34.92	78.94	54.57	139.59	141.62	106, 284[g]
Velocity of sound (m/sec)	–	1282	307	1101	357	–	1246	1294	279, 285
Viscosity, μ									
N-sec/m² × 10³	–	0.0256	0.00074	0.0133	0.00111	(0.0035)	0.00834	0.00875	285
Centipoise[h]	–	0.0256	0.00074	0.0133	0.00111	(0.0035)	0.00834	0.00875	285
Thermal conductivity (mW/cm-K), k	9.0[282]	0.74	0.105	1.00	0.165	i	1.739	1.842	279, 280
Prandtl number, $N_{pr} = \mu C_p/k$	–	2.27	0.742	1.30	0.821	–	0.680	0.681	f
Dielectric constant, ϵ	1.287	1.253	1.00039	1.231	1.0040	1.0937	1.000271	1.000253	283
Index of refraction, $n = \sqrt{\epsilon j}$	1.134	1.119	1.000196	1.1093	1.0020	1.0458	1.000136	1.000126	k
Surface tension (N/m) × 10³	–	3.00	–	1.94	–	0	–	–	121
Equiv. vol./vol. liquid at NBT	0.8184	0.9190	546.3	1	53.30	2.357	789.3	847.1	l

Note: Gas constant[1] – R = 62, 362.7 cm³-mm Hg/mol-K; Molecular weight[107] = 2.01594; Mole = gram mole.

[a] Values in parentheses are estimates.

[b] Apply to all items on the line except for specific values separately referenced in parentheses.

[c] These temperatures given in Reference 106 are based on the IPTS-1968 temperature scale. To compare with the corresponding temperatures given in Reference 1 for parahydrogen which are based on the NBS-1955 temperature scale, 0.01 K should be subtracted, i.e., use 13.947, 20.380, and 33.8 respectively for the triple point, normal boiling point, and critical point when determining property differences between normal and parahydrogen.

[d] Compressibility factor calculated from the tabulated data using the gas constant, R, from Reference 1.

[e] Heats of fusion and vaporization calculated from enthalpy differences and as reported in References 106, 114, 284.

[f] Calculated from property values given in this table.

[g] Base point (zero values) for enthalpy, internal energy, and entropy are 0 K for the ideal gas at 1 atm pressure.

[h] Units for poise are g/cm-sec.

[i] Anomalously large.

[j] Long wavelengths.

[k] Index of refraction calculated from the dielectric constant data given in Reference 283, in good agreement with the values reported in Reference 110.

[l] Ratio of specific volumes reported in References 1 and 278 or 106 and 284.

Table 4.6
REFRACTIVE INDEX OF SATURATED LIQUID PARAHYDROGEN AT THREE WAVELENGTHS

Temperature, K	Wavelength 4358 Å	5462 Å	6939 Å
15	1.1222	1.1204	1.1191
16	1.1208	1.1189	1.1177
17	1.1192	1.1174	1.1162
18	1.1176	1.1158	1.1146
19	1.1159	1.1141	1.1129
20	1.1141	1.1124	1.1112
21	1.1123	1.1106	1.1095
22	1.1103	1.1086	1.1075
23	1.1082	1.1065	1.1054
24	1.1059	1.1042	1.1032
25	1.1034	1.1018	1.1008
26	1.1007	1.0991	1.0981
27	1.0977	1.0962	1.0952
28	1.0944	1.0929	1.0920
29	1.0906	1.0892	1.0883
30	1.0862	1.0849	1.0840

Table 4.7
JOULE-THOMSON INVERSION CURVE OF PARAHYDROGEN

Temperature		Pressure		Density	
K	R	atm	psia	$mol/cm^3 \times 10^3$	lb_m/ft^3
28	50.4	9.87	145.1	30.06	3.783
29	52.2	15.05	221.2	29.90	3.763
30	54.0	20.08	295.1	29.73	3.742
31	55.8	25.01	367.6	29.56	3.720
32	57.6	29.85	438.7	29.40	3.700
34	61.2	39.16	575.5	29.05	3.656
36	64.8	48.06	706.3	28.70	3.612
40	72.0	64.59	949.2	27.99	3.523
50	90.0	98.93	1454	26.16	3.292
60	108.0	124.4	1828	24.30	3.058
80	144.0	153.5	2256	20.58	2.590
100	180.0	161.4	2372	17.04	2.145
120	216.0	162.1	2353	14.12	1.777
140	252.0	140.5	2064	10.86	1.367
160	288.0	102.2	1502	7.176	0.9031
180	324.0	50.97	749.1	3.321	0.4179
200	360.0	0.54	8.6	0.036	0.0045

Table 4.8

LATENT HEAT OF SUBLIMATION OF PARAHYDROGEN

Temperature, K	Heat of sublimation, cal/mol
13.813	244.90
13.000	242.30
12.000	238.71
11.000	234.76
10.000	230.55
9.000	226.12
8.000	221.53
7.000	216.82
6.000	212.02
5.000	207.15
4.000	202.24
3.000	197.30
2.000	192.34
1.000	187.37

Table 4.9

VAPOR PRESSURE OF PARAHYDROGEN BELOW THE TRIPLE POINT

Temperature, K	Pressure, mm Hg
13.813	52.89
13.000	30.13
12.000	13.78
11.000	5.567
10.000	1.917
9.000	5.343×10^{-1}
8.000	1.118×10^{-1}
7.000	1.561×10^{-2}
6.000	1.198×10^{-3}
5.000	3.570×10^{-5}
4.000	2.080×10^{-7}
3.000	4.832×10^{-11}
2.000	3.985×10^{-18}
1.000	8.255×10^{-39}

Table 4.10

ORTHO-PARA CONCENTRATIONS AT EQUILIBRIUM

Temperature, K	Percentage in para form for H_2
10	99.9999
20	99.821
20.39	99.789
23.57	–
30	97.021
33.10	95.034
40	88.727
50	77.054
60	65.569
70	55.991
80	48.537
90	42.882
100	38.620
120	32.959
150	28.603
200	25.974
250	25.264
298.16	25.075
300	25.072
350	25.019
400	25.005
500	25.000

Table 4.11

HEATS OF CONVERSION FROM NORMAL TO PARAHYDROGEN

Temperature, K	cal/mol
10	253.9865
20	253.987
20.39	253.986
30	253.986
33.1	253.986
40	253.976
50	253.845
60	253.212
70	251.400
80	247.623
90	241.275
100	232.079
120	205.857
150	155.385
200	78.91
250	33.98
298.16	13.76
300	13.28

Table 4.12
THERMODYNAMIC PROPERTIES OF NORMAL HYDROGEN

0.01 MPA ISOBAR

TEMPERATURE DEG. K	VOLUME CU M/KG	ISOTHERM DERIVATIVE CU M-MPA/KG	ISOCHORE DERIVATIVE MPA/K	INTERNAL ENERGY KJ/KG-K	ENTHALPY KJ/KG	ENTROPY KJ/KG-K	CV KJ / KG-K	CP KJ / KG-K	VELOCITY OF SOUND M/S
* 13.804	0.01298	1.1628	0.9233	218.2	218.4	14.081	4.67	6.38	1260
14	0.01301	1.1302	0.9233	219.5	219.7	14.175	4.72	6.50	1248
* 14.430	0.01307	1.1173	0.9196	222.4	222.5	14.374	4.80	6.67	1245
* 14.430	5.83386	0.0571	0.0007	614.8	673.2	45.613	6.22	10.59	312
15	6.07554	0.0596	0.0007	618.4	679.2	46.023	6.22	10.56	318
16	6.49796	0.0640	0.0006	624.8	689.7	46.703	6.21	10.52	329
17	6.91887	0.0683	0.0006	631.0	700.2	47.339	6.20	10.49	340
18	7.33861	0.0725	0.0006	637.3	710.7	47.938	6.20	10.46	350
19	7.75743	0.0768	0.0005	643.6	721.2	48.503	6.20	10.44	360
20	8.17550	0.0810	0.0005	649.8	731.6	49.038	6.20	10.43	369
21	8.59295	0.0852	0.0005	656.1	742.0	49.548	6.20	10.42	379
22	9.00988	0.0895	0.0005	662.3	752.4	50.033	6.20	10.41	388
23	9.42636	0.0937	0.0004	668.6	762.8	50.496	6.20	10.40	396
24	9.84245	0.0979	0.0004	674.8	773.2	50.938	6.19	10.39	405
25	10.25822	0.1021	0.0004	681.0	783.6	51.362	6.19	10.38	414
26	10.67369	0.1062	0.0004	687.3	794.0	51.767	6.19	10.38	422
27	11.08891	0.1104	0.0004	693.5	804.4	52.159	6.19	10.37	430
28	11.50391	0.1146	0.0004	699.7	814.7	52.535	6.19	10.37	438
29	11.91870	0.1188	0.0003	705.9	825.1	52.899	6.19	10.36	446
30	12.33332	0.1229	0.0003	712.1	835.5	53.250	6.19	10.36	453
31	12.74778	0.1271	0.0003	718.3	845.8	53.590	6.19	10.35	461
32	13.16210	0.1313	0.0003	724.5	856.2	53.919	6.19	10.35	468
33	13.57629	0.1354	0.0003	730.7	866.5	54.237	6.19	10.35	476
34	13.99036	0.1396	0.0003	737.0	876.9	54.546	6.19	10.35	483
35	14.40432	0.1437	0.0003	743.2	887.2	54.846	6.19	10.34	490
36	14.81819	0.1479	0.0003	749.4	897.6	55.138	6.19	10.34	497
37	15.23198	0.1520	0.0003	755.6	907.9	55.421	6.19	10.34	504
38	15.64568	0.1562	0.0003	761.8	918.2	55.697	6.19	10.34	511
39	16.05931	0.1603	0.0003	768.0	928.6	55.966	6.19	10.34	517
40	16.47288	0.1645	0.0003	774.2	938.9	56.227	6.19	10.34	524
42	17.29983	0.1728	0.0002	786.6	959.6	56.732	6.19	10.34	537
44	18.12657	0.1811	0.0002	799.0	980.3	57.212	6.20	10.34	550
46	18.95313	0.1893	0.0002	811.4	1001.0	57.672	6.20	10.34	562
48	19.77953	0.1976	0.0002	823.9	1021.6	58.112	6.20	10.34	574
50	20.60580	0.2059	0.0002	836.3	1042.3	58.534	6.21	10.35	586
52	21.43196	0.2142	0.0002	848.7	1063.0	58.940	6.22	10.35	597
54	22.25800	0.2224	0.0002	861.2	1083.7	59.331	6.23	10.36	608
56	23.08396	0.2307	0.0002	873.6	1104.5	59.709	6.24	10.37	619
58	23.90983	0.2390	0.0002	886.1	1125.2	60.073	6.25	10.38	630
60	24.73564	0.2472	0.0002	898.6	1146.0	60.425	6.27	10.40	641
65	26.79987	0.2679	0.0002	930.1	1198.1	61.258	6.31	10.45	666
70	28.86380	0.2886	0.0001	961.8	1250.5	62.035	6.38	10.51	689
75	30.92747	0.3092	0.0001	994.0	1303.2	62.763	6.46	10.59	712
80	32.99095	0.3299	0.0001	1026.5	1356.4	63.449	6.56	10.69	733
85	35.05427	0.3505	0.0001	1059.6	1410.2	64.101	6.67	10.80	753
90	37.11746	0.3711	0.0001	1093.3	1464.5	64.722	6.80	10.92	772
95	39.18054	0.3918	0.0001	1127.6	1519.4	65.316	6.93	11.06	791
100	41.24353	0.4124	0.0001	1162.6	1575.0	65.886	7.07	11.20	808
120	49.49480	0.4950	0.0001	1309.8	1804.8	67.976	7.66	11.78	873
140	57.74534	0.5775	0.0001	1468.7	2046.2	69.836	8.22	12.34	931
160	65.99544	0.6600	0.0001	1638.0	2298.0	71.516	8.70	12.82	986
180	74.24525	0.7425	0.0001	1816.0	2558.4	73.049	9.09	13.21	1039
200	82.49487	0.8250	0.0001	2001.0	2826.0	74.459	9.40	13.53	1089
220	90.74435	0.9075	0.0000	2191.7	3099.1	75.760	9.65	13.78	1138
240	98.99373	0.9900	0.0000	2386.7	3376.7	76.968	9.85	13.97	1185
260	107.24303	1.0725	0.0000	2585.2	3657.6	78.093	9.99	14.12	1231
280	115.49227	1.1550	0.0000	2786.3	3941.2	79.143	10.10	14.23	1275
300	123.73215	1.2374	0.0000	2989.2	4226.6	80.127	10.18	14.31	1319
350	144.35340	1.4436	0.0000	3501.8	4945.3	82.343	10.30	14.42	1422
400	164.97452	1.6498	0.0000	4018.3	5668.0	84.273	10.35	14.47	1519
450	185.59556	1.8560	0.0000	4536.5	6392.5	85.980	10.37	14.49	1611
500	206.21654	2.0622	0.0000	5055.6	7117.8	87.507	10.39	14.51	1697
550	226.83749	2.2685	0.0000	5575.3	7843.7	88.891	10.40	14.53	1780
600	247.45841	2.4747	0.0000	6096.1	8570.7	90.153	10.42	14.55	1858
700	288.70020	2.8871	0.0000	7141.0	10028.0	92.403	10.48	14.60	2006
800	329.94195	3.2995	0.0000	8193.5	11492.9	94.361	10.57	14.69	2142
900	371.18367	3.7119	0.0000	9256.9	12968.7	96.097	10.70	14.82	2268
1000	412.42538	4.1243	0.0000	10334.0	14458.2	97.664	10.86	14.98	2386
1200	494.90891	4.9492	0.0000	12544.2	17493.3	100.412	11.24	15.36	2601
1400	577.39658	5.7740	0.0000	14838.4	20612.3	102.822	11.70	15.82	2795
1600	659.93251	6.5988	0.0000	17233.3	23832.7	104.978	12.30	16.44	2969
1800	742.78144	7.4237	0.0000	19799.6	27227.4	106.971	13.52	17.71	3119
2000	826.97486	8.2485	0.0000	22778.4	31048.1	108.991	16.56	20.99	3232
2500	1072.23177	10.3106	0.0000	37496.2	48218.6	116.495	49.17	57.67	3474
3000	1538.60332	12.3727	0.0000	88375.8	103761.9	136.424	145.32	179.51	3904

* TWO-PHASE BOUNDARY

Table 4.12 (continued)
THERMODYNAMIC PROPERTIES OF NORMAL HYDROGEN

0.01 MPA ISOBAR

TEMPERATURE DEG. K	DENSITY KG/CU M	$V(DH/DV)_P$ KJ/KG	$V(DP/DU)_V$ MPA-CU M/KJ $\times 10^2$	$-V(DP/DV)_T$ MPA	$(DV/DT)_P/V$ 1/DEG. K	THERMAL CONDUCTIVITY W/K-M $\times 10^3$	VISCOSITY KG/M-S $\times 10^7$	THERMAL DIFFUSIVITY SQ M/HR	DIELECTRIC CONSTANT	PRANDTL NUMBER
* 13.804	77.0167	618.8	0.25650	89.5531	0.01031	72.63	254.83	0.00053	1.25158	2.2385
14	76.8435	611.8	0.25478	86.8501	0.01063	74.61	247.76	0.00054	1.25097	2.1598
* 14.430	76.4900	620.0	0.25021	85.4586	0.01076	78.59	234.09	0.00055	1.24972	1.9873
* 14.430	0.1714	145.5	0.06684	0.0098	0.07279	12.77	7.74	0.02532	1.00052	0.6420
15	0.1646	151.5	0.06684	0.0098	0.06968	13.01	8.07	0.02696	1.00050	0.6546
16	0.1539	162.1	0.06684	0.0098	0.06488	13.45	8.64	0.02992	1.00046	0.6756
17	0.1445	172.6	0.06683	0.0099	0.06074	13.89	9.21	0.03300	1.00044	0.6949
18	0.1363	183.1	0.06681	0.0099	0.05713	14.40	9.77	0.03635	1.00041	0.7099
19	0.1289	193.6	0.06680	0.0099	0.05394	14.90	10.32	0.03985	1.00039	0.7235
20	0.1223	204.1	0.06679	0.0099	0.05109	15.53	10.87	0.04382	1.00037	0.7303
21	0.1164	214.6	0.06678	0.0099	0.04855	16.32	11.41	0.04846	1.00035	0.7286
22	0.1110	225.0	0.06677	0.0099	0.04625	17.11	11.95	0.05334	1.00033	0.7267
23	0.1061	235.4	0.06676	0.0099	0.04416	17.91	12.48	0.05844	1.00032	0.7247
24	0.1016	245.9	0.06675	0.0099	0.04226	18.70	13.01	0.06378	1.00031	0.7225
25	0.0975	256.3	0.06675	0.0099	0.04051	19.49	13.52	0.06934	1.00029	0.7202
26	0.0937	266.7	0.06674	0.0100	0.03891	20.28	14.04	0.07510	1.00028	0.7181
27	0.0902	277.1	0.06674	0.0100	0.03743	21.07	14.54	0.08109	1.00027	0.7159
28	0.0869	287.4	0.06673	0.0100	0.03606	21.85	15.04	0.08730	1.00026	0.7136
29	0.0839	297.8	0.06673	0.0100	0.03479	22.64	15.54	0.09374	1.00025	0.7112
30	0.0811	308.2	0.06672	0.0100	0.03361	23.42	16.03	0.10040	1.00024	0.7088
31	0.0784	318.6	0.06672	0.0100	0.03251	24.12	16.51	0.10690	1.00024	0.7089
32	0.0760	328.9	0.06672	0.0100	0.03147	24.82	16.99	0.11360	1.00023	0.7087
33	0.0737	339.3	0.06672	0.0100	0.03050	25.51	17.47	0.12049	1.00022	0.7085
34	0.0715	349.6	0.06672	0.0100	0.02959	26.22	17.93	0.12762	1.00022	0.7078
35	0.0694	360.0	0.06672	0.0100	0.02873	26.92	18.40	0.13496	1.00021	0.7070
36	0.0675	370.4	0.06670	0.0100	0.02793	27.63	18.86	0.14248	1.00020	0.7061
37	0.0657	380.8	0.06669	0.0100	0.02716	28.33	19.31	0.15021	1.00020	0.7051
38	0.0639	391.1	0.06669	0.0100	0.02644	29.03	19.76	0.15815	1.00019	0.7039
39	0.0623	401.5	0.06668	0.0100	0.02575	29.74	20.21	0.16627	1.00019	0.7027
40	0.0607	411.9	0.06667	0.0100	0.02510	30.44	20.65	0.17460	1.00018	0.7015
42	0.0578	432.6	0.06665	0.0100	0.02390	31.83	21.53	0.19177	1.00017	0.6990
44	0.0552	453.4	0.06662	0.0100	0.02280	33.22	22.38	0.20971	1.00017	0.6964
46	0.0528	474.2	0.06659	0.0100	0.02180	34.61	23.22	0.22839	1.00016	0.6938
48	0.0506	495.1	0.06655	0.0100	0.02089	35.99	24.05	0.24780	1.00015	0.6910
50	0.0485	516.1	0.06649	0.0100	0.02005	37.36	24.86	0.26790	1.00015	0.6884
52	0.0467	537.2	0.06640	0.0100	0.01927	38.73	25.66	0.28864	1.00014	0.6859
54	0.0449	558.4	0.06630	0.0100	0.01855	40.09	26.45	0.31003	1.00014	0.6836
56	0.0433	579.7	0.06618	0.0100	0.01789	41.43	27.22	0.33197	1.00013	0.6815
58	0.0418	601.3	0.06603	0.0100	0.01727	42.75	27.99	0.35440	1.00013	0.6798
60	0.0404	623.0	0.06587	0.0100	0.01669	44.11	28.74	0.37778	1.00012	0.6775
65	0.0373	678.2	0.06535	0.0100	0.01540	47.57	30.58	0.43934	1.00011	0.6716
70	0.0346	735.0	0.06467	0.0100	0.01430	50.68	32.36	0.50105	1.00010	0.6712
75	0.0323	793.7	0.06385	0.0100	0.01334	53.66	34.09	0.56408	1.00010	0.6729
80	0.0303	854.4	0.06291	0.0100	0.01251	55.99	35.77	0.62219	1.00009	0.6829
85	0.0285	917.4	0.06185	0.0100	0.01177	58.96	37.41	0.68903	1.00009	0.6852
90	0.0269	982.7	0.06070	0.0100	0.01112	61.64	39.01	0.75391	1.00008	0.6914
95	0.0255	1049.9	0.05955	0.0100	0.01053	64.09	40.57	0.81772	1.00008	0.6998
100	0.0242	1119.3	0.05835	0.0100	0.01000	76.11	45.74	1.00922	1.00007	0.6729
120	0.0202	1413.9	0.05387	0.0100	0.00833	94.27	54.09	1.42538	1.00006	0.6761
140	0.0173	1728.1	0.05019	0.0100	0.00714	107.65	59.20	1.81288	1.00005	0.6788
160	0.0152	2051.4	0.04743	0.0100	0.00625	118.63	63.00	2.19814	1.00005	0.6809
180	0.0135	2378.4	0.04539	0.0100	0.00556	128.13	66.17	2.59175	1.00004	0.6824
200	0.0121	2705.4	0.04387	0.0100	0.00500	136.58	69.01	2.99867	1.00004	0.6834
220	0.0110	3031.2	0.04273	0.0100	0.00455	144.29	71.65	3.42120	1.00003	0.6842
240	0.0101	3353.1	0.04189	0.0100	0.00417	151.36	74.18	3.86109	1.00003	0.6847
260	0.0093	3670.9	0.04127	0.0100	0.00385	157.94	76.63	4.31898	1.00003	0.6850
280	0.0087	3984.0	0.04082	0.0100	0.00357	164.11	79.03	4.79584	1.00003	0.6851
300	0.0081	4292.7	0.04050	0.0100	0.00333	169.97	81.39	5.29127	1.00002	0.6852
350	0.0069	5048.8	0.04004	0.0100	0.00286	183.62	87.21	6.61534	1.00002	0.6851
400	0.0061	5790.0	0.03985	0.0100	0.00250	196.44	92.95	8.06046	1.00002	0.6849
450	0.0054	6522.9	0.03977	0.0100	0.00222	208.91	98.62	9.62527	1.00002	0.6846
500	0.0048	7257.1	0.03970	0.0100	0.00200	221.06	104.23	11.30738	1.00001	0.6843
550	0.0044	7990.6	0.03964	0.0100	0.00182	233.14	109.79	13.10500	1.00001	0.6841
600	0.0040	8728.4	0.03957	0.0100	0.00167	245.19	115.28	15.01522	1.00001	0.6839
700	0.0035	10222.3	0.03936	0.0100	0.00143	269.26	126.07	19.16391	1.00001	0.6837
800	0.0030	11754.6	0.03902	0.0100	0.00125	293.59	136.61	23.73441	1.00001	0.6837
900	0.0027	13338.6	0.03856	0.0100	0.00111	318.41	146.90	28.70892	1.00001	0.6837
1000	0.0024	14983.1	0.03798	0.0100	0.00100	440.56	156.95	43.65775	1.00001	0.5338
1200	0.0020	18437.5	0.03669	0.0100	0.00083	511.07	176.42	59.26423	1.00001	0.5304
1400	0.0017	22150.2	0.03526	0.0100	0.00071	584.47	195.11	76.78772	1.00001	0.5282
1600	0.0015	26303.1	0.03352	0.0100	0.00063	668.83	213.14	96.64986	1.00000	0.5239
1800	0.0013	31869.2	0.03053	0.0100	0.00056	802.94	230.62	121.20117	1.00000	0.5088
2000	0.0012	41867.6	0.02497	0.0100	0.00050	1102.50	247.74	156.38992	1.00000	0.4716
2500	0.0009	138643.2	0.00872	0.0096	0.00042	4540.27	292.93	303.88519	1.00000	0.3721
3000	0.0006	433048.8	0.00353	0.0080	0.00041	15311.58	346.27	472.46613	1.00000	0.4059

* TWO-PHASE BOUNDARY

Table 4.12 (continued)
THERMODYNAMIC PROPERTIES OF NORMAL HYDROGEN

0.02 MPA ISOBAR

TEMPERATURE DEG. K	VOLUME CU M/KG	ISOTHERM DERIVATIVE CU M-MPA/KG	ISOCHORE DERIVATIVE MPA/K	INTERNAL ENERGY KJ/KG-K	ENTHALPY KJ/KG	ENTROPY KJ/KG-K	CV KJ / KG-K	CP KJ / KG-K	VELOCITY OF SOUND M/S
* 13.807	0.01298	1.1634	0.9234	218.2	218.5	14.082	4.67	6.38	1260
14	0.01301	1.1318	0.9234	219.5	219.8	14.173	4.72	6.50	1249
15	0.01315	1.0325	0.9076	226.2	226.5	14.636	4.92	6.99	1211
* 15.832	0.01329	1.0154	0.8919	232.2	232.5	15.026	5.09	7.28	1205
* 15.832	3.15931	0.0610	0.0013	622.2	685.4	43.633	6.27	10.81	325
16	3.19578	0.0618	0.0013	623.3	687.2	43.747	6.26	10.80	326
17	3.41127	0.0663	0.0012	629.7	697.9	44.399	6.24	10.71	337
18	3.62529	0.0708	0.0011	636.1	708.6	45.009	6.23	10.64	348
19	3.83822	0.0752	0.0011	642.4	719.2	45.583	6.22	10.59	358
20	4.05030	0.0795	0.0010	648.8	729.8	46.126	6.21	10.56	368
21	4.26168	0.0839	0.0010	655.1	740.3	46.641	6.21	10.53	377
22	4.47250	0.0882	0.0009	661.4	750.8	47.131	6.21	10.51	386
23	4.68285	0.0925	0.0009	667.7	761.3	47.598	6.21	10.49	395
24	4.89279	0.0967	0.0008	674.0	771.8	48.044	6.20	10.47	404
25	5.10238	0.1010	0.0008	680.2	782.3	48.471	6.20	10.46	413
26	5.31167	0.1052	0.0008	686.5	792.7	48.879	6.20	10.44	421
27	5.52070	0.1095	0.0008	692.8	803.2	49.274	6.20	10.43	429
28	5.72950	0.1137	0.0007	699.0	813.6	49.652	6.20	10.42	437
29	5.93809	0.1179	0.0007	705.3	824.0	50.018	6.20	10.41	445
30	6.14651	0.1221	0.0007	711.5	834.4	50.371	6.20	10.41	453
31	6.35476	0.1263	0.0007	717.7	844.8	50.712	6.20	10.40	460
32	6.56286	0.1305	0.0006	724.0	855.2	51.042	6.20	10.39	468
33	6.77084	0.1347	0.0006	730.2	865.6	51.361	6.20	10.39	475
34	6.97869	0.1389	0.0006	736.4	876.0	51.672	6.19	10.38	483
35	7.18644	0.1431	0.0006	742.6	886.4	51.973	6.19	10.38	490
36	7.39410	0.1473	0.0006	748.9	896.8	52.265	6.19	10.37	497
37	7.60166	0.1515	0.0005	755.1	907.1	52.549	6.20	10.37	504
38	7.80915	0.1556	0.0005	761.3	917.5	52.826	6.19	10.37	510
39	8.01656	0.1598	0.0005	767.5	927.9	53.095	6.20	10.37	517
40	8.22390	0.1640	0.0005	773.8	938.2	53.358	6.20	10.36	524
42	8.63841	0.1723	0.0005	786.2	959.0	53.863	6.20	10.36	537
44	9.05271	0.1806	0.0005	798.6	979.7	54.345	6.20	10.36	549
46	9.46683	0.1889	0.0004	811.1	1000.4	54.805	6.20	10.36	562
48	9.88078	0.1973	0.0004	823.5	1021.1	55.246	6.20	10.36	574
50	10.29461	0.2056	0.0004	835.9	1041.8	55.669	6.21	10.36	586
52	10.70831	0.2139	0.0004	848.4	1062.5	56.075	6.22	10.37	597
54	11.12192	0.2222	0.0004	860.9	1083.3	56.467	6.23	10.37	608
56	11.53542	0.2304	0.0004	873.3	1104.0	56.845	6.24	10.38	619
58	11.94885	0.2387	0.0003	885.8	1124.8	57.209	6.25	10.39	630
60	12.36221	0.2470	0.0003	898.4	1145.6	57.562	6.27	10.41	641
65	13.39533	0.2677	0.0003	929.9	1197.8	58.395	6.32	10.45	666
70	14.42813	0.2884	0.0003	961.6	1250.2	59.173	6.38	10.52	689
75	15.46069	0.3091	0.0003	993.8	1303.0	59.901	6.46	10.60	712
80	16.49305	0.3298	0.0003	1026.4	1356.2	60.588	6.56	10.69	733
85	17.52525	0.3504	0.0002	1059.4	1409.9	61.240	6.67	10.80	753
90	18.55732	0.3711	0.0002	1093.1	1464.3	61.862	6.80	10.93	772
95	19.58928	0.3917	0.0002	1127.5	1519.2	62.456	6.93	11.06	791
100	20.62115	0.4124	0.0002	1162.5	1574.9	63.026	7.07	11.20	808
120	24.74793	0.4950	0.0002	1309.7	1804.7	65.117	7.66	11.79	873
140	28.87400	0.5775	0.0001	1468.6	2046.1	66.977	8.22	12.35	931
160	32.99962	0.6601	0.0001	1637.9	2297.9	68.656	8.70	12.82	987
180	37.12496	0.7426	0.0001	1815.9	2558.4	70.190	9.09	13.21	1039
200	41.25010	0.8251	0.0001	2001.0	2826.0	71.600	9.40	13.53	1090
220	45.37510	0.9076	0.0001	2191.6	3099.1	72.901	9.65	13.78	1138
240	49.50000	0.9901	0.0001	2386.7	3376.7	74.109	9.85	13.97	1185
260	53.62482	1.0726	0.0001	2585.2	3657.7	75.234	9.99	14.12	1231
280	57.74958	1.1551	0.0001	2786.2	3941.2	76.284	10.10	14.23	1275
300	61.86966	1.2375	0.0001	2989.2	4226.6	77.268	10.18	14.31	1319
350	72.18050	1.4438	0.0001	3501.8	4945.4	79.484	10.30	14.42	1422
400	82.49121	1.6500	0.0001	4018.2	5668.1	81.415	10.35	14.47	1519
450	92.80183	1.8562	0.0000	4536.5	6392.6	83.121	10.37	14.49	1611
500	103.11240	2.0624	0.0000	5055.6	7117.8	84.648	10.39	14.51	1697
550	113.42293	2.2686	0.0000	5575.3	7843.8	86.033	10.40	14.53	1780
600	123.73343	2.4748	0.0000	6096.1	8570.8	87.295	10.42	14.55	1859
700	144.35440	2.8873	0.0000	7141.0	10028.1	89.544	10.48	14.60	2006
800	164.97531	3.2997	0.0000	8193.5	11493.0	91.502	10.57	14.69	2142
900	185.59621	3.7121	0.0000	9256.9	12968.8	93.238	10.70	14.82	2268
1000	206.21708	4.1245	0.0000	10333.9	14458.3	94.806	10.86	14.98	2386
1200	247.45886	4.9494	0.0000	12544.2	17493.4	97.553	11.24	15.36	2601
1400	288.70208	5.7742	0.0000	14837.9	20611.9	99.963	11.69	15.81	2795
1600	329.96236	6.5990	0.0000	17227.8	23827.1	102.116	12.25	16.38	2971
1800	371.33330	7.4239	0.0000	19763.5	27190.1	104.090	13.22	17.40	3125
2000	413.17959	8.2487	0.0000	22614.2	30877.7	106.041	15.49	19.82	3249
2500	530.09558	10.3108	0.0000	34967.0	45568.9	112.480	38.93	46.02	3488
3000	726.79296	12.3729	0.0000	73552.5	88088.4	127.747	113.04	137.57	3875

* TWO-PHASE BOUNDARY

Table 4.12 (continued)
THERMODYNAMIC PROPERTIES OF NORMAL HYDROGEN

0.02 MPA ISOBAR

TEMPERATURE DEG. K	DENSITY KG/CU M	$V(DH/DV)_P$ KJ/KG	$V(DP/DU)_V$ MPA-CU M/KJ X 10^2	$-V(DP/DV)_T$ MPA	$(DV/DT)_P/V$ 1/DEG. K	THERMAL CONDUCTIVITY W/K-M X 10^3	VISCOSITY KG/M-S X 10^7	THERMAL DIFFUSIVITY SQ M/HR	DIELECTRIC CONSTANT	PRANDTL NUMBER
* 13.807	77.0224	619.1	0.25649	89.6087	0.01031	72.67	254.89	0.00053	1.25160	2.2378
14	76.8522	612.4	0.25479	86.9799	0.01062	74.62	247.94	0.00054	1.25100	2.1602
15	76.0247	604.6	0.24263	78.4985	0.01156	83.06	218.31	0.00056	1.24809	1.8373
* 15.832	75.2560	623.4	0.23300	76.4147	0.01167	88.03	197.78	0.00058	1.24539	1.6349
* 15.832	0.3165	157.7	0.06680	0.0193	0.06860	13.56	8.58	0.01426	1.00095	0.6841
16	0.3129	159.4	0.06682	0.0193	0.06771	13.63	8.67	0.01452	1.00094	0.6870
17	0.2931	170.1	0.06687	0.0194	0.06293	14.04	9.24	0.01611	1.00088	0.7043
18	0.2758	180.8	0.06689	0.0195	0.05887	14.52	9.80	0.01781	1.00083	0.7178
19	0.2605	191.4	0.06689	0.0196	0.05535	15.02	10.35	0.01958	1.00079	0.7304
20	0.2469	202.0	0.06689	0.0196	0.05227	15.63	10.90	0.02158	1.00074	0.7363
21	0.2346	212.6	0.06688	0.0197	0.04953	16.41	11.44	0.02391	1.00071	0.7339
22	0.2236	223.2	0.06686	0.0197	0.04708	17.20	11.98	0.02636	1.00067	0.7315
23	0.2135	233.7	0.06685	0.0197	0.04487	17.99	12.50	0.02892	1.00064	0.7290
24	0.2044	244.2	0.06684	0.0198	0.04287	18.78	13.03	0.03159	1.00062	0.7264
25	0.1960	254.7	0.06683	0.0198	0.04105	19.57	13.55	0.03438	1.00059	0.7238
26	0.1883	265.2	0.06682	0.0198	0.03938	20.35	14.06	0.03726	1.00057	0.7214
27	0.1811	275.7	0.06681	0.0198	0.03784	21.13	14.56	0.04026	1.00055	0.7189
28	0.1745	286.2	0.06680	0.0198	0.03642	21.91	15.06	0.04337	1.00053	0.7164
29	0.1684	296.6	0.06679	0.0199	0.03511	22.70	15.56	0.04659	1.00051	0.7138
30	0.1627	307.0	0.06679	0.0199	0.03389	23.48	16.05	0.04992	1.00049	0.7112
31	0.1574	317.5	0.06678	0.0199	0.03276	24.18	16.53	0.05318	1.00047	0.7111
32	0.1524	327.9	0.06678	0.0199	0.03170	24.87	17.01	0.05654	1.00046	0.7108
33	0.1477	338.3	0.06677	0.0199	0.03071	25.57	17.48	0.05999	1.00045	0.7103
34	0.1433	348.7	0.06677	0.0199	0.02978	26.27	17.95	0.06357	1.00043	0.7095
35	0.1392	359.1	0.06677	0.0199	0.02890	26.97	18.42	0.06724	1.00042	0.7086
36	0.1352	369.5	0.06675	0.0199	0.02808	27.68	18.88	0.07101	1.00041	0.7076
37	0.1316	379.9	0.06674	0.0199	0.02730	28.38	19.33	0.07488	1.00040	0.7064
38	0.1281	390.3	0.06674	0.0199	0.02656	29.08	19.78	0.07885	1.00039	0.7052
39	0.1247	400.7	0.06673	0.0199	0.02587	29.78	20.23	0.08292	1.00038	0.7040
40	0.1216	411.1	0.06671	0.0199	0.02521	30.49	20.67	0.08709	1.00037	0.7026
42	0.1158	431.9	0.06669	0.0199	0.02398	31.88	21.54	0.09569	1.00035	0.7000
44	0.1105	452.8	0.06666	0.0200	0.02288	33.27	22.40	0.10467	1.00033	0.6973
46	0.1056	473.6	0.06663	0.0200	0.02187	34.65	23.24	0.11402	1.00032	0.6945
48	0.1012	494.5	0.06658	0.0200	0.02094	36.03	24.06	0.12373	1.00030	0.6917
50	0.0971	515.6	0.06652	0.0200	0.02010	37.40	24.87	0.13379	1.00029	0.6890
52	0.0934	536.7	0.06643	0.0200	0.01931	38.77	25.67	0.14417	1.00028	0.6865
54	0.0899	557.9	0.06633	0.0200	0.01859	40.12	26.46	0.15487	1.00027	0.6841
56	0.0867	579.3	0.06621	0.0200	0.01792	41.46	27.24	0.16585	1.00026	0.6820
58	0.0837	600.9	0.06606	0.0200	0.01730	42.79	28.00	0.17707	1.00025	0.6802
60	0.0809	622.6	0.06590	0.0200	0.01672	44.15	28.75	0.18877	1.00024	0.6779
65	0.0747	677.9	0.06538	0.0200	0.01542	47.60	30.59	0.21957	1.00022	0.6719
70	0.0693	734.8	0.06470	0.0200	0.01431	50.71	32.37	0.25044	1.00021	0.6714
75	0.0647	793.5	0.06387	0.0200	0.01336	53.69	34.10	0.28196	1.00019	0.6731
80	0.0606	854.2	0.06293	0.0200	0.01252	56.01	35.78	0.31104	1.00018	0.6830
85	0.0571	917.2	0.06187	0.0200	0.01178	58.99	37.42	0.34448	1.00017	0.6853
90	0.0539	982.6	0.06072	0.0200	0.01112	61.66	39.02	0.37693	1.00016	0.6915
95	0.0510	1049.7	0.05957	0.0200	0.01054	64.12	40.58	0.40886	1.00015	0.6999
100	0.0485	1119.2	0.05837	0.0200	0.01001	76.12	45.74	0.50451	1.00015	0.6730
120	0.0404	1413.9	0.05388	0.0200	0.00834	94.28	54.09	0.71261	1.00012	0.6762
140	0.0346	1728.1	0.05020	0.0200	0.00714	107.66	59.20	0.90640	1.00010	0.6789
160	0.0303	2051.4	0.04744	0.0200	0.00625	118.63	63.00	1.09907	1.00009	0.6809
180	0.0269	2378.5	0.04539	0.0200	0.00556	128.13	66.17	1.29591	1.00008	0.6824
200	0.0242	2705.5	0.04388	0.0200	0.00500	136.58	69.01	1.49940	1.00007	0.6835
220	0.0220	3031.4	0.04274	0.0200	0.00455	144.29	71.65	1.71069	1.00007	0.6842
240	0.0202	3353.3	0.04190	0.0200	0.00417	151.37	74.18	1.93066	1.00006	0.6847
260	0.0186	3671.1	0.04128	0.0200	0.00385	157.94	76.63	2.15962	1.00006	0.6850
280	0.0173	3984.3	0.04083	0.0200	0.00357	164.12	79.03	2.39807	1.00005	0.6851
300	0.0162	4292.9	0.04050	0.0200	0.00333	169.97	81.39	2.64581	1.00005	0.6852
350	0.0139	5049.0	0.04005	0.0200	0.00286	183.63	87.21	3.30790	1.00004	0.6851
400	0.0121	5790.2	0.03985	0.0200	0.00250	196.45	92.95	4.03052	1.00004	0.6849
450	0.0108	6523.2	0.03977	0.0200	0.00222	208.82	98.62	4.81297	1.00003	0.6846
500	0.0097	7257.4	0.03970	0.0200	0.00200	221.07	104.24	5.65408	1.00003	0.6843
550	0.0088	7990.9	0.03964	0.0200	0.00182	233.15	109.79	6.55295	1.00003	0.6841
600	0.0081	8728.7	0.03957	0.0200	0.00167	245.20	115.28	7.50813	1.00002	0.6839
700	0.0069	10222.6	0.03936	0.0200	0.00143	269.27	126.07	9.58260	1.00002	0.6837
800	0.0061	11754.9	0.03902	0.0200	0.00125	293.61	136.61	11.86800	1.00002	0.6837
900	0.0054	13338.9	0.03856	0.0200	0.00111	318.42	146.90	14.35538	1.00002	0.6837
1000	0.0048	14983.4	0.03798	0.0200	0.00100	440.56	156.96	21.82932	1.00001	0.5338
1200	0.0040	18437.3	0.03669	0.0200	0.00083	511.05	176.43	29.63228	1.00001	0.5304
1400	0.0035	22140.8	0.03528	0.0200	0.00071	584.08	195.12	38.38558	1.00001	0.5283
1600	0.0030	26208.2	0.03368	0.0200	0.00063	665.06	213.15	48.22769	1.00001	0.5250
1800	0.0027	31308.9	0.03120	0.0200	0.00056	781.06	230.62	60.00541	1.00001	0.5138
2000	0.0024	39572.9	0.02668	0.0200	0.00050	1013.92	247.68	76.08398	1.00001	0.4842
2500	0.0019	111902.3	0.01089	0.0195	0.00041	3527.69	291.64	146.26990	1.00001	0.3805
3000	0.0014	351285.0	0.00429	0.0170	0.00039	12245.90	345.55	232.91357	1.00000	0.3882

* TWO-PHASE BOUNDARY

Table 4.12 (continued)
THERMODYNAMIC PROPERTIES OF NORMAL HYDROGEN

0.04 MPA ISOBAR

TEMPERATURE DEG. K	VOLUME CU M/KG	ISOTHERM DERIVATIVE CU M-MPA/KG	ISOCHORE DERIVATIVE MPA/K	INTERNAL ENERGY KJ/KG-K	ENTHALPY KJ/KG	ENTROPY KJ/KG-K	CV KJ / KG-K	CP KJ / KG-K	VELOCITY OF SOUND M/S
* 13.814	0.01298	1.1647	0.9237	218.2	218.8	14.082	4.68	6.38	1261
14	0.01301	1.1349	0.9236	219.5	220.0	14.171	4.72	6.50	1250
15	0.01315	1.0346	0.9083	226.2	226.7	14.633	4.92	6.99	1212
16	0.01331	0.9689	0.8905	233.4	233.9	15.099	5.12	7.44	1187
17	0.01348	0.9053	0.8754	241.1	241.6	15.566	5.30	7.92	1163
* 17.497	0.01357	0.8806	0.8682	245.1	245.6	15.799	5.39	8.15	1154
* 17.497	1.71035	0.0646	0.0025	630.1	698.5	41.681	6.34	11.20	338
18	1.76648	0.0670	0.0024	633.4	704.1	41.997	6.32	11.10	343
19	1.87705	0.0718	0.0022	640.0	715.1	42.593	6.28	10.96	354
20	1.98652	0.0764	0.0021	646.6	726.0	43.153	6.26	10.86	364
21	2.09514	0.0810	0.0020	653.0	736.8	43.682	6.25	10.79	374
22	2.20310	0.0855	0.0019	659.5	747.6	44.184	6.24	10.73	384
23	2.31052	0.0900	0.0018	665.9	758.3	44.660	6.23	10.68	393
24	2.41749	0.0944	0.0017	672.3	769.0	45.114	6.22	10.64	402
25	2.52409	0.0988	0.0017	678.6	779.6	45.547	6.22	10.61	411
26	2.63035	0.1032	0.0016	685.0	790.2	45.961	6.22	10.58	419
27	2.73634	0.1076	0.0015	691.3	800.8	46.361	6.21	10.56	427
28	2.84208	0.1119	0.0015	697.6	811.3	46.743	6.21	10.54	436
29	2.94761	0.1162	0.0014	703.9	821.8	47.113	6.21	10.52	444
30	3.05295	0.1205	0.0014	710.2	832.4	47.469	6.21	10.51	452
31	3.15812	0.1248	0.0013	716.5	842.8	47.814	6.21	10.49	459
32	3.26314	0.1291	0.0013	722.8	853.3	48.146	6.21	10.48	467
33	3.36802	0.1333	0.0012	729.1	863.8	48.469	6.20	10.47	474
34	3.47278	0.1376	0.0012	735.3	874.3	48.781	6.20	10.46	482
35	3.57744	0.1418	0.0012	741.6	884.7	49.084	6.20	10.45	489
36	3.68199	0.1461	0.0011	747.9	895.2	49.379	6.20	10.44	496
37	3.78646	0.1503	0.0011	754.1	905.6	49.665	6.20	10.43	503
38	3.89084	0.1545	0.0011	760.4	916.0	49.943	6.20	10.42	510
39	3.99515	0.1588	0.0010	766.6	926.4	50.213	6.20	10.42	516
40	4.09939	0.1630	0.0010	772.9	936.9	50.477	6.20	10.41	523
42	4.30768	0.1714	0.0010	785.4	957.7	50.985	6.20	10.40	536
44	4.51576	0.1798	0.0009	797.9	978.5	51.468	6.20	10.40	549
46	4.72366	0.1882	0.0009	810.3	999.3	51.930	6.20	10.39	561
48	4.93140	0.1965	0.0008	822.8	1020.1	52.373	6.21	10.39	574
50	5.13901	0.2049	0.0008	835.3	1040.8	52.797	6.21	10.39	585
52	5.34649	0.2132	0.0008	847.8	1061.6	53.204	6.22	10.39	597
54	5.55387	0.2216	0.0007	860.3	1082.4	53.597	6.23	10.40	608
56	5.76116	0.2299	0.0007	872.8	1103.2	53.976	6.24	10.41	619
58	5.96837	0.2383	0.0007	885.3	1124.0	54.341	6.25	10.41	630
60	6.17550	0.2466	0.0007	897.8	1144.9	54.694	6.27	10.43	640
65	6.69306	0.2674	0.0006	929.4	1197.1	55.529	6.32	10.47	666
70	7.21030	0.2881	0.0006	961.2	1249.6	56.308	6.38	10.53	689
75	7.72730	0.3088	0.0005	993.4	1302.5	57.037	6.46	10.61	712
80	8.24410	0.3296	0.0005	1026.0	1355.7	57.725	6.56	10.70	733
85	8.76074	0.3503	0.0005	1059.1	1409.5	58.378	6.67	10.81	753
90	9.27725	0.3710	0.0004	1092.8	1463.9	58.999	6.80	10.94	773
95	9.79365	0.3917	0.0004	1127.2	1518.9	59.594	6.93	11.07	791
100	10.30996	0.4123	0.0004	1162.2	1574.6	60.165	7.07	11.21	808
120	12.37451	0.4950	0.0003	1309.5	1804.4	62.256	7.66	11.79	873
140	14.43833	0.5776	0.0003	1468.4	2046.0	64.117	8.22	12.35	932
160	16.50171	0.6602	0.0003	1637.8	2297.9	65.796	8.70	12.82	987
180	18.56481	0.7428	0.0002	1815.8	2558.4	67.331	9.09	13.22	1039
200	20.62771	0.8253	0.0002	2000.8	2826.0	68.740	9.40	13.53	1090
220	22.69048	0.9079	0.0002	2191.5	3099.1	70.042	9.65	13.78	1138
240	24.75314	0.9904	0.0002	2386.6	3376.7	71.250	9.85	13.97	1185
260	26.81572	1.0729	0.0002	2585.1	3657.7	72.375	9.99	14.12	1231
280	28.87824	1.1554	0.0001	2786.2	3941.3	73.425	10.10	14.23	1276
300	30.93842	1.2378	0.0001	2989.2	4226.7	74.409	10.18	14.31	1319
350	36.09405	1.4441	0.0001	3501.7	4945.5	76.626	10.30	14.43	1422
400	41.24955	1.6503	0.0001	4018.2	5668.2	78.556	10.35	14.47	1519
450	46.40496	1.8565	0.0001	4536.5	6392.7	80.262	10.37	14.50	1611
500	51.56032	2.0627	0.0001	5055.6	7118.0	81.790	10.39	14.51	1698
550	56.71565	2.2690	0.0001	5575.3	7843.9	83.174	10.40	14.53	1780
600	61.87095	2.4752	0.0001	6096.1	8570.9	84.436	10.42	14.55	1859
700	72.18149	2.8876	0.0001	7141.0	10028.2	86.686	10.48	14.60	2006
800	82.49200	3.3000	0.0001	8193.5	11493.2	88.643	10.57	14.69	2142
900	92.80248	3.7124	0.0000	9256.9	12969.0	90.380	10.70	14.82	2268
1000	103.11294	4.1249	0.0000	10333.9	14458.5	91.947	10.86	14.98	2386
1200	123.73385	4.9497	0.0000	12544.2	17493.6	94.695	11.24	15.36	2601
1400	144.35526	5.7745	0.0000	14837.5	20611.7	97.104	11.68	15.81	2795
1600	164.98269	6.5994	0.0000	17223.9	23823.3	99.254	12.21	16.34	2972
1800	185.64924	7.4242	0.0000	19737.9	27163.9	101.215	13.02	17.18	3130
2000	206.48386	8.2490	0.0000	22498.0	30757.4	103.118	14.72	19.00	3262
2500	262.92037	10.3111	0.0000	33176.9	43693.8	108.803	31.61	37.77	3507
3000	347.85821	12.3732	0.0000	62713.5	76627.9	120.644	86.94	104.62	3853

* TWO-PHASE BOUNDARY

Table 4.12 (continued)
THERMODYNAMIC PROPERTIES OF NORMAL HYDROGEN

0.04 MPA ISOBAR

TEMPERATURE DEG. K	DENSITY KG/CU M	$V(\partial H/\partial V)_P$ KJ/KG	$V(\partial P/\partial U)_V$ MPA-CU M/KJ X 10^2	$-V(\partial P/\partial V)_T$ MPA	$(\partial V/\partial T)_P/V$ 1/DEG. K	THERMAL CONDUCTIVITY W/K-M X 10^3	VISCOSITY KG/M-S X 10^7	THERMAL DIFFUSIVITY SQ M/HR	DIELECTRIC CONSTANT	PRANDTL NUMBER
* 13.814	77.0340	619.8	0.25646	89.7200	0.01029	72.76	255.02	0.00053	1.25164	2.2364
14	76.8696	613.6	0.25483	87.2391	0.01059	74.63	248.30	0.00054	1.25106	2.1611
15	76.0433	605.3	0.24280	78.6773	0.01154	83.08	218.62	0.00056	1.24816	1.8387
16	75.1384	608.0	0.23158	72.8003	0.01223	88.91	194.63	0.00057	1.24498	1.6280
17	74.1849	607.4	0.22254	67.1585	0.01304	92.73	175.22	0.00057	1.24164	1.4961
* 17.497	73.6915	608.9	0.21861	64.8894	0.01338	94.22	166.89	0.00056	1.23991	1.4431
* 17.497	0.5847	170.9	0.06672	0.0378	0.06552	14.61	9.58	0.00803	1.00176	0.7338
18	0.5661	176.3	0.06682	0.0379	0.06297	14.83	9.86	0.00849	1.00171	0.7381
19	0.5328	187.1	0.06694	0.0382	0.05858	15.27	10.41	0.00941	1.00161	0.7470
20	0.5034	197.9	0.06700	0.0385	0.05487	15.86	10.95	0.01044	1.00152	0.7503
21	0.4773	208.7	0.06702	0.0387	0.05167	16.62	11.49	0.01162	1.00144	0.7460
22	0.4539	219.5	0.06702	0.0388	0.04887	17.39	12.03	0.01285	1.00137	0.7420
23	0.4328	230.3	0.06701	0.0389	0.04639	18.16	12.55	0.01414	1.00130	0.7383
24	0.4137	241.0	0.06700	0.0391	0.04416	18.94	13.07	0.01549	1.00125	0.7348
25	0.3962	251.7	0.06698	0.0392	0.04216	19.72	13.59	0.01689	1.00119	0.7313
26	0.3802	262.4	0.06697	0.0392	0.04034	20.49	14.10	0.01834	1.00115	0.7282
27	0.3655	273.0	0.06695	0.0393	0.03869	21.27	14.61	0.01984	1.00110	0.7252
28	0.3519	283.6	0.06694	0.0394	0.03717	22.04	15.10	0.02140	1.00106	0.7222
29	0.3393	294.2	0.06692	0.0394	0.03577	22.82	15.60	0.02301	1.00102	0.7192
30	0.3276	304.7	0.06691	0.0395	0.03447	23.59	16.09	0.02468	1.00099	0.7163
31	0.3166	315.3	0.06690	0.0395	0.03328	24.29	16.57	0.02632	1.00095	0.7157
32	0.3065	325.8	0.06689	0.0396	0.03216	24.98	17.05	0.02801	1.00092	0.7150
33	0.2969	336.3	0.06688	0.0396	0.03112	25.68	17.52	0.02974	1.00089	0.7142
34	0.2880	346.8	0.06687	0.0396	0.03015	26.38	17.99	0.03154	1.00087	0.7131
35	0.2795	357.3	0.06687	0.0396	0.02924	27.08	18.45	0.03338	1.00084	0.7118
36	0.2716	367.8	0.06685	0.0397	0.02838	27.78	18.91	0.03527	1.00082	0.7106
37	0.2641	378.3	0.06684	0.0397	0.02758	28.48	19.36	0.03722	1.00080	0.7092
38	0.2570	388.7	0.06683	0.0397	0.02682	29.18	19.81	0.03921	1.00077	0.7078
39	0.2503	399.2	0.06682	0.0397	0.02610	29.88	20.26	0.04125	1.00075	0.7064
40	0.2439	409.7	0.06680	0.0398	0.02542	30.58	20.70	0.04334	1.00074	0.7049
42	0.2321	430.6	0.06678	0.0398	0.02416	31.97	21.57	0.04765	1.00070	0.7020
44	0.2214	451.5	0.06675	0.0398	0.02303	33.35	22.42	0.05215	1.00067	0.6991
46	0.2117	472.5	0.06671	0.0398	0.02200	34.73	23.26	0.05683	1.00064	0.6961
48	0.2028	493.5	0.06666	0.0399	0.02106	36.11	24.09	0.06170	1.00061	0.6931
50	0.1946	514.6	0.06659	0.0399	0.02019	37.48	24.90	0.06673	1.00059	0.6903
52	0.1870	535.8	0.06650	0.0399	0.01940	38.84	25.70	0.07193	1.00056	0.6876
54	0.1801	557.1	0.06640	0.0399	0.01866	40.19	26.48	0.07729	1.00054	0.6851
56	0.1736	578.5	0.06627	0.0399	0.01799	41.53	27.26	0.08279	1.00052	0.6829
58	0.1676	600.1	0.06613	0.0399	0.01735	42.85	28.02	0.08841	1.00050	0.6811
60	0.1619	621.9	0.06596	0.0399	0.01677	44.21	28.78	0.09426	1.00049	0.6787
65	0.1494	677.3	0.06543	0.0399	0.01546	47.66	30.61	0.10968	1.00045	0.6725
70	0.1387	734.2	0.06475	0.0400	0.01434	50.77	32.39	0.12514	1.00042	0.6719
75	0.1294	793.0	0.06392	0.0400	0.01338	53.74	34.12	0.14091	1.00039	0.6736
80	0.1213	853.8	0.06297	0.0400	0.01253	56.06	35.80	0.15546	1.00037	0.6834
85	0.1141	916.9	0.06191	0.0400	0.01179	59.03	37.43	0.17220	1.00034	0.6856
90	0.1078	982.3	0.06076	0.0400	0.01113	61.71	39.03	0.18845	1.00032	0.6917
95	0.1021	1049.5	0.05960	0.0400	0.01054	64.16	40.59	0.20442	1.00031	0.7001
100	0.0970	1119.0	0.05840	0.0400	0.01001	76.13	45.74	0.25215	1.00029	0.6732
120	0.0808	1413.8	0.05390	0.0400	0.00834	94.29	54.08	0.35622	1.00024	0.6763
140	0.0693	1728.1	0.05022	0.0400	0.00715	107.66	59.20	0.45316	1.00021	0.6790
160	0.0606	2051.5	0.04746	0.0400	0.00625	118.64	63.00	0.54954	1.00018	0.6810
180	0.0539	2378.7	0.04541	0.0400	0.00556	128.14	66.17	0.64799	1.00016	0.6825
200	0.0485	2705.8	0.04389	0.0400	0.00500	136.59	69.01	0.74976	1.00015	0.6835
220	0.0441	3031.7	0.04275	0.0400	0.00455	144.30	71.66	0.85543	1.00013	0.6842
240	0.0404	3353.6	0.04191	0.0400	0.00417	151.37	74.18	0.96544	1.00012	0.6847
260	0.0373	3671.4	0.04129	0.0400	0.00385	157.95	76.63	1.07995	1.00011	0.6850
280	0.0346	3984.7	0.04084	0.0400	0.00357	164.13	79.03	1.19919	1.00010	0.6852
300	0.0323	4293.4	0.04051	0.0400	0.00333	169.98	81.40	1.32308	1.00010	0.6852
350	0.0277	5049.5	0.04005	0.0400	0.00286	183.64	87.22	1.65418	1.00008	0.6851
400	0.0242	5790.7	0.03986	0.0400	0.00250	196.46	92.96	2.01554	1.00007	0.6849
450	0.0215	6523.7	0.03978	0.0400	0.00222	208.83	98.63	2.40682	1.00006	0.6846
500	0.0194	7258.0	0.03970	0.0400	0.00200	221.09	104.24	2.82743	1.00006	0.6843
550	0.0176	7991.5	0.03965	0.0400	0.00182	233.17	109.80	3.27693	1.00005	0.6841
600	0.0162	8729.3	0.03958	0.0400	0.00167	245.22	115.29	3.75458	1.00005	0.6839
700	0.0139	10223.2	0.03936	0.0400	0.00143	269.29	126.08	4.79195	1.00004	0.6837
800	0.0121	11755.5	0.03903	0.0400	0.00125	293.63	136.62	5.93479	1.00004	0.6837
900	0.0108	13339.5	0.03856	0.0400	0.00111	318.45	146.92	7.17862	1.00003	0.6837
1000	0.0097	14984.0	0.03798	0.0400	0.00100	440.56	156.97	10.91511	1.00003	0.5338
1200	0.0081	18437.7	0.03670	0.0400	0.00083	511.03	176.44	14.81653	1.00002	0.5305
1400	0.0069	22134.7	0.03530	0.0400	0.00071	583.80	195.14	19.19024	1.00002	0.5284
1600	0.0061	26141.5	0.03379	0.0400	0.00062	662.39	213.17	24.07967	1.00002	0.5258
1800	0.0054	30913.0	0.03169	0.0400	0.00056	765.58	230.63	29.78615	1,00002	0.5175
2000	0.0048	37948.7	0.02805	0.0400	0.00050	951.20	247.66	37.21792	1.00001	0.4946
2500	0.0038	92571.8	0.01331	0.0392	0.00041	2795.55	290.67	70.06111	1.00001	0.3927
3000	0.0029	279107.0	0.00533	0.0356	0.00037	9541.08	342.92	114.20189	1.00001	0.3760

* TWO-PHASE BOUNDARY

Table 4.12 (continued)
THERMODYNAMIC PROPERTIES OF NORMAL HYDROGEN

0.06 MPA ISOBAR

TEMPERATURE	VOLUME	ISOTHERM DERIVATIVE	ISOCHORE DERIVATIVE	INTERNAL ENERGY	ENTHALPY	ENTROPY	CV	CP	VELOCITY OF SOUND
DEG. K	CU M/KG	CU M-MPA/KG	MPA/K	KJ/KG-K	KJ/KG	KJ/KG-K	KJ /	KG-K	M/S
* 13.821	0.01298	1.1660	0.9239	218.2	219.0	14.083	4.68	6.38	1261
14	0.01301	1.1380	0.9239	219.4	220.2	14.168	4.71	6.49	1252
15	0.01315	1.0367	0.9090	226.1	226.9	14.630	4.92	6.99	1213
16	0.01331	0.9709	0.8912	233.3	234.1	15.096	5.12	7.43	1188
17	0.01348	0.9073	0.8761	241.0	241.8	15.562	5.30	7.91	1164
18	0.01366	0.8460	0.8616	249.2	250.0	16.030	5.47	8.42	1141
* 18.621	0.01378	0.8147	0.8524	254.5	255.3	16.321	5.56	8.72	1130
* 18.621	1.19270	0.0663	0.0036	634.8	706.3	40.545	6.41	11.53	345
19	1.22174	0.0682	0.0035	637.4	710.7	40.777	6.38	11.43	350
20	1.29743	0.0732	0.0033	644.2	722.0	41.358	6.33	11.23	360
21	1.37208	0.0780	0.0031	650.9	733.2	41.904	6.29	11.09	371
22	1.44595	0.0828	0.0029	657.5	744.2	42.418	6.27	10.98	381
23	1.51921	0.0875	0.0028	664.0	755.2	42.904	6.26	10.90	390
24	1.59196	0.0921	0.0026	670.5	766.0	43.367	6.25	10.83	400
25	1.66430	0.0966	0.0025	677.0	776.8	43.808	6.24	10.78	409
26	1.73629	0.1011	0.0024	683.4	787.6	44.228	6.23	10.73	417
27	1.80798	0.1056	0.0023	689.8	798.3	44.633	6.23	10.70	426
28	1.87941	0.1101	0.0022	696.2	809.0	45.020	6.23	10.66	434
29	1.95062	0.1145	0.0021	702.6	819.6	45.394	6.22	10.63	442
30	2.02162	0.1189	0.0021	709.0	830.2	45.754	6.22	10.61	450
31	2.09245	0.1232	0.0020	715.3	840.8	46.101	6.22	10.59	458
32	2.16313	0.1276	0.0019	721.6	851.4	46.437	6.22	10.57	466
33	2.23367	0.1319	0.0019	727.9	862.0	46.762	6.21	10.55	473
34	2.30408	0.1363	0.0018	734.3	872.5	47.077	6.21	10.53	481
35	2.37437	0.1406	0.0018	740.6	883.0	47.382	6.21	10.52	488
36	2.44457	0.1449	0.0017	746.9	893.5	47.678	6.21	10.50	495
37	2.51468	0.1492	0.0017	753.2	904.0	47.966	6.21	10.49	502
38	2.58470	0.1534	0.0016	759.5	914.5	48.246	6.21	10.48	509
39	2.65465	0.1577	0.0016	765.7	925.0	48.518	6.21	10.47	516
40	2.72452	0.1620	0.0015	772.0	935.5	48.783	6.21	10.46	523
42	2.86409	0.1705	0.0014	784.6	956.4	49.293	6.21	10.45	536
44	3.00344	0.1789	0.0014	797.1	977.3	49.779	6.21	10.44	549
46	3.14260	0.1874	0.0013	809.6	998.2	50.242	6.21	10.43	561
48	3.28161	0.1958	0.0013	822.1	1019.0	50.686	6.21	10.42	573
50	3.42047	0.2042	0.0012	834.6	1039.8	51.111	6.22	10.42	585
52	3.55922	0.2126	0.0012	847.1	1060.7	51.520	6.22	10.42	597
54	3.69786	0.2210	0.0011	859.7	1081.5	51.914	6.23	10.42	608
56	3.83641	0.2294	0.0011	872.2	1102.4	52.293	6.24	10.43	619
58	3.97487	0.2378	0.0010	884.7	1123.2	52.660	6.26	10.44	630
60	4.11327	0.2461	0.0010	897.3	1144.1	53.013	6.27	10.45	640
65	4.45897	0.2670	0.0009	928.9	1196.4	53.849	6.32	10.49	666
70	4.80437	0.2878	0.0009	960.8	1249.0	54.629	6.38	10.55	690
75	5.14951	0.3086	0.0008	993.0	1301.9	55.359	6.46	10.62	712
80	5.49446	0.3294	0.0008	1025.6	1355.3	56.048	6.56	10.71	733
85	5.83924	0.3501	0.0007	1058.8	1409.1	56.701	6.67	10.82	754
90	6.18390	0.3709	0.0007	1092.5	1463.5	57.324	6.80	10.94	773
95	6.52845	0.3916	0.0006	1126.8	1518.6	57.918	6.93	11.07	791
100	6.87291	0.4123	0.0006	1161.9	1574.3	58.489	7.07	11.21	809
120	8.25003	0.4951	0.0005	1309.2	1804.2	60.582	7.66	11.80	873
140	9.62644	0.5777	0.0004	1468.3	2045.8	62.443	8.22	12.35	932
160	11.00241	0.6604	0.0004	1637.6	2297.8	64.123	8.70	12.83	987
180	12.37810	0.7430	0.0003	1815.7	2558.3	65.658	9.09	13.22	1039
200	13.75359	0.8255	0.0003	2000.7	2825.9	67.067	9.40	13.53	1090
220	15.12893	0.9081	0.0003	2191.4	3099.1	68.369	9.65	13.78	1139
240	16.50418	0.9906	0.0003	2386.5	3376.8	69.577	9.85	13.97	1186
260	17.87935	1.0732	0.0002	2585.0	3657.8	70.702	9.99	14.12	1231
280	19.25446	1.1557	0.0002	2786.1	3941.3	71.753	10.10	14.23	1276
300	20.62801	1.2381	0.0002	2989.1	4226.8	72.736	10.18	14.31	1319
350	24.06523	1.4444	0.0002	3501.7	4945.6	74.953	10.30	14.43	1422
400	27.50233	1.6506	0.0002	4018.2	5668.3	76.884	10.35	14.48	1519
450	30.93934	1.8568	0.0001	4536.5	6392.8	78.590	10.37	14.50	1611
500	34.37630	2.0631	0.0001	5055.5	7118.1	80.117	10.39	14.51	1698
550	37.81322	2.2693	0.0001	5575.3	7844.1	81.502	10.40	14.53	1780
600	41.25012	2.4755	0.0001	6096.1	8571.1	82.763	10.42	14.55	1859
700	48.12386	2.8879	0.0001	7140.9	10028.4	85.013	10.48	14.60	2006
800	54.99756	3.3004	0.0001	8193.5	11493.4	86.971	10.57	14.69	2142
900	61.87123	3.7128	0.0001	9256.8	12969.1	88.707	10.70	14.82	2268
1000	68.74489	4.1252	0.0001	10333.9	14458.6	90.275	10.86	14.98	2386
1200	82.49218	4.9501	0.0001	12544.2	17493.7	93.022	11.24	15.36	2601
1400	96.23973	5.7749	0.0000	14837.4	20611.7	95.431	11.68	15.81	2795
1600	109.99056	6.5997	0.0000	17222.2	23821.7	97.581	12.19	16.32	2972
1800	123.76268	7.4246	0.0000	19726.6	27152.3	99.536	12.93	17.08	3132
2000	137.62628	8.2494	0.0000	22446.6	30704.2	101.417	14.39	18.63	3268
2500	174.65337	10.3115	0.0000	32383.6	42862.8	106.768	28.35	34.10	3519
3000	227.24916	12.3736	0.0000	57839.5	71474.4	117.065	74.57	89.35	3845

* TWO-PHASE BOUNDARY

Table 4.12 (continued)
THERMODYNAMIC PROPERTIES OF NORMAL HYDROGEN

0.06 MPA ISOBAR

TEMPERATURE DEG. K	DENSITY KG/CU M	$V(DH/DV)_P$ KJ/KG	$V(DP/DU)_V$ MPA-CU M/KJ X 10^2	$-V(DP/DV)_T$ MPA	$(DV/DT)_P/V$ 1/DEG. K	THERMAL CONDUCTIVITY W/K-M X 10^3	VISCOSITY KG/M-S X 10^7	THERMAL DIFFUSIVITY SQ M/HR	DIELECTRIC CONSTANT	PRANDTL NUMBER
* 13.821	77.0455	620.4	0.25644	89.8314	0.01028	72.84	255.14	0.00053	1.25168	2.2350
14	76.8869	614.7	0.25486	87.4981	0.01056	74.65	248.65	0.00054	1.25112	2.1620
15	76.0618	606.0	0.24296	78.8559	0.01153	83.11	218.92	0.00056	1.24822	1.8401
16	75.1582	608.7	0.23173	72.9692	0.01221	88.95	194.90	0.00057	1.24505	1.6289
17	74.2067	608.1	0.22269	67.3243	0.01301	92.77	175.47	0.00057	1.24172	1.4968
18	73.2096	605.0	0.21517	61.9320	0.01391	95.46	159.47	0.00056	1.23823	1.4062
* 18.621	72.5645	604.7	0.21108	59.1157	0.01442	96.52	150.87	0.00055	1.23598	1.3628
* 18.621	0.8384	178.9	0.06666	0.0556	0.06446	15.43	10.26	0.00574	1.00253	0.7667
19	0.8185	183.1	0.06678	0.0558	0.06246	15.58	10.47	0.00599	1.00247	0.7681
20	0.7708	194.0	0.06699	0.0564	0.05790	16.11	11.01	0.00670	1.00232	0.7673
21	0.7288	205.0	0.06709	0.0569	0.05410	16.84	11.55	0.00750	1.00220	0.7600
22	0.6916	215.9	0.06714	0.0573	0.05086	17.59	12.08	0.00834	1.00209	0.7539
23	0.6582	226.9	0.06715	0.0576	0.04804	18.35	12.60	0.00921	1.00198	0.7486
24	0.6282	237.8	0.06714	0.0578	0.04556	19.11	13.12	0.01011	1.00189	0.7438
25	0.6009	248.6	0.06713	0.0581	0.04335	19.88	13.64	0.01105	1.00181	0.7394
26	0.5759	259.5	0.06711	0.0582	0.04137	20.65	14.15	0.01202	1.00174	0.7355
27	0.5531	270.3	0.06709	0.0584	0.03958	21.41	14.65	0.01303	1.00167	0.7318
28	0.5321	281.0	0.06707	0.0586	0.03794	22.18	15.15	0.01407	1.00160	0.7283
29	0.5127	291.7	0.06705	0.0587	0.03645	22.94	15.64	0.01515	1.00155	0.7249
30	0.4947	302.4	0.06703	0.0588	0.03508	23.71	16.13	0.01627	1.00149	0.7215
31	0.4779	313.1	0.06702	0.0589	0.03381	24.40	16.61	0.01737	1.00144	0.7204
32	0.4623	323.7	0.06701	0.0590	0.03264	25.10	17.08	0.01850	1.00139	0.7193
33	0.4477	334.3	0.06699	0.0591	0.03155	25.79	17.56	0.01966	1.00135	0.7181
34	0.4340	344.9	0.06698	0.0591	0.03053	26.48	18.02	0.02086	1.00131	0.7167
35	0.4212	355.5	0.06697	0.0592	0.02958	27.18	18.49	0.02209	1.00127	0.7152
36	0.4091	366.0	0.06695	0.0593	0.02870	27.88	18.94	0.02336	1.00123	0.7137
37	0.3977	376.6	0.06693	0.0593	0.02786	28.58	19.40	0.02466	1.00120	0.7121
38	0.3869	387.1	0.06692	0.0594	0.02707	29.28	19.85	0.02599	1.00117	0.7105
39	0.3767	397.7	0.06691	0.0594	0.02633	29.98	20.29	0.02736	1.00114	0.7089
40	0.3670	408.2	0.06689	0.0594	0.02563	30.67	20.73	0.02875	1.00111	0.7072
42	0.3492	429.2	0.06686	0.0595	0.02434	32.06	21.60	0.03163	1.00105	0.7040
44	0.3330	450.3	0.06683	0.0596	0.02318	33.44	22.45	0.03464	1.00100	0.7009
46	0.3182	471.3	0.06678	0.0596	0.02213	34.82	23.29	0.03777	1.00096	0.6977
48	0.3047	492.4	0.06673	0.0597	0.02117	36.19	24.12	0.04102	1.00092	0.6946
50	0.2924	513.6	0.06666	0.0597	0.02029	37.56	24.93	0.04438	1.00088	0.6915
52	0.2810	534.8	0.06657	0.0597	0.01948	38.92	25.72	0.04785	1.00085	0.6888
54	0.2704	556.2	0.06646	0.0598	0.01874	40.27	26.51	0.05143	1.00081	0.6862
56	0.2607	577.7	0.06633	0.0598	0.01805	41.60	27.28	0.05510	1.00079	0.6838
58	0.2516	599.4	0.06619	0.0598	0.01741	42.92	28.05	0.05885	1.00076	0.6819
60	0.2431	621.2	0.06602	0.0598	0.01682	44.28	28.80	0.06276	1.00073	0.6794
65	0.2243	676.6	0.06549	0.0599	0.01550	47.72	30.63	0.07305	1.00068	0.6731
70	0.2081	733.7	0.06480	0.0599	0.01437	50.83	32.41	0.08337	1.00063	0.6724
75	0.1942	792.6	0.06397	0.0599	0.01340	53.79	34.14	0.09389	1.00059	0.6740
80	0.1820	853.4	0.06302	0.0599	0.01255	56.11	35.82	0.10361	1.00055	0.6838
85	0.1713	916.5	0.06195	0.0600	0.01181	59.08	37.45	0.11477	1.00052	0.6859
90	0.1617	982.0	0.06080	0.0600	0.01114	61.75	39.05	0.12562	1.00049	0.6920
95	0.1532	1049.2	0.05964	0.0600	0.01055	64.21	40.61	0.13628	1.00046	0.7003
100	0.1455	1118.8	0.05844	0.0600	0.01002	76.15	45.74	0.16803	1.00044	0.6735
120	0.1212	1413.7	0.05392	0.0600	0.00834	94.29	54.08	0.23743	1.00037	0.6765
140	0.1039	1728.1	0.05024	0.0600	0.00715	107.67	59.20	0.30208	1.00031	0.6791
160	0.0909	2051.7	0.04747	0.0600	0.00625	118.64	63.00	0.36636	1.00027	0.6811
180	0.0808	2378.9	0.04542	0.0600	0.00556	128.15	66.17	0.43202	1.00024	0.6825
200	0.0727	2706.0	0.04390	0.0600	0.00500	136.60	69.01	0.49988	1.00022	0.6836
220	0.0661	3032.0	0.04275	0.0600	0.00455	144.31	71.66	0.57035	1.00020	0.6843
240	0.0606	3354.0	0.04191	0.0600	0.00417	151.38	74.18	0.64370	1.00018	0.6847
260	0.0559	3671.8	0.04129	0.0600	0.00385	157.96	76.63	0.72005	1.00017	0.6850
280	0.0519	3985.1	0.04084	0.0600	0.00357	164.14	79.04	0.79956	1.00016	0.6852
300	0.0485	4293.8	0.04052	0.0600	0.00333	169.99	81.40	0.88218	1.00015	0.6852
350	0.0416	5050.0	0.04006	0.0600	0.00286	183.65	87.22	1.10294	1.00013	0.6851
400	0.0364	5791.3	0.03986	0.0600	0.00250	196.48	92.96	1.34388	1.00011	0.6849
450	0.0323	6524.3	0.03978	0.0600	0.00222	208.85	98.63	1.60477	1.00010	0.6846
500	0.0291	7258.5	0.03971	0.0600	0.00200	221.10	104.25	1.88522	1.00009	0.6843
550	0.0264	7992.1	0.03965	0.0600	0.00182	233.19	109.80	2.18492	1.00008	0.6841
600	0.0242	8729.9	0.03958	0.0600	0.00167	245.23	115.30	2.50340	1.00007	0.6839
700	0.0208	10223.8	0.03937	0.0600	0.00143	269.32	126.09	3.19506	1.00006	0.6837
800	0.0182	11756.1	0.03903	0.0600	0.00125	293.65	136.64	3.95705	1.00005	0.6837
900	0.0162	13340.2	0.03856	0.0600	0.00111	318.48	146.93	4.78638	1.00005	0.6837
1000	0.0145	14984.6	0.03799	0.0600	0.00100	440.56	156.99	7.27704	1.00004	0.5339
1200	0.0121	18438.2	0.03670	0.0600	0.00083	511.03	176.46	9.87799	1.00004	0.5305
1400	0.0104	22132.3	0.03531	0.0600	0.00071	583.68	195.16	12.79295	1.00003	0.5285
1600	0.0091	26112.3	0.03384	0.0600	0.00062	661.21	213.19	16.04321	1.00003	0.5262
1800	0.0081	30738.0	0.03192	0.0600	0.00056	758.72	230.65	19.79253	1.00002	0.5192
2000	0.0073	37229.0	0.02870	0.0599	0.00050	923.40	247.66	24.55330	1.00002	0.4997
2500	0.0057	83897.9	0.01479	0.0590	0.00041	2466.96	290.23	45.48040	1.00002	0.4012
3000	0.0044	243245.4	0.00610	0.0544	0.00037	8196.95	341.07	75.05428	1.00001	0.3718

* TWO-PHASE BOUNDARY

Table 4.12 (continued)

THERMODYNAMIC PROPERTIES OF NORMAL HYDROGEN

0.08 MPA ISOBAR

TEMPERATURE	VOLUME	ISOTHERM DERIVATIVE	ISOCHORE DERIVATIVE	INTERNAL ENERGY	ENTHALPY	ENTROPY	CV	CP	VELOCITY OF SOUND
DEG. K	CU M/KG	CU M-MPA/KG	MPA/K	KJ/KG-K	KJ/KG	KJ/KG-K	KJ /	KG-K	M/S
* 13.828	0.01298	1.1672	0.9241	218.3	219.3	14.083	4.68	6.38	1262
14	0.01300	1.1411	0.9241	219.4	220.4	14.165	4.71	6.49	1253
15	0.01314	1.0388	0.9097	226.1	227.1	14.627	4.92	6.98	1214
16	0.01330	0.9729	0.8919	233.3	234.3	15.093	5.12	7.43	1189
17	0.01347	0.9092	0.8768	241.0	242.0	15.559	5.30	7.91	1165
18	0.01366	0.8480	0.8623	249.1	250.2	16.026	5.47	8.41	1142
19	0.01385	0.7833	0.8470	257.8	258.9	16.495	5.62	8.96	1117
* 19.496	0.01396	0.7643	0.8384	262.3	263.4	16.729	5.69	9.18	1111
* 19.496	0.92223	0.0671	0.0047	638.1	711.9	39.738	6.46	11.84	351
20	0.95189	0.0698	0.0045	641.7	717.8	40.039	6.41	11.68	356
21	1.00981	0.0750	0.0042	648.6	729.4	40.604	6.35	11.44	367
22	1.06682	0.0800	0.0040	655.4	740.7	41.133	6.32	11.27	378
23	1.12311	0.0849	0.0038	662.1	751.9	41.631	6.29	11.14	388
24	1.17885	0.0897	0.0036	668.7	763.0	42.103	6.27	11.04	397
25	1.23413	0.0944	0.0034	675.3	774.0	42.552	6.26	10.96	406
26	1.28903	0.0990	0.0033	681.8	784.9	42.979	6.25	10.89	415
27	1.34361	0.1036	0.0031	688.3	795.8	43.389	6.25	10.84	424
28	1.39792	0.1082	0.0030	694.8	806.6	43.781	6.24	10.79	433
29	1.45199	0.1127	0.0029	701.2	817.4	44.160	6.24	10.75	441
30	1.50585	0.1172	0.0028	707.7	828.1	44.524	6.23	10.72	449
31	1.55953	0.1217	0.0027	714.1	838.8	44.874	6.23	10.68	457
32	1.61305	0.1261	0.0026	720.4	849.5	45.213	6.23	10.66	465
33	1.66642	0.1305	0.0025	726.8	860.1	45.541	6.22	10.63	472
34	1.71967	0.1349	0.0024	733.2	870.7	45.858	6.22	10.61	480
35	1.77280	0.1393	0.0024	739.5	881.3	46.165	6.22	10.59	487
36	1.82582	0.1437	0.0023	745.9	891.9	46.463	6.22	10.57	494
37	1.87875	0.1480	0.0022	752.2	902.5	46.753	6.22	10.55	501
38	1.93160	0.1523	0.0022	758.5	913.0	47.034	6.21	10.54	508
39	1.98437	0.1567	0.0021	764.8	923.6	47.308	6.21	10.53	515
40	2.03707	0.1610	0.0020	771.1	934.1	47.574	6.21	10.52	522
42	2.14228	0.1695	0.0019	783.7	955.1	48.087	6.21	10.50	535
44	2.24726	0.1781	0.0018	796.3	976.1	48.574	6.21	10.48	548
46	2.35206	0.1866	0.0018	808.9	997.0	49.040	6.21	10.47	561
48	2.45670	0.1951	0.0017	821.4	1018.0	49.485	6.21	10.46	573
50	2.56120	0.2036	0.0016	834.0	1038.9	49.912	6.22	10.45	585
52	2.66558	0.2120	0.0016	846.5	1059.7	50.321	6.23	10.45	596
54	2.76985	0.2205	0.0015	859.1	1080.6	50.716	6.23	10.45	608
56	2.87403	0.2289	0.0014	871.6	1101.5	51.096	6.24	10.45	619
58	2.97813	0.2373	0.0014	884.2	1122.4	51.464	6.26	10.46	630
60	3.08215	0.2457	0.0013	896.8	1143.4	51.818	6.27	10.47	640
65	3.34194	0.2666	0.0012	928.4	1195.8	52.656	6.32	10.50	666
70	3.60140	0.2875	0.0011	960.3	1248.4	53.437	6.38	10.56	690
75	3.86062	0.3084	0.0011	992.6	1301.4	54.168	6.47	10.63	712
80	4.11964	0.3292	0.0010	1025.2	1354.8	54.857	6.56	10.72	733
85	4.37850	0.3500	0.0009	1058.4	1408.7	55.511	6.67	10.83	754
90	4.63723	0.3708	0.0009	1092.2	1463.1	56.134	6.80	10.95	773
95	4.89585	0.3915	0.0008	1126.5	1518.2	56.729	6.93	11.08	791
100	5.15438	0.4123	0.0008	1161.6	1574.0	57.300	7.07	11.22	809
120	6.18780	0.4951	0.0007	1309.0	1804.0	59.393	7.66	11.80	873
140	7.22050	0.5779	0.0006	1468.1	2045.7	61.255	8.22	12.35	932
160	8.25276	0.6605	0.0005	1637.5	2297.7	62.936	8.70	12.83	987
180	9.28474	0.7432	0.0004	1815.5	2558.3	64.470	9.09	13.22	1040
200	10.31652	0.8257	0.0004	2000.6	2825.9	65.880	9.40	13.53	1090
220	11.34817	0.9083	0.0004	2191.3	3099.2	67.182	9.65	13.78	1139
240	12.37971	0.9909	0.0003	2386.4	3376.8	68.390	9.85	13.97	1186
260	13.41117	1.0734	0.0003	2584.9	3657.8	69.516	9.99	14.12	1232
280	14.44257	1.1560	0.0003	2786.0	3941.4	70.566	10.10	14.23	1276
300	15.47280	1.2384	0.0003	2989.0	4226.9	71.550	10.18	14.31	1319
350	18.05083	1.4447	0.0002	3501.6	4945.7	73.767	10.30	14.43	1422
400	20.62872	1.6509	0.0002	4018.1	5668.4	75.697	10.35	14.48	1519
450	23.20653	1.8572	0.0002	4536.4	6393.0	77.403	10.37	14.50	1611
500	25.78429	2.0634	0.0002	5055.5	7118.3	78.931	10.39	14.51	1698
550	28.36201	2.2696	0.0001	5575.2	7844.2	80.315	10.40	14.53	1780
600	30.93970	2.4759	0.0001	6096.0	8571.2	81.577	10.42	14.55	1859
700	36.09504	2.8883	0.0001	7140.9	10028.5	83.827	10.48	14.60	2006
800	41.25034	3.3007	0.0001	8193.5	11493.5	85.785	10.57	14.69	2142
900	46.40561	3.7131	0.0001	9256.8	12969.3	87.521	10.70	14.82	2268
1000	51.56086	4.1256	0.0001	10333.9	14458.8	89.088	10.86	14.98	2386
1200	61.87135	4.9504	0.0001	12544.2	17493.9	91.836	11.24	15.36	2601
1400	72.18199	5.7753	0.0001	14837.3	20611.8	94.245	11.68	15.81	2795
1600	82.49476	6.6001	0.0001	17221.2	23820.8	96.394	12.18	16.31	2973
1800	92.82135	7.4249	0.0000	19719.8	27145.5	98.346	12.87	17.02	3133
2000	103.20737	8.2498	0.0000	22415.9	30672.5	100.213	14.19	18.42	3272
2500	130.71063	10.3119	0.0000	31910.6	42367.5	105.365	26.40	31.92	3529
3000	168.34401	12.3739	0.0000	54917.2	68384.7	114.736	66.97	80.08	3841

* TWO-PHASE BOUNDARY

Table 4.12 (continued)
THERMODYNAMIC PROPERTIES OF NORMAL HYDROGEN

0.08 MPA ISOBAR

TEMPERATURE DEG. K	DENSITY KG/CU M	$V(DH/DV)_P$ KJ/KG	$V(DP/DU)_V$ MPA-CU M/KJ X 10^2	$-V(DP/DV)_T$ MPA	$(DV/DT)_P/V$ 1/DEG. K	THERMAL CONDUCTIVITY W/K-M X 10^3	VISCOSITY KG/M-S X 10^7	THERMAL DIFFUSIVITY SQ M/HR	DIELECTRIC CONSTANT	PRANDTL NUMBER
* 13.828	77.0570	621.0	0.25641	89.9428	0.01027	72.92	255.26	0.00053	1.25172	2.2336
14	76.9043	615.9	0.25489	87.7567	0.01053	74.67	249.01	0.00054	1.25118	2.1630
15	76.0803	606.6	0.24313	79.0342	0.01151	83.13	219.23	0.00056	1.24829	1.8415
16	75.1781	609.4	0.23189	73.1378	0.01219	88.98	195.17	0.00057	1.24512	1.6299
17	74.2283	608.8	0.22283	67.4898	0.01299	92.81	175.71	0.00057	1.24179	1.4974
18	73.2332	605.8	0.21531	62.1031	0.01388	95.51	159.70	0.00056	1.23831	1.4065
19	72.1879	598.0	0.20884	56.5441	0.01498	97.04	146.25	0.00054	1.23466	1.3500
* 19.496	71.6402	599.5	0.20583	54.7548	0.01531	97.43	140.27	0.00053	1.23276	1.3216
* 19.496	1.0843	184.6	0.06665	0.0727	0.06416	16.11	10.80	0.00452	1.00327	0.7938
20	1.0505	190.1	0.06684	0.0733	0.06144	16.41	11.07	0.00481	1.00317	0.7879
21	0.9903	201.2	0.06708	0.0742	0.05686	17.10	11.60	0.00543	1.00299	0.7764
22	0.9374	212.4	0.06720	0.0750	0.05307	17.81	12.13	0.00607	1.00283	0.7675
23	0.8904	223.5	0.06726	0.0756	0.04986	18.55	12.65	0.00673	1.00269	0.7601
24	0.8483	234.6	0.06727	0.0761	0.04707	19.30	13.17	0.00742	1.00256	0.7538
25	0.8103	245.6	0.06726	0.0765	0.04463	20.05	13.69	0.00813	1.00244	0.7481
26	0.7758	256.6	0.06725	0.0768	0.04246	20.80	14.19	0.00886	1.00234	0.7433
27	0.7443	267.5	0.06722	0.0771	0.04051	21.56	14.69	0.00962	1.00224	0.7388
28	0.7153	278.4	0.06720	0.0774	0.03876	22.31	15.19	0.01041	1.00216	0.7347
29	0.6887	289.3	0.06718	0.0776	0.03716	23.07	15.68	0.01122	1.00208	0.7307
30	0.6641	300.1	0.06715	0.0778	0.03570	23.83	16.17	0.01206	1.00200	0.7270
31	0.6412	310.9	0.06713	0.0780	0.03436	24.52	16.65	0.01289	1.00193	0.7253
32	0.6199	321.6	0.06712	0.0782	0.03313	25.21	17.12	0.01374	1.00187	0.7237
33	0.6001	332.3	0.06710	0.0783	0.03199	25.90	17.59	0.01462	1.00181	0.7222
34	0.5815	343.0	0.06709	0.0785	0.03093	26.60	18.06	0.01552	1.00175	0.7204
35	0.5641	353.7	0.06707	0.0786	0.02994	27.29	18.52	0.01645	1.00170	0.7186
36	0.5477	364.3	0.06705	0.0787	0.02901	27.99	18.98	0.01740	1.00165	0.7168
37	0.5323	375.0	0.06703	0.0788	0.02815	28.68	19.43	0.01838	1.00160	0.7150
38	0.5177	385.6	0.06702	0.0789	0.02734	29.38	19.88	0.01938	1.00156	0.7132
39	0.5039	396.2	0.06700	0.0789	0.02657	30.08	20.32	0.02041	1.00152	0.7114
40	0.4909	406.8	0.06698	0.0790	0.02585	30.77	20.76	0.02146	1.00148	0.7095
42	0.4668	427.9	0.06695	0.0791	0.02453	32.15	21.63	0.02363	1.00141	0.7061
44	0.4450	449.0	0.06691	0.0792	0.02334	33.53	22.48	0.02588	1.00134	0.7027
46	0.4252	470.2	0.06686	0.0793	0.02226	34.90	23.32	0.02824	1.00128	0.6993
48	0.4070	491.3	0.06681	0.0794	0.02128	36.27	24.14	0.03068	1.00123	0.6960
50	0.3904	512.6	0.06674	0.0795	0.02039	37.64	24.95	0.03321	1.00118	0.6928
52	0.3752	533.9	0.06664	0.0795	0.01957	38.99	25.75	0.03581	1.00113	0.6899
54	0.3610	555.3	0.06653	0.0796	0.01881	40.34	26.53	0.03850	1.00109	0.6872
56	0.3479	576.9	0.06640	0.0796	0.01812	41.68	27.31	0.04126	1.00105	0.6848
58	0.3358	598.6	0.06625	0.0797	0.01747	42.99	28.07	0.04408	1.00101	0.6828
60	0.3244	620.5	0.06608	0.0797	0.01687	44.35	28.82	0.04701	1.00098	0.6802
65	0.2992	676.0	0.06554	0.0798	0.01554	47.79	30.65	0.05474	1.00090	0.6737
70	0.2777	733.2	0.06485	0.0798	0.01440	50.89	32.43	0.06248	1.00084	0.6729
75	0.2590	792.1	0.06402	0.0799	0.01342	53.84	34.16	0.07038	1.00078	0.6745
80	0.2427	853.1	0.06306	0.0799	0.01257	56.16	35.83	0.07768	1.00073	0.6841
85	0.2284	916.2	0.06199	0.0799	0.01182	59.13	37.47	0.08606	1.00069	0.6862
90	0.2156	981.7	0.06083	0.0800	0.01116	61.80	39.06	0.09420	1.00065	0.6922
95	0.2043	1049.0	0.05967	0.0800	0.01056	64.25	40.62	0.10221	1.00062	0.7005
100	0.1940	1118.6	0.05847	0.0800	0.01003	76.16	45.74	0.12597	1.00058	0.6737
120	0.1616	1413.7	0.05395	0.0800	0.00835	94.30	54.08	0.17803	1.00049	0.6767
140	0.1385	1728.2	0.05026	0.0800	0.00715	107.68	59.20	0.22654	1.00042	0.6792
160	0.1212	2051.8	0.04749	0.0800	0.00625	118.65	63.00	0.27477	1.00037	0.6812
180	0.1077	2379.1	0.04543	0.0800	0.00556	128.15	66.17	0.32403	1.00032	0.6826
200	0.0969	2706.3	0.04391	0.0800	0.00500	136.61	69.01	0.37494	1.00029	0.6836
220	0.0881	3032.3	0.04276	0.0800	0.00455	144.32	71.66	0.42781	1.00027	0.6843
240	0.0808	3354.3	0.04192	0.0800	0.00417	151.39	74.19	0.48283	1.00024	0.6848
260	0.0746	3672.2	0.04130	0.0800	0.00385	157.97	76.64	0.54011	1.00022	0.6851
280	0.0692	3985.5	0.04085	0.0800	0.00357	164.15	79.04	0.59975	1.00021	0.6852
300	0.0646	4294.2	0.04052	0.0800	0.00333	170.00	81.40	0.66172	1.00019	0.6852
350	0.0554	5050.5	0.04006	0.0800	0.00286	183.66	87.23	0.82732	1.00017	0.6851
400	0.0485	5791.8	0.03987	0.0800	0.00250	196.49	92.97	1.00805	1.00015	0.6849
450	0.0431	6524.8	0.03978	0.0800	0.00222	208.86	98.64	1.20375	1.00013	0.6846
500	0.0388	7259.1	0.03971	0.0800	0.00200	221.12	104.26	1.41411	1.00012	0.6843
550	0.0353	7992.6	0.03965	0.0800	0.00182	233.20	109.81	1.63892	1.00011	0.6841
600	0.0323	8730.5	0.03958	0.0800	0.00167	245.25	115.31	1.87781	1.00010	0.6839
700	0.0277	10224.4	0.03937	0.0800	0.00143	269.34	126.10	2.39662	1.00008	0.6837
800	0.0242	11756.8	0.03903	0.0800	0.00125	293.68	136.65	2.96818	1.00007	0.6837
900	0.0215	13340.8	0.03857	0.0800	0.00111	318.50	146.94	3.59025	1.00006	0.6837
1000	0.0194	14985.3	0.03799	0.0800	0.00100	440.56	157.00	5.45800	1.00006	0.5339
1200	0.0162	18438.8	0.03670	0.0800	0.00083	511.03	176.47	7.40874	1.00005	0.5306
1400	0.0139	22131.2	0.03531	0.0800	0.00071	583.61	195.18	9.59459	1.00004	0.5286
1600	0.0121	26095.2	0.03387	0.0800	0.00062	660.51	213.21	12.02811	1.00004	0.5264
1800	0.0108	30634.0	0.03205	0.0800	0.00056	754.63	230.67	14.81525	1.00003	0.5203
2000	0.0097	36800.1	0.02910	0.0799	0.00050	906.81	247.67	18.29582	1.00003	0.5030
2500	0.0077	78695.3	0.01585	0.0789	0.00041	2269.86	289.97	33.46093	1.00002	0.4078
3000	0.0059	220733.0	0.00670	0.0735	0.00036	7352.97	339.76	55.64640	1.00002	0.3700

* TWO-PHASE BOUNDARY

Table 4.12 (continued)
THERMODYNAMIC PROPERTIES OF NORMAL HYDROGEN

0.10 MPA ISOBAR

TEMPERATURE DEG. K	VOLUME CU M/KG	ISOTHERM DERIVATIVE CU M-MPA/KG	ISOCHORE DERIVATIVE MPA/K	INTERNAL ENERGY KJ/KG-K	ENTHALPY KJ/KG	ENTROPY KJ/KG-K	CV KJ /	CP KG-K	VELOCITY OF SOUND M/S
* 13.834	0.01298	1.1685	0.9244	218.3	219.6	14.084	4.68	6.38	1262
14	0.01300	1.1442	0.9243	219.4	220.7	14.162	4.71	6.48	1254
15	0.01314	1.0409	0.9104	226.0	227.3	14.624	4.92	6.98	1216
16	0.01330	0.9748	0.8925	233.2	234.6	15.090	5.12	7.43	1190
17	0.01347	0.9112	0.8774	240.9	242.2	15.556	5.30	7.91	1166
18	0.01365	0.8501	0.8630	249.0	250.4	16.022	5.47	8.41	1143
19	0.01385	0.7853	0.8476	257.7	259.1	16.490	5.62	8.95	1119
20	0.01406	0.7221	0.8304	266.9	268.3	16.963	5.75	9.53	1094
* 20.224	0.01412	0.7122	0.8261	269.1	270.5	17.072	5.77	9.64	1090
* 20.224	0.75454	0.0675	0.0057	640.6	716.0	39.109	6.50	12.14	355
21	0.79178	0.0717	0.0054	646.2	725.3	39.562	6.43	11.86	364
22	0.83883	0.0770	0.0051	653.2	737.1	40.108	6.37	11.60	375
23	0.88507	0.0822	0.0048	660.1	748.6	40.620	6.33	11.41	385
24	0.93068	0.0872	0.0046	666.8	759.9	41.102	6.30	11.27	395
25	0.97579	0.0921	0.0043	673.5	771.1	41.560	6.29	11.16	404
26	1.02048	0.0969	0.0041	680.2	782.2	41.994	6.27	11.07	413
27	1.06484	0.1016	0.0040	686.8	793.3	42.411	6.26	10.99	422
28	1.10890	0.1063	0.0038	693.3	804.2	42.808	6.26	10.93	431
29	1.15271	0.1110	0.0036	699.8	815.1	43.191	6.25	10.87	439
30	1.19630	0.1156	0.0035	706.3	826.0	43.559	6.25	10.83	448
31	1.23970	0.1201	0.0034	712.8	836.8	43.913	6.24	10.79	456
32	1.28294	0.1246	0.0033	719.2	847.5	44.255	6.24	10.75	463
33	1.32603	0.1291	0.0032	725.7	858.3	44.585	6.23	10.72	471
34	1.36898	0.1336	0.0031	732.1	869.0	44.905	6.23	10.69	479
35	1.41181	0.1380	0.0030	738.4	879.6	45.214	6.23	10.66	486
36	1.45454	0.1425	0.0029	744.8	890.3	45.515	6.22	10.64	493
37	1.49717	0.1469	0.0028	751.2	900.9	45.806	6.22	10.62	501
38	1.53972	0.1512	0.0027	757.6	911.5	46.089	6.22	10.60	508
39	1.58218	0.1556	0.0026	763.9	922.1	46.364	6.22	10.58	515
40	1.62458	0.1600	0.0026	770.2	932.7	46.632	6.22	10.57	521
42	1.70918	0.1686	0.0024	782.9	953.8	47.146	6.22	10.54	535
44	1.79355	0.1772	0.0023	795.5	974.9	47.636	6.22	10.52	548
46	1.87774	0.1858	0.0022	808.1	995.9	48.103	6.22	10.50	560
48	1.96176	0.1944	0.0021	820.7	1016.9	48.550	6.22	10.49	573
50	2.04564	0.2029	0.0020	833.3	1037.9	48.978	6.22	10.48	585
52	2.12940	0.2114	0.0020	845.9	1058.8	49.389	6.23	10.48	596
54	2.21305	0.2199	0.0019	858.5	1079.8	49.784	6.24	10.47	608
56	2.29661	0.2284	0.0018	871.0	1100.7	50.166	6.25	10.47	619
58	2.38009	0.2368	0.0017	883.6	1121.6	50.534	6.26	10.48	630
60	2.46349	0.2452	0.0017	896.3	1142.6	50.889	6.27	10.49	640
65	2.67172	0.2663	0.0016	927.9	1195.1	51.728	6.32	10.52	666
70	2.87963	0.2872	0.0014	959.9	1247.9	52.510	6.39	10.57	690
75	3.08729	0.3081	0.0013	992.2	1300.9	53.242	6.47	10.64	712
80	3.29475	0.3290	0.0013	1024.9	1354.3	53.932	6.56	10.73	734
85	3.50206	0.3498	0.0012	1058.1	1408.3	54.586	6.67	10.84	754
90	3.70923	0.3706	0.0011	1091.8	1462.8	55.210	6.80	10.96	773
95	3.91630	0.3914	0.0011	1126.2	1517.9	55.805	6.93	11.09	791
100	4.12327	0.4122	0.0010	1161.3	1573.6	56.377	7.07	11.22	809
120	4.95046	0.4952	0.0008	1308.8	1803.8	58.471	7.66	11.80	873
140	5.77694	0.5780	0.0007	1467.9	2045.6	60.334	8.22	12.36	932
160	6.60298	0.6607	0.0006	1637.3	2297.6	62.014	8.70	12.83	987
180	7.42873	0.7433	0.0006	1815.4	2558.3	63.549	9.09	13.22	1040
200	8.25429	0.8260	0.0005	2000.5	2825.9	64.959	9.40	13.53	1090
220	9.07970	0.9086	0.0005	2191.2	3099.2	66.261	9.65	13.78	1139
240	9.90502	0.9911	0.0004	2386.3	3376.8	67.470	9.85	13.97	1186
260	10.73026	1.0737	0.0004	2584.9	3657.9	68.595	9.99	14.12	1232
280	11.55544	1.1562	0.0004	2785.9	3941.5	69.645	10.10	14.23	1276
300	12.37968	1.2387	0.0003	2989.0	4226.9	70.629	10.18	14.31	1319
350	14.44218	1.4450	0.0003	3501.6	4945.8	72.846	10.30	14.43	1423
400	16.50455	1.6512	0.0003	4018.1	5668.5	74.777	10.35	14.48	1520
450	18.56684	1.8575	0.0002	4536.4	6393.1	76.483	10.37	14.50	1611
500	20.62908	2.0637	0.0002	5055.5	7118.4	78.010	10.39	14.51	1698
550	22.69128	2.2700	0.0002	5575.2	7844.3	79.395	10.40	14.53	1780
600	24.75345	2.4762	0.0002	6096.0	8571.4	80.657	10.42	14.55	1859
700	28.87775	2.8886	0.0001	7140.9	10028.7	82.906	10.48	14.60	2006
800	33.00201	3.3011	0.0001	8193.5	11493.7	84.864	10.57	14.69	2142
900	37.12624	3.7135	0.0001	9256.8	12969.4	86.601	10.70	14.82	2268
1000	41.25045	4.1259	0.0001	10333.9	14459.0	88.168	10.86	14.98	2386
1200	49.49885	4.9500	0.0001	12544.2	17494.1	90.916	11.24	15.36	2601
1400	57.74736	5.7756	0.0001	14837.2	20611.9	93.324	11.68	15.81	2796
1600	65.99738	6.6004	0.0001	17220.5	23820.2	95.473	12.17	16.30	2973
1800	74.25729	7.4253	0.0001	19715.2	27140.9	97.422	12.83	16.98	3134
2000	82.55973	8.2501	0.0001	22395.0	30651.0	99.281	14.05	18.27	3275
2500	104.41653	10.3122	0.0000	31587.8	42029.5	104.298	25.06	30.43	3536
3000	133.52968	12.3743	0.0000	52916.7	66269.7	113.034	61.69	73.70	3839

* TWO-PHASE BOUNDARY

Table 4.12 (continued)
THERMODYNAMIC PROPERTIES OF NORMAL HYDROGEN

0.10 MPA ISOBAR

TEMPERATURE DEG. K	DENSITY KG/CU M	$V(DH/DV)_P$ KJ/KG	$V(DP/DU)_V$ MPA-CU M/KJ X 10^2	$-V(DP/DV)_T$ MPA	$(DV/DT)_P/V$ 1/DEG. K	THERMAL CONDUCTIVITY W/K-M X 10^3	VISCOSITY KG/M-S X 10^7	THERMAL DIFFUSIVITY SQ M/HR	DIELECTRIC CONSTANT	PRANDTL NUMBER
* 13.834	77.0685	621.7	0.25639	90.0543	0.01026	73.01	255.39	0.00053	1.25176	2.2323
14	76.9216	617.1	0.25492	88.0151	0.01050	74.68	249.37	0.00054	1.25124	2.1639
15	76.0988	607.3	0.24330	79.2124	0.01149	83.15	219.54	0.00056	1.24835	1.8428
16	75.1978	610.0	0.23204	73.3062	0.01218	89.01	195.44	0.00057	1.24519	1.6308
17	74.2500	609.5	0.22298	67.6550	0.01297	92.85	175.96	0.00057	1.24187	1.4981
18	73.2568	606.6	0.21545	62.2739	0.01386	95.55	159.92	0.00056	1.23840	1.4069
19	72.2131	598.8	0.20897	56.7088	0.01495	97.09	146.46	0.00054	1.23475	1.3501
20	71.1039	588.9	0.20322	51.3422	0.01617	98.44	134.88	0.00052	1.23089	1.3051
* 20.224	70.8360	588.4	0.20198	50.4463	0.01638	98.84	132.43	0.00052	1.22996	1.2911
* 20.224	1.3253	188.9	0.06666	0.0894	0.06427	16.89	11.25	0.00378	1.00400	0.8085
21	1.2630	197.5	0.06698	0.0906	0.06004	17.38	11.66	0.00418	1.00381	0.7956
22	1.1921	208.8	0.06722	0.0918	0.05556	18.06	12.19	0.00470	1.00360	0.7829
23	1.1299	220.1	0.06733	0.0929	0.05186	18.76	12.71	0.00524	1.00341	0.7729
24	1.0745	231.3	0.06738	0.0937	0.04872	19.49	13.22	0.00579	1.00324	0.7646
25	1.0248	242.5	0.06739	0.0944	0.04600	20.23	13.73	0.00637	1.00309	0.7576
26	0.9799	253.7	0.06738	0.0950	0.04362	20.97	14.24	0.00696	1.00296	0.7516
27	0.9391	264.8	0.06736	0.0955	0.04151	21.71	14.74	0.00757	1.00283	0.7463
28	0.9018	275.9	0.06733	0.0959	0.03961	22.46	15.23	0.00820	1.00272	0.7414
29	0.8675	286.9	0.06730	0.0963	0.03791	23.20	15.72	0.00885	1.00262	0.7369
30	0.8359	297.8	0.06727	0.0966	0.03636	23.95	16.21	0.00953	1.00252	0.7326
31	0.8066	308.7	0.06725	0.0969	0.03494	24.64	16.69	0.01020	1.00243	0.7304
32	0.7795	319.5	0.06723	0.0971	0.03364	25.33	17.16	0.01088	1.00235	0.7283
33	0.7541	330.4	0.06720	0.0974	0.03244	26.02	17.63	0.01159	1.00227	0.7263
34	0.7305	341.1	0.06719	0.0976	0.03133	26.71	18.10	0.01232	1.00220	0.7242
35	0.7083	351.9	0.06718	0.0978	0.03030	27.40	18.56	0.01306	1.00214	0.7221
36	0.6875	362.6	0.06715	0.0979	0.02934	28.09	19.01	0.01383	1.00207	0.7200
37	0.6679	373.3	0.06713	0.0981	0.02844	28.79	19.47	0.01461	1.00201	0.7180
38	0.6495	384.0	0.06711	0.0982	0.02760	29.48	19.91	0.01542	1.00196	0.7159
39	0.6320	394.6	0.06709	0.0983	0.02682	30.17	20.36	0.01624	1.00191	0.7139
40	0.6155	405.3	0.06707	0.0985	0.02607	30.87	20.79	0.01708	1.00186	0.7119
42	0.5851	426.6	0.06704	0.0987	0.02471	32.24	21.66	0.01882	1.00176	0.7082
44	0.5576	447.8	0.06699	0.0988	0.02349	33.62	22.51	0.02063	1.00168	0.7045
46	0.5326	469.0	0.06694	0.0990	0.02239	34.99	23.35	0.02252	1.00161	0.7009
48	0.5097	490.3	0.06688	0.0991	0.02140	36.35	24.17	0.02447	1.00154	0.6974
50	0.4888	511.6	0.06681	0.0992	0.02049	37.71	24.98	0.02650	1.00147	0.6941
52	0.4696	533.0	0.06671	0.0993	0.01965	39.07	25.77	0.02859	1.00142	0.6911
54	0.4519	554.5	0.06660	0.0994	0.01889	40.41	26.56	0.03074	1.00136	0.6882
56	0.4354	576.1	0.06646	0.0994	0.01818	41.75	27.33	0.03295	1.00131	0.6857
58	0.4202	597.9	0.06631	0.0995	0.01753	43.06	28.09	0.03521	1.00127	0.6836
60	0.4059	619.8	0.06614	0.0995	0.01692	44.41	28.84	0.03756	1.00122	0.6810
65	0.3743	675.4	0.06560	0.0997	0.01557	47.85	30.67	0.04375	1.00113	0.6743
70	0.3473	732.7	0.06490	0.0997	0.01443	50.95	32.45	0.04995	1.00105	0.6734
75	0.3239	791.7	0.06406	0.0998	0.01345	53.90	34.17	0.05627	1.00098	0.6749
80	0.3035	852.7	0.06310	0.0999	0.01259	56.21	35.85	0.06212	1.00091	0.6845
85	0.2855	915.9	0.06203	0.0999	0.01183	59.18	37.48	0.06884	1.00086	0.6865
90	0.2696	981.4	0.06087	0.0999	0.01117	61.85	39.08	0.07535	1.00081	0.6925
95	0.2553	1048.7	0.05971	0.1000	0.01057	64.30	40.63	0.08176	1.00077	0.7007
100	0.2425	1118.3	0.05850	0.1000	0.01004	76.18	45.74	0.10074	1.00073	0.6739
120	0.2020	1413.6	0.05397	0.1000	0.00835	94.31	54.08	0.14239	1.00061	0.6768
140	0.1731	1728.2	0.05028	0.1000	0.00715	107.69	59.19	0.18122	1.00052	0.6793
160	0.1514	2051.9	0.04750	0.1001	0.00625	118.66	63.00	0.21982	1.00046	0.6812
180	0.1346	2379.3	0.04544	0.1001	0.00556	128.16	66.17	0.25924	1.00041	0.6826
200	0.1211	2706.5	0.04392	0.1001	0.00500	136.62	69.01	0.29998	1.00037	0.6836
220	0.1101	3032.5	0.04277	0.1001	0.00454	144.33	71.66	0.34228	1.00033	0.6843
240	0.1010	3354.6	0.04193	0.1001	0.00417	151.40	74.19	0.38631	1.00030	0.6848
260	0.0932	3672.6	0.04131	0.1001	0.00385	157.97	76.64	0.43214	1.00028	0.6851
280	0.0865	3985.9	0.04086	0.1001	0.00357	164.16	79.04	0.47986	1.00026	0.6852
300	0.0808	4294.7	0.04053	0.1001	0.00333	170.01	81.41	0.52945	1.00024	0.6853
350	0.0692	5051.0	0.04007	0.1001	0.00286	183.67	87.23	0.66195	1.00021	0.6851
400	0.0606	5792.3	0.03987	0.1000	0.00250	196.50	92.97	0.80656	1.00018	0.6849
450	0.0539	6525.3	0.03979	0.1000	0.00222	208.87	98.65	0.96314	1.00016	0.6846
500	0.0485	7259.6	0.03971	0.1000	0.00200	221.13	104.26	1.13145	1.00015	0.6843
550	0.0441	7993.2	0.03966	0.1000	0.00182	233.22	109.82	1.31132	1.00013	0.6841
600	0.0404	8731.0	0.03958	0.1000	0.00167	245.27	115.31	1.50245	1.00012	0.6839
700	0.0346	10225.0	0.03937	0.1000	0.00143	269.36	126.11	1.91755	1.00010	0.6837
800	0.0303	11757.4	0.03903	0.1000	0.00125	293.70	136.66	2.37486	1.00009	0.6837
900	0.0269	13341.4	0.03857	0.1000	0.00111	318.53	146.95	2.87258	1.00008	0.6837
1000	0.0242	14985.9	0.03799	0.1000	0.00100	440.56	157.01	4.36658	1.00007	0.5340
1200	0.0202	18439.4	0.03670	0.1000	0.00083	511.02	176.49	5.92719	1.00006	0.5306
1400	0.0173	22130.6	0.03531	0.1000	0.00071	583.56	195.19	7.67568	1.00005	0.5287
1600	0.0152	26083.8	0.03389	0.1000	0.00062	660.02	213.23	9.62022	1.00005	0.5266
1800	0.0135	30563.2	0.03215	0.1000	0.00056	751.84	230.69	11.83628	1.00004	0.5210
2000	0.0121	36507.6	0.02938	0.0999	0.00050	895.49	247.69	14.57039	1.00004	0.5052
2500	0.0096	75131.5	0.01667	0.0988	0.00041	2134.82	289.81	26.37134	1.00003	0.4131
3000	0.0075	204885.4	0.00721	0.0927	0.00036	6758.73	338.77	44.08577	1.00002	0.3694

* TWO-PHASE BOUNDARY

Table 4.12 (continued)
THERMODYNAMIC PROPERTIES OF NORMAL HYDROGEN

0.101325 MPA ISOBAR

TEMPERATURE DEG. K	VOLUME CU M/KG	ISOTHERM DERIVATIVE CU M-MPA/KG	ISOCHORE DERIVATIVE MPA/K	INTERNAL ENERGY KJ/KG-K	ENTHALPY KJ/KG	ENTROPY KJ/KG-K	CV KJ /	CP KG-K	VELOCITY OF SOUND M/S
* 13.835	0.01298	1.1686	0.9244	218.3	219.6	14.084	4.68	6.38	1263
14	0.01300	1.1444	0.9243	219.3	220.7	14.162	4.71	6.48	1254
15	0.01314	1.0411	0.9105	226.0	227.4	14.624	4.92	6.98	1216
16	0.01330	0.9750	0.8926	233.2	234.6	15.089	5.12	7.43	1190
17	0.01347	0.9113	0.8775	240.9	242.3	15.555	5.30	7.90	1166
18	0.01365	0.8502	0.8630	249.0	250.4	16.022	5.47	8.41	1143
19	0.01385	0.7854	0.8477	257.7	259.1	16.490	5.62	8.95	1119
20	0.01406	0.7222	0.8305	266.9	268.3	16.963	5.75	9.52	1094
* 20.268	0.01413	0.7093	0.8253	269.5	270.9	17.093	5.78	9.66	1089
* 20.268	0.74739	0.0675	0.0058	640.7	716.5	39.082	6.50	12.15	355
21	0.78035	0.0715	0.0055	646.0	725.1	39.499	6.44	11.89	363
22	0.82689	0.0768	0.0052	653.0	736.8	40.047	6.37	11.62	374
23	0.87260	0.0820	0.0049	659.9	748.3	40.559	6.33	11.43	385
24	0.91769	0.0870	0.0046	666.7	759.7	41.043	6.31	11.29	395
25	0.96226	0.0919	0.0044	673.4	770.9	41.501	6.29	11.17	404
26	1.00643	0.0968	0.0042	680.1	782.0	41.935	6.27	11.08	413
27	1.05025	0.1015	0.0040	686.7	793.1	42.352	6.26	11.00	422
28	1.09378	0.1062	0.0039	693.2	804.1	42.750	6.26	10.94	431
29	1.13705	0.1109	0.0037	699.7	815.0	43.134	6.25	10.88	439
30	1.18011	0.1154	0.0036	706.2	825.8	43.501	6.25	10.83	448
31	1.22297	0.1200	0.0034	712.7	836.6	43.856	6.24	10.79	456
32	1.26567	0.1245	0.0033	719.2	847.4	44.198	6.24	10.76	463
33	1.30822	0.1290	0.0032	725.6	858.1	44.529	6.23	10.72	471
34	1.35063	0.1335	0.0031	732.0	868.8	44.848	6.23	10.69	479
35	1.39293	0.1379	0.0030	738.4	879.5	45.158	6.23	10.67	486
36	1.43512	0.1424	0.0029	744.8	890.2	45.458	6.22	10.64	493
37	1.47721	0.1468	0.0028	751.1	900.8	45.750	6.22	10.62	501
38	1.51922	0.1512	0.0027	757.5	911.4	46.033	6.22	10.60	508
39	1.56115	0.1555	0.0027	763.8	922.0	46.308	6.22	10.59	515
40	1.60300	0.1599	0.0026	770.2	932.6	46.576	6.22	10.57	521
42	1.68652	0.1686	0.0025	782.8	953.7	47.091	6.22	10.54	535
44	1.76982	0.1772	0.0024	795.5	974.8	47.581	6.22	10.52	548
46	1.85292	0.1858	0.0022	808.1	995.8	48.048	6.22	10.51	560
48	1.93587	0.1943	0.0021	820.7	1016.8	48.495	6.22	10.49	573
50	2.01867	0.2029	0.0021	833.2	1037.8	48.923	6.22	10.48	585
52	2.10135	0.2114	0.0020	845.8	1058.7	49.334	6.23	10.48	596
54	2.18393	0.2199	0.0019	858.4	1079.7	49.729	6.24	10.47	608
56	2.26641	0.2283	0.0018	871.0	1100.6	50.111	6.25	10.48	619
58	2.34880	0.2368	0.0018	883.6	1121.6	50.479	6.26	10.48	630
60	2.43113	0.2452	0.0017	896.2	1142.6	50.834	6.27	10.49	640
65	2.63666	0.2662	0.0016	927.9	1195.1	51.673	6.32	10.52	666
70	2.84187	0.2872	0.0015	959.9	1247.8	52.455	6.39	10.57	690
75	3.04684	0.3081	0.0014	992.1	1300.9	53.187	6.47	10.65	712
80	3.25161	0.3290	0.0013	1024.8	1354.3	53.877	6.56	10.73	734
85	3.45621	0.3498	0.0012	1058.0	1408.2	54.532	6.67	10.84	754
90	3.66069	0.3706	0.0011	1091.8	1462.7	55.155	6.80	10.96	773
95	3.86506	0.3914	0.0011	1126.2	1517.9	55.751	6.93	11.09	791
100	4.06933	0.4122	0.0010	1161.3	1573.6	56.322	7.07	11.23	809
120	4.88574	0.4952	0.0008	1308.8	1803.8	58.417	7.66	11.80	873
140	5.70143	0.5780	0.0007	1467.9	2045.6	60.279	8.22	12.36	932
160	6.51668	0.6607	0.0006	1637.3	2297.6	61.960	8.70	12.83	987
180	7.33165	0.7434	0.0006	1815.4	2558.3	63.495	9.09	13.22	1040
200	8.14642	0.8260	0.0005	2000.5	2825.9	64.905	9.40	13.53	1090
220	8.96105	0.9086	0.0005	2191.2	3099.2	66.207	9.65	13.78	1139
240	9.77558	0.9911	0.0004	2386.3	3376.8	67.415	9.85	13.97	1186
260	10.59003	1.0737	0.0004	2584.8	3657.9	68.541	9.99	14.12	1232
280	11.40442	1.1563	0.0004	2785.9	3941.5	69.591	10.10	14.23	1276
300	12.21788	1.2387	0.0003	2989.0	4226.9	70.575	10.18	14.31	1319
350	14.25342	1.4450	0.0003	3501.6	4945.8	72.792	10.30	14.43	1423
400	16.28883	1.6513	0.0003	4018.1	5668.6	74.722	10.35	14.48	1520
450	18.32416	1.8575	0.0002	4536.4	6393.1	76.428	10.37	14.50	1611
500	20.35943	2.0638	0.0002	5055.5	7118.4	77.956	10.39	14.51	1698
550	22.39466	2.2700	0.0002	5575.2	7844.4	79.340	10.40	14.53	1780
600	24.42987	2.4762	0.0002	6096.0	8571.4	80.602	10.42	14.55	1859
700	28.50024	2.8887	0.0001	7140.9	10028.7	82.852	10.48	14.60	2006
800	32.57056	3.3011	0.0001	8193.5	11493.7	84.810	10.57	14.69	2142
900	36.64086	3.7135	0.0001	9256.8	12969.5	86.546	10.70	14.82	2268
1000	40.71114	4.1259	0.0001	10333.9	14459.0	88.114	10.86	14.98	2386
1200	48.85168	4.9508	0.0001	12544.2	17494.1	90.861	11.24	15.36	2601
1400	56.99233	5.7756	0.0001	14837.2	20611.9	93.270	11.68	15.81	2796
1600	65.13445	6.6005	0.0001	17220.5	23820.2	95.419	12.17	16.30	2973
1800	73.28628	7.4253	0.0001	19714.9	27140.7	97.368	12.83	16.98	3134
2000	81.47980	8.2501	0.0001	22393.8	30649.8	99.226	14.04	18.26	3275
2500	103.04279	10.3122	0.0000	31569.9	42010.7	104.235	24.99	30.35	3536
3000	131.72066	12.3743	0.0000	52805.4	66152.0	112.936	61.40	73.34	3839

* TWO-PHASE BOUNDARY

Table 4.12 (continued)
THERMODYNAMIC PROPERTIES OF NORMAL HYDROGEN

0.101325 MPA ISOBAR

TEMPERATURE DEG. K	DENSITY KG/CU M	$V(DH/DV)_P$ KJ/KG	$V(DP/DU)_V$ MPA-CU M/KJ X 10^2	$-V(DP/DV)_T$ MPA	$(DV/DT)_P/V$ 1/DEG. K	THERMAL CONDUCTIVITY W/K-M X 10^3	VISCOSITY KG/M-S X 10^7	THERMAL DIFFUSIVITY SQ M/HR	DIELECTRIC CONSTANT	PRANDTL NUMBER
* 13.835	77.0693	621.7	0.25638	90.0617	0.01026	73.01	255.40	0.00053	1.25176	2.2322
14	76.9227	617.2	0.25493	88.0322	0.01050	74.68	249.39	0.00054	1.25125	2.1640
15	76.1000	607.3	0.24331	79.2241	0.01149	83.16	219.56	0.00056	1.24835	1.8429
16	75.1991	610.1	0.23205	73.3174	0.01217	89.01	195.46	0.00057	1.24519	1.6309
17	74.2514	609.6	0.22299	67.6660	0.01297	92.85	175.98	0.00057	1.24187	1.4981
18	73.2583	606.7	0.21546	62.2852	0.01386	95.56	159.94	0.00056	1.23840	1.4069
19	72.2148	598.9	0.20898	56.7197	0.01494	97.10	146.47	0.00054	1.23476	1.3501
20	71.1057	589.0	0.20322	51.3532	0.01617	98.44	134.89	0.00052	1.23089	1.3051
* 20.268	70.7859	587.9	0.20175	50.2056	0.01644	98.92	131.98	0.00052	1.22978	1.2893
* 20.268	1.3380	189.1	0.06667	0.0903	0.06424	16.94	11.28	0.00375	1.00404	0.8091
21	1.2815	197.3	0.06697	0.0916	0.06027	17.40	11.67	0.00411	1.00387	0.7970
22	1.2094	208.6	0.06721	0.0929	0.05573	18.08	12.19	0.00463	1.00365	0.7840
23	1.1460	219.8	0.06734	0.0940	0.05200	18.78	12.71	0.00516	1.00346	0.7738
24	1.0897	231.1	0.06739	0.0948	0.04883	19.50	13.23	0.00571	1.00329	0.7654
25	1.0392	242.3	0.06740	0.0955	0.04610	20.24	13.74	0.00628	1.00313	0.7582
26	0.9936	253.5	0.06739	0.0961	0.04370	20.98	14.24	0.00686	1.00300	0.7522
27	0.9522	264.6	0.06736	0.0967	0.04157	21.72	14.74	0.00746	1.00287	0.7468
28	0.9143	275.7	0.06734	0.0971	0.03967	22.47	15.24	0.00809	1.00276	0.7418
29	0.8795	286.7	0.06731	0.0975	0.03796	23.21	15.73	0.00873	1.00265	0.7373
30	0.8474	297.7	0.06728	0.0978	0.03640	23.96	16.21	0.00940	1.00256	0.7330
31	0.8177	308.6	0.06726	0.0981	0.03498	24.65	16.69	0.01006	1.00247	0.7307
32	0.7901	319.4	0.06724	0.0984	0.03367	25.34	17.16	0.01073	1.00238	0.7286
33	0.7644	330.2	0.06721	0.0986	0.03247	26.02	17.63	0.01143	1.00230	0.7266
34	0.7404	341.0	0.06720	0.0988	0.03136	26.72	18.10	0.01215	1.00223	0.7245
35	0.7179	351.7	0.06718	0.0990	0.03032	27.41	18.56	0.01289	1.00216	0.7223
36	0.6968	362.5	0.06716	0.0992	0.02936	28.10	19.02	0.01364	1.00210	0.7202
37	0.6770	373.2	0.06714	0.0994	0.02846	28.79	19.47	0.01442	1.00204	0.7182
38	0.6582	383.9	0.06712	0.0995	0.02762	29.49	19.91	0.01521	1.00198	0.7161
39	0.6406	394.5	0.06710	0.0996	0.02683	30.18	20.36	0.01602	1.00193	0.7141
40	0.6238	405.2	0.06708	0.0997	0.02609	30.87	20.80	0.01685	1.00188	0.7121
42	0.5929	426.5	0.06704	0.0999	0.02472	32.25	21.66	0.01857	1.00179	0.7083
44	0.5650	447.7	0.06700	0.1001	0.02350	33.62	22.51	0.02036	1.00170	0.7046
46	0.5397	469.0	0.06695	0.1003	0.02240	34.99	23.35	0.02222	1.00163	0.7010
48	0.5166	490.2	0.06689	0.1004	0.02140	36.36	24.17	0.02415	1.00156	0.6975
50	0.4954	511.5	0.06681	0.1005	0.02049	37.72	24.98	0.02615	1.00149	0.6942
52	0.4759	532.9	0.06671	0.1006	0.01966	39.07	25.78	0.02821	1.00143	0.6911
54	0.4579	554.4	0.06660	0.1007	0.01889	40.42	26.56	0.03034	1.00138	0.6883
56	0.4412	576.0	0.06647	0.1007	0.01819	41.75	27.33	0.03252	1.00133	0.6858
58	0.4257	597.8	0.06631	0.1008	0.01753	43.06	28.09	0.03475	1.00128	0.6837
60	0.4113	619.7	0.06614	0.1009	0.01692	44.42	28.84	0.03707	1.00124	0.6810
65	0.3793	675.4	0.06560	0.1010	0.01558	47.85	30.68	0.04318	1.00114	0.6744
70	0.3519	732.6	0.06490	0.1011	0.01443	50.95	32.45	0.04930	1.00106	0.6734
75	0.3282	791.7	0.06407	0.1011	0.01345	53.90	34.17	0.05554	1.00099	0.6750
80	0.3075	852.6	0.06311	0.1012	0.01259	56.22	35.85	0.06131	1.00093	0.6845
85	0.2893	915.8	0.06204	0.1012	0.01184	59.18	37.48	0.06793	1.00087	0.6865
90	0.2732	981.4	0.06088	0.1012	0.01117	61.85	39.08	0.07437	1.00082	0.6925
95	0.2587	1048.7	0.05971	0.1013	0.01057	64.30	40.64	0.08070	1.00078	0.7007
100	0.2457	1118.3	0.05850	0.1013	0.01004	76.18	45.74	0.09942	1.00074	0.6740
120	0.2047	1413.6	0.05397	0.1013	0.00835	94.31	54.08	0.14053	1.00062	0.6768
140	0.1754	1728.2	0.05028	0.1014	0.00715	107.69	59.19	0.17885	1.00053	0.6793
160	0.1535	2051.9	0.04750	0.1014	0.00625	118.66	63.00	0.21694	1.00046	0.6812
180	0.1364	2379.3	0.04544	0.1014	0.00556	128.16	66.17	0.25585	1.00041	0.6826
200	0.1228	2706.5	0.04392	0.1014	0.00500	136.62	69.01	0.29606	1.00037	0.6836
220	0.1116	3032.6	0.04277	0.1014	0.00454	144.33	71.66	0.33781	1.00034	0.6843
240	0.1023	3354.7	0.04193	0.1014	0.00417	151.40	74.19	0.38126	1.00031	0.6848
260	0.0944	3672.6	0.04131	0.1014	0.00385	157.98	76.64	0.42649	1.00028	0.6851
280	0.0877	3985.9	0.04086	0.1014	0.00357	164.16	79.04	0.47359	1.00026	0.6852
300	0.0818	4294.7	0.04053	0.1014	0.00333	170.01	81.41	0.52253	1.00025	0.6853
350	0.0702	5051.0	0.04007	0.1014	0.00286	183.67	87.23	0.65330	1.00021	0.6851
400	0.0614	5792.3	0.03987	0.1014	0.00250	196.50	92.97	0.79602	1.00018	0.6849
450	0.0546	6525.4	0.03979	0.1014	0.00222	208.88	98.65	0.95055	1.00016	0.6846
500	0.0491	7259.7	0.03971	0.1014	0.00200	221.13	104.26	1.11666	1.00015	0.6843
550	0.0447	7993.2	0.03966	0.1014	0.00182	233.22	109.82	1.29418	1.00013	0.6841
600	0.0409	8731.1	0.03958	0.1014	0.00167	245.27	115.31	1.48282	1.00012	0.6839
700	0.0351	10225.1	0.03937	0.1014	0.00143	269.36	126.11	1.89250	1.00011	0.6837
800	0.0307	11757.4	0.03903	0.1014	0.00125	293.70	136.66	2.34383	1.00009	0.6837
900	0.0273	13341.4	0.03857	0.1013	0.00111	318.53	146.95	2.83504	1.00008	0.6837
1000	0.0246	14985.9	0.03799	0.1013	0.00100	440.56	157.01	4.30949	1.00007	0.5340
1200	0.0205	18439.4	0.03670	0.1013	0.00083	511.02	176.49	5.84969	1.00006	0.5306
1400	0.0175	22130.6	0.03531	0.1013	0.00071	583.55	195.19	7.57531	1.00005	0.5287
1600	0.0154	26083.1	0.03389	0.1013	0.00062	660.00	213.23	9.49429	1.00005	0.5266
1800	0.0136	30559.3	0.03215	0.1013	0.00056	751.69	230.69	11.68063	1.00004	0.5211
2000	0.0123	36491.4	0.02940	0.1013	0.00050	894.86	247.69	14.37620	1.00004	0.5054
2500	0.0097	74933.1	0.01671	0.1001	0.00040	2127.30	289.80	26.00356	1.00003	0.4134
3000	0.0076	203[illegible]33.2	0.00725	0.0939	0.00036	6725.27	338.71	43.48332	1.00002	0.3694

* TWO-PHASE BOUNDARY

Table 4.12 (continued)
THERMODYNAMIC PROPERTIES OF NORMAL HYDROGEN

0.15 MPA ISOBAR

TEMPERATURE DEG. K	VOLUME CU M/KG	ISOTHERM DERIVATIVE CU M-MPA/KG	ISOCHORE DERIVATIVE MPA/K	INTERNAL ENERGY KJ/KG-K	ENTHALPY KJ/KG	ENTROPY KJ/KG-K	CV KJ / KG-K	CP KJ / KG-K	VELOCITY OF SOUND M/S
* 13.851	0.01297	1.1717	0.9250	218.3	220.2	14.085	4.68	6.38	1264
14	0.01299	1.1520	0.9249	219.3	221.2	14.156	4.71	6.47	1257
15	0.01313	1.0461	0.9122	225.9	227.9	14.617	4.92	6.97	1218
16	0.01329	0.9798	0.8942	233.1	235.1	15.082	5.11	7.42	1192
17	0.01346	0.9161	0.8791	240.7	242.8	15.547	5.30	7.89	1168
18	0.01364	0.8552	0.8646	248.9	250.9	16.013	5.46	8.39	1146
19	0.01384	0.7903	0.8493	257.5	259.6	16.480	5.61	8.93	1121
20	0.01405	0.7272	0.8321	266.7	268.8	16.952	5.74	9.50	1097
21	0.01429	0.6608	0.8128	276.5	278.6	17.433	5.86	10.14	1070
* 21.672	0.01447	0.6202	0.7984	283.4	285.6	17.761	5.92	10.58	1053
* 21.672	0.52255	0.0673	0.0084	644.7	723.1	37.959	6.59	12.84	362
22	0.53312	0.0692	0.0082	647.2	727.2	38.146	6.55	12.67	366
23	0.56639	0.0751	0.0077	654.7	739.6	38.700	6.46	12.25	378
24	0.59879	0.0807	0.0072	661.9	751.7	39.214	6.40	11.95	388
25	0.63055	0.0861	0.0068	669.0	763.5	39.697	6.36	11.73	399
26	0.66180	0.0914	0.0065	675.9	775.2	40.152	6.33	11.55	408
27	0.69263	0.0965	0.0062	682.8	786.7	40.586	6.31	11.42	418
28	0.72313	0.1015	0.0059	689.6	798.0	40.997	6.30	11.30	427
29	0.75333	0.1065	0.0056	696.3	809.3	41.393	6.29	11.21	436
30	0.78329	0.1113	0.0054	703.0	820.4	41.771	6.28	11.13	444
31	0.81304	0.1161	0.0052	709.6	831.5	42.135	6.27	11.06	453
32	0.84260	0.1209	0.0050	716.2	842.6	42.485	6.26	11.00	461
33	0.87200	0.1256	0.0048	722.7	853.5	42.823	6.26	10.94	469
34	0.90126	0.1302	0.0047	729.3	864.4	43.149	6.25	10.90	476
35	0.93039	0.1348	0.0045	735.8	875.3	43.464	6.25	10.85	484
36	0.95940	0.1394	0.0044	742.2	886.2	43.770	6.24	10.82	491
37	0.98832	0.1439	0.0043	748.7	897.0	44.066	6.24	10.78	499
38	1.01714	0.1485	0.0041	755.2	907.7	44.353	6.24	10.75	506
39	1.04588	0.1530	0.0040	761.6	918.5	44.632	6.23	10.73	513
40	1.07455	0.1574	0.0039	768.0	929.2	44.903	6.23	10.70	520
42	1.13168	0.1663	0.0037	780.8	950.6	45.424	6.23	10.66	534
44	1.18858	0.1751	0.0035	793.6	971.8	45.919	6.23	10.63	547
46	1.24528	0.1839	0.0034	806.3	993.1	46.391	6.23	10.60	559
48	1.30182	0.1926	0.0032	819.0	1014.2	46.841	6.23	10.57	572
50	1.35822	0.2013	0.0031	831.6	1035.4	47.272	6.23	10.56	584
52	1.41449	0.2099	0.0029	844.3	1056.5	47.686	6.24	10.55	596
54	1.47065	0.2185	0.0028	856.9	1077.5	48.084	6.24	10.54	607
56	1.52672	0.2271	0.0027	869.6	1098.6	48.468	6.25	10.53	618
58	1.58270	0.2356	0.0026	882.3	1119.7	48.838	6.26	10.53	629
60	1.63861	0.2441	0.0025	894.9	1140.7	49.195	6.28	10.54	640
65	1.77810	0.2654	0.0023	926.8	1193.5	50.037	6.33	10.56	666
70	1.91727	0.2865	0.0022	958.8	1246.4	50.822	6.39	10.61	690
75	2.05619	0.3075	0.0020	991.2	1299.6	51.556	6.47	10.67	712
80	2.19491	0.3285	0.0019	1023.9	1353.2	52.248	6.56	10.76	734
85	2.33348	0.3495	0.0018	1057.2	1407.2	52.904	6.68	10.86	754
90	2.47191	0.3704	0.0017	1091.0	1461.8	53.528	6.80	10.98	773
95	2.61023	0.3913	0.0016	1125.5	1517.0	54.125	6.93	11.10	792
100	2.74846	0.4121	0.0015	1160.6	1572.9	54.697	7.07	11.24	809
120	3.30069	0.4953	0.0013	1308.2	1803.3	56.794	7.66	11.81	874
140	3.85219	0.5782	0.0011	1467.4	2045.2	58.658	8.22	12.37	933
160	4.40326	0.6611	0.0009	1636.9	2297.4	60.340	8.70	12.84	988
180	4.95405	0.7438	0.0008	1815.0	2558.2	61.875	9.09	13.23	1040
200	5.50464	0.8265	0.0008	2000.2	2825.9	63.286	9.40	13.54	1091
220	6.05509	0.9091	0.0007	2191.0	3099.2	64.588	9.65	13.79	1139
240	6.60544	0.9918	0.0006	2386.1	3376.9	65.796	9.85	13.98	1186
260	7.15571	1.0744	0.0006	2584.6	3658.0	66.922	9.99	14.12	1232
280	7.70593	1.1569	0.0005	2785.8	3941.6	67.972	10.10	14.23	1277
300	8.25551	1.2394	0.0005	2988.8	4227.1	68.956	10.18	14.31	1320
350	9.63066	1.4457	0.0004	3501.5	4946.1	71.174	10.30	14.43	1423
400	11.00567	1.6520	0.0004	4018.0	5668.8	73.104	10.35	14.48	1520
450	12.38060	1.8583	0.0003	4536.3	6393.4	74.810	10.37	14.50	1612
500	13.75547	2.0646	0.0003	5055.4	7118.7	76.338	10.39	14.51	1698
550	15.13031	2.2708	0.0003	5575.2	7844.7	77.722	10.40	14.53	1781
600	16.50512	2.4770	0.0003	6096.0	8571.7	78.984	10.42	14.55	1859
700	19.25470	2.8895	0.0002	7140.9	10029.1	81.234	10.48	14.60	2007
800	22.00423	3.3019	0.0002	8193.4	11494.1	83.192	10.57	14.69	2143
900	24.75374	3.7144	0.0002	9256.8	12969.9	84.928	10.70	14.82	2269
1000	27.50323	4.1268	0.0002	10333.9	14459.4	86.496	10.86	14.98	2386
1200	33.00218	4.9517	0.0001	12544.2	17494.5	89.243	11.24	15.36	2602
1400	38.50118	5.7765	0.0001	14837.1	20612.3	91.652	11.68	15.80	2796
1600	44.00101	6.6013	0.0001	17219.4	23819.5	93.800	12.16	16.29	2973
1800	49.50621	7.4262	0.0001	19708.0	27133.9	95.746	12.78	16.92	3136
2000	55.03455	8.2510	0.0001	22362.5	30617.7	97.591	13.83	18.04	3279
2500	69.45464	10.3131	0.0001	31085.9	41504.1	102.396	22.98	28.11	3549
3000	87.83039	12.3752	0.0001	49798.0	62972.6	110.142	53.35	63.69	3839

* TWO-PHASE BOUNDARY

Table 4.12 (continued)
THERMODYNAMIC PROPERTIES OF NORMAL HYDROGEN

0.15 MPA ISOBAR

TEMPERATURE DEG. K	DENSITY KG/CU M	$V(DH/DV)_P$ KJ/KG	$V(DP/DU)_V$ MPA-CU M/KJ X 10^2	$-V(DP/DV)_T$ MPA	$(DV/DT)_P/V$ 1/DEG. K	THERMAL CONDUCTIVITY W/K-M X 10^3	VISCOSITY KG/M-S X 10^7	THERMAL DIFFUSIVITY SQ M/HR	DIELECTRIC CONSTANT	PRANDTL NUMBER
* 13.851	77.0972	623.3	0.25632	90.3332	0.01024	73.22	255.70	0.00054	1.25186	2.2289
14	76.9648	620.0	0.25501	88.6598	0.01043	74.72	250.26	0.00054	1.25139	2.1663
15	76.1448	609.0	0.24371	79.6566	0.01145	83.21	220.31	0.00056	1.24851	1.8463
16	75.2472	611.7	0.23242	73.7262	0.01213	89.09	196.11	0.00057	1.24536	1.6332
17	74.3038	611.3	0.22334	68.0668	0.01292	92.95	176.57	0.00057	1.24206	1.4998
18	73.3154	608.6	0.21579	62.6992	0.01379	95.67	160.49	0.00056	1.23860	1.4078
19	72.2760	600.8	0.20930	57.1192	0.01487	97.23	146.98	0.00054	1.23497	1.3505
20	71.1714	591.1	0.20352	51.7551	0.01608	98.59	135.36	0.00052	1.23112	1.3048
21	69.9823	577.2	0.19830	46.2470	0.01758	100.14	125.13	0.00051	1.22699	1.2675
* 21.672	69.1286	568.4	0.19499	42.8767	0.01862	100.73	118.87	0.00050	1.22403	1.2490
* 21.672	1.9137	196.1	0.06680	0.1287	0.06547	18.61	12.17	0.00273	1.00578	0.8398
22	1.8757	199.9	0.06699	0.1299	0.06337	18.79	12.34	0.00285	1.00566	0.8319
23	1.7656	211.6	0.06737	0.1326	0.05790	19.39	12.85	0.00323	1.00533	0.8119
24	1.6700	223.2	0.06757	0.1348	0.05353	20.04	13.36	0.00361	1.00504	0.7967
25	1.5859	234.9	0.06765	0.1366	0.04993	20.72	13.86	0.00401	1.00479	0.7847
26	1.5110	246.5	0.06768	0.1381	0.04688	21.41	14.36	0.00442	1.00456	0.7751
27	1.4438	258.0	0.06767	0.1394	0.04426	22.12	14.86	0.00483	1.00436	0.7669
28	1.3829	269.4	0.06764	0.1404	0.04196	22.83	15.35	0.00526	1.00417	0.7598
29	1.3274	280.8	0.06760	0.1413	0.03992	23.55	15.83	0.00570	1.00401	0.7535
30	1.2767	292.0	0.06757	0.1421	0.03810	24.28	16.31	0.00615	1.00385	0.7478
31	1.2300	303.2	0.06753	0.1428	0.03647	24.96	16.79	0.00661	1.00371	0.7439
32	1.1868	314.4	0.06750	0.1434	0.03498	25.64	17.26	0.00707	1.00358	0.7405
33	1.1468	325.4	0.06747	0.1440	0.03363	26.31	17.73	0.00755	1.00346	0.7373
34	1.1096	336.4	0.06745	0.1445	0.03239	27.00	18.19	0.00804	1.00335	0.7342
35	1.0748	347.4	0.06743	0.1449	0.03124	27.68	18.65	0.00854	1.00324	0.7312
36	1.0423	358.3	0.06740	0.1453	0.03019	28.37	19.10	0.00906	1.00314	0.7284
37	1.0118	369.2	0.06738	0.1456	0.02921	29.05	19.55	0.00959	1.00305	0.7257
38	0.9831	380.0	0.06735	0.1460	0.02829	29.74	20.00	0.01013	1.00297	0.7230
39	0.9561	390.9	0.06733	0.1462	0.02744	30.43	20.44	0.01068	1.00288	0.7205
40	0.9306	401.7	0.06730	0.1465	0.02664	31.11	20.88	0.01125	1.00281	0.7180
42	0.8836	423.2	0.06725	0.1470	0.02519	32.48	21.74	0.01241	1.00266	0.7135
44	0.8413	444.7	0.06720	0.1473	0.02389	33.84	22.58	0.01363	1.00254	0.7091
46	0.8030	466.2	0.06714	0.1477	0.02273	35.20	23.42	0.01489	1.00242	0.7050
48	0.7682	487.6	0.06707	0.1479	0.02169	36.56	24.24	0.01620	1.00232	0.7010
50	0.7363	509.1	0.06699	0.1482	0.02074	37.91	25.04	0.01756	1.00222	0.6973
52	0.7070	530.7	0.06688	0.1484	0.01987	39.26	25.84	0.01896	1.00213	0.6940
54	0.6800	552.3	0.06676	0.1486	0.01908	40.60	26.62	0.02040	1.00205	0.6908
56	0.6550	574.1	0.06662	0.1487	0.01835	41.93	27.39	0.02188	1.00197	0.6881
58	0.6318	596.0	0.06647	0.1489	0.01767	43.23	28.15	0.02339	1.00190	0.6858
60	0.6103	618.0	0.06629	0.1490	0.01705	44.58	28.90	0.02496	1.00184	0.6829
65	0.5624	673.9	0.06574	0.1492	0.01567	48.01	30.73	0.02910	1.00170	0.6759
70	0.5216	731.4	0.06503	0.1494	0.01450	51.10	32.50	0.03325	1.00157	0.6747
75	0.4863	790.6	0.06418	0.1496	0.01350	54.03	34.22	0.03747	1.00147	0.6761
80	0.4556	851.7	0.06321	0.1497	0.01263	56.34	35.89	0.04138	1.00137	0.6854
85	0.4285	915.0	0.06213	0.1498	0.01187	59.30	37.52	0.04587	1.00129	0.6873
90	0.4045	980.7	0.06097	0.1498	0.01120	61.96	39.12	0.05022	1.00122	0.6931
95	0.3831	1048.1	0.05979	0.1499	0.01059	64.41	40.67	0.05451	1.00115	0.7012
100	0.3638	1117.8	0.05858	0.1499	0.01006	76.21	45.74	0.06709	1.00110	0.6745
120	0.3030	1413.4	0.05403	0.1501	0.00836	94.33	54.07	0.09488	1.00091	0.6772
140	0.2596	1728.3	0.05032	0.1501	0.00715	107.70	59.19	0.12079	1.00078	0.6796
160	0.2271	2052.2	0.04754	0.1501	0.00626	118.68	63.00	0.14655	1.00068	0.6814
180	0.2019	2379.7	0.04548	0.1501	0.00556	128.18	66.17	0.17285	1.00061	0.6828
200	0.1817	2707.1	0.04395	0.1501	0.00500	136.64	69.02	0.20003	1.00055	0.6838
220	0.1652	3033.3	0.04280	0.1501	0.00454	144.35	71.67	0.22825	1.00050	0.6844
240	0.1514	3355.5	0.04195	0.1501	0.00417	151.42	74.20	0.25762	1.00046	0.6849
260	0.1397	3673.5	0.04133	0.1501	0.00384	158.00	76.65	0.28818	1.00042	0.6851
280	0.1298	3986.9	0.04087	0.1501	0.00357	164.18	79.05	0.32001	1.00039	0.6853
300	0.1211	4295.8	0.04054	0.1501	0.00333	170.04	81.42	0.35308	1.00036	0.6853
350	0.1038	5052.2	0.04008	0.1501	0.00286	183.70	87.24	0.44145	1.00031	0.6852
400	0.0909	5793.6	0.03988	0.1501	0.00250	196.53	92.98	0.53789	1.00027	0.6849
450	0.0808	6526.7	0.03980	0.1501	0.00222	208.91	98.66	0.64232	1.00024	0.6846
500	0.0727	7261.0	0.03972	0.1501	0.00200	221.17	104.28	0.75456	1.00022	0.6843
550	0.0661	7994.6	0.03967	0.1501	0.00182	233.26	109.84	0.87451	1.00020	0.6841
600	0.0606	8732.5	0.03959	0.1501	0.00167	245.32	115.33	1.00198	1.00018	0.6839
700	0.0519	10226.5	0.03938	0.1501	0.00143	269.41	126.14	1.27880	1.00016	0.6837
800	0.0454	11758.9	0.03904	0.1501	0.00125	293.76	136.68	1.58377	1.00014	0.6837
900	0.0404	13342.9	0.03857	0.1501	0.00111	318.60	146.98	1.91568	1.00012	0.6837
1000	0.0364	14987.5	0.03799	0.1500	0.00100	440.56	157.05	2.91135	1.00011	0.5341
1200	0.0303	18440.9	0.03671	0.1500	0.00083	511.02	176.53	3.95179	1.00009	0.5307
1400	0.0260	22130.4	0.03532	0.1500	0.00071	583.48	195.24	5.11727	1.00008	0.5288
1600	0.0227	26066.6	0.03392	0.1500	0.00062	659.28	213.28	6.41128	1.00007	0.5269
1800	0.0202	30453.8	0.03229	0.1500	0.00056	747.50	230.74	7.87443	1.00006	0.5222
2000	0.0182	36053.5	0.02984	0.1499	0.00050	877.89	247.74	9.64366	1.00005	0.5090
2500	0.0144	69569.8	0.01813	0.1485	0.00040	1924.03	289.57	17.11322	1.00004	0.4231
3000	0.0114	179472.6	0.00823	0.1409	0.00035	5805.59	337.08	28.82253	1.00003	0.3698

* TWO-PHASE BOUNDARY

Table 4.12 (continued)
THERMODYNAMIC PROPERTIES OF NORMAL HYDROGEN

0.20 MPA ISOBAR

TEMPERATURE DEG. K	VOLUME CU M/KG	ISOTHERM DERIVATIVE CU M-MPA/KG	ISOCHORE DERIVATIVE MPA/K	INTERNAL ENERGY KJ/KG-K	ENTHALPY KJ/KG	ENTROPY KJ/KG-K	CV KJ / KG-K	CP KJ / KG-K	VELOCITY OF SOUND M/S
* 13.868	0.01297	1.1749	0.9255	218.3	220.9	14.086	4.68	6.38	1265
14	0.01299	1.1597	0.9255	219.2	221.8	14.149	4.71	6.46	1260
15	0.01312	1.0513	0.9140	225.8	228.4	14.609	4.91	6.97	1221
16	0.01328	0.9847	0.8959	233.0	235.6	15.074	5.11	7.41	1195
17	0.01345	0.9209	0.8807	240.6	243.3	15.538	5.29	7.88	1171
18	0.01363	0.8603	0.8662	248.7	251.4	16.003	5.46	8.38	1149
19	0.01382	0.7953	0.8509	257.3	260.1	16.470	5.61	8.92	1124
20	0.01404	0.7323	0.8337	266.4	269.3	16.941	5.74	9.48	1100
21	0.01427	0.6661	0.8146	276.2	279.0	17.421	5.85	10.12	1073
22	0.01454	0.5995	0.7928	286.6	289.5	17.908	5.95	10.82	1044
* 22.805	0.01478	0.5396	0.7723	295.6	298.5	18.310	6.01	11.52	1017
* 22.805	0.40014	0.0661	0.0112	647.0	727.1	37.114	6.66	13.55	367
23	0.40532	0.0674	0.0110	648.6	729.7	37.228	6.63	13.41	369
24	0.43157	0.0738	0.0102	656.5	742.8	37.785	6.52	12.84	381
25	0.45696	0.0799	0.0096	664.0	755.4	38.301	6.45	12.44	392
26	0.48170	0.0856	0.0090	671.4	767.7	38.781	6.40	12.15	403
27	0.50594	0.0912	0.0086	678.5	779.7	39.236	6.37	11.92	413
28	0.52976	0.0966	0.0081	685.6	791.6	39.664	6.35	11.74	423
29	0.55326	0.1018	0.0078	692.6	803.2	40.074	6.33	11.59	432
30	0.57647	0.1070	0.0074	699.4	814.7	40.465	6.32	11.46	441
31	0.59944	0.1121	0.0071	706.3	826.1	40.839	6.30	11.36	449
32	0.62222	0.1170	0.0069	713.0	837.5	41.198	6.29	11.27	458
33	0.64481	0.1219	0.0066	719.7	848.7	41.543	6.29	11.19	466
34	0.66725	0.1268	0.0064	726.4	859.8	41.877	6.28	11.12	474
35	0.68955	0.1316	0.0062	733.0	870.9	42.198	6.27	11.06	482
36	0.71173	0.1363	0.0060	739.6	882.0	42.509	6.26	11.00	489
37	0.73380	0.1410	0.0058	746.2	892.9	42.810	6.26	10.96	497
38	0.75578	0.1457	0.0056	752.7	903.9	43.102	6.25	10.91	504
39	0.77766	0.1503	0.0054	759.2	914.8	43.385	6.25	10.88	511
40	0.79948	0.1549	0.0053	765.7	925.6	43.660	6.25	10.84	518
42	0.84289	0.1640	0.0050	778.7	947.3	44.187	6.24	10.78	532
44	0.88607	0.1730	0.0047	791.6	968.8	44.687	6.24	10.73	546
46	0.92904	0.1819	0.0045	804.4	990.2	45.163	6.24	10.69	559
48	0.97185	0.1908	0.0043	817.2	1011.6	45.618	6.24	10.66	571
50	1.01450	0.1996	0.0041	830.0	1032.9	46.052	6.24	10.64	583
52	1.05704	0.2084	0.0040	842.7	1054.1	46.469	6.24	10.62	595
54	1.09946	0.2171	0.0038	855.4	1075.3	46.870	6.25	10.60	607
56	1.14178	0.2258	0.0037	868.1	1096.5	47.256	6.26	10.59	618
58	1.18402	0.2344	0.0035	880.9	1117.7	47.627	6.27	10.59	629
60	1.22618	0.2430	0.0034	893.6	1138.9	47.986	6.28	10.59	640
65	1.33130	0.2645	0.0031	925.6	1191.8	48.833	6.33	10.60	666
70	1.43610	0.2858	0.0029	957.7	1244.9	49.620	6.39	10.64	690
75	1.54066	0.3069	0.0027	990.2	1298.3	50.357	6.47	10.70	712
80	1.64501	0.3281	0.0025	1023.0	1352.0	51.050	6.57	10.78	734
85	1.74920	0.3491	0.0024	1056.3	1406.2	51.707	6.68	10.88	754
90	1.85326	0.3701	0.0022	1090.2	1460.9	52.333	6.80	11.00	774
95	1.95721	0.3911	0.0021	1124.7	1516.2	52.930	6.93	11.12	792
100	2.06107	0.4120	0.0020	1159.9	1572.1	53.504	7.07	11.26	810
120	2.47580	0.4954	0.0017	1307.7	1802.8	55.603	7.66	11.82	874
140	2.88982	0.5785	0.0014	1466.9	2044.9	57.468	8.22	12.37	933
160	3.30341	0.6615	0.0013	1636.5	2297.2	59.151	8.70	12.84	988
180	3.71671	0.7443	0.0011	1814.7	2558.1	60.687	9.09	13.23	1041
200	4.12982	0.8270	0.0010	1999.9	2825.9	62.098	9.40	13.54	1091
220	4.54279	0.9097	0.0009	2190.7	3099.3	63.400	9.65	13.79	1140
240	4.95565	0.9924	0.0008	2385.9	3377.0	64.609	9.85	13.98	1187
260	5.36844	1.0750	0.0008	2584.4	3658.1	65.735	9.99	14.13	1233
280	5.78117	1.1576	0.0007	2785.6	3941.8	66.785	10.10	14.23	1277
300	6.19343	1.2401	0.0007	2988.7	4227.3	67.769	10.19	14.31	1320
350	7.22489	1.4465	0.0006	3501.3	4946.3	69.987	10.30	14.43	1423
400	8.25622	1.6528	0.0005	4017.9	5669.1	71.917	10.35	14.48	1520
450	9.28747	1.8591	0.0004	4536.2	6393.7	73.624	10.37	14.50	1612
500	10.31867	2.0654	0.0004	5055.4	7119.1	75.151	10.39	14.52	1699
550	11.34982	2.2716	0.0004	5575.1	7845.1	76.536	10.40	14.53	1781
600	12.38096	2.4779	0.0003	6095.9	8572.1	77.798	10.42	14.55	1860
700	14.44317	2.8904	0.0003	7140.8	10029.5	80.048	10.48	14.60	2007
800	16.50534	3.3028	0.0003	8193.4	11494.5	82.006	10.57	14.69	2143
900	18.56749	3.7152	0.0002	9256.8	12970.3	83.742	10.70	14.82	2269
1000	20.62962	4.1277	0.0002	10333.9	14459.8	85.309	10.86	14.98	2386
1200	24.75385	4.9525	0.0002	12544.1	17494.9	88.057	11.24	15.36	2602
1400	28.87810	5.7774	0.0001	14837.0	20612.6	90.466	11.68	15.80	2796
1600	33.00288	6.6022	0.0001	17218.7	23819.3	92.613	12.15	16.28	2974
1800	37.13116	7.4271	0.0001	19703.7	27130.0	94.557	12.74	16.88	3137
2000	41.27447	8.2519	0.0001	22343.1	30598.0	96.393	13.71	17.90	3282
2500	52.02197	10.3140	0.0001	30786.7	41191.1	101.073	21.74	26.73	3559
3000	65.34074	12.3761	0.0001	47934.6	61002.7	108.228	48.29	57.68	3840

* TWO-PHASE BOUNDARY

Table 4.12 (continued)
THERMODYNAMIC PROPERTIES OF NORMAL HYDROGEN

0.20 MPA ISOBAR

TEMPERATURE DEG. K	DENSITY KG/CU M	$V(\partial H/\partial V)_P$ KJ/KG	$V(\partial P/\partial U)_V$ MPA-CU M/KJ X 10^2	$-V(\partial P/\partial V)_T$ MPA	$(\partial V/\partial T)_P/V$ 1/DEG. K	THERMAL CONDUCTIVITY W/K-M X 10^3	VISCOSITY KG/M-S X 10^7	THERMAL DIFFUSIVITY SQ M/HR	DIELECTRIC CONSTANT	PRANDTL NUMBER
* 13.868	77.1259	624.9	0.25626	90.6124	0.01021	73.43	256.01	0.00054	1.25196	2.2255
14	77.0078	622.9	0.25509	89.3027	0.01036	74.76	251.16	0.00054	1.25155	2.1687
15	76.1907	610.6	0.24412	80.0995	0.01141	83.27	221.08	0.00056	1.24867	1.8499
16	75.2962	613.4	0.23280	74.1448	0.01208	89.17	196.79	0.00058	1.24553	1.6356
17	74.3574	613.0	0.22370	68.4766	0.01286	93.05	177.19	0.00057	1.24224	1.5014
18	73.3736	610.5	0.21613	63.1221	0.01372	95.79	161.05	0.00056	1.23880	1.4087
19	72.3384	602.9	0.20963	57.5276	0.01479	97.36	147.49	0.00054	1.23519	1.3510
20	71.2386	593.3	0.20383	52.1662	0.01598	98.74	135.84	0.00053	1.23136	1.3046
21	70.0571	579.6	0.19860	46.6655	0.01746	100.31	125.59	0.00051	1.22725	1.2666
22	68.7873	563.1	0.19370	41.2384	0.01923	101.11	116.47	0.00049	1.22284	1.2469
* 22.805	67.6762	544.6	0.18973	36.5158	0.02115	101.20	109.69	0.00047	1.21900	1.2486
* 22.805	2.4991	200.5	0.06703	0.1652	0.06756	20.09	12.91	0.00214	1.00755	0.8709
23	2.4672	202.9	0.06715	0.1663	0.06610	20.18	13.01	0.00220	1.00745	0.8646
24	2.3171	215.0	0.06761	0.1710	0.05973	20.70	13.51	0.00250	1.00700	0.8380
25	2.1884	227.1	0.06784	0.1747	0.05478	21.29	14.00	0.00282	1.00661	0.8183
26	2.0760	239.1	0.06793	0.1778	0.05079	21.92	14.50	0.00313	1.00627	0.8032
27	1.9765	251.1	0.06795	0.1802	0.04747	22.58	14.98	0.00345	1.00597	0.7910
28	1.8876	262.9	0.06793	0.1823	0.04464	23.25	15.47	0.00378	1.00570	0.7808
29	1.8075	274.6	0.06790	0.1841	0.04219	23.94	15.95	0.00411	1.00546	0.7722
30	1.7347	286.3	0.06785	0.1856	0.04005	24.63	16.43	0.00446	1.00524	0.7646
31	1.6682	297.8	0.06781	0.1869	0.03815	25.30	16.90	0.00481	1.00504	0.7588
32	1.6072	309.2	0.06777	0.1881	0.03645	25.96	17.37	0.00516	1.00485	0.7537
33	1.5508	320.5	0.06773	0.1891	0.03491	26.63	17.83	0.00552	1.00468	0.7492
34	1.4987	331.7	0.06771	0.1900	0.03352	27.30	18.29	0.00590	1.00452	0.7448
35	1.4502	342.9	0.06769	0.1908	0.03225	27.98	18.74	0.00628	1.00438	0.7408
36	1.4050	354.0	0.06766	0.1915	0.03108	28.65	19.19	0.00667	1.00424	0.7372
37	1.3628	365.1	0.06762	0.1922	0.03001	29.33	19.64	0.00707	1.00411	0.7337
38	1.3231	376.1	0.06760	0.1927	0.02902	30.01	20.08	0.00748	1.00399	0.7304
39	1.2859	387.1	0.06757	0.1933	0.02809	30.69	20.52	0.00790	1.00388	0.7273
40	1.2508	398.1	0.06753	0.1937	0.02723	31.37	20.96	0.00833	1.00377	0.7243
42	1.1864	419.9	0.06748	0.1946	0.02568	32.72	21.82	0.00921	1.00358	0.7189
44	1.1286	441.7	0.06741	0.1953	0.02430	34.07	22.66	0.01012	1.00340	0.7139
46	1.0764	463.4	0.06734	0.1958	0.02308	35.42	23.49	0.01108	1.00325	0.7092
48	1.0290	485.0	0.06727	0.1963	0.02198	36.77	24.31	0.01207	1.00310	0.7048
50	0.9857	506.7	0.06717	0.1968	0.02099	38.12	25.11	0.01309	1.00297	0.7007
52	0.9460	528.4	0.06706	0.1971	0.02009	39.46	25.90	0.01414	1.00285	0.6969
54	0.9095	550.2	0.06693	0.1975	0.01927	40.79	26.68	0.01523	1.00274	0.6935
56	0.8758	572.1	0.06679	0.1977	0.01851	42.11	27.45	0.01634	1.00264	0.6905
58	0.8446	594.1	0.06662	0.1980	0.01782	43.41	28.21	0.01748	1.00255	0.6879
60	0.8155	616.3	0.06644	0.1982	0.01718	44.75	28.95	0.01866	1.00246	0.6849
65	0.7511	672.5	0.06587	0.1986	0.01577	48.17	30.78	0.02177	1.00226	0.6775
70	0.6963	730.1	0.06515	0.1990	0.01458	51.24	32.55	0.02489	1.00210	0.6759
75	0.6491	789.5	0.06430	0.1992	0.01356	54.16	34.26	0.02806	1.00196	0.6772
80	0.6079	850.8	0.06332	0.1994	0.01268	56.46	35.93	0.03101	1.00183	0.6864
85	0.5717	914.2	0.06224	0.1996	0.01190	59.42	37.56	0.03438	1.00172	0.6880
90	0.5396	980.0	0.06106	0.1997	0.01122	62.08	39.15	0.03766	1.00163	0.6937
95	0.5109	1047.5	0.05988	0.1998	0.01062	64.52	40.71	0.04088	1.00154	0.7016
100	0.4852	1117.3	0.05867	0.1999	0.01007	76.25	45.74	0.05027	1.00146	0.6751
120	0.4039	1413.2	0.05409	0.2001	0.00837	94.35	54.06	0.07112	1.00122	0.6776
140	0.3460	1728.4	0.05037	0.2002	0.00716	107.72	59.19	0.09058	1.00104	0.6798
160	0.3027	2052.4	0.04758	0.2002	0.00626	118.70	63.00	0.10991	1.00091	0.6816
180	0.2691	2380.2	0.04551	0.2003	0.00556	128.20	66.18	0.12966	1.00081	0.6829
200	0.2421	2707.7	0.04398	0.2003	0.00500	136.66	69.02	0.15006	1.00073	0.6839
220	0.2201	3034.0	0.04282	0.2003	0.00454	144.37	71.67	0.17123	1.00066	0.6845
240	0.2018	3356.4	0.04197	0.2003	0.00417	151.44	74.20	0.19327	1.00061	0.6849
260	0.1863	3674.5	0.04135	0.2002	0.00384	158.02	76.65	0.21620	1.00056	0.6852
280	0.1730	3987.9	0.04089	0.2002	0.00357	164.20	79.06	0.24009	1.00052	0.6853
300	0.1615	4296.9	0.04056	0.2002	0.00333	170.06	81.42	0.26490	1.00049	0.6853
350	0.1384	5053.5	0.04009	0.2002	0.00286	183.73	87.25	0.33121	1.00042	0.6852
400	0.1211	5794.9	0.03989	0.2002	0.00250	196.56	93.00	0.40356	1.00036	0.6849
450	0.1077	6528.1	0.03981	0.2002	0.00222	208.94	98.67	0.48191	1.00032	0.6846
500	0.0969	7262.4	0.03973	0.2002	0.00200	221.21	104.29	0.56612	1.00029	0.6844
550	0.0881	7996.1	0.03967	0.2001	0.00182	233.30	109.86	0.65611	1.00027	0.6841
600	0.0808	8733.9	0.03960	0.2001	0.00167	245.36	115.35	0.75174	1.00024	0.6839
700	0.0692	10228.0	0.03938	0.2001	0.00143	269.47	126.16	0.95942	1.00021	0.6837
800	0.0606	11760.4	0.03904	0.2001	0.00125	293.82	136.71	1.18822	1.00018	0.6837
900	0.0539	13344.5	0.03858	0.2001	0.00111	318.67	147.01	1.43723	1.00016	0.6837
1000	0.0485	14989.1	0.03800	0.2001	0.00100	440.56	157.08	2.18374	1.00015	0.5342
1200	0.0404	18442.5	0.03671	0.2001	0.00083	511.02	176.57	2.96410	1.00012	0.5308
1400	0.0346	22130.9	0.03533	0.2001	0.00071	583.43	195.28	3.83814	1.00010	0.5290
1600	0.0303	26057.1	0.03394	0.2001	0.00062	658.83	213.33	4.80761	1.00009	0.5272
1800	0.0269	30389.4	0.03238	0.2000	0.00056	744.92	230.80	5.89858	1.00008	0.5230
2000	0.0242	35783.4	0.03011	0.1999	0.00050	867.40	247.79	7.20098	1.00007	0.5113
2500	0.0192	66242.4	0.01914	0.1983	0.00040	1797.89	289.46	12.59695	1.00006	0.4303
3000	0.0153	163878.2	0.00902	0.1894	0.00035	5220.51	336.01	21.28967	1.00005	0.3713

* TWO-PHASE BOUNDARY

Table 4.12 (continued)
THERMODYNAMIC PROPERTIES OF NORMAL HYDROGEN

0.25 MPA ISOBAR

TEMPERATURE DEG. K	VOLUME CU M/KG	ISOTHERM DERIVATIVE CU M-MPA/KG	ISOCHORE DERIVATIVE MPA/K	INTERNAL ENERGY KJ/KG-K	ENTHALPY KJ/KG	ENTROPY KJ/KG-K	CV KJ / KG-K	CP KJ / KG-K	VELOCITY OF SOUND M/S
* 13.885	0.01296	1.1781	0.9261	218.3	221.6	14.087	4.69	6.38	1267
14	0.01298	1.1673	0.9261	219.1	222.3	14.142	4.71	6.44	1264
15	0.01312	1.0565	0.9158	225.7	229.0	14.602	4.91	6.96	1224
16	0.01327	0.9896	0.8975	232.9	236.2	15.066	5.11	7.40	1197
17	0.01344	0.9257	0.8824	240.5	243.8	15.530	5.29	7.87	1174
18	0.01362	0.8653	0.8678	248.5	251.9	15.994	5.46	8.37	1151
19	0.01381	0.8002	0.8526	257.1	260.6	16.460	5.61	8.90	1127
20	0.01402	0.7373	0.8354	266.2	269.7	16.930	5.74	9.46	1103
21	0.01426	0.6713	0.8163	275.9	279.5	17.408	5.85	10.09	1076
22	0.01452	0.6051	0.7946	286.3	289.9	17.894	5.95	10.79	1048
23	0.01482	0.5340	0.7690	297.4	301.1	18.392	6.03	11.62	1015
* 23.751	0.01507	0.4836	0.7464	306.3	310.1	18.776	6.08	12.29	989
* 23.751	0.32422	0.0645	0.0140	648.3	729.4	36.440	6.73	14.28	370
24	0.32992	0.0663	0.0137	650.4	732.9	36.588	6.69	14.05	373
25	0.35185	0.0731	0.0127	658.6	746.6	37.147	6.56	13.36	386
26	0.37292	0.0795	0.0119	666.5	759.7	37.659	6.49	12.87	397
27	0.39336	0.0856	0.0112	674.1	772.4	38.139	6.44	12.52	408
28	0.41331	0.0914	0.0106	681.4	784.8	38.587	6.40	12.24	418
29	0.43286	0.0971	0.0100	688.7	796.9	39.014	6.38	12.02	428
30	0.45210	0.1026	0.0096	695.8	808.8	39.418	6.36	11.84	437
31	0.47106	0.1079	0.0092	702.8	820.6	39.804	6.34	11.69	446
32	0.48980	0.1131	0.0088	709.8	832.2	40.173	6.33	11.57	455
33	0.50834	0.1183	0.0084	716.6	843.7	40.527	6.31	11.46	463
34	0.52671	0.1233	0.0081	723.4	855.1	40.868	6.30	11.36	471
35	0.54494	0.1283	0.0078	730.2	866.4	41.196	6.29	11.28	479
36	0.56303	0.1332	0.0076	736.9	877.7	41.513	6.28	11.20	487
37	0.58101	0.1380	0.0073	743.6	888.9	41.819	6.28	11.14	495
38	0.59889	0.1429	0.0071	750.2	900.0	42.115	6.27	11.08	502
39	0.61668	0.1476	0.0069	756.9	911.0	42.403	6.26	11.03	510
40	0.63439	0.1523	0.0067	763.4	922.0	42.681	6.26	10.99	517
42	0.66959	0.1617	0.0063	776.5	943.9	43.215	6.25	10.91	531
44	0.70454	0.1709	0.0060	789.6	965.7	43.721	6.25	10.85	545
46	0.73928	0.1800	0.0057	802.5	987.3	44.202	6.25	10.79	558
48	0.77385	0.1890	0.0054	815.4	1008.9	44.660	6.24	10.75	570
50	0.80827	0.1980	0.0052	828.3	1030.3	45.098	6.25	10.72	583
52	0.84256	0.2069	0.0050	841.1	1051.7	45.518	6.25	10.69	595
54	0.87674	0.2157	0.0048	853.9	1073.1	45.921	6.26	10.67	606
56	0.91082	0.2245	0.0046	866.7	1094.4	46.309	6.26	10.65	618
58	0.94481	0.2332	0.0044	879.5	1115.7	46.683	6.28	10.64	629
60	0.97873	0.2419	0.0043	892.3	1137.0	47.044	6.29	10.64	640
65	1.06323	0.2636	0.0039	924.4	1190.2	47.894	6.33	10.64	666
70	1.14741	0.2850	0.0036	956.6	1243.5	48.684	6.40	10.68	690
75	1.23134	0.3064	0.0034	989.1	1297.0	49.423	6.47	10.73	713
80	1.31507	0.3276	0.0032	1022.1	1350.9	50.118	6.57	10.81	734
85	1.39864	0.3488	0.0030	1055.5	1405.1	50.777	6.68	10.91	755
90	1.48207	0.3699	0.0028	1089.4	1459.9	51.404	6.80	11.02	774
95	1.56540	0.3909	0.0027	1124.0	1515.3	52.002	6.93	11.14	792
100	1.64864	0.4119	0.0025	1159.2	1571.3	52.577	7.08	11.27	810
120	1.98088	0.4955	0.0021	1307.1	1802.3	54.678	7.66	11.84	875
140	2.31241	0.5788	0.0018	1466.5	2044.6	56.545	8.22	12.38	933
160	2.64350	0.6618	0.0016	1636.1	2297.0	58.228	8.70	12.85	989
180	2.97431	0.7448	0.0014	1814.4	2558.0	59.765	9.09	13.23	1041
200	3.30493	0.8276	0.0013	1999.6	2825.8	61.176	9.40	13.54	1092
220	3.63540	0.9103	0.0011	2190.4	3099.3	62.479	9.66	13.79	1140
240	3.96578	0.9930	0.0010	2385.7	3377.1	63.688	9.85	13.98	1187
260	4.29608	1.0757	0.0010	2584.2	3658.3	64.813	10.00	14.13	1233
280	4.62632	1.1583	0.0009	2785.4	3942.0	65.864	10.11	14.24	1277
300	4.95618	1.2408	0.0008	2988.5	4227.5	66.849	10.19	14.31	1321
350	5.78144	1.4473	0.0007	3501.2	4946.6	69.066	10.30	14.43	1424
400	6.60656	1.6536	0.0006	4017.8	5669.4	70.997	10.35	14.48	1521
450	7.43160	1.8599	0.0006	4536.2	6394.1	72.703	10.37	14.50	1612
500	8.25658	2.0662	0.0005	5055.3	7119.4	74.231	10.39	14.52	1699
550	9.08153	2.2725	0.0005	5575.0	7845.4	75.615	10.40	14.53	1781
600	9.90646	2.4787	0.0004	6095.9	8572.5	76.877	10.42	14.55	1860
700	11.55625	2.8912	0.0004	7140.8	10029.9	79.127	10.48	14.60	2007
800	13.20601	3.3037	0.0003	8193.4	11494.9	81.085	10.57	14.69	2143
900	14.85574	3.7161	0.0003	9256.8	12970.7	82.821	10.70	14.82	2269
1000	16.50545	4.1286	0.0003	10333.9	14460.2	84.389	10.86	14.98	2387
1200	19.80485	4.9534	0.0002	12544.1	17495.3	87.136	11.24	15.36	2602
1400	23.10426	5.7783	0.0002	14837.0	20613.0	89.545	11.68	15.80	2796
1600	26.40404	6.6031	0.0002	17218.3	23819.3	91.693	12.15	16.28	2974
1800	29.70632	7.4280	0.0001	19700.8	27127.4	93.635	12.72	16.86	3138
2000	33.01935	8.2528	0.0001	22329.8	30584.7	95.466	13.62	17.80	3284
2500	41.58048	10.3149	0.0001	30582.5	40977.6	100.059	20.89	25.79	3566
3000	51.98234	12.3770	0.0001	46661.3	59656.9	106.810	44.80	53.57	3842

* TWO-PHASE BOUNDARY

Table 4.12 (continued)
THERMODYNAMIC PROPERTIES OF NORMAL HYDROGEN

0.25 MPA ISOBAR

TEMPERATURE DEG. K	DENSITY KG/CU M	$V(DH/DV)_P$ KJ/KG	$V(DP/DU)_V$ MPA-CU M/KJ X 10^2	$-V(DP/DV)_T$ MPA	$(DV/DT)_P/V$ 1/DEG. K	THERMAL CONDUCTIVITY W/K-M X 10^3	VISCOSITY KG/M-S X 10^7	THERMAL DIFFUSIVITY SQ M/HR	DIELECTRIC CONSTANT	PRANDTL NUMBER
* 13.885	77.1545	626.5	0.25619	90.8919	0.01019	73.63	256.32	0.00054	1.25206	2.2222
14	77.0506	625.7	0.25517	89.9438	0.01030	74.79	252.05	0.00054	1.25170	2.1711
15	76.2363	612.3	0.24452	80.5409	0.01137	83.33	221.85	0.00057	1.24883	1.8534
16	75.3451	615.0	0.23317	74.5620	0.01204	89.24	197.46	0.00058	1.24571	1.6381
17	74.4107	614.8	0.22406	68.8847	0.01281	93.14	177.80	0.00057	1.24243	1.5032
18	73.4314	612.5	0.21646	63.5428	0.01366	95.90	161.61	0.00056	1.23901	1.4097
19	72.4005	604.9	0.20995	57.9339	0.01472	97.49	148.01	0.00054	1.23541	1.3514
20	71.3054	595.5	0.20413	52.5755	0.01589	98.89	136.31	0.00053	1.23159	1.3044
21	70.1314	582.0	0.19889	47.0814	0.01734	100.49	126.05	0.00051	1.22751	1.2657
22	68.8698	565.8	0.19399	41.6725	0.01907	101.30	116.92	0.00049	1.22313	1.2451
23	67.4908	544.5	0.18907	36.0382	0.02134	101.38	108.58	0.00047	1.21836	1.2443
* 23.751	66.3708	528.3	0.18510	32.0967	0.02325	100.99	102.75	0.00045	1.21449	1.2500
* 23.751	3.0843	203.3	0.06730	0.1988	0.07022	21.44	13.56	0.00175	1.00932	0.9028
24	3.0310	206.5	0.06747	0.2009	0.06806	21.53	13.68	0.00182	1.00916	0.8928
25	2.8422	219.1	0.06792	0.2078	0.06097	21.98	14.16	0.00208	1.00859	0.8607
26	2.6816	231.6	0.06813	0.2132	0.05558	22.51	14.64	0.00235	1.00810	0.8374
27	2.5422	244.1	0.06820	0.2176	0.05129	23.09	15.12	0.00261	1.00768	0.8195
28	2.4195	256.3	0.06821	0.2212	0.04776	23.71	15.60	0.00288	1.00731	0.8051
29	2.3102	268.5	0.06818	0.2243	0.04478	24.35	16.07	0.00316	1.00698	0.7934
30	2.2119	280.4	0.06813	0.2268	0.04223	25.01	16.54	0.00344	1.00668	0.7834
31	2.1229	292.3	0.06808	0.2290	0.04000	25.66	17.01	0.00372	1.00641	0.7752
32	2.0417	304.0	0.06804	0.2310	0.03804	26.31	17.47	0.00401	1.00616	0.7681
33	1.9672	315.6	0.06798	0.2326	0.03630	26.96	17.93	0.00431	1.00594	0.7620
34	1.8986	327.0	0.06796	0.2341	0.03473	27.62	18.39	0.00461	1.00573	0.7563
35	1.8351	338.4	0.06794	0.2354	0.03332	28.28	18.84	0.00492	1.00554	0.7512
36	1.7761	349.7	0.06791	0.2366	0.03203	28.95	19.29	0.00524	1.00536	0.7465
37	1.7211	361.0	0.06787	0.2376	0.03086	29.62	19.73	0.00556	1.00520	0.7422
38	1.6697	372.2	0.06784	0.2385	0.02978	30.28	20.17	0.00589	1.00504	0.7382
39	1.6216	383.4	0.06781	0.2394	0.02878	30.96	20.61	0.00623	1.00489	0.7345
40	1.5763	394.5	0.06777	0.2401	0.02785	31.63	21.04	0.00657	1.00476	0.7309
42	1.4935	416.6	0.06770	0.2415	0.02618	32.97	21.90	0.00728	1.00451	0.7246
44	1.4194	438.6	0.06762	0.2426	0.02473	34.31	22.74	0.00802	1.00428	0.7188
46	1.3527	460.6	0.06754	0.2435	0.02344	35.65	23.56	0.00879	1.00408	0.7135
48	1.2922	482.4	0.06746	0.2443	0.02228	36.98	24.38	0.00958	1.00390	0.7086
50	1.2372	504.3	0.06736	0.2450	0.02125	38.32	25.18	0.01041	1.00373	0.7040
52	1.1869	526.2	0.06724	0.2455	0.02031	39.65	25.97	0.01125	1.00358	0.6999
54	1.1406	548.2	0.06710	0.2460	0.01946	40.98	26.74	0.01212	1.00344	0.6962
56	1.0979	570.2	0.06695	0.2465	0.01868	42.29	27.51	0.01302	1.00331	0.6929
58	1.0584	592.3	0.06678	0.2469	0.01797	43.59	28.26	0.01393	1.00319	0.6901
60	1.0217	614.6	0.06659	0.2472	0.01731	44.92	29.01	0.01488	1.00308	0.6869
65	0.9405	671.0	0.06601	0.2479	0.01586	48.33	30.83	0.01738	1.00284	0.6790
70	0.8715	728.9	0.06528	0.2484	0.01465	51.39	32.59	0.01988	1.00263	0.6772
75	0.8121	788.4	0.06442	0.2488	0.01361	54.29	34.31	0.02242	1.00245	0.6783
80	0.7604	849.8	0.06343	0.2491	0.01272	56.59	35.98	0.02478	1.00229	0.6873
85	0.7150	913.4	0.06234	0.2494	0.01194	59.54	37.60	0.02749	1.00216	0.6887
90	0.6747	979.3	0.06116	0.2496	0.01125	62.20	39.19	0.03012	1.00203	0.6943
95	0.6388	1046.9	0.05997	0.2497	0.01064	64.64	40.74	0.03270	1.00193	0.7021
100	0.6066	1116.8	0.05875	0.2498	0.01009	76.29	45.74	0.04017	1.00183	0.6757
120	0.5048	1413.1	0.05415	0.2502	0.00838	94.37	54.06	0.05686	1.00152	0.6779
140	0.4324	1728.4	0.05042	0.2503	0.00716	107.74	59.19	0.07245	1.00130	0.6801
160	0.3783	2052.7	0.04762	0.2504	0.00626	118.72	63.00	0.08793	1.00114	0.6818
180	0.3362	2380.6	0.04554	0.2504	0.00556	128.22	66.18	0.10374	1.00101	0.6831
200	0.3026	2708.3	0.04400	0.2504	0.00500	136.68	69.03	0.12007	1.00091	0.6840
220	0.2751	3034.8	0.04285	0.2504	0.00454	144.39	71.68	0.13702	1.00083	0.6846
240	0.2522	3357.2	0.04199	0.2504	0.00416	151.46	74.21	0.15466	1.00076	0.6850
260	0.2328	3675.4	0.04137	0.2504	0.00384	158.04	76.66	0.17302	1.00070	0.6853
280	0.2162	3989.0	0.04091	0.2504	0.00357	164.22	79.07	0.19213	1.00065	0.6854
300	0.2018	4298.0	0.04057	0.2504	0.00333	170.08	81.43	0.21199	1.00061	0.6854
350	0.1730	5054.7	0.04010	0.2503	0.00285	183.75	87.26	0.26506	1.00052	0.6852
400	0.1514	5796.2	0.03990	0.2503	0.00250	196.59	93.01	0.32296	1.00046	0.6849
450	0.1346	6529.4	0.03982	0.2503	0.00222	208.98	98.69	0.38566	1.00041	0.6846
500	0.1211	7263.8	0.03974	0.2503	0.00200	221.25	104.31	0.45305	1.00036	0.6844
550	0.1101	7997.5	0.03968	0.2502	0.00182	233.35	109.87	0.52507	1.00033	0.6841
600	0.1009	8735.4	0.03961	0.2502	0.00167	245.41	115.37	0.60160	1.00030	0.6839
700	0.0865	10229.5	0.03939	0.2502	0.00143	269.52	126.18	0.76780	1.00026	0.6837
800	0.0757	11761.9	0.03905	0.2502	0.00125	293.89	136.74	0.95089	1.00023	0.6836
900	0.0673	13346.0	0.03858	0.2501	0.00111	318.73	147.04	1.15016	1.00020	0.6837
1000	0.0606	14990.6	0.03800	0.2501	0.00100	440.56	157.11	1.74717	1.00018	0.5343
1200	0.0505	18444.1	0.03671	0.2501	0.00083	511.02	176.61	2.37148	1.00015	0.5310
1400	0.0433	22131.9	0.03533	0.2501	0.00071	583.40	195.33	3.07069	1.00013	0.5291
1600	0.0379	26051.2	0.03396	0.2501	0.00062	658.53	213.38	3.84570	1.00011	0.5274
1800	0.0337	30346.0	0.03244	0.2500	0.00056	743.15	230.85	4.71498	1.00010	0.5236
2000	0.0303	35599.7	0.03030	0.2499	0.00050	860.23	247.84	5.74331	1.00009	0.5130
2500	0.0240	63967.0	0.01990	0.2481	0.00040	1711.60	289.40	9.93600	1.00007	0.4360
3000	0.0192	153046.9	0.00967	0.2381	0.00035	4814.04	335.26	16.81836	1.00006	0.3730

* TWO-PHASE BOUNDARY

Table 4.12 (continued)
THERMODYNAMIC PROPERTIES OF NORMAL HYDROGEN

0.30 MPA ISOBAR

TEMPERATURE DEG. K	VOLUME CU M/KG	ISOTHERM DERIVATIVE CU M-MPA/KG	ISOCHORE DERIVATIVE MPA/K	INTERNAL ENERGY KJ/KG-K	ENTHALPY KJ/KG	ENTROPY KJ/KG-K	CV KJ / KG-K	CP KJ / KG-K	VELOCITY OF SOUND M/S
* 13.901	0.01296	1.1812	0.9267	218.3	222.2	14.088	4.69	6.38	1268
14	0.01297	1.1750	0.9266	219.0	222.9	14.136	4.71	6.43	1267
15	0.01311	1.0616	0.9176	225.6	229.5	14.595	4.91	6.96	1226
16	0.01326	0.9945	0.8992	232.7	236.7	15.059	5.11	7.40	1200
17	0.01343	0.9305	0.8840	240.3	244.4	15.521	5.29	7.86	1176
18	0.01361	0.8704	0.8694	248.4	252.5	15.985	5.46	8.35	1154
19	0.01380	0.8051	0.8542	256.9	261.1	16.450	5.61	8.89	1130
20	0.01401	0.7424	0.8370	266.0	270.2	16.919	5.74	9.44	1105
21	0.01424	0.6765	0.8180	275.7	280.0	17.396	5.85	10.06	1079
22	0.01450	0.6106	0.7964	286.0	290.4	17.880	5.95	10.75	1051
23	0.01480	0.5414	0.7710	297.1	301.5	18.377	6.02	11.55	1019
24	0.01513	0.4708	0.7406	309.0	313.5	18.888	6.09	12.49	983
* 24.571	0.01535	0.4345	0.7207	316.2	320.8	19.188	6.12	13.04	962
* 24.571	0.27220	0.0625	0.0168	648.8	730.5	35.872	6.78	15.05	372
25	0.28074	0.0658	0.0162	652.7	736.9	36.130	6.71	14.59	378
26	0.29965	0.0730	0.0150	661.2	751.1	36.684	6.59	13.80	391
27	0.31775	0.0797	0.0140	669.2	764.6	37.195	6.52	13.25	403
28	0.33525	0.0861	0.0132	677.0	777.6	37.667	6.47	12.84	413
29	0.35227	0.0921	0.0125	684.6	790.3	38.113	6.43	12.52	424
30	0.36892	0.0980	0.0119	692.0	802.7	38.533	6.40	12.27	433
31	0.38526	0.1036	0.0113	699.2	814.8	38.931	6.38	12.06	443
32	0.40135	0.1091	0.0108	706.4	826.8	39.312	6.36	11.89	452
33	0.41722	0.1145	0.0104	713.5	838.6	39.676	6.35	11.75	460
34	0.43291	0.1198	0.0100	720.4	850.3	40.024	6.33	11.62	469
35	0.44844	0.1250	0.0096	727.3	861.9	40.360	6.31	11.51	477
36	0.46382	0.1300	0.0093	734.2	873.3	40.683	6.30	11.41	485
37	0.47909	0.1351	0.0090	741.0	884.7	40.995	6.29	11.33	493
38	0.49425	0.1400	0.0087	747.7	896.0	41.296	6.29	11.26	501
39	0.50932	0.1449	0.0084	754.4	907.2	41.588	6.28	11.20	508
40	0.52429	0.1498	0.0081	761.1	918.4	41.870	6.27	11.14	516
42	0.55403	0.1593	0.0077	774.4	940.6	42.411	6.27	11.04	530
44	0.58350	0.1688	0.0073	787.5	962.6	42.922	6.26	10.96	544
46	0.61277	0.1781	0.0069	800.6	984.4	43.408	6.26	10.90	557
48	0.64185	0.1873	0.0066	813.6	1006.2	43.871	6.25	10.84	570
50	0.67078	0.1964	0.0063	826.6	1027.8	44.312	6.25	10.80	582
52	0.69958	0.2054	0.0060	839.5	1049.4	44.735	6.26	10.76	594
54	0.72827	0.2143	0.0058	852.4	1070.9	45.141	6.26	10.73	606
56	0.75685	0.2232	0.0056	865.2	1092.3	45.531	6.27	10.71	618
58	0.78535	0.2321	0.0054	878.1	1113.7	45.907	6.28	10.70	629
60	0.81376	0.2409	0.0052	891.0	1135.1	46.270	6.29	10.69	640
65	0.88453	0.2627	0.0047	923.2	1188.5	47.123	6.34	10.69	666
70	0.95496	0.2843	0.0044	955.5	1242.0	47.917	6.40	10.71	690
75	1.02514	0.3058	0.0041	988.1	1295.7	48.658	6.48	10.76	713
80	1.09512	0.3272	0.0038	1021.2	1349.7	49.355	6.57	10.84	734
85	1.16494	0.3484	0.0036	1054.6	1404.1	50.015	6.68	10.93	755
90	1.23463	0.3696	0.0034	1088.6	1459.0	50.643	6.81	11.04	774
95	1.30420	0.3907	0.0032	1123.2	1514.5	51.242	6.94	11.16	793
100	1.37369	0.4118	0.0030	1158.5	1570.6	51.818	7.08	11.29	810
120	1.65094	0.4956	0.0025	1306.5	1801.8	53.921	7.67	11.85	875
140	1.92747	0.5791	0.0022	1466.0	2044.3	55.789	8.22	12.39	934
160	2.20356	0.6622	0.0019	1635.7	2296.8	57.474	8.70	12.85	989
180	2.47938	0.7452	0.0017	1814.0	2557.9	59.011	9.09	13.24	1042
200	2.75500	0.8281	0.0015	1999.3	2825.8	60.422	9.40	13.55	1092
220	3.03048	0.9109	0.0014	2190.2	3099.3	61.726	9.66	13.79	1141
240	3.30587	0.9937	0.0013	2385.4	3377.2	62.935	9.85	13.98	1188
260	3.58117	1.0763	0.0012	2584.0	3658.4	64.061	10.00	14.13	1233
280	3.85642	1.1590	0.0011	2785.2	3942.1	65.112	10.11	14.24	1278
300	4.13135	1.2416	0.0010	2988.3	4227.8	66.096	10.19	14.32	1321
350	4.81913	1.4480	0.0009	3501.1	4946.8	68.314	10.30	14.43	1424
400	5.50678	1.6544	0.0008	4017.7	5669.7	70.245	10.35	14.48	1521
450	6.19435	1.8607	0.0007	4536.1	6394.4	71.951	10.37	14.50	1613
500	6.88186	2.0670	0.0006	5055.2	7119.8	73.479	10.39	14.52	1699
550	7.56934	2.2733	0.0005	5575.0	7845.8	74.863	10.40	14.53	1782
600	8.25679	2.4796	0.0005	6095.8	8572.9	76.125	10.42	14.55	1860
700	9.63164	2.8921	0.0004	7140.8	10030.3	78.375	10.48	14.60	2008
800	11.00645	3.3045	0.0004	8193.4	11495.3	80.333	10.57	14.69	2143
900	12.38124	3.7170	0.0003	9256.7	12971.1	82.069	10.70	14.82	2269
1000	13.75601	4.1294	0.0003	10333.8	14460.6	83.637	10.86	14.98	2387
1200	16.50551	4.9543	0.0003	12544.1	17495.8	86.384	11.24	15.36	2602
1400	19.25502	5.7792	0.0002	14836.9	20613.4	88.793	11.68	15.80	2796
1600	22.00482	6.6040	0.0002	17217.9	23819.4	90.940	12.15	16.27	2974
1800	24.75651	7.4289	0.0002	19698.7	27125.6	92.881	12.70	16.84	3138
2000	27.51638	8.2537	0.0002	22320.1	30575.0	94.708	13.56	17.73	3286
2500	34.62797	10.3158	0.0001	30431.8	40820.2	99.238	20.26	25.09	3572
3000	43.14034	12.3779	0.0001	45720.7	58662.8	105.690	42.21	50.52	3844

* TWO-PHASE BOUNDARY

Table 4.12 (continued)
THERMODYNAMIC PROPERTIES OF NORMAL HYDROGEN

0.30 MPA ISOBAR

TEMPERATURE DEG. K	DENSITY KG/CU M	$V(DH/DV)_P$ KJ/KG	$V(DP/DU)_V$ MPA-CU M/KJ X 10^2	$-V(DP/DV)_T$ MPA	$(DV/DT)_P/V$ 1/DEG. K	THERMAL CONDUCTIVITY W/K-M X 10^3	VISCOSITY KG/M-S X 10^7	THERMAL DIFFUSIVITY SQ M/HR	DIELECTRIC CONSTANT	PRANDTL NUMBER
* 13.901	77.1830	628.1	0.25612	91.1717	0.01016	73.84	256.63	0.00054	1.25216	2.2189
14	77.0933	628.6	0.25525	90.5832	0.01023	74.83	252.95	0.00054	1.25185	2.1737
15	76.2817	613.9	0.24492	80.9809	0.01133	83.39	222.62	0.00057	1.24899	1.8570
16	75.3938	616.7	0.23354	74.9777	0.01199	89.32	198.14	0.00058	1.24588	1.6405
17	74.4636	616.5	0.22441	69.2910	0.01276	93.24	178.41	0.00057	1.24262	1.5049
18	73.4888	614.4	0.21680	63.9613	0.01359	96.02	162.17	0.00056	1.23921	1.4107
19	72.4621	606.9	0.21028	58.3382	0.01464	97.62	148.52	0.00055	1.23562	1.3519
20	71.3718	597.7	0.20443	52.9831	0.01580	99.03	136.79	0.00053	1.23182	1.3042
21	70.2051	584.3	0.19917	47.4947	0.01722	100.66	126.51	0.00051	1.22776	1.2649
22	68.9515	568.4	0.19426	42.1036	0.01892	101.50	117.36	0.00049	1.22341	1.2434
23	67.5828	548.3	0.18937	36.5865	0.02107	101.60	109.02	0.00047	1.21868	1.2398
24	66.0882	524.8	0.18403	31.1119	0.02380	101.03	101.33	0.00044	1.21352	1.2529
* 24.571	65.1597	512.3	0.18067	28.3138	0.02545	100.39	97.14	0.00043	1.21032	1.2616
* 24.571	3.6738	205.1	0.06760	0.2296	0.07336	22.71	14.13	0.00148	1.01111	0.9363
25	3.5621	210.8	0.06787	0.2344	0.06921	22.83	14.33	0.00158	1.01077	0.9161
26	3.3372	224.0	0.06825	0.2437	0.06162	23.20	14.80	0.00181	1.01009	0.8801
27	3.1471	236.9	0.06842	0.2509	0.05592	23.69	15.27	0.00205	1.00951	0.8537
28	2.9829	249.7	0.06846	0.2568	0.05142	24.23	15.74	0.00228	1.00902	0.8336
29	2.8387	262.2	0.06844	0.2616	0.04775	24.81	16.20	0.00251	1.00858	0.8177
30	2.7106	274.6	0.06840	0.2656	0.04469	25.42	16.67	0.00275	1.00819	0.8045
31	2.5956	286.8	0.06834	0.2690	0.04207	26.05	17.13	0.00299	1.00784	0.7934
32	2.4916	298.8	0.06829	0.2719	0.03980	26.68	17.59	0.00324	1.00753	0.7839
33	2.3968	310.7	0.06822	0.2745	0.03781	27.31	18.04	0.00349	1.00724	0.7759
34	2.3100	322.4	0.06822	0.2767	0.03604	27.95	18.49	0.00375	1.00698	0.7686
35	2.2300	333.9	0.06820	0.2786	0.03446	28.60	18.94	0.00401	1.00673	0.7621
36	2.1560	345.5	0.06816	0.2804	0.03304	29.26	19.39	0.00428	1.00651	0.7564
37	2.0873	356.9	0.06813	0.2819	0.03175	29.91	19.83	0.00455	1.00630	0.7512
38	2.0233	368.3	0.06809	0.2833	0.03057	30.57	20.26	0.00483	1.00611	0.7464
39	1.9634	379.6	0.06805	0.2846	0.02949	31.23	20.70	0.00512	1.00593	0.7419
40	1.9073	390.9	0.06801	0.2857	0.02849	31.90	21.13	0.00541	1.00576	0.7378
42	1.8050	413.3	0.06793	0.2876	0.02671	33.22	21.98	0.00600	1.00545	0.7304
44	1.7138	435.6	0.06784	0.2892	0.02516	34.54	22.81	0.00662	1.00517	0.7239
46	1.6319	457.8	0.06775	0.2906	0.02380	35.87	23.64	0.00726	1.00493	0.7179
48	1.5580	479.8	0.06766	0.2917	0.02259	37.20	24.45	0.00793	1.00470	0.7124
50	1.4908	501.9	0.06755	0.2927	0.02151	38.53	25.25	0.00862	1.00450	0.7075
52	1.4294	524.0	0.06742	0.2936	0.02054	39.85	26.03	0.00933	1.00431	0.7030
54	1.3731	546.1	0.06728	0.2943	0.01966	41.17	26.81	0.01006	1.00414	0.6989
56	1.3213	568.3	0.06711	0.2949	0.01885	42.48	27.57	0.01080	1.00399	0.6954
58	1.2733	590.5	0.06694	0.2955	0.01812	43.77	28.32	0.01157	1.00384	0.6923
60	1.2289	612.9	0.06674	0.2960	0.01744	45.10	29.07	0.01236	1.00371	0.6889
65	1.1305	669.6	0.06615	0.2970	0.01596	48.49	30.88	0.01445	1.00341	0.6806
70	1.0472	727.7	0.06541	0.2977	0.01472	51.54	32.64	0.01654	1.00316	0.6784
75	0.9755	787.4	0.06454	0.2983	0.01367	54.42	34.36	0.01866	1.00294	0.6794
80	0.9131	848.9	0.06354	0.2987	0.01276	56.72	36.02	0.02064	1.00275	0.6882
85	0.8584	912.6	0.06244	0.2991	0.01197	59.67	37.64	0.02290	1.00259	0.6895
90	0.8100	978.6	0.06125	0.2994	0.01128	62.32	39.23	0.02509	1.00244	0.6949
95	0.7668	1046.3	0.06006	0.2996	0.01066	64.75	40.78	0.02725	1.00231	0.7026
100	0.7280	1116.3	0.05883	0.2998	0.01011	76.33	45.74	0.03345	1.00219	0.6763
120	0.6057	1412.9	0.05421	0.3002	0.00838	94.40	54.05	0.04736	1.00183	0.6783
140	0.5188	1728.5	0.05046	0.3004	0.00717	107.76	59.18	0.06036	1.00156	0.6803
160	0.4538	2053.0	0.04766	0.3005	0.00626	118.74	63.00	0.07328	1.00137	0.6820
180	0.4033	2381.1	0.04557	0.3006	0.00556	128.24	66.18	0.08646	1.00122	0.6832
200	0.3630	2709.0	0.04403	0.3006	0.00500	136.70	69.03	0.10008	1.00109	0.6841
220	0.3300	3035.5	0.04287	0.3006	0.00454	144.41	71.68	0.11421	1.00099	0.6847
240	0.3025	3358.1	0.04202	0.3006	0.00416	151.48	74.21	0.12892	1.00091	0.6851
260	0.2792	3676.4	0.04138	0.3006	0.00384	158.06	76.67	0.14423	1.00084	0.6853
280	0.2593	3990.0	0.04092	0.3005	0.00357	164.25	79.07	0.16016	1.00078	0.6854
300	0.2421	4299.1	0.04059	0.3005	0.00333	170.11	81.44	0.17672	1.00073	0.6854
350	0.2075	5055.9	0.04012	0.3005	0.00285	183.78	87.27	0.22096	1.00063	0.6852
400	0.1816	5797.5	0.03991	0.3004	0.00250	196.62	93.02	0.26923	1.00055	0.6849
450	0.1614	6530.8	0.03983	0.3004	0.00222	209.01	98.70	0.32150	1.00049	0.6846
500	0.1453	7265.2	0.03975	0.3004	0.00200	221.29	104.33	0.37768	1.00044	0.6844
550	0.1321	7998.9	0.03969	0.3003	0.00182	233.39	109.89	0.43771	1.00040	0.6841
600	0.1211	8736.9	0.03961	0.3003	0.00167	245.45	115.39	0.50151	1.00036	0.6839
700	0.1038	10231.0	0.03940	0.3003	0.00143	269.57	126.21	0.64005	1.00031	0.6837
800	0.0909	11763.5	0.03906	0.3002	0.00125	293.95	136.77	0.79267	1.00027	0.6836
900	0.0808	13347.6	0.03859	0.3002	0.00111	318.80	147.07	0.95878	1.00024	0.6837
1000	0.0727	14992.2	0.03801	0.3002	0.00100	440.56	157.15	1.45613	1.00022	0.5344
1200	0.0606	18445.7	0.03672	0.3002	0.00083	511.01	176.65	1.97641	1.00018	0.5311
1400	0.0519	22133.0	0.03533	0.3001	0.00071	583.38	195.38	2.55906	1.00016	0.5292
1600	0.0454	26047.2	0.03397	0.3001	0.00062	658.30	213.43	3.20457	1.00014	0.5276
1800	0.0404	30314.4	0.03249	0.3001	0.00056	741.85	230.91	3.92682	1.00012	0.5241
2000	0.0363	35464.6	0.03045	0.3000	0.00050	854.94	247.90	4.77531	1.00011	0.5142
2500	0.0289	62285.1	0.02051	0.2979	0.00040	1647.79	289.38	8.18729	1.00009	0.4406
3000	0.0232	144953.9	0.01022	0.2869	0.00035	4510.27	334.70	13.86503	1.00007	0.3749

* TWO-PHASE BOUNDARY

Table 4.12 (continued)
THERMODYNAMIC PROPERTIES OF NORMAL HYDROGEN

0.35 MPA ISOBAR

TEMPERATURE DEG. K	VOLUME CU M/KG	ISOTHERM DERIVATIVE CU M-MPA/KG	ISOCHORE DERIVATIVE MPA/K	INTERNAL ENERGY KJ/KG-K	ENTHALPY KJ/KG	ENTROPY KJ/KG-K	CV KJ / KG-K	CP KJ / KG-K	VELOCITY OF SOUND M/S
* 13.918	0.01295	1.1844	0.9273	218.3	222.9	14.090	4.69	6.39	1270
14	0.01296	1.1826	0.9272	218.9	223.4	14.129	4.71	6.42	1270
15	0.01310	1.0667	0.9194	225.5	230.1	14.588	4.91	6.95	1229
16	0.01326	0.9993	0.9009	232.6	237.3	15.051	5.10	7.39	1203
17	0.01342	0.9353	0.8856	240.2	244.9	15.513	5.29	7.86	1179
18	0.01360	0.8753	0.8710	248.2	253.0	15.976	5.45	8.34	1157
19	0.01379	0.8100	0.8559	256.7	261.6	16.440	5.60	8.87	1132
20	0.01400	0.7473	0.8386	265.8	270.7	16.909	5.73	9.42	1108
21	0.01423	0.6817	0.8197	275.4	280.4	17.384	5.85	10.04	1082
22	0.01449	0.6161	0.7982	285.7	290.8	17.867	5.94	10.72	1054
23	0.01478	0.5473	0.7730	296.7	301.9	18.361	6.02	11.51	1023
24	0.01511	0.4770	0.7430	308.6	313.9	18.870	6.09	12.43	987
25	0.01549	0.4056	0.7075	321.4	326.8	19.397	6.14	13.55	946
* 25.299	0.01562	0.3858	0.6949	325.5	330.9	19.562	6.16	13.89	933
* 25.299	0.23414	0.0603	0.0198	648.8	730.7	35.376	6.84	15.88	374
26	0.24651	0.0660	0.0186	655.3	741.6	35.797	6.72	15.02	384
27	0.26317	0.0735	0.0172	664.0	756.2	36.348	6.61	14.16	397
28	0.27906	0.0805	0.0161	672.3	770.0	36.850	6.54	13.56	408
29	0.29438	0.0870	0.0151	680.3	783.3	37.318	6.49	13.11	419
30	0.30925	0.0933	0.0143	688.0	796.2	37.756	6.45	12.76	429
31	0.32378	0.0993	0.0136	695.5	808.9	38.170	6.42	12.48	439
32	0.33801	0.1051	0.0130	702.9	821.2	38.562	6.40	12.26	449
33	0.35201	0.1107	0.0124	710.2	833.4	38.937	6.38	12.06	458
34	0.36580	0.1162	0.0119	717.3	845.4	39.294	6.36	11.90	466
35	0.37942	0.1216	0.0114	724.4	857.2	39.637	6.34	11.76	475
36	0.39289	0.1269	0.0110	731.4	868.9	39.967	6.33	11.64	483
37	0.40623	0.1321	0.0106	738.3	880.5	40.285	6.31	11.54	491
38	0.41946	0.1372	0.0103	745.2	892.0	40.591	6.30	11.45	499
39	0.43259	0.1422	0.0099	752.0	903.4	40.888	6.30	11.37	507
40	0.44562	0.1472	0.0096	758.7	914.7	41.174	6.29	11.30	514
42	0.47146	0.1570	0.0091	772.2	937.2	41.722	6.28	11.18	529
44	0.49704	0.1666	0.0086	785.5	959.4	42.240	6.27	11.08	543
46	0.52239	0.1761	0.0082	798.7	981.5	42.730	6.26	11.00	556
48	0.54756	0.1855	0.0078	811.8	1003.5	43.197	6.26	10.93	569
50	0.57257	0.1947	0.0074	824.9	1025.3	43.642	6.26	10.88	582
52	0.59745	0.2039	0.0071	837.9	1047.0	44.068	6.26	10.84	594
54	0.62222	0.2130	0.0068	850.8	1068.6	44.477	6.27	10.80	606
56	0.64688	0.2220	0.0065	863.8	1090.2	44.870	6.28	10.78	617
58	0.67145	0.2309	0.0063	876.7	1111.7	45.248	6.29	10.75	629
60	0.69594	0.2398	0.0061	889.6	1133.2	45.612	6.30	10.74	639
65	0.75689	0.2618	0.0056	921.9	1186.9	46.469	6.34	10.73	666
70	0.81750	0.2836	0.0051	954.4	1240.5	47.266	6.40	10.75	690
75	0.87786	0.3052	0.0048	987.1	1294.4	48.009	6.48	10.79	713
80	0.93802	0.3267	0.0045	1020.2	1348.5	48.707	6.57	10.86	735
85	0.99802	0.3481	0.0042	1053.8	1403.1	49.369	6.68	10.95	755
90	1.05788	0.3694	0.0039	1087.8	1458.1	49.998	6.81	11.06	775
95	1.11764	0.3906	0.0037	1122.5	1513.6	50.599	6.94	11.17	793
100	1.17730	0.4117	0.0035	1157.8	1569.8	51.175	7.08	11.30	811
120	1.41525	0.4958	0.0029	1306.0	1801.3	53.281	7.67	11.86	876
140	1.65251	0.5794	0.0025	1465.6	2043.9	55.150	8.22	12.40	934
160	1.88933	0.6626	0.0022	1635.4	2296.6	56.835	8.70	12.86	990
180	2.12586	0.7457	0.0020	1813.7	2557.8	58.373	9.09	13.24	1042
200	2.36220	0.8287	0.0018	1999.0	2825.8	59.785	9.41	13.55	1093
220	2.59840	0.9115	0.0016	2189.9	3099.4	61.089	9.66	13.80	1141
240	2.83450	0.9943	0.0015	2385.2	3377.3	62.298	9.85	13.99	1188
260	3.07053	1.0770	0.0013	2583.8	3658.5	63.424	10.00	14.13	1234
280	3.30649	1.1597	0.0013	2785.0	3942.3	64.475	10.11	14.24	1278
300	3.54219	1.2423	0.0012	2988.2	4228.0	65.460	10.19	14.32	1321
350	4.13177	1.4488	0.0010	3501.0	4947.1	67.678	10.30	14.43	1425
400	4.72123	1.6552	0.0009	4017.6	5670.0	69.609	10.35	14.48	1522
450	5.31060	1.8615	0.0008	4536.0	6394.7	71.315	10.37	14.50	1613
500	5.89992	2.0679	0.0007	5055.2	7120.1	72.843	10.39	14.52	1700
550	6.48920	2.2742	0.0006	5574.9	7846.2	74.227	10.40	14.53	1782
600	7.07846	2.4804	0.0006	6095.8	8573.2	75.489	10.42	14.55	1861
700	8.25692	2.8929	0.0005	7140.7	10030.7	77.739	10.48	14.60	2008
800	9.43534	3.3054	0.0004	8193.3	11495.7	79.697	10.57	14.69	2144
900	10.61374	3.7179	0.0004	9256.7	12971.5	81.434	10.70	14.82	2270
1000	11.79212	4.1303	0.0004	10333.8	14461.1	83.001	10.86	14.98	2387
1200	14.14885	4.9552	0.0003	12544.1	17496.2	85.749	11.24	15.36	2603
1400	16.50557	5.7801	0.0003	14836.9	20613.8	88.157	11.68	15.80	2797
1600	18.86252	6.6049	0.0002	17217.7	23813.6	90.304	12.14	16.27	2975
1800	21.22097	7.4298	0.0002	19697.0	27124.3	92.244	12.69	16.82	3139
2000	23.58592	8.2546	0.0002	22312.5	30567.5	94.068	13.51	17.68	3287
2500	29.66646	10.3167	0.0001	30314.6	40697.9	98.549	19.78	24.55	3577
3000	36.85890	12.3788	0.0001	44989.2	57889.9	104.769	40.19	48.15	3846

* TWO-PHASE BOUNDARY

Table 4.12 (continued)
THERMODYNAMIC PROPERTIES OF NORMAL HYDROGEN

0.35 MPA ISOBAR

TEMPERATURE DEG. K	DENSITY KG/CU M	$V(DH/DV)_P$ KJ/KG	$V(DP/DU)_V$ MPA-CU M/KJ X 10^2	$-V(DP/DV)_T$ MPA	$(DV/DT)_P/V$ 1/DEG. K	THERMAL CONDUCTIVITY W/K-M X 10^3	VISCOSITY KG/M-S X 10^7	THERMAL DIFFUSIVITY SQ M/HR	DIELECTRIC CONSTANT	PRANDTL NUMBER
* 13.918	77.2115	629.7	0.25606	91.4518	0.01014	74.05	256.94	0.00054	1.25226	2.2156
14	77.1358	631.5	0.25533	91.2209	0.01016	74.87	253.85	0.00054	1.25200	2.1762
15	76.3270	615.5	0.24531	81.4195	0.01129	83.45	223.39	0.00057	1.24915	1.8606
16	75.4422	618.3	0.23391	75.3921	0.01195	89.40	198.82	0.00058	1.24605	1.6430
17	74.5163	618.2	0.22476	69.6955	0.01271	93.34	179.02	0.00057	1.24280	1.5066
18	73.5458	616.3	0.21712	64.3776	0.01353	96.13	162.73	0.00056	1.23941	1.4117
19	72.5233	608.8	0.21059	58.7405	0.01457	97.75	149.03	0.00055	1.23583	1.3524
20	71.4378	599.8	0.20472	53.3890	0.01571	99.18	137.27	0.00053	1.23205	1.3041
21	70.2782	586.7	0.19946	47.9054	0.01711	100.82	126.97	0.00051	1.22802	1.2641
22	69.0324	571.1	0.19454	42.5317	0.01877	101.69	117.81	0.00049	1.22369	1.2417
23	67.6752	551.3	0.18966	37.0390	0.02087	101.82	109.47	0.00047	1.21899	1.2370
24	66.1935	528.1	0.18438	31.5729	0.02353	101.28	101.78	0.00044	1.21388	1.2489
25	64.5558	501.3	0.17842	26.1848	0.02702	100.08	94.56	0.00041	1.20824	1.2799
* 25.299	64.0123	493.5	0.17627	24.6966	0.02814	99.54	92.41	0.00040	1.20637	1.2891
* 25.299	4.2710	206.2	0.06790	0.2575	0.07699	23.94	14.66	0.00127	1.01293	0.9724
26	4.0566	216.0	0.06830	0.2678	0.06953	24.06	14.98	0.00142	1.01228	0.9349
27	3.7998	229.6	0.06858	0.2793	0.06168	24.39	15.43	0.00163	1.01150	0.8958
28	3.5834	242.9	0.06868	0.2883	0.05581	24.83	15.88	0.00184	1.01084	0.8674
29	3.3970	255.9	0.06869	0.2956	0.05122	25.33	16.34	0.00205	1.01027	0.8457
30	3.2336	268.7	0.06865	0.3016	0.04749	25.88	16.80	0.00226	1.00978	0.8284
31	3.0885	281.3	0.06860	0.3066	0.04438	26.47	17.25	0.00247	1.00934	0.8136
32	2.9585	293.6	0.06854	0.3108	0.04174	27.07	17.70	0.00269	1.00894	0.8014
33	2.8408	305.8	0.06846	0.3145	0.03945	27.69	18.15	0.00291	1.00859	0.7911
34	2.7337	317.7	0.06847	0.3177	0.03745	28.31	18.60	0.00313	1.00826	0.7819
35	2.6356	329.5	0.06845	0.3205	0.03569	28.94	19.05	0.00336	1.00796	0.7739
36	2.5452	341.2	0.06842	0.3229	0.03411	29.58	19.49	0.00359	1.00769	0.7669
37	2.4616	352.9	0.06838	0.3251	0.03269	30.22	19.92	0.00383	1.00744	0.7606
38	2.3840	364.4	0.06834	0.3270	0.03141	30.87	20.36	0.00407	1.00720	0.7549
39	2.3117	375.9	0.06830	0.3288	0.03023	31.52	20.79	0.00432	1.00698	0.7497
40	2.2441	387.4	0.06825	0.3303	0.02916	32.17	21.22	0.00457	1.00678	0.7449
42	2.1211	410.1	0.06816	0.3330	0.02725	33.48	22.06	0.00508	1.00641	0.7365
44	2.0119	432.6	0.06806	0.3353	0.02561	34.79	22.89	0.00562	1.00607	0.7291
46	1.9143	455.0	0.06796	0.3372	0.02417	36.11	23.71	0.00617	1.00578	0.7224
48	1.8263	477.3	0.06786	0.3387	0.02291	37.42	24.52	0.00675	1.00551	0.7164
50	1.7465	499.5	0.06774	0.3401	0.02178	38.74	25.31	0.00734	1.00527	0.7109
52	1.6738	521.8	0.06760	0.3413	0.02077	40.05	26.10	0.00795	1.00505	0.7061
54	1.6072	544.1	0.06745	0.3423	0.01986	41.36	26.87	0.00858	1.00485	0.7017
56	1.5459	566.4	0.06728	0.3431	0.01902	42.66	27.63	0.00922	1.00467	0.6979
58	1.4893	588.8	0.06710	0.3439	0.01827	43.95	28.38	0.00988	1.00449	0.6946
60	1.4369	611.2	0.06690	0.3446	0.01757	45.27	29.12	0.01056	1.00434	0.6909
65	1.3212	668.2	0.06629	0.3459	0.01606	48.65	30.94	0.01236	1.00399	0.6822
70	1.2232	726.5	0.06554	0.3469	0.01480	51.70	32.69	0.01416	1.00369	0.6797
75	1.1391	786.3	0.06466	0.3477	0.01373	54.56	34.40	0.01597	1.00344	0.6806
80	1.0661	848.0	0.06366	0.3483	0.01281	56.84	36.06	0.01767	1.00322	0.6891
85	1.0020	911.9	0.06254	0.3488	0.01201	59.79	37.69	0.01962	1.00302	0.6902
90	0.9453	977.9	0.06135	0.3492	0.01131	62.43	39.27	0.02150	1.00285	0.6955
95	0.8947	1045.8	0.06015	0.3495	0.01068	64.87	40.82	0.02336	1.00270	0.7031
100	0.8494	1115.8	0.05892	0.3497	0.01013	76.37	45.74	0.02864	1.00256	0.6769
120	0.7066	1412.8	0.05427	0.3503	0.00839	94.42	54.05	0.04058	1.00213	0.6787
140	0.6051	1728.6	0.05051	0.3506	0.00717	107.78	59.18	0.05173	1.00182	0.6806
160	0.5293	2053.3	0.04769	0.3507	0.00626	118.76	63.00	0.06281	1.00160	0.6822
180	0.4704	2381.6	0.04560	0.3508	0.00556	128.26	66.18	0.07412	1.00142	0.6833
200	0.4233	2709.6	0.04406	0.3508	0.00500	136.72	69.03	0.08580	1.00128	0.6842
220	0.3849	3036.3	0.04289	0.3508	0.00454	144.43	71.69	0.09792	1.00116	0.6848
240	0.3528	3358.9	0.04204	0.3508	0.00416	151.50	74.22	0.11054	1.00106	0.6852
260	0.3257	3677.3	0.04140	0.3508	0.00384	158.08	76.67	0.12366	1.00098	0.6854
280	0.3024	3991.0	0.04094	0.3507	0.00357	164.27	79.08	0.13733	1.00091	0.6855
300	0.2823	4300.3	0.04061	0.3507	0.00333	170.13	81.45	0.15153	1.00085	0.6855
350	0.2420	5057.2	0.04013	0.3506	0.00285	183.81	87.28	0.18946	1.00073	0.6853
400	0.2118	5798.8	0.03993	0.3506	0.00250	196.66	93.03	0.23085	1.00064	0.6850
450	0.1883	6532.1	0.03984	0.3505	0.00222	209.05	98.72	0.27567	1.00057	0.6846
500	0.1695	7266.6	0.03976	0.3505	0.00200	221.32	104.34	0.32384	1.00051	0.6844
550	0.1541	8000.3	0.03970	0.3505	0.00182	233.43	109.91	0.37531	1.00046	0.6841
600	0.1413	8738.3	0.03962	0.3504	0.00166	245.50	115.41	0.43001	1.00043	0.6839
700	0.1211	10232.5	0.03940	0.3504	0.00143	269.63	126.23	0.54880	1.00036	0.6837
800	0.1060	11765.0	0.03906	0.3503	0.00125	294.01	136.79	0.67966	1.00032	0.6836
900	0.0942	13349.1	0.03859	0.3503	0.00111	318.87	147.11	0.82208	1.00028	0.6837
1000	0.0848	14993.8	0.03801	0.3503	0.00100	440.56	157.18	1.24824	1.00026	0.5346
1200	0.0707	18447.3	0.03672	0.3502	0.00083	511.01	176.69	1.69421	1.00021	0.5312
1400	0.0606	22134.2	0.03534	0.3502	0.00071	583.36	195.42	2.19362	1.00018	0.5293
1600	0.0530	26044.6	0.03398	0.3502	0.00062	658.13	213.48	2.74669	1.00016	0.5278
1800	0.0471	30290.2	0.03253	0.3501	0.00056	740.84	230.97	3.36433	1.00014	0.5245
2000	0.0424	35359.9	0.03056	0.3500	0.00050	850.83	247.96	4.08594	1.00013	0.5153
2500	0.0337	60976.7	0.02100	0.3478	0.00040	1598.13	289.38	6.95285	1.00010	0.4445
3000	0.0271	138607.2	0.01070	0.3358	0.00035	4271.99	334.27	11.77275	1.00008	0.3768

* TWO-PHASE BOUNDARY

Table 4.12 (continued)
THERMODYNAMIC PROPERTIES OF NORMAL HYDROGEN

0.40 MPA ISOBAR

TEMPERATURE DEG. K	VOLUME CU M/KG	ISOTHERM DERIVATIVE CU M-MPA/KG	ISOCHORE DERIVATIVE MPA/K	INTERNAL ENERGY KJ/KG-K	ENTHALPY KJ/KG	ENTROPY KJ/KG-K	CV KJ / KG-K	CP KJ / KG-K	VELOCITY OF SOUND M/S
* 13.935	0.01295	1.1876	0.9279	218.4	223.5	14.091	4.69	6.39	1271
14	0.01296	1.1902	0.9278	218.8	224.0	14.122	4.71	6.41	1273
15	0.01309	1.0718	0.9211	225.4	230.6	14.581	4.91	6.94	1231
16	0.01325	1.0042	0.9025	232.5	237.8	15.043	5.10	7.38	1205
17	0.01341	0.9401	0.8872	240.0	245.4	15.505	5.29	7.85	1181
18	0.01358	0.8826	0.8729	248.0	253.5	15.965	5.45	8.32	1160
19	0.01378	0.8191	0.8576	256.5	262.0	16.429	5.60	8.84	1137
20	0.01399	0.7523	0.8402	265.6	271.2	16.898	5.73	9.40	1111
21	0.01421	0.6868	0.8213	275.2	280.9	17.372	5.84	10.01	1085
22	0.01447	0.6215	0.8000	285.4	291.2	17.853	5.94	10.68	1057
23	0.01476	0.5532	0.7751	296.4	302.3	18.346	6.02	11.46	1026
24	0.01508	0.4831	0.7454	308.2	314.2	18.852	6.09	12.37	991
25	0.01546	0.4122	0.7103	320.9	327.0	19.377	6.14	13.46	950
* 25.957	0.01590	0.3437	0.6694	334.2	340.6	19.906	6.19	14.74	905
* 25.957	0.20495	0.0579	0.0229	648.3	730.3	34.929	6.89	16.79	376
26	0.20572	0.0583	0.0228	648.7	731.0	34.958	6.88	16.71	376
27	0.22161	0.0668	0.0208	658.4	747.0	35.561	6.72	15.34	390
28	0.23648	0.0745	0.0193	667.3	761.9	36.100	6.62	14.44	403
29	0.25063	0.0817	0.0180	675.7	776.0	36.596	6.56	13.80	415
30	0.26426	0.0884	0.0170	683.8	789.5	37.055	6.51	13.33	425
31	0.27747	0.0948	0.0160	691.7	802.6	37.486	6.47	12.96	436
32	0.29036	0.1009	0.0153	699.3	815.4	37.893	6.44	12.66	445
33	0.30298	0.1068	0.0145	706.8	828.0	38.279	6.42	12.41	455
34	0.31537	0.1126	0.0139	714.1	840.3	38.646	6.39	12.20	464
35	0.32758	0.1182	0.0133	721.4	852.4	38.997	6.36	12.03	473
36	0.33963	0.1237	0.0128	728.5	864.4	39.334	6.35	11.88	481
37	0.35154	0.1290	0.0124	735.6	876.2	39.658	6.33	11.75	489
38	0.36332	0.1343	0.0119	742.6	887.9	39.970	6.32	11.64	497
39	0.37500	0.1395	0.0115	749.5	899.5	40.271	6.31	11.55	505
40	0.38659	0.1446	0.0112	756.4	911.0	40.563	6.30	11.46	513
42	0.40952	0.1547	0.0105	770.0	933.8	41.118	6.29	11.32	528
44	0.43217	0.1645	0.0099	783.4	956.3	41.641	6.28	11.20	542
46	0.45460	0.1742	0.0094	796.7	978.6	42.137	6.27	11.11	555
48	0.47684	0.1837	0.0089	810.0	1000.7	42.608	6.27	11.03	568
50	0.49892	0.1931	0.0085	823.1	1022.7	43.057	6.27	10.96	581
52	0.52086	0.2024	0.0082	836.2	1044.6	43.486	6.27	10.91	593
54	0.54268	0.2116	0.0078	849.3	1066.4	43.897	6.28	10.87	605
56	0.56440	0.2207	0.0075	862.3	1088.1	44.293	6.28	10.84	617
58	0.58603	0.2297	0.0072	875.3	1109.7	44.673	6.29	10.81	628
60	0.60758	0.2387	0.0070	888.3	1131.3	45.039	6.30	10.79	639
65	0.66116	0.2610	0.0064	920.7	1185.2	45.900	6.35	10.77	666
70	0.71441	0.2829	0.0059	953.3	1239.1	46.699	6.41	10.78	690
75	0.76741	0.3047	0.0055	986.1	1293.1	47.444	6.48	10.82	713
80	0.82020	0.3263	0.0051	1019.3	1347.4	48.145	6.58	10.89	735
85	0.87283	0.3478	0.0048	1052.9	1402.0	48.808	6.69	10.97	755
90	0.92533	0.3691	0.0045	1087.0	1457.1	49.439	6.81	11.08	775
95	0.97772	0.3904	0.0043	1121.7	1512.8	50.040	6.94	11.19	793
100	1.03002	0.4116	0.0041	1157.1	1569.1	50.617	7.08	11.32	811
120	1.23851	0.4959	0.0034	1305.4	1800.8	52.726	7.67	11.87	876
140	1.44630	0.5796	0.0029	1465.1	2043.6	54.596	8.23	12.40	935
160	1.65365	0.6630	0.0025	1635.0	2296.4	56.282	8.70	12.86	990
180	1.86072	0.7462	0.0022	1813.4	2557.7	57.821	9.09	13.25	1043
200	2.06760	0.8292	0.0020	1998.7	2825.8	59.233	9.41	13.55	1093
220	2.27434	0.9121	0.0018	2189.7	3099.4	60.537	9.66	13.80	1142
240	2.48098	0.9949	0.0017	2385.0	3377.4	61.746	9.85	13.99	1189
260	2.68754	1.0777	0.0015	2583.6	3658.7	62.873	10.00	14.13	1234
280	2.89405	1.1604	0.0014	2784.9	3942.5	63.924	10.11	14.24	1279
300	3.10031	1.2430	0.0013	2988.0	4228.2	64.909	10.19	14.32	1322
350	3.61625	1.4496	0.0011	3500.9	4947.4	67.127	10.30	14.43	1425
400	4.13206	1.6560	0.0010	4017.5	5670.3	69.058	10.35	14.48	1522
450	4.64779	1.8624	0.0009	4535.9	6395.0	70.764	10.37	14.50	1613
500	5.16346	2.0687	0.0008	5055.1	7120.5	72.292	10.39	14.52	1700
550	5.67910	2.2750	0.0007	5574.9	7846.5	73.677	10.40	14.53	1782
600	6.19471	2.4813	0.0007	6095.7	8573.6	74.939	10.42	14.55	1861
700	7.22588	2.8938	0.0006	7140.7	10031.0	77.189	10.48	14.60	2008
800	8.25701	3.3063	0.0005	8193.3	11496.1	79.147	10.57	14.69	2144
900	9.28811	3.7187	0.0004	9256.7	12971.9	80.883	10.70	14.82	2270
1000	10.31920	4.1312	0.0004	10333.8	14461.5	82.450	10.86	14.98	2388
1200	12.38135	4.9561	0.0003	12544.1	17496.6	85.198	11.24	15.36	2603
1400	14.44349	5.7809	0.0003	14836.9	20614.2	87.607	11.68	15.80	2797
1600	16.50580	6.6058	0.0003	17217.5	23819.8	89.754	12.14	16.27	2975
1800	18.56935	7.4306	0.0002	19695.6	27123.4	91.693	12.68	16.81	3139
2000	20.63822	8.2555	0.0002	22306.3	30561.6	93.514	13.47	17.64	3288
2500	25.94803	10.3176	0.0002	30220.2	40599.4	97.955	19.38	24.11	3581
3000	32.16812	12.3797	0.0001	44399.4	57266.7	103.987	38.56	46.24	3848

* TWO-PHASE BOUNDARY

Table 4.12 (continued)
THERMODYNAMIC PROPERTIES OF NORMAL HYDROGEN

0.40 MPA ISOBAR

TEMPERATURE DEG. K	DENSITY KG/CU M	$V(DH/DV)_P$ KJ/KG	$V(DP/DU)_V$ MPA-CU M/KJ X 10^2	$-V(DP/DV)_T$ MPA	$(DV/DT)_P/V$ 1/DEG. K	THERMAL CONDUCTIVITY W/K-M X 10^3	VISCOSITY KG/M-S X 10^7	THERMAL DIFFUSIVITY SQ M/HR	DIELECTRIC CONSTANT	PRANDTL NUMBER
* 13.935	77.2399	631.3	0.25599	91.7322	0.01011	74.25	257.26	0.00054	1.25236	2.2124
14	77.1782	634.3	0.25541	91.8568	0.01010	74.91	254.75	0.00055	1.25215	2.1789
15	76.3720	617.1	0.24570	81.8567	0.01125	83.51	224.16	0.00057	1.24931	1.8642
16	75.4904	619.9	0.23427	75.8051	0.01191	89.48	199.49	0.00058	1.24621	1.6455
17	74.5686	619.9	0.22511	70.0983	0.01266	93.43	179.63	0.00057	1.24298	1.5084
18	73.6116	619.2	0.21750	64.9667	0.01344	96.26	163.38	0.00057	1.23964	1.4120
19	72.5911	612.8	0.21094	59.4563	0.01442	97.89	149.60	0.00055	1.23607	1.3508
20	71.5035	602.0	0.20502	53.7931	0.01562	99.33	137.75	0.00053	1.23228	1.3040
21	70.3507	589.0	0.19974	48.3137	0.01700	100.99	127.43	0.00052	1.22827	1.2634
22	69.1127	573.7	0.19481	42.9569	0.01862	101.88	118.25	0.00050	1.22397	1.2401
23	67.7665	554.3	0.18995	37.4872	0.02068	102.04	109.91	0.00047	1.21931	1.2344
24	66.2974	531.4	0.18472	32.0293	0.02327	101.53	102.22	0.00045	1.21424	1.2450
25	64.6759	505.0	0.17883	26.6581	0.02664	100.37	95.02	0.00042	1.20865	1.2739
* 25.957	62.9088	476.2	0.17182	21.6193	0.03096	98.50	88.30	0.00038	1.20259	1.3216
* 25.957	4.8790	206.9	0.06820	0.2826	0.08113	25.16	15.16	0.00111	1.01478	1.0116
26	4.8610	207.6	0.06823	0.2835	0.08050	25.15	15.18	0.00111	1.01472	1.0083
27	4.5124	222.0	0.06869	0.3016	0.06910	25.23	15.61	0.00131	1.01366	0.9489
28	4.2287	235.9	0.06888	0.3152	0.06120	25.51	16.05	0.00150	1.01280	0.9083
29	3.9899	249.5	0.06892	0.3258	0.05532	25.91	16.49	0.00169	1.01207	0.8787
30	3.7842	262.8	0.06889	0.3344	0.05072	26.38	16.94	0.00188	1.01145	0.8557
31	3.6040	275.7	0.06884	0.3415	0.04699	26.93	17.38	0.00208	1.01090	0.8364
32	3.4440	288.4	0.06878	0.3475	0.04389	27.50	17.83	0.00227	1.01042	0.8207
33	3.3005	300.9	0.06870	0.3526	0.04125	28.08	18.27	0.00247	1.00998	0.8077
34	3.1708	313.0	0.06871	0.3570	0.03898	28.68	18.71	0.00267	1.00959	0.7963
35	3.0527	325.1	0.06871	0.3608	0.03700	29.29	19.15	0.00287	1.00923	0.7865
36	2.9444	337.0	0.06868	0.3641	0.03525	29.91	19.59	0.00308	1.00890	0.7781
37	2.8447	348.8	0.06864	0.3670	0.03369	30.54	20.02	0.00329	1.00860	0.7706
38	2.7524	360.6	0.06860	0.3697	0.03229	31.17	20.45	0.00350	1.00832	0.7639
39	2.6666	372.2	0.06855	0.3720	0.03102	31.81	20.88	0.00372	1.00806	0.7579
40	2.5867	383.8	0.06850	0.3741	0.02986	32.46	21.31	0.00394	1.00782	0.7523
42	2.4419	406.8	0.06840	0.3777	0.02781	33.74	22.15	0.00440	1.00738	0.7427
44	2.3139	429.7	0.06828	0.3807	0.02607	35.04	22.97	0.00487	1.00699	0.7344
46	2.1997	452.3	0.06817	0.3832	0.02456	36.34	23.79	0.00536	1.00664	0.7270
48	2.0971	474.8	0.06806	0.3853	0.02323	37.65	24.59	0.00586	1.00633	0.7204
50	2.0043	497.2	0.06793	0.3871	0.02205	38.96	25.38	0.00638	1.00605	0.7145
52	1.9199	519.6	0.06778	0.3886	0.02100	40.26	26.16	0.00692	1.00580	0.7093
54	1.8427	542.0	0.06762	0.3899	0.02006	41.56	26.93	0.00747	1.00556	0.7045
56	1.7718	564.5	0.06745	0.3910	0.01920	42.85	27.69	0.00803	1.00535	0.7004
58	1.7064	587.0	0.06726	0.3920	0.01842	44.13	28.44	0.00861	1.00515	0.6969
60	1.6459	609.6	0.06705	0.3929	0.01770	45.45	29.18	0.00921	1.00497	0.6930
65	1.5125	666.8	0.06644	0.3947	0.01615	48.81	30.99	0.01079	1.00456	0.6838
70	1.3998	725.3	0.06567	0.3960	0.01487	51.85	32.74	0.01237	1.00422	0.6810
75	1.3031	785.3	0.06478	0.3970	0.01378	54.69	34.45	0.01396	1.00393	0.6817
80	1.2192	847.1	0.06377	0.3978	0.01285	56.97	36.11	0.01545	1.00368	0.6900
85	1.1457	911.1	0.06265	0.3984	0.01204	59.91	37.73	0.01716	1.00346	0.6910
90	1.0807	977.3	0.06144	0.3989	0.01133	62.55	39.31	0.01881	1.00326	0.6961
95	1.0228	1045.2	0.06024	0.3993	0.01071	64.98	40.85	0.02044	1.00308	0.7036
100	0.9709	1115.4	0.05900	0.3996	0.01015	76.41	45.75	0.02504	1.00293	0.6775
120	0.8074	1412.7	0.05433	0.4004	0.00840	94.44	54.05	0.03549	1.00243	0.6791
140	0.6914	1728.7	0.05056	0.4008	0.00717	107.80	59.18	0.04526	1.00208	0.6808
160	0.6047	2053.6	0.04773	0.4010	0.00626	118.78	63.00	0.05496	1.00182	0.6823
180	0.5374	2382.1	0.04564	0.4010	0.00556	128.28	66.19	0.06487	1.00162	0.6835
200	0.4837	2710.2	0.04409	0.4010	0.00500	136.74	69.04	0.07509	1.00146	0.6843
220	0.4397	3037.0	0.04292	0.4010	0.00454	144.45	71.69	0.08571	1.00133	0.6849
240	0.4031	3359.8	0.04206	0.4010	0.00416	151.53	74.23	0.09675	1.00121	0.6852
260	0.3721	3678.3	0.04142	0.4010	0.00384	158.11	76.68	0.10824	1.00112	0.6854
280	0.3455	3992.1	0.04096	0.4010	0.00357	164.29	79.09	0.12020	1.00104	0.6855
300	0.3225	4301.4	0.04062	0.4009	0.00333	170.16	81.46	0.13263	1.00097	0.6855
350	0.2765	5058.4	0.04014	0.4008	0.00285	183.84	87.29	0.16583	1.00083	0.6853
400	0.2420	5800.1	0.03994	0.4008	0.00250	196.69	93.05	0.20207	1.00073	0.6850
450	0.2152	6533.5	0.03985	0.4007	0.00222	209.08	98.73	0.24129	1.00065	0.6846
500	0.1937	7268.0	0.03977	0.4006	0.00200	221.36	104.36	0.28345	1.00058	0.6844
550	0.1761	8001.8	0.03971	0.4006	0.00182	233.47	109.93	0.32851	1.00053	0.6841
600	0.1614	8739.8	0.03963	0.4005	0.00166	245.55	115.43	0.37639	1.00049	0.6839
700	0.1384	10234.0	0.03941	0.4005	0.00143	269.68	126.25	0.48036	1.00042	0.6837
800	0.1211	11766.5	0.03907	0.4004	0.00125	294.07	136.82	0.59490	1.00036	0.6836
900	0.1077	13350.7	0.03860	0.4004	0.00111	318.94	147.14	0.71955	1.00032	0.6837
1000	0.0969	14995.4	0.03801	0.4003	0.00100	440.56	157.21	1.09232	1.00029	0.5347
1200	0.0808	18448.9	0.03672	0.4003	0.00083	511.01	176.73	1.48256	1.00024	0.5313
1400	0.0692	22135.5	0.03534	0.4002	0.00071	583.35	195.47	1.91955	1.00021	0.5295
1600	0.0606	26042.8	0.03399	0.4002	0.00062	657.99	213.53	2.40333	1.00018	0.5279
1800	0.0539	30271.1	0.03256	0.4002	0.00056	740.02	231.02	2.94275	1.00016	0.5248
2000	0.0485	35276.0	0.03065	0.4000	0.00050	847.52	248.02	3.57013	1.00015	0.5161
2500	0.0385	59921.5	0.02142	0.3976	0.00040	1558.07	289.39	6.03619	1.00012	0.4478
3000	0.0311	133455.3	0.01112	0.3848	0.00035	4078.54	333.93	10.21507	1.00009	0.3786

* TWO-PHASE BOUNDARY

Table 4.12 (continued)
THERMODYNAMIC PROPERTIES OF NORMAL HYDROGEN

0.50 MPA ISOBAR

TEMPERATURE DEG. K	VOLUME CU M/KG	ISOTHERM DERIVATIVE CU M-MPA/KG	ISOCHORE DERIVATIVE MPA/K	INTERNAL ENERGY KJ/KG-K	ENTHALPY KJ/KG	ENTROPY KJ/KG-K	CV KJ /	CP KG-K	VELOCITY OF SOUND M/S
* 13.968	0.01294	1.1940	0.9290	218.4	224.9	14.093	4.70	6.39	1274
14	0.01294	1.2053	0.9289	218.6	225.1	14.109	4.70	6.38	1279
15	0.01308	1.0819	0.9246	225.2	231.7	14.567	4.91	6.93	1237
16	0.01323	1.0138	0.9058	232.3	238.9	15.028	5.10	7.37	1210
17	0.01339	0.9559	0.8908	239.7	246.4	15.486	5.28	7.81	1189
18	0.01356	0.8927	0.8759	247.7	254.5	15.947	5.45	8.29	1166
19	0.01375	0.8295	0.8607	256.2	263.1	16.409	5.60	8.81	1142
20	0.01396	0.7653	0.8436	265.1	272.1	16.875	5.73	9.35	1118
21	0.01419	0.6968	0.8246	274.7	281.8	17.348	5.84	9.96	1090
22	0.01444	0.6323	0.8035	284.9	292.1	17.827	5.94	10.62	1063
23	0.01472	0.5647	0.7790	295.7	303.1	18.316	6.02	11.37	1033
24	0.01504	0.4952	0.7500	307.3	314.9	18.818	6.08	12.25	998
25	0.01541	0.4251	0.7158	319.9	327.6	19.337	6.14	13.29	959
26	0.01584	0.3527	0.6746	333.6	341.5	19.882	6.19	14.61	912
27	0.01638	0.2793	0.6263	348.8	357.0	20.464	6.25	16.43	856
* 27.116	0.01645	0.2724	0.6200	350.7	358.9	20.537	6.26	16.62	850
* 27.116	0.16290	0.0528	0.0296	646.2	727.6	34.141	7.00	18.92	378
28	0.17542	0.0614	0.0270	655.7	743.4	34.714	6.84	17.03	391
29	0.18839	0.0701	0.0247	665.6	759.8	35.288	6.72	15.69	405
30	0.20054	0.0780	0.0229	674.7	775.0	35.803	6.64	14.78	417
31	0.21210	0.0854	0.0215	683.3	789.4	36.276	6.58	14.12	428
32	0.22323	0.0923	0.0203	691.6	803.2	36.717	6.53	13.62	439
33	0.23401	0.0988	0.0192	699.7	816.7	37.130	6.50	13.23	449
34	0.24452	0.1051	0.0183	707.5	829.7	37.520	6.45	12.90	458
35	0.25480	0.1112	0.0174	715.1	842.5	37.890	6.42	12.63	468
36	0.26489	0.1171	0.0167	722.6	855.0	38.243	6.39	12.41	477
37	0.27482	0.1229	0.0160	729.9	867.4	38.581	6.37	12.23	485
38	0.28462	0.1285	0.0154	737.2	879.5	38.905	6.36	12.07	494
39	0.29430	0.1340	0.0149	744.4	891.5	39.216	6.34	11.93	502
40	0.30387	0.1394	0.0144	751.4	903.4	39.517	6.33	11.81	510
42	0.32275	0.1500	0.0135	765.4	926.8	40.088	6.32	11.61	525
44	0.34133	0.1603	0.0127	779.2	949.9	40.624	6.30	11.46	540
46	0.35967	0.1703	0.0120	792.8	972.7	41.131	6.29	11.33	554
48	0.37782	0.1802	0.0114	806.3	995.2	41.611	6.29	11.22	567
50	0.39579	0.1899	0.0108	819.7	1017.6	42.067	6.28	11.14	580
52	0.41363	0.1994	0.0104	832.9	1039.8	42.503	6.29	11.07	593
54	0.43134	0.2089	0.0099	846.2	1061.8	42.919	6.29	11.01	605
56	0.44894	0.2182	0.0095	859.3	1083.8	43.320	6.29	10.97	617
58	0.46645	0.2275	0.0091	872.5	1105.7	43.704	6.30	10.93	628
60	0.48388	0.2366	0.0088	885.6	1127.5	44.074	6.31	10.90	639
65	0.52716	0.2593	0.0080	918.3	1181.9	44.943	6.35	10.86	666
70	0.57010	0.2816	0.0074	951.1	1236.2	45.748	6.41	10.86	690
75	0.61279	0.3036	0.0069	984.1	1290.5	46.498	6.49	10.88	714
80	0.65527	0.3254	0.0064	1017.4	1345.1	47.202	6.58	10.94	736
85	0.69759	0.3471	0.0060	1051.2	1400.0	47.868	6.69	11.02	756
90	0.73977	0.3687	0.0057	1085.4	1455.3	48.501	6.81	11.12	776
95	0.78185	0.3901	0.0054	1120.2	1511.1	49.104	6.94	11.22	794
100	0.82383	0.4115	0.0051	1155.6	1567.6	49.683	7.08	11.35	812
120	0.99107	0.4962	0.0042	1304.3	1799.8	51.796	7.67	11.89	877
140	1.15760	0.5802	0.0036	1464.2	2043.0	53.669	8.23	12.42	936
160	1.32371	0.6638	0.0031	1634.2	2296.0	55.357	8.70	12.88	991
180	1.48953	0.7472	0.0028	1812.7	2557.5	56.897	9.09	13.26	1044
200	1.65516	0.8303	0.0025	1998.2	2825.7	58.310	9.41	13.56	1094
220	1.82065	0.9133	0.0023	2189.2	3099.5	59.614	9.66	13.80	1143
240	1.98605	0.9962	0.0021	2384.5	3377.5	60.824	9.85	13.99	1190
260	2.15136	1.0790	0.0019	2583.2	3658.9	61.951	10.00	14.14	1235
280	2.31662	1.1618	0.0018	2784.5	3942.8	63.002	10.11	14.24	1280
300	2.48169	1.2445	0.0017	2987.7	4228.6	63.987	10.19	14.32	1323
350	2.89453	1.4511	0.0014	3500.6	4947.9	66.206	10.30	14.43	1426
400	3.30723	1.6576	0.0013	4017.3	5670.9	68.137	10.35	14.48	1523
450	3.71985	1.8640	0.0011	4535.8	6395.7	69.844	10.37	14.50	1614
500	4.13242	2.0703	0.0010	5055.0	7121.2	71.371	10.39	14.52	1701
550	4.54495	2.2767	0.0009	5574.8	7847.2	72.756	10.40	14.53	1783
600	4.95746	2.4830	0.0008	6095.6	8574.4	74.018	10.42	14.55	1862
700	5.78242	2.8955	0.0007	7140.6	10031.8	76.268	10.48	14.60	2009
800	6.60734	3.3080	0.0006	8193.2	11496.9	78.226	10.57	14.69	2145
900	7.43224	3.7205	0.0006	9256.6	12972.8	79.963	10.70	14.82	2270
1000	8.25712	4.1330	0.0005	10333.8	14462.3	81.530	10.86	14.98	2388
1200	9.90685	4.9578	0.0004	12544.0	17497.5	84.278	11.24	15.36	2603
1400	11.55656	5.7827	0.0004	14836.8	20615.1	86.686	11.68	15.80	2797
1600	13.20641	6.6076	0.0003	17217.1	23820.3	88.833	12.14	16.26	2975
1800	14.85713	7.4324	0.0003	19693.5	27122.1	90.771	12.66	16.79	3140
2000	16.51165	8.2573	0.0003	22297.0	30552.8	92.589	13.40	17.57	3290
2500	20.74647	10.3194	0.0002	30075.8	40449.0	96.969	18.78	23.44	3587
3000	25.63286	12.3815	0.0002	43497.1	56313.6	102.715	36.04	43.31	3853

* TWO-PHASE BOUNDARY

Table 4.12 (continued)
THERMODYNAMIC PROPERTIES OF NORMAL HYDROGEN

0.50 MPA ISOBAR

TEMPERATURE DEG. K	DENSITY KG/CU M	V(DH/DV)$_P$ KJ/KG	V(DP/DU)$_V$ MPA-CU M/KJ X 10^2	-V(DP/DV)$_T$ MPA	(DV/DT)$_P$/V 1/DEG. K	THERMAL CONDUCTIVITY W/K-M X 10^3	VISCOSITY KG/M-S X 10^7	THERMAL DIFFUSIVITY SQ M/HR	DIELECTRIC CONSTANT	PRANDTL NUMBER
* 13.968	77.2965	634.5	0.25586	92.2941	0.01007	74.66	257.88	0.00054	1.25256	2.2061
14	77.2625	639.9	0.25557	93.1235	0.00998	74.98	256.56	0.00055	1.25244	2.1842
15	76.4616	620.4	0.24647	82.7270	0.01118	83.62	225.71	0.00057	1.24962	1.8716
16	75.5862	623.1	0.23499	76.6271	0.01182	89.63	200.85	0.00058	1.24655	1.6506
17	74.6876	626.0	0.22585	71.3910	0.01248	93.65	181.04	0.00058	1.24340	1.5039
18	73.7238	623.1	0.21812	65.8110	0.01331	96.49	164.50	0.00057	1.24003	1.4140
19	72.7116	617.0	0.21153	60.3138	0.01427	98.15	150.63	0.00055	1.23649	1.3514
20	71.6412	607.7	0.20563	54.8242	0.01539	99.63	138.76	0.00054	1.23276	1.3022
21	70.4941	593.6	0.20030	49.1229	0.01679	101.32	128.35	0.00052	1.22877	1.2621
22	69.2711	578.8	0.19535	43.7989	0.01834	102.25	119.14	0.00050	1.22452	1.2372
23	67.9460	560.1	0.19052	38.3717	0.02030	102.47	110.79	0.00048	1.21993	1.2294
24	66.5013	537.8	0.18538	32.9286	0.02278	102.02	103.11	0.00045	1.21494	1.2379
25	64.9112	512.3	0.17963	27.5905	0.02594	100.93	95.91	0.00042	1.20946	1.2629
26	63.1156	482.2	0.17263	22.2624	0.03030	99.09	88.98	0.00039	1.20330	1.3120
27	61.0558	447.2	0.16402	17.0512	0.03673	96.50	82.17	0.00035	1.19625	1.3988
* 27.116	60.7811	443.9	0.16287	16.5596	0.03744	96.11	81.33	0.00034	1.19531	1.4064
* 27.116	6.1386	207.3	0.06874	0.3240	0.09124	27.63	16.10	0.00086	1.01862	1.1024
28	5.7006	221.3	0.06912	0.3503	0.07695	27.33	16.44	0.00101	1.01728	1.0239
29	5.3082	236.3	0.06929	0.3723	0.06639	27.35	16.84	0.00118	1.01609	0.9656
30	4.9866	250.6	0.06931	0.3891	0.05897	27.58	17.25	0.00135	1.01511	0.9246
31	4.7147	264.5	0.06928	0.4024	0.05340	27.99	17.67	0.00151	1.01428	0.8918
32	4.4797	278.0	0.06923	0.4133	0.04901	28.46	18.10	0.00168	1.01356	0.8663
33	4.2733	291.1	0.06913	0.4223	0.04544	28.97	18.53	0.00184	1.01293	0.8461
34	4.0897	303.7	0.06919	0.4300	0.04247	29.50	18.96	0.00201	1.01238	0.8290
35	3.9247	316.2	0.06922	0.4365	0.03995	30.05	19.38	0.00218	1.01188	0.8147
36	3.7752	328.6	0.06920	0.4422	0.03778	30.63	19.81	0.00235	1.01142	0.8028
37	3.6387	340.8	0.06916	0.4471	0.03588	31.22	20.23	0.00253	1.01101	0.7924
38	3.5135	352.9	0.06912	0.4515	0.03420	31.82	20.66	0.00270	1.01063	0.7834
39	3.3979	364.9	0.06906	0.4554	0.03270	32.43	21.08	0.00288	1.01028	0.7754
40	3.2909	376.9	0.06900	0.4588	0.03134	33.05	21.49	0.00306	1.00995	0.7682
42	3.0984	400.5	0.06888	0.4647	0.02900	34.29	22.32	0.00343	1.00937	0.7559
44	2.9297	423.8	0.06874	0.4696	0.02703	35.56	23.14	0.00381	1.00885	0.7456
46	2.7803	446.9	0.06861	0.4736	0.02535	36.83	23.95	0.00421	1.00840	0.7366
48	2.6468	469.8	0.06847	0.4769	0.02389	38.11	24.74	0.00462	1.00800	0.7288
50	2.5266	492.6	0.06832	0.4797	0.02261	39.39	25.53	0.00504	1.00763	0.7218
52	2.4176	515.4	0.06815	0.4822	0.02148	40.68	26.30	0.00547	1.00730	0.7157
54	2.3184	538.1	0.06798	0.4843	0.02046	41.96	27.07	0.00592	1.00700	0.7103
56	2.2275	560.8	0.06779	0.4861	0.01955	43.24	27.82	0.00637	1.00673	0.7055
58	2.1438	583.6	0.06758	0.4876	0.01872	44.50	28.56	0.00684	1.00647	0.7015
60	2.0666	606.4	0.06737	0.4890	0.01797	45.80	29.30	0.00732	1.00624	0.6971
65	1.8970	664.1	0.06672	0.4918	0.01635	49.14	31.10	0.00859	1.00573	0.6870
70	1.7541	722.9	0.06593	0.4939	0.01502	52.16	32.84	0.00986	1.00529	0.6835
75	1.6319	783.3	0.06502	0.4954	0.01389	54.96	34.54	0.01114	1.00493	0.6840
80	1.5261	845.4	0.06399	0.4967	0.01294	57.23	36.19	0.01234	1.00461	0.6919
85	1.4335	909.6	0.06285	0.4976	0.01211	60.16	37.81	0.01371	1.00433	0.6924
90	1.3518	976.0	0.06164	0.4984	0.01139	62.79	39.39	0.01504	1.00408	0.6972
95	1.2790	1044.1	0.06042	0.4990	0.01075	65.21	40.93	0.01635	1.00386	0.7045
100	1.2138	1114.4	0.05917	0.4995	0.01018	76.49	45.75	0.01999	1.00366	0.6787
120	1.0090	1412.4	0.05445	0.5007	0.00842	94.49	54.04	0.02836	1.00304	0.6798
140	0.8639	1729.0	0.05065	0.5012	0.00718	107.84	59.17	0.03619	1.00261	0.6814
160	0.7555	2054.3	0.04781	0.5015	0.00627	118.82	63.00	0.04397	1.00228	0.6827
180	0.6714	2383.0	0.04570	0.5016	0.00556	128.32	66.19	0.05191	1.00202	0.6838
200	0.6042	2711.5	0.04414	0.5016	0.00500	136.78	69.05	0.06010	1.00182	0.6845
220	0.5493	3038.5	0.04296	0.5016	0.00454	144.49	71.70	0.06860	1.00166	0.6850
240	0.5035	3361.5	0.04210	0.5016	0.00416	151.57	74.24	0.07744	1.00152	0.6854
260	0.4648	3680.2	0.04146	0.5016	0.00384	158.15	76.69	0.08664	1.00140	0.6855
280	0.4317	3994.1	0.04099	0.5015	0.00357	164.34	79.10	0.09622	1.00130	0.6856
300	0.4030	4303.6	0.04065	0.5015	0.00333	170.20	81.48	0.10618	1.00121	0.6856
350	0.3455	5060.9	0.04017	0.5013	0.00285	183.89	87.32	0.13276	1.00104	0.6853
400	0.3024	5802.8	0.03996	0.5012	0.00250	196.75	93.07	0.16177	1.00091	0.6850
450	0.2688	6536.2	0.03987	0.5011	0.00222	209.15	98.76	0.19317	1.00081	0.6847
500	0.2420	7270.8	0.03978	0.5010	0.00200	221.44	104.39	0.22692	1.00073	0.6844
550	0.2200	8004.6	0.03972	0.5009	0.00182	233.56	109.96	0.26299	1.00066	0.6841
600	0.2017	8742.7	0.03964	0.5009	0.00166	245.64	115.47	0.30132	1.00061	0.6840
700	0.1729	10237.0	0.03942	0.5007	0.00143	269.79	126.30	0.38455	1.00052	0.6837
800	0.1513	11769.5	0.03908	0.5007	0.00125	294.19	136.87	0.47623	1.00046	0.6836
900	0.1345	13353.8	0.03860	0.5006	0.00111	319.07	147.20	0.57602	1.00041	0.6837
1000	0.1211	14998.5	0.03802	0.5005	0.00100	440.56	157.28	0.87403	1.00036	0.5349
1200	0.1009	18452.1	0.03673	0.5004	0.00083	511.01	176.81	1.18626	1.00030	0.5316
1400	0.0865	22138.4	0.03535	0.5004	0.00071	583.32	195.56	1.53585	1.00026	0.5297
1600	0.0757	26040.8	0.03400	0.5003	0.00062	657.77	213.64	1.92269	1.00023	0.5283
1800	0.0673	30242.7	0.03260	0.5003	0.00056	738.77	231.13	2.35300	1.00020	0.5254
2000	0.0606	35148.4	0.03079	0.5001	0.00050	842.45	248.14	2.84995	1.00018	0.5175
2500	0.0482	58307.1	0.02209	0.4974	0.00040	1496.73	289.44	4.76813	1.00015	0.4534
3000	0.0390	125516.5	0.01185	0.4830	0.00035	3780.36	333.43	8.05486	1.00012	0.3820

* TWO-PHASE BOUNDARY

Table 4.12 (continued)
THERMODYNAMIC PROPERTIES OF NORMAL HYDROGEN

0.60 MPA ISOBAR

TEMPERATURE DEG. K	VOLUME CU M/KG	ISOTHERM DERIVATIVE CU M-MPA/KG	ISOCHORE DERIVATIVE MPA/K	INTERNAL ENERGY KJ/KG-K	ENTHALPY KJ/KG	ENTROPY KJ/KG-K	CV KJ / KG-K	CP KJ / KG-K	VELOCITY OF SOUND M/S
* 14.001	0.01293	1.2078	0.9302	218.4	226.2	14.096	4.70	6.38	1280
15	0.01306	1.0920	0.9281	225.0	232.8	14.553	4.90	6.92	1242
16	0.01321	1.0321	0.9095	232.0	239.9	15.010	5.10	7.33	1219
17	0.01337	0.9660	0.8938	239.5	247.5	15.469	5.28	7.79	1194
18	0.01354	0.9027	0.8790	247.4	255.5	15.929	5.44	8.27	1171
19	0.01373	0.8398	0.8638	255.8	264.1	16.390	5.59	8.77	1148
20	0.01393	0.7755	0.8468	264.7	273.1	16.854	5.72	9.31	1123
21	0.01416	0.7107	0.8280	274.2	282.7	17.324	5.84	9.90	1098
22	0.01440	0.6428	0.8069	284.3	292.9	17.801	5.93	10.56	1069
23	0.01468	0.5760	0.7829	295.0	303.9	18.286	6.01	11.29	1040
24	0.01499	0.5082	0.7545	306.5	315.5	18.783	6.08	12.12	1007
25	0.01535	0.4376	0.7211	318.9	328.2	19.298	6.14	13.14	968
26	0.01577	0.3665	0.6813	332.4	341.9	19.835	6.19	14.38	923
27	0.01629	0.2933	0.6346	347.3	357.0	20.407	6.25	16.08	869
28	0.01694	0.2216	0.5810	364.1	374.2	21.031	6.32	18.56	807
* 28.123	0.01704	0.2189	0.5735	366.3	376.6	21.114	6.34	18.60	802
* 28.123	0.13376	0.0472	0.0369	642.7	722.9	33.437	7.13	21.64	378
29	0.14530	0.0570	0.0332	653.5	740.6	34.057	6.95	18.80	393
30	0.15699	0.0666	0.0302	664.2	758.4	34.660	6.81	16.93	407
31	0.16776	0.0752	0.0279	674.0	774.7	35.194	6.71	15.73	420
32	0.17792	0.0831	0.0260	683.2	790.0	35.679	6.64	14.88	431
33	0.18761	0.0905	0.0244	692.0	804.6	36.128	6.59	14.25	442
34	0.19695	0.0974	0.0231	700.4	818.5	36.546	6.53	13.74	453
35	0.20602	0.1041	0.0219	708.5	832.1	36.938	6.48	13.35	463
36	0.21486	0.1104	0.0209	716.4	845.3	37.310	6.45	13.03	472
37	0.22352	0.1166	0.0200	724.1	858.2	37.663	6.42	12.77	482
38	0.23202	0.1226	0.0192	731.6	870.8	38.001	6.40	12.55	490
39	0.24038	0.1285	0.0185	739.0	883.3	38.324	6.38	12.36	499
40	0.24864	0.1342	0.0178	746.4	895.6	38.635	6.36	12.20	507
42	0.26485	0.1453	0.0166	760.8	919.7	39.223	6.34	11.93	523
44	0.28074	0.1560	0.0156	774.9	943.3	39.773	6.32	11.73	538
46	0.29637	0.1665	0.0147	788.8	966.6	40.291	6.31	11.56	552
48	0.31179	0.1767	0.0139	802.5	989.6	40.780	6.30	11.43	566
50	0.32704	0.1867	0.0132	816.1	1012.4	41.245	6.30	11.32	579
52	0.34214	0.1965	0.0126	829.6	1034.9	41.687	6.30	11.23	592
54	0.35712	0.2062	0.0121	843.0	1057.3	42.109	6.30	11.16	604
56	0.37198	0.2158	0.0115	856.4	1079.5	42.515	6.31	11.10	616
58	0.38675	0.2252	0.0111	869.6	1101.7	42.903	6.31	11.05	628
60	0.40143	0.2346	0.0107	882.9	1123.7	43.277	6.32	11.01	639
65	0.43784	0.2576	0.0097	915.9	1178.6	44.154	6.36	10.94	666
70	0.47391	0.2802	0.0090	948.9	1233.3	44.965	6.42	10.93	691
75	0.50972	0.3025	0.0083	982.1	1287.9	45.719	6.49	10.94	714
80	0.54533	0.3246	0.0078	1015.6	1342.8	46.427	6.58	10.99	736
85	0.58077	0.3465	0.0073	1049.4	1397.9	47.096	6.69	11.06	757
90	0.61608	0.3682	0.0068	1083.8	1453.4	47.731	6.82	11.16	776
95	0.65128	0.3899	0.0065	1118.7	1509.5	48.337	6.94	11.26	795
100	0.68639	0.4113	0.0061	1154.2	1566.1	48.917	7.08	11.38	813
120	0.82612	0.4965	0.0051	1303.2	1798.8	51.035	7.67	11.91	878
140	0.96515	0.5808	0.0043	1463.2	2042.3	52.911	8.23	12.43	937
160	1.10375	0.6646	0.0038	1633.4	2295.7	54.600	8.70	12.89	992
180	1.24207	0.7481	0.0034	1812.0	2557.3	56.141	9.10	13.26	1044
200	1.38020	0.8314	0.0030	1997.6	2825.7	57.555	9.41	13.57	1095
220	1.51820	0.9145	0.0027	2188.7	3099.6	58.860	9.66	13.81	1143
240	1.65609	0.9975	0.0025	2384.1	3377.7	60.070	9.85	14.00	1190
260	1.79391	1.0804	0.0023	2582.8	3659.2	61.197	10.00	14.14	1236
280	1.93167	1.1632	0.0021	2784.1	3943.1	62.249	10.11	14.25	1280
300	2.06928	1.2459	0.0020	2987.4	4229.0	63.234	10.19	14.32	1324
350	2.41338	1.4526	0.0017	3500.4	4948.4	65.453	10.30	14.44	1427
400	2.75734	1.6592	0.0015	4017.1	5671.5	67.384	10.35	14.48	1523
450	3.10122	1.8656	0.0013	4535.6	6396.3	69.091	10.37	14.50	1615
500	3.44506	2.0720	0.0012	5054.8	7121.9	70.619	10.39	14.52	1701
550	3.78885	2.2783	0.0011	5574.7	7848.0	72.004	10.40	14.53	1784
600	4.13262	2.4847	0.0010	6095.5	8575.1	73.266	10.42	14.55	1862
700	4.82011	2.8972	0.0009	7140.6	10032.6	75.516	10.48	14.60	2009
800	5.50756	3.3097	0.0008	8193.2	11497.7	77.474	10.57	14.69	2145
900	6.19499	3.7222	0.0007	9256.6	12973.6	79.211	10.70	14.82	2271
1000	6.88239	4.1347	0.0006	10333.7	14463.2	80.778	10.86	14.98	2389
1200	8.25718	4.9596	0.0005	12544.0	17498.3	83.526	11.24	15.36	2604
1400	9.63195	5.7845	0.0004	14836.8	20615.9	85.934	11.68	15.80	2798
1600	11.00681	6.6094	0.0004	17216.9	23821.0	88.081	12.14	16.26	2976
1800	12.38234	7.4342	0.0003	19692.0	27121.4	90.018	12.65	16.78	3141
2000	13.76076	8.2591	0.0003	22290.0	30546.5	91.833	13.36	17.52	3291
2500	17.28176	10.3212	0.0002	29969.2	40338.2	96.168	18.34	22.95	3593
3000	21.29856	12.3833	0.0002	42830.8	55610.0	101.702	34.18	41.14	3857

* TWO-PHASE BOUNDARY

Table 4.12 (continued)
THERMODYNAMIC PROPERTIES OF NORMAL HYDROGEN

0.60 MPA ISOBAR

TEMPERATURE DEG. K	DENSITY KG/CU M	$V(DH/DV)_P$ KJ/KG	$V(DP/DU)_V$ MPA-CU M/KJ X 10^2	$-V(DP/DV)_T$ MPA	$(DV/DT)_P/V$ 1/DEG. K	THERMAL CONDUCTIVITY W/K-M X 10^3	VISCOSITY KG/M-S X 10^7	THERMAL DIFFUSIVITY SQ M/HR	DIELECTRIC CONSTANT	PRANDTL NUMBER
* 14.001	77.3529	640.6	0.25573	93.4243	0.00996	75.07	258.50	0.00055	1.25276	2.1963
15	76.5504	623.6	0.24723	83.5919	0.01110	83.73	227.26	0.00057	1.24994	1.8790
16	75.7022	630.0	0.23579	78.1349	0.01164	89.81	202.51	0.00058	1.24696	1.6534
17	74.7902	629.7	0.22649	72.2485	0.01237	93.84	182.26	0.00058	1.24376	1.5130
18	73.8345	627.0	0.21873	66.6472	0.01319	96.71	165.61	0.00057	1.24041	1.4160
19	72.8306	621.3	0.21212	61.1633	0.01412	98.41	151.65	0.00055	1.23691	1.3522
20	71.7697	612.1	0.20621	55.6607	0.01521	99.92	139.71	0.00054	1.23321	1.3020
21	70.6402	600.0	0.20087	50.2035	0.01649	101.66	129.29	0.00052	1.22928	1.2585
22	69.4268	583.9	0.19588	44.6299	0.01808	102.62	120.02	0.00050	1.22506	1.2345
23	68.1216	565.8	0.19108	39.2410	0.01995	102.89	111.66	0.00048	1.22054	1.2249
24	66.7023	544.7	0.18601	33.8976	0.02226	102.51	103.99	0.00046	1.21563	1.2298
25	65.1405	519.3	0.18040	28.5044	0.02530	101.48	96.80	0.00043	1.21025	1.2531
26	63.3921	490.4	0.17372	23.2325	0.02933	99.75	89.93	0.00039	1.20424	1.2967
27	61.3956	456.4	0.16546	18.0098	0.03524	97.28	83.19	0.00035	1.19741	1.3751
28	59.0308	417.9	0.15564	13.0801	0.04442	93.92	76.36	0.00031	1.18935	1.5092
* 28.123	58.6975	416.7	0.15422	12.8515	0.04462	93.42	75.48	0.00031	1.18821	1.5024
* 28.123	7.4761	207.0	0.06917	0.3528	0.10454	30.27	17.01	0.00067	1.02271	1.2160
29	6.8825	222.0	0.06951	0.3925	0.08467	29.46	17.28	0.00082	1.02089	1.1027
30	6.3700	238.0	0.06964	0.4245	0.07115	29.18	17.64	0.00097	1.01933	1.0232
31	5.9608	253.0	0.06965	0.4484	0.06216	29.32	18.02	0.00113	1.01808	0.9662
32	5.6207	267.4	0.06962	0.4671	0.05565	29.62	18.41	0.00128	1.01704	0.9249
33	5.3301	281.3	0.06952	0.4822	0.05064	30.00	18.81	0.00142	1.01615	0.8935
34	5.0773	294.4	0.06966	0.4946	0.04667	30.43	19.22	0.00157	1.01538	0.8682
35	4.8539	307.4	0.06972	0.5051	0.04342	30.91	19.63	0.00172	1.01470	0.8478
36	4.6542	320.2	0.06972	0.5140	0.04069	31.42	20.05	0.00187	1.01409	0.8312
37	4.4739	332.8	0.06970	0.5217	0.03836	31.96	20.46	0.00201	1.01355	0.8172
38	4.3100	345.3	0.06966	0.5285	0.03633	32.52	20.87	0.00217	1.01305	0.8052
39	4.1600	357.7	0.06960	0.5344	0.03455	33.10	21.28	0.00232	1.01259	0.7947
40	4.0220	370.0	0.06953	0.5397	0.03297	33.69	21.69	0.00247	1.01217	0.7855
42	3.7758	394.2	0.06938	0.5486	0.03028	34.88	22.51	0.00279	1.01142	0.7701
44	3.5621	418.0	0.06922	0.5558	0.02806	36.10	23.31	0.00311	1.01077	0.7575
46	3.3742	441.6	0.06906	0.5617	0.02619	37.33	24.11	0.00344	1.01020	0.7467
48	3.2072	464.9	0.06889	0.5667	0.02458	38.59	24.90	0.00379	1.00970	0.7375
50	3.0577	488.1	0.06872	0.5708	0.02319	39.85	25.67	0.00414	1.00924	0.7294
52	2.9228	511.2	0.06853	0.5744	0.02197	41.11	26.44	0.00451	1.00883	0.7224
54	2.8002	534.3	0.06834	0.5774	0.02088	42.37	27.20	0.00488	1.00846	0.7162
56	2.6883	557.3	0.06813	0.5800	0.01991	43.63	27.95	0.00527	1.00812	0.7107
58	2.5857	580.3	0.06791	0.5823	0.01904	44.87	28.69	0.00566	1.00781	0.7061
60	2.4911	603.3	0.06768	0.5843	0.01824	46.17	29.42	0.00606	1.00753	0.7013
65	2.2839	661.5	0.06701	0.5883	0.01655	49.48	31.21	0.00713	1.00690	0.6903
70	2.1101	720.7	0.06620	0.5913	0.01516	52.47	32.94	0.00819	1.00637	0.6861
75	1.9618	781.4	0.06526	0.5935	0.01401	55.24	34.63	0.00926	1.00592	0.6863
80	1.8338	843.8	0.06421	0.5953	0.01303	57.48	36.28	0.01027	1.00554	0.6937
85	1.7219	908.2	0.06306	0.5966	0.01218	60.41	37.89	0.01142	1.00520	0.6939
90	1.6232	974.8	0.06183	0.5977	0.01144	63.03	39.46	0.01253	1.00490	0.6984
95	1.5354	1043.1	0.06060	0.5986	0.01079	65.44	41.00	0.01363	1.00463	0.7055
100	1.4569	1113.6	0.05933	0.5993	0.01022	76.58	45.76	0.01663	1.00440	0.6799
120	1.2105	1412.2	0.05457	0.6010	0.00843	94.54	54.03	0.02361	1.00365	0.6806
140	1.0361	1729.2	0.05074	0.6018	0.00719	107.89	59.17	0.03015	1.00313	0.6819
160	0.9060	2054.9	0.04788	0.6022	0.00627	118.86	63.00	0.03665	1.00273	0.6831
180	0.8051	2384.0	0.04576	0.6023	0.00556	128.36	66.20	0.04327	1.00243	0.6840
200	0.7245	2712.8	0.04420	0.6024	0.00500	136.82	69.05	0.05011	1.00218	0.6847
220	0.6587	3040.1	0.04301	0.6024	0.00454	144.53	71.71	0.05720	1.00199	0.6852
240	0.6038	3363.3	0.04214	0.6023	0.00416	151.61	74.25	0.06458	1.00182	0.6855
260	0.5574	3682.1	0.04150	0.6022	0.00384	158.19	76.71	0.07225	1.00168	0.6857
280	0.5177	3996.2	0.04103	0.6022	0.00357	164.39	79.12	0.08024	1.00156	0.6857
300	0.4833	4305.9	0.04068	0.6021	0.00333	170.25	81.49	0.08854	1.00146	0.6856
350	0.4144	5063.3	0.04020	0.6019	0.00285	183.95	87.34	0.11071	1.00125	0.6854
400	0.3627	5805.4	0.03998	0.6017	0.00249	196.81	93.10	0.13490	1.00109	0.6850
450	0.3225	6539.0	0.03989	0.6016	0.00222	209.22	98.79	0.16109	1.00097	0.6847
500	0.2903	7273.6	0.03980	0.6014	0.00200	221.52	104.43	0.18923	1.00087	0.6844
550	0.2639	8007.5	0.03974	0.6013	0.00181	233.64	110.00	0.21931	1.00080	0.6841
600	0.2420	8745.6	0.03966	0.6012	0.00166	245.73	115.51	0.25127	1.00073	0.6840
700	0.2075	10239.9	0.03943	0.6011	0.00143	269.89	126.35	0.32067	1.00063	0.6837
800	0.1816	11772.6	0.03909	0.6009	0.00125	294.31	136.93	0.39712	1.00055	0.6836
900	0.1614	13356.9	0.03861	0.6008	0.00111	319.21	147.26	0.48033	1.00049	0.6837
1000	0.1453	15001.7	0.03803	0.6008	0.00100	440.56	157.35	0.72851	1.00044	0.5351
1200	0.1211	18455.4	0.03674	0.6006	0.00083	511.01	176.89	0.98872	1.00036	0.5318
1400	0.1038	22141.3	0.03535	0.6006	0.00071	583.31	195.65	1.28006	1.00031	0.5300
1600	0.0909	26040.3	0.03401	0.6005	0.00062	657.61	213.74	1.60232	1.00027	0.5286
1800	0.0808	30222.7	0.03264	0.6004	0.00056	737.85	231.25	1.96015	1.00024	0.5259
2000	0.0727	35055.2	0.03090	0.6002	0.00050	838.71	248.26	2.37121	1.00022	0.5187
2500	0.0579	57115.1	0.02262	0.5972	0.00040	1451.41	289.52	3.93420	1.00017	0.4578
3000	0.0470	119609.2	0.01246	0.5814	0.00034	3558.42	333.10	6.63130	1.00014	0.3851

* TWO-PHASE BOUNDARY

Table 4.12 (continued)
THERMODYNAMIC PROPERTIES OF NORMAL HYDROGEN

0.70 MPA ISOBAR

TEMPERATURE DEG. K	VOLUME CU M/KG	ISOTHERM DERIVATIVE CU M-MPA/KG	ISOCHORE DERIVATIVE MPA/K	INTERNAL ENERGY KJ/KG-K	ENTHALPY KJ/KG	ENTROPY KJ/KG-K	CV KJ / KG-K	CP	VELOCITY OF SOUND M/S
* 14.034	0.01292	1.2145	0.9313	218.5	227.5	14.098	4.71	6.38	1283
15	0.01305	1.1212	0.9307	224.8	233.9	14.539	4.90	6.88	1254
16	0.01319	1.0426	0.9125	231.8	241.0	14.995	5.09	7.32	1224
17	0.01335	0.9761	0.8968	239.2	248.5	15.453	5.27	7.77	1199
18	0.01352	0.9125	0.8820	247.1	256.6	15.911	5.44	8.24	1176
19	0.01371	0.8500	0.8668	255.5	265.1	16.371	5.59	8.74	1153
20	0.01391	0.7857	0.8500	264.3	274.1	16.833	5.72	9.28	1129
21	0.01413	0.7213	0.8313	273.7	283.6	17.301	5.83	9.85	1104
22	0.01437	0.6554	0.8104	283.7	293.8	17.774	5.93	10.48	1076
23	0.01464	0.5871	0.7866	294.4	304.7	18.257	6.01	11.21	1046
24	0.01495	0.5200	0.7588	305.8	316.2	18.751	6.08	12.02	1014
25	0.01530	0.4498	0.7262	318.0	328.7	19.260	6.13	13.00	976
26	0.01571	0.3816	0.6878	331.3	342.3	19.789	6.18	14.14	934
27	0.01620	0.3102	0.6429	345.8	357.1	20.350	6.24	15.68	883
28	0.01682	0.2364	0.5913	362.1	373.9	20.959	6.31	18.03	822
29	0.01765	0.1660	0.5302	381.2	393.6	21.649	6.41	21.71	750
* 29.019	0.01767	0.1659	0.5291	381.6	393.9	21.662	6.41	21.70	749
* 29.019	0.11211	0.0412	0.0451	638.0	716.5	32.781	7.28	25.29	378
30	0.12448	0.0537	0.0395	651.7	738.8	33.540	7.04	20.53	396
31	0.13519	0.0641	0.0356	663.4	758.0	34.170	6.89	18.10	411
32	0.14493	0.0733	0.0327	673.9	775.3	34.719	6.78	16.59	424
33	0.15402	0.0817	0.0304	683.6	791.4	35.215	6.70	15.56	435
34	0.16265	0.0894	0.0285	692.7	806.6	35.667	6.61	14.79	447
35	0.17092	0.0967	0.0269	701.4	821.1	36.087	6.55	14.20	458
36	0.17893	0.1036	0.0255	709.8	835.0	36.481	6.50	13.75	468
37	0.18672	0.1102	0.0243	717.9	848.6	36.853	6.46	13.38	478
38	0.19432	0.1166	0.0232	725.8	861.8	37.206	6.43	13.08	487
39	0.20178	0.1229	0.0223	733.5	874.8	37.542	6.41	12.83	496
40	0.20911	0.1289	0.0214	741.1	887.5	37.864	6.39	12.62	504
42	0.22344	0.1406	0.0199	756.0	912.4	38.471	6.37	12.28	521
44	0.23742	0.1518	0.0186	770.5	936.7	39.036	6.35	12.02	536
46	0.25114	0.1626	0.0175	784.7	960.5	39.566	6.33	11.81	551
48	0.26463	0.1732	0.0166	798.7	984.0	40.065	6.32	11.64	565
50	0.27793	0.1835	0.0157	812.6	1007.1	40.537	6.32	11.51	578
52	0.29109	0.1936	0.0149	826.2	1030.0	40.986	6.31	11.40	591
54	0.30411	0.2036	0.0143	839.8	1052.7	41.415	6.31	11.31	604
56	0.31701	0.2133	0.0136	853.3	1075.2	41.825	6.32	11.23	616
58	0.32982	0.2230	0.0131	866.8	1097.6	42.219	6.32	11.17	628
60	0.34255	0.2325	0.0126	880.1	1119.9	42.596	6.33	11.12	639
65	0.37406	0.2560	0.0115	913.4	1175.3	43.481	6.37	11.03	666
70	0.40522	0.2789	0.0105	946.7	1230.4	44.298	6.42	11.00	691
75	0.43612	0.3015	0.0098	980.1	1285.4	45.057	6.50	11.01	715
80	0.46581	0.3238	0.0091	1013.7	1340.5	45.768	6.59	11.04	737
85	0.49734	0.3459	0.0085	1047.7	1395.9	46.440	6.70	11.11	757
90	0.52774	0.3678	0.0080	1082.2	1451.6	47.078	6.82	11.19	777
95	0.55802	0.3896	0.0076	1117.2	1507.8	47.685	6.95	11.29	796
100	0.58822	0.4112	0.0072	1152.8	1564.6	48.267	7.09	11.41	814
120	0.70831	0.4968	0.0059	1302.1	1797.9	50.390	7.67	11.93	879
140	0.82769	0.5814	0.0051	1462.3	2041.7	52.269	8.23	12.45	938
160	0.94664	0.6655	0.0044	1632.6	2295.3	53.960	8.71	12.90	993
180	1.06532	0.7491	0.0039	1811.4	2557.1	55.502	9.10	13.27	1045
200	1.18381	0.8325	0.0035	1997.0	2825.7	56.916	9.41	13.57	1096
220	1.30216	0.9157	0.0032	2188.1	3099.7	58.222	9.66	13.82	1144
240	1.42041	0.9988	0.0029	2383.6	3377.9	59.433	9.85	14.00	1191
260	1.53859	1.0817	0.0027	2582.4	3659.5	60.560	10.00	14.14	1237
280	1.65671	1.1646	0.0025	2783.8	3943.5	61.612	10.11	14.25	1281
300	1.77470	1.2474	0.0023	2987.1	4229.4	62.598	10.19	14.33	1324
350	2.06970	1.4541	0.0020	3500.1	4948.9	64.817	10.30	14.44	1427
400	2.36456	1.6607	0.0018	4016.9	5672.1	66.748	10.35	14.48	1524
450	2.65935	1.8672	0.0016	4535.5	6397.0	68.455	10.37	14.50	1616
500	2.95408	2.0737	0.0014	5054.7	7122.6	69.983	10.39	14.52	1702
550	3.24878	2.2800	0.0013	5574.5	7848.7	71.368	10.40	14.53	1784
600	3.54346	2.4863	0.0012	6095.5	8575.9	72.630	10.42	14.55	1863
700	4.13275	2.8989	0.0010	7140.5	10033.4	74.880	10.48	14.60	2010
800	4.72201	3.3115	0.0009	8193.1	11498.5	76.838	10.57	14.69	2146
900	5.31123	3.7240	0.0008	9256.5	12974.4	78.575	10.70	14.82	2272
1000	5.90045	4.1365	0.0007	10333.7	14464.0	80.142	10.86	14.98	2389
1200	7.07884	4.9614	0.0006	12544.0	17499.2	82.890	11.24	15.36	2604
1400	8.25722	5.7863	0.0005	14836.7	20616.8	85.299	11.68	15.80	2798
1600	9.43567	6.6111	0.0004	17216.7	23821.7	87.445	12.13	16.26	2976
1800	10.61465	7.4360	0.0004	19690.8	27121.1	89.382	12.64	16.77	3141
2000	11.79592	8.2608	0.0004	22284.7	30541.8	91.194	13.32	17.48	3292
2500	14.80859	10.3230	0.0003	29886.3	40252.3	95.494	17.99	22.57	3597
3000	18.21479	12.3851	0.0002	42312.9	55063.2	100.864	32.73	39.46	3860

* TWO-PHASE BOUNDARY

THERMODYNAMIC PROPERTIES OF NORMAL HYDROGEN

0.70 MPA ISOBAR

TEMPERATURE DEG. K	DENSITY KG/CU M	$V(DH/DV)_P$ KJ/KG	$V(DP/DU)_V$ MPA-CU M/KJ X 10^2	$-V(DP/DV)_T$ MPA	$(DV/DT)_P/V$ 1/DEG. K	THERMAL CONDUCTIVITY W/K-M X 10^3	VISCOSITY KG/M-S X 10^7	THERMAL DIFFUSIVITY SQ M/HR	DIELECTRIC CONSTANT	PRANDTL NUMBER
* 14.034	77.4091	644.0	0.25560	94.0105	0.00991	75.48	259.12	0.00055	1.25296	2.1901
15	76.6375	634.7	0.24774	85.9246	0.01083	83.85	228.80	0.00057	1.25024	1.8761
16	75.7962	633.6	0.23643	79.0247	0.01155	89.96	203.88	0.00058	1.24729	1.6580
17	74.8918	633.3	0.22711	73.0991	0.01227	94.03	183.49	0.00058	1.24412	1.5163
18	73.9440	630.7	0.21933	67.4756	0.01307	96.93	166.73	0.00057	1.24080	1.4182
19	72.9482	625.4	0.21269	62.0050	0.01398	98.66	152.68	0.00056	1.23732	1.3530
20	71.8967	616.4	0.20678	56.4899	0.01505	100.21	140.66	0.00054	1.23365	1.3020
21	70.7792	604.8	0.20142	51.0545	0.01628	101.99	130.19	0.00053	1.22976	1.2571
22	69.5856	589.9	0.19642	45.6088	0.01777	103.00	120.92	0.00051	1.22561	1.2306
23	68.2935	571.3	0.19162	40.0958	0.01962	103.31	112.52	0.00049	1.22113	1.2208
24	66.8966	550.9	0.18662	34.7861	0.02181	102.98	104.85	0.00046	1.21630	1.2235
25	65.3638	526.1	0.18114	29.4004	0.02470	102.02	97.68	0.00043	1.21102	1.2442
26	63.6599	499.3	0.17471	24.2906	0.02832	100.38	90.86	0.00040	1.20516	1.2796
27	61.7337	467.0	0.16688	19.1479	0.03357	98.07	84.23	0.00036	1.19857	1.3467
28	59.4604	428.5	0.15755	14.0551	0.04207	94.89	77.51	0.00032	1.19081	1.4724
29	56.6553	385.1	0.14604	9.4055	0.05638	90.57	70.44	0.00027	1.18129	1.6885
* 29.019	56.5997	385.0	0.14582	9.3871	0.05636	90.49	70.31	0.00027	1.18110	1.6862
* 29.019	8.9195	206.1	0.06948	0.3678	0.12267	33.25	17.94	0.00053	1.02714	1.3638
30	8.0336	224.4	0.06982	0.4315	0.09151	31.56	18.14	0.00069	1.02442	1.1802
31	7.3969	241.0	0.06993	0.4744	0.07508	31.10	18.44	0.00084	1.02247	1.0728
32	6.8999	256.5	0.06995	0.5059	0.06468	31.06	18.78	0.00098	1.02095	1.0031
33	6.4925	271.4	0.06986	0.5302	0.05733	31.23	19.15	0.00111	1.01970	0.9539
34	6.1483	285.1	0.07009	0.5496	0.05186	31.51	19.53	0.00125	1.01865	0.9163
35	5.8506	298.6	0.07022	0.5655	0.04757	31.88	19.91	0.00138	1.01774	0.8873
36	5.5888	311.9	0.07026	0.5789	0.04409	32.31	20.31	0.00151	1.01694	0.8643
37	5.3557	324.9	0.07025	0.5904	0.04119	32.78	20.71	0.00165	1.01623	0.8454
38	5.1461	337.8	0.07021	0.6002	0.03873	33.28	21.10	0.00178	1.01559	0.8296
39	4.9559	350.6	0.07015	0.6088	0.03661	33.81	21.50	0.00191	1.01501	0.8162
40	4.7823	363.2	0.07007	0.6164	0.03476	34.36	21.90	0.00205	1.01448	0.8046
42	4.4755	388.0	0.06990	0.6291	0.03165	35.49	22.70	0.00232	1.01355	0.7855
44	4.2119	412.4	0.06971	0.6393	0.02914	36.66	23.49	0.00261	1.01275	0.7701
46	3.9819	436.4	0.06952	0.6476	0.02706	37.86	24.28	0.00290	1.01205	0.7574
48	3.7789	460.2	0.06933	0.6545	0.02530	39.08	25.06	0.00320	1.01143	0.7466
50	3.5980	483.8	0.06913	0.6603	0.02379	40.31	25.82	0.00350	1.01088	0.7373
52	3.4354	507.2	0.06892	0.6652	0.02247	41.55	26.58	0.00382	1.01039	0.7293
54	3.2883	530.5	0.06871	0.6694	0.02131	42.79	27.34	0.00414	1.00994	0.7223
56	3.1544	553.8	0.06848	0.6730	0.02028	44.03	28.08	0.00447	1.00954	0.7161
58	3.0319	577.0	0.06825	0.6761	0.01935	45.26	28.81	0.00481	1.00916	0.7109
60	2.9193	600.3	0.06800	0.6788	0.01852	46.54	29.54	0.00516	1.00882	0.7056
65	2.6734	658.9	0.06730	0.6843	0.01674	49.82	31.32	0.00608	1.00808	0.6936
70	2.4678	718.5	0.06647	0.6883	0.01531	52.78	33.05	0.00700	1.00746	0.6887
75	2.2929	779.6	0.06551	0.6914	0.01412	55.51	34.73	0.00792	1.00693	0.6886
80	2.1422	842.2	0.06444	0.6937	0.01311	57.74	36.37	0.00879	1.00647	0.6956
85	2.0107	906.9	0.06327	0.6956	0.01225	60.66	37.97	0.00978	1.00607	0.6954
90	1.8949	973.7	0.06202	0.6970	0.01150	63.27	39.54	0.01074	1.00572	0.6996
95	1.7920	1042.1	0.06078	0.6982	0.01084	65.67	41.08	0.01168	1.00541	0.7064
100	1.7000	1112.8	0.05950	0.6991	0.01025	76.66	45.76	0.01423	1.00513	0.6811
120	1.4118	1412.1	0.05469	0.7014	0.00845	94.59	54.03	0.02022	1.00426	0.6813
140	1.2082	1729.5	0.05084	0.7025	0.00720	107.93	59.17	0.02584	1.00364	0.6824
160	1.0564	2055.6	0.04796	0.7030	0.00627	118.90	63.00	0.03142	1.00319	0.6834
180	0.9387	2385.0	0.04583	0.7032	0.00556	128.40	66.20	0.03710	1.00283	0.6843
200	0.8447	2714.1	0.04425	0.7032	0.00500	136.86	69.06	0.04297	1.00255	0.6849
220	0.7680	3041.6	0.04306	0.7032	0.00454	144.57	71.72	0.04906	1.00232	0.6854
240	0.7040	3365.0	0.04218	0.7031	0.00416	151.65	74.26	0.05538	1.00212	0.6857
260	0.6499	3684.0	0.04154	0.7030	0.00384	158.24	76.72	0.06197	1.00196	0.6858
280	0.6036	3998.3	0.04106	0.7029	0.00356	164.43	79.14	0.06882	1.00182	0.6858
300	0.5635	4308.1	0.04071	0.7029	0.00333	170.30	81.51	0.07594	1.00170	0.6857
350	0.4832	5065.8	0.04022	0.7026	0.00285	184.00	87.36	0.09496	1.00146	0.6854
400	0.4229	5808.0	0.04000	0.7023	0.00249	196.87	93.12	0.11571	1.00127	0.6851
450	0.3760	6541.7	0.03991	0.7021	0.00222	209.29	98.82	0.13817	1.00113	0.6847
500	0.3385	7276.4	0.03982	0.7020	0.00200	221.59	104.46	0.16231	1.00102	0.6844
550	0.3078	8010.4	0.03975	0.7018	0.00181	233.73	110.04	0.18811	1.00093	0.6842
600	0.2822	8748.5	0.03967	0.7017	0.00166	245.82	115.55	0.21552	1.00085	0.6840
700	0.2420	10242.9	0.03944	0.7015	0.00143	270.00	126.40	0.27505	1.00073	0.6837
800	0.2118	11775.6	0.03910	0.7013	0.00125	294.43	136.98	0.34062	1.00064	0.6836
900	0.1883	13360.0	0.03862	0.7012	0.00111	319.35	147.32	0.41198	1.00057	0.6837
1000	0.1695	15004.8	0.03804	0.7010	0.00100	440.56	157.42	0.62457	1.00051	0.5354
1200	0.1413	18458.6	0.03674	0.7009	0.00083	511.01	176.97	0.84762	1.00043	0.5321
1400	0.1211	22144.4	0.03536	0.7008	0.00071	583.30	195.74	1.09735	1.00036	0.5302
1600	0.1060	26040.7	0.03402	0.7007	0.00062	657.49	213.84	1.37350	1.00032	0.5289
1800	0.0942	30208.0	0.03267	0.7005	0.00056	737.13	231.36	1.67972	1.00028	0.5263
2000	0.0848	34983.6	0.03099	0.7003	0.00050	835.81	248.38	2.03000	1.00026	0.5196
2500	0.0675	55188.8	0.02305	0.6971	0.00040	1416.15	289.62	3.34507	1.00020	0.4616
3000	0.0549	114990.9	0.01299	0.6799	0.00034	3384.84	332.87	5.62467	1.00017	0.3881

* TWO-PHASE BOUNDARY

Table 4.12 (continued)
THERMODYNAMIC PROPERTIES OF NORMAL HYDROGEN

0.80 MPA ISOBAR

TEMPERATURE DEG. K	VOLUME CU M/KG	ISOTHERM DERIVATIVE CU M-MPA/KG	ISOCHORE DERIVATIVE MPA/K	INTERNAL ENERGY KJ/KG-K	ENTHALPY KJ/KG	ENTROPY KJ/KG-K	CV KJ /	CP KG-K	VELOCITY OF SOUND M/S
* 14.067	0.01291	1.2211	0.9325	218.5	228.8	14.100	4.71	6.38	1286
15	0.01303	1.1342	0.9319	224.6	235.0	14.525	4.90	6.85	1259
16	0.01318	1.0530	0.9155	231.5	242.1	14.980	5.09	7.30	1229
17	0.01333	0.9860	0.8998	238.9	249.6	15.437	5.27	7.75	1204
18	0.01350	0.9223	0.8850	246.8	257.6	15.894	5.43	8.22	1181
19	0.01369	0.8601	0.8698	255.1	266.1	16.352	5.58	8.71	1159
20	0.01388	0.7958	0.8531	263.9	275.1	16.812	5.71	9.24	1134
21	0.01410	0.7318	0.8346	273.3	284.6	17.278	5.83	9.80	1109
22	0.01434	0.6662	0.8139	283.2	294.7	17.749	5.93	10.42	1082
23	0.01461	0.5980	0.7903	293.8	305.5	18.229	6.01	11.13	1053
24	0.01491	0.5315	0.7631	305.0	317.0	18.718	6.08	11.92	1021
25	0.01525	0.4638	0.7313	317.1	329.3	19.223	6.13	12.83	985
26	0.01565	0.3950	0.6939	330.2	342.7	19.746	6.18	13.94	944
27	0.01612	0.3242	0.6505	344.4	357.3	20.298	6.24	15.39	895
28	0.01670	0.2545	0.6010	360.3	373.7	20.890	6.30	17.39	838
29	0.01747	0.1831	0.5432	378.6	392.5	21.553	6.39	20.66	769
* 29.828	0.01837	0.1248	0.4856	396.8	411.5	22.196	6.49	25.51	700
* 29.828	0.09519	0.0349	0.0545	631.9	708.1	32.145	7.46	30.49	378
30	0.09783	0.0379	0.0526	635.3	713.6	32.330	7.38	28.36	382
31	0.10956	0.0516	0.0455	650.7	738.3	33.141	7.12	22.06	400
32	0.11945	0.0627	0.0408	663.2	758.7	33.790	6.95	19.10	415
33	0.12833	0.0723	0.0374	674.3	777.0	34.350	6.84	17.33	428
34	0.13656	0.0810	0.0347	684.4	793.6	34.849	6.71	16.11	441
35	0.14434	0.0890	0.0325	693.8	809.3	35.303	6.62	15.25	453
36	0.15178	0.0966	0.0306	702.8	824.2	35.723	6.56	14.60	464
37	0.15896	0.1037	0.0290	711.4	838.6	36.116	6.51	14.10	474
38	0.16594	0.1106	0.0276	719.7	852.5	36.487	6.48	13.70	484
39	0.17273	0.1172	0.0264	727.8	866.0	36.838	6.45	13.37	493
40	0.17939	0.1236	0.0253	735.7	879.2	37.173	6.42	13.09	502
42	0.19234	0.1358	0.0234	751.1	905.0	37.801	6.39	12.66	519
44	0.20491	0.1475	0.0218	766.0	929.9	38.381	6.37	12.33	534
46	0.21720	0.1588	0.0205	780.6	954.3	38.923	6.35	12.07	549
48	0.22925	0.1697	0.0193	794.9	978.3	39.433	6.34	11.87	564
50	0.24110	0.1804	0.0183	808.9	1001.8	39.914	6.33	11.71	577
52	0.25280	0.1908	0.0173	822.8	1025.1	40.370	6.33	11.57	591
54	0.26436	0.2010	0.0165	836.6	1048.1	40.805	6.33	11.46	603
56	0.27580	0.2110	0.0158	850.3	1070.9	41.221	6.33	11.37	616
58	0.28715	0.2208	0.0151	863.9	1093.6	41.619	6.33	11.29	627
60	0.29840	0.2305	0.0145	877.4	1116.1	42.000	6.34	11.23	639
65	0.32623	0.2544	0.0132	911.0	1172.0	42.893	6.38	11.12	666
70	0.35372	0.2777	0.0121	944.5	1227.5	43.716	6.43	11.07	691
75	0.38093	0.3005	0.0112	978.0	1282.8	44.479	6.50	11.07	715
80	0.40794	0.3231	0.0105	1011.9	1338.2	45.195	6.59	11.10	737
85	0.43478	0.3454	0.0098	1046.0	1393.8	45.869	6.70	11.15	758
90	0.46149	0.3675	0.0092	1080.6	1449.8	46.510	6.82	11.23	778
95	0.48810	0.3894	0.0087	1115.7	1506.2	47.119	6.95	11.33	797
100	0.51461	0.4112	0.0082	1151.4	1563.1	47.702	7.09	11.44	815
120	0.61995	0.4972	0.0068	1300.9	1796.9	49.830	7.68	11.95	880
140	0.72459	0.5821	0.0058	1461.4	2041.1	51.711	8.23	12.46	939
160	0.82881	0.6663	0.0050	1631.9	2294.9	53.404	8.71	12.91	994
180	0.93276	0.7501	0.0045	1810.7	2556.9	54.947	9.10	13.28	1046
200	1.03651	0.8336	0.0040	1996.4	2825.6	56.363	9.41	13.58	1097
220	1.14013	0.9169	0.0037	2187.6	3099.7	57.669	9.66	13.82	1145
240	1.24366	1.0000	0.0033	2383.2	3378.1	58.880	9.85	14.01	1192
260	1.34710	1.0831	0.0031	2582.0	3659.7	60.008	10.00	14.15	1238
280	1.45049	1.1660	0.0029	2783.4	3943.8	61.060	10.11	14.25	1282
300	1.55377	1.2488	0.0027	2986.8	4229.8	62.046	10.19	14.33	1325
350	1.81194	1.4557	0.0023	3499.9	4949.5	64.265	10.30	14.44	1428
400	2.06998	1.6623	0.0020	4016.7	5672.7	66.197	10.35	14.48	1525
450	2.32794	1.8689	0.0018	4535.3	6397.7	67.904	10.37	14.50	1616
500	2.58586	2.0753	0.0016	5054.6	7123.2	69.432	10.39	14.52	1703
550	2.84373	2.2817	0.0015	5574.4	7849.4	70.817	10.40	14.53	1785
600	3.10158	2.4880	0.0013	6095.4	8576.6	72.079	10.42	14.55	1864
700	3.61723	2.9007	0.0011	7140.4	10034.2	74.330	10.48	14.61	2011
800	4.13284	3.3132	0.0010	8193.1	11499.3	76.288	10.57	14.69	2146
900	4.64842	3.7257	0.0009	9256.5	12975.2	78.024	10.70	14.82	2272
1000	5.16399	4.1382	0.0008	10333.6	14464.8	79.591	10.86	14.98	2390
1200	6.19509	4.9631	0.0007	12544.0	17500.0	82.339	11.24	15.36	2605
1400	7.22618	5.7880	0.0006	14836.7	20617.6	84.748	11.68	15.80	2799
1600	8.25732	6.6129	0.0005	17216.5	23822.4	86.894	12.13	16.26	2977
1800	9.28889	7.4378	0.0004	19689.8	27121.0	88.831	12.63	16.76	3142
2000	10.32234	8.2626	0.0004	22280.3	30538.2	90.641	13.30	17.45	3293
2500	12.95466	10.3247	0.0003	29819.5	40183.3	94.913	17.71	22.26	3601
3000	15.90915	12.3869	0.0003	41895.2	54622.5	100.150	31.55	38.10	3863

* TWO-PHASE BOUNDARY

Table 4.12 (continued)
THERMODYNAMIC PROPERTIES OF NORMAL HYDROGEN

0.80 MPA ISOBAR

TEMPERATURE DEG. K	DENSITY KG/CU M	$V(\partial H/\partial V)_P$ KJ/KG	$V(\partial P/\partial U)_V$ MPA-CU M/KJ X 10^2	$-V(\partial P/\partial V)_T$ MPA	$(\partial V/\partial T)_P/V$ 1/DEG. K	THERMAL CONDUCTIVITY W/K-M X 10^3	VISCOSITY KG/M-S X 10^7	THERMAL DIFFUSIVITY SQ M/HR	DIELECTRIC CONSTANT	PRANDTL NUMBER
* 14.067	77.4651	647.3	0.25548	94.5954	0.00986	75.88	259.74	0.00055	1.25315	2.1840
15	76.7248	639.7	0.24789	87.0229	0.01071	83.96	230.36	0.00058	1.25055	1.8798
16	75.8894	637.2	0.23706	79.9090	0.01146	90.11	205.24	0.00059	1.24761	1.6627
17	74.9923	637.0	0.22772	73.9430	0.01217	94.21	184.71	0.00058	1.24447	1.5196
18	74.0521	634.5	0.21992	68.2963	0.01296	97.15	167.84	0.00057	1.24118	1.4204
19	73.0644	629.5	0.21325	62.8392	0.01384	98.91	153.70	0.00056	1.23772	1.3540
20	72.0222	620.7	0.20734	57.3120	0.01489	100.49	141.61	0.00054	1.23409	1.3021
21	70.9163	609.5	0.20196	51.8962	0.01608	102.31	131.10	0.00053	1.23024	1.2559
22	69.7362	595.0	0.19695	46.4549	0.01752	103.37	121.79	0.00051	1.22613	1.2282
23	68.4619	576.7	0.19215	40.9370	0.01931	103.72	113.38	0.00049	1.22172	1.2171
24	67.0865	556.9	0.18721	35.6589	0.02140	103.44	105.70	0.00047	1.21696	1.2177
25	65.5817	533.8	0.18185	30.4154	0.02404	102.55	98.55	0.00044	1.21177	1.2332
26	63.9171	507.2	0.17564	25.2504	0.02748	101.00	91.76	0.00041	1.20605	1.2664
27	62.0450	475.9	0.16816	20.1166	0.03234	98.80	85.20	0.00037	1.19963	1.3271
28	59.8722	440.8	0.15929	15.2350	0.03945	95.82	78.64	0.00033	1.19221	1.4271
29	57.2336	398.5	0.14856	10.4781	0.05184	91.83	71.81	0.00028	1.18324	1.6155
* 29.828	54.4335	356.9	0.13744	6.7925	0.07148	87.32	65.59	0.00023	1.17378	1.9164
* 29.828	10.5055	205.0	0.06963	0.3666	0.14877	36.90	18.93	0.00041	1.03202	1.5642
30	10.2218	208.9	0.06974	0.3877	0.13577	36.00	18.91	0.00045	1.03114	1.4895
31	9.1271	228.2	0.07006	0.4710	0.09664	33.73	19.00	0.00060	1.02777	1.2426
32	8.3720	245.4	0.07019	0.5246	0.07784	32.98	19.24	0.00074	1.02545	1.1145
33	7.7923	261.3	0.07012	0.5634	0.06630	32.75	19.54	0.00087	1.02368	1.0339
34	7.3227	275.8	0.07050	0.5930	0.05844	32.79	19.88	0.00100	1.02224	0.9770
35	6.9280	289.8	0.07071	0.6167	0.05262	32.99	20.23	0.00112	1.02103	0.9353
36	6.5883	303.6	0.07080	0.6361	0.04810	33.30	20.60	0.00125	1.01999	0.9034
37	6.2907	317.1	0.07081	0.6525	0.04447	33.68	20.98	0.00137	1.01908	0.8781
38	6.0264	330.4	0.07078	0.6664	0.04146	34.12	21.36	0.00149	1.01828	0.8574
39	5.7893	343.5	0.07072	0.6784	0.03891	34.59	21.74	0.00161	1.01755	0.8402
40	5.5746	356.5	0.07063	0.6889	0.03672	35.09	22.13	0.00173	1.01690	0.8256
42	5.1992	381.9	0.07044	0.7062	0.03314	36.14	22.91	0.00198	1.01575	0.8021
44	4.8801	406.9	0.07022	0.7200	0.03030	37.26	23.68	0.00223	1.01478	0.7836
46	4.6041	431.4	0.06999	0.7312	0.02799	38.41	24.45	0.00249	1.01394	0.7687
48	4.3621	455.6	0.06978	0.7405	0.02605	39.59	25.22	0.00275	1.01320	0.7561
50	4.1476	479.5	0.06955	0.7482	0.02441	40.79	25.98	0.00302	1.01255	0.7455
52	3.9557	503.3	0.06932	0.7547	0.02299	42.00	26.73	0.00330	1.01197	0.7364
54	3.7827	526.9	0.06908	0.7602	0.02175	43.22	27.48	0.00359	1.01144	0.7285
56	3.6258	550.5	0.06883	0.7649	0.02065	44.44	28.21	0.00388	1.01097	0.7216
58	3.4826	573.9	0.06859	0.7690	0.01967	45.65	28.94	0.00418	1.01053	0.7158
60	3.3512	597.4	0.06833	0.7726	0.01880	46.91	29.66	0.00449	1.01013	0.7100
65	3.0653	656.5	0.06760	0.7797	0.01694	50.16	31.43	0.00530	1.00927	0.6969
70	2.8271	716.5	0.06673	0.7850	0.01546	53.10	33.15	0.00611	1.00854	0.6913
75	2.6251	777.8	0.06575	0.7889	0.01423	55.79	34.82	0.00691	1.00793	0.6909
80	2.4513	840.7	0.06467	0.7920	0.01320	58.01	36.46	0.00768	1.00741	0.6974
85	2.3000	905.6	0.06348	0.7944	0.01232	60.91	38.06	0.00855	1.00695	0.6969
90	2.1669	972.6	0.06222	0.7962	0.01155	63.51	39.62	0.00939	1.00654	0.7008
95	2.0488	1041.2	0.06096	0.7978	0.01088	65.91	41.15	0.01022	1.00619	0.7073
100	1.9432	1112.0	0.05967	0.7990	0.01029	76.75	45.77	0.01243	1.00587	0.6823
120	1.6130	1412.0	0.05481	0.8019	0.00846	94.64	54.02	0.01768	1.00487	0.6821
140	1.3801	1729.9	0.05093	0.8033	0.00720	107.98	59.17	0.02260	1.00416	0.6829
160	1.2065	2056.3	0.04804	0.8039	0.00628	118.95	63.01	0.02749	1.00364	0.6838
180	1.0721	2386.1	0.04589	0.8042	0.00557	128.45	66.21	0.03248	1.00323	0.6846
200	0.9648	2715.4	0.04430	0.8042	0.00500	136.90	69.07	0.03762	1.00291	0.6852
220	0.8771	3043.2	0.04311	0.8042	0.00454	144.62	71.73	0.04295	1.00265	0.6856
240	0.8041	3366.8	0.04223	0.8041	0.00416	151.70	74.28	0.04849	1.00242	0.6858
260	0.7423	3686.0	0.04157	0.8040	0.00384	158.28	76.74	0.05426	1.00224	0.6859
280	0.6894	4000.4	0.04110	0.8039	0.00356	164.48	79.15	0.06026	1.00208	0.6859
300	0.6436	4310.4	0.04075	0.8037	0.00332	170.35	81.53	0.06650	1.00194	0.6858
350	0.5519	5068.3	0.04025	0.8034	0.00285	184.05	87.38	0.08315	1.00166	0.6855
400	0.4831	5810.6	0.04003	0.8031	0.00249	196.93	93.15	0.10132	1.00146	0.6851
450	0.4296	6544.4	0.03993	0.8028	0.00222	209.36	98.85	0.12099	1.00129	0.6847
500	0.3867	7279.3	0.03984	0.8026	0.00199	221.67	104.49	0.14212	1.00117	0.6844
550	0.3517	8013.2	0.03977	0.8024	0.00181	233.81	110.07	0.16471	1.00106	0.6842
600	0.3224	8751.4	0.03969	0.8022	0.00166	245.91	115.59	0.18871	1.00097	0.6840
700	0.2765	10245.9	0.03946	0.8019	0.00143	270.11	126.44	0.24083	1.00083	0.6837
800	0.2420	11778.7	0.03911	0.8017	0.00125	294.55	137.04	0.29824	1.00073	0.6836
900	0.2151	13363.1	0.03863	0.8015	0.00111	319.48	147.38	0.36071	1.00065	0.6837
1000	0.1936	15008.0	0.03805	0.8014	0.00100	440.56	157.48	0.54661	1.00058	0.5356
1200	0.1614	18461.9	0.03675	0.8011	0.00083	511.01	177.05	0.74180	1.00049	0.5323
1400	0.1384	22147.5	0.03537	0.8010	0.00071	583.29	195.83	0.96032	1.00042	0.5305
1600	0.1211	26041.6	0.03403	0.8009	0.00062	657.39	213.94	1.20191	1.00036	0.5291
1800	0.1077	30196.8	0.03269	0.8007	0.00056	736.56	231.47	1.46950	1.00032	0.5267
2000	0.0969	34926.6	0.03105	0.8005	0.00050	833.46	248.50	1.77454	1.00029	0.5204
2500	0.0772	55442.4	0.02340	0.7970	0.00040	1387.70	289.72	2.90726	1.00023	0.4648
3000	0.0629	111251.3	0.01345	0.7786	0.00034	3244.24	332.71	4.87643	1.00019	0.3908

* TWO-PHASE BOUNDARY

Table 4.12 (continued)
THERMODYNAMIC PROPERTIES OF NORMAL HYDROGEN

0.90 MPA ISOBAR

TEMPERATURE DEG. K	VOLUME CU M/KG	ISOTHERM DERIVATIVE CU M-MPA/KG	ISOCHORE DERIVATIVE MPA/K	INTERNAL ENERGY KJ/KG-K	ENTHALPY KJ/KG	ENTROPY KJ/KG-K	CV KJ /	CP KG-K	VELOCITY OF SOUND M/S
* 14.100	0.01290	1.2278	0.9336	218.6	230.2	14.103	4.72	6.38	1289
15	0.01302	1.1472	0.9331	224.4	236.1	14.511	4.90	6.83	1265
16	0.01316	1.0633	0.9185	231.3	243.2	14.965	5.09	7.28	1234
17	0.01332	0.9959	0.9027	238.7	250.7	15.421	5.27	7.73	1209
18	0.01348	0.9319	0.8879	246.5	258.6	15.877	5.43	8.20	1186
19	0.01367	0.8700	0.8728	254.8	267.1	16.333	5.58	8.68	1164
20	0.01386	0.8057	0.8562	263.6	276.0	16.792	5.71	9.20	1140
21	0.01407	0.7421	0.8377	272.8	285.5	17.256	5.82	9.76	1115
22	0.01431	0.6767	0.8172	282.7	295.6	17.724	5.92	10.37	1088
23	0.01457	0.6109	0.7941	293.1	306.3	18.200	6.00	11.04	1060
24	0.01486	0.5428	0.7672	304.3	317.7	18.687	6.07	11.82	1028
25	0.01520	0.4762	0.7360	316.3	330.0	19.187	6.13	12.70	993
26	0.01558	0.4081	0.6998	329.1	343.2	19.703	6.18	13.76	953
27	0.01604	0.3377	0.6579	343.1	357.6	20.247	6.23	15.13	906
28	0.01660	0.2691	0.6099	358.6	373.5	20.826	6.29	16.96	851
29	0.01731	0.2011	0.5551	376.2	391.7	21.465	6.37	19.69	788
30	0.01833	0.1312	0.4891	397.3	413.8	22.211	6.48	24.86	709
* 30.567	0.01918	0.0889	0.4424	412.2	429.5	22.730	6.58	31.34	650
* 30.567	0.08139	0.0281	0.0656	624.3	697.6	31.506	7.67	38.60	376
31	0.08768	0.0365	0.0595	634.0	713.0	32.005	7.46	30.60	387
32	0.09864	0.0507	0.0511	650.5	739.3	32.841	7.18	23.21	405
33	0.10775	0.0622	0.0457	663.7	760.7	33.502	7.00	19.86	420
34	0.11588	0.0721	0.0418	675.2	779.5	34.063	6.83	17.87	434
35	0.12339	0.0811	0.0387	685.6	796.7	34.561	6.71	16.56	447
36	0.13047	0.0893	0.0362	695.4	812.8	35.014	6.63	15.64	459
37	0.13723	0.0971	0.0342	704.5	828.1	35.433	6.57	14.94	470
38	0.14374	0.1044	0.0324	713.3	842.7	35.824	6.52	14.40	480
39	0.15006	0.1114	0.0308	721.8	856.9	36.192	6.48	13.97	490
40	0.15620	0.1182	0.0294	730.1	870.7	36.541	6.46	13.61	499
42	0.16811	0.1311	0.0271	746.0	897.3	37.191	6.42	13.06	517
44	0.17961	0.1433	0.0252	761.4	923.0	37.789	6.39	12.66	533
46	0.19079	0.1550	0.0235	776.3	948.0	38.344	6.37	12.35	548
48	0.20173	0.1663	0.0221	790.9	972.5	38.865	6.36	12.11	563
50	0.21246	0.1773	0.0209	805.3	996.5	39.355	6.35	11.91	577
52	0.22303	0.1880	0.0198	819.4	1020.1	39.818	6.34	11.75	590
54	0.23345	0.1984	0.0189	833.4	1043.5	40.260	6.34	11.62	603
56	0.24376	0.2086	0.0180	847.2	1066.6	40.681	6.34	11.51	615
58	0.25396	0.2187	0.0172	861.0	1089.5	41.083	6.34	11.42	627
60	0.26408	0.2286	0.0165	874.6	1112.3	41.469	6.35	11.34	639
65	0.28905	0.2528	0.0150	908.5	1168.7	42.370	6.38	11.21	666
70	0.31367	0.2764	0.0138	942.3	1224.6	43.199	6.44	11.15	692
75	0.33802	0.2996	0.0127	976.0	1280.2	43.967	6.51	11.13	716
80	0.36216	0.3224	0.0118	1010.0	1335.9	44.686	6.60	11.15	738
85	0.38614	0.3448	0.0111	1044.3	1391.8	45.364	6.70	11.20	759
90	0.40998	0.3671	0.0104	1079.0	1448.0	46.006	6.82	11.27	779
95	0.43371	0.3892	0.0098	1114.2	1504.5	46.618	6.95	11.36	798
100	0.45736	0.4111	0.0093	1150.0	1561.6	47.203	7.09	11.47	815
120	0.55124	0.4975	0.0077	1299.8	1795.9	49.335	7.68	11.97	881
140	0.64442	0.5827	0.0065	1460.5	2040.5	51.219	8.23	12.48	940
160	0.73717	0.6671	0.0057	1631.1	2294.5	52.914	8.71	12.92	995
180	0.82966	0.7511	0.0050	1810.0	2556.7	54.458	9.10	13.29	1047
200	0.92195	0.8347	0.0045	1995.8	2825.6	55.874	9.41	13.59	1098
220	1.01411	0.9181	0.0041	2187.1	3099.8	57.181	9.66	13.83	1146
240	1.10618	1.0013	0.0038	2382.7	3378.3	58.393	9.85	14.01	1193
260	1.19817	1.0844	0.0035	2581.6	3660.0	59.520	10.00	14.15	1239
280	1.29010	1.1674	0.0032	2783.1	3944.2	60.573	10.11	14.26	1283
300	1.38193	1.2503	0.0030	2986.5	4230.2	61.559	10.19	14.33	1326
350	1.61146	1.4572	0.0026	3499.7	4950.0	63.779	10.30	14.44	1429
400	1.84086	1.6639	0.0023	4016.6	5673.3	65.711	10.35	14.49	1526
450	2.07019	1.8705	0.0020	4535.1	6398.3	67.418	10.37	14.50	1617
500	2.29945	2.0770	0.0018	5054.4	7123.9	68.946	10.39	14.52	1704
550	2.52869	2.2834	0.0016	5574.3	7850.1	70.331	10.40	14.53	1786
600	2.75790	2.4897	0.0015	6095.3	8577.4	71.593	10.42	14.55	1864
700	3.21627	2.9024	0.0013	7140.3	10035.0	73.844	10.48	14.61	2011
800	3.67460	3.3149	0.0011	8193.0	11500.1	75.802	10.57	14.69	2147
900	4.13290	3.7275	0.0010	9256.4	12976.1	77.538	10.70	14.82	2273
1000	4.59118	4.1400	0.0009	10333.6	14465.7	79.106	10.86	14.98	2390
1200	5.50773	4.9649	0.0008	12543.9	17500.9	81.853	11.24	15.36	2605
1400	6.42425	5.7898	0.0006	14836.7	20618.5	84.262	11.68	15.80	2799
1600	7.34082	6.6147	0.0006	17216.4	23823.1	86.408	12.13	16.26	2977
1800	8.25775	7.4396	0.0005	19689.0	27121.0	88.344	12.62	16.75	3142
2000	9.17625	8.2644	0.0005	22276.7	30535.4	90.153	13.27	17.43	3294
2500	11.51332	10.3265	0.0004	29764.2	40126.2	94.402	17.48	22.01	3604
3000	14.12044	12.3886	0.0003	41549.2	54257.6	99.529	30.58	36.98	3867

* TWO-PHASE BOUNDARY

Table 4.12 (continued)
THERMODYNAMIC PROPERTIES OF NORMAL HYDROGEN

0.90 MPA ISOBAR

TEMPERATURE DEG. K	DENSITY KG/CU M	$V(DH/DV)_P$ KJ/KG	$V(DP/DU)_V$ MPA-CU M/KJ X 10^2	$-V(DP/DV)_T$ MPA	$(DV/DT)_P/V$ 1/DEG. K	THERMAL CONDUCTIVITY W/K-M X 10^3	VISCOSITY KG/M-S X 10^7	THERMAL DIFFUSIVITY SQ M/HR	DIELECTRIC CONSTANT	PRANDTL NUMBER
* 14.100	77.5208	650.6	0.25535	95.1791	0.00981	76.29	260.36	0.00056	1.25335	2.1781
15	76.8115	644.7	0.24804	88.1146	0.01059	84.07	231.93	0.00058	1.25085	1.8836
16	75.9818	640.8	0.23767	80.7877	0.01137	90.26	206.61	0.00059	1.24794	1.6675
17	75.0917	640.5	0.22831	74.7803	0.01207	94.39	185.94	0.00059	1.24482	1.5231
18	74.1591	638.2	0.22049	69.1097	0.01285	97.36	168.95	0.00058	1.24155	1.4227
19	73.1793	633.5	0.21380	63.6661	0.01371	99.15	154.72	0.00056	1.23812	1.3551
20	72.1462	624.9	0.20788	58.1271	0.01473	100.77	142.56	0.00055	1.23452	1.3023
21	71.0514	614.1	0.20249	52.7287	0.01589	102.62	132.00	0.00053	1.23071	1.2550
22	69.8844	599.9	0.19747	47.2907	0.01728	103.72	122.66	0.00052	1.22665	1.2260
23	68.6311	583.2	0.19269	41.9288	0.01894	104.13	114.25	0.00049	1.22230	1.2118
24	67.2723	562.8	0.18778	36.5169	0.02101	103.90	106.55	0.00047	1.21760	1.2125
25	65.7941	540.6	0.18251	31.3292	0.02349	103.07	99.40	0.00044	1.21250	1.2249
26	64.1659	514.8	0.17651	26.1854	0.02673	101.60	92.65	0.00041	1.20690	1.2546
27	62.3439	484.3	0.16936	21.0554	0.03124	99.51	86.15	0.00038	1.20065	1.3101
28	60.2507	450.8	0.16085	16.2152	0.03761	96.69	79.70	0.00034	1.19350	1.3975
29	57.7615	412.0	0.15082	11.6174	0.04778	92.98	73.10	0.00029	1.18504	1.5477
30	54.5650	363.7	0.13827	7.1562	0.06834	87.88	65.89	0.00023	1.17422	1.8639
* 30.567	52.1365	328.3	0.12886	4.6342	0.09546	85.67	61.14	0.00019	1.16605	2.2367
* 30.567	12.2868	203.6	0.06963	0.3458	0.18963	41.99	20.06	0.00032	1.03752	1.8447
31	11.4051	213.9	0.06993	0.4158	0.14310	38.62	19.89	0.00040	1.03479	1.5758
32	10.1382	233.6	0.07028	0.5144	0.09938	35.76	19.87	0.00055	1.03088	1.2896
33	9.2804	251.0	0.07028	0.5773	0.07913	34.73	20.04	0.00068	1.02824	1.1459
34	8.6298	266.3	0.07086	0.6224	0.06711	34.35	20.30	0.00080	1.02625	1.0562
35	8.1045	281.1	0.07119	0.6569	0.05894	34.30	20.60	0.00092	1.02463	0.9950
36	7.6647	295.4	0.07134	0.6846	0.05293	34.44	20.93	0.00103	1.02329	0.9504
37	7.2871	309.4	0.07139	0.7073	0.04829	34.69	21.28	0.00115	1.02213	0.9163
38	6.9569	323.1	0.07138	0.7264	0.04457	35.03	21.64	0.00126	1.02112	0.8893
39	6.6642	336.6	0.07131	0.7427	0.04149	35.43	22.00	0.00137	1.02022	0.8673
40	6.4019	349.9	0.07122	0.7568	0.03890	35.87	22.37	0.00148	1.01942	0.8489
42	5.9485	376.0	0.07100	0.7799	0.03474	36.83	23.12	0.00171	1.01804	0.8201
44	5.5677	401.5	0.07074	0.7980	0.03153	37.88	23.88	0.00193	1.01688	0.7981
46	5.2414	426.5	0.07049	0.8126	0.02896	38.98	24.64	0.00217	1.01588	0.7806
48	4.9572	451.1	0.07024	0.8245	0.02684	40.12	25.39	0.00241	1.01502	0.7661
50	4.7068	475.4	0.06998	0.8344	0.02505	41.29	26.14	0.00265	1.01425	0.7540
52	4.4838	499.5	0.06972	0.8428	0.02352	42.47	26.88	0.00290	1.01358	0.7438
54	4.2835	523.4	0.06946	0.8499	0.02220	43.66	27.62	0.00316	1.01297	0.7349
56	4.1025	547.2	0.06919	0.8559	0.02103	44.86	28.35	0.00342	1.01242	0.7272
58	3.9376	570.9	0.06893	0.8611	0.02000	46.05	29.07	0.00369	1.01191	0.7208
60	3.7868	594.6	0.06865	0.8656	0.01908	47.29	29.79	0.00396	1.01146	0.7144
65	3.4596	654.1	0.06789	0.8746	0.01714	50.51	31.54	0.00469	1.01046	0.7003
70	3.1881	714.5	0.06701	0.8813	0.01560	53.42	33.25	0.00541	1.00964	0.6940
75	2.9584	776.1	0.06600	0.8863	0.01434	56.07	34.92	0.00613	1.00894	0.6932
80	2.7612	839.3	0.06490	0.8901	0.01328	58.27	36.55	0.00681	1.00834	0.6993
85	2.5898	904.4	0.06369	0.8931	0.01238	61.16	38.14	0.00759	1.00782	0.6983
90	2.4391	971.6	0.06241	0.8954	0.01160	63.76	39.70	0.00835	1.00737	0.7019
95	2.3057	1040.4	0.06114	0.8973	0.01092	66.14	41.23	0.00909	1.00696	0.7083
100	2.1865	1111.3	0.05984	0.8988	0.01032	76.84	45.78	0.01103	1.00660	0.6834
120	1.8141	1411.9	0.05493	0.9025	0.00848	94.69	54.02	0.01570	1.00548	0.6828
140	1.5518	1730.2	0.05102	0.9042	0.00721	108.02	59.17	0.02009	1.00468	0.6833
160	1.3565	2057.0	0.04811	0.9050	0.00628	118.99	63.01	0.02444	1.00409	0.6841
180	1.2053	2387.1	0.04595	0.9053	0.00557	128.49	66.22	0.02888	1.00364	0.6848
200	1.0847	2716.7	0.04436	0.9054	0.00500	136.95	69.08	0.03345	1.00327	0.6854
220	0.9861	3044.7	0.04315	0.9053	0.00454	144.66	71.75	0.03820	1.00297	0.6857
240	0.9040	3368.5	0.04227	0.9052	0.00416	151.74	74.29	0.04313	1.00273	0.6859
260	0.8346	3687.9	0.04161	0.9051	0.00384	158.33	76.75	0.04826	1.00252	0.6860
280	0.7751	4002.5	0.04113	0.9049	0.00356	164.53	79.17	0.05360	1.00234	0.6860
300	0.7236	4312.7	0.04078	0.9047	0.00332	170.40	81.55	0.05915	1.00218	0.6859
350	0.6206	5070.8	0.04027	0.9043	0.00285	184.11	87.40	0.07396	1.00187	0.6855
400	0.5432	5813.3	0.04005	0.9039	0.00249	197.00	93.17	0.09012	1.00164	0.6851
450	0.4830	6547.2	0.03995	0.9035	0.00222	209.43	98.88	0.10762	1.00146	0.6847
500	0.4349	7282.1	0.03985	0.9032	0.00199	221.75	104.52	0.12642	1.00131	0.6844
550	0.3955	8016.1	0.03979	0.9030	0.00181	233.89	110.11	0.14651	1.00119	0.6842
600	0.3626	8754.3	0.03970	0.9028	0.00166	246.01	115.63	0.16786	1.00109	0.6840
700	0.3109	10248.9	0.03947	0.9024	0.00143	270.21	126.49	0.21421	1.00094	0.6837
800	0.2721	11781.7	0.03912	0.9021	0.00125	294.68	137.09	0.26527	1.00082	0.6836
900	0.2420	13366.2	0.03864	0.9019	0.00111	319.62	147.44	0.32084	1.00073	0.6837
1000	0.2178	15011.1	0.03806	0.9017	0.00100	440.56	157.55	0.48597	1.00066	0.5358
1200	0.1816	18465.1	0.03676	0.9014	0.00083	511.01	177.13	0.65949	1.00055	0.5325
1400	0.1557	22150.7	0.03537	0.9012	0.00071	583.28	195.92	0.85374	1.00047	0.5307
1600	0.1362	26043.0	0.03404	0.9011	0.00062	657.31	214.04	1.06846	1.00041	0.5294
1800	0.1211	30188.1	0.03271	0.9009	0.00055	736.08	231.58	1.30606	1.00036	0.5271
2000	0.1090	34880.0	0.03111	0.9006	0.00050	831.52	248.62	1.57614	1.00033	0.5211
2500	0.0869	54824.5	0.02371	0.8969	0.00040	1364.13	289.83	2.56941	1.00026	0.4675
3000	0.0708	108142.4	0.01385	0.8774	0.00034	3127.32	332.61	4.29912	1.00021	0.3933

* TWO-PHASE BOUNDARY

Table 4.12 (continued)
THERMODYNAMIC PROPERTIES OF NORMAL HYDROGEN

1.00 MPA ISOBAR

TEMPERATURE DEG. K	VOLUME CU M/KG	ISOTHERM DERIVATIVE CU M-MPA/KG	ISOCHORE DERIVATIVE MPA/K	INTERNAL ENERGY KJ/KG-K	ENTHALPY KJ/KG	ENTROPY KJ/KG-K	CV KJ / KG-K	CP	VELOCITY OF SOUND M/S
* 14.133	0.01289	1.2344	0.9347	218.6	231.5	14.105	4.72	6.38	1292
15	0.01300	1.1600	0.9343	224.2	237.2	14.498	4.90	6.80	1270
16	0.01315	1.0735	0.9214	231.1	244.2	14.951	5.08	7.27	1239
17	0.01330	1.0056	0.9056	238.4	251.7	15.405	5.26	7.71	1214
18	0.01347	0.9414	0.8908	246.2	259.7	15.860	5.43	8.18	1191
19	0.01364	0.8798	0.8757	254.5	268.1	16.315	5.57	8.66	1169
20	0.01384	0.8155	0.8593	263.2	277.0	16.772	5.70	9.17	1145
21	0.01405	0.7523	0.8409	272.4	286.4	17.234	5.82	9.71	1121
22	0.01428	0.6871	0.8205	282.2	296.5	17.700	5.92	10.31	1094
23	0.01454	0.6219	0.7976	292.6	307.1	18.173	6.00	10.97	1066
24	0.01482	0.5539	0.7713	303.6	318.4	18.656	6.07	11.74	1035
25	0.01515	0.4883	0.7407	315.4	330.6	19.151	6.13	12.58	1001
26	0.01553	0.4207	0.7056	328.1	343.7	19.662	6.18	13.59	962
27	0.01597	0.3531	0.6648	341.9	357.9	20.198	6.23	14.84	917
28	0.01650	0.2850	0.6185	357.0	373.5	20.765	6.29	16.52	865
29	0.01717	0.2166	0.5658	374.0	391.2	21.386	6.36	19.00	804
30	0.01809	0.1485	0.5041	394.0	412.1	22.094	6.45	23.26	731
31	0.01957	0.0784	0.4243	419.9	439.5	22.993	6.63	33.89	633
* 31.248	0.02016	0.0594	0.3982	428.5	448.6	23.287	6.70	40.59	600
* 31.248	0.06965	0.0213	0.0788	614.7	684.4	30.834	7.93	52.16	374
32	0.08043	0.0367	0.0654	634.1	714.6	31.790	7.50	31.66	393
33	0.09052	0.0511	0.0561	651.4	741.9	32.631	7.22	23.86	411
34	0.09887	0.0627	0.0502	664.9	763.8	33.285	6.97	20.32	427
35	0.10632	0.0727	0.0459	676.7	783.0	33.842	6.81	18.27	442
36	0.11320	0.0818	0.0426	687.3	800.5	34.337	6.70	16.91	454
37	0.11969	0.0903	0.0398	697.3	817.0	34.786	6.62	15.94	466
38	0.12587	0.0982	0.0375	706.6	832.5	35.202	6.56	15.21	477
39	0.13183	0.1057	0.0356	715.6	847.4	35.589	6.52	14.65	487
40	0.13759	0.1128	0.0339	724.3	861.8	35.954	6.49	14.19	497
42	0.14869	0.1264	0.0310	740.8	889.5	36.629	6.44	13.51	515
44	0.15934	0.1391	0.0287	756.7	916.0	37.245	6.41	13.01	531
46	0.16965	0.1513	0.0267	772.0	941.6	37.815	6.39	12.64	547
48	0.17971	0.1629	0.0251	786.9	966.6	38.346	6.37	12.35	562
50	0.18954	0.1742	0.0236	801.5	991.1	38.846	6.36	12.12	576
52	0.19921	0.1852	0.0224	815.9	1015.1	39.317	6.36	11.93	590
54	0.20873	0.1959	0.0213	830.1	1038.8	39.765	6.35	11.78	603
56	0.21813	0.2063	0.0203	844.2	1062.3	40.192	6.35	11.65	615
58	0.22742	0.2166	0.0194	858.0	1085.5	40.599	6.36	11.55	627
60	0.23663	0.2267	0.0185	871.9	1108.5	40.989	6.36	11.46	639
65	0.25931	0.2513	0.0168	906.0	1165.4	41.898	6.39	11.31	667
70	0.28164	0.2752	0.0154	940.0	1221.7	42.733	6.44	11.22	692
75	0.30370	0.2986	0.0142	974.0	1277.7	43.506	6.51	11.19	716
80	0.32554	0.3217	0.0132	1008.1	1333.7	44.229	6.60	11.20	739
85	0.34723	0.3444	0.0123	1042.5	1389.8	44.909	6.71	11.24	760
90	0.36878	0.3668	0.0116	1077.4	1446.1	45.554	6.83	11.31	780
95	0.39022	0.3890	0.0109	1112.7	1502.9	46.168	6.96	11.40	798
100	0.41157	0.4111	0.0103	1148.6	1560.1	46.754	7.09	11.50	816
120	0.49627	0.4979	0.0085	1298.7	1795.0	48.892	7.68	11.99	882
140	0.58028	0.5833	0.0073	1459.6	2039.9	50.778	8.24	12.49	941
160	0.66386	0.6680	0.0063	1630.3	2294.2	52.474	8.71	12.93	996
180	0.74718	0.7521	0.0056	1809.4	2556.6	54.020	9.10	13.30	1048
200	0.83031	0.8358	0.0050	1995.3	2825.6	55.437	9.41	13.59	1099
220	0.91330	0.9193	0.0046	2186.6	3099.9	56.744	9.66	13.83	1147
240	0.99620	1.0026	0.0042	2382.3	3378.5	57.956	9.85	14.02	1194
260	1.07902	1.0858	0.0039	2581.2	3660.3	59.084	10.00	14.16	1240
280	1.16179	1.1688	0.0036	2782.7	3944.5	60.137	10.11	14.26	1284
300	1.24446	1.2517	0.0033	2986.2	4230.7	61.124	10.19	14.33	1327
350	1.45108	1.4588	0.0029	3499.4	4950.5	63.344	10.30	14.44	1430
400	1.65757	1.6655	0.0025	4016.4	5673.9	65.276	10.35	14.49	1527
450	1.86398	1.8721	0.0022	4535.0	6399.0	66.983	10.37	14.50	1618
500	2.07033	2.0786	0.0020	5054.3	7124.6	68.511	10.39	14.52	1704
550	2.27666	2.2851	0.0018	5574.2	7850.9	69.896	10.41	14.53	1786
600	2.48295	2.4914	0.0017	6095.2	8578.1	71.159	10.42	14.55	1865
700	2.89550	2.9041	0.0014	7140.3	10035.8	73.409	10.48	14.61	2012
800	3.30800	3.3167	0.0013	8193.0	11501.0	75.367	10.57	14.69	2147
900	3.72048	3.7292	0.0011	9256.4	12976.9	77.104	10.70	14.82	2273
1000	4.13294	4.1417	0.0010	10333.6	14466.5	78.671	10.86	14.98	2391
1200	4.95784	4.9667	0.0008	12543.9	17501.7	81.419	11.24	15.36	2606
1400	5.78271	5.7916	0.0007	14836.6	20619.3	83.828	11.68	15.80	2800
1600	6.60763	6.6165	0.0006	17216.3	23823.9	85.974	12.13	16.26	2978
1800	7.43284	7.4413	0.0006	19688.4	27121.2	87.909	12.62	16.75	3143
2000	8.25941	8.2662	0.0005	22273.7	30533.1	89.717	13.25	17.41	3295
2500	10.36066	10.3283	0.0004	29717.4	40078.1	93.946	17.29	21.79	3607
3000	12.69252	12.3904	0.0003	41256.5	53949.0	98.980	29.75	36.03	3869

* TWO-PHASE BOUNDARY

Table 4.12 (continued)
THERMODYNAMIC PROPERTIES OF NORMAL HYDROGEN

1.00 MPA ISOBAR

TEMPERATURE DEG. K	DENSITY KG/CU M	V(DH/DV)$_P$ KJ/KG	V(DP/DU)$_V$ MPA-CU M/KJ X 10^2	-V(DP/DV)$_T$ MPA	(DV/DT)$_P$/V 1/DEG. K	THERMAL CONDUCTIVITY W/K-M X 10^3	VISCOSITY KG/M-S X 10^7	THERMAL DIFFUSIVITY SQ M/HR	DIELECTRIC CONSTANT	PRANDTL NUMBER
* 14.133	77.5763	654.0	0.25522	95.7614	0.00976	76.69	260.99	0.00056	1.25355	2.1723
15	76.8975	649.7	0.24818	89.1999	0.01047	84.18	233.50	0.00058	1.25116	1.8875
16	76.0734	644.3	0.23827	81.6612	0.01128	90.41	207.98	0.00059	1.24826	1.6724
17	75.1902	644.1	0.22890	75.6113	0.01198	94.58	187.17	0.00059	1.24516	1.5266
18	74.2648	641.8	0.22106	69.9161	0.01274	97.58	170.06	0.00058	1.24192	1.4251
19	73.2929	637.5	0.21433	64.4858	0.01358	99.40	155.74	0.00056	1.23852	1.3564
20	72.2689	629.1	0.20842	58.9352	0.01458	101.04	143.51	0.00055	1.23495	1.3027
21	71.1846	618.6	0.20301	53.5525	0.01570	102.94	132.90	0.00054	1.23117	1.2541
22	70.0303	604.8	0.19798	48.1167	0.01705	104.07	123.52	0.00052	1.22716	1.2241
23	68.7931	588.6	0.19320	42.7839	0.01864	104.53	115.09	0.00050	1.22286	1.2083
24	67.4541	568.5	0.18834	37.3608	0.02064	104.35	107.38	0.00047	1.21823	1.2077
25	66.0015	547.2	0.18315	32.2261	0.02298	103.58	100.25	0.00045	1.21322	1.2173
26	64.4069	522.0	0.17735	27.0974	0.02604	102.18	93.53	0.00042	1.20773	1.2442
27	62.6325	493.8	0.17046	22.1181	0.03006	100.19	87.08	0.00039	1.20164	1.2900
28	60.6157	461.3	0.16232	17.2762	0.03580	97.53	80.74	0.00035	1.19475	1.3671
29	58.2362	423.5	0.15281	12.6128	0.04486	94.04	74.28	0.00031	1.18665	1.5007
30	55.2770	378.6	0.14128	8.2063	0.06142	89.37	67.41	0.00025	1.17662	1.7544
31	51.0906	320.0	0.12530	4.0067	0.10589	85.69	59.29	0.00018	1.16254	2.3450
* 31.248	49.6094	300.4	0.11970	2.9465	0.13513	86.22	56.77	0.00015	1.15759	2.6729
* 31.248	14.3567	202.2	0.06924	0.3054	0.25798	49.56	21.45	0.00024	1.04393	2.2578
32	12.4337	220.6	0.07011	0.4557	0.14350	40.71	20.84	0.00037	1.03797	1.6202
33	11.0478	240.2	0.07028	0.5646	0.09932	37.54	20.72	0.00051	1.03369	1.3169
34	10.1147	256.8	0.07117	0.6340	0.07916	36.34	20.83	0.00064	1.03081	1.1650
35	9.4056	272.3	0.07166	0.6842	0.06708	35.87	21.05	0.00075	1.02863	1.0718
36	8.8336	287.3	0.07190	0.7230	0.05886	35.75	21.32	0.00086	1.02687	1.0082
37	8.3550	301.8	0.07199	0.7541	0.05282	35.83	21.62	0.00097	1.02540	0.9618
38	7.9444	315.9	0.07200	0.7798	0.04815	36.05	21.95	0.00107	1.02414	0.9263
39	7.5856	329.8	0.07193	0.8014	0.04440	36.35	22.29	0.00118	1.02304	0.8981
40	7.2677	343.5	0.07183	0.8199	0.04132	36.72	22.64	0.00128	1.02207	0.8751
42	6.7254	370.2	0.07158	0.8499	0.03648	37.57	23.36	0.00149	1.02041	0.8398
44	6.2759	396.2	0.07129	0.8731	0.03284	38.53	24.09	0.00170	1.01904	0.8136
46	5.8943	421.7	0.07099	0.8917	0.02998	39.58	24.83	0.00191	1.01787	0.7931
48	5.5647	446.7	0.07071	0.9067	0.02765	40.67	25.57	0.00213	1.01687	0.7766
50	5.2758	471.4	0.07042	0.9192	0.02571	41.80	26.31	0.00235	1.01599	0.7629
52	5.0198	495.8	0.07013	0.9296	0.02407	42.95	27.04	0.00258	1.01521	0.7514
54	4.7908	520.0	0.06985	0.9384	0.02265	44.11	27.77	0.00281	1.01451	0.7415
56	4.5844	544.1	0.06955	0.9459	0.02141	45.29	28.49	0.00305	1.01388	0.7330
58	4.3971	568.0	0.06928	0.9524	0.02033	46.46	29.21	0.00329	1.01331	0.7259
60	4.2261	591.9	0.06899	0.9580	0.01936	47.68	29.91	0.00354	1.01279	0.7189
65	3.8564	651.8	0.06819	0.9691	0.01734	50.86	31.66	0.00420	1.01167	0.7037
70	3.5506	712.5	0.06728	0.9772	0.01575	53.74	33.36	0.00486	1.01074	0.6966
75	3.2928	774.4	0.06625	0.9834	0.01445	56.35	35.02	0.00551	1.00996	0.6955
80	3.0718	837.9	0.06513	0.9881	0.01337	58.54	36.64	0.00612	1.00929	0.7011
85	2.8800	903.2	0.06391	0.9917	0.01245	61.42	38.23	0.00683	1.00870	0.6998
90	2.7117	970.6	0.06261	0.9946	0.01166	64.00	39.78	0.00751	1.00819	0.7031
95	2.5627	1039.5	0.06132	0.9969	0.01096	66.38	41.30	0.00818	1.00774	0.7092
100	2.4297	1110.6	0.06001	0.9988	0.01035	76.93	45.80	0.00991	1.00734	0.6846
120	2.0150	1411.8	0.05505	1.0032	0.00849	94.75	54.01	0.01412	1.00608	0.6835
140	1.7233	1730.6	0.05112	1.0053	0.00722	108.07	59.17	0.01807	1.00520	0.6838
160	1.5063	2057.8	0.04819	1.0062	0.00628	119.03	63.01	0.02200	1.00455	0.6845
180	1.3384	2388.2	0.04602	1.0066	0.00557	128.53	66.22	0.02600	1.00404	0.6851
200	1.2044	2718.0	0.04441	1.0067	0.00500	136.99	69.09	0.03012	1.00363	0.6856
220	1.0949	3046.3	0.04320	1.0066	0.00454	144.70	71.76	0.03440	1.00330	0.6859
240	1.0038	3370.3	0.04231	1.0064	0.00416	151.79	74.30	0.03884	1.00303	0.6861
260	0.9268	3689.9	0.04165	1.0062	0.00384	158.37	76.77	0.04346	1.00280	0.6861
280	0.8607	4004.6	0.04116	1.0060	0.00356	164.57	79.18	0.04827	1.00260	0.6861
300	0.8036	4315.0	0.04081	1.0059	0.00332	170.45	81.56	0.05327	1.00242	0.6860
350	0.6891	5073.3	0.04030	1.0053	0.00285	184.16	87.42	0.06661	1.00208	0.6856
400	0.6033	5815.9	0.04007	1.0048	0.00249	197.06	93.20	0.08117	1.00182	0.6852
450	0.5365	6549.9	0.03997	1.0044	0.00221	209.50	98.91	0.09692	1.00162	0.6848
500	0.4830	7284.9	0.03987	1.0040	0.00199	221.82	104.56	0.11386	1.00146	0.6844
550	0.4392	8019.0	0.03980	1.0037	0.00181	233.98	110.15	0.13195	1.00132	0.6842
600	0.4027	8757.2	0.03972	1.0034	0.00166	246.10	115.67	0.15117	1.00121	0.6840
700	0.3454	10251.9	0.03948	1.0030	0.00142	270.32	126.54	0.19292	1.00104	0.6837
800	0.3023	11784.7	0.03913	1.0026	0.00125	294.80	137.14	0.23890	1.00091	0.6836
900	0.2688	13369.3	0.03865	1.0023	0.00111	319.75	147.50	0.28895	1.00081	0.6837
1000	0.2420	15014.3	0.03806	1.0021	0.00100	440.56	157.62	0.43747	1.00073	0.5361
1200	0.2017	18468.4	0.03676	1.0018	0.00083	511.01	177.21	0.59364	1.00061	0.5328
1400	0.1729	22153.9	0.03538	1.0015	0.00071	583.27	196.01	0.76848	1.00052	0.5310
1600	0.1513	26044.8	0.03404	1.0013	0.00062	657.24	214.15	0.96170	1.00046	0.5297
1800	0.1345	30181.3	0.03273	1.0011	0.00055	735.67	231.70	1.17535	1.00041	0.5275
2000	0.1211	34841.2	0.03116	1.0008	0.00050	829.88	248.74	1.41760	1.00036	0.5217
2500	0.0965	54302.1	0.02397	0.9969	0.00040	1344.18	289.95	2.30096	1.00029	0.4700
3000	0.0788	105504.7	0.01422	0.9762	0.00034	3028.08	332.54	3.84065	1.00024	0.3956

* TWO-PHASE BOUNDARY

Table 4.12 (continued)
THERMODYNAMIC PROPERTIES OF NORMAL HYDROGEN

1.292800 MPA ISOBAR

TEMPERATURE DEG. K	VOLUME CU M/KG	ISOTHERM DERIVATIVE CU M-MPA/KG	ISOCHORE DERIVATIVE MPA/K	INTERNAL ENERGY KJ/KG-K	ENTHALPY KJ/KG	ENTROPY KJ/KG-K	CV KJ /	CP KG-K	VELOCITY OF SOUND M/S
* 14.229	0.01286	1.2537	0.9367	218.7	235.4	14.112	4.73	6.38	1300
15	0.01296	1.1970	0.9378	223.7	240.4	14.459	4.89	6.74	1285
16	0.01310	1.1028	0.9299	230.5	247.4	14.909	5.08	7.23	1253
17	0.01325	1.0336	0.9139	237.7	254.8	15.360	5.25	7.66	1228
18	0.01341	0.9688	0.8991	245.4	262.7	15.811	5.42	8.12	1205
19	0.01358	0.9102	0.8842	253.5	271.1	16.261	5.56	8.57	1184
20	0.01377	0.8467	0.8680	262.1	279.9	16.714	5.69	9.07	1161
21	0.01397	0.7813	0.8499	271.2	289.2	17.170	5.81	9.60	1136
22	0.01420	0.7166	0.8300	280.7	299.1	17.630	5.91	10.17	1111
23	0.01444	0.6530	0.8077	290.9	309.6	18.096	5.99	10.79	1084
24	0.01471	0.5882	0.7826	301.7	320.7	18.569	6.06	11.47	1055
25	0.01502	0.5232	0.7537	313.1	332.6	19.053	6.12	12.25	1023
26	0.01537	0.4570	0.7209	325.4	345.2	19.549	6.17	13.16	987
27	0.01577	0.3916	0.6835	338.5	358.9	20.065	6.22	14.23	946
28	0.01624	0.3269	0.6411	352.8	373.8	20.604	6.27	15.56	901
29	0.01682	0.2603	0.5941	368.5	390.2	21.181	6.33	17.45	847
30	0.01755	0.1976	0.5409	386.1	408.8	21.812	6.40	20.08	787
31	0.01856	0.1349	0.4787	407.0	430.9	22.538	6.50	24.64	715
32	0.02020	0.0718	0.3996	434.0	460.1	23.464	6.68	35.71	620
* 32.976	0.03128	-0.0001	0.1927	521.6	562.0	26.576	9.88	-14658.08	348
* 32.976	0.03241	-0.0001	0.1868	527.0	568.9	26.786	9.80	-13261.84	351
33	0.03934	0.0025	0.1548	557.4	608.3	27.980	9.44	501.14	363
34	0.06012	0.0298	0.0915	621.0	698.7	30.694	7.72	42.28	404
35	0.06941	0.0456	0.0760	643.1	732.8	31.684	7.23	28.61	425
36	0.07672	0.0584	0.0669	659.2	758.4	32.405	6.98	23.22	441
37	0.08307	0.0694	0.0607	672.7	780.1	32.999	6.82	20.35	455
38	0.08886	0.0793	0.0559	684.6	799.5	33.516	6.71	18.54	468
39	0.09425	0.0884	0.0521	695.5	817.3	33.980	6.64	17.27	480
40	0.09935	0.0969	0.0489	705.7	834.1	34.405	6.59	16.34	490
42	0.10894	0.1126	0.0439	724.6	865.4	35.169	6.51	15.06	510
44	0.11795	0.1270	0.0401	742.2	894.6	35.848	6.47	14.20	528
46	0.12654	0.1405	0.0369	758.8	922.4	36.465	6.44	13.60	545
48	0.13482	0.1533	0.0344	774.8	949.1	37.034	6.42	13.14	560
50	0.14286	0.1655	0.0322	790.3	975.0	37.563	6.40	12.79	575
52	0.15070	0.1773	0.0303	805.5	1000.3	38.059	6.40	12.51	589
54	0.15839	0.1888	0.0287	820.3	1025.1	38.527	6.39	12.28	602
56	0.16594	0.1999	0.0272	835.0	1049.5	38.971	6.39	12.09	615
58	0.17338	0.2107	0.0259	849.4	1073.5	39.393	6.39	11.94	628
60	0.18072	0.2214	0.0247	863.6	1097.3	39.795	6.39	11.81	640
65	0.19875	0.2471	0.0223	898.8	1155.7	40.729	6.41	11.58	668
70	0.21640	0.2720	0.0203	933.5	1213.2	41.583	6.46	11.44	694
75	0.23379	0.2961	0.0187	968.0	1270.3	42.370	6.53	11.38	718
80	0.25096	0.3198	0.0173	1002.7	1327.1	43.103	6.61	11.35	741
85	0.26796	0.3431	0.0162	1037.5	1383.9	43.793	6.72	11.37	762
90	0.28483	0.3660	0.0152	1072.7	1440.9	44.445	6.84	11.43	782
95	0.30159	0.3886	0.0143	1108.3	1498.2	45.064	6.96	11.50	801
100	0.31826	0.4111	0.0135	1144.4	1555.9	45.655	7.10	11.59	819
120	0.38425	0.4990	0.0111	1295.5	1792.2	47.806	7.69	12.05	884
140	0.44956	0.5853	0.0094	1456.9	2038.1	49.701	8.24	12.53	943
160	0.51445	0.6705	0.0082	1628.1	2293.1	51.402	8.71	12.96	999
180	0.57907	0.7551	0.0073	1807.5	2556.1	52.950	9.10	13.32	1051
200	0.64351	0.8392	0.0065	1993.6	2825.5	54.369	9.42	13.61	1101
220	0.70781	0.9229	0.0059	2185.2	3100.2	55.678	9.66	13.85	1150
240	0.77202	1.0064	0.0054	2381.0	3379.1	56.892	9.86	14.03	1197
260	0.83616	1.0897	0.0050	2580.1	3661.1	58.021	10.00	14.17	1242
280	0.90024	1.1729	0.0046	2781.7	3945.5	59.074	10.11	14.27	1286
300	0.96425	1.2560	0.0043	2985.3	4231.9	60.061	10.19	14.34	1329
350	1.12416	1.4633	0.0037	3498.7	4952.1	62.282	10.31	14.45	1432
400	1.28395	1.6702	0.0032	4015.8	5675.7	64.215	10.35	14.49	1529
450	1.44365	1.8769	0.0029	4534.5	6400.9	65.923	10.37	14.51	1620
500	1.60330	2.0835	0.0026	5053.9	7126.7	67.452	10.39	14.52	1706
550	1.76292	2.2900	0.0024	5573.9	7853.0	68.837	10.41	14.53	1788
600	1.92252	2.4964	0.0022	6094.9	8580.3	70.099	10.42	14.55	1867
700	2.24165	2.9091	0.0018	7140.1	10038.1	72.350	10.48	14.61	2014
800	2.56075	3.3218	0.0016	8192.8	11503.3	74.308	10.57	14.70	2149
900	2.87982	3.7343	0.0014	9256.3	12979.3	76.044	10.70	14.82	2275
1000	3.19888	4.1469	0.0013	10333.5	14469.0	77.612	10.86	14.98	2392
1200	3.83696	4.9718	0.0011	12543.8	17504.2	80.360	11.24	15.36	2607
1400	4.47502	5.7968	0.0009	14836.6	20621.9	82.768	11.68	15.80	2801
1600	5.11310	6.6217	0.0008	17216.0	23826.2	84.914	12.13	16.25	2979
1800	5.75139	7.4465	0.0007	19686.8	27122.2	86.849	12.60	16.74	3144
2000	6.39059	8.2714	0.0006	22266.9	30528.7	88.654	13.21	17.36	3297
2500	8.01231	10.3335	0.0005	29613.1	39971.5	92.839	16.85	21.31	3614
3000	9.79099	12.3957	0.0004	40604.3	53262.1	97.666	27.91	33.90	3877

* TWO-PHASE BOUNDARY

Table 4.12 (continued)
THERMODYNAMIC PROPERTIES OF NORMAL HYDROGEN

1.292800 MPA ISOBAR

TEMPERATURE DEG. K	DENSITY KG/CU M	$V(DH/DV)_P$ KJ/KG	$V(DP/DU)_V$ MPA-CU M/KJ X 10^2	$-V(DP/DV)_T$ MPA	$(DV/DT)_P/V$ 1/DEG. K	THERMAL CONDUCTIVITY W/K-M X 10^3	VISCOSITY KG/M-S X 10^7	THERMAL DIFFUSIVITY SQ M/HR	DIELECTRIC CONSTANT	PRANDTL NUMBER
* 14.229	77.7376	664.1	0.25449	97.4607	0.00961	77.85	262.81	0.00056	1.25411	2.1545
15	77.1455	663.8	0.24860	92.3420	0.01016	84.50	238.11	0.00058	1.25203	1.8998
16	76.3375	654.5	0.23996	84.1885	0.01105	90.83	212.01	0.00059	1.24919	1.6873
17	75.4732	654.2	0.23054	78.0093	0.01172	95.10	190.76	0.00059	1.24615	1.5374
18	74.5678	652.2	0.22264	72.2386	0.01245	98.19	173.31	0.00058	1.24298	1.4327
19	73.6240	649.8	0.21587	67.0094	0.01319	100.11	158.77	0.00057	1.23968	1.3597
20	72.6196	642.4	0.20992	61.4904	0.01412	101.84	146.28	0.00056	1.23617	1.3025
21	71.5645	631.5	0.20447	55.9166	0.01520	103.83	135.51	0.00054	1.23249	1.2526
22	70.4452	618.6	0.19942	50.4817	0.01644	105.08	126.02	0.00053	1.22860	1.2197
23	69.2507	603.8	0.19462	45.2201	0.01786	105.66	117.53	0.00051	1.22445	1.1997
24	67.9678	586.1	0.18987	39.9775	0.01958	105.62	109.81	0.00049	1.22001	1.1929
25	66.5794	566.0	0.18490	34.8312	0.02164	105.01	102.67	0.00046	1.21521	1.1974
26	65.0753	542.6	0.17949	29.7379	0.02424	103.82	96.02	0.00044	1.21003	1.2167
27	63.4185	517.0	0.17329	24.8318	0.02753	102.08	89.69	0.00041	1.20434	1.2502
28	61.5718	488.4	0.16604	20.1288	0.03185	99.77	83.56	0.00037	1.19801	1.3029
29	59.4606	454.6	0.15784	15.4766	0.03838	96.80	77.47	0.00034	1.19081	1.3965
30	56.9770	418.0	0.14833	11.2605	0.04803	93.03	71.26	0.00029	1.18237	1.5379
31	53.8752	374.2	0.13670	7.2696	0.06585	87.74	64.60	0.00024	1.17190	1.8140
32	49.5003	317.8	0.12084	3.5564	0.11236	85.51	56.75	0.00017	1.15723	2.3699
* 32.976	31.9741	198.6	0.06098	-0.0026	-73.80653	-5850.18	35.91	0.00004	1.09971	8.9969
* 32.976	30.8504	199.5	0.06181	-0.0028	-66.46463	-5289.75	34.93	0.00005	1.09609	8.7585
33	25.4217	204.8	0.06449	0.0633	2.44661	241.60	30.45	0.00007	1.07873	6.3152
34	15.6336	228.8	0.07130	0.4952	0.18481	50.08	24.13	0.00026	1.05103	2.0376
35	14.4079	247.1	0.07298	0.6567	0.11577	43.73	23.27	0.00038	1.04409	1.5227
36	13.0352	264.2	0.07358	0.7615	0.08789	41.38	23.04	0.00049	1.03983	1.2928
37	12.0376	280.4	0.07389	0.8358	0.07258	40.31	23.04	0.00059	1.03675	1.1633
38	11.2542	296.0	0.07399	0.8927	0.06263	39.81	23.16	0.00069	1.03432	1.0784
39	10.6104	311.0	0.07394	0.9380	0.05554	39.63	23.36	0.00078	1.03234	1.0182
40	10.0653	325.7	0.07381	0.9753	0.05017	39.64	23.60	0.00087	1.03066	0.9729
42	9.1792	354.2	0.07345	1.0334	0.04250	40.01	24.17	0.00104	1.02793	0.9093
44	8.4784	381.8	0.07302	1.0767	0.03720	40.66	24.80	0.00122	1.02578	0.8662
46	7.9028	408.6	0.07259	1.1102	0.03328	41.47	25.46	0.00139	1.02402	0.8346
48	7.4174	434.8	0.07218	1.1370	0.03023	42.39	26.14	0.00157	1.02253	0.8104
50	7.0001	460.5	0.07178	1.1588	0.02777	43.38	26.83	0.00174	1.02125	0.7910
52	6.6357	485.8	0.07140	1.1768	0.02575	44.42	27.53	0.00193	1.02014	0.7752
54	6.3136	510.9	0.07103	1.1919	0.02404	45.50	28.23	0.00211	1.01915	0.7620
56	6.0264	535.6	0.07065	1.2046	0.02258	46.59	28.92	0.00230	1.01828	0.7506
58	5.7678	560.2	0.07032	1.2154	0.02131	47.69	29.61	0.00249	1.01749	0.7413
60	5.5334	584.6	0.06998	1.2248	0.02020	48.86	30.30	0.00269	1.01677	0.7325
65	5.0315	645.7	0.06909	1.2433	0.01793	51.92	32.01	0.00321	1.01524	0.7139
70	4.6210	707.4	0.06809	1.2567	0.01618	54.71	33.68	0.00372	1.01399	0.7044
75	4.2774	770.1	0.06699	1.2667	0.01477	57.20	35.31	0.00423	1.01295	0.7022
80	3.9848	834.2	0.06581	1.2743	0.01361	59.33	36.91	0.00472	1.01206	0.7065
85	3.7319	900.1	0.06453	1.2802	0.01264	62.17	38.48	0.00527	1.01129	0.7040
90	3.5109	968.0	0.06319	1.2849	0.01180	64.73	40.02	0.00581	1.01062	0.7065
95	3.3158	1037.4	0.06186	1.2886	0.01108	67.07	41.53	0.00633	1.01003	0.7119
100	3.1421	1109.0	0.06051	1.2916	0.01045	77.21	45.84	0.00763	1.00950	0.6880
120	2.6025	1411.9	0.05540	1.2987	0.00853	94.92	54.01	0.01090	1.00786	0.6856
140	2.2244	1731.8	0.05139	1.3020	0.00724	108.22	59.17	0.01397	1.00672	0.6853
160	1.9438	2060.1	0.04841	1.3034	0.00629	119.17	63.03	0.01703	1.00587	0.6855
180	1.7269	2391.4	0.04620	1.3039	0.00557	128.67	66.25	0.02013	1.00521	0.6859
200	1.5540	2722.0	0.04457	1.3040	0.00500	137.12	69.12	0.02334	1.00469	0.6862
220	1.4128	3051.0	0.04334	1.3039	0.00454	144.83	71.79	0.02665	1.00426	0.6864
240	1.2953	3375.6	0.04243	1.3036	0.00416	151.92	74.34	0.03010	1.00391	0.6865
260	1.1959	3695.6	0.04176	1.3033	0.00383	158.51	76.81	0.03368	1.00361	0.6865
280	1.1108	4010.8	0.04126	1.3029	0.00356	164.71	79.23	0.03741	1.00335	0.6864
300	1.0371	4321.7	0.04090	1.3026	0.00332	170.59	81.62	0.04129	1.00313	0.6862
350	0.8896	5080.6	0.04038	1.3017	0.00284	184.33	87.49	0.05163	1.00268	0.6857
400	0.7788	5823.6	0.04014	1.3008	0.00249	197.24	93.27	0.06292	1.00235	0.6852
450	0.6927	6557.9	0.04002	1.3001	0.00221	209.70	98.99	0.07513	1.00209	0.6848
500	0.6237	7293.1	0.03992	1.2995	0.00199	222.05	104.65	0.08825	1.00188	0.6845
550	0.5672	8027.4	0.03985	1.2990	0.00181	234.23	110.25	0.10227	1.00171	0.6842
600	0.5202	8765.8	0.03976	1.2985	0.00166	246.37	115.79	0.11717	1.00157	0.6840
700	0.4461	10260.6	0.03952	1.2978	0.00142	270.63	126.68	0.14952	1.00134	0.6837
800	0.3905	11793.7	0.03916	1.2972	0.00125	295.15	137.30	0.18515	1.00118	0.6836
900	0.3472	13378.3	0.03868	1.2967	0.00111	320.15	147.68	0.22393	1.00105	0.6837
1000	0.3126	15023.5	0.03809	1.2963	0.00100	440.56	157.81	0.33859	1.00094	0.5368
1200	0.2606	18477.9	0.03678	1.2958	0.00083	511.01	177.44	0.45942	1.00079	0.5335
1400	0.2235	22163.3	0.03539	1.2954	0.00071	583.25	196.28	0.59469	1.00067	0.5317
1600	0.1956	26051.0	0.03406	1.2950	0.00062	657.08	214.45	0.74412	1.00059	0.5305
1800	0.1739	30168.7	0.03277	1.2947	0.00055	734.77	232.03	0.90905	1.00052	0.5285
2000	0.1565	34757.2	0.03128	1.2943	0.00050	826.22	249.10	1.09503	1.00047	0.5234
2500	0.1248	53140.5	0.02459	1.2897	0.00040	1299.70	290.31	1.75944	1.00038	0.4759
3000	0.1021	93603.9	0.01512	1.2660	0.00034	2805.93	332.51	2.91717	1.00031	0.4018

* TWO-PHASE BOUNDARY

Table 4.12 (continued)
THERMODYNAMIC PROPERTIES OF NORMAL HYDROGEN

1.50 MPA ISOBAR

TEMPERATURE DEG. K	VOLUME CU M/KG	ISOTHERM DERIVATIVE CU M-MPA/KG	ISOCHORE DERIVATIVE MPA/K	INTERNAL ENERGY KJ/KG-K	ENTHALPY KJ/KG	ENTROPY KJ/KG-K	CV KJ /	CP KG-K	VELOCITY OF SOUND M/S
* 14.296	0.01285	1.2672	0.9378	218.8	238.1	14.117	4.74	6.38	1306
15	0.01293	1.2227	0.9402	223.3	242.7	14.432	4.89	6.70	1295
16	0.01307	1.1232	0.9358	230.0	249.6	14.879	5.07	7.20	1263
17	0.01321	1.0597	0.9197	237.2	257.0	15.327	5.25	7.62	1240
18	0.01337	0.9938	0.9049	244.8	264.9	15.775	5.41	8.06	1217
19	0.01354	0.9302	0.8898	252.9	273.2	16.224	5.56	8.52	1194
20	0.01372	0.8675	0.8738	261.4	282.0	16.674	5.69	9.00	1172
21	0.01392	0.8044	0.8561	270.3	291.2	17.127	5.80	9.51	1148
22	0.01414	0.7407	0.8365	279.8	301.0	17.583	5.90	10.06	1123
23	0.01437	0.6765	0.8147	289.8	311.3	18.042	5.99	10.65	1097
24	0.01464	0.6103	0.7902	300.4	322.3	18.511	6.06	11.32	1068
25	0.01493	0.5466	0.7624	311.6	334.0	18.987	6.12	12.05	1037
26	0.01526	0.4806	0.7310	323.6	346.5	19.474	6.17	12.91	1003
27	0.01564	0.4190	0.6953	336.4	359.8	19.978	6.22	13.84	966
28	0.01608	0.3544	0.6555	350.1	374.2	20.501	6.26	15.04	923
29	0.01661	0.2909	0.6113	365.1	390.0	21.055	6.32	16.59	874
30	0.01726	0.2295	0.5622	381.7	407.6	21.650	6.38	18.68	820
31	0.01810	0.1686	0.5067	400.6	427.7	22.311	6.45	21.92	757
32	0.01930	0.1106	0.4417	423.2	452.2	23.086	6.57	27.61	682
33	0.02140	0.0534	0.3575	453.8	485.9	24.124	6.80	42.98	581
34	0.03016	0.0078	0.2048	525.2	570.5	26.640	8.76	174.96	395
35	0.04884	0.0245	0.1167	604.2	677.5	29.754	7.70	54.06	415
36	0.05798	0.0408	0.0942	631.9	718.9	30.922	7.27	33.56	434
37	0.06494	0.0542	0.0817	650.9	748.3	31.728	7.02	26.25	450
38	0.07082	0.0656	0.0737	666.0	772.2	32.366	6.84	22.59	466
39	0.07614	0.0761	0.0668	679.1	793.3	32.914	6.73	20.00	475
40	0.08105	0.0857	0.0620	690.9	812.5	33.399	6.66	18.44	487
42	0.09006	0.1030	0.0546	712.1	847.2	34.246	6.57	16.44	508
44	0.09836	0.1186	0.0492	731.2	878.7	34.980	6.51	15.21	526
46	0.10618	0.1331	0.0450	749.0	908.3	35.636	6.48	14.37	544
48	0.11365	0.1467	0.0416	765.9	936.4	36.235	6.45	13.77	560
50	0.12085	0.1597	0.0388	782.2	963.4	36.787	6.43	13.31	575
52	0.12785	0.1721	0.0363	797.9	989.7	37.301	6.42	12.95	589
54	0.13467	0.1840	0.0343	813.3	1015.3	37.785	6.42	12.66	603
56	0.14135	0.1956	0.0324	828.4	1040.4	38.242	6.41	12.42	616
58	0.14792	0.2068	0.0308	843.1	1065.0	38.675	6.41	12.23	628
60	0.15439	0.2178	0.0294	857.7	1089.3	39.086	6.41	12.07	640
65	0.17022	0.2443	0.0263	893.5	1148.9	40.039	6.43	11.78	669
70	0.18567	0.2698	0.0239	928.8	1207.3	40.905	6.47	11.60	695
75	0.20084	0.2945	0.0220	963.8	1265.1	41.702	6.54	11.51	720
80	0.21580	0.3186	0.0203	998.8	1322.5	42.443	6.62	11.46	743
85	0.23060	0.3423	0.0190	1033.9	1379.8	43.139	6.73	11.47	764
90	0.24526	0.3655	0.0177	1069.3	1437.2	43.796	6.84	11.51	784
95	0.25981	0.3885	0.0167	1105.2	1494.9	44.419	6.97	11.57	803
100	0.27427	0.4112	0.0158	1141.5	1552.9	45.014	7.10	11.65	821
120	0.33142	0.4999	0.0129	1293.2	1790.3	47.175	7.69	12.09	886
140	0.38790	0.5868	0.0110	1455.0	2036.9	49.075	8.25	12.56	945
160	0.44397	0.6724	0.0095	1626.5	2292.4	50.779	8.72	12.98	1001
180	0.49977	0.7572	0.0084	1806.1	2555.8	52.329	9.11	13.34	1053
200	0.55538	0.8415	0.0076	1992.4	2825.5	53.751	9.42	13.63	1103
220	0.61087	0.9255	0.0069	2184.1	3100.4	55.060	9.67	13.86	1152
240	0.66627	1.0091	0.0063	2380.1	3379.5	56.275	9.86	14.04	1199
260	0.72159	1.0926	0.0058	2579.3	3661.7	57.405	10.00	14.17	1244
280	0.77685	1.1759	0.0054	2780.9	3946.2	58.459	10.11	14.28	1288
300	0.83206	1.2591	0.0050	2984.7	4232.8	59.446	10.19	14.35	1331
350	0.96994	1.4665	0.0043	3498.3	4953.2	61.668	10.31	14.45	1434
400	1.10769	1.6735	0.0038	4015.4	5677.0	63.601	10.35	14.49	1530
450	1.24536	1.8803	0.0033	4534.2	6402.2	65.309	10.37	14.51	1622
500	1.38297	2.0869	0.0030	5053.7	7128.1	66.838	10.39	14.52	1708
550	1.52056	2.2934	0.0027	5573.7	7854.5	68.223	10.41	14.54	1790
600	1.65812	2.4999	0.0025	6094.7	8581.9	69.486	10.42	14.55	1868
700	1.93319	2.9127	0.0021	7139.9	10039.7	71.736	10.48	14.61	2015
800	2.20822	3.3253	0.0019	8192.7	11505.0	73.695	10.57	14.70	2150
900	2.48322	3.7379	0.0017	9256.2	12981.0	75.431	10.70	14.82	2276
1000	2.75821	4.1505	0.0015	10333.4	14470.7	76.999	10.86	14.98	2393
1200	3.30816	4.9755	0.0013	12543.8	17506.0	79.746	11.24	15.36	2608
1400	3.85809	5.8004	0.0011	14836.5	20623.6	82.155	11.68	15.80	2802
1600	4.40804	6.6253	0.0009	17215.9	23827.9	84.301	12.13	16.25	2980
1800	4.95814	7.4502	0.0008	19686.0	27123.3	86.236	12.60	16.73	3145
2000	5.50898	8.2751	0.0008	22263.3	30526.8	88.039	13.18	17.33	3298
2500	6.90505	10.3372	0.0006	29558.6	39916.2	92.201	16.62	21.06	3617
3000	8.42675	12.3993	0.0005	40263.2	52903.4	96.920	26.94	32.79	3881

* TWO-PHASE BOUNDARY

Table 4.12 (continued)
THERMODYNAMIC PROPERTIES OF NORMAL HYDROGEN

1.50 MPA ISOBAR

TEMPERATURE DEG. K	DENSITY KG/CU M	$V(DH/DV)_P$ KJ/KG	$V(DP/DU)_V$ MPA-CU M/KJ X 10^2	$-V(DP/DV)_T$ MPA	$(DV/DT)_P/V$ 1/DEG. K	THERMAL CONDUCTIVITY W/K-M X 10^3	VISCOSITY KG/M-S X 10^7	THERMAL DIFFUSIVITY SQ M/HR	DIELECTRIC CONSTANT	PRANDTL NUMBER
* 14.296	77.8506	671.3	0.25391	98.6543	0.00951	78.66	264.09	0.00057	1.25451	2.1424
15	77.3176	673.7	0.24889	94.5343	0.00995	84.72	241.40	0.00059	1.25264	1.9091
16	76.5207	661.5	0.24110	85.9509	0.01089	91.13	214.88	0.00060	1.24983	1.6983
17	75.6762	664.0	0.23165	80.1912	0.01147	95.47	193.41	0.00060	1.24687	1.5429
18	74.7864	662.0	0.22371	74.3219	0.01218	98.64	175.72	0.00059	1.24375	1.4359
19	73.8480	657.9	0.21687	68.6960	0.01295	100.60	160.87	0.00058	1.24046	1.3627
20	72.8612	651.2	0.21090	63.2062	0.01382	102.39	148.24	0.00056	1.23701	1.3034
21	71.8281	641.8	0.20547	57.7812	0.01482	104.46	137.37	0.00055	1.23341	1.2505
22	70.7307	629.8	0.20040	52.3925	0.01597	105.78	127.79	0.00054	1.22959	1.2148
23	69.5672	615.2	0.19561	47.0594	0.01731	106.44	119.26	0.00052	1.22555	1.1933
24	68.3128	597.3	0.19089	41.6902	0.01895	106.48	111.48	0.00050	1.22120	1.1853
25	66.9679	578.4	0.18604	36.6072	0.02083	105.98	104.35	0.00047	1.21655	1.1861
26	65.5158	555.8	0.18085	31.4849	0.02322	104.91	97.72	0.00045	1.21154	1.2021
27	63.9263	533.2	0.17498	26.7877	0.02596	103.32	91.43	0.00042	1.20608	1.2247
28	62.1792	505.8	0.16830	22.0382	0.02974	101.22	85.42	0.00039	1.20009	1.2696
29	60.2136	475.4	0.16072	17.5136	0.03491	98.55	79.52	0.00036	1.19338	1.3391
30	57.9478	441.9	0.15217	13.2993	0.04227	95.19	73.59	0.00032	1.18567	1.4441
31	55.2427	403.0	0.14212	9.3138	0.05440	90.61	67.45	0.00027	1.17651	1.6321
32	51.8031	358.1	0.12979	5.7282	0.07711	86.46	60.78	0.00022	1.16493	1.9410
33	46.7258	300.0	0.11255	2.4954	0.14328	85.30	52.63	0.00015	1.14799	2.6518
34	33.1607	221.1	0.07051	0.2588	0.79145	120.16	37.31	0.00007	1.10354	5.4323
35	20.4766	232.5	0.07404	0.5017	0.23253	59.12	27.31	0.00019	1.06308	2.4976
36	17.2459	251.0	0.07514	0.7044	0.13372	48.97	25.44	0.00030	1.05294	1.7437
37	15.3982	268.1	0.07559	0.8349	0.09789	45.37	24.75	0.00040	1.04717	1.4320
38	14.1209	284.3	0.07625	0.9270	0.07946	43.71	24.50	0.00049	1.04320	1.2663
39	13.1343	299.1	0.07558	0.9995	0.06687	42.71	24.46	0.00059	1.04014	1.1456
40	12.3383	314.5	0.07542	1.0570	0.05862	42.26	24.54	0.00067	1.03768	1.0704
42	11.1031	344.1	0.07494	1.1436	0.04777	42.07	24.90	0.00083	1.03386	0.9728
44	10.1664	372.6	0.07437	1.2062	0.04081	42.38	25.40	0.00099	1.03097	0.9115
46	9.4180	400.3	0.07382	1.2538	0.03591	42.97	25.98	0.00114	1.02867	0.8690
48	8.7991	427.2	0.07330	1.2912	0.03223	43.73	26.60	0.00130	1.02677	0.8375
50	8.2746	453.6	0.07280	1.3213	0.02934	44.60	27.25	0.00146	1.02516	0.8130
52	7.8219	479.5	0.07234	1.3460	0.02700	45.54	27.91	0.00162	1.02377	0.7936
54	7.4254	505.1	0.07190	1.3665	0.02507	46.54	28.58	0.00178	1.02255	0.7775
56	7.0744	530.3	0.07146	1.3836	0.02342	47.56	29.25	0.00195	1.02148	0.7639
58	6.7603	555.2	0.07108	1.3983	0.02202	48.61	29.92	0.00212	1.02052	0.7528
60	6.4771	580.0	0.07071	1.4109	0.02081	49.73	30.60	0.00229	1.01965	0.7425
65	5.8748	641.9	0.06974	1.4355	0.01835	52.70	32.26	0.00274	1.01781	0.7212
70	5.3859	704.2	0.06868	1.4533	0.01648	55.41	33.91	0.00319	1.01632	0.7100
75	4.9790	767.4	0.06753	1.4665	0.01499	57.81	35.52	0.00363	1.01508	0.7071
80	4.6339	831.9	0.06630	1.4766	0.01378	59.89	37.11	0.00406	1.01403	0.7103
85	4.3366	898.2	0.06498	1.4843	0.01277	62.71	38.66	0.00454	1.01313	0.7070
90	4.0774	966.4	0.06360	1.4905	0.01191	65.24	40.19	0.00501	1.01234	0.7088
95	3.8490	1036.2	0.06224	1.4953	0.01116	67.57	41.69	0.00546	1.01165	0.7137
100	3.6461	1108.0	0.06086	1.4992	0.01052	77.42	45.87	0.00656	1.01103	0.6904
120	3.0173	1412.1	0.05565	1.5085	0.00856	95.05	54.01	0.00938	1.00912	0.6871
140	2.5780	1732.9	0.05159	1.5126	0.00725	108.33	59.18	0.01204	1.00779	0.6863
160	2.2524	2061.8	0.04857	1.5145	0.00630	119.27	63.04	0.01468	1.00680	0.6863
180	2.0009	2393.7	0.04633	1.5151	0.00557	128.76	66.27	0.01737	1.00604	0.6864
200	1.8006	2724.9	0.04468	1.5152	0.00500	137.21	69.14	0.02013	1.00544	0.6866
220	1.6370	3054.3	0.04344	1.5150	0.00454	144.93	71.82	0.02300	1.00494	0.6868
240	1.5009	3379.3	0.04252	1.5146	0.00415	152.01	74.37	0.02597	1.00453	0.6868
260	1.3858	3699.8	0.04184	1.5141	0.00383	158.60	76.84	0.02907	1.00418	0.6867
280	1.2872	4015.2	0.04134	1.5136	0.00356	164.81	79.27	0.03229	1.00388	0.6866
300	1.2018	4326.4	0.04096	1.5132	0.00332	170.69	81.65	0.03564	1.00363	0.6863
350	1.0310	5085.8	0.04043	1.5119	0.00284	184.44	87.53	0.04457	1.00311	0.6858
400	0.9028	5829.1	0.04018	1.5108	0.00249	197.37	93.32	0.05430	1.00272	0.6853
450	0.8030	6563.6	0.04006	1.5098	0.00221	209.84	99.05	0.06484	1.00242	0.6848
500	0.7231	7298.9	0.03996	1.5090	0.00199	222.21	104.72	0.07617	1.00218	0.6845
550	0.6577	8033.3	0.03988	1.5083	0.00181	234.40	110.33	0.08827	1.00198	0.6842
600	0.6031	8771.8	0.03979	1.5077	0.00166	246.56	115.87	0.10113	1.00182	0.6840
700	0.5173	10266.8	0.03954	1.5067	0.00142	270.85	126.77	0.12905	1.00156	0.6837
800	0.4529	11800.0	0.03918	1.5059	0.00125	295.40	137.41	0.15980	1.00137	0.6836
900	0.4027	13384.8	0.03870	1.5053	0.00111	320.43	147.80	0.19326	1.00121	0.6837
1000	0.3626	15030.0	0.03811	1.5048	0.00100	440.56	157.95	0.29194	1.00109	0.5372
1200	0.3023	18484.7	0.03680	1.5040	0.00083	511.01	177.60	0.39611	1.00091	0.5340
1400	0.2592	22170.1	0.03541	1.5034	0.00071	583.24	196.47	0.51270	1.00078	0.5322
1600	0.2269	26056.2	0.03408	1.5030	0.00062	657.00	214.66	0.64148	1.00068	0.5310
1800	0.2017	30164.1	0.03280	1.5026	0.00055	734.30	232.26	0.78348	1.00061	0.5291
2000	0.1815	34715.3	0.03134	1.5021	0.00050	824.31	249.36	0.94314	1.00055	0.5243
2500	0.1448	52535.0	0.02492	1.4971	0.00040	1276.43	290.58	1.50695	1.00044	0.4793
3000	0.1187	96506.1	0.01564	1.4714	0.00034	2689.20	332.58	2.48770	1.00036	0.4056

* TWO-PHASE BOUNDARY

Table 4.12 (continued)
THERMODYNAMIC PROPERTIES OF NORMAL HYDROGEN

2.00 MPA ISOBAR

TEMPERATURE DEG. K	VOLUME CU M/KG	ISOTHERM DERIVATIVE CU M-MPA/KG	ISOCHORE DERIVATIVE MPA/K	INTERNAL ENERGY KJ/KG-K	ENTHALPY KJ/KG	ENTROPY KJ/KG-K	CV KJ /	CP KG-K	VELOCITY OF SOUND M/S
* 14.457	0.01280	1.2994	0.9407	219.1	244.7	14.128	4.77	6.38	1319
15	0.01287	1.2650	0.9449	222.5	248.2	14.369	4.88	6.63	1311
16	0.01300	1.1850	0.9435	229.1	255.1	14.811	5.06	7.09	1288
17	0.01313	1.1096	0.9322	236.1	262.4	15.254	5.23	7.53	1264
18	0.01328	1.0414	0.9177	243.6	270.1	15.697	5.39	7.96	1240
19	0.01345	0.9770	0.9029	251.4	278.3	16.139	5.54	8.41	1218
20	0.01362	0.9157	0.8873	259.7	286.9	16.582	5.67	8.86	1196
21	0.01381	0.8527	0.8704	268.4	296.0	17.027	5.79	9.34	1173
22	0.01401	0.7908	0.8514	277.6	305.6	17.474	5.89	9.85	1150
23	0.01423	0.7273	0.8307	287.3	315.7	17.923	5.97	10.39	1125
24	0.01447	0.6647	0.8075	297.5	326.4	18.378	6.05	10.98	1098
25	0.01474	0.6021	0.7817	308.2	337.7	18.839	6.11	11.63	1070
26	0.01504	0.5392	0.7529	319.6	349.7	19.307	6.16	12.35	1039
27	0.01537	0.4780	0.7209	331.7	362.4	19.788	6.21	13.15	1006
28	0.01575	0.4174	0.6856	344.5	376.0	20.281	6.26	14.08	969
29	0.01619	0.3585	0.6467	358.2	390.6	20.794	6.30	15.17	929
30	0.01671	0.2994	0.6046	373.1	406.5	21.331	6.34	16.57	884
31	0.01733	0.2429	0.5588	389.2	423.9	21.902	6.40	18.37	835
32	0.01812	0.1881	0.5085	407.2	443.5	22.523	6.46	20.89	780
33	0.01915	0.1373	0.4528	427.7	466.0	23.218	6.54	24.60	719
34	0.02062	0.0915	0.3897	452.1	493.4	24.033	6.67	30.65	649
35	0.02303	0.0526	0.3166	483.2	529.2	25.073	6.89	42.21	568
36	0.02752	0.0300	0.2395	525.1	580.1	26.505	7.30	59.51	494
37	0.03460	0.0290	0.1764	571.0	640.2	28.152	7.32	54.86	466
38	0.04137	0.0388	0.1415	604.2	687.0	29.401	7.23	40.85	468
39	0.04711	0.0505	0.1203	628.7	722.9	30.335	7.06	31.86	477
40	0.05206	0.0618	0.1061	647.8	752.0	31.071	6.91	26.65	488
42	0.06054	0.0824	0.0877	678.0	799.0	32.220	6.72	21.09	509
44	0.06776	0.1002	0.0762	702.1	837.7	33.119	6.60	18.32	527
46	0.07446	0.1169	0.0680	723.6	872.5	33.893	6.55	16.63	545
48	0.08071	0.1323	0.0617	743.1	904.6	34.576	6.52	15.51	561
50	0.08664	0.1468	0.0567	761.5	934.7	35.192	6.50	14.71	576
52	0.09232	0.1606	0.0526	778.9	963.5	35.757	6.49	14.11	591
54	0.09781	0.1737	0.0491	795.7	991.3	36.281	6.48	13.65	605
56	0.10314	0.1863	0.0461	812.0	1018.3	36.772	6.47	13.27	618
58	0.10835	0.1984	0.0435	827.8	1044.5	37.233	6.46	12.96	631
60	0.11345	0.2102	0.0413	843.3	1070.2	37.668	6.46	12.72	643
65	0.12584	0.2385	0.0367	880.9	1132.6	38.665	6.47	12.27	673
70	0.13784	0.2654	0.0331	917.5	1193.2	39.565	6.51	11.99	700
75	0.14956	0.2913	0.0302	953.6	1252.7	40.386	6.57	11.82	724
80	0.16106	0.3164	0.0278	989.4	1311.6	41.146	6.64	11.73	747
85	0.17240	0.3409	0.0258	1025.3	1370.1	41.856	6.74	11.69	769
90	0.18360	0.3650	0.0241	1061.4	1428.6	42.525	6.86	11.70	789
95	0.19470	0.3886	0.0227	1097.8	1487.2	43.158	6.98	11.74	808
100	0.20571	0.4119	0.0214	1134.6	1546.0	43.761	7.12	11.80	826
120	0.24906	0.5024	0.0174	1287.7	1785.9	45.945	7.70	12.19	892
140	0.29176	0.5905	0.0147	1450.6	2034.1	47.857	8.25	12.63	951
160	0.33405	0.6770	0.0128	1622.7	2290.8	49.570	8.72	13.04	1006
180	0.37609	0.7625	0.0113	1802.9	2555.0	51.126	9.11	13.38	1058
200	0.41794	0.8474	0.0101	1989.6	2825.5	52.550	9.42	13.66	1108
220	0.45968	0.9317	0.0092	2181.7	3101.0	53.863	9.67	13.88	1157
240	0.50131	1.0157	0.0084	2377.9	3380.5	55.079	9.86	14.06	1203
260	0.54288	1.0995	0.0077	2577.3	3663.1	56.211	10.01	14.19	1249
280	0.58439	1.1830	0.0072	2779.2	3948.0	57.266	10.12	14.29	1293
300	0.62587	1.2665	0.0067	2983.2	4234.9	58.255	10.20	14.36	1336
350	0.72937	1.4743	0.0057	3497.1	4955.8	60.478	10.31	14.46	1438
400	0.83275	1.6815	0.0050	4014.5	5680.0	62.412	10.36	14.50	1534
450	0.93605	1.8885	0.0045	4533.5	6405.5	64.121	10.38	14.51	1625
500	1.03929	2.0952	0.0040	5053.0	7131.6	65.650	10.39	14.53	1711
550	1.14251	2.3019	0.0036	5573.1	7858.2	67.036	10.41	14.54	1793
600	1.24570	2.5084	0.0033	6094.2	8585.6	68.298	10.43	14.56	1871
700	1.45203	2.9213	0.0029	7139.6	10043.6	70.549	10.48	14.61	2018
800	1.65832	3.3340	0.0025	8192.4	11509.1	72.508	10.57	14.70	2153
900	1.86459	3.7466	0.0022	9256.0	12985.1	74.244	10.70	14.82	2279
1000	2.07084	4.1592	0.0020	10333.2	14474.9	75.812	10.86	14.98	2396
1200	2.48332	4.9843	0.0017	12543.6	17510.3	78.560	11.24	15.36	2610
1400	2.89578	5.8092	0.0014	14836.4	20627.9	80.969	11.68	15.80	2804
1600	3.30824	6.6342	0.0013	17215.6	23832.1	83.114	12.13	16.25	2982
1800	3.72080	7.4591	0.0011	19684.6	27126.2	85.048	12.59	16.72	3147
2000	4.13384	8.2840	0.0010	22257.2	30524.9	86.849	13.14	17.29	3301
2500	5.17877	10.3461	0.0008	29464.0	39821.5	90.971	16.23	20.62	3625
3000	6.30535	12.4082	0.0007	39671.0	52281.7	95.502	25.26	30.87	3890

* TWO-PHASE BOUNDARY

Table 4.12 (continued)
THERMODYNAMIC PROPERTIES OF NORMAL HYDROGEN

2.00 MPA ISOBAR

TEMPERATURE DEG. K	DENSITY KG/CU M	$V(DH/DV)_P$ KJ/KG	$V(DP/DU)_V$ MPA-CU M/KJ X 10^2	$-V(DP/DV)_T$ MPA	$(DV/DT)_P/V$ 1/DEG. K	THERMAL CONDUCTIVITY W/K-M X 10^3	VISCOSITY KG/M-S X 10^7	THERMAL DIFFUSIVITY SQ M/HR	DIELECTRIC CONSTANT	PRANDTL NUMBER
* 14.457	78.1136	688.5	0.25262	101.5121	0.00927	80.58	267.20	0.00058	1.25546	2.1155
15	77.7130	689.7	0.24935	98.3063	0.00961	85.23	249.22	0.00060	1.25403	1.9386
16	76.9464	685.5	0.24215	91.1801	0.01035	91.81	221.79	0.00061	1.25133	1.7135
17	76.1334	682.5	0.23391	84.4778	0.01104	96.32	199.58	0.00060	1.24847	1.5606
18	75.2732	680.3	0.22598	78.3875	0.01171	99.64	181.27	0.00060	1.24545	1.4489
19	74.3695	676.5	0.21912	72.6572	0.01243	101.74	165.95	0.00059	1.24229	1.3714
20	73.4224	671.4	0.21311	67.2294	0.01320	103.67	152.96	0.00057	1.23897	1.3073
21	72.4296	662.9	0.20771	61.7632	0.01409	105.90	141.77	0.00056	1.23551	1.2507
22	71.3823	652.7	0.20261	56.4471	0.01508	107.38	131.98	0.00055	1.23186	1.2100
23	70.2758	639.5	0.19784	51.1094	0.01625	108.22	123.28	0.00053	1.22801	1.1840
24	69.0976	624.6	0.19317	45.9266	0.01758	108.47	115.43	0.00051	1.22392	1.1686
25	67.8400	607.5	0.18853	40.8465	0.01914	108.19	108.27	0.00049	1.21956	1.1634
26	66.4945	588.0	0.18367	35.8550	0.02100	107.37	101.65	0.00047	1.21492	1.1689
27	65.0436	567.2	0.17842	31.0906	0.02319	106.09	95.46	0.00045	1.20992	1.1832
28	63.4743	544.2	0.17267	26.4928	0.02588	104.38	89.61	0.00042	1.20453	1.2090
29	61.7543	519.4	0.16625	22.1397	0.02921	102.20	83.98	0.00039	1.19864	1.2466
30	59.8446	491.1	0.15926	17.9155	0.03375	99.54	78.48	0.00036	1.19212	1.3066
31	57.6930	460.7	0.15143	14.0151	0.03987	95.97	73.03	0.00033	1.18480	1.3977
32	55.2006	426.7	0.14259	10.3835	0.04897	91.79	67.50	0.00029	1.17637	1.5365
33	52.2285	389.7	0.13250	7.1714	0.06313	87.82	61.76	0.00025	1.16636	1.7303
34	48.4961	349.1	0.12051	4.4390	0.08778	84.91	55.59	0.00021	1.15388	2.0066
35	43.4287	304.9	0.10583	2.2862	0.13846	82.44	48.63	0.00016	1.13707	2.4902
36	36.3370	270.5	0.09027	1.0885	0.22002	80.21	40.96	0.00013	1.11384	3.0388
37	28.9056	260.5	0.08341	0.8378	0.21056	70.15	34.76	0.00016	1.08985	2.7180
38	24.1693	270.3	0.08099	0.9366	0.15110	60.34	31.40	0.00022	1.07475	2.1255
39	21.2266	283.9	0.08030	1.0722	0.11220	54.69	29.55	0.00029	1.06544	1.7216
40	19.2072	298.3	0.07987	1.1871	0.08935	51.56	28.51	0.00036	1.05909	1.4741
42	16.5171	327.3	0.07909	1.3614	0.06444	48.54	27.62	0.00050	1.05066	1.2000
44	14.7578	355.3	0.07821	1.4784	0.05155	47.48	27.48	0.00063	1.04518	1.0599
46	13.4303	384.1	0.07721	1.5698	0.04329	47.22	27.66	0.00076	1.04106	0.9739
48	12.3903	412.4	0.07631	1.6397	0.03762	47.42	28.01	0.00089	1.03784	0.9165
50	11.5426	440.0	0.07551	1.6949	0.03344	47.88	28.47	0.00101	1.03522	0.8751
52	10.8320	467.0	0.07479	1.7393	0.03022	48.51	29.00	0.00114	1.03302	0.8438
54	10.2239	493.6	0.07414	1.7759	0.02765	49.26	29.56	0.00127	1.03115	0.8191
56	9.6956	519.7	0.07350	1.8060	0.02553	50.08	30.15	0.00140	1.02952	0.7987
58	9.2297	545.4	0.07301	1.8315	0.02377	50.96	30.76	0.00153	1.02809	0.7825
60	8.8146	570.9	0.07253	1.8533	0.02228	51.94	31.37	0.00167	1.02681	0.7681
65	7.9467	634.3	0.07135	1.8955	0.01935	54.65	32.93	0.00202	1.02415	0.7396
70	7.2548	697.8	0.07013	1.9256	0.01719	57.17	34.50	0.00237	1.02203	0.7238
75	6.6864	762.1	0.06884	1.9479	0.01551	59.32	36.06	0.00270	1.02029	0.7187
80	6.2088	827.6	0.06750	1.9647	0.01417	61.30	37.60	0.00303	1.01883	0.7194
85	5.8005	894.7	0.06608	1.9776	0.01307	64.05	39.12	0.00340	1.01759	0.7141
90	5.4465	963.6	0.06462	1.9877	0.01214	66.52	40.61	0.00376	1.01651	0.7144
95	5.1361	1034.0	0.06318	1.9958	0.01135	68.78	42.08	0.00411	1.01556	0.7181
100	4.8612	1106.5	0.06173	2.0021	0.01066	77.96	45.98	0.00489	1.01472	0.6960
120	4.0150	1413.2	0.05625	2.0171	0.00863	95.39	54.04	0.00702	1.01215	0.6906
140	3.4275	1735.8	0.05205	2.0238	0.00728	108.61	59.21	0.00903	1.01037	0.6886
160	2.9935	2066.4	0.04895	2.0266	0.00631	119.53	63.09	0.01103	1.00905	0.6879
180	2.6590	2399.7	0.04665	2.0275	0.00558	129.01	66.32	0.01306	1.00803	0.6877
200	2.3927	2732.1	0.04496	2.0275	0.00500	137.45	69.20	0.01514	1.00723	0.6877
220	2.1754	3062.6	0.04367	2.0270	0.00453	145.16	71.88	0.01730	1.00657	0.6876
240	1.9948	3388.6	0.04273	2.0262	0.00415	152.24	74.44	0.01954	1.00602	0.6875
260	1.8420	3709.8	0.04203	2.0253	0.00383	158.84	76.92	0.02187	1.00556	0.6873
280	1.7112	4026.0	0.04151	2.0243	0.00355	165.05	79.35	0.02430	1.00517	0.6871
300	1.5978	4338.1	0.04112	2.0236	0.00331	170.94	81.75	0.02682	1.00482	0.6867
350	1.3710	5098.4	0.04056	2.0213	0.00284	184.72	87.64	0.03354	1.00414	0.6860
400	1.2008	5842.4	0.04029	2.0192	0.00248	197.68	93.45	0.04087	1.00362	0.6854
450	1.0683	6577.4	0.04016	2.0175	0.00221	210.19	99.20	0.04880	1.00322	0.6849
500	0.9622	7313.1	0.04005	2.0160	0.00199	222.59	104.88	0.05733	1.00290	0.6845
550	0.8753	8047.7	0.03996	2.0147	0.00181	234.82	110.51	0.06643	1.00264	0.6842
600	0.8028	8786.4	0.03986	2.0136	0.00166	247.01	116.07	0.07610	1.00242	0.6840
700	0.6887	10281.7	0.03960	2.0119	0.00142	271.38	127.01	0.09711	1.00208	0.6837
800	0.6030	11815.2	0.03924	2.0105	0.00124	296.01	137.68	0.12024	1.00182	0.6836
900	0.5363	13400.2	0.03874	2.0094	0.00111	321.11	148.11	0.14541	1.00162	0.6837
1000	0.4829	15045.8	0.03815	2.0085	0.00100	440.56	158.29	0.21918	1.00146	0.5384
1200	0.4027	18500.9	0.03683	2.0071	0.00083	511.01	178.00	0.29734	1.00121	0.5352
1400	0.3453	22186.6	0.03543	2.0061	0.00071	583.23	196.92	0.38481	1.00104	0.5335
1600	0.3023	26070.0	0.03411	2.0053	0.00062	656.86	215.17	0.48139	1.00091	0.5323
1800	0.2688	30161.0	0.03284	2.0047	0.00055	733.48	232.83	0.58771	1.00081	0.5306
2000	0.2419	34647.6	0.03145	2.0039	0.00050	820.98	249.97	0.70663	1.00073	0.5264
2500	0.1931	51488.5	0.02553	1.9978	0.00040	1235.99	291.26	1.11761	1.00058	0.4859
3000	0.1586	91109.1	0.01664	1.9679	0.00034	2485.57	332.94	1.82796	1.00048	0.4134

* TWO-PHASE BOUNDARY

Table 4.12 (continued)
THERMODYNAMIC PROPERTIES OF NORMAL HYDROGEN

2.50 MPA ISOBAR

TEMPERATURE DEG. K	VOLUME CU M/KG	ISOTHERM DERIVATIVE CU M-MPA/KG	ISOCHORE DERIVATIVE MPA/K	INTERNAL ENERGY KJ/KG-K	ENTHALPY KJ/KG	ENTROPY KJ/KG-K	CV KJ / KG-K	CP KJ / KG-K	VELOCITY OF SOUND M/S
* 14.615	0.01276	1.3311	0.9438	219.3	251.2	14.140	4.79	6.38	1332
15	0.01280	1.3125	0.9471	221.7	253.7	14.309	4.87	6.55	1329
16	0.01293	1.2382	0.9495	228.2	260.5	14.745	5.05	7.00	1310
17	0.01306	1.1575	0.9442	235.1	267.7	15.183	5.23	7.46	1285
18	0.01320	1.0930	0.9296	242.4	275.4	15.621	5.38	7.86	1264
19	0.01336	1.0271	0.9152	250.1	283.5	16.058	5.53	8.29	1241
20	0.01352	0.9641	0.8999	258.2	292.0	16.495	5.66	8.73	1220
21	0.01370	0.9029	0.8836	266.7	300.9	16.932	5.77	9.18	1198
22	0.01389	0.8411	0.8655	275.6	310.3	17.371	5.87	9.65	1176
23	0.01410	0.7795	0.8456	285.0	320.2	17.811	5.96	10.16	1152
24	0.01432	0.7170	0.8236	294.9	330.7	18.255	6.04	10.70	1127
25	0.01457	0.6554	0.7993	305.2	341.7	18.704	6.11	11.28	1100
26	0.01484	0.5935	0.7725	316.2	353.3	19.157	6.16	11.92	1072
27	0.01514	0.5348	0.7431	327.7	365.5	19.620	6.21	12.60	1042
28	0.01548	0.4768	0.7109	339.8	378.5	20.091	6.25	13.37	1009
29	0.01586	0.4193	0.6761	352.6	392.3	20.576	6.29	14.25	974
30	0.01630	0.3619	0.6386	366.3	407.0	21.075	6.33	15.31	935
31	0.01680	0.3088	0.5984	380.9	422.9	21.596	6.37	16.52	895
32	0.01740	0.2566	0.5555	396.7	440.2	22.144	6.42	18.07	850
33	0.01813	0.2071	0.5095	413.9	459.2	22.730	6.47	20.06	802
34	0.01904	0.1628	0.4605	432.9	480.5	23.364	6.53	22.59	750
35	0.02022	0.1226	0.4081	454.2	504.8	24.068	6.61	26.06	695
36	0.02184	0.0888	0.3529	478.5	533.1	24.867	6.73	30.79	638
37	0.02410	0.0645	0.2985	506.5	566.7	25.788	6.88	36.57	586
38	0.02724	0.0509	0.2453	537.2	605.3	26.816	7.05	40.38	540
39	0.03114	0.0477	0.2036	568.1	645.9	27.872	7.14	39.98	517
40	0.03530	0.0523	0.1728	595.1	683.4	28.820	7.05	35.51	513
42	0.04305	0.0690	0.1338	637.8	745.4	30.336	6.90	27.09	521
44	0.04977	0.0872	0.1111	669.8	794.3	31.472	6.76	22.18	535
46	0.05576	0.1048	0.0962	696.1	835.5	32.389	6.66	19.28	551
48	0.06125	0.1213	0.0855	719.0	872.1	33.168	6.60	17.44	566
50	0.06631	0.1365	0.0774	739.6	905.4	33.849	6.57	16.22	581
52	0.07120	0.1512	0.0713	759.0	937.0	34.469	6.55	15.41	597
54	0.07589	0.1653	0.0660	777.4	967.1	35.037	6.54	14.72	610
56	0.08040	0.1787	0.0614	795.2	996.2	35.566	6.53	14.17	623
58	0.08478	0.1916	0.0576	812.1	1024.1	36.056	6.51	13.74	636
60	0.08905	0.2042	0.0544	828.6	1051.2	36.515	6.50	13.39	648
65	0.09936	0.2339	0.0478	868.1	1116.5	37.560	6.51	12.78	678
70	0.10928	0.2621	0.0428	906.1	1179.3	38.492	6.54	12.39	705
75	0.11891	0.2890	0.0389	943.3	1240.6	39.338	6.59	12.14	730
80	0.12832	0.3151	0.0357	980.1	1300.9	40.116	6.67	11.99	753
85	0.13758	0.3403	0.0330	1016.7	1360.7	40.841	6.76	11.92	774
90	0.14670	0.3650	0.0308	1053.4	1420.2	41.522	6.87	11.89	795
95	0.15572	0.3893	0.0288	1090.4	1479.7	42.165	6.99	11.90	814
100	0.16465	0.4131	0.0271	1127.6	1539.3	42.776	7.13	11.95	832
120	0.19970	0.5052	0.0220	1282.3	1781.6	44.982	7.72	12.29	897
140	0.23411	0.5944	0.0185	1446.1	2031.4	46.907	8.26	12.70	956
160	0.26813	0.6818	0.0161	1619.0	2289.3	48.627	8.73	13.09	1011
180	0.30190	0.7680	0.0142	1799.7	2554.4	50.188	9.12	13.42	1063
200	0.33550	0.8533	0.0127	1986.8	2825.6	51.617	9.43	13.69	1113
220	0.36897	0.9381	0.0115	2179.2	3101.6	52.932	9.68	13.91	1161
240	0.40235	1.0225	0.0105	2375.7	3381.6	54.150	9.87	14.08	1208
260	0.43566	1.1065	0.0097	2575.4	3664.6	55.283	10.01	14.21	1253
280	0.46892	1.1902	0.0090	2777.5	3949.8	56.340	10.12	14.31	1297
300	0.50216	1.2740	0.0084	2981.7	4237.1	57.330	10.20	14.37	1340
350	0.58503	1.4821	0.0072	3495.9	4958.5	59.555	10.31	14.47	1442
400	0.66779	1.6896	0.0063	4013.6	5683.0	61.490	10.36	14.51	1538
450	0.75046	1.8967	0.0056	4532.7	6408.9	63.199	10.38	14.52	1629
500	0.83309	2.1036	0.0050	5052.4	7135.1	64.729	10.39	14.53	1715
550	0.91568	2.3103	0.0046	5572.6	7861.8	66.114	10.41	14.54	1797
600	0.99825	2.5169	0.0042	6093.8	8589.4	67.377	10.43	14.56	1875
700	1.16334	2.9299	0.0036	7139.2	10047.6	69.628	10.48	14.61	2021
800	1.32839	3.3427	0.0031	8192.1	11513.1	71.587	10.57	14.70	2156
900	1.49341	3.7553	0.0028	9255.7	12989.3	73.324	10.70	14.82	2281
1000	1.65842	4.1680	0.0025	10333.0	14479.1	74.891	10.86	14.99	2398
1200	1.98841	4.9931	0.0021	12543.5	17514.6	77.639	11.24	15.36	2613
1400	2.31838	5.8180	0.0018	14836.3	20632.3	80.048	11.68	15.80	2806
1600	2.64836	6.6430	0.0016	17215.4	23836.3	82.194	12.12	16.25	2984
1800	2.97840	7.4679	0.0014	19683.7	27129.7	84.128	12.58	16.71	3149
2000	3.30878	8.2928	0.0013	22253.0	30524.9	85.926	13.12	17.26	3303
2500	4.14356	10.3550	0.0010	29399.4	39758.3	90.022	15.96	20.32	3630
3000	5.03682	12.4171	0.0008	39266.7	51858.8	94.424	24.11	29.55	3898

* TWO-PHASE BOUNDARY

Table 4.12 (continued)
THERMODYNAMIC PROPERTIES OF NORMAL HYDROGEN

2.50 MPA ISOBAR

TEMPERATURE DEG. K	DENSITY KG/CU M	$V(DH/DV)_P$ KJ/KG	$V(DP/DU)_V$ MPA-CU M/KJ X 10^2	$-V(DP/DV)_T$ MPA	$(DV/DT)_P/V$ 1/DEG. K	THERMAL CONDUCTIVITY W/K-M X 10^3	VISCOSITY KG/M-S X 10^7	THERMAL DIFFUSIVITY SQ M/HR	DIELECTRIC CONSTANT	PRANDTL NUMBER
* 14.615	78.3836	705.3	0.25148	104.3378	0.00905	82.46	270.30	0.00059	1.25639	2.0913
15	78.1041	708.6	0.24920	102.5141	0.00924	85.73	257.35	0.00060	1.25541	1.9652
16	77.3614	706.3	0.24281	95.7904	0.00991	92.48	228.86	0.00061	1.25279	1.7326
17	76.5720	700.2	0.23596	88.6329	0.01065	97.14	205.81	0.00061	1.25001	1.5803
18	75.7402	700.3	0.22800	82.7863	0.01123	100.60	186.86	0.00061	1.24709	1.4608
19	74.8684	696.8	0.22111	76.9010	0.01190	102.83	171.05	0.00060	1.24403	1.3793
20	73.9536	691.6	0.21508	71.2964	0.01262	104.90	157.65	0.00058	1.24083	1.3118
21	72.9991	684.8	0.20968	65.9100	0.01341	107.28	146.15	0.00058	1.23750	1.2507
22	71.9949	675.5	0.20462	60.5553	0.01429	108.91	136.11	0.00056	1.23399	1.2066
23	70.9403	664.1	0.19987	55.2969	0.01529	109.91	127.25	0.00055	1.23032	1.1758
24	69.8221	650.3	0.19526	50.0652	0.01645	110.33	119.28	0.00053	1.22643	1.1565
25	68.6372	634.8	0.19073	44.9873	0.01777	110.24	112.05	0.00051	1.22232	1.1464
26	67.3769	617.1	0.18610	39.9903	0.01932	109.64	105.41	0.00049	1.21796	1.1460
27	66.0317	599.0	0.18121	35.3142	0.02104	108.61	99.24	0.00047	1.21332	1.1516
28	64.5913	579.1	0.17599	30.7942	0.02309	107.17	93.46	0.00045	1.20836	1.1658
29	63.0401	557.1	0.17040	26.4313	0.02558	105.35	87.98	0.00042	1.20304	1.1900
30	61.3589	532.4	0.16439	22.2056	0.02876	103.15	82.72	0.00040	1.19728	1.2278
31	59.5118	507.4	0.15780	18.3786	0.03256	100.15	77.61	0.00037	1.19098	1.2803
32	57.4635	479.7	0.15062	14.7479	0.03767	96.76	72.58	0.00034	1.18402	1.3554
33	55.1554	449.9	0.14281	11.4246	0.04460	92.91	67.56	0.00030	1.17621	1.4590
34	52.5200	419.4	0.13426	8.5493	0.05386	89.17	62.51	0.00027	1.16734	1.5835
35	49.4439	387.1	0.12483	6.0624	0.06732	86.23	57.37	0.00024	1.15704	1.7338
36	45.7965	355.0	0.11454	4.0682	0.08675	83.05	52.11	0.00021	1.14490	1.9320
37	41.4871	327.9	0.10459	2.6760	0.11154	79.63	46.84	0.00019	1.13068	2.1516
38	36.7123	307.6	0.09479	1.8688	0.13127	75.29	42.00	0.00018	1.11506	2.2526
39	32.1093	301.0	0.08888	1.5331	0.13282	70.08	38.12	0.00020	1.10014	2.1748
40	28.3274	304.4	0.08646	1.4810	0.11665	64.70	35.39	0.00023	1.08800	1.9429
42	23.2312	324.5	0.08350	1.6025	0.08349	57.50	32.25	0.00033	1.07177	1.5194
44	20.0928	349.8	0.08183	1.7516	0.06342	54.04	30.78	0.00044	1.06187	1.2636
46	17.9350	376.8	0.08048	1.8787	0.05118	52.41	30.17	0.00055	1.05510	1.1099
48	16.3274	404.2	0.07928	1.9804	0.04315	51.74	30.01	0.00065	1.05007	1.0119
50	15.0802	431.2	0.07819	2.0578	0.03762	51.64	30.14	0.00076	1.04618	0.9470
52	14.0443	459.1	0.07751	2.1241	0.03356	51.87	30.42	0.00086	1.04296	0.9035
54	13.1774	486.2	0.07659	2.1785	0.03028	52.29	30.80	0.00097	1.04027	0.8672
56	12.4383	512.7	0.07569	2.2229	0.02764	52.85	31.25	0.00108	1.03799	0.8378
58	11.7954	538.8	0.07506	2.2606	0.02550	53.51	31.75	0.00119	1.03600	0.8152
60	11.2294	564.7	0.07446	2.2926	0.02372	54.32	32.28	0.00130	1.03425	0.7959
65	10.0642	629.1	0.07303	2.3545	0.02031	56.71	33.69	0.00159	1.03066	0.7589
70	9.1511	693.6	0.07162	2.3985	0.01786	59.01	35.15	0.00187	1.02785	0.7379
75	8.4099	758.7	0.07019	2.4308	0.01600	60.89	36.63	0.00215	1.02557	0.7304
80	7.7927	824.9	0.06872	2.4551	0.01454	62.75	38.12	0.00242	1.02368	0.7285
85	7.2686	892.6	0.06720	2.4738	0.01335	65.42	39.59	0.00272	1.02207	0.7211
90	6.8166	962.1	0.06565	2.4883	0.01236	67.82	41.05	0.00301	1.02069	0.7198
95	6.4219	1033.1	0.06413	2.4998	0.01152	70.02	42.49	0.00330	1.01948	0.7223
100	6.0735	1106.0	0.06261	2.5089	0.01080	78.55	46.12	0.00390	1.01842	0.7014
120	5.0075	1415.1	0.05685	2.5300	0.00868	95.77	54.10	0.00560	1.01517	0.6940
140	4.2715	1739.4	0.05252	2.5391	0.00730	108.93	59.26	0.00723	1.01293	0.6909
160	3.7295	2071.4	0.04932	2.5428	0.00632	119.82	63.14	0.00884	1.01128	0.6896
180	3.3124	2406.1	0.04697	2.5438	0.00558	129.27	66.38	0.01047	1.01002	0.6890
200	2.9807	2739.7	0.04522	2.5435	0.00500	137.70	69.27	0.01215	1.00901	0.6887
220	2.7103	3071.2	0.04391	2.5425	0.00453	145.40	71.96	0.01388	1.00819	0.6884
240	2.4854	3398.0	0.04294	2.5412	0.00414	152.48	74.52	0.01569	1.00751	0.6881
260	2.2954	3720.0	0.04221	2.5397	0.00382	159.08	77.00	0.01756	1.00693	0.6878
280	2.1326	4036.9	0.04168	2.5382	0.00354	165.29	79.44	0.01951	1.00644	0.6875
300	1.9914	4349.8	0.04127	2.5371	0.00330	171.19	81.84	0.02153	1.00601	0.6871
350	1.7093	5111.2	0.04069	2.5333	0.00283	185.00	87.75	0.02693	1.00516	0.6863
400	1.4975	5855.8	0.04040	2.5301	0.00248	197.99	93.58	0.03281	1.00452	0.6856
450	1.3325	6591.2	0.04026	2.5274	0.00220	210.54	99.34	0.03918	1.00402	0.6850
500	1.2004	7327.2	0.04013	2.5250	0.00198	222.97	105.05	0.04602	1.00362	0.6846
550	1.0921	8062.1	0.04004	2.5230	0.00180	235.24	110.69	0.05333	1.00329	0.6842
600	1.0018	8801.0	0.03993	2.5213	0.00165	247.47	116.27	0.06109	1.00302	0.6840
700	0.8596	10296.7	0.03966	2.5185	0.00142	271.91	127.24	0.07795	1.00259	0.6837
800	0.7528	11830.4	0.03929	2.5163	0.00124	296.61	137.95	0.09651	1.00227	0.6836
900	0.6696	13415.7	0.03879	2.5146	0.00110	321.79	148.41	0.11671	1.00202	0.6837
1000	0.6030	15061.5	0.03819	2.5132	0.00099	440.56	158.63	0.17552	1.00182	0.5396
1200	0.5029	18517.1	0.03686	2.5111	0.00083	511.01	178.40	0.23807	1.00152	0.5364
1400	0.4313	22203.1	0.03546	2.5095	0.00071	583.22	197.38	0.30807	1.00130	0.5347
1600	0.3776	26084.9	0.03413	2.5083	0.00062	656.77	215.68	0.38534	1.00114	0.5336
1800	0.3358	30164.5	0.03288	2.5074	0.00055	732.93	233.39	0.47031	1.00101	0.5321
2000	0.3022	34607.2	0.03153	2.5063	0.00050	818.72	250.59	0.56500	1.00091	0.5283
2500	0.2413	50780.1	0.02597	2.4990	0.00040	1208.37	291.96	0.88707	1.00073	0.4910
3000	0.1985	87415.3	0.01741	2.4653	0.00034	2345.90	333.45	1.43955	1.00060	0.4200

* TWO-PHASE BOUNDARY

Table 4.12 (continued)
THERMODYNAMIC PROPERTIES OF NORMAL HYDROGEN

3.00 MPA ISOBAR

TEMPERATURE	VOLUME	ISOTHERM DERIVATIVE	ISOCHORE DERIVATIVE	INTERNAL ENERGY	ENTHALPY	ENTROPY	CV	CP	VELOCITY OF SOUND
DEG. K	CU M/KG	CU M-MPA/KG	MPA/K	KJ/KG-K	KJ/KG	KJ/KG-K	KJ /	KG-K	M/S
* 14.772	0.01272	1.3623	0.9470	219.6	257.8	14.151	4.81	6.38	1345
15	0.01274	1.3585	0.9492	221.0	259.2	14.250	4.86	6.47	1345
16	0.01286	1.2794	0.9544	227.4	266.0	14.684	5.05	6.93	1326
17	0.01299	1.2075	0.9510	234.1	273.1	15.116	5.22	7.37	1306
18	0.01313	1.1408	0.9404	241.3	280.7	15.549	5.37	7.78	1285
19	0.01327	1.0731	0.9266	248.8	288.6	15.980	5.52	8.20	1262
20	0.01343	1.0093	0.9118	256.8	297.0	16.411	5.65	8.62	1241
21	0.01360	0.9488	0.8960	265.1	305.9	16.843	5.76	9.05	1221
22	0.01378	0.8875	0.8787	273.8	315.1	17.275	5.86	9.50	1199
23	0.01397	0.8271	0.8595	282.9	324.9	17.707	5.95	9.97	1177
24	0.01419	0.7655	0.8386	292.5	335.1	18.142	6.03	10.47	1153
25	0.01442	0.7053	0.8155	302.5	345.8	18.579	6.10	11.00	1128
26	0.01467	0.6461	0.7904	313.1	357.1	19.019	6.16	11.57	1102
27	0.01494	0.5884	0.7629	324.1	368.9	19.468	6.21	12.17	1074
28	0.01525	0.5315	0.7331	335.7	381.4	19.921	6.25	12.84	1044
29	0.01559	0.4762	0.7011	347.9	394.7	20.385	6.29	13.57	1013
30	0.01597	0.4222	0.6670	360.7	408.6	20.859	6.33	14.39	980
31	0.01640	0.3688	0.6309	374.3	423.5	21.345	6.36	15.36	943
32	0.01689	0.3184	0.5926	388.7	439.4	21.850	6.40	16.47	905
33	0.01747	0.2698	0.5524	404.1	456.5	22.377	6.44	17.82	864
34	0.01815	0.2258	0.5101	420.6	475.0	22.931	6.48	19.38	822
35	0.01896	0.1858	0.4661	438.4	495.3	23.519	6.53	21.25	778
36	0.01997	0.1498	0.4207	457.8	517.7	24.149	6.59	23.55	732
37	0.02123	0.1196	0.3747	478.8	542.5	24.830	6.66	26.23	686
38	0.02281	0.0967	0.3293	501.7	570.1	25.565	6.74	28.92	644
39	0.02479	0.0809	0.2866	525.9	600.3	26.349	6.83	31.19	607
40	0.02717	0.0718	0.2490	550.7	632.2	27.157	6.91	32.43	580
42	0.03273	0.0711	0.1916	597.0	695.2	28.693	6.91	30.15	557
44	0.03841	0.0830	0.1546	635.6	750.9	29.990	6.84	25.55	557
46	0.04369	0.0984	0.1306	667.1	798.1	31.041	6.76	21.97	566
48	0.04856	0.1144	0.1138	693.8	839.5	31.921	6.69	19.52	578
50	0.05311	0.1299	0.1015	717.4	876.7	32.682	6.64	17.83	590
52	0.05740	0.1449	0.0920	738.9	911.1	33.356	6.61	16.62	604
54	0.06150	0.1593	0.0844	758.9	943.4	33.966	6.59	15.72	617
56	0.06543	0.1732	0.0782	778.1	974.4	34.530	6.58	15.04	629
58	0.06925	0.1867	0.0731	796.2	1004.0	35.050	6.56	14.52	643
60	0.07295	0.1998	0.0686	813.7	1032.5	35.534	6.55	14.08	655
65	0.08185	0.2307	0.0598	855.2	1100.8	36.626	6.54	13.28	684
70	0.09035	0.2599	0.0532	894.8	1165.9	37.591	6.56	12.78	711
75	0.09858	0.2877	0.0480	933.1	1228.9	38.460	6.61	12.46	736
80	0.10659	0.3145	0.0439	970.9	1290.6	39.258	6.69	12.25	759
85	0.11445	0.3405	0.0405	1008.2	1351.6	39.997	6.78	12.13	781
90	0.12217	0.3658	0.0376	1045.6	1412.1	40.689	6.89	12.08	801
95	0.12980	0.3905	0.0351	1083.0	1472.4	41.341	7.01	12.07	820
100	0.13733	0.4149	0.0330	1120.8	1532.8	41.960	7.14	12.09	838
120	0.16683	0.5085	0.0266	1277.0	1777.5	44.188	7.73	12.38	902
140	0.19571	0.5987	0.0224	1441.8	2028.9	46.125	8.27	12.77	961
160	0.22421	0.6868	0.0194	1615.3	2287.9	47.853	8.74	13.13	1016
180	0.25246	0.7736	0.0171	1796.5	2553.9	49.419	9.13	13.46	1068
200	0.28054	0.8595	0.0153	1984.1	2825.7	50.851	9.43	13.72	1118
220	0.30851	0.9446	0.0139	2176.8	3102.3	52.169	9.68	13.93	1166
240	0.33638	1.0293	0.0127	2373.6	3382.7	53.390	9.87	14.10	1213
260	0.36419	1.1135	0.0117	2573.5	3666.1	54.524	10.02	14.23	1258
280	0.39195	1.1975	0.0108	2775.8	3951.6	55.582	10.12	14.32	1301
300	0.41969	1.2816	0.0101	2980.3	4239.3	56.573	10.20	14.38	1344
350	0.48881	1.4900	0.0086	3494.8	4961.2	58.800	10.31	14.48	1446
400	0.55781	1.6977	0.0075	4012.7	5686.1	60.736	10.36	14.51	1542
450	0.62674	1.9049	0.0067	4532.0	6412.2	62.445	10.38	14.52	1633
500	0.69561	2.1119	0.0060	5051.8	7138.6	63.975	10.40	14.53	1718
550	0.76446	2.3187	0.0055	5572.1	7865.5	65.362	10.41	14.54	1800
600	0.83328	2.5254	0.0050	6093.4	8593.2	66.625	10.43	14.56	1878
700	0.97087	2.9385	0.0043	7138.9	10051.5	68.876	10.48	14.61	2024
800	1.10842	3.3513	0.0038	8191.9	11517.2	70.835	10.57	14.70	2159
900	1.24595	3.7640	0.0033	9255.5	12993.4	72.572	10.70	14.82	2284
1000	1.38347	4.1767	0.0030	10332.9	14483.3	74.139	10.86	14.99	2401
1200	1.65847	5.0018	0.0025	12543.4	17518.8	76.887	11.24	15.36	2615
1400	1.93345	5.8268	0.0021	14836.2	20636.6	79.296	11.68	15.80	2808
1600	2.20843	6.6518	0.0019	17215.2	23840.5	81.442	12.12	16.25	2986
1800	2.48346	7.4767	0.0017	19682.9	27133.3	83.375	12.57	16.70	3151
2000	2.75875	8.3016	0.0015	22249.8	30526.1	85.173	13.10	17.24	3305
2500	3.45369	10.3638	0.0012	29351.7	39712.7	89.248	15.76	20.10	3635
3000	4.19314	12.4259	0.0010	38968.3	51547.7	93.555	23.26	28.58	3904

* TWO-PHASE BOUNDARY

Table 4.12 (continued)
THERMODYNAMIC PROPERTIES OF NORMAL HYDROGEN

3.00 MPA ISOBAR

TEMPERATURE DEG. K	DENSITY KG/CU M	$V(DH/DV)_P$ KJ/KG	$V(DP/DU)_V$ MPA-CU M/KJ X 10^2	$-V(DP/DV)_T$ MPA	$(DV/DT)_P/V$ 1/DEG. K	THERMAL CONDUCTIVITY W/K-M X 10^3	VISCOSITY KG/M-S X 10^7	THERMAL DIFFUSIVITY SQ M/HR	DIELECTRIC CONSTANT	PRANDTL NUMBER
* 14.772	78.6428	721.9	0.25043	107.1338	0.00884	84.29	273.40	0.00060	1.25731	2.0698
15	78.4822	726.8	0.24911	106.6179	0.00890	86.21	265.60	0.00061	1.25674	1.9933
16	77.7515	722.4	0.24323	99.4716	0.00959	93.11	235.84	0.00062	1.25416	1.7554
17	76.9926	720.1	0.23670	92.9694	0.01023	97.93	212.07	0.00062	1.25149	1.5952
18	76.1868	718.9	0.22967	86.9158	0.01082	101.52	192.47	0.00062	1.24866	1.4746
19	75.3438	715.1	0.22288	80.8495	0.01146	103.89	176.15	0.00061	1.24570	1.3896
20	74.4611	710.3	0.21687	75.1511	0.01213	106.09	162.34	0.00060	1.24261	1.3188
21	73.5386	704.5	0.21146	69.7716	0.01284	108.60	150.50	0.00059	1.23938	1.2539
22	72.5731	696.3	0.20646	64.4117	0.01364	110.37	140.21	0.00058	1.23601	1.2066
23	71.5604	686.3	0.20172	59.1907	0.01452	111.51	131.14	0.00056	1.23248	1.1719
24	70.4955	673.7	0.19717	53.9663	0.01554	112.09	123.04	0.00055	1.22877	1.1492
25	69.3718	659.9	0.19271	48.9308	0.01667	112.17	115.72	0.00053	1.22487	1.1347
26	68.1839	644.6	0.18824	44.0539	0.01794	111.75	109.03	0.00051	1.22075	1.1284
27	66.9234	628.3	0.18359	39.3788	0.01937	110.92	102.85	0.00049	1.21640	1.1286
28	65.5849	610.4	0.17873	34.8590	0.02103	109.72	97.09	0.00047	1.21178	1.1361
29	64.1571	591.2	0.17362	30.5492	0.02295	108.16	91.68	0.00045	1.20687	1.1499
30	62.6303	570.4	0.16832	26.4435	0.02523	106.27	86.54	0.00042	1.20163	1.1717
31	60.9904	547.6	0.16255	22.4936	0.02805	103.67	81.63	0.00040	1.19603	1.2093
32	59.2057	523.9	0.15637	18.8488	0.03144	100.76	76.87	0.00037	1.18994	1.2564
33	57.2569	498.4	0.14982	15.4471	0.03576	97.54	72.21	0.00034	1.18332	1.3195
34	55.1099	472.8	0.14282	12.4434	0.04099	94.03	67.64	0.00032	1.17606	1.3942
35	52.7287	446.6	0.13541	9.7964	0.04758	90.55	63.12	0.00029	1.16804	1.4813
36	50.0718	420.0	0.12756	7.5024	0.05608	87.76	58.66	0.00027	1.15913	1.5741
37	47.1064	394.4	0.11943	5.6338	0.06651	84.75	54.27	0.00025	1.14925	1.6801
38	43.8311	372.3	0.11138	4.2386	0.07769	81.46	50.05	0.00023	1.13840	1.7771
39	40.3347	354.9	0.10399	3.2613	0.08789	77.97	46.15	0.00022	1.12689	1.8462
40	36.8045	343.9	0.09794	2.6409	0.09429	74.35	42.76	0.00022	1.11536	1.8649
42	30.5512	341.7	0.09069	2.1711	0.08823	67.11	37.90	0.00026	1.09513	1.7026
44	26.0341	356.9	0.08678	2.1601	0.07158	61.71	35.12	0.00033	1.08067	1.4541
46	22.8887	378.9	0.08442	2.2517	0.05799	58.52	33.56	0.00042	1.07069	1.2601
48	20.5912	403.7	0.08265	2.3549	0.04834	56.76	32.72	0.00051	1.06344	1.1249
50	18.8289	429.5	0.08118	2.4449	0.04152	55.89	32.34	0.00060	1.05790	1.0318
52	17.4203	456.0	0.07989	2.5248	0.03644	55.55	32.25	0.00069	1.05349	0.9649
54	16.2609	482.6	0.07876	2.5911	0.03258	55.58	32.37	0.00078	1.04986	0.9157
56	15.2835	509.2	0.07778	2.6476	0.02953	55.84	32.62	0.00087	1.04682	0.8783
58	14.4407	535.6	0.07715	2.6965	0.02711	56.26	32.95	0.00097	1.04420	0.8503
60	13.7071	561.7	0.07648	2.7380	0.02507	56.87	33.36	0.00106	1.04192	0.8259
65	12.2174	626.5	0.07478	2.8188	0.02120	58.89	34.55	0.00131	1.03730	0.7791
70	11.0676	691.4	0.07317	2.8764	0.01848	60.92	35.87	0.00155	1.03375	0.7524
75	10.1443	757.0	0.07157	2.9187	0.01645	62.53	37.26	0.00178	1.03090	0.7423
80	9.3815	823.7	0.06997	2.9505	0.01487	64.25	38.67	0.00201	1.02856	0.7375
85	8.7377	891.9	0.06834	2.9749	0.01360	66.83	40.09	0.00227	1.02658	0.7280
90	8.1851	961.9	0.06669	2.9939	0.01256	69.15	41.51	0.00252	1.02488	0.7251
95	7.7044	1033.2	0.06509	3.0089	0.01168	71.28	42.92	0.00276	1.02341	0.7264
100	7.2815	1106.6	0.06350	3.0208	0.01092	79.19	46.29	0.00324	1.02211	0.7066
120	5.9941	1417.8	0.05745	3.0477	0.00873	96.20	54.18	0.00467	1.01818	0.6972
140	5.1097	1743.6	0.05299	3.0591	0.00732	109.28	59.33	0.00603	1.01548	0.6931
160	4.4602	2076.9	0.04970	3.0635	0.00632	120.12	63.21	0.00738	1.01350	0.6912
180	3.9610	2412.8	0.04728	3.0643	0.00558	129.54	66.45	0.00875	1.01199	0.6902
200	3.5645	2747.5	0.04549	3.0635	0.00499	137.95	69.35	0.01016	1.01078	0.6896
220	3.2414	3080.0	0.04414	3.0618	0.00452	145.64	72.03	0.01161	1.00980	0.6892
240	2.9728	3407.7	0.04314	3.0598	0.00414	152.72	74.60	0.01312	1.00899	0.6888
260	2.7458	3730.4	0.04240	3.0575	0.00381	159.32	77.09	0.01468	1.00830	0.6884
280	2.5514	4047.9	0.04184	3.0552	0.00354	165.54	79.53	0.01631	1.00771	0.6880
300	2.3827	4361.7	0.04142	3.0536	0.00330	171.45	81.94	0.01801	1.00720	0.6874
350	2.0458	5124.0	0.04081	3.0481	0.00283	185.28	87.86	0.02252	1.00618	0.6865
400	1.7927	5869.3	0.04051	3.0434	0.00247	198.30	93.70	0.02744	1.00541	0.6857
450	1.5956	6605.1	0.04036	3.0394	0.00220	210.88	99.49	0.03276	1.00482	0.6851
500	1.4376	7341.4	0.04022	3.0360	0.00198	223.35	105.21	0.03848	1.00434	0.6846
550	1.3081	8076.5	0.04012	3.0331	0.00180	235.66	110.87	0.04459	1.00395	0.6843
600	1.2001	8815.6	0.04000	3.0307	0.00165	247.93	116.47	0.05108	1.00362	0.6840
700	1.0300	10311.6	0.03973	3.0266	0.00142	272.44	127.48	0.06517	1.00311	0.6837
800	0.9022	11845.6	0.03934	3.0235	0.00124	297.22	138.22	0.08069	1.00272	0.6836
900	0.8026	13431.2	0.03884	3.0210	0.00110	322.47	148.71	0.09757	1.00242	0.6837
1000	0.7228	15077.2	0.03823	3.0190	0.00099	440.56	158.96	0.14642	1.00218	0.5407
1200	0.6030	18533.4	0.03689	3.0159	0.00083	511.01	178.80	0.19857	1.00182	0.5376
1400	0.5172	22219.7	0.03549	3.0137	0.00071	583.21	197.84	0.25692	1.00156	0.5360
1600	0.4528	26100.4	0.03416	3.0120	0.00062	656.69	216.20	0.32132	1.00136	0.5349
1800	0.4027	30171.7	0.03292	3.0106	0.00055	732.51	233.96	0.39207	1.00121	0.5335
2000	0.3625	34582.2	0.03160	3.0092	0.00050	817.04	251.21	0.47072	1.00109	0.5300
2500	0.2835	50262.1	0.02630	3.0008	0.00040	1187.98	292.67	0.73486	1.00087	0.4952
3000	0.2385	84685.3	0.01803	2.9634	0.00034	2242.45	334.06	1.18453	1.00072	0.4257

* TWO-PHASE BOUNDARY

Table 4.12 (continued)
THERMODYNAMIC PROPERTIES OF NORMAL HYDROGEN

3.50 MPA ISOBAR

TEMPERATURE DEG. K	VOLUME CU M/KG	ISOTHERM DERIVATIVE CU M-MPA/KG	ISOCHORE DERIVATIVE MPA/K	INTERNAL ENERGY KJ/KG-K	ENTHALPY KJ/KG	ENTROPY KJ/KG-K	CV KJ /	CP KG-K	VELOCITY OF SOUND M/S
* 14.927	0.01267	1.3929	0.9496	219.9	264.3	14.163	4.83	6.38	1357
15	0.01268	1.3951	0.9505	220.4	264.8	14.194	4.84	6.41	1358
16	0.01280	1.3237	0.9576	226.6	271.4	14.623	5.04	6.85	1342
17	0.01292	1.2538	0.9572	233.2	278.5	15.051	5.21	7.28	1324
18	0.01305	1.1850	0.9506	240.3	286.0	15.479	5.37	7.71	1304
19	0.01319	1.1224	0.9370	247.7	293.9	15.906	5.51	8.10	1284
20	0.01334	1.0577	0.9228	255.5	302.2	16.332	5.64	8.50	1263
21	0.01350	0.9958	0.9075	263.6	310.9	16.758	5.75	8.92	1243
22	0.01368	0.9357	0.8908	272.1	320.0	17.184	5.86	9.35	1222
23	0.01386	0.8759	0.8726	281.0	329.6	17.609	5.95	9.79	1201
24	0.01406	0.8161	0.8525	290.4	339.6	18.035	6.03	10.25	1178
25	0.01427	0.7561	0.8307	300.1	350.1	18.463	6.10	10.74	1154
26	0.01451	0.6973	0.8068	310.3	361.1	18.893	6.16	11.26	1130
27	0.01476	0.6399	0.7809	320.9	372.6	19.329	6.21	11.82	1104
28	0.01504	0.5828	0.7530	332.1	384.7	19.768	6.26	12.42	1076
29	0.01535	0.5285	0.7232	343.7	397.5	20.216	6.30	13.06	1047
30	0.01569	0.4773	0.6917	355.9	410.9	20.669	6.33	13.73	1017
31	0.01606	0.4258	0.6585	368.7	424.9	21.132	6.36	14.51	985
32	0.01649	0.3763	0.6236	382.2	439.9	21.606	6.40	15.39	951
33	0.01697	0.3294	0.5872	396.4	455.8	22.095	6.43	16.37	916
34	0.01752	0.2847	0.5494	411.4	472.7	22.600	6.46	17.53	879
35	0.01816	0.2442	0.5103	427.3	490.8	23.126	6.50	18.81	841
36	0.01891	0.2077	0.4703	444.2	510.4	23.676	6.53	20.24	802
37	0.01979	0.1756	0.4300	462.1	531.4	24.253	6.58	21.83	763
38	0.02084	0.1481	0.3899	481.2	554.1	24.858	6.62	23.56	726
39	0.02209	0.1257	0.3508	501.2	578.5	25.493	6.67	25.31	690
40	0.02356	0.1099	0.3140	522.1	604.5	26.151	6.73	26.64	660
42	0.02717	0.0926	0.2509	564.4	659.5	27.491	6.81	27.89	616
44	0.03140	0.0917	0.2035	604.3	714.2	28.763	6.82	26.41	596
46	0.03578	0.1007	0.1699	638.6	763.8	29.866	6.80	23.69	592
48	0.04000	0.1135	0.1462	668.6	808.6	30.819	6.75	21.20	597
50	0.04401	0.1278	0.1288	694.9	848.9	31.643	6.70	19.27	606
52	0.04782	0.1421	0.1156	718.6	885.9	32.369	6.66	17.84	617
54	0.05145	0.1564	0.1052	740.4	920.5	33.021	6.64	16.75	628
56	0.05495	0.1703	0.0967	761.1	953.4	33.620	6.62	15.91	640
58	0.05832	0.1839	0.0897	780.4	984.5	34.167	6.60	15.24	652
60	0.06159	0.1970	0.0838	798.8	1014.4	34.674	6.58	14.71	663
65	0.06947	0.2289	0.0725	842.4	1085.6	35.812	6.58	13.77	692
70	0.07695	0.2588	0.0640	883.5	1152.8	36.809	6.59	13.16	719
75	0.08415	0.2874	0.0576	923.0	1217.6	37.703	6.64	12.76	743
80	0.09115	0.3148	0.0524	961.7	1280.7	38.518	6.70	12.50	766
85	0.09799	0.3413	0.0482	999.8	1342.8	39.271	6.79	12.35	788
90	0.10472	0.3672	0.0447	1037.8	1404.3	39.975	6.90	12.26	808
95	0.11134	0.3924	0.0416	1075.8	1465.5	40.636	7.02	12.22	827
100	0.11788	0.4172	0.0390	1114.0	1526.6	41.262	7.15	12.23	845
120	0.14339	0.5120	0.0313	1271.8	1773.6	43.512	7.74	12.47	908
140	0.16830	0.6032	0.0263	1437.5	2026.5	45.461	8.28	12.83	967
160	0.19285	0.6921	0.0227	1611.7	2286.6	47.196	8.75	13.18	1021
180	0.21716	0.7794	0.0200	1793.4	2553.4	48.767	9.13	13.49	1073
200	0.24130	0.8657	0.0179	1981.4	2825.9	50.203	9.44	13.75	1123
220	0.26533	0.9512	0.0162	2174.4	3103.1	51.523	9.69	13.96	1171
240	0.28927	1.0361	0.0148	2371.5	3383.9	52.745	9.88	14.12	1217
260	0.31314	1.1206	0.0136	2571.6	3667.6	53.881	10.02	14.24	1262
280	0.33697	1.2048	0.0126	2774.1	3953.5	54.940	10.13	14.33	1306
300	0.36079	1.2892	0.0118	2978.8	4241.6	55.933	10.21	14.40	1349
350	0.42009	1.4979	0.0101	3493.7	4964.0	58.161	10.32	14.48	1450
400	0.47926	1.7058	0.0088	4011.8	5689.2	60.098	10.36	14.52	1546
450	0.53837	1.9132	0.0078	4531.2	6415.5	61.808	10.38	14.53	1636
500	0.59742	2.1203	0.0070	5051.2	7142.2	63.338	10.40	14.54	1722
550	0.65644	2.3272	0.0064	5571.6	7869.1	64.725	10.41	14.55	1803
600	0.71544	2.5339	0.0058	6092.9	8597.0	65.988	10.43	14.56	1881
700	0.83339	2.9470	0.0050	7138.6	10055.4	68.240	10.48	14.61	2027
800	0.95131	3.3600	0.0044	8191.6	11521.2	70.199	10.57	14.70	2162
900	1.06920	3.7727	0.0039	9255.3	12997.5	71.936	10.70	14.83	2287
1000	1.18707	4.1854	0.0035	10332.7	14487.5	73.503	10.86	14.99	2403
1200	1.42280	5.0106	0.0029	12543.3	17523.1	76.251	11.24	15.37	2617
1400	1.65850	5.8356	0.0025	14836.1	20640.9	78.660	11.68	15.80	2810
1600	1.89420	6.6606	0.0022	17215.1	23844.8	80.806	12.12	16.25	2988
1800	2.12994	7.4855	0.0019	19682.4	27137.2	82.739	12.57	16.70	3153
2000	2.36588	8.3104	0.0018	22247.4	30528.0	84.536	13.08	17.22	3308
2500	2.96107	10.3726	0.0014	29314.6	39678.3	88.595	15.60	19.93	3639
3000	3.59160	12.4347	0.0012	38736.3	51306.9	92.829	22.59	27.82	3910

* TWO-PHASE BOUNDARY

Table 4.12 (continued)
THERMODYNAMIC PROPERTIES OF NORMAL HYDROGEN

3.50 MPA ISOBAR

TEMPERATURE DEG. K	DENSITY KG/CU M	$V(DH/DV)_P$ KJ/KG	$V(DP/DU)_V$ MPA-CU M/KJ X 10^2	$-V(DP/DV)_T$ MPA	$(DV/DT)_P/V$ 1/DEG. K	THERMAL CONDUCTIVITY W/K-M X 10^3	VISCOSITY KG/M-S X 10^7	THERMAL DIFFUSIVITY SQ M/HR	DIELECTRIC CONSTANT	PRANDTL NUMBER
* 14.927	78.8975	738.4	0.24927	109.8930	0.00864	86.07	276.50	0.00062	1.25820	2.0499
15	78.8440	741.3	0.24890	109.9949	0.00864	86.68	273.88	0.00062	1.25802	2.0242
16	78.1394	740.2	0.24329	103.4356	0.00926	93.74	243.10	0.00063	1.25553	1.7771
17	77.3997	738.6	0.23732	97.0466	0.00986	98.70	218.42	0.00063	1.25292	1.6121
18	76.6149	736.0	0.23117	90.7901	0.01047	102.42	198.10	0.00062	1.25016	1.4905
19	75.7943	735.1	0.22439	85.0704	0.01101	104.90	181.21	0.00062	1.24728	1.3987
20	74.9428	730.5	0.21842	79.2684	0.01164	107.22	167.01	0.00061	1.24429	1.3246
21	74.0524	724.8	0.21303	73.7426	0.01231	109.86	154.84	0.00060	1.24118	1.2571
22	73.1206	717.8	0.20806	68.4215	0.01302	111.77	144.27	0.00059	1.23792	1.2063
23	72.1477	708.8	0.20339	63.1958	0.01381	113.04	134.99	0.00058	1.23452	1.1687
24	71.1269	697.9	0.19890	58.0441	0.01469	113.76	126.73	0.00056	1.23097	1.1420
25	70.0562	685.1	0.19453	52.9693	0.01568	113.99	119.30	0.00055	1.22725	1.1244
26	68.9286	671.0	0.19016	48.0622	0.01679	113.73	112.54	0.00053	1.22333	1.1146
27	67.7393	655.9	0.18568	43.3471	0.01802	113.08	106.33	0.00051	1.21922	1.1111
28	66.4831	639.0	0.18105	38.7445	0.01944	112.07	100.56	0.00049	1.21488	1.1145
29	65.1537	621.7	0.17629	34.4343	0.02100	110.73	95.18	0.00047	1.21030	1.1223
30	63.7444	604.0	0.17143	30.4265	0.02273	109.09	90.10	0.00045	1.20545	1.1340
31	62.2492	584.0	0.16626	26.5052	0.02485	106.77	85.30	0.00043	1.20033	1.1593
32	60.6468	563.1	0.16077	22.8203	0.02733	104.19	80.69	0.00040	1.19485	1.1916
33	58.9266	541.4	0.15501	19.4131	0.03025	101.38	76.24	0.00038	1.18899	1.2313
34	57.0738	518.4	0.14900	16.2475	0.03382	98.37	71.93	0.00035	1.18270	1.2818
35	55.0661	495.5	0.14267	13.4449	0.03795	95.11	67.73	0.00033	1.17591	1.3392
36	52.8898	472.7	0.13611	10.9834	0.04282	91.89	63.65	0.00031	1.16858	1.4022
37	50.5307	450.6	0.12936	8.8738	0.04845	89.25	59.69	0.00029	1.16067	1.4599
38	47.9855	429.5	0.12265	7.1055	0.05487	86.47	55.87	0.00028	1.15217	1.5226
39	45.2729	410.4	0.11611	5.6889	0.06167	83.58	52.26	0.00026	1.14317	1.5827
40	42.4430	395.9	0.11001	4.6658	0.06730	80.54	48.93	0.00026	1.13382	1.6190
42	36.8054	378.8	0.10013	3.4086	0.07362	74.54	43.44	0.00026	1.11536	1.6255
44	31.8437	379.1	0.09373	2.9211	0.06968	69.09	39.67	0.00030	1.09929	1.5165
46	27.9523	392.2	0.08944	2.8134	0.06040	64.94	37.31	0.00035	1.08680	1.3608
48	25.0017	411.6	0.08668	2.8385	0.05150	62.20	35.85	0.00042	1.07739	1.2218
50	22.7238	434.4	0.08461	2.9033	0.04437	60.53	34.97	0.00050	1.07017	1.1133
52	20.9139	458.7	0.08292	2.9720	0.03889	59.60	34.48	0.00058	1.06445	1.0321
54	19.4351	484.1	0.08149	3.0397	0.03460	59.15	34.27	0.00065	1.05980	0.9703
56	18.1994	509.8	0.08027	3.0997	0.03120	59.05	34.26	0.00073	1.05593	0.9228
58	17.1469	535.4	0.07930	3.1526	0.02846	59.18	34.40	0.00082	1.05263	0.8857
60	16.2361	561.1	0.07843	3.1993	0.02621	59.56	34.64	0.00090	1.04979	0.8553
65	14.3943	626.2	0.07655	3.2950	0.02199	61.16	35.54	0.00111	1.04405	0.8000
70	12.9961	691.2	0.07475	3.3640	0.01904	62.91	36.68	0.00132	1.03971	0.7674
75	11.8836	757.1	0.07299	3.4150	0.01686	64.21	37.95	0.00152	1.03627	0.7542
80	10.9709	824.0	0.07125	3.4537	0.01517	65.78	39.27	0.00173	1.03345	0.7465
85	10.2046	892.5	0.06950	3.4833	0.01383	68.26	40.63	0.00195	1.03109	0.7348
90	9.5497	962.8	0.06775	3.5064	0.01273	70.50	41.99	0.00217	1.02907	0.7303
95	8.9818	1034.5	0.06606	3.5245	0.01182	72.57	43.36	0.00238	1.02733	0.7303
100	8.4835	1108.1	0.06439	3.5390	0.01103	79.88	46.50	0.00277	1.02580	0.7117
120	6.9740	1421.3	0.05804	3.5710	0.00877	96.67	54.29	0.00400	1.02117	0.7004
140	5.9417	1748.3	0.05345	3.5842	0.00734	109.65	59.42	0.00518	1.01802	0.6952
160	5.1854	2082.9	0.05008	3.5888	0.00633	120.45	63.30	0.00634	1.01571	0.6928
180	4.6050	2419.9	0.04759	3.5892	0.00558	129.84	66.53	0.00752	1.01394	0.6914
200	4.1442	2755.6	0.04576	3.5876	0.00499	138.23	69.43	0.00873	1.01254	0.6906
220	3.7689	3089.0	0.04438	3.5850	0.00452	145.90	72.12	0.00998	1.01140	0.6900
240	3.4570	3417.5	0.04335	3.5819	0.00413	152.97	74.69	0.01128	1.01045	0.6894
260	3.1934	3741.0	0.04258	3.5787	0.00381	159.57	77.18	0.01263	1.00965	0.6889
280	2.9676	4059.1	0.04201	3.5754	0.00353	165.79	79.63	0.01403	1.00897	0.6884
300	2.7717	4373.8	0.04158	3.5733	0.00329	171.71	82.03	0.01549	1.00838	0.6878
350	2.3805	5136.9	0.04094	3.5657	0.00282	185.56	87.97	0.01937	1.00719	0.6867
400	2.0865	5882.8	0.04062	3.5592	0.00247	198.61	93.83	0.02360	1.00630	0.6859
450	1.8575	6619.0	0.04045	3.5537	0.00219	211.23	99.63	0.02818	1.00561	0.6852
500	1.6739	7355.7	0.04031	3.5490	0.00198	223.73	105.37	0.03310	1.00505	0.6847
550	1.5234	8091.0	0.04019	3.5451	0.00180	236.08	111.05	0.03835	1.00460	0.6843
600	1.3977	8830.3	0.04007	3.5417	0.00165	248.38	116.67	0.04393	1.00422	0.6840
700	1.1999	10326.6	0.03979	3.5362	0.00142	272.97	127.71	0.05605	1.00362	0.6837
800	1.0512	11860.8	0.03939	3.5319	0.00124	297.82	138.49	0.06939	1.00317	0.6836
900	0.9353	13446.6	0.03888	3.5285	0.00110	323.15	149.02	0.08390	1.00282	0.6837
1000	0.8424	15092.9	0.03827	3.5258	0.00099	440.56	159.30	0.12563	1.00254	0.5419
1200	0.7028	18549.6	0.03693	3.5216	0.00083	511.01	179.20	0.17035	1.00212	0.5388
1400	0.6030	22236.3	0.03552	3.5186	0.00071	583.21	198.30	0.22038	1.00182	0.5372
1600	0.5279	26116.2	0.03418	3.5163	0.00062	656.64	216.71	0.27559	1.00159	0.5362
1800	0.4695	30181.3	0.03295	3.5144	0.00055	732.19	234.53	0.33620	1.00142	0.5349
2000	0.4227	34566.7	0.03165	3.5126	0.00050	815.74	251.83	0.40344	1.00127	0.5317
2500	0.3377	49863.7	0.02657	3.5030	0.00040	1172.12	293.40	0.62697	1.00102	0.4988
3000	0.2784	82563.1	0.01855	3.4622	0.00034	2161.83	334.72	1.00469	1.00084	0.4308

* TWO-PHASE BOUNDARY

Table 4.12 (continued)
THERMODYNAMIC PROPERTIES OF NORMAL HYDROGEN

4.00 MPA ISOBAR

TEMPERATURE DEG. K	VOLUME CU M/KG	ISOTHERM DERIVATIVE CU M-MPA/KG	ISOCHORE DERIVATIVE MPA/K	INTERNAL ENERGY KJ/KG-K	ENTHALPY KJ/KG	ENTROPY KJ/KG-K	CV KJ / KG-K	CP KJ / KG-K	VELOCITY OF SOUND M/S
* 15.080	0.01263	1.4112	0.9524	220.2	270.8	14.174	4.85	6.39	1364
16	0.01274	1.3667	0.9608	225.9	276.8	14.564	5.03	6.78	1358
17	0.01286	1.2950	0.9624	232.4	283.8	14.988	5.20	7.21	1340
18	0.01298	1.2281	0.9572	239.3	291.3	15.412	5.36	7.62	1322
19	0.01312	1.1684	0.9466	246.6	299.1	15.835	5.50	8.01	1304
20	0.01326	1.1021	0.9333	254.2	307.3	16.256	5.63	8.41	1283
21	0.01341	1.0393	0.9185	262.2	315.9	16.677	5.74	8.81	1263
22	0.01358	0.9817	0.9023	270.6	324.9	17.097	5.85	9.21	1244
23	0.01375	0.9227	0.8848	279.3	334.3	17.516	5.94	9.63	1223
24	0.01394	0.8638	0.8656	288.4	344.2	17.935	6.02	10.07	1202
25	0.01414	0.8051	0.8448	297.9	354.5	18.355	6.09	10.52	1179
26	0.01436	0.7467	0.8221	307.8	365.2	18.776	6.15	11.01	1156
27	0.01460	0.6877	0.7975	318.1	376.5	19.201	6.21	11.53	1130
28	0.01486	0.6334	0.7712	328.9	388.3	19.628	6.26	12.06	1105
29	0.01514	0.5804	0.7431	340.1	400.6	20.062	6.30	12.62	1078
30	0.01545	0.5290	0.7135	351.8	413.5	20.500	6.33	13.22	1051
31	0.01578	0.4784	0.6827	363.9	427.1	20.944	6.37	13.89	1022
32	0.01616	0.4311	0.6504	376.7	441.3	21.396	6.40	14.60	992
33	0.01657	0.3833	0.6169	390.0	456.3	21.857	6.43	15.43	959
34	0.01704	0.3410	0.5822	404.0	472.1	22.330	6.45	16.27	927
35	0.01757	0.2997	0.5466	418.6	488.9	22.816	6.48	17.25	893
36	0.01817	0.2623	0.5102	434.0	506.7	23.317	6.51	18.31	859
37	0.01886	0.2287	0.4736	450.1	525.6	23.834	6.54	19.45	825
38	0.01965	0.1993	0.4371	467.0	545.6	24.368	6.57	20.64	791
39	0.02056	0.1742	0.4012	484.6	566.8	24.920	6.60	21.83	759
40	0.02161	0.1532	0.3666	502.8	589.3	25.488	6.63	23.01	729
42	0.02413	0.1259	0.3036	540.6	637.1	26.653	6.70	24.61	680
44	0.02719	0.1135	0.2519	578.1	686.9	27.811	6.75	24.94	648
46	0.03060	0.1131	0.2117	612.9	735.3	28.887	6.78	23.85	631
48	0.03412	0.1204	0.1813	644.7	781.2	29.864	6.76	22.02	626
50	0.03759	0.1314	0.1586	673.1	823.5	30.728	6.73	20.27	629
52	0.04095	0.1439	0.1413	698.7	862.5	31.493	6.70	18.80	635
54	0.04418	0.1571	0.1277	722.2	898.9	32.180	6.68	17.62	644
56	0.04730	0.1702	0.1168	744.2	933.4	32.809	6.66	16.69	653
58	0.05032	0.1836	0.1078	764.7	966.0	33.381	6.63	15.93	664
60	0.05325	0.1967	0.1003	784.2	997.2	33.9[illegible]	6.62	15.31	675
65	0.06027	0.2283	0.0859	829.7	1070.8	35.087	6.60	14.22	701
70	0.06699	0.2590	0.0754	872.3	1140.3	36.118	6.62	13.51	727
75	0.07341	0.2880	0.0675	913.0	1206.7	37.034	6.66	13.06	751
80	0.07964	0.3159	0.0612	952.6	1271.1	37.866	6.72	12.75	774
85	0.08572	0.3430	0.0561	991.5	1334.3	38.633	6.81	12.55	795
90	0.09168	0.3692	0.0519	1030.1	1396.8	39.348	6.92	12.44	815
95	0.09754	0.3949	0.0483	1068.6	1458.8	40.018	7.03	12.37	834
100	0.10333	0.4200	0.0452	1107.3	1520.6	40.651	7.16	12.36	852
120	0.12584	0.5160	0.0361	1266.6	1769.9	42.922	7.75	12.56	914
140	0.14777	0.6080	0.0303	1433.2	2024.3	44.882	8.29	12.89	972
160	0.16935	0.6975	0.0261	1608.1	2285.5	46.624	8.75	13.23	1027
180	0.19069	0.7854	0.0230	1790.3	2553.1	48.200	9.14	13.53	1078
200	0.21188	0.8721	0.0205	1978.7	2826.2	49.639	9.44	13.78	1128
220	0.23295	0.9579	0.0186	2172.1	3103.9	50.962	9.69	13.98	1176
240	0.25394	1.0431	0.0169	2369.4	3385.1	52.186	9.88	14.14	1222
260	0.27486	1.1278	0.0156	2569.7	3669.2	53.324	10.02	14.26	1267
280	0.29574	1.2122	0.0144	2772.4	3955.4	54.384	10.13	14.35	1310
300	0.31662	1.2969	0.0135	2977.4	4243.9	55.377	10.21	14.41	1353
350	0.36854	1.5058	0.0115	3492.6	4966.7	57.607	10.32	14.49	1454
400	0.42035	1.7139	0.0100	4010.9	5692.3	59.545	10.36	14.52	1550
450	0.47209	1.9215	0.0089	4530.5	6418.9	61.256	10.38	14.53	1640
500	0.52378	2.1287	0.0080	5050.6	7145.7	62.787	10.40	14.54	1725
550	0.57543	2.3356	0.0073	5571.1	7872.8	64.173	10.41	14.55	1807
600	0.62706	2.5424	0.0067	6092.5	8600.7	65.437	10.43	14.56	1884
700	0.73029	2.9556	0.0057	7138.2	10059.4	67.689	10.48	14.61	2030
800	0.83347	3.3686	0.0050	8191.4	11525.3	69.648	10.57	14.70	2164
900	0.93663	3.7814	0.0044	9255.2	13001.7	71.385	10.70	14.83	2289
1000	1.03977	4.1941	0.0040	10332.6	14491.6	72.952	10.86	14.99	2406
1200	1.24604	5.0193	0.0033	12543.2	17527.4	75.701	11.24	15.37	2619
1400	1.45228	5.8443	0.0029	14836.1	20645.2	78.110	11.68	15.80	2812
1600	1.65852	6.6693	0.0025	17215.0	23849.1	80.255	12.12	16.25	2990
1800	1.86479	7.4942	0.0022	19681.9	27141.1	82.188	12.57	16.70	3155
2000	2.07123	8.3192	0.0020	22245.5	30530.4	83.984	13.07	17.21	3310
2500	2.59169	10.3813	0.0016	29284.7	39651.5	88.031	15.48	19.79	3643
3000	3.14109	12.4435	0.0013	38549.3	51113.7	92.205	22.06	27.21	3915

* TWO-PHASE BOUNDARY

Table 4.12 (continued)
THERMODYNAMIC PROPERTIES OF NORMAL HYDROGEN

4.00 MPA ISOBAR

TEMPERATURE DEG. K	DENSITY KG/CU M	$V(DH/DV)_P$ KJ/KG	$V(DP/DU)_V$ MPA-CU M/KJ X 10^2	$-V(DP/DV)_T$ MPA	$(DV/DT)_P/V$ 1/DEG. K	THERMAL CONDUCTIVITY W/K-M X 10^3	VISCOSITY KG/M-S X 10^7	THERMAL DIFFUSIVITY SQ M/HR	DIELECTRIC CONSTANT	PRANDTL NUMBER
* 15.080	79.1477	750.0	0.24825	111.6967	0.00853	87.81	279.60	0.00062	1.25909	2.0363
16	78.5151	757.3	0.24339	107.3080	0.00895	94.35	250.48	0.00064	1.25686	1.8001
17	77.7893	755.0	0.23774	100.7400	0.00955	99.43	224.78	0.00064	1.25430	1.6305
18	77.0297	753.5	0.23179	94.5996	0.01012	103.29	203.80	0.00063	1.25162	1.5044
19	76.2330	753.7	0.22567	89.0699	0.01063	105.88	186.36	0.00062	1.24882	1.4097
20	75.4072	748.8	0.21984	83.1052	0.01123	108.33	171.72	0.00061	1.24592	1.3330
21	74.5439	743.3	0.21449	77.4730	0.01186	111.09	159.18	0.00061	1.24290	1.2627
22	73.6412	738.1	0.20952	72.2903	0.01248	113.11	148.31	0.00060	1.23974	1.2080
23	72.7031	730.2	0.20492	67.0801	0.01319	114.51	138.81	0.00059	1.23646	1.1676
24	71.7214	720.6	0.20047	61.9556	0.01397	115.35	130.37	0.00058	1.23304	1.1378
25	70.6965	709.1	0.19619	56.9159	0.01484	115.72	122.82	0.00056	1.22947	1.1170
26	69.6206	696.1	0.19190	51.9873	0.01581	115.60	115.96	0.00054	1.22573	1.1041
27	68.4918	681.1	0.18756	47.1047	0.01693	115.10	109.69	0.00052	1.22182	1.0989
28	67.3048	666.8	0.18310	42.6332	0.01809	114.26	103.90	0.00051	1.21771	1.0968
29	66.0553	651.3	0.17857	38.3405	0.01938	113.10	98.51	0.00049	1.21340	1.0995
30	64.7413	634.6	0.17400	34.2491	0.02083	111.67	93.47	0.00047	1.20888	1.1067
31	63.3553	616.7	0.16925	30.3107	0.02252	109.56	88.72	0.00045	1.20412	1.1247
32	61.8867	598.7	0.16427	26.6805	0.02438	107.24	84.19	0.00043	1.19909	1.1459
33	60.3319	578.3	0.15910	23.1272	0.02667	104.73	79.88	0.00041	1.19378	1.1766
34	58.6740	559.2	0.15372	20.0072	0.02910	102.05	75.72	0.00038	1.18813	1.2073
35	56.9093	538.5	0.14815	17.0574	0.03204	99.19	71.72	0.00036	1.18215	1.2475
36	55.0254	518.0	0.14238	14.4332	0.03535	96.17	67.85	0.00034	1.17577	1.2919
37	53.0180	498.0	0.13651	12.1255	0.03906	93.18	64.12	0.00033	1.16901	1.3386
38	50.8854	478.9	0.13066	10.1427	0.04309	90.71	60.55	0.00031	1.16186	1.3775
39	48.6348	461.2	0.12493	8.4737	0.04735	88.15	57.15	0.00030	1.15434	1.4157
40	46.2851	445.1	0.11937	7.0906	0.05170	85.53	53.96	0.00029	1.14652	1.4518
42	41.4430	422.8	0.10935	5.2174	0.05819	80.20	48.39	0.00028	1.13053	1.4846
44	36.7771	413.2	0.10154	4.1750	0.06034	75.23	44.10	0.00030	1.11527	1.4616
46	32.6747	416.4	0.09553	3.6962	0.05727	70.95	41.07	0.00033	1.10197	1.3804
48	29.3046	428.7	0.09147	3.5296	0.05137	67.63	39.07	0.00038	1.09113	1.2720
50	26.6002	446.5	0.08856	3.4940	0.04540	65.33	37.76	0.00044	1.08248	1.1715
52	24.4207	467.6	0.08631	3.5143	0.04020	63.82	36.91	0.00050	1.07554	1.0873
54	22.6331	490.6	0.08449	3.5550	0.03592	62.91	36.40	0.00057	1.06988	1.0196
56	21.1407	514.5	0.08297	3.5988	0.03244	62.44	36.13	0.00064	1.06517	0.9659
58	19.8723	539.1	0.08179	3.6482	0.02955	62.25	36.05	0.00071	1.06118	0.9223
60	18.7786	564.0	0.08072	3.6932	0.02715	62.38	36.10	0.00078	1.05774	0.8864
65	16.5932	627.6	0.07839	3.7886	0.02266	63.53	36.67	0.00097	1.05090	0.8208
70	14.9280	692.9	0.07632	3.8661	0.01950	64.95	37.59	0.00116	1.04571	0.7821
75	13.6221	758.7	0.07442	3.9234	0.01721	65.95	38.70	0.00133	1.04165	0.7662
80	12.5567	825.7	0.07254	3.9672	0.01544	67.35	39.92	0.00151	1.03835	0.7556
85	11.6663	894.4	0.07067	4.0010	0.01403	69.73	41.20	0.00171	1.03560	0.7416
90	10.9079	964.8	0.06882	4.0274	0.01289	71.88	42.51	0.00191	1.03326	0.7354
95	10.2521	1036.7	0.06704	4.0481	0.01194	73.87	43.83	0.00210	1.03123	0.7342
100	9.6782	1110.6	0.06530	4.0646	0.01113	80.62	46.74	0.00243	1.02947	0.7165
120	7.9469	1425.4	0.05864	4.1004	0.00881	97.18	54.43	0.00351	1.02415	0.7034
140	6.7673	1753.6	0.05391	4.1147	0.00735	110.06	59.53	0.00454	1.02054	0.6972
160	5.9050	2089.3	0.05045	4.1190	0.00633	120.79	63.39	0.00557	1.01791	0.6943
180	5.2440	2427.4	0.04790	4.1185	0.00557	130.14	66.63	0.00660	1.01589	0.6926
200	4.7197	2764.0	0.04603	4.1158	0.00498	138.51	69.52	0.00767	1.01429	0.6915
220	4.2927	3098.3	0.04461	4.1120	0.00451	146.17	72.21	0.00877	1.01299	0.6907
240	3.9379	3427.6	0.04356	4.1077	0.00413	153.23	74.78	0.00991	1.01192	0.6901
260	3.6382	3751.7	0.04277	4.1033	0.00380	159.82	77.27	0.01109	1.01100	0.6894
280	3.3814	4070.4	0.04218	4.0989	0.00353	166.05	79.72	0.01232	1.01023	0.6889
300	3.1584	4385.9	0.04173	4.0961	0.00329	171.97	82.13	0.01360	1.00955	0.6882
350	2.7134	5149.9	0.04107	4.0859	0.00281	185.84	88.09	0.01701	1.00820	0.6869
400	2.3790	5896.4	0.04073	4.0774	0.00246	198.92	93.96	0.02073	1.00719	0.6860
450	2.1183	6633.0	0.04055	4.0702	0.00219	211.57	99.78	0.02475	1.00640	0.6853
500	1.9092	7369.9	0.04039	4.0641	0.00197	224.11	105.54	0.02906	1.00576	0.6847
550	1.7378	8105.5	0.04027	4.0589	0.00179	236.49	111.24	0.03367	1.00525	0.6843
600	1.5947	8845.0	0.04014	4.0544	0.00165	248.83	116.87	0.03857	1.00481	0.6840
700	1.3693	10341.5	0.03985	4.0472	0.00141	273.50	127.95	0.04920	1.00413	0.6837
800	1.1998	11876.0	0.03944	4.0417	0.00124	298.42	138.76	0.06091	1.00362	0.6836
900	1.0677	13462.1	0.03893	4.0372	0.00110	323.83	149.32	0.07365	1.00322	0.6836
1000	0.9617	15108.6	0.03831	4.0336	0.00099	440.56	159.63	0.11004	1.00290	0.5430
1200	0.8025	18565.8	0.03696	4.0282	0.00083	511.01	179.59	0.14918	1.00242	0.5400
1400	0.6886	22252.9	0.03554	4.0242	0.00071	583.20	198.76	0.19297	1.00208	0.5385
1600	0.6029	26132.4	0.03421	4.0212	0.00062	656.59	217.23	0.24129	1.00182	0.5375
1800	0.5363	30192.4	0.03298	4.0188	0.00055	731.94	235.10	0.29431	1.00162	0.5363
2000	0.4828	34557.8	0.03170	4.0165	0.00050	814.69	252.45	0.35302	1.00146	0.5332
2500	0.3858	49546.3	0.02679	4.0056	0.00040	1159.33	294.12	0.54656	1.00116	0.5021
3000	0.3184	80853.3	0.01899	3.9615	0.00034	2096.71	335.42	0.87127	1.00096	0.4353

* TWO-PHASE BOUNDARY

Table 4.12 (continued)
THERMODYNAMIC PROPERTIES OF NORMAL HYDROGEN

5.00 MPA ISOBAR

TEMPERATURE	VOLUME	ISOTHERM DERIVATIVE	ISOCHORE DERIVATIVE	INTERNAL ENERGY	ENTHALPY	ENTROPY	CV	CP	VELOCITY OF SOUND
DEG. K	CU M/KG	CU M-MPA/KG	MPA/K	KJ/KG-K	KJ/KG	KJ/KG-K	KJ /	KG-K	M/S
* 15.382	0.01256	1.4746	0.9589	220.9	283.7	14.196	4.88	6.40	1390
16	0.01262	1.4419	0.9656	224.6	287.8	14.454	5.01	6.66	1384
17	0.01273	1.3771	0.9705	231.0	294.6	14.870	5.19	7.07	1370
18	0.01285	1.3108	0.9690	237.6	301.9	15.286	5.35	7.48	1354
19	0.01298	1.2480	0.9623	244.7	309.6	15.700	5.49	7.87	1337
20	0.01311	1.1937	0.9514	252.1	317.6	16.113	5.62	8.22	1322
21	0.01325	1.1291	0.9383	259.8	326.0	16.525	5.73	8.61	1302
22	0.01340	1.0703	0.9233	267.8	334.8	16.935	5.84	8.98	1284
23	0.01356	1.0116	0.9071	276.2	344.0	17.343	5.93	9.37	1264
24	0.01373	0.9542	0.8895	284.9	353.6	17.750	6.01	9.76	1245
25	0.01391	0.8976	0.8703	294.0	363.5	18.157	6.08	10.17	1225
26	0.01411	0.8409	0.8497	303.4	373.9	18.562	6.15	10.59	1204
27	0.01432	0.7852	0.8274	313.1	384.7	18.970	6.21	11.03	1181
28	0.01454	0.7307	0.8035	323.3	396.0	19.379	6.26	11.49	1158
29	0.01478	0.6782	0.7782	333.8	407.7	19.791	6.31	11.97	1134
30	0.01504	0.6262	0.7516	344.7	419.9	20.205	6.34	12.47	1109
31	0.01533	0.5776	0.7238	356.0	432.6	20.621	6.38	12.98	1084
32	0.01563	0.5319	0.6951	367.7	445.9	21.042	6.41	13.51	1059
33	0.01597	0.4873	0.6657	379.9	459.7	21.467	6.44	14.09	1033
34	0.01633	0.4434	0.6355	392.4	474.1	21.896	6.46	14.72	1005
35	0.01673	0.4034	0.6045	405.4	489.1	22.332	6.48	15.36	977
36	0.01717	0.3645	0.5731	419.0	504.8	22.775	6.51	16.07	949
37	0.01766	0.3303	0.5413	432.9	521.2	23.225	6.53	16.76	921
38	0.01820	0.2981	0.5095	447.4	538.4	23.682	6.55	17.50	893
39	0.01879	0.2690	0.4779	462.3	556.2	24.146	6.56	18.25	865
40	0.01945	0.2443	0.4470	477.6	574.9	24.617	6.57	18.95	839
42	0.02098	0.2041	0.3883	509.2	614.2	25.575	6.60	20.27	791
44	0.02281	0.1771	0.3356	541.7	655.7	26.542	6.64	21.20	752
46	0.02491	0.1615	0.2903	573.5	698.1	27.483	6.70	21.60	721
48	0.02724	0.1555	0.2525	604.9	741.1	28.398	6.73	21.34	702
50	0.02970	0.1565	0.2217	634.6	783.1	29.256	6.74	20.60	691
52	0.03222	0.1620	0.1967	662.2	823.3	30.045	6.74	19.63	687
54	0.03473	0.1706	0.1766	687.9	861.6	30.767	6.72	18.64	688
56	0.03721	0.1808	0.1604	712.1	898.2	31.433	6.71	17.74	691
58	0.03964	0.1920	0.1470	734.6	932.8	32.042	6.68	16.94	697
60	0.04202	0.2035	0.1358	755.9	966.0	32.604	6.66	16.27	705
65	0.04776	0.2333	0.1147	805.1	1043.9	33.851	6.64	15.00	726
70	0.05325	0.2631	0.0997	850.5	1116.7	34.930	6.65	14.16	748
75	0.05853	0.2918	0.0886	893.3	1186.0	35.887	6.69	13.60	770
80	0.06365	0.3204	0.0799	934.7	1252.9	36.751	6.76	13.21	791
85	0.06867	0.3483	0.0728	975.2	1318.5	37.547	6.84	12.94	812
90	0.07355	0.3752	0.0670	1015.0	1382.7	38.281	6.94	12.77	831
95	0.07834	0.4014	0.0621	1054.6	1446.3	38.968	7.05	12.66	849
100	0.08306	0.4271	0.0580	1094.2	1509.4	39.616	7.18	12.61	866
120	0.10133	0.5249	0.0459	1256.4	1763.1	41.926	7.78	12.73	927
140	0.11907	0.6184	0.0383	1424.9	2020.2	43.908	8.31	13.01	984
160	0.13648	0.7090	0.0329	1601.1	2283.5	45.664	8.77	13.32	1038
180	0.15367	0.7977	0.0289	1784.3	2552.6	47.249	9.15	13.60	1089
200	0.17071	0.8851	0.0258	1973.4	2827.0	48.694	9.46	13.83	1138
220	0.18764	0.9716	0.0233	2167.4	3105.6	50.022	9.70	14.03	1185
240	0.20449	1.0573	0.0213	2365.3	3387.7	51.250	9.89	14.18	1231
260	0.22128	1.1424	0.0196	2566.1	3672.5	52.390	10.03	14.29	1276
280	0.23802	1.2271	0.0181	2769.1	3959.2	53.452	10.14	14.38	1319
300	0.25478	1.3124	0.0169	2974.6	4248.5	54.448	10.22	14.43	1362
350	0.29639	1.5219	0.0144	3490.4	4972.3	56.681	10.32	14.51	1462
400	0.33788	1.7303	0.0126	4009.2	5698.6	58.620	10.37	14.53	1557
450	0.37930	1.9381	0.0112	4529.1	6425.6	60.332	10.38	14.54	1647
500	0.42067	2.1455	0.0100	5049.4	7152.8	61.864	10.40	14.55	1732
550	0.46201	2.3525	0.0091	5570.1	7880.2	63.251	10.41	14.55	1813
600	0.50333	2.5594	0.0083	6091.7	8608.3	64.515	10.43	14.57	1891
700	0.58593	2.9728	0.0072	7137.6	10067.3	66.767	10.48	14.62	2036
800	0.66849	3.3858	0.0063	8190.9	11533.4	68.727	10.57	14.70	2170
900	0.75103	3.7987	0.0056	9254.8	13009.9	70.464	10.70	14.83	2294
1000	0.83355	4.2114	0.0050	10332.3	14500.0	72.032	10.86	14.99	2411
1200	0.99857	5.0367	0.0042	12543.0	17535.9	74.780	11.24	15.37	2624
1400	1.16357	5.8618	0.0036	14835.9	20653.8	77.189	11.68	15.80	2816
1600	1.32856	6.6867	0.0031	17214.8	23857.6	79.335	12.12	16.25	2994
1800	1.49358	7.5117	0.0028	19681.2	27149.1	81.267	12.56	16.69	3159
2000	1.65871	8.3366	0.0025	22242.5	30536.1	83.062	13.05	17.19	3314
2500	2.07467	10.3988	0.0020	29239.1	39612.4	87.089	15.29	19.58	3649
3000	2.51137	12.4609	0.0017	38263.4	50820.3	91.173	21.24	26.28	3924

* TWO-PHASE BOUNDARY

Table 4.12 (continued)
THERMODYNAMIC PROPERTIES OF NORMAL HYDROGEN

5.00 MPA ISOBAR

TEMPERATURE DEG. K	DENSITY KG/CU M	$V(DH/DV)_P$ KJ/KG	$V(DP/DU)_V$ MPA-CU M/KJ X 10^2	$-V(DP/DV)_T$ MPA	$(DV/DT)_P/V$ 1/DEG. K	THERMAL CONDUCTIVITY W/K-M X 10^3	VISCOSITY KG/M-S X 10^7	THERMAL DIFFUSIVITY SQ M/HR	DIELECTRIC CONSTANT	PRANDTL NUMBER
* 15.382	79.6358	783.3	0.24654	117.4329	0.00817	91.17	285.79	0.00064	1.26081	2.0050
16	79.2264	787.3	0.24343	114.2398	0.00845	95.52	265.42	0.00065	1.25937	1.8493
17	78.5408	788.2	0.23818	108.1565	0.00897	100.86	237.88	0.00065	1.25695	1.6680
18	77.8155	787.1	0.23283	101.9981	0.00950	104.95	215.35	0.00065	1.25439	1.5345
19	77.0622	786.1	0.22740	96.1733	0.01001	107.77	196.75	0.00064	1.25174	1.4359
20	76.2761	787.1	0.22203	91.0477	0.01045	110.42	181.10	0.00063	1.24897	1.3490
21	75.4646	781.7	0.21690	85.2096	0.01101	113.42	167.84	0.00063	1.24612	1.2738
22	74.6146	777.0	0.21204	79.8570	0.01156	115.66	156.35	0.00062	1.24314	1.2145
23	73.7364	770.4	0.20750	74.5884	0.01216	117.28	146.37	0.00061	1.24007	1.1694
24	72.8222	762.8	0.20319	69.4868	0.01280	118.35	137.55	0.00060	1.23688	1.1348
25	71.8715	753.7	0.19904	64.5099	0.01349	118.95	129.69	0.00059	1.23356	1.1087
26	70.8805	743.0	0.19495	59.6058	0.01425	119.08	122.61	0.00057	1.23011	1.0906
27	69.8496	731.3	0.19081	54.8474	0.01509	118.84	116.19	0.00056	1.22653	1.0786
28	68.7720	718.7	0.18661	50.2511	0.01599	118.27	110.29	0.00054	1.22279	1.0717
29	67.6476	705.4	0.18239	45.8796	0.01696	117.41	104.84	0.00052	1.21890	1.0684
30	66.4752	690.6	0.17822	41.6295	0.01805	116.30	99.78	0.00051	1.21485	1.0697
31	65.2503	676.0	0.17392	37.6892	0.01920	114.51	95.05	0.00049	1.21063	1.0775
32	63.9697	661.5	0.16954	34.0273	0.02043	112.55	90.59	0.00047	1.20623	1.0876
33	62.6318	645.8	0.16514	30.5213	0.02181	110.43	86.38	0.00045	1.20164	1.1018
34	61.2349	628.9	0.16063	27.1527	0.02341	108.22	82.39	0.00043	1.19686	1.1207
35	59.7690	612.6	0.15598	24.1108	0.02507	105.87	78.58	0.00042	1.19186	1.1400
36	58.2355	595.3	0.15124	21.2278	0.02700	103.41	74.94	0.00040	1.18664	1.1647
37	56.6288	579.2	0.14640	18.7024	0.02894	100.85	71.46	0.00038	1.18120	1.1877
38	54.9544	562.8	0.14161	16.3795	0.03110	98.22	68.14	0.00037	1.17554	1.2143
39	53.2122	546.8	0.13691	14.3153	0.03339	95.61	64.97	0.00035	1.16967	1.2406
40	51.4071	532.5	0.13227	12.5594	0.03559	93.44	61.98	0.00035	1.16360	1.2570
42	47.6583	507.6	0.12338	9.7251	0.03993	89.04	56.55	0.00033	1.15109	1.2872
44	43.8420	490.4	0.11527	7.7629	0.04323	84.74	51.94	0.00033	1.13844	1.2995
46	40.1370	482.4	0.10789	6.4812	0.04479	80.80	48.24	0.00034	1.12625	1.2898
48	36.7078	482.5	0.10215	5.7089	0.04424	77.32	45.39	0.00036	1.11504	1.2530
50	33.6647	489.5	0.09763	5.2680	0.04208	74.45	43.31	0.00039	1.10517	1.1983
52	31.0368	501.8	0.09406	5.0278	0.03912	72.20	41.84	0.00043	1.09669	1.1375
54	28.7934	518.3	0.09122	4.9125	0.03596	70.55	40.81	0.00047	1.08949	1.0781
56	26.8763	537.3	0.08896	4.8583	0.03301	69.39	40.11	0.00052	1.08336	1.0252
58	25.2271	558.1	0.08717	4.8432	0.03034	68.61	39.64	0.00058	1.07811	0.9786
60	23.7978	580.1	0.08565	4.8435	0.02804	68.24	39.37	0.00063	1.07357	0.9385
65	20.9367	638.9	0.08246	4.8841	0.02348	68.44	39.24	0.00078	1.06453	0.8603
70	18.7801	701.4	0.07981	4.9402	0.02019	69.20	39.67	0.00094	1.05775	0.8118
75	17.0856	765.6	0.07743	4.9860	0.01776	69.56	40.43	0.00108	1.05244	0.7904
80	15.7118	832.6	0.07524	5.0342	0.01587	70.61	41.39	0.00122	1.04815	0.7743
85	14.5616	901.5	0.07309	5.0718	0.01435	72.74	42.47	0.00139	1.04457	0.7555
90	13.5961	972.0	0.07098	5.1010	0.01313	74.69	43.63	0.00155	1.04157	0.7456
95	12.7650	1044.0	0.06902	5.1242	0.01213	76.52	44.83	0.00170	1.03900	0.7419
100	12.0401	1118.1	0.06712	5.1426	0.01128	82.25	47.31	0.00195	1.03675	0.7255
120	9.8688	1435.7	0.05983	5.1798	0.00887	98.33	54.78	0.00282	1.03005	0.7091
140	8.3984	1765.6	0.05483	5.1932	0.00737	110.97	59.80	0.00366	1.02554	0.7012
160	7.3272	2103.3	0.05119	5.1949	0.00633	121.55	63.63	0.00448	1.02225	0.6972
180	6.5075	2443.2	0.04852	5.1912	0.00557	130.81	66.84	0.00532	1.01975	0.6949
200	5.8579	2781.6	0.04656	5.1850	0.00497	139.10	69.72	0.00618	1.01776	0.6933
220	5.3293	3117.4	0.04507	5.1778	0.00450	146.72	72.40	0.00707	1.01615	0.6922
240	4.8902	3448.1	0.04397	5.1703	0.00411	153.76	74.97	0.00798	1.01481	0.6913
260	4.5191	3773.5	0.04314	5.1628	0.00379	160.35	77.47	0.00894	1.01368	0.6905
280	4.2013	4093.3	0.04251	5.1556	0.00351	166.57	79.92	0.00993	1.01272	0.6898
300	3.9249	4410.6	0.04203	5.1512	0.00327	172.50	82.34	0.01096	1.01188	0.6889
350	3.3740	5176.2	0.04132	5.1348	0.00280	186.41	88.32	0.01371	1.01020	0.6874
400	2.9596	5923.7	0.04095	5.1212	0.00245	199.55	94.22	0.01670	1.00895	0.6863
450	2.6365	6661.1	0.04074	5.1097	0.00218	212.26	100.07	0.01994	1.00797	0.6854
500	2.3772	7398.6	0.04056	5.1001	0.00197	224.87	105.86	0.02341	1.00718	0.6848
550	2.1644	8134.6	0.04042	5.0919	0.00179	237.32	111.60	0.02712	1.00654	0.6844
600	1.9868	8874.4	0.04028	5.0849	0.00164	249.74	117.27	0.03106	1.00600	0.6840
700	1.7067	10371.5	0.03996	5.0736	0.00141	274.56	128.42	0.03962	1.00515	0.6837
800	1.4959	11906.4	0.03955	5.0649	0.00123	299.63	139.30	0.04904	1.00451	0.6835
900	1.3315	13492.9	0.03902	5.0580	0.00110	325.18	149.93	0.05930	1.00402	0.6836
1000	1.1997	15140.0	0.03839	5.0524	0.00099	440.56	160.30	0.08821	1.00362	0.5454
1200	1.0014	18598.1	0.03702	5.0439	0.00083	511.01	180.39	0.11955	1.00302	0.5425
1400	0.8594	22286.1	0.03560	5.0377	0.00071	583.20	199.68	0.15461	1.00259	0.5410
1600	0.7527	26165.0	0.03425	5.0331	0.00062	656.53	218.26	0.19328	1.00227	0.5401
1800	0.6695	30217.6	0.03303	5.0293	0.00055	731.54	236.25	0.23568	1.00202	0.5390
2000	0.6029	34552.6	0.03178	5.0259	0.00050	813.09	253.70	0.28249	1.00182	0.5363
2500	0.4820	49069.9	0.02714	5.0123	0.00040	1139.77	295.60	0.43478	1.00145	0.5078
3000	0.3982	78243.2	0.01970	4.9618	0.00034	1996.91	336.92	0.68695	1.00120	0.4434

* TWO-PHASE BOUNDARY

Table 4.12 (continued)
THERMODYNAMIC PROPERTIES OF NORMAL HYDROGEN

6.00 MPA ISOBAR

TEMPERATURE DEG. K	VOLUME CU M/KG	ISOTHERM DERIVATIVE CU M-MPA/KG	ISOCHORE DERIVATIVE MPA/K	INTERNAL ENERGY KJ/KG-K	ENTHALPY KJ/KG	ENTROPY KJ/KG-K	CV KJ/KG-K	CP KJ/KG-K	VELOCITY OF SOUND M/S
* 15.678	0.01248	1.5352	0.9665	221.7	296.6	14.218	4.92	6.40	1414
16	0.01252	1.5184	0.9702	223.6	298.7	14.350	4.98	6.54	1411
17	0.01262	1.4521	0.9773	229.7	305.4	14.759	5.17	6.95	1397
18	0.01273	1.3891	0.9784	236.2	312.6	15.167	5.33	7.34	1383
19	0.01285	1.3275	0.9746	243.0	320.1	15.575	5.48	7.72	1368
20	0.01297	1.2678	0.9665	250.2	328.0	15.980	5.61	8.09	1352
21	0.01310	1.2171	0.9551	257.6	336.3	16.385	5.72	8.43	1339
22	0.01324	1.1548	0.9421	265.4	344.9	16.786	5.83	8.79	1320
23	0.01339	1.0975	0.9270	273.5	353.9	17.185	5.92	9.15	1302
24	0.01355	1.0404	0.9107	281.9	363.2	17.582	6.00	9.52	1284
25	0.01371	0.9847	0.8930	290.6	372.9	17.978	6.08	9.89	1265
26	0.01389	0.9301	0.8740	299.6	383.0	18.371	6.15	10.26	1246
27	0.01407	0.8757	0.8536	309.0	393.4	18.766	6.21	10.66	1226
28	0.01427	0.8223	0.8318	318.6	404.3	19.160	6.26	11.06	1205
29	0.01449	0.7702	0.8086	328.6	415.5	19.556	6.31	11.48	1183
30	0.01471	0.7202	0.7842	338.9	427.2	19.952	6.35	11.90	1161
31	0.01496	0.6722	0.7588	349.6	439.3	20.349	6.39	12.33	1139
32	0.01522	0.6260	0.7325	360.6	451.9	20.748	6.42	12.78	1116
33	0.01550	0.5802	0.7055	371.9	464.9	21.148	6.45	13.25	1092
34	0.01580	0.5386	0.6782	383.6	478.4	21.551	6.48	13.73	1068
35	0.01613	0.4990	0.6504	395.6	492.4	21.956	6.50	14.22	1045
36	0.01648	0.4609	0.6222	407.9	506.8	22.364	6.52	14.74	1021
37	0.01687	0.4244	0.5936	420.7	521.8	22.775	6.54	15.28	996
38	0.01728	0.3915	0.5651	433.7	537.4	23.190	6.55	15.81	972
39	0.01773	0.3604	0.5364	447.1	553.5	23.608	6.56	16.35	948
40	0.01822	0.3333	0.5082	460.8	570.1	24.028	6.57	16.86	925
42	0.01931	0.2862	0.4536	488.9	604.8	24.875	6.59	17.85	881
44	0.02058	0.2504	0.4027	517.9	641.4	25.725	6.61	18.69	841
46	0.02204	0.2252	0.3568	546.7	678.9	26.558	6.67	19.30	807
48	0.02366	0.2085	0.3165	575.9	717.8	27.387	6.70	19.61	781
50	0.02542	0.2003	0.2818	604.5	757.0	28.187	6.73	19.54	763
52	0.02728	0.1977	0.2523	632.1	795.8	28.948	6.74	19.20	750
54	0.02921	0.1995	0.2273	658.5	833.7	29.663	6.74	18.67	743
56	0.03116	0.2049	0.2063	683.7	870.7	30.336	6.74	18.03	741
58	0.03311	0.2122	0.1887	707.4	906.0	30.957	6.72	17.38	741
60	0.03505	0.2211	0.1738	729.9	940.2	31.536	6.70	16.78	744
65	0.03981	0.2460	0.1456	782.0	1020.8	32.826	6.68	15.55	757
70	0.04442	0.2737	0.1256	829.7	1096.2	33.944	6.69	14.65	774
75	0.04887	0.3015	0.1109	874.5	1167.8	34.932	6.72	14.03	793
80	0.05321	0.3291	0.0994	917.5	1236.8	35.822	6.78	13.59	812
85	0.05743	0.3565	0.0903	959.3	1303.8	36.637	6.86	13.28	831
90	0.06156	0.3834	0.0829	1000.3	1369.7	37.390	6.96	13.08	848
95	0.06562	0.4098	0.0766	1040.9	1434.7	38.092	7.07	12.94	866
100	0.06965	0.4362	0.0713	1081.4	1499.3	38.755	7.20	12.85	882
120	0.08506	0.5350	0.0560	1246.6	1757.0	41.103	7.80	12.89	940
140	0.09999	0.6296	0.0464	1416.8	2016.7	43.105	8.33	13.12	996
160	0.11460	0.7212	0.0398	1594.2	2281.9	44.873	8.79	13.40	1049
180	0.12901	0.8107	0.0349	1778.4	2552.5	46.467	9.16	13.66	1099
200	0.14328	0.8987	0.0311	1968.3	2828.0	47.919	9.47	13.89	1148
220	0.15745	0.9856	0.0281	2162.9	3107.6	49.251	9.71	14.07	1195
240	0.17154	1.0718	0.0256	2361.3	3390.5	50.482	9.90	14.21	1241
260	0.18557	1.1573	0.0235	2562.5	3675.9	51.625	10.04	14.32	1285
280	0.19955	1.2423	0.0218	2765.9	3963.2	52.689	10.15	14.40	1328
300	0.21357	1.3282	0.0203	2971.8	4253.2	53.688	10.22	14.45	1370
350	0.24829	1.5381	0.0173	3488.3	4978.0	55.923	10.33	14.52	1471
400	0.28290	1.7469	0.0151	4007.5	5704.9	57.864	10.37	14.54	1565
450	0.31744	1.9548	0.0134	4527.7	6432.4	59.577	10.39	14.55	1655
500	0.35194	2.1623	0.0120	5048.3	7159.9	61.110	10.40	14.55	1739
550	0.38640	2.3695	0.0109	5569.2	7887.6	62.497	10.41	14.56	1820
600	0.42085	2.5765	0.0100	6090.9	8615.9	63.762	10.43	14.57	1897
700	0.48969	2.9900	0.0086	7137.1	10075.2	66.015	10.48	14.62	2042
800	0.55850	3.4031	0.0075	8190.5	11541.5	67.974	10.57	14.70	2175
900	0.62729	3.8160	0.0067	9254.5	13018.2	69.711	10.70	14.83	2300
1000	0.69606	4.2287	0.0060	10332.0	14508.4	71.279	10.86	14.99	2416
1200	0.83359	5.0540	0.0050	12542.9	17544.4	74.028	11.24	15.37	2628
1400	0.97109	5.8791	0.0043	14835.8	20662.4	76.437	11.68	15.80	2821
1600	1.10859	6.7041	0.0038	17214.7	23866.2	78.583	12.12	16.25	2998
1800	1.24610	7.5290	0.0033	19680.8	27157.4	80.515	12.56	16.69	3163
2000	1.38370	8.3539	0.0030	22240.3	30542.5	82.309	13.04	17.17	3317
2500	1.73009	10.4161	0.0024	29205.4	39585.9	86.322	15.15	19.42	3654
3000	2.09227	12.4783	0.0020	38052.4	50606.0	90.339	20.64	25.59	3932

* TWO-PHASE BOUNDARY

Table 4.12 (continued)
THERMODYNAMIC PROPERTIES OF NORMAL HYDROGEN

6.00 MPA ISOBAR

TEMPERATURE DEG. K	DENSITY KG/CU M	$V(DH/DV)_P$ KJ/KG	$V(DP/DU)_V$ MPA-CU M/KJ X 10^2	$-V(DP/DV)_T$ MPA	$(DV/DT)_P/V$ 1/DEG. K	THERMAL CONDUCTIVITY W/K-M X 10^3	VISCOSITY KG/M-S X 10^7	THERMAL DIFFUSIVITY SQ M/HR	DIELECTRIC CONSTANT	PRANDTL NUMBER
* 15.678	80.1084	815.0	0.24531	122.9816	0.00786	94.38	291.98	0.00066	1.26248	1.9813
16	79.9026	817.4	0.24368	121.3254	0.00800	96.62	280.92	0.00067	1.26176	1.9004
17	79.2468	818.3	0.23857	115.0744	0.00849	102.22	251.31	0.00067	1.25944	1.7087
18	78.5592	819.2	0.23347	109.1289	0.00897	106.54	227.24	0.00066	1.25701	1.5666
19	77.8398	819.0	0.22846	103.3346	0.00943	109.56	207.34	0.00066	1.25447	1.4619
20	77.0922	818.0	0.22351	97.7382	0.00989	112.41	190.68	0.00065	1.25184	1.3721
21	76.3144	819.4	0.21868	92.8849	0.01028	115.60	176.53	0.00065	1.24911	1.2867
22	75.5170	813.9	0.21407	87.2047	0.01080	118.06	164.45	0.00064	1.24631	1.2247
23	74.6827	809.0	0.20964	81.9628	0.01131	119.87	153.88	0.00063	1.24338	1.1746
24	73.8252	802.5	0.20543	76.8050	0.01186	121.15	144.65	0.00062	1.24038	1.1361
25	72.9346	795.0	0.20141	71.8215	0.01243	121.95	136.45	0.00061	1.23727	1.1061
26	72.0101	786.6	0.19746	66.9741	0.01305	122.28	129.09	0.00060	1.23404	1.0836
27	71.0541	776.9	0.19351	62.2237	0.01372	122.26	122.45	0.00058	1.23072	1.0675
28	70.0607	766.3	0.18952	57.6127	0.01444	121.91	116.40	0.00057	1.22726	1.0564
29	69.0310	754.8	0.18553	53.1672	0.01521	121.28	110.84	0.00055	1.22369	1.0492
30	67.9619	742.7	0.18162	48.9441	0.01602	120.41	105.70	0.00054	1.21999	1.0447
31	66.8519	730.3	0.17760	44.9390	0.01688	118.86	100.92	0.00052	1.21615	1.0470
32	65.7014	717.4	0.17355	41.1262	0.01781	117.16	96.46	0.00050	1.21218	1.0520
33	64.5097	703.2	0.16949	37.4312	0.01885	115.32	92.27	0.00049	1.20808	1.0604
34	63.2732	689.9	0.16549	34.0776	0.01990	113.39	88.31	0.00047	1.20384	1.0694
35	61.9922	676.4	0.16143	30.9365	0.02103	111.36	84.57	0.00045	1.19945	1.0800
36	60.6643	662.3	0.15728	27.9606	0.02225	109.25	81.02	0.00044	1.19491	1.0930
37	59.2908	647.7	0.15309	25.1656	0.02359	107.05	77.65	0.00043	1.19023	1.1082
38	57.8698	633.8	0.14897	22.6546	0.02494	104.80	74.43	0.00041	1.18540	1.1229
39	56.4047	619.6	0.14490	20.3267	0.02639	102.50	71.38	0.00040	1.18044	1.1388
40	54.8979	606.9	0.14089	18.2964	0.02778	100.18	68.49	0.00039	1.17534	1.1523
42	51.7818	583.1	0.13298	14.8206	0.03061	95.95	63.18	0.00037	1.16486	1.1752
44	48.5797	564.4	0.12542	12.1657	0.03310	92.15	58.53	0.00037	1.15415	1.1869
46	45.3752	552.6	0.11793	10.2169	0.03492	88.53	54.57	0.00036	1.14351	1.1896
48	42.2687	546.1	0.11168	8.8143	0.03591	85.23	51.32	0.00037	1.13325	1.1808
50	39.3388	546.3	0.10643	7.8782	0.03577	82.29	48.71	0.00039	1.12363	1.1568
52	36.6521	551.5	0.10207	7.2446	0.03482	79.79	46.70	0.00041	1.11486	1.1240
54	34.2378	561.2	0.09842	6.8315	0.03327	77.76	45.19	0.00044	1.10702	1.0849
56	32.0953	574.7	0.09542	6.5765	0.03137	76.15	44.08	0.00047	1.10010	1.0434
58	30.2035	590.5	0.09300	6.4089	0.02944	74.92	43.27	0.00051	1.09401	1.0040
60	28.5293	608.7	0.09094	6.3073	0.02756	74.12	42.71	0.00056	1.08864	0.9666
65	25.1194	660.2	0.08678	6.1803	0.02355	73.45	42.01	0.00068	1.07776	0.8892
70	22.5144	718.5	0.08347	6.1615	0.02039	73.55	41.98	0.00080	1.06951	0.8363
75	20.4609	780.4	0.08060	6.1681	0.01797	73.31	42.37	0.00092	1.06303	0.8108
80	18.7948	845.2	0.07801	6.1862	0.01608	73.97	43.05	0.00104	1.05779	0.7907
85	17.4128	912.7	0.07558	6.2078	0.01455	75.85	43.91	0.00118	1.05346	0.7688
90	16.2435	982.6	0.07326	6.2279	0.01331	77.59	44.89	0.00132	1.04981	0.7565
95	15.2387	1054.3	0.07110	6.2454	0.01227	79.23	45.96	0.00145	1.04668	0.7505
100	14.3578	1128.7	0.06898	6.2622	0.01138	84.07	48.01	0.00164	1.04394	0.7337
120	11.7562	1448.3	0.06100	6.2901	0.00890	99.64	55.24	0.00237	1.03588	0.7143
140	10.0012	1779.5	0.05575	6.2970	0.00737	111.99	60.15	0.00307	1.03046	0.7048
160	8.7257	2118.7	0.05193	6.2927	0.00633	122.40	63.92	0.00377	1.02654	0.6999
180	7.7511	2460.2	0.04914	6.2834	0.00555	131.53	67.09	0.00447	1.02355	0.6970
200	6.9792	2800.1	0.04708	6.2720	0.00496	139.75	69.95	0.00519	1.02119	0.6951
220	6.3512	3137.4	0.04553	6.2600	0.00449	147.31	72.62	0.00593	1.01927	0.6937
240	5.8296	3469.3	0.04437	6.2480	0.00410	154.32	75.18	0.00670	1.01768	0.6925
260	5.3888	3795.9	0.04350	6.2364	0.00377	160.89	77.68	0.00750	1.01633	0.6915
280	5.0112	4116.7	0.04284	6.2255	0.00350	167.11	80.13	0.00834	1.01518	0.6906
300	4.6823	4435.8	0.04234	6.2192	0.00326	173.04	82.56	0.00920	1.01418	0.6896
350	4.0275	5202.9	0.04157	6.1949	0.00279	186.98	88.55	0.01151	1.01219	0.6878
400	3.5348	5951.3	0.04116	6.1748	0.00244	200.17	94.48	0.01402	1.01069	0.6865
450	3.1502	6689.4	0.04093	6.1581	0.00217	212.95	100.36	0.01673	1.00952	0.6856
500	2.8414	7427.3	0.04073	6.1441	0.00196	225.63	106.19	0.01964	1.00859	0.6849
550	2.5880	8163.7	0.04058	6.1323	0.00178	238.15	111.96	0.02276	1.00782	0.6844
600	2.3762	8903.8	0.04042	6.1222	0.00164	250.64	117.67	0.02606	1.00718	0.6841
700	2.0421	10401.4	0.04008	6.1058	0.00141	275.61	128.88	0.03324	1.00617	0.6836
800	1.7905	11936.8	0.03965	6.0932	0.00123	300.83	139.84	0.04113	1.00541	0.6835
900	1.5942	13523.8	0.03911	6.0833	0.00110	326.53	150.53	0.04973	1.00481	0.6836
1000	1.4367	15171.3	0.03847	6.0752	0.00099	440.56	160.98	0.07365	1.00434	0.5477
1200	1.1996	18630.4	0.03709	6.0630	0.00082	511.01	181.19	0.09979	1.00362	0.5449
1400	1.0298	22319.3	0.03565	6.0541	0.00071	583.19	200.60	0.12903	1.00311	0.5435
1600	0.9020	26198.0	0.03430	6.0474	0.00062	656.48	219.30	0.16127	1.00272	0.5427
1800	0.8025	30245.4	0.03307	6.0421	0.00055	731.25	237.40	0.19659	1.00242	0.5417
2000	0.7227	34558.1	0.03184	6.0374	0.00050	811.91	254.96	0.23553	1.00218	0.5392
2500	0.5780	48728.1	0.02741	6.0206	0.00040	1125.33	297.09	0.36084	1.00174	0.5128
3000	0.4780	76323.3	0.02028	5.9640	0.00034	1923.06	338.48	0.56595	1.00144	0.4505

* TWO-PHASE BOUNDARY

Table 4.12 (continued)
THERMODYNAMIC PROPERTIES OF NORMAL HYDROGEN

7.00 MPA ISOBAR

TEMPERATURE DEG. K	VOLUME CU M/KG	ISOTHERM DERIVATIVE CU M-MPA/KG	ISOCHORE DERIVATIVE MPA/K	INTERNAL ENERGY KJ/KG-K	ENTHALPY KJ/KG	ENTROPY KJ/KG-K	CV KJ / KG-K	CP KJ / KG-K	VELOCITY OF SOUND M/S
* 15.968	0.01241	1.5924	0.9744	222.5	309.4	14.240	4.95	6.42	1437
16	0.01242	1.5910	0.9748	222.7	309.6	14.253	4.96	6.43	1437
17	0.01251	1.5282	0.9838	228.6	316.2	14.656	5.15	6.84	1424
18	0.01262	1.4642	0.9869	234.9	323.2	15.057	5.32	7.22	1410
19	0.01273	1.4027	0.9850	241.6	330.7	15.457	5.47	7.60	1396
20	0.01284	1.3450	0.9792	248.5	338.4	15.856	5.60	7.95	1382
21	0.01297	1.2874	0.9699	255.8	346.6	16.254	5.72	8.30	1367
22	0.01310	1.2395	0.9580	263.3	355.0	16.648	5.82	8.62	1354
23	0.01323	1.1794	0.9449	271.2	363.8	17.039	5.91	8.96	1337
24	0.01338	1.1240	0.9297	279.3	372.9	17.428	6.00	9.30	1320
25	0.01353	1.0684	0.9134	287.7	382.4	17.815	6.08	9.65	1303
26	0.01369	1.0143	0.8957	296.4	392.2	18.198	6.15	10.00	1285
27	0.01386	0.9618	0.8768	305.4	402.4	18.583	6.21	10.36	1267
28	0.01404	0.9094	0.8568	314.7	413.0	18.965	6.27	10.72	1247
29	0.01423	0.8584	0.8355	324.2	423.9	19.348	6.32	11.10	1228
30	0.01444	0.8083	0.8129	334.1	435.1	19.730	6.36	11.47	1207
31	0.01465	0.7604	0.7894	344.2	446.8	20.112	6.40	11.86	1187
32	0.01488	0.7147	0.7650	354.7	458.8	20.495	6.44	12.24	1166
33	0.01513	0.6708	0.7400	365.4	471.3	20.877	6.47	12.63	1145
34	0.01539	0.6287	0.7144	376.4	484.1	21.261	6.50	13.03	1123
35	0.01567	0.5885	0.6887	387.7	497.3	21.644	6.52	13.44	1102
36	0.01596	0.5506	0.6628	399.3	511.0	22.029	6.54	13.86	1080
37	0.01628	0.5134	0.6367	411.1	525.1	22.415	6.56	14.30	1058
38	0.01662	0.4800	0.6104	423.2	539.6	22.802	6.57	14.72	1037
39	0.01698	0.4485	0.5841	435.6	554.5	23.190	6.58	15.14	1016
40	0.01737	0.4194	0.5579	448.3	569.9	23.578	6.59	15.55	995
42	0.01823	0.3681	0.5067	474.2	601.8	24.356	6.60	16.33	955
44	0.01920	0.3261	0.4581	500.7	635.2	25.132	6.61	17.05	917
46	0.02030	0.2935	0.4130	527.3	669.4	25.892	6.67	17.68	882
48	0.02152	0.2706	0.3721	554.5	705.1	26.653	6.70	18.07	854
50	0.02285	0.2544	0.3357	581.6	741.5	27.396	6.73	18.28	831
52	0.02427	0.2442	0.3036	608.3	778.1	28.114	6.75	18.31	814
54	0.02576	0.2400	0.2758	634.2	814.6	28.802	6.76	18.11	802
56	0.02731	0.2398	0.2516	659.6	850.7	29.461	6.76	17.78	794
58	0.02889	0.2428	0.2307	683.7	885.9	30.078	6.74	17.35	790
60	0.03048	0.2480	0.2127	706.8	920.1	30.658	6.73	16.89	789
65	0.03445	0.2667	0.1776	760.7	1001.8	31.965	6.71	15.83	793
70	0.03836	0.2901	0.1526	810.3	1078.8	33.106	6.71	14.99	805
75	0.04218	0.3157	0.1341	856.7	1152.0	34.117	6.75	14.35	819
80	0.04592	0.3420	0.1198	901.1	1222.6	35.028	6.81	13.89	835
85	0.04957	0.3686	0.1085	944.1	1291.1	35.860	6.89	13.55	852
90	0.05315	0.3949	0.0992	986.2	1358.3	36.628	6.99	13.33	868
95	0.05667	0.4211	0.0915	1027.8	1424.4	37.343	7.09	13.17	884
100	0.06013	0.4469	0.0850	1069.0	1490.0	38.015	7.22	13.07	900
120	0.07350	0.5464	0.0662	1237.1	1751.6	40.400	7.83	13.03	954
140	0.08640	0.6418	0.0547	1408.9	2013.7	42.420	8.35	13.23	1008
160	0.09901	0.7340	0.0468	1587.6	2280.7	44.201	8.80	13.48	1060
180	0.11143	0.8241	0.0410	1772.7	2552.7	45.803	9.18	13.73	1110
200	0.12371	0.9126	0.0365	1963.3	2829.3	47.260	9.48	13.94	1158
220	0.13590	1.0000	0.0329	2158.5	3109.8	48.597	9.72	14.11	1205
240	0.14801	1.0865	0.0300	2357.4	3393.4	49.831	9.91	14.25	1250
260	0.16007	1.1724	0.0275	2559.0	3679.5	50.976	10.05	14.35	1294
280	0.17208	1.2577	0.0255	2762.8	3967.3	52.043	10.15	14.43	1337
300	0.18414	1.3443	0.0237	2969.1	4258.1	53.043	10.23	14.48	1379
350	0.21394	1.5545	0.0202	3486.2	4983.8	55.281	10.33	14.54	1479
400	0.24363	1.7635	0.0176	4005.8	5711.2	57.225	10.38	14.56	1573
450	0.27326	1.9716	0.0156	4526.4	6439.2	58.939	10.39	14.55	1662
500	0.30284	2.1793	0.0141	5047.2	7167.1	60.472	10.40	14.56	1746
550	0.33239	2.3865	0.0128	5568.3	7895.0	61.860	10.42	14.56	1827
600	0.36192	2.5936	0.0117	6090.1	8623.6	63.125	10.43	14.57	1903
700	0.42095	3.0071	0.0100	7136.5	10083.1	65.378	10.49	14.62	2048
800	0.47993	3.4203	0.0088	8190.1	11549.6	67.338	10.57	14.71	2181
900	0.53890	3.8332	0.0078	9254.2	13026.5	69.075	10.70	14.83	2305
1000	0.59785	4.2460	0.0070	10331.8	14516.8	70.643	10.86	14.99	2421
1200	0.71574	5.0713	0.0058	12542.7	17552.9	73.392	11.24	15.37	2633
1400	0.83360	5.8964	0.0050	14835.8	20671.0	75.801	11.68	15.80	2825
1600	0.95146	6.7213	0.0044	17214.6	23874.8	77.947	12.12	16.25	3001
1800	1.06933	7.5463	0.0039	19680.4	27165.7	79.879	12.56	16.68	3166
2000	1.18726	8.3712	0.0035	22238.7	30549.5	81.672	13.02	17.16	3321
2500	1.48400	10.4334	0.0028	29179.3	39567.3	85.674	15.04	19.30	3659
3000	1.79328	12.4955	0.0023	37888.4	50441.4	89.639	20.17	25.06	3938

* TWO-PHASE BOUNDARY

Table 4.12 (continued)
THERMODYNAMIC PROPERTIES OF NORMAL HYDROGEN

7.00 MPA ISOBAR

TEMPERATURE DEG. K	DENSITY KG/CU M	$V(DH/DV)_P$ KJ/KG	$V(DP/DU)_V$ MPA-CU M/KJ X 10^2	$-V(DP/DV)_T$ MPA	$(DV/DT)_P/V$ 1/DEG. K	THERMAL CONDUCTIVITY W/K-M X 10^3	VISCOSITY KG/M-S X 10^7	THERMAL DIFFUSIVITY SQ M/HR	DIELECTRIC CONSTANT	PRANDTL NUMBER
* 15.968	80.5666	844.9	0.24430	128.2933	0.00760	97.46	298.16	0.00068	1.26411	1.9633
16	80.5475	845.3	0.24413	128.1504	0.00761	97.68	297.03	0.00068	1.26404	1.9553
17	79.9143	848.5	0.23909	122.1251	0.00806	103.51	265.12	0.00068	1.26180	1.7507
18	79.2588	849.6	0.23410	116.0489	0.00850	108.05	239.40	0.00068	1.25948	1.6008
19	78.5724	850.1	0.22924	110.2150	0.00894	111.26	218.17	0.00067	1.25706	1.4898
20	77.8601	850.4	0.22459	104.7199	0.00935	114.31	200.45	0.00066	1.25455	1.3945
21	77.1160	849.3	0.22001	99.2809	0.00977	117.69	185.39	0.00066	1.25193	1.3069
22	76.3500	851.1	0.21557	94.6381	0.01012	120.32	172.53	0.00066	1.24923	1.2354
23	75.5655	845.5	0.21141	89.1232	0.01060	122.33	161.46	0.00065	1.24648	1.1831
24	74.7481	840.7	0.20730	84.0149	0.01107	123.78	151.72	0.00064	1.24361	1.1404
25	73.9086	834.2	0.20338	78.9602	0.01157	124.76	143.14	0.00063	1.24067	1.1072
26	73.0392	827.1	0.19956	74.0861	0.01209	125.27	135.47	0.00062	1.23764	1.0814
27	72.1426	819.5	0.19575	69.3888	0.01264	125.42	128.58	0.00060	1.23451	1.0617
28	71.2151	810.7	0.19194	64.7641	0.01323	125.26	122.32	0.00059	1.23128	1.0473
29	70.2596	801.1	0.18816	60.3127	0.01385	124.82	116.61	0.00058	1.22795	1.0366
30	69.2709	790.3	0.18445	55.9937	0.01452	124.14	111.36	0.00056	1.22452	1.0292
31	68.2493	779.4	0.18065	51.8953	0.01521	122.79	106.49	0.00055	1.22098	1.0283
32	67.1955	768.5	0.17683	48.0249	0.01593	121.28	101.97	0.00053	1.21734	1.0292
33	66.1095	757.0	0.17304	44.3451	0.01669	119.65	97.75	0.00052	1.21359	1.0320
34	64.9888	745.2	0.16925	40.8590	0.01749	117.93	93.78	0.00050	1.20973	1.0362
35	63.8360	733.2	0.16551	37.5682	0.01833	116.13	90.04	0.00049	1.20577	1.0422
36	62.6474	721.3	0.16175	34.4949	0.01921	114.25	86.51	0.00047	1.20169	1.0494
37	61.4259	708.4	0.15802	31.5377	0.02019	112.30	83.17	0.00046	1.19751	1.0593
38	60.1709	696.5	0.15431	28.8802	0.02113	110.30	80.00	0.00045	1.19323	1.0676
39	58.8836	684.5	0.15071	26.4106	0.02212	108.27	77.00	0.00044	1.18885	1.0765
40	57.5659	672.8	0.14712	24.1448	0.02311	106.21	74.14	0.00043	1.18437	1.0852
42	54.8551	650.9	0.13999	20.1923	0.02509	102.05	68.90	0.00041	1.17520	1.1027
44	52.0711	632.3	0.13299	16.9828	0.02697	98.29	64.24	0.00040	1.16583	1.1148
46	49.2584	619.0	0.12577	14.4590	0.02856	94.97	60.18	0.00039	1.15642	1.1204
48	46.4718	610.8	0.11947	12.5758	0.02958	91.83	56.70	0.00039	1.14714	1.1157
50	43.7729	606.7	0.11397	11.1370	0.03014	88.97	53.79	0.00040	1.13821	1.1054
52	41.2091	606.8	0.10920	10.0635	0.03017	86.44	51.41	0.00041	1.12976	1.0890
54	38.8158	612.0	0.10514	9.3166	0.02960	84.24	49.51	0.00043	1.12192	1.0646
56	36.6163	620.6	0.10169	8.7809	0.02865	82.40	48.02	0.00046	1.11475	1.0363
58	34.6166	632.2	0.09883	8.4036	0.02745	80.88	46.88	0.00048	1.10825	1.0058
60	32.8099	646.4	0.09632	8.1382	0.02613	79.79	46.02	0.00052	1.10241	0.9744
65	29.0283	690.1	0.09122	7.7421	0.02294	78.41	44.77	0.00061	1.09024	0.9041
70	26.0692	742.6	0.08720	7.5635	0.02018	77.90	44.34	0.00072	1.08079	0.8531
75	23.7060	800.9	0.08383	7.4838	0.01792	77.12	44.42	0.00082	1.07328	0.8267
80	21.7779	863.3	0.08084	7.4491	0.01609	77.39	44.82	0.00092	1.06718	0.8044
85	20.1741	929.2	0.07807	7.4365	0.01459	79.01	45.46	0.00104	1.06212	0.7799
90	18.8139	997.8	0.07550	7.4304	0.01336	80.52	46.26	0.00116	1.05785	0.7656
95	17.6464	1068.8	0.07312	7.4312	0.01232	81.98	47.17	0.00127	1.05419	0.7576
100	16.6298	1142.1	0.07086	7.4325	0.01144	86.05	48.83	0.00143	1.05102	0.7414
120	13.6060	1463.0	0.06216	7.4343	0.00891	101.10	55.79	0.00205	1.04160	0.7191
140	11.5745	1795.0	0.05665	7.4281	0.00737	113.13	60.57	0.00266	1.03531	0.7082
160	10.1003	2135.4	0.05266	7.4137	0.00631	123.32	64.26	0.00326	1.03077	0.7025
180	8.9746	2478.3	0.04975	7.3960	0.00554	132.31	67.38	0.00387	1.02730	0.6991
200	8.0834	2819.5	0.04760	7.3773	0.00494	140.43	70.21	0.00449	1.02457	0.6968
220	7.3585	3158.0	0.04598	7.3588	0.00447	147.94	72.86	0.00513	1.02235	0.6950
240	6.7563	3491.1	0.04477	7.3410	0.00408	154.91	75.41	0.00579	1.02051	0.6937
260	6.2474	3818.7	0.04386	7.3243	0.00376	161.45	77.90	0.00648	1.01895	0.6925
280	5.8112	4140.4	0.04317	7.3087	0.00348	167.66	80.35	0.00720	1.01762	0.6915
300	5.4307	4461.4	0.04264	7.3002	0.00324	173.59	82.78	0.00795	1.01646	0.6903
350	4.6743	5229.8	0.04182	7.2663	0.00278	187.56	88.79	0.00994	1.01415	0.6883
400	4.1046	5979.2	0.04138	7.2384	0.00243	200.80	94.75	0.01210	1.01242	0.6868
450	3.6596	6717.8	0.04111	7.2154	0.00217	213.64	100.66	0.01444	1.01107	0.6858
500	3.3021	7456.3	0.04090	7.1961	0.00195	226.39	106.52	0.01695	1.00998	0.6850
550	3.0085	8193.0	0.04073	7.1799	0.00178	238.98	112.32	0.01964	1.00909	0.6845
600	2.7630	8933.4	0.04056	7.1661	0.00163	251.54	118.06	0.02249	1.00835	0.6841
700	2.3756	10431.4	0.04020	7.1437	0.00140	276.65	129.35	0.02867	1.00718	0.6836
800	2.0836	11967.2	0.03975	7.1266	0.00123	302.02	140.37	0.03548	1.00629	0.6835
900	1.8556	13554.6	0.03919	7.1130	0.00109	327.88	151.14	0.04289	1.00560	0.6836
1000	1.6726	15202.5	0.03854	7.1020	0.00099	440.56	161.65	0.06326	1.00505	0.5500
1200	1.3972	18662.6	0.03715	7.0854	0.00082	511.01	181.99	0.08568	1.00422	0.5473
1400	1.1996	22352.4	0.03570	7.0733	0.00071	583.19	201.52	0.11076	1.00362	0.5460
1600	1.0510	26231.2	0.03434	7.0643	0.00062	656.44	220.34	0.13841	1.00317	0.5453
1800	0.9352	30274.7	0.03312	7.0570	0.00055	731.02	238.55	0.16868	1.00282	0.5444
2000	0.8423	34570.3	0.03190	7.0508	0.00050	810.99	256.23	0.20200	1.00254	0.5422
2500	0.6739	48470.9	0.02763	7.0306	0.00040	1114.10	298.60	0.30835	1.00203	0.5174
3000	0.5576	74837.9	0.02075	6.9679	0.00033	1865.56	340.11	0.48061	1.00168	0.4568

* TWO-PHASE BOUNDARY

Table 4.12 (continued)
THERMODYNAMIC PROPERTIES OF NORMAL HYDROGEN

8.00 MPA ISOBAR

TEMPERATURE DEG. K	VOLUME CU M/KG	ISOTHERM DERIVATIVE CU M-MPA/KG	ISOCHORE DERIVATIVE MPA/K	INTERNAL ENERGY KJ/KG-K	ENTHALPY KJ/KG	ENTROPY KJ/KG-K	CV KJ / KG-K	CP KJ / KG-K	VELOCITY OF SOUND M/S
* 16.253	0.01234	1.6420	1.010	223.3	322.0	14.257	4.99	6.53	1465
17	0.01241	1.6009	0.9901	227.7	327.0	14.557	5.13	6.73	1450
18	0.01251	1.5400	0.9949	233.8	333.9	14.953	5.30	7.11	1437
19	0.01262	1.4780	0.9947	240.3	341.2	15.347	5.46	7.48	1424
20	0.01273	1.4176	0.9904	247.1	348.9	15.740	5.59	7.83	1409
21	0.01284	1.3626	0.9829	254.1	356.9	16.131	5.71	8.16	1396
22	0.01296	1.3074	0.9726	261.5	365.2	16.519	5.82	8.49	1382
23	0.01309	1.2612	0.9603	269.1	373.9	16.904	5.91	8.79	1370
24	0.01323	1.2048	0.9469	277.0	382.8	17.286	6.00	9.12	1354
25	0.01337	1.1497	0.9317	285.2	392.1	17.664	6.07	9.45	1337
26	0.01352	1.0957	0.9153	293.6	401.7	18.039	6.14	9.78	1320
27	0.01367	1.0434	0.8978	302.3	411.7	18.415	6.21	10.11	1303
28	0.01384	0.9928	0.8791	311.2	421.9	18.787	6.27	10.44	1286
29	0.01401	0.9423	0.8593	320.4	432.5	19.160	6.33	10.79	1268
30	0.01420	0.8935	0.8385	329.9	443.5	19.531	6.37	11.13	1249
31	0.01439	0.8453	0.8166	339.7	454.8	19.902	6.41	11.48	1230
32	0.01460	0.7993	0.7939	349.7	466.4	20.271	6.45	11.83	1210
33	0.01481	0.7555	0.7704	359.9	478.4	20.640	6.48	12.17	1191
34	0.01504	0.7134	0.7465	370.4	490.7	21.009	6.51	12.52	1171
35	0.01529	0.6734	0.7222	381.1	503.4	21.376	6.54	12.87	1151
36	0.01554	0.6345	0.6978	392.1	516.5	21.744	6.56	13.24	1131
37	0.01582	0.5986	0.6734	403.4	529.9	22.112	6.58	13.59	1112
38	0.01611	0.5644	0.6489	414.8	543.7	22.479	6.60	13.95	1093
39	0.01641	0.5326	0.6244	426.5	557.8	22.846	6.60	14.29	1074
40	0.01674	0.5024	0.5998	438.3	572.3	23.212	6.61	14.64	1055
42	0.01745	0.4474	0.5516	462.6	602.2	23.943	6.62	15.31	1017
44	0.01824	0.4026	0.5050	487.5	633.5	24.669	6.63	15.91	983
46	0.01912	0.3647	0.4611	512.4	665.3	25.376	6.68	16.49	949
48	0.02009	0.3360	0.4206	538.0	698.7	26.087	6.71	16.91	920
50	0.02114	0.3135	0.3835	563.7	732.8	26.784	6.74	17.22	895
52	0.02227	0.2983	0.3503	589.3	767.4	27.462	6.76	17.36	875
54	0.02346	0.2882	0.3206	614.5	802.2	28.118	6.77	17.38	860
56	0.02471	0.2824	0.2943	639.4	837.1	28.754	6.78	17.27	848
58	0.02600	0.2806	0.2712	663.4	871.5	29.357	6.77	17.05	841
60	0.02732	0.2819	0.2508	686.7	905.3	29.930	6.76	16.75	836
65	0.03068	0.2941	0.2100	741.6	987.0	31.237	6.74	15.91	833
70	0.03403	0.3128	0.1802	792.4	1064.6	32.388	6.74	15.16	839
75	0.03734	0.3348	0.1580	840.1	1138.8	33.413	6.77	14.57	849
80	0.04059	0.3590	0.1408	885.7	1210.4	34.337	6.83	14.11	861
85	0.04380	0.3840	0.1271	929.7	1280.1	35.182	6.91	13.77	875
90	0.04694	0.4096	0.1160	972.7	1348.3	35.962	7.01	13.53	889
95	0.05004	0.4349	0.1068	1015.1	1415.5	36.689	7.11	13.36	904
100	0.05309	0.4605	0.0991	1057.1	1481.9	37.370	7.24	13.24	918
120	0.06494	0.5604	0.0771	1227.8	1747.3	39.789	7.85	13.22	971
140	0.07623	0.6547	0.0632	1401.3	2011.2	41.823	8.37	13.33	1021
160	0.08733	0.7475	0.0539	1581.2	2279.8	43.615	8.82	13.56	1072
180	0.09825	0.8380	0.0471	1767.1	2553.2	45.225	9.19	13.79	1121
200	0.10904	0.9270	0.0419	1958.5	2830.8	46.688	9.49	13.99	1169
220	0.11974	1.0148	0.0377	2154.2	3112.2	48.028	9.73	14.15	1215
240	0.13037	1.1016	0.0343	2353.6	3396.5	49.266	9.91	14.28	1260
260	0.14095	1.1877	0.0315	2555.6	3683.2	50.413	10.05	14.38	1303
280	0.15148	1.2733	0.0292	2759.7	3971.6	51.482	10.16	14.45	1346
300	0.16207	1.3605	0.0271	2966.5	4263.0	52.485	10.24	14.50	1388
350	0.18818	1.5711	0.0231	3484.2	4989.6	54.725	10.34	14.55	1487
400	0.21418	1.7803	0.0202	4004.2	5717.7	56.670	10.38	14.57	1581
450	0.24012	1.9885	0.0179	4525.1	6446.1	58.385	10.39	14.56	1669
500	0.26602	2.1963	0.0161	5046.2	7174.3	59.919	10.41	14.56	1753
550	0.29188	2.4036	0.0146	5567.4	7902.4	61.308	10.42	14.57	1833
600	0.31773	2.6107	0.0134	6089.4	8631.2	62.573	10.43	14.58	1910
700	0.36939	3.0243	0.0114	7136.0	10091.1	64.826	10.49	14.62	2054
800	0.42101	3.4375	0.0100	8189.7	11557.7	66.786	10.57	14.71	2186
900	0.47261	3.8504	0.0089	9253.9	13034.7	68.524	10.70	14.83	2310
1000	0.52419	4.2632	0.0080	10331.6	14525.1	70.092	10.86	14.99	2426
1200	0.62735	5.0885	0.0067	12542.6	17561.4	72.841	11.24	15.37	2637
1400	0.73048	5.9135	0.0057	14835.7	20679.6	75.250	11.68	15.80	2829
1600	0.83361	6.7385	0.0050	17214.6	23883.4	77.396	12.12	16.25	3005
1800	0.93674	7.5634	0.0044	19680.1	27174.1	79.328	12.55	16.68	3170
2000	1.03993	8.3883	0.0040	22237.3	30556.8	81.120	13.02	17.15	3325
2500	1.29946	10.4505	0.0032	29158.2	39553.9	85.114	14.95	19.21	3664
3000	1.56927	12.5126	0.0027	37756.2	50310.4	89.037	19.79	24.63	3944

* TWO-PHASE BOUNDARY

Table 4.12 (continued)
THERMODYNAMIC PROPERTIES OF NORMAL HYDROGEN

8.00 MPA ISOBAR

TEMPERATURE DEG. K	DENSITY KG/CU M	$V(\partial H/\partial V)_P$ KJ/KG	$V(\partial P/\partial U)_V$ MPA-CU M/KJ X 10^2	$-V(\partial P/\partial V)_T$ MPA	$-(\partial V/\partial T)_P/V$ 1/DEG. K	THERMAL CONDUCTIVITY W/K-M X 10^3	VISCOSITY KG/M-S X 10^7	THERMAL DIFFUSIVITY SQ M/HR	DIELECTRIC CONSTANT	PRANDTL NUMBER
* 16.253	81.0262	860.5	0.24956	133.0474	0.00759	100.44	304.73	0.00068	1.26573	1.9811
17	80.5578	876.8	0.23972	128.9649	0.00768	104.76	279.55	0.00070	1.26407	1.7963
18	79.9223	879.9	0.23480	123.0814	0.00808	109.49	251.89	0.00069	1.26183	1.6363
19	79.2669	880.9	0.23004	117.1587	0.00849	112.90	229.28	0.00069	1.25951	1.5190
20	78.5812	880.8	0.22547	111.3942	0.00889	116.11	210.37	0.00068	1.25709	1.4188
21	77.8729	881.4	0.22110	106.1125	0.00926	119.69	194.42	0.00068	1.25459	1.3262
22	77.1386	880.4	0.21681	100.8483	0.00964	122.49	180.78	0.00067	1.25201	1.2531
23	76.3826	882.0	0.21271	96.3300	0.00997	124.65	169.02	0.00067	1.24935	1.1922
24	75.6049	877.4	0.20886	91.0901	0.01040	126.26	158.77	0.00066	1.24662	1.1470
25	74.8090	872.1	0.20506	86.0107	0.01083	127.40	149.79	0.00065	1.24383	1.1107
26	73.9871	865.9	0.20134	81.0675	0.01129	128.07	141.79	0.00064	1.24095	1.0823
27	73.1404	859.3	0.19765	76.3177	0.01176	128.38	134.61	0.00063	1.23799	1.0599
28	72.2675	852.4	0.19397	71.7474	0.01225	128.38	128.13	0.00061	1.23494	1.0424
29	71.3703	844.2	0.19035	67.2527	0.01278	128.11	122.23	0.00060	1.23182	1.0292
30	70.4457	835.3	0.18684	62.9424	0.01332	127.59	116.83	0.00059	1.22860	1.0189
31	69.4946	825.6	0.18324	58.7436	0.01390	126.39	111.85	0.00057	1.22530	1.0156
32	68.5164	815.8	0.17960	54.7641	0.01450	125.04	107.23	0.00056	1.22191	1.0141
33	67.5114	805.9	0.17600	51.0066	0.01510	123.57	102.94	0.00054	1.21843	1.0140
34	66.4798	795.6	0.17242	47.4290	0.01574	122.03	98.93	0.00053	1.21487	1.0151
35	65.4218	785.2	0.16886	44.0567	0.01639	120.40	95.16	0.00051	1.21122	1.0174
36	64.3373	774.3	0.16529	40.8199	0.01709	118.70	91.62	0.00050	1.20749	1.0216
37	63.2256	764.0	0.16182	37.8475	0.01779	116.93	88.27	0.00049	1.20367	1.0261
38	62.0887	753.4	0.15844	35.0456	0.01852	115.11	85.10	0.00048	1.19978	1.0312
39	60.9269	742.9	0.15516	32.4475	0.01924	113.26	82.10	0.00047	1.19581	1.0362
40	59.7429	732.4	0.15188	30.0121	0.01999	111.39	79.26	0.00046	1.19177	1.0415
42	57.3140	712.0	0.14537	25.6429	0.02151	107.60	74.03	0.00044	1.18352	1.0536
44	54.8247	695.3	0.13886	22.0737	0.02288	103.86	69.35	0.00043	1.17510	1.0622
46	52.3052	682.0	0.13194	19.0754	0.02418	100.50	65.22	0.00042	1.16662	1.0700
48	49.7865	672.5	0.12581	16.7280	0.02514	97.54	61.61	0.00042	1.15818	1.0680
50	47.3102	666.0	0.12027	14.8314	0.02586	94.79	58.51	0.00042	1.14993	1.0631
52	44.9111	664.2	0.11535	13.3977	0.02614	92.26	55.88	0.00043	1.14197	1.0516
54	42.6222	665.7	0.11104	12.2818	0.02610	90.03	53.70	0.00044	1.13441	1.0364
56	40.4657	670.5	0.10732	11.4269	0.02576	88.08	51.91	0.00045	1.12732	1.0178
58	38.4552	678.4	0.10418	10.7920	0.02513	86.40	50.47	0.00047	1.12074	0.9958
60	36.5979	689.2	0.10140	10.3163	0.02431	85.11	49.34	0.00050	1.11469	0.9712
65	32.5995	726.4	0.09559	9.5875	0.02190	83.19	47.52	0.00058	1.10173	0.9087
70	29.3897	773.2	0.09094	9.1934	0.01960	82.18	46.69	0.00066	1.09140	0.8613
75	25.7843	826.9	0.08707	8.9679	0.01762	80.94	46.48	0.00075	1.08307	0.8365
80	24.6339	886.2	0.08368	8.8430	0.01592	80.83	46.64	0.00084	1.07622	0.8142
85	22.8327	949.8	0.08060	8.7688	0.01450	82.18	47.08	0.00094	1.07051	0.7889
90	21.3028	1017.1	0.07774	8.7248	0.01330	83.48	47.70	0.00104	1.06568	0.7730
95	19.9830	1086.6	0.07513	8.6903	0.01229	84.74	48.46	0.00114	1.06152	0.7639
100	18.8344	1159.4	0.07268	8.6726	0.01142	88.17	49.75	0.00127	1.05792	0.7471
120	15.3987	1479.6	0.06378	8.6301	0.00893	102.69	56.44	0.00182	1.04718	0.7266
140	13.1174	1812.0	0.05754	8.5880	0.00735	114.37	61.06	0.00236	1.04009	0.7114
160	11.4504	2153.3	0.05338	8.5589	0.00630	124.32	64.65	0.00288	1.03493	0.7050
180	10.1779	2497.3	0.05035	8.5295	0.00552	133.15	67.71	0.00342	1.03101	0.7011
200	9.1706	2839.7	0.04812	8.5012	0.00492	141.16	70.49	0.00396	1.02791	0.6984
220	8.3512	3179.3	0.04643	8.4745	0.00445	148.60	73.12	0.00453	1.02539	0.6964
240	7.6703	3513.5	0.04517	8.4496	0.00407	155.52	75.65	0.00511	1.02330	0.6948
260	7.0948	3842.0	0.04422	8.4266	0.00374	162.03	78.13	0.00572	1.02154	0.6935
280	6.6015	4164.6	0.04350	8.4053	0.00347	168.23	80.58	0.00635	1.02003	0.6923
300	6.1701	4487.4	0.04293	8.3944	0.00323	174.15	83.01	0.00701	1.01872	0.6910
350	5.3141	5257.0	0.04207	8.3490	0.00277	188.14	89.03	0.00876	1.01610	0.6887
400	4.6690	6007.2	0.04159	8.3119	0.00242	201.43	95.02	0.01066	1.01414	0.6871
450	4.1646	6746.4	0.04130	8.2815	0.00216	214.32	100.95	0.01272	1.01260	0.6859
500	3.7592	7485.3	0.04106	8.2561	0.00195	227.14	106.85	0.01493	1.01137	0.6851
550	3.4260	8222.3	0.04088	8.2348	0.00177	239.80	112.68	0.01730	1.01036	0.6845
600	3.1473	8963.0	0.04070	8.2166	0.00163	252.43	118.46	0.01981	1.00951	0.6841
700	2.7072	10461.5	0.04031	8.1873	0.00140	277.70	129.82	0.02525	1.00818	0.6836
800	2.3753	11997.6	0.03985	8.1649	0.00123	303.22	140.91	0.03125	1.00717	0.6835
900	2.1159	13585.4	0.03928	8.1471	0.00109	329.22	151.74	0.03777	1.00639	0.6836
1000	1.9077	15233.8	0.03862	8.1328	0.00098	440.56	162.32	0.05546	1.00576	0.5523
1200	1.5940	18694.7	0.03721	8.1111	0.00082	511.01	182.80	0.07510	1.00481	0.5497
1400	1.3690	22385.4	0.03575	8.0954	0.00071	583.18	202.45	0.09705	1.00413	0.5485
1600	1.1996	26264.5	0.03439	8.0836	0.00062	656.41	221.39	0.12126	1.00362	0.5479
1800	1.0675	30305.0	0.03316	8.0742	0.00055	730.84	239.71	0.14775	1.00322	0.5471
2000	0.9616	34587.0	0.03196	8.0662	0.00050	810.25	257.50	0.17687	1.00290	0.5450
2500	0.7696	48270.9	0.02781	8.0422	0.00040	1105.05	300.12	0.26917	1.00232	0.5216
3000	0.6372	73647.0	0.02115	7.9735	0.00033	1819.14	341.77	0.41728	1.00192	0.4627

* TWO-PHASE BOUNDARY

Table 4.12 (continued)
THERMODYNAMIC PROPERTIES OF NORMAL HYDROGEN

10.00 MPA ISOBAR

TEMPERATURE DEG. K	VOLUME CU M/KG	ISOTHERM DERIVATIVE CU M-MPA/KG	ISOCHORE DERIVATIVE MPA/K	INTERNAL ENERGY KJ/KG-K	ENTHALPY KJ/KG	ENTROPY KJ/KG-K	CV KJ /	CP KG-K	VELOCITY OF SOUND M/S
* 16.808	0.01221	1.7491	1.024	225.1	347.2	14.297	5.07	6.58	1506
17	0.01223	1.7383	1.024	226.2	348.5	14.372	5.11	6.64	1503
18	0.01232	1.6821	1.021	232.1	355.3	14.761	5.27	6.97	1491
19	0.01241	1.6243	1.012	238.3	362.4	15.144	5.42	7.27	1476
20	0.01251	1.5672	1.011	244.7	369.8	15.526	5.57	7.61	1464
21	0.01262	1.5088	1.006	251.5	377.6	15.906	5.69	7.93	1450
22	0.01272	1.4503	0.9983	258.5	385.7	16.284	5.80	8.25	1436
23	0.01284	1.3999	0.9884	265.7	394.1	16.657	5.90	8.55	1424
24	0.01296	1.3481	0.9764	273.2	402.8	17.027	5.99	8.84	1410
25	0.01308	1.3034	0.9632	281.0	411.8	17.394	6.07	9.12	1399
26	0.01321	1.2515	0.9495	288.9	421.1	17.756	6.14	9.41	1385
27	0.01335	1.2002	0.9342	297.1	430.6	18.117	6.21	9.71	1370
28	0.01349	1.1494	0.9179	305.6	440.5	18.474	6.28	10.01	1354
29	0.01364	1.1003	0.9005	314.3	450.6	18.831	6.34	10.31	1338
30	0.01379	1.0532	0.8822	323.1	461.1	19.185	6.38	10.60	1322
31	0.01396	1.0079	0.8630	332.3	471.8	19.538	6.43	10.89	1307
32	0.01413	0.9630	0.8431	341.6	482.9	19.888	6.47	11.19	1290
33	0.01431	0.9196	0.8225	351.1	494.2	20.237	6.51	11.48	1273
34	0.01449	0.8772	0.8012	360.9	505.8	20.584	6.54	11.77	1256
35	0.01469	0.8367	0.7794	370.8	517.7	20.929	6.57	12.06	1239
36	0.01490	0.7978	0.7574	381.0	529.9	21.273	6.60	12.34	1222
37	0.01511	0.7606	0.7352	391.3	542.4	21.615	6.62	12.63	1204
38	0.01534	0.7254	0.7129	401.8	555.2	21.956	6.64	12.90	1187
39	0.01557	0.6901	0.6906	412.5	568.2	22.294	6.65	13.19	1170
40	0.01582	0.6588	0.6686	423.3	581.5	22.631	6.66	13.45	1154
42	0.01635	0.6014	0.6249	445.4	609.0	23.300	6.67	13.96	1122
44	0.01693	0.5501	0.5821	468.0	637.4	23.961	6.69	14.46	1091
46	0.01757	0.5076	0.5408	490.6	666.2	24.601	6.73	14.91	1060
48	0.01825	0.4709	0.5016	513.9	696.4	25.244	6.76	15.30	1032
50	0.01899	0.4405	0.4649	537.5	727.4	25.876	6.79	15.63	1007
52	0.01977	0.4159	0.4308	561.2	758.9	26.494	6.81	15.88	985
54	0.02061	0.3968	0.3996	584.8	790.9	27.097	6.82	16.05	966
56	0.02148	0.3825	0.3711	608.5	823.3	27.688	6.83	16.13	951
58	0.02239	0.3723	0.3452	631.6	855.6	28.255	6.82	16.13	938
60	0.02334	0.3655	0.3218	654.4	887.8	28.801	6.81	16.07	929
65	0.02580	0.3615	0.2729	709.3	967.4	30.074	6.80	15.72	914
70	0.02835	0.3702	0.2352	761.3	1044.8	31.222	6.80	15.21	910
75	0.03090	0.3855	0.2062	810.6	1119.6	32.255	6.83	14.73	912
80	0.03345	0.4039	0.1835	857.8	1192.2	33.192	6.88	14.34	918
85	0.03598	0.4247	0.1653	903.3	1263.1	34.052	6.95	14.03	926
90	0.03848	0.4468	0.1505	947.8	1332.7	34.847	7.05	13.81	935
95	0.04096	0.4702	0.1382	991.6	1401.2	35.588	7.15	13.63	946
100	0.04342	0.4938	0.1278	1034.9	1469.0	36.284	7.27	13.51	958
120	0.05297	0.5907	0.0987	1209.6	1739.4	38.748	7.89	13.44	1003
140	0.06221	0.6861	0.0807	1386.7	2008.7	40.824	8.41	13.55	1052
160	0.07104	0.7761	0.0682	1568.8	2279.2	42.629	8.85	13.69	1096
180	0.07985	0.8673	0.0595	1756.5	2554.9	44.253	9.22	13.90	1144
200	0.08854	0.9568	0.0528	1949.1	2834.5	45.726	9.51	14.07	1190
220	0.09715	1.0451	0.0475	2146.0	3117.5	47.074	9.75	14.23	1235
240	0.10570	1.1324	0.0432	2346.2	3403.2	48.317	9.93	14.35	1279
260	0.11419	1.2190	0.0396	2549.0	3691.0	49.470	10.07	14.44	1322
280	0.12265	1.3049	0.0366	2753.8	3980.3	50.542	10.17	14.50	1364
300	0.13119	1.3936	0.0340	2961.4	4273.2	51.549	10.25	14.54	1406
350	0.15212	1.6046	0.0290	3480.2	5001.4	53.794	10.35	14.58	1503
400	0.17295	1.8141	0.0252	4001.1	5730.7	55.742	10.39	14.59	1596
450	0.19373	2.0226	0.0224	4522.6	6459.9	57.460	10.40	14.58	1684
500	0.21446	2.2304	0.0201	5044.1	7188.9	58.995	10.41	14.58	1767
550	0.23517	2.4378	0.0182	5565.7	7917.4	60.384	10.42	14.58	1846
600	0.25586	2.6449	0.0167	6088.0	8646.6	61.650	10.44	14.59	1922
700	0.29720	3.0586	0.0143	7135.0	10107.0	63.904	10.49	14.63	2065
800	0.33850	3.4718	0.0125	8189.0	11574.0	65.865	10.58	14.71	2197
900	0.37979	3.8847	0.0111	9253.4	13051.3	67.603	10.70	14.83	2320
1000	0.42106	4.2974	0.0100	10331.2	14541.9	69.171	10.86	14.99	2435
1200	0.50359	5.1227	0.0083	12542.5	17578.4	71.920	11.24	15.37	2646
1400	0.58610	5.9477	0.0071	14835.7	20696.7	74.330	11.68	15.80	2837
1600	0.66860	6.7726	0.0062	17214.6	23900.6	76.476	12.12	16.24	3013
1800	0.75111	7.5975	0.0056	19679.8	27190.9	78.407	12.55	16.68	3177
2000	0.83366	8.4223	0.0050	22235.5	30572.0	80.199	13.00	17.14	3332
2500	1.04113	10.4844	0.0040	29126.2	39537.5	84.179	14.81	19.06	3672
3000	1.25599	12.5464	0.0033	37554.2	50114.1	88.037	19.21	23.97	3955

* TWO-PHASE BOUNDARY

Table 4.12 (continued)
THERMODYNAMIC PROPERTIES OF NORMAL HYDROGEN

10.00 MPA ISOBAR

TEMPERATURE DEG. K	DENSITY KG/CU M	$V(DH/DV)_P$ KJ/KG	$V(DP/DU)_V$ MPA-CU M/KJ X 10^2	$-V(DP/DV)_T$ MPA	$(DV/DT)_P/V$ 1/DEG. K	THERMAL CONDUCTIVITY W/K-M X 10^3	VISCOSITY KG/M-S X 10^7	THERMAL DIFFUSIVITY SQ M/HR	DIELECTRIC CONSTANT	PRANDTL NUMBER
* 16.808	81.8729	919.9	0.24645	143.2003	0.00715	106.03	316.91	0.00071	1.26874	1.9661
17	81.7600	922.0	0.24510	142.1202	0.00720	107.12	309.94	0.00071	1.26834	1.9216
18	81.1621	931.5	0.23860	136.5241	0.00748	112.22	278.08	0.00071	1.26621	1.7262
19	80.5558	939.9	0.23167	130.8463	0.00774	115.98	252.38	0.00071	1.26407	1.5825
20	79.9218	942.6	0.22727	125.2542	0.00807	119.53	230.97	0.00071	1.26182	1.4701
21	79.2664	943.0	0.22304	119.5946	0.00841	123.45	212.99	0.00071	1.25951	1.3687
22	78.5887	941.9	0.21893	113.9783	0.00876	126.57	197.68	0.00070	1.25711	1.2884
23	77.8936	942.9	0.21502	109.0437	0.00906	129.04	184.56	0.00070	1.25466	1.2223
24	77.1767	941.9	0.21121	104.0384	0.00939	130.93	173.15	0.00069	1.25214	1.1691
25	76.4428	942.9	0.20758	99.6355	0.00967	132.34	163.18	0.00068	1.24956	1.1240
26	75.6908	939.1	0.20418	94.7233	0.01002	133.27	154.38	0.00067	1.24692	1.0904
27	74.9239	934.7	0.20070	89.9250	0.01039	133.85	146.57	0.00066	1.24423	1.0633
28	74.1366	929.4	0.19723	85.2159	0.01077	134.11	139.55	0.00065	1.24147	1.0417
29	73.3292	923.7	0.19384	80.6838	0.01116	134.10	133.20	0.00064	1.23865	1.0241
30	72.5015	917.6	0.19059	76.3574	0.01155	133.85	127.43	0.00063	1.23576	1.0093
31	71.6523	911.6	0.18729	72.2213	0.01195	132.88	122.13	0.00061	1.23280	1.0011
32	70.7876	904.5	0.18399	68.1673	0.01237	131.78	117.27	0.00060	1.22979	0.9956
33	69.9029	897.1	0.18072	64.2854	0.01279	130.55	112.77	0.00059	1.22671	0.9915
34	68.9988	889.1	0.17747	60.5267	0.01324	129.26	108.59	0.00057	1.22358	0.9887
35	68.0764	881.0	0.17421	56.9585	0.01368	127.89	104.70	0.00056	1.22038	0.9870
36	67.1359	872.8	0.17096	53.5628	0.01414	126.44	101.05	0.00055	1.21713	0.9863
37	66.1783	864.4	0.16775	50.3323	0.01461	124.93	97.62	0.00054	1.21383	0.9866
38	65.2037	856.0	0.16466	47.2967	0.01507	123.38	94.40	0.00053	1.21047	0.9871
39	64.2133	846.2	0.16172	44.3160	0.01558	121.79	91.36	0.00052	1.20706	0.9892
40	63.2056	837.8	0.15886	41.6392	0.01606	120.17	88.48	0.00051	1.20360	0.9904
42	61.1498	821.8	0.15317	36.7751	0.01699	116.87	83.17	0.00049	1.19657	0.9938
44	59.0530	806.9	0.14743	32.4879	0.01792	113.58	78.42	0.00048	1.18942	0.9980
46	56.9271	796.7	0.14110	28.8948	0.01872	110.38	74.16	0.00047	1.18221	1.0018
48	54.7942	787.3	0.13536	25.8049	0.01944	107.32	70.37	0.00046	1.17499	1.0035
50	52.6712	780.0	0.13004	23.2021	0.02004	104.56	67.02	0.00046	1.16785	1.0018
52	50.5770	775.3	0.12515	21.0339	0.02048	102.15	64.09	0.00046	1.16082	0.9962
54	48.5318	773.3	0.12071	19.2554	0.02075	99.95	61.53	0.00046	1.15399	0.9879
56	46.5526	773.9	0.11676	17.8052	0.02084	97.94	59.34	0.00047	1.14741	0.9772
58	44.6539	776.8	0.11334	16.6230	0.02077	96.13	57.47	0.00048	1.14112	0.9645
60	42.8456	782.3	0.11023	15.6610	0.02055	94.65	55.91	0.00049	1.13515	0.9494
65	38.7532	806.9	0.10355	14.0104	0.01948	92.11	53.06	0.00054	1.12172	0.9055
70	35.2794	844.5	0.09801	13.0614	0.01801	90.32	51.41	0.00061	1.11040	0.8656
75	32.3603	891.0	0.09331	12.4759	0.01653	88.42	50.57	0.00067	1.10095	0.8423
80	29.8978	943.6	0.08922	12.0748	0.01520	87.63	50.28	0.00074	1.09303	0.8227
85	27.7951	1002.0	0.08555	11.8046	0.01401	88.51	50.34	0.00082	1.08629	0.7981
90	25.9853	1065.0	0.08218	11.6098	0.01296	89.39	50.65	0.00090	1.08052	0.7823
95	24.4121	1132.0	0.07914	11.4780	0.01204	90.28	51.15	0.00098	1.07552	0.7721
100	23.0330	1202.1	0.07630	11.3739	0.01124	92.74	51.83	0.00107	1.07115	0.7551
120	18.8773	1518.3	0.06630	11.1500	0.00885	106.22	57.97	0.00151	1.05805	0.7336
140	16.0754	1851.5	0.05974	11.0295	0.00732	117.12	62.22	0.00194	1.04929	0.7198
160	14.0768	2192.1	0.05479	10.9245	0.00625	126.53	65.56	0.00236	1.04306	0.7096
180	12.5242	2537.8	0.05153	10.8616	0.00548	134.99	68.46	0.00279	1.03825	0.7048
200	11.2944	2882.1	0.04913	10.8066	0.00488	142.74	71.14	0.00323	1.03445	0.7015
220	10.2934	3223.6	0.04732	10.7578	0.00441	150.00	73.69	0.00369	1.03136	0.6990
240	9.4610	3559.5	0.04596	10.7141	0.00403	156.81	76.18	0.00416	1.02880	0.6970
260	8.7570	3889.7	0.04492	10.6747	0.00371	163.25	78.63	0.00465	1.02664	0.6953
280	8.1531	4213.8	0.04414	10.6392	0.00344	169.41	81.07	0.00516	1.02478	0.6939
300	7.6226	4540.6	0.04352	10.6229	0.00320	175.31	83.49	0.00570	1.02316	0.6923
350	6.5738	5312.3	0.04255	10.5486	0.00274	189.32	89.53	0.00711	1.01995	0.6895
400	5.7819	6063.9	0.04201	10.4889	0.00241	202.69	95.56	0.00865	1.01753	0.6876
450	5.1619	6804.1	0.04167	10.4403	0.00214	215.70	101.55	0.01032	1.01564	0.6863
500	4.6628	7543.7	0.04139	10.4000	0.00193	228.64	107.50	0.01211	1.01412	0.6853
550	4.2523	8281.3	0.04117	10.3662	0.00176	241.44	113.40	0.01402	1.01287	0.6846
600	3.9084	9022.4	0.04096	10.3375	0.00162	254.21	119.25	0.01605	1.01183	0.6842
700	3.3648	10521.6	0.04054	10.2915	0.00139	279.77	130.75	0.02046	1.01017	0.6836
800	2.9542	12058.4	0.04004	10.2563	0.00122	305.60	141.98	0.02532	1.00893	0.6834
900	2.6330	13646.9	0.03945	10.2286	0.00109	331.91	152.95	0.03059	1.00796	0.6835
1000	2.3749	15296.1	0.03877	10.2062	0.00098	440.56	163.66	0.04454	1.00717	0.5569
1200	1.9857	18758.8	0.03733	10.1723	0.00082	511.00	184.40	0.06028	1.00600	0.5546
1400	1.7062	22451.3	0.03585	10.1479	0.00070	583.18	204.31	0.07787	1.00515	0.5536
1600	1.4957	26331.2	0.03447	10.1295	0.00062	656.36	223.49	0.09725	1.00451	0.5531
1800	1.3314	30367.5	0.03324	10.1150	0.00055	730.56	242.05	0.11845	1.00402	0.5525
2000	1.1995	34629.0	0.03205	10.1029	0.00049	809.11	260.06	0.14171	1.00362	0.5508
2500	0.9605	47982.5	0.02811	10.0701	0.00040	1091.21	303.19	0.21462	1.00290	0.5295
3000	0.7962	71842.6	0.02179	9.9893	0.00033	1748.06	345.18	0.32975	1.00240	0.4733

* TWO-PHASE BOUNDARY

Table 4.12 (continued)
THERMODYNAMIC PROPERTIES OF NORMAL HYDROGEN

12.00 MPA ISOBAR

TEMPERATURE DEG. K	VOLUME CU M/KG	ISOTHERM DERIVATIVE CU M-MPA/KG	ISOCHORE DERIVATIVE MPA/K	INTERNAL ENERGY KJ/KG-K	ENTHALPY KJ/KG	ENTROPY KJ/KG-K	CV KJ /	CP KG-K	VELOCITY OF SOUND M/S
* 17.346	0.01210	1.8521	1.038	227.1	372.2	14.336	5.15	6.63	1544
18	0.01215	1.8168	1.037	230.8	376.6	14.585	5.25	6.83	1537
19	0.01224	1.7629	1.035	236.8	383.6	14.962	5.41	7.13	1525
20	0.01233	1.7091	1.031	243.0	390.9	15.336	5.54	7.43	1514
21	0.01242	1.6510	1.027	249.4	398.4	15.705	5.67	7.74	1501
22	0.01252	1.5973	1.021	256.1	406.3	16.072	5.79	8.04	1489
23	0.01262	1.5426	1.013	263.1	414.5	16.436	5.89	8.33	1477
24	0.01272	1.4862	1.003	270.2	422.9	16.797	5.98	8.61	1463
25	0.01284	1.4391	0.9916	277.7	431.7	17.153	6.07	8.88	1451
26	0.01295	1.3900	0.9786	285.3	440.7	17.505	6.14	9.15	1439
27	0.01307	1.3454	0.9647	293.1	450.0	17.856	6.21	9.41	1427
28	0.01320	1.2977	0.9507	301.2	459.5	18.202	6.28	9.68	1414
29	0.01333	1.2500	0.9353	309.4	469.3	18.547	6.34	9.95	1400
30	0.01346	1.2023	0.9191	317.9	479.4	18.888	6.40	10.22	1386
31	0.01360	1.1560	0.9019	326.5	489.8	19.227	6.45	10.48	1371
32	0.01375	1.1136	0.8840	335.3	500.4	19.564	6.49	10.74	1357
33	0.01391	1.0708	0.8655	344.4	511.2	19.899	6.53	11.00	1343
34	0.01406	1.0288	0.8465	353.6	522.4	20.231	6.57	11.25	1328
35	0.01423	0.9889	0.8269	363.0	533.7	20.561	6.60	11.50	1313
36	0.01440	0.9495	0.8069	372.5	545.4	20.889	6.63	11.75	1297
37	0.01458	0.9134	0.7866	382.2	557.2	21.214	6.66	11.99	1282
38	0.01477	0.8776	0.7660	392.1	569.3	21.537	6.68	12.22	1267
39	0.01496	0.8431	0.7455	402.1	581.7	21.857	6.69	12.45	1252
40	0.01517	0.8101	0.7249	412.2	594.2	22.175	6.70	12.67	1238
42	0.01559	0.7470	0.6841	432.9	620.0	22.803	6.72	13.12	1208
44	0.01606	0.6919	0.6441	454.0	646.7	23.423	6.74	13.54	1179
46	0.01655	0.6457	0.6052	475.0	673.5	24.022	6.79	13.94	1151
48	0.01708	0.6042	0.5677	496.8	701.8	24.622	6.82	14.29	1125
50	0.01765	0.5690	0.5319	518.9	730.7	25.212	6.84	14.59	1101
52	0.01825	0.5385	0.4982	541.2	760.2	25.790	6.86	14.84	1079
54	0.01888	0.5129	0.4667	563.5	790.1	26.354	6.87	15.05	1060
56	0.01955	0.4916	0.4373	586.0	820.5	26.910	6.88	15.20	1042
58	0.02024	0.4753	0.4101	608.2	851.1	27.445	6.88	15.28	1028
60	0.02096	0.4623	0.3851	630.2	881.7	27.964	6.87	15.33	1016
65	0.02285	0.4435	0.3313	684.0	958.3	29.188	6.86	15.26	993
70	0.02485	0.4390	0.2881	735.9	1034.0	30.312	6.86	15.03	981
75	0.02689	0.4458	0.2537	785.7	1108.4	31.338	6.89	14.71	976
80	0.02895	0.4593	0.2260	833.8	1181.2	32.277	6.93	14.39	976
85	0.03101	0.4755	0.2037	880.3	1252.4	33.141	7.00	14.13	980
90	0.03307	0.4942	0.1853	925.8	1322.5	33.944	7.09	13.93	985
95	0.03511	0.5142	0.1700	970.5	1391.3	34.692	7.19	13.78	992
100	0.03714	0.5351	0.1571	1014.7	1460.4	35.396	7.31	13.68	1000
120	0.04511	0.6260	0.1208	1192.8	1734.1	37.891	7.92	13.61	1037
140	0.05285	0.7197	0.0984	1372.5	2006.7	39.992	8.44	13.70	1081
160	0.06040	0.8125	0.0832	1557.0	2281.8	41.827	8.88	13.85	1126
180	0.06761	0.8980	0.0723	1746.4	2557.7	43.453	9.24	14.03	1167
200	0.07490	0.9879	0.0638	1940.3	2839.0	44.935	9.53	14.16	1211
220	0.08211	1.0765	0.0573	2138.1	3123.4	46.290	9.77	14.30	1255
240	0.08926	1.1642	0.0521	2339.2	3410.3	47.539	9.95	14.40	1298
260	0.09637	1.2510	0.0477	2542.7	3699.2	48.696	10.09	14.49	1340
280	0.10344	1.3372	0.0441	2748.1	3989.4	49.771	10.19	14.54	1382
300	0.11061	1.4275	0.0409	2956.5	4283.8	50.782	10.27	14.57	1424
350	0.12809	1.6388	0.0348	3476.5	5013.5	53.033	10.36	14.61	1520
400	0.14547	1.8484	0.0303	3998.2	5743.9	54.983	10.40	14.60	1611
450	0.16280	2.0569	0.0269	4520.3	6473.9	56.703	10.41	14.59	1698
500	0.18009	2.2648	0.0241	5042.2	7203.4	58.239	10.42	14.59	1781
550	0.19736	2.4722	0.0219	5564.2	7932.5	59.629	10.43	14.58	1860
600	0.21461	2.6793	0.0201	6086.7	8662.0	60.896	10.44	14.59	1935
700	0.24906	3.0929	0.0172	7134.2	10122.9	63.151	10.49	14.63	2077
800	0.28349	3.5060	0.0150	8188.4	11590.3	65.112	10.58	14.71	2208
900	0.31790	3.9188	0.0133	9253.0	13067.8	66.850	10.70	14.84	2330
1000	0.35230	4.3315	0.0120	10331.0	14558.6	68.419	10.87	14.99	2445
1200	0.42107	5.1566	0.0100	12542.5	17595.4	71.168	11.24	15.37	2655
1400	0.48983	5.9815	0.0086	14835.8	20713.9	73.578	11.68	15.80	2845
1600	0.55859	6.8063	0.0075	17214.8	23917.3	75.723	12.12	16.24	3020
1800	0.62734	7.6311	0.0067	19679.7	27207.8	77.655	12.55	16.67	3184
2000	0.69613	8.4559	0.0060	22234.2	30587.7	79.446	12.99	17.12	3338
2500	0.86894	10.5178	0.0048	29102.7	39530.0	83.416	14.72	18.95	3680
3000	1.04737	12.5798	0.0040	37405.3	49973.3	87.227	18.73	23.48	3965

* TWO-PHASE BOUNDARY

Table 4.12 (continued)
THERMODYNAMIC PROPERTIES OF NORMAL HYDROGEN

12.00 MPA ISOBAR

TEMPERATURE DEG. K	DENSITY KG/CU M	$V(\partial H/\partial V)_P$ KJ/KG	$V(\partial P/\partial U)_V$ MPA-CU M/KJ $\times 10^{-2}$	$-V(\partial P/\partial V)_T$ MPA	$(\partial V/\partial T)_P/V$ 1/DEG. K	THERMAL CONDUCTIVITY W/K-M $\times 10^{3}$	VISCOSITY KG/M-S $\times 10^{7}$	THERMAL DIFFUSIVITY SQ M/HR	DIELECTRIC CONSTANT	PRANDTL NUMBER
* 17.346	82.6756	977.1	0.24393	153.1232	0.00678	111.38	329.05	0.00073	1.27159	1.9572
18	82.3055	984.2	0.23990	149.5335	0.00694	114.78	306.11	0.00074	1.27027	1.8211
19	81.7264	993.1	0.23428	144.0750	0.00713	118.83	276.63	0.00073	1.26822	1.6609
20	81.1312	999.7	0.22926	138.6583	0.00744	122.68	252.39	0.00073	1.26611	1.5293
21	80.5315	1001.9	0.22488	132.9611	0.00772	126.94	232.38	0.00073	1.26398	1.4161
22	79.9020	1004.2	0.22092	127.6311	0.00800	130.37	215.25	0.00073	1.26175	1.3267
23	79.2534	1004.6	0.21711	122.2533	0.00829	133.10	200.58	0.00073	1.25946	1.2547
24	78.5874	1002.8	0.21339	116.7973	0.00859	135.25	187.90	0.00072	1.25711	1.1966
25	77.9078	1004.0	0.20984	112.1188	0.00884	136.90	176.88	0.00071	1.25471	1.1472
26	77.2094	1003.1	0.20637	107.3243	0.00912	138.06	167.16	0.00070	1.25225	1.1075
27	76.4976	1003.4	0.20295	102.9186	0.00937	138.95	158.57	0.00069	1.24975	1.0742
28	75.7724	1001.0	0.19973	98.3287	0.00967	139.32	150.91	0.00068	1.24720	1.0483
29	75.0316	997.5	0.19654	93.7885	0.00997	139.52	144.03	0.00067	1.24461	1.0269
30	74.2761	992.6	0.19348	89.3044	0.01029	139.48	137.80	0.00066	1.24196	1.0093
31	73.5041	987.6	0.19037	84.9711	0.01061	138.71	132.13	0.00065	1.23926	0.9936
32	72.7171	983.7	0.18727	80.9802	0.01092	137.78	126.94	0.00064	1.23651	0.9894
33	71.9154	978.3	0.18424	77.0050	0.01124	136.73	122.16	0.00062	1.23371	0.9825
34	71.1002	972.4	0.18126	73.1453	0.01157	135.52	117.75	0.00061	1.23088	0.9770
35	70.2703	966.5	0.17828	69.4894	0.01190	134.42	113.66	0.00060	1.22799	0.9725
36	69.4279	960.1	0.17529	65.9188	0.01224	133.15	109.84	0.00059	1.22507	0.9695
37	68.5720	954.5	0.17232	62.6369	0.01256	131.82	106.28	0.00058	1.22210	0.9603
38	67.7045	947.8	0.16945	59.4184	0.01289	130.44	102.93	0.00057	1.21910	0.9643
39	66.8261	940.7	0.16673	56.3441	0.01323	129.03	99.79	0.00056	1.21606	0.9626
40	65.9356	933.5	0.16407	53.4161	0.01357	127.59	96.83	0.00055	1.21299	0.9615
42	64.1275	918.5	0.15878	47.9026	0.01423	124.62	91.39	0.00053	1.20677	0.9620
44	62.2808	905.8	0.15351	43.0902	0.01495	121.63	86.50	0.00052	1.20044	0.9629
46	60.4123	898.3	0.14761	39.0098	0.01551	118.69	82.11	0.00051	1.19405	0.9640
48	58.5352	890.3	0.14225	35.3675	0.01605	115.86	78.17	0.00050	1.18766	0.9643
50	56.6572	884.2	0.13721	32.2400	0.01650	113.15	74.64	0.00049	1.18129	0.9623
52	54.7950	879.2	0.13251	29.5076	0.01688	110.60	71.49	0.00049	1.17500	0.9596
54	52.9589	876.0	0.12817	27.1606	0.01718	108.27	68.71	0.00049	1.16881	0.9551
56	51.1592	874.4	0.12423	25.1500	0.01739	106.28	66.24	0.00049	1.16277	0.9477
58	49.4050	875.2	0.12073	23.4846	0.01746	104.43	64.09	0.00050	1.15691	0.9380
60	47.7036	877.8	0.11751	22.0559	0.01746	102.89	62.21	0.00051	1.15126	0.9268
65	43.7550	894.0	0.11037	19.4053	0.01707	100.07	58.59	0.00054	1.13815	0.8936
70	40.2475	922.1	0.10430	17.6700	0.01631	97.84	56.20	0.00058	1.12601	0.8636
75	37.1910	961.7	0.09903	16.5807	0.01530	95.52	54.74	0.00063	1.11662	0.8430
80	34.5429	1010.0	0.09439	15.6651	0.01425	94.18	53.93	0.00068	1.10801	0.8240
85	32.2500	1064.0	0.09021	15.3363	0.01328	94.06	53.61	0.00075	1.10060	0.8002
90	30.2433	1123.4	0.08640	14.9467	0.01240	95.17	53.61	0.00081	1.09414	0.7846
95	28.4826	1186.7	0.08299	14.6446	0.01161	95.74	53.85	0.00088	1.08849	0.7751
100	26.9253	1254.0	0.07982	14.4084	0.01091	97.57	54.15	0.00095	1.08352	0.7590
120	22.1672	1564.0	0.06878	13.8769	0.00870	110.11	59.77	0.00131	1.06841	0.7388
140	19.9214	1892.2	0.06161	13.6180	0.00722	120.20	63.60	0.00167	1.05819	0.7248
160	16.5567	2240.4	0.05656	13.4531	0.00619	129.00	66.65	0.00203	1.05079	0.7156
180	14.7907	2578.2	0.05287	13.2823	0.00544	137.02	69.34	0.00238	1.04528	0.7100
200	13.3517	2926.7	0.05012	13.1900	0.00484	144.47	71.88	0.00275	1.04081	0.7044
220	12.1790	3269.7	0.04818	13.1110	0.00437	151.52	74.35	0.00313	1.03718	0.7014
240	11.2029	3607.1	0.04673	13.0422	0.00399	158.19	76.77	0.00353	1.03417	0.6991
260	10.3766	3938.7	0.04562	12.9814	0.00369	164.54	79.19	0.00394	1.03162	0.6971
280	9.6672	4264.0	0.04477	12.9272	0.00341	170.64	81.60	0.00437	1.02943	0.6955
300	9.0407	4595.0	0.04410	12.9053	0.00317	176.51	84.01	0.00482	1.02751	0.6936
350	7.8071	5360.6	0.04303	12.7940	0.00272	190.52	90.05	0.00601	1.02372	0.6903
400	6.8741	6121.3	0.04242	12.7058	0.00239	203.95	96.11	0.00731	1.02087	0.6881
450	6.1424	6862.3	0.04203	12.6343	0.00213	217.07	102.15	0.00872	1.01863	0.6866
500	5.5527	7602.5	0.04171	12.5755	0.00192	230.13	108.16	0.01023	1.01683	0.6855
550	5.0659	8340.5	0.04146	12.5264	0.00175	243.06	114.12	0.01184	1.01535	0.6848
600	4.6597	9081.9	0.04123	12.4847	0.00161	255.98	120.03	0.01355	1.01411	0.6842
700	4.0150	10581.8	0.04076	12.4180	0.00138	281.84	131.67	0.01727	1.01215	0.6836
800	3.5275	12119.2	0.04023	12.3672	0.00121	307.97	143.04	0.02136	1.01067	0.6834
900	3.1455	13708.4	0.03962	12.3272	0.00109	334.59	154.15	0.02581	1.00951	0.6835
1000	2.8385	15358.3	0.03892	12.2950	0.00098	440.56	165.01	0.03727	1.00858	0.5616
1200	2.3749	18822.6	0.03745	12.2463	0.00082	511.00	186.02	0.05040	1.00717	0.5595
1400	2.0415	22516.8	0.03595	12.2113	0.00070	583.17	206.18	0.06508	1.00616	0.5587
1600	1.7902	26397.8	0.03456	12.1849	0.00062	656.32	225.61	0.08125	1.00540	0.5584
1800	1.5940	30431.4	0.03332	12.1642	0.00055	730.36	244.41	0.09892	1.00481	0.5580
2000	1.4365	34678.1	0.03214	12.1471	0.00049	808.28	262.64	0.11828	1.00433	0.5565
2500	1.1508	47788.9	0.02834	12.1042	0.00040	1080.98	306.32	0.17848	1.00347	0.5369
3000	0.9548	70529.5	0.02231	12.0108	0.00033	1695.49	348.70	0.27223	1.00288	0.4830

* TWO-PHASE BOUNDARY

Table 4.12 (continued)
THERMODYNAMIC PROPERTIES OF NORMAL HYDROGEN

14.00 MPA ISOBAR

TEMPERATURE DEG. K	VOLUME CU M/KG	ISOTHERM DERIVATIVE CU M-MPA/KG	ISOCHORE DERIVATIVE MPA/K	INTERNAL ENERGY KJ/KG-K	ENTHALPY KJ/KG	ENTROPY KJ/KG-K	CV KJ /	CP KG-K	VELOCITY OF SOUND M/S
* 17.869	0.01198	1.9517	1.052	229.2	396.9	14.374	5.21	6.67	1580
18	0.01199	1.9449	1.052	229.9	397.8	14.423	5.24	6.71	1579
19	0.01207	1.8929	1.051	235.6	404.7	14.793	5.39	7.00	1569
20	0.01216	1.8408	1.049	241.6	411.8	15.160	5.53	7.29	1559
21	0.01224	1.7889	1.045	247.9	419.3	15.524	5.65	7.57	1548
22	0.01233	1.7372	1.039	254.3	427.0	15.883	5.77	7.85	1537
23	0.01242	1.6804	1.036	261.0	434.9	16.237	5.87	8.14	1526
24	0.01252	1.6297	1.027	267.9	443.2	16.589	5.97	8.41	1515
25	0.01262	1.5784	1.017	275.0	451.7	16.937	6.06	8.67	1503
26	0.01273	1.5244	1.006	282.4	460.5	17.281	6.14	8.93	1489
27	0.01283	1.4799	0.9933	289.9	469.6	17.623	6.21	9.18	1479
28	0.01295	1.4333	0.9796	297.6	478.9	17.960	6.28	9.43	1466
29	0.01306	1.3879	0.9652	305.6	488.4	18.296	6.35	9.67	1454
30	0.01318	1.3439	0.9509	313.7	498.2	18.627	6.40	9.91	1442
31	0.01331	1.2995	0.9355	321.9	508.3	18.956	6.46	10.16	1430
32	0.01344	1.2548	0.9194	330.4	518.5	19.283	6.51	10.40	1416
33	0.01357	1.2134	0.9025	339.0	529.1	19.606	6.55	10.63	1403
34	0.01371	1.1721	0.8851	347.8	539.8	19.927	6.59	10.86	1390
35	0.01386	1.1320	0.8672	356.8	550.8	20.246	6.62	11.09	1377
36	0.01401	1.0925	0.8490	365.9	562.0	20.561	6.66	11.32	1363
37	0.01416	1.0568	0.8303	375.1	573.4	20.874	6.69	11.53	1350
38	0.01432	1.0212	0.8114	384.5	585.1	21.185	6.71	11.74	1336
39	0.01449	0.9864	0.7923	394.0	596.9	21.492	6.72	11.94	1323
40	0.01466	0.9529	0.7731	403.7	608.9	21.797	6.74	12.13	1310
42	0.01502	0.8909	0.7347	423.3	633.6	22.398	6.76	12.50	1284
44	0.01541	0.8333	0.6969	443.3	658.9	22.988	6.78	12.87	1257
46	0.01582	0.7799	0.6598	463.1	684.5	23.556	6.84	13.26	1230
48	0.01625	0.7351	0.6238	483.8	711.4	24.127	6.87	13.58	1205
50	0.01671	0.6957	0.5891	504.8	738.8	24.688	6.90	13.86	1183
52	0.01720	0.6606	0.5560	526.0	766.8	25.237	6.92	14.11	1161
54	0.01771	0.6313	0.5246	547.3	795.3	25.774	6.93	14.31	1142
56	0.01824	0.6055	0.4951	568.9	824.3	26.303	6.94	14.48	1124
58	0.01880	0.5834	0.4674	590.2	853.4	26.813	6.93	14.61	1109
60	0.01937	0.5657	0.4416	611.5	882.7	27.310	6.92	14.69	1095
65	0.02089	0.5344	0.3847	663.9	956.4	28.489	6.92	14.77	1068
70	0.02250	0.5198	0.3378	715.1	1030.2	29.582	6.92	14.70	1051
75	0.02418	0.5164	0.2992	764.8	1103.3	30.592	6.94	14.54	1040
80	0.02589	0.5221	0.2676	813.2	1175.6	31.524	6.99	14.34	1035
85	0.02761	0.5335	0.2415	860.3	1246.8	32.388	7.05	14.13	1034
90	0.02933	0.5479	0.2199	906.3	1317.0	33.192	7.14	13.97	1036
95	0.03106	0.5648	0.2018	951.7	1386.5	33.943	7.24	13.84	1039
100	0.03278	0.5833	0.1865	996.6	1455.5	34.651	7.35	13.76	1045
120	0.03958	0.6666	0.1431	1177.2	1731.3	37.164	7.96	13.73	1073
140	0.04623	0.7567	0.1163	1359.2	2006.4	39.284	8.47	13.82	1111
160	0.05272	0.8476	0.0980	1545.6	2283.8	41.135	8.92	13.96	1152
180	0.05909	0.9381	0.0849	1736.8	2564.1	42.786	9.28	14.10	1194
200	0.06537	1.0274	0.0749	1932.1	2847.2	44.278	9.56	14.23	1236
220	0.07138	1.1089	0.0672	2130.7	3130.0	45.625	9.79	14.36	1275
240	0.07753	1.1967	0.0610	2332.5	3418.0	46.879	9.97	14.46	1318
260	0.08365	1.2837	0.0559	2536.7	3707.8	48.039	10.10	14.53	1359
280	0.08973	1.3701	0.0516	2742.7	3998.9	49.117	10.20	14.58	1399
300	0.09592	1.4620	0.0479	2951.8	4294.7	50.132	10.28	14.61	1441
350	0.11093	1.6734	0.0407	3472.9	5025.9	52.388	10.37	14.63	1536
400	0.12585	1.8830	0.0354	3995.4	5757.3	54.341	10.40	14.62	1627
450	0.14071	2.0915	0.0314	4518.1	6488.1	56.062	10.41	14.60	1713
500	0.15554	2.2993	0.0282	5040.5	7218.1	57.600	10.42	14.60	1794
550	0.17035	2.5067	0.0256	5562.8	7947.6	58.991	10.43	14.59	1873
600	0.18514	2.7137	0.0234	6085.6	8677.5	60.258	10.45	14.60	1947
700	0.21468	3.1271	0.0200	7133.4	10138.9	62.514	10.50	14.64	2088
800	0.24419	3.5401	0.0175	8188.0	11606.6	64.475	10.58	14.72	2219
900	0.27369	3.9528	0.0156	9252.7	13084.4	66.214	10.71	14.84	2340
1000	0.30317	4.3654	0.0140	10330.9	14575.3	67.783	10.87	15.00	2454
1200	0.36213	5.1903	0.0117	12542.6	17612.3	70.532	11.25	15.37	2663
1400	0.42106	6.0150	0.0100	14836.0	20730.9	72.942	11.68	15.80	2853
1600	0.47999	6.8397	0.0087	17215.0	23934.9	75.087	12.12	16.24	3028
1800	0.53893	7.6644	0.0078	19679.8	27224.8	77.019	12.55	16.67	3191
2000	0.59788	8.4891	0.0070	22233.4	30603.8	78.809	12.99	17.12	3345
2500	0.74595	10.5509	0.0056	29084.7	39527.9	82.772	14.64	18.86	3687
3000	0.89849	12.6127	0.0047	37289.7	49868.5	86.546	18.44	23.11	3973

* TWO-PHASE BOUNDARY

Table 4.12 (continued)
THERMODYNAMIC PROPERTIES OF NORMAL HYDROGEN

14.00 MPA ISOBAR

TEMPERATURE DEG. K	DENSITY KG/CU M	$V(DH/DV)_P$ KJ/KG	$V(DP/DU)_V$ MPA-CU M/KJ X 10^2	$-V(DP/DV)_T$ MPA	$(DV/DT)_P/V$ 1/DEG. K	THERMAL CONDUCTIVITY W/K-M X 10^3	VISCOSITY KG/M-S X 10^7	THERMAL DIFFUSIVITY SQ M/HR	DIELECTRIC CONSTANT	PRANDTL NUMBER
* 17.869	83.4399	1032.3	0.24187	162.8525	0.00646	116.52	341.17	0.00075	1.27431	1.9531
18	83.3689	1033.9	0.24110	162.1439	0.00649	117.20	336.21	0.00075	1.27405	1.9250
19	82.8207	1044.6	0.23561	156.7682	0.00671	121.54	302.68	0.00075	1.27210	1.7445
20	82.2581	1052.9	0.23071	151.4246	0.00693	125.67	275.26	0.00075	1.27010	1.5972
21	81.6813	1059.2	0.22626	146.1222	0.00715	130.19	252.51	0.00076	1.26806	1.4687
22	81.0903	1063.9	0.22214	140.8714	0.00738	133.89	233.37	0.00076	1.26596	1.3678
23	80.4940	1063.1	0.21902	135.2625	0.00766	136.91	217.21	0.00075	1.26385	1.2913
24	79.8708	1065.2	0.21544	130.1661	0.00789	139.29	203.13	0.00075	1.26164	1.2261
25	79.2334	1065.7	0.21197	125.0644	0.00814	141.16	190.92	0.00074	1.25939	1.1725
26	78.5812	1063.8	0.20856	119.7910	0.00840	142.53	180.21	0.00073	1.25709	1.1295
27	77.9182	1065.5	0.20518	115.3118	0.00861	143.51	170.79	0.00072	1.25475	1.0922
28	77.2394	1065.2	0.20185	110.7064	0.00885	144.16	162.39	0.00071	1.25236	1.0617
29	76.5489	1064.3	0.19865	106.2418	0.00909	144.53	154.88	0.00070	1.24993	1.0362
30	75.8487	1062.5	0.19578	101.9367	0.00933	144.66	148.14	0.00069	1.24747	1.0150
31	75.1334	1059.9	0.19285	97.6386	0.00958	144.03	142.01	0.00068	1.24496	1.0012
32	74.4075	1056.1	0.18992	93.3670	0.00985	143.25	136.44	0.00067	1.24242	0.9905
33	73.6684	1053.0	0.18705	89.3899	0.01010	142.34	131.33	0.00065	1.23983	0.9809
34	72.9202	1049.0	0.18422	85.4730	0.01036	141.37	126.64	0.00064	1.23722	0.9731
35	72.1585	1044.6	0.18142	81.6823	0.01062	140.31	122.29	0.00063	1.23456	0.9667
36	71.3871	1039.7	0.17864	77.9869	0.01089	139.17	118.26	0.00062	1.23187	0.9618
37	70.6058	1036.0	0.17589	74.6181	0.01113	137.97	114.51	0.00061	1.22916	0.9568
38	69.8145	1031.1	0.17324	71.2911	0.01138	136.71	111.01	0.00060	1.22641	0.9529
39	69.0158	1025.5	0.17071	68.0748	0.01164	135.43	107.73	0.00059	1.22364	0.9495
40	68.2087	1019.8	0.16823	64.9973	0.01189	134.11	104.65	0.00058	1.22084	0.9465
42	66.5716	1009.1	0.16328	59.3071	0.01239	131.38	99.00	0.00057	1.21518	0.9421
44	64.9100	998.8	0.15830	54.0908	0.01288	128.62	93.96	0.00055	1.20946	0.9400
46	63.2267	990.9	0.15266	49.3116	0.01338	125.88	89.42	0.00054	1.20368	0.9418
48	61.5326	984.8	0.14756	45.2351	0.01379	123.21	85.33	0.00053	1.19788	0.9406
50	59.8340	979.6	0.14277	41.6279	0.01415	120.63	81.64	0.00052	1.19208	0.9383
52	58.1427	975.1	0.13826	38.4099	0.01448	118.17	78.32	0.00052	1.18633	0.9355
54	56.4686	972.6	0.13405	35.6485	0.01472	115.86	75.35	0.00052	1.18065	0.9307
56	54.8188	970.8	0.13021	33.1922	0.01492	113.71	72.69	0.00052	1.17508	0.9256
58	53.1998	969.8	0.12677	31.0358	0.01506	111.68	70.31	0.00052	1.16962	0.9196
60	51.6176	971.2	0.12354	29.1990	0.01512	110.12	68.21	0.00052	1.16431	0.9098
65	47.8611	982.2	0.11623	25.5765	0.01504	107.17	63.98	0.00055	1.15176	0.8820
70	44.4354	1005.4	0.10983	23.0981	0.01462	104.68	60.99	0.00058	1.14040	0.8566
75	41.3625	1038.3	0.10419	21.3608	0.01401	102.18	58.97	0.00061	1.13027	0.8394
80	38.6321	1080.8	0.09913	20.1711	0.01327	100.39	57.67	0.00065	1.12132	0.8237
85	36.2217	1131.0	0.09454	19.3230	0.01250	100.55	56.93	0.00071	1.11346	0.8003
90	34.0900	1186.9	0.09036	18.6790	0.01177	100.75	56.60	0.00076	1.10655	0.7849
95	32.1975	1247.5	0.08661	18.1847	0.01110	101.04	56.57	0.00082	1.10043	0.7751
100	30.5058	1312.8	0.08315	17.7952	0.01048	102.49	56.59	0.00088	1.09498	0.7595
120	25.2655	1616.0	0.07119	16.8430	0.00850	114.24	61.75	0.00119	1.07823	0.7420
140	21.6318	1945.3	0.06343	16.3678	0.00710	123.51	65.14	0.00149	1.06672	0.7287
160	18.9674	2289.0	0.05798	16.0775	0.00610	131.67	67.87	0.00179	1.05833	0.7195
180	16.9222	2637.9	0.05407	15.8753	0.00535	139.22	70.34	0.00210	1.05193	0.7125
200	15.2983	2985.5	0.05120	15.7172	0.00476	146.34	72.72	0.00242	1.04686	0.7068
220	14.0096	3317.3	0.04903	15.5346	0.00433	153.15	75.07	0.00274	1.04286	0.7038
240	12.8974	3656.0	0.04748	15.4341	0.00395	159.65	77.42	0.00308	1.03941	0.7011
260	11.9548	3988.7	0.04630	15.3466	0.00364	165.89	79.78	0.00344	1.03649	0.6989
280	11.1448	4315.2	0.04539	15.2694	0.00338	171.92	82.16	0.00381	1.03399	0.6970
300	10.4251	4650.6	0.04467	15.2420	0.00314	177.75	84.55	0.00420	1.03177	0.6949
350	9.0149	5425.6	0.04351	15.0854	0.00270	191.74	90.58	0.00523	1.02743	0.6912
400	7.9461	6179.4	0.04282	14.9625	0.00237	205.23	96.66	0.00636	1.02415	0.6887
450	7.1067	6921.0	0.04238	14.8637	0.00211	218.43	102.75	0.00758	1.02158	0.6869
500	6.4291	7661.6	0.04202	14.7826	0.00191	231.61	108.81	0.00889	1.01951	0.6857
550	5.8704	8400.0	0.04174	14.7150	0.00174	244.68	114.84	0.01028	1.01780	0.6849
600	5.4014	9141.7	0.04148	14.6579	0.00160	257.74	120.82	0.01177	1.01637	0.6843
700	4.6581	10642.0	0.04098	14.5667	0.00138	283.89	132.59	0.01499	1.01411	0.6836
800	4.0951	12180.0	0.04042	14.4973	0.00121	310.32	144.11	0.01854	1.01239	0.6834
900	3.6538	13769.7	0.03978	14.4428	0.00108	337.26	155.36	0.02240	1.01105	0.6835
1000	3.2984	15420.3	0.03906	14.3990	0.00097	440.56	166.35	0.03207	1.00997	0.5662
1200	2.7615	18886.2	0.03756	14.3328	0.00081	511.00	187.63	0.04334	1.00834	0.5644
1400	2.3749	22582.1	0.03604	14.2853	0.00070	583.17	208.06	0.05594	1.00717	0.5638
1600	2.0834	26464.2	0.03464	14.2496	0.00061	656.29	227.74	0.06981	1.00629	0.5637
1800	1.8555	30496.1	0.03340	14.2216	0.00055	730.19	246.78	0.08497	1.00560	0.5635
2000	1.6726	34731.7	0.03222	14.1986	0.00049	807.63	265.26	0.10156	1.00505	0.5622
2500	1.3406	47654.5	0.02853	14.1443	0.00040	1073.04	309.48	0.15278	1.00404	0.5440
3000	1.1130	69525.3	0.02273	14.0377	0.00033	1654.59	352.30	0.23163	1.00336	0.4920

* TWO-PHASE BOUNDARY

Table 4.12 (continued)
THERMODYNAMIC PROPERTIES OF NORMAL HYDROGEN

16.00 MPA ISOBAR

TEMPERATURE DEG. K	VOLUME CU M/KG	ISOTHERM DERIVATIVE CU M-MPA/KG	ISOCHORE DERIVATIVE MPA/K	INTERNAL ENERGY KJ/KG-K	ENTHALPY KJ/KG	ENTROPY KJ/KG-K	CV KJ / KG-K	CP KJ / KG-K	VELOCITY OF SOUND M/S
* 18.378	0.01188	2.0484	1.066	231.4	421.5	14.411	5.28	6.71	1615
19	0.01193	2.0171	1.066	234.9	425.7	14.637	5.37	6.89	1609
20	0.01200	1.9667	1.065	240.7	432.8	14.998	5.51	7.17	1600
21	0.01208	1.9163	1.062	246.7	440.1	15.356	5.64	7.44	1591
22	0.01217	1.8661	1.058	253.0	447.6	15.709	5.75	7.70	1581
23	0.01225	1.8161	1.052	259.5	455.5	16.057	5.86	7.96	1571
24	0.01234	1.7664	1.045	266.1	463.6	16.402	5.96	8.22	1561
25	0.01243	1.7119	1.041	273.0	471.9	16.741	6.05	8.49	1550
26	0.01253	1.6642	1.031	280.1	480.5	17.077	6.13	8.74	1540
27	0.01262	1.6158	1.020	287.4	489.4	17.412	6.21	8.98	1528
28	0.01273	1.5644	1.007	294.8	498.5	17.742	6.28	9.22	1515
29	0.01283	1.5218	0.9937	302.5	507.8	18.070	6.35	9.45	1505
30	0.01294	1.4775	0.9796	310.3	517.3	18.394	6.41	9.67	1493
31	0.01306	1.4313	0.9650	318.2	527.1	18.715	6.46	9.90	1481
32	0.01317	1.3906	0.9504	326.4	537.1	19.032	6.52	10.12	1470
33	0.01329	1.3492	0.9351	334.7	547.4	19.347	6.56	10.34	1458
34	0.01342	1.3072	0.9192	343.1	557.8	19.659	6.61	10.56	1446
35	0.01355	1.2689	0.9026	351.7	568.5	19.969	6.64	10.77	1434
36	0.01368	1.2299	0.8857	360.5	579.4	20.275	6.68	10.97	1422
37	0.01381	1.1921	0.8684	369.4	590.4	20.579	6.71	11.18	1409
38	0.01396	1.1571	0.8508	378.4	601.7	20.879	6.74	11.37	1397
39	0.01410	1.1222	0.8330	387.5	613.2	21.177	6.75	11.55	1385
40	0.01425	1.0890	0.8151	396.8	624.8	21.472	6.77	11.73	1373
42	0.01456	1.0265	0.7789	415.6	648.6	22.052	6.79	12.06	1350
44	0.01490	0.9676	0.7429	434.7	673.1	22.620	6.82	12.39	1326
46	0.01525	0.9144	0.7075	453.7	697.7	23.166	6.88	12.73	1301
48	0.01562	0.8633	0.6729	473.6	723.4	23.715	6.92	13.05	1277
50	0.01601	0.8176	0.6391	493.7	749.8	24.254	6.95	13.34	1253
52	0.01641	0.7824	0.6068	514.1	776.7	24.781	6.97	13.56	1234
54	0.01684	0.7466	0.5758	534.6	804.0	25.297	6.98	13.78	1214
56	0.01729	0.7195	0.5463	555.4	832.0	25.806	6.99	13.93	1198
58	0.01775	0.6931	0.5184	576.0	860.0	26.298	6.98	14.07	1182
60	0.01823	0.6719	0.4922	596.6	888.2	26.776	6.98	14.17	1168
65	0.01949	0.6301	0.4336	647.7	959.5	27.916	6.97	14.34	1138
70	0.02083	0.6056	0.3840	698.0	1031.3	28.981	6.97	14.37	1117
75	0.02224	0.5943	0.3425	747.3	1103.1	29.971	7.00	14.31	1103
80	0.02368	0.5913	0.3076	795.6	1174.4	30.892	7.04	14.22	1093
85	0.02515	0.5968	0.2785	842.8	1245.2	31.750	7.10	14.09	1088
90	0.02663	0.6071	0.2539	889.3	1315.3	32.552	7.19	13.96	1086
95	0.02811	0.6208	0.2332	935.1	1384.8	33.304	7.28	13.86	1087
100	0.02959	0.6361	0.2155	980.4	1453.9	34.012	7.39	13.79	1089
120	0.03550	0.7105	0.1654	1163.0	1730.9	36.537	7.99	13.81	1108
140	0.04131	0.7966	0.1342	1346.8	2007.7	38.670	8.51	13.91	1141
160	0.04700	0.8856	0.1130	1534.9	2287.0	40.533	8.95	14.05	1179
180	0.05259	0.9744	0.0977	1727.5	2569.0	42.195	9.31	14.19	1219
200	0.05810	1.0633	0.0861	1924.1	2853.7	43.695	9.59	14.30	1259
220	0.06353	1.1504	0.0770	2123.9	3140.4	45.061	9.82	14.39	1299
240	0.06875	1.2299	0.0700	2326.2	3426.1	46.305	9.99	14.50	1336
260	0.07411	1.3170	0.0641	2531.0	3716.8	47.468	10.12	14.58	1377
280	0.07945	1.4035	0.0592	2737.5	4008.7	48.550	10.22	14.62	1417
300	0.08491	1.4973	0.0548	2947.3	4305.9	49.569	10.30	14.64	1459
350	0.09806	1.7085	0.0466	3469.5	5038.5	51.828	10.38	14.65	1553
400	0.11113	1.9180	0.0405	3992.8	5770.8	53.784	10.41	14.64	1642
450	0.12415	2.1264	0.0359	4516.0	6502.3	55.507	10.42	14.62	1727
500	0.13713	2.3341	0.0322	5038.8	7232.9	57.045	10.43	14.61	1808
550	0.15009	2.5413	0.0292	5561.4	7962.8	58.437	10.44	14.60	1885
600	0.16303	2.7482	0.0268	6084.5	8693.0	59.705	10.45	14.60	1960
700	0.18889	3.1614	0.0229	7132.7	10154.9	61.962	10.50	14.64	2100
800	0.21471	3.5742	0.0200	8187.6	11623.0	63.924	10.58	14.72	2229
900	0.24052	3.9867	0.0178	9252.5	13100.9	65.663	10.71	14.84	2350
1000	0.26632	4.3991	0.0160	10330.8	14592.0	67.232	10.87	15.00	2464
1200	0.31791	5.2237	0.0133	12542.7	17629.2	69.981	11.25	15.37	2672
1400	0.36948	6.0483	0.0114	14836.3	20748.0	72.391	11.68	15.80	2861
1600	0.42104	6.8728	0.0100	17215.3	23952.0	74.537	12.12	16.24	3035
1800	0.47261	7.6974	0.0089	19680.0	27241.8	76.468	12.55	16.67	3198
2000	0.52419	8.5219	0.0080	22232.9	30620.0	78.258	12.98	17.11	3352
2500	0.65371	10.5835	0.0064	29070.3	39529.6	82.214	14.58	18.79	3694
3000	0.78689	12.6451	0.0053	37196.8	49787.0	85.959	18.18	22.80	3981

* TWO-PHASE BOUNDARY

Table 4.12 (continued)
THERMODYNAMIC PROPERTIES OF NORMAL HYDROGEN

16.00 MPA ISOBAR

TEMPERATURE DEG. K	DENSITY KG/CU M	$V(DH/DV)_P$ KJ/KG	$V(DP/DU)_V$ MPA-CU M/KJ X 10^2	$-V(DP/DV)_T$ MPA	$(DV/DT)_P/V$ 1/DEG. K	THERMAL CONDUCTIVITY W/K-M X 10^3	VISCOSITY KG/M-S X 10^7	THERMAL DIFFUSIVITY SQ M/HR	DIELECTRIC CONSTANT	PRANDTL NUMBER
* 18.378	84.1703	1085.8	0.24014	172.4170	0.00618	121.39	353.30	0.00077	1.27691	1.9543
19	83.8438	1093.2	0.23686	169.1204	0.00630	124.11	330.57	0.00077	1.27574	1.8357
20	83.3088	1103.2	0.23204	163.8434	0.00650	128.51	299.59	0.00077	1.27384	1.6714
21	82.7610	1111.2	0.22767	158.5978	0.00670	133.30	273.99	0.00078	1.27189	1.5293
22	82.2006	1117.5	0.22362	153.3933	0.00689	137.24	252.55	0.00078	1.26990	1.4178
23	81.6276	1122.3	0.21989	148.2402	0.00710	140.47	234.38	0.00078	1.26787	1.3287
24	81.0422	1125.8	0.21637	143.1502	0.00730	143.08	218.82	0.00077	1.26579	1.2567
25	80.4489	1123.8	0.21388	137.7179	0.00756	145.17	205.41	0.00076	1.26369	1.2016
26	79.8365	1126.0	0.21059	132.8603	0.00776	146.73	193.63	0.00076	1.26152	1.1529
27	79.2100	1127.0	0.20728	127.9866	0.00797	147.88	183.25	0.00075	1.25931	1.1125
28	78.5728	1125.7	0.20401	122.9176	0.00819	148.69	174.06	0.00074	1.25706	1.0796
29	77.9254	1127.5	0.20084	118.5882	0.00838	149.22	165.88	0.00073	1.25478	1.0503
30	77.2670	1127.1	0.19786	114.1641	0.00858	149.50	158.54	0.00072	1.25246	1.0257
31	76.5983	1125.0	0.19489	109.6379	0.00880	148.99	151.91	0.00071	1.25011	1.0095
32	75.9211	1124.4	0.19212	105.5768	0.00900	148.33	145.91	0.00069	1.24773	0.9957
33	75.2310	1122.6	0.18939	101.4996	0.00921	147.53	140.41	0.00068	1.24530	0.9843
34	74.5334	1119.4	0.18671	97.4322	0.00943	146.67	135.39	0.00067	1.24286	0.9748
35	73.8266	1117.4	0.18404	93.6801	0.00964	145.72	130.76	0.00066	1.24039	0.9662
36	73.1104	1114.2	0.18137	89.9185	0.00985	144.68	126.48	0.00065	1.23788	0.9594
37	72.3861	1110.7	0.17876	86.2894	0.01006	143.58	122.51	0.00064	1.23536	0.9537
38	71.6542	1107.6	0.17627	82.9097	0.01026	142.43	118.81	0.00063	1.23280	0.9482
39	70.9152	1103.4	0.17391	79.5832	0.01047	141.24	115.36	0.00062	1.23023	0.9434
40	70.1694	1099.3	0.17160	76.4173	0.01067	140.01	112.12	0.00061	1.22764	0.9390
42	68.6604	1091.3	0.16696	70.4824	0.01105	137.45	106.22	0.00060	1.22240	0.9321
44	67.1339	1083.3	0.16225	64.9599	0.01144	134.85	100.98	0.00058	1.21712	0.9277
46	65.5916	1079.2	0.15683	59.9771	0.01180	132.27	96.27	0.00057	1.21180	0.9266
48	64.0397	1072.6	0.15193	55.2845	0.01217	129.73	92.03	0.00056	1.20647	0.9262
50	62.4792	1066.6	0.14729	51.0822	0.01251	127.25	88.19	0.00055	1.20112	0.9249
52	60.9229	1065.3	0.14295	47.6678	0.01273	124.87	84.72	0.00054	1.19580	0.9200
54	59.3772	1061.2	0.13887	44.3290	0.01299	122.62	81.59	0.00054	1.19053	0.9171
56	57.8466	1061.3	0.13512	41.6200	0.01313	120.49	78.75	0.00054	1.18532	0.9105
58	56.3397	1059.8	0.13176	39.0511	0.01328	118.50	76.21	0.00054	1.18022	0.9048
60	54.8590	1060.9	0.12859	36.8625	0.01335	116.80	73.92	0.00054	1.17521	0.8966
65	51.3040	1068.9	0.12127	32.3268	0.01341	113.58	69.21	0.00056	1.16326	0.8734
70	47.9986	1087.9	0.11472	29.0695	0.01321	110.93	65.71	0.00058	1.15222	0.8514
75	44.9729	1117.1	0.10883	26.7252	0.01281	108.36	63.20	0.00061	1.14217	0.8349
80	42.2333	1154.1	0.10348	24.9716	0.01232	106.25	61.47	0.00064	1.13313	0.8226
85	39.7686	1200.5	0.09859	23.7322	0.01173	106.16	60.33	0.00068	1.12504	0.8006
90	37.5586	1253.7	0.09408	22.8006	0.01114	106.11	59.66	0.00073	1.11782	0.7850
95	35.5736	1312.3	0.09002	22.0838	0.01056	106.15	59.34	0.00078	1.11136	0.7746
100	33.7906	1375.1	0.08628	21.4951	0.01003	107.38	59.08	0.00083	1.10558	0.7586
120	28.1711	1671.3	0.07349	20.0154	0.00826	118.51	63.85	0.00110	1.08750	0.7442
140	24.2091	1998.3	0.06519	19.2857	0.00696	127.01	66.82	0.00136	1.07487	0.7318
160	21.2755	2341.5	0.05937	18.8405	0.00600	134.51	69.21	0.00162	1.06559	0.7228
180	19.0141	2690.3	0.05520	18.5275	0.00527	141.56	71.43	0.00189	1.05848	0.7158
200	17.2120	3039.8	0.05213	18.3010	0.00470	148.32	73.63	0.00217	1.05283	0.7098
220	15.7408	3384.8	0.04983	18.1087	0.00425	154.87	75.85	0.00246	1.04824	0.7050
240	14.5460	3707.4	0.04817	17.8896	0.00391	161.19	78.12	0.00275	1.04452	0.7027
260	13.4930	4039.6	0.04697	17.7700	0.00361	167.30	80.41	0.00306	1.04125	0.7006
280	12.5871	4367.0	0.04600	17.6655	0.00335	173.24	82.75	0.00339	1.03845	0.6985
300	11.7769	4707.3	0.04523	17.6334	0.00311	179.02	85.12	0.00374	1.03594	0.6961
350	10.1977	5483.5	0.04397	17.4229	0.00267	192.98	91.13	0.00465	1.03107	0.6920
400	8.9985	6238.0	0.04322	17.2590	0.00235	206.50	97.23	0.00564	1.02738	0.6892
450	8.0551	6980.1	0.04273	17.1280	0.00209	219.79	103.35	0.00672	1.02448	0.6872
500	7.2924	7721.1	0.04233	17.0209	0.00189	233.09	109.47	0.00788	1.02215	0.6859
550	6.6628	8459.7	0.04202	16.9319	0.00173	246.28	115.55	0.00911	1.02022	0.6850
600	6.1338	9201.7	0.04174	16.8568	0.00159	259.48	121.60	0.01043	1.01860	0.6843
700	5.2942	10702.4	0.04119	16.7372	0.00137	285.92	133.51	0.01328	1.01604	0.6836
800	4.6574	12240.7	0.04060	16.6464	0.00120	312.67	145.16	0.01642	1.01410	0.6834
900	4.1576	13831.0	0.03994	16.5752	0.00107	339.93	156.56	0.01984	1.01258	0.6834
1000	3.7548	15482.3	0.03920	16.5179	0.00097	440.56	167.69	0.02817	1.01136	0.5708
1200	3.1456	18949.6	0.03768	16.4316	0.00081	511.00	189.25	0.03805	1.00951	0.5693
1400	2.7065	22647.1	0.03614	16.3698	0.00070	583.17	209.95	0.04908	1.00818	0.5690
1600	2.3751	26530.4	0.03472	16.3233	0.00061	656.27	229.89	0.06124	1.00717	0.5691
1800	2.1159	30561.1	0.03347	16.2870	0.00055	730.07	249.18	0.07451	1.00639	0.5690
2000	1.9077	34788.0	0.03229	16.2573	0.00049	807.10	267.90	0.08902	1.00576	0.5679
2500	1.5297	47560.1	0.02869	16.1900	0.00040	1066.63	312.69	0.13358	1.00462	0.5509
3000	1.2708	68730.0	0.02308	16.0698	0.00033	1621.58	355.97	0.20147	1.00383	0.5005

* TWO-PHASE BOUNDARY

Table 4.12 (continued)
THERMODYNAMIC PROPERTIES OF NORMAL HYDROGEN

18.00 MPA ISOBAR

TEMPERATURE DEG. K	VOLUME CU M/KG	ISOTHERM DERIVATIVE CU M-MPA/KG	ISOCHORE DERIVATIVE MPA/K	INTERNAL ENERGY KJ/KG-K	ENTHALPY KJ/KG	ENTROPY KJ/KG-K	CV KJ / KG-K	CP KJ / KG-K	VELOCITY OF SOUND M/S
* 18.874	0.01178	2.1426	1.080	233.7	445.8	14.447	5.33	6.76	1648
19	0.01179	2.1364	1.080	234.4	446.7	14.492	5.35	6.79	1647
20	0.01186	2.0875	1.080	240.1	453.6	14.847	5.49	7.06	1639
21	0.01194	2.0386	1.078	245.9	460.8	15.200	5.62	7.32	1630
22	0.01201	1.9897	1.075	252.0	468.3	15.547	5.74	7.58	1621
23	0.01209	1.9409	1.070	258.3	476.0	15.890	5.85	7.83	1612
24	0.01217	1.8923	1.064	264.8	483.9	16.228	5.95	8.07	1603
25	0.01226	1.8441	1.057	271.4	492.1	16.563	6.04	8.31	1593
26	0.01235	1.7964	1.049	278.3	500.5	16.892	6.12	8.55	1584
27	0.01244	1.7451	1.043	285.3	509.2	17.219	6.20	8.81	1574
28	0.01253	1.7001	1.032	292.6	518.1	17.542	6.28	9.04	1564
29	0.01263	1.6542	1.020	300.0	527.3	17.864	6.35	9.26	1553
30	0.01273	1.6057	1.007	307.5	536.6	18.182	6.41	9.48	1541
31	0.01283	1.5650	0.9933	315.3	546.2	18.495	6.47	9.69	1531
32	0.01294	1.5226	0.9789	323.1	556.0	18.807	6.52	9.89	1520
33	0.01305	1.4807	0.9642	331.2	566.0	19.114	6.57	10.10	1508
34	0.01316	1.4380	0.9496	339.3	576.2	19.419	6.62	10.31	1497
35	0.01328	1.3992	0.9343	347.7	586.6	19.721	6.66	10.51	1486
36	0.01339	1.3617	0.9186	356.1	597.2	20.020	6.70	10.70	1475
37	0.01352	1.3241	0.9024	364.7	608.0	20.316	6.73	10.89	1464
38	0.01364	1.2873	0.8859	373.4	619.0	20.609	6.76	11.07	1452
39	0.01377	1.2532	0.8692	382.2	630.2	20.898	6.78	11.24	1441
40	0.01391	1.2192	0.8523	391.2	641.5	21.185	6.80	11.41	1430
42	0.01419	1.1548	0.8181	409.3	664.6	21.749	6.83	11.72	1408
44	0.01448	1.0969	0.7838	427.8	688.4	22.301	6.86	12.02	1387
46	0.01478	1.0427	0.7499	446.1	712.2	22.830	6.91	12.34	1364
48	0.01511	0.9914	0.7165	465.3	737.2	23.362	6.96	12.63	1342
50	0.01544	0.9454	0.6840	484.7	762.7	23.882	6.99	12.89	1321
52	0.01580	0.9019	0.6523	504.4	788.7	24.393	7.01	13.13	1300
54	0.01617	0.8648	0.6217	524.2	815.2	24.892	7.03	13.34	1281
56	0.01655	0.8327	0.5924	544.4	842.3	25.386	7.04	13.50	1264
58	0.01694	0.8033	0.5646	564.4	869.4	25.863	7.03	13.64	1248
60	0.01735	0.7794	0.5381	584.4	896.8	26.327	7.03	13.74	1234
65	0.01843	0.7289	0.4784	634.3	966.1	27.434	7.02	13.95	1204
70	0.01958	0.6962	0.4270	683.7	1036.1	28.473	7.02	14.05	1180
75	0.02078	0.6762	0.3832	732.4	1106.4	29.443	7.05	14.08	1162
80	0.02202	0.6677	0.3459	780.4	1176.7	30.351	7.09	14.04	1150
85	0.02329	0.6663	0.3142	827.6	1246.8	31.201	7.15	13.98	1141
90	0.02457	0.6711	0.2872	874.2	1316.6	31.998	7.23	13.91	1136
95	0.02587	0.6806	0.2641	920.3	1386.0	32.749	7.33	13.84	1134
100	0.02717	0.6927	0.2442	965.9	1455.0	33.457	7.43	13.79	1134
120	0.03237	0.7587	0.1876	1149.9	1732.5	35.986	8.02	13.86	1145
140	0.03751	0.8391	0.1523	1335.2	2010.5	38.128	8.54	13.98	1172
160	0.04258	0.9252	0.1281	1524.8	2291.2	40.001	8.98	14.12	1206
180	0.04756	1.0128	0.1106	1718.8	2574.8	41.672	9.34	14.26	1243
200	0.05246	1.1000	0.0974	1916.6	2860.9	43.179	9.63	14.37	1281
220	0.05730	1.1874	0.0870	2117.5	3148.9	44.551	9.85	14.45	1320
240	0.06208	1.2728	0.0786	2320.8	3438.3	45.811	10.02	14.51	1358
260	0.06670	1.3507	0.0720	2525.8	3726.4	46.965	10.14	14.58	1394
280	0.07145	1.4373	0.0667	2732.6	4018.7	48.048	10.23	14.66	1435
300	0.07635	1.5332	0.0618	2943.1	4317.4	49.071	10.31	14.67	1477
350	0.08806	1.7441	0.0524	3466.3	5051.3	51.334	10.39	14.67	1569
400	0.09968	1.9533	0.0456	3990.3	5784.6	53.292	10.42	14.65	1657
450	0.11126	2.1615	0.0404	4514.1	6516.7	55.017	10.43	14.63	1741
500	0.12281	2.3690	0.0362	5037.3	7247.8	56.556	10.44	14.61	1821
550	0.13433	2.5760	0.0329	5560.2	7978.2	57.949	10.44	14.61	1898
600	0.14584	2.7827	0.0301	6083.6	8708.6	59.217	10.45	14.61	1972
700	0.16882	3.1956	0.0258	7132.2	10171.0	61.475	10.50	14.64	2111
800	0.19178	3.6082	0.0225	8187.3	11639.3	63.437	10.59	14.72	2240
900	0.21472	4.0205	0.0200	9252.5	13117.5	65.176	10.71	14.84	2360
1000	0.23766	4.4327	0.0180	10330.9	14608.7	66.745	10.87	15.00	2473
1200	0.28351	5.2570	0.0150	12543.0	17646.1	69.495	11.25	15.37	2680
1400	0.32935	6.0812	0.0128	14836.7	20765.0	71.905	11.68	15.80	2868
1600	0.37518	6.9056	0.0112	17215.8	23969.1	74.051	12.12	16.24	3042
1800	0.42102	7.7300	0.0100	19680.4	27258.7	75.982	12.55	16.67	3205
2000	0.46687	8.5544	0.0090	22232.7	30636.3	77.772	12.98	17.10	3358
2500	0.58196	10.6157	0.0072	29058.5	39533.8	81.723	14.53	18.73	3700
3000	0.70014	12.6772	0.0060	37119.9	49722.4	85.443	17.95	22.55	3989

* TWO-PHASE BOUNDARY

Table 4.12 (continued)
THERMODYNAMIC PROPERTIES OF NORMAL HYDROGEN

18.00 MPA ISOBAR

TEMPERATURE DEG. K	DENSITY KG/CU M	$V(DH/DV)_P$ KJ/KG	$V(DP/DU)_V$ MPA-CU M/KJ X 10^2	$-V(DP/DV)_T$ MPA	$(DV/DT)_P/V$ 1/DEG. K	THERMAL CONDUCTIVITY W/K-M X 10^3	VISCOSITY KG/M-S X 10^7	THERMAL DIFFUSIVITY SQ M/HR	DIELECTRIC CONSTANT	PRANDTL NUMBER
* 18.874	84.8704	1137.7	0.23871	181.8403	0.00594	126.03	365.44	0.00079	1.27940	1.9595
19	84.8070	1139.4	0.23807	181.1834	0.00596	126.57	360.52	0.00079	1.27918	1.9347
20	84.2955	1151.0	0.23330	175.9684	0.00614	131.23	325.55	0.00079	1.27735	1.7520
21	83.7725	1160.5	0.22897	170.7764	0.00631	136.27	296.78	0.00080	1.27549	1.5953
22	83.2381	1168.4	0.22497	165.6158	0.00649	140.45	272.79	0.00080	1.27359	1.4725
23	82.6924	1174.6	0.22129	160.4964	0.00667	143.88	252.54	0.00080	1.27165	1.3744
24	82.1357	1179.6	0.21783	155.4285	0.00685	146.69	235.26	0.00080	1.26967	1.2951
25	81.5680	1183.3	0.21458	150.4231	0.00703	148.95	220.37	0.00079	1.26765	1.2300
26	80.9895	1186.2	0.21145	145.4917	0.00721	150.69	207.42	0.00078	1.26560	1.1769
27	80.4001	1184.7	0.20915	140.3040	0.00744	152.01	196.07	0.00077	1.26352	1.1362
28	79.7993	1187.5	0.20598	135.6647	0.00761	152.38	186.04	0.00076	1.26139	1.0988
29	79.1848	1168.9	0.20288	130.9867	0.00779	153.64	177.07	0.00075	1.25922	1.0670
30	78.5637	1187.5	0.19997	126.1467	0.00798	154.05	169.09	0.00074	1.25703	1.0405
31	77.9345	1189.4	0.19703	121.9677	0.00814	153.66	161.92	0.00073	1.25481	1.0207
32	77.2941	1189.5	0.19415	117.6904	0.00832	153.09	155.41	0.00072	1.25255	1.0044
33	76.6471	1188.7	0.19140	113.4883	0.00850	152.39	149.51	0.00071	1.25023	0.9908
34	75.9926	1186.4	0.18882	109.2804	0.00869	151.63	144.12	0.00070	1.24798	0.9798
35	75.3265	1185.2	0.18627	105.3932	0.00886	150.76	139.15	0.00069	1.24564	0.9698
36	74.6550	1184.1	0.18374	101.6578	0.00904	149.32	134.59	0.00068	1.24329	0.9612
37	73.9771	1182.0	0.18123	97.9564	0.00921	148.80	130.37	0.00066	1.24091	0.9541
38	73.2918	1179.1	0.17883	94.3467	0.00939	147.72	126.45	0.00066	1.23852	0.9478
39	72.6014	1176.6	0.17658	90.9873	0.00955	146.60	122.80	0.00065	1.23611	0.9415
40	71.9051	1173.2	0.17439	87.6682	0.00972	145.45	119.39	0.00064	1.23368	0.9362
42	70.4961	1166.7	0.17000	81.4092	0.01005	143.02	113.19	0.00062	1.22877	0.9279
44	69.0729	1161.9	0.16554	75.7670	0.01035	140.54	107.70	0.00061	1.22383	0.9212
46	67.6380	1160.3	0.16033	70.5258	0.01063	138.06	102.80	0.00060	1.21887	0.9186
48	66.1964	1156.6	0.15563	65.6252	0.01092	135.62	98.39	0.00058	1.21389	0.9162
50	64.7504	1153.5	0.15117	61.2150	0.01117	133.24	94.41	0.00057	1.20891	0.9133
52	63.3028	1149.6	0.14693	57.0946	0.01142	130.93	90.79	0.00057	1.20394	0.9107
54	61.8615	1147.6	0.14296	53.5005	0.01162	128.72	87.51	0.00056	1.19900	0.9067
56	60.4295	1146.8	0.13929	50.3192	0.01177	126.61	84.53	0.00056	1.19411	0.9013
58	59.0159	1145.5	0.13602	47.4072	0.01191	124.63	81.83	0.00056	1.18930	0.8957
60	57.6214	1146.8	0.13287	44.9082	0.01198	122.93	79.38	0.00056	1.18456	0.8874
65	54.2514	1153.3	0.12563	39.5450	0.01210	119.63	74.26	0.00057	1.17315	0.8660
70	51.0762	1170.1	0.11904	35.5587	0.01201	116.67	70.34	0.00059	1.16251	0.8472
75	48.1267	1195.6	0.11299	32.5426	0.01177	114.10	67.41	0.00061	1.15264	0.8317
80	45.4153	1230.7	0.10745	30.3250	0.01141	111.75	65.28	0.00063	1.14364	0.8201
85	42.9418	1273.1	0.10233	28.6114	0.01098	111.48	63.78	0.00067	1.13547	0.7999
90	40.6944	1322.9	0.09758	27.3084	0.01052	111.23	62.78	0.00071	1.12807	0.7853
95	38.6540	1378.8	0.09326	26.3060	0.01004	111.06	62.18	0.00075	1.12139	0.7749
100	36.8044	1439.6	0.08927	25.4928	0.00958	112.19	61.57	0.00080	1.11536	0.7570
120	30.8955	1731.1	0.07569	23.4407	0.00800	122.82	66.03	0.00103	1.09624	0.7449
140	26.6567	2053.8	0.06691	22.3681	0.00681	130.61	68.60	0.00126	1.08266	0.7343
160	23.4862	2395.9	0.06074	21.7295	0.00590	137.48	70.64	0.00149	1.07258	0.7257
180	21.0273	2745.6	0.05630	21.2970	0.00519	144.02	72.60	0.00173	1.06481	0.7187
200	19.0621	3094.9	0.05305	20.9689	0.00464	150.40	74.60	0.00198	1.05863	0.7128
220	17.4520	3443.1	0.05060	20.7219	0.00420	156.67	76.69	0.00224	1.05359	0.7075
240	16.1070	3782.9	0.04874	20.5002	0.00384	162.79	78.86	0.00251	1.04938	0.7030
260	14.9925	4100.5	0.04736	20.2510	0.00355	168.77	81.09	0.00278	1.04591	0.7004
280	13.9953	4419.3	0.04660	20.1147	0.00332	174.61	83.38	0.00306	1.04281	0.7000
300	13.0969	4764.8	0.04579	20.0796	0.00308	180.33	85.71	0.00338	1.04002	0.6974
350	11.3563	5541.9	0.04443	19.8062	0.00265	194.23	91.69	0.00420	1.03464	0.6928
400	10.0318	6297.0	0.04361	19.5952	0.00233	207.79	97.80	0.00509	1.03056	0.6897
450	8.9880	7039.5	0.04307	19.4272	0.00208	221.16	103.96	0.00606	1.02735	0.6876
500	8.1430	7780.8	0.04264	19.2904	0.00188	234.55	110.12	0.00710	1.02475	0.6861
550	7.4445	8519.7	0.04229	19.1769	0.00171	247.87	116.27	0.00821	1.02261	0.6851
600	6.8570	9261.8	0.04198	19.0812	0.00158	261.21	122.37	0.00939	1.02081	0.6844
700	5.9234	10762.8	0.04140	18.9292	0.00135	287.95	134.42	0.01195	1.01796	0.6836
800	5.2143	12301.5	0.04078	18.8141	0.00120	315.00	146.22	0.01477	1.01580	0.6833
900	4.6572	13892.2	0.04010	18.7239	0.00107	342.58	157.76	0.01784	1.01410	0.6834
1000	4.2077	15544.1	0.03934	18.6515	0.00096	440.56	169.04	0.02513	1.01274	0.5754
1200	3.5272	19012.8	0.03779	18.5425	0.00081	511.00	190.88	0.03393	1.01067	0.5742
1400	3.0363	22711.8	0.03623	18.4645	0.00070	583.17	211.85	0.04375	1.00918	0.5741
1600	2.6654	26596.3	0.03479	18.4059	0.00061	656.25	232.06	0.05456	1.00805	0.5744
1800	2.3752	30626.3	0.03354	18.3602	0.00054	729.96	251.61	0.06637	1.00717	0.5746
2000	2.1419	34846.2	0.03236	18.3229	0.00049	806.67	270.57	0.07927	1.00647	0.5737
2500	1.7183	47494.0	0.02883	18.2413	0.00039	1061.32	315.94	0.11868	1.00519	0.5577
3000	1.4283	68083.8	0.02339	18.1067	0.00033	1594.22	359.72	0.17820	1.00431	0.5088

* TWO-PHASE BOUNDARY

Table 4.12 (continued)
THERMODYNAMIC PROPERTIES OF NORMAL HYDROGEN

20.00 MPA ISOBAR

TEMPERATURE DEG. K	VOLUME CU M/KG	ISOTHERM DERIVATIVE CU M-MPA/KG	ISOCHORE DERIVATIVE MPA/K	INTERNAL ENERGY KJ/KG-K	ENTHALPY KJ/KG	ENTROPY KJ/KG-K	CV KJ / KG-K	CP KJ / KG-K	VELOCITY OF SOUND M/S
* 19.360	0.01169	2.2344	1.094	236.2	470.0	14.482	5.38	6.80	1680
20	0.01173	2.2040	1.094	239.7	474.4	14.706	5.47	6.97	1675
21	0.01180	2.1564	1.093	245.4	481.5	15.054	5.60	7.22	1667
22	0.01187	2.1087	1.090	251.3	488.8	15.396	5.72	7.47	1659
23	0.01195	2.0611	1.087	257.4	496.4	15.734	5.83	7.71	1651
24	0.01203	2.0136	1.082	263.7	504.2	16.068	5.93	7.95	1642
25	0.01210	1.9664	1.075	270.2	512.3	16.397	6.03	8.18	1634
26	0.01219	1.9195	1.068	276.9	520.6	16.720	6.11	8.41	1625
27	0.01227	1.8731	1.060	283.8	529.1	17.042	6.20	8.63	1615
28	0.01236	1.8274	1.051	290.8	537.9	17.359	6.28	8.86	1606
29	0.01245	1.7797	1.044	298.0	546.9	17.675	6.35	9.10	1597
30	0.01254	1.7369	1.032	305.3	556.1	17.987	6.41	9.30	1588
31	0.01263	1.6937	1.019	312.8	565.5	18.295	6.47	9.51	1577
32	0.01273	1.6480	1.006	320.5	575.1	18.600	6.53	9.71	1566
33	0.01283	1.6088	0.9921	328.3	584.9	18.902	6.58	9.90	1556
34	0.01293	1.5685	0.9777	336.2	594.9	19.201	6.63	10.09	1546
35	0.01304	1.5277	0.9630	344.3	605.1	19.496	6.67	10.28	1535
36	0.01315	1.4861	0.9483	352.5	615.4	19.789	6.71	10.48	1523
37	0.01326	1.4494	0.9331	360.8	626.0	20.078	6.75	10.66	1513
38	0.01337	1.4146	0.9177	369.3	636.8	20.365	6.78	10.82	1503
39	0.01349	1.3793	0.9019	377.8	647.7	20.648	6.80	10.99	1493
40	0.01361	1.3444	0.8859	386.5	658.7	20.928	6.82	11.15	1482
42	0.01386	1.2803	0.8533	404.1	681.3	21.479	6.85	11.44	1462
44	0.01412	1.2210	0.8206	422.0	704.5	22.018	6.89	11.73	1442
46	0.01440	1.1649	0.7880	439.8	727.7	22.534	6.95	12.03	1420
48	0.01468	1.1145	0.7558	458.4	752.1	23.051	6.99	12.30	1400
50	0.01498	1.0671	0.7242	477.2	776.9	23.558	7.03	12.54	1380
52	0.01529	1.0229	0.6935	496.3	802.2	24.055	7.05	12.77	1361
54	0.01562	0.9819	0.6636	515.6	828.0	24.541	7.07	12.98	1342
56	0.01595	0.9463	0.6346	535.3	854.3	25.021	7.08	13.15	1325
58	0.01630	0.9139	0.6067	554.8	880.8	25.486	7.08	13.29	1310
60	0.01666	0.8863	0.5803	574.3	907.5	25.938	7.07	13.40	1296
65	0.01760	0.8298	0.5197	623.1	975.1	27.018	7.07	13.62	1265
70	0.01860	0.7893	0.4671	671.6	1043.6	28.034	7.07	13.76	1239
75	0.01964	0.7626	0.4216	719.7	1112.6	28.987	7.09	13.84	1220
80	0.02073	0.7461	0.3823	767.3	1181.9	29.881	7.13	13.87	1204
85	0.02184	0.7398	0.3486	814.4	1251.1	30.721	7.20	13.85	1193
90	0.02297	0.7396	0.3195	860.9	1320.3	31.513	7.28	13.83	1186
95	0.02412	0.7444	0.2943	907.1	1389.4	32.260	7.37	13.80	1181
100	0.02527	0.7532	0.2725	953.0	1458.3	32.966	7.48	13.77	1178
120	0.02990	0.8101	0.2096	1138.0	1735.9	35.496	8.06	13.88	1181
140	0.03451	0.8842	0.1702	1324.5	2014.7	37.645	8.57	14.03	1203
160	0.03906	0.9669	0.1432	1515.4	2296.6	39.525	9.01	14.19	1234
180	0.04355	1.0521	0.1235	1710.5	2581.4	41.203	9.37	14.32	1268
200	0.04796	1.1390	0.1087	1909.4	2868.7	42.717	9.66	14.43	1304
220	0.05233	1.2245	0.0970	2111.4	3157.9	44.095	9.88	14.51	1341
240	0.05664	1.3105	0.0876	2315.6	3448.5	45.359	10.05	14.56	1378
260	0.06091	1.3950	0.0800	2521.6	3739.8	46.526	10.17	14.59	1415
280	0.06514	1.4793	0.0735	2728.6	4031.5	47.606	10.26	14.60	1451
300	0.06951	1.5696	0.0688	2939.0	4329.2	48.624	10.33	14.70	1495
350	0.08005	1.7800	0.0583	3463.2	5064.3	50.892	10.41	14.69	1585
400	0.09053	1.9889	0.0507	3987.9	5798.4	52.852	10.43	14.67	1672
450	0.10095	2.1968	0.0449	4512.2	6531.3	54.578	10.44	14.64	1755
500	0.11135	2.4040	0.0403	5035.9	7262.8	56.119	10.44	14.62	1835
550	0.12172	2.6108	0.0365	5559.1	7993.5	57.512	10.45	14.61	1911
600	0.13208	2.8173	0.0334	6082.8	8724.3	58.781	10.46	14.61	1984
700	0.15277	3.2299	0.0286	7131.7	10187.0	61.039	10.51	14.65	2122
800	0.17343	3.6420	0.0250	8187.1	11655.7	63.002	10.59	14.72	2250
900	0.19408	4.0541	0.0222	9252.5	13134.0	64.741	10.71	14.84	2370
1000	0.21472	4.4660	0.0200	10331.0	14625.4	66.310	10.87	15.00	2482
1200	0.25599	5.2899	0.0167	12543.3	17663.0	69.060	11.25	15.37	2689
1400	0.29724	6.1139	0.0143	14837.2	20782.0	71.470	11.68	15.80	2876
1600	0.33849	6.9380	0.0125	17216.3	23986.1	73.616	12.12	16.24	3049
1800	0.37974	7.7622	0.0111	19680.9	27275.7	75.547	12.55	16.67	3211
2000	0.42100	8.5865	0.0100	22232.7	30652.7	77.337	12.97	17.10	3364
2500	0.52456	10.6475	0.0080	29048.8	39540.0	81.283	14.48	18.69	3706
3000	0.63076	12.7087	0.0067	37055.1	49670.3	84.983	17.77	22.33	3996

* TWO-PHASE BOUNDARY

Table 4.12 (continued)
THERMODYNAMIC PROPERTIES OF NORMAL HYDROGEN

20.00 MPA ISOBAR

TEMPERATURE DEG. K	DENSITY KG/CU M	$V(DH/DV)_P$ KJ/KG	$V(DP/DU)_V$ MPA-CU M/KJ X 10^2	$-V(DP/DV)_T$ MPA	$(DV/DT)_P/V$ 1/DEG. K	THERMAL CONDUCTIVITY W/K-M X 10^3	VISCOSITY KG/M-S X 10^7	THERMAL DIFFUSIVITY SQ M/HR	DIELECTRIC CONSTANT	PRANDTL NUMBER
* 19.360	85.5433	1188.4	0.23748	191.1415	0.00572	130.46	377.62	0.00081	1.28181	1.9680
20	85.2277	1196.5	0.23452	187.8410	0.00582	133.83	353.32	0.00081	1.28068	1.8393
21	84.7262	1207.5	0.23021	182.7004	0.00598	139.12	321.04	0.00082	1.27889	1.6668
22	84.2142	1216.9	0.22624	177.5827	0.00614	143.52	294.22	0.00082	1.27706	1.5318
23	83.6921	1224.6	0.22258	172.4970	0.00630	147.16	271.67	0.00082	1.27520	1.4241
24	83.1599	1231.0	0.21916	167.4527	0.00646	150.15	252.50	0.00082	1.27331	1.3370
25	82.6179	1236.0	0.21594	162.4598	0.00662	152.58	236.03	0.00081	1.27138	1.2657
26	82.0662	1240.2	0.21286	157.5289	0.00678	154.47	221.76	0.00081	1.26942	1.2072
27	81.5050	1243.8	0.20984	152.6707	0.00694	155.94	209.29	0.00080	1.26743	1.1589
28	80.9346	1246.8	0.20687	147.8964	0.00710	157.06	198.32	0.00079	1.26541	1.1185
29	80.3501	1246.2	0.20469	142.9994	0.00730	157.84	188.54	0.00078	1.26334	1.0868
30	79.7603	1248.6	0.20186	138.5345	0.00745	158.37	179.85	0.00077	1.26125	1.0564
31	79.1617	1250.3	0.19900	134.0749	0.00760	158.07	172.05	0.00076	1.25914	1.0347
32	78.5555	1249.8	0.19620	129.4586	0.00777	157.59	165.02	0.00074	1.25700	1.0171
33	77.9419	1251.6	0.19346	125.3936	0.00791	156.98	158.65	0.00073	1.25483	1.0008
34	77.3220	1251.9	0.19082	121.2815	0.00806	156.30	152.86	0.00072	1.25265	0.9870
35	76.6946	1251.0	0.18826	117.1683	0.00822	155.51	147.55	0.00071	1.25044	0.9756
36	76.0619	1248.7	0.18579	113.0376	0.00839	154.64	142.68	0.00070	1.24822	0.9666
37	75.4195	1248.2	0.18338	109.3153	0.00854	153.68	138.17	0.00069	1.24597	0.9579
38	74.7739	1247.5	0.18109	105.7719	0.00868	152.68	134.00	0.00068	1.24370	0.9500
39	74.1217	1245.4	0.17893	102.2326	0.00882	151.62	130.13	0.00067	1.24142	0.9430
40	73.4653	1242.7	0.17682	98.7645	0.00897	150.52	126.52	0.00066	1.23912	0.9369
42	72.1398	1238.5	0.17261	92.3572	0.00924	148.20	119.99	0.00065	1.23450	0.9266
44	70.8011	1235.4	0.16832	86.4488	0.00949	145.81	114.23	0.00063	1.22984	0.9186
46	69.4518	1235.3	0.16330	80.9077	0.00974	143.42	109.10	0.00062	1.22515	0.9152
48	68.0970	1234.8	0.15876	75.8970	0.00996	141.05	104.50	0.00061	1.22045	0.9110
50	66.7399	1233.4	0.15447	71.2198	0.01017	138.73	100.35	0.00060	1.21576	0.9073
52	65.3829	1231.7	0.15040	66.8821	0.01037	136.48	96.59	0.00059	1.21109	0.9039
54	64.0296	1229.6	0.14656	62.8680	0.01056	134.31	93.17	0.00058	1.20643	0.9004
56	62.6829	1228.9	0.14295	59.3154	0.01070	132.23	90.06	0.00058	1.20181	0.8954
58	61.3466	1227.7	0.13971	56.0628	0.01082	130.25	87.22	0.00058	1.19724	0.8897
60	60.0282	1228.7	0.13665	53.2057	0.01091	128.55	84.63	0.00058	1.19274	0.8823
65	56.8213	1235.5	0.12944	47.1485	0.01102	125.23	79.15	0.00058	1.18185	0.8608
70	53.7734	1250.6	0.12286	42.4454	0.01100	122.23	74.87	0.00059	1.17155	0.8430
75	50.9103	1274.2	0.11675	38.8224	0.01086	119.46	71.57	0.00061	1.16194	0.8291
80	48.2479	1305.7	0.11107	36.0001	0.01062	116.91	69.08	0.00063	1.15305	0.8194
85	45.7912	1346.4	0.10578	33.8776	0.01029	116.51	67.25	0.00066	1.14488	0.7996
90	43.5352	1393.9	0.10084	32.2001	0.00992	116.11	65.95	0.00069	1.13742	0.7855
95	41.4668	1447.1	0.09632	30.8672	0.00953	115.78	65.07	0.00073	1.13061	0.7755
100	39.5742	1506.3	0.09212	29.8080	0.00914	116.87	64.04	0.00077	1.12440	0.7546
120	33.4477	1793.6	0.07778	27.0964	0.00774	127.13	68.24	0.00099	1.10447	0.7447
140	28.9793	2112.2	0.06855	25.6237	0.00664	134.29	70.44	0.00119	1.09008	0.7359
160	25.6015	2452.6	0.06207	24.7531	0.00578	140.53	72.14	0.00139	1.07930	0.7282
180	22.9647	2801.5	0.05739	24.1615	0.00511	146.57	73.83	0.00160	1.07093	0.7215
200	20.8486	3153.1	0.05396	23.7458	0.00458	152.56	75.64	0.00183	1.06425	0.7154
220	19.1108	3500.9	0.05136	23.4010	0.00415	158.55	77.58	0.00206	1.05878	0.7101
240	17.6549	3844.4	0.04940	23.1363	0.00379	164.46	79.64	0.00230	1.05422	0.7052
260	16.4175	4179.8	0.04788	22.9019	0.00349	170.28	81.79	0.00256	1.05035	0.7009
280	15.3509	4508.9	0.04669	22.7079	0.00324	176.02	84.03	0.00283	1.04703	0.6968
300	14.3862	4823.0	0.04633	22.5803	0.00305	181.67	86.33	0.00309	1.04403	0.6986
350	12.4914	5600.9	0.04488	22.2353	0.00262	195.50	92.27	0.00383	1.03815	0.6936
400	11.0465	6356.5	0.04400	21.9708	0.00231	209.07	98.38	0.00465	1.03368	0.6902
450	9.9058	7099.3	0.04341	21.7611	0.00206	222.51	104.57	0.00552	1.03017	0.6879
500	8.9811	7840.9	0.04294	21.5907	0.00186	236.01	110.78	0.00647	1.02732	0.6864
550	8.2157	8579.8	0.04256	21.4496	0.00170	249.46	116.98	0.00748	1.02497	0.6852
600	7.5713	9322.1	0.04223	21.3309	0.00157	262.92	123.15	0.00855	1.02300	0.6845
700	6.5460	10823.3	0.04160	21.1426	0.00135	289.96	135.33	0.01089	1.01986	0.6836
800	5.7660	12362.2	0.04096	21.0002	0.00119	317.33	147.27	0.01346	1.01748	0.6833
900	5.1525	13953.4	0.04025	20.8888	0.00106	345.23	158.95	0.01625	1.01561	0.6834
1000	4.6572	15605.7	0.03948	20.7994	0.00096	440.56	170.38	0.02270	1.01410	0.5801
1200	3.9065	19075.7	0.03789	20.6651	0.00081	511.00	192.51	0.03063	1.01182	0.5791
1400	3.3643	22776.3	0.03631	20.5690	0.00069	583.17	213.76	0.03948	1.01017	0.5793
1600	2.9543	26662.0	0.03487	20.4971	0.00061	656.24	234.24	0.04923	1.00893	0.5798
1800	2.6334	30691.6	0.03360	20.4409	0.00054	729.87	254.05	0.05986	1.00796	0.5802
2000	2.3753	34905.7	0.03243	20.3953	0.00049	806.30	273.27	0.07147	1.00717	0.5795
2500	1.9064	47448.9	0.02895	20.2979	0.00039	1056.83	319.24	0.10680	1.00576	0.5645
3000	1.5854	67548.4	0.02365	20.1484	0.00033	1571.05	363.54	0.15972	1.00478	0.5168

* TWO-PHASE BOUNDARY

Table 4.12 (continued)
THERMODYNAMIC PROPERTIES OF NORMAL HYDROGEN

22.00 MPA ISOBAR

TEMPERATURE DEG. K	VOLUME CU M/KG	ISOTHERM DERIVATIVE CU M-MPA/KG	ISOCHORE DERIVATIVE MPA/K	INTERNAL ENERGY KJ/KG-K	ENTHALPY KJ/KG	ENTROPY KJ/KG-K	CV KJ / KG-K	CP KJ / KG-K	VELOCITY OF SOUND M/S
* 19.835	0.01160	2.3243	1.107	238.7	493.9	14.516	5.43	6.84	1711
20	0.01161	2.3167	1.107	239.6	495.1	14.574	5.45	6.88	1710
21	0.01168	2.2703	1.107	245.2	502.1	14.917	5.59	7.13	1702
22	0.01175	2.2238	1.105	250.9	509.3	15.255	5.71	7.37	1695
23	0.01182	2.1773	1.102	256.9	516.8	15.588	5.82	7.61	1687
24	0.01189	2.1308	1.098	263.0	524.6	15.917	5.92	7.84	1680
25	0.01196	2.0846	1.093	269.4	532.5	16.242	6.02	8.07	1672
26	0.01204	2.0385	1.086	275.9	540.7	16.561	6.11	8.29	1663
27	0.01212	1.9928	1.079	282.5	549.1	16.878	6.19	8.50	1655
28	0.01220	1.9476	1.070	289.4	557.7	17.190	6.27	8.72	1646
29	0.01228	1.9030	1.061	296.4	566.5	17.500	6.34	8.93	1637
30	0.01236	1.8591	1.051	303.5	575.5	17.806	6.41	9.14	1628
31	0.01245	1.8157	1.043	310.9	584.8	18.110	6.47	9.35	1620
32	0.01254	1.7748	1.031	318.3	594.3	18.410	6.53	9.55	1611
33	0.01264	1.7338	1.018	325.9	603.9	18.707	6.58	9.73	1601
34	0.01273	1.6912	1.004	333.7	613.7	19.001	6.63	9.92	1590
35	0.01283	1.6534	0.9903	341.5	623.7	19.291	6.68	10.09	1581
36	0.01293	1.6150	0.9760	349.5	633.9	19.578	6.72	10.27	1571
37	0.01303	1.5756	0.9615	357.6	644.3	19.862	6.76	10.45	1560
38	0.01314	1.5350	0.9468	365.8	654.8	20.143	6.79	10.62	1549
39	0.01324	1.5001	0.9318	374.2	665.5	20.421	6.82	10.78	1540
40	0.01335	1.4676	0.9167	382.6	676.4	20.695	6.84	10.92	1531
42	0.01358	1.4030	0.8856	399.7	698.5	21.235	6.88	11.21	1512
44	0.01382	1.3406	0.8542	417.2	721.2	21.763	6.91	11.49	1492
46	0.01407	1.2848	0.8227	434.4	743.9	22.267	6.98	11.77	1472
48	0.01433	1.2332	0.7915	452.6	767.7	22.774	7.02	12.03	1453
50	0.01459	1.1839	0.7609	470.9	792.0	23.270	7.06	12.27	1434
52	0.01487	1.1394	0.7309	489.6	816.8	23.755	7.09	12.48	1416
54	0.01516	1.0982	0.7017	508.4	841.9	24.230	7.11	12.68	1399
56	0.01546	1.0601	0.6734	527.6	867.7	24.700	7.12	12.85	1383
58	0.01577	1.0243	0.6458	546.6	893.6	25.153	7.12	12.99	1367
60	0.01609	0.9927	0.6193	565.7	919.7	25.596	7.12	13.12	1353
65	0.01692	0.9299	0.5583	613.6	985.8	26.653	7.11	13.35	1321
70	0.01780	0.8831	0.5046	661.3	1053.0	27.649	7.11	13.51	1295
75	0.01873	0.8498	0.4577	708.8	1120.8	28.585	7.14	13.62	1274
80	0.01969	0.8291	0.4170	756.0	1189.1	29.467	7.18	13.68	1257
85	0.02068	0.8156	0.3814	802.7	1257.6	30.298	7.24	13.72	1243
90	0.02168	0.8110	0.3506	849.2	1326.2	31.083	7.32	13.73	1234
95	0.02271	0.8116	0.3237	895.3	1394.9	31.825	7.41	13.73	1226
100	0.02374	0.8163	0.3002	941.3	1463.6	32.529	7.52	13.74	1222
120	0.02790	0.8643	0.2314	1127.1	1741.0	35.057	8.09	13.88	1218
140	0.03207	0.9309	0.1880	1314.6	2020.1	37.208	8.60	14.07	1234
160	0.03620	1.0099	0.1582	1506.5	2302.8	39.095	9.04	14.24	1261
180	0.04028	1.0936	0.1365	1702.7	2588.7	40.779	9.40	14.38	1293
200	0.04430	1.1784	0.1200	1902.6	2877.1	42.298	9.69	14.48	1327
220	0.04827	1.2638	0.1071	2105.5	3167.4	43.681	9.91	14.56	1362
240	0.05219	1.3483	0.0966	2310.7	3459.0	44.950	10.08	14.61	1398
260	0.05608	1.4326	0.0881	2517.5	3751.2	46.120	10.20	14.64	1433
280	0.05993	1.5155	0.0810	2725.4	4043.8	47.204	10.29	14.64	1468
300	0.06392	1.6365	0.0758	2935.1	4341.2	48.220	10.34	14.73	1513
350	0.07351	1.8164	0.0642	3460.2	5077.5	50.491	10.42	14.71	1602
400	0.08303	2.0248	0.0558	3985.7	5812.4	52.454	10.44	14.68	1687
450	0.09252	2.2323	0.0494	4510.5	6545.9	54.181	10.44	14.65	1770
500	0.10197	2.4392	0.0443	5034.6	7277.9	55.723	10.45	14.63	1848
550	0.11140	2.6457	0.0402	5558.1	8009.0	57.117	10.45	14.62	1924
600	0.12082	2.8520	0.0368	6082.0	8740.0	58.386	10.46	14.62	1996
700	0.13963	3.2641	0.0315	7131.4	10203.2	60.645	10.51	14.65	2133
800	0.15841	3.6759	0.0275	8187.0	11672.1	62.608	10.59	14.73	2260
900	0.17718	4.0876	0.0244	9252.5	13150.6	64.348	10.72	14.84	2379
1000	0.19595	4.4993	0.0220	10331.3	14642.1	65.917	10.88	15.00	2491
1200	0.23346	5.3227	0.0183	12543.8	17679.9	68.667	11.25	15.37	2697
1400	0.27096	6.1463	0.0157	14837.7	20798.9	71.077	11.68	15.80	2883
1600	0.30846	6.9702	0.0137	17217.0	24003.1	73.223	12.12	16.24	3056
1800	0.34596	7.7941	0.0122	19681.5	27292.6	75.154	12.55	16.67	3218
2000	0.38347	8.6182	0.0110	22232.9	30669.2	76.943	12.97	17.10	3370
2500	0.47759	10.6788	0.0088	29040.7	39547.7	80.886	14.45	18.64	3712
3000	0.57401	12.7398	0.0073	36999.6	49627.8	84.567	17.60	22.15	4003

* TWO-PHASE BOUNDARY

Table 4.12 (continued)
THERMODYNAMIC PROPERTIES OF NORMAL HYDROGEN

22.00 MPA ISOBAR

TEMPERATURE DEG. K	DENSITY KG/CU M	V(DH/DV)$_P$ KJ/KG	V(DP/DU)$_V$ MPA-CU M/KJ X 10^2	-V(DP/DV)$_T$ MPA	(DV/DT)$_P$/V 1/DEG. K	THERMAL CONDUCTIVITY W/K-M X 10^3	VISCOSITY KG/M-S X 10^7	THERMAL DIFFUSIVITY SQ M/HR	DIELECTRIC CONSTANT	PRANDTL NUMBER
* 19.835	86.1915	1237.7	0.23646	200.3367	0.00553	135.33	389.82	0.00083	1.28412	1.9701
20	86.1126	1240.0	0.23571	199.4935	0.00555	136.34	383.11	0.00083	1.28384	1.9336
21	85.6298	1252.5	0.23141	194.4029	0.00569	141.86	346.91	0.00084	1.28211	1.7439
22	85.1375	1263.3	0.22745	189.3277	0.00584	146.47	316.97	0.00084	1.28036	1.5959
23	84.6359	1272.4	0.22380	184.2761	0.00598	150.31	291.88	0.00084	1.27857	1.4779
24	84.1251	1280.1	0.22039	179.2570	0.00613	153.47	270.62	0.00084	1.27675	1.3826
25	83.6054	1286.5	0.21719	174.2797	0.00627	156.07	252.43	0.00083	1.27489	1.3045
26	83.0769	1291.9	0.21414	169.3539	0.00641	158.10	236.71	0.00083	1.27301	1.2405
27	82.5398	1296.7	0.21115	164.4893	0.00656	159.71	223.02	0.00082	1.27110	1.1875
28	81.9943	1300.9	0.20822	159.6962	0.00670	160.95	211.00	0.00081	1.26917	1.1432
29	81.4407	1304.3	0.20543	154.9848	0.00685	161.87	200.38	0.00080	1.26720	1.1057
30	80.8791	1306.6	0.20282	150.3656	0.00699	162.51	190.94	0.00079	1.26521	1.0734
31	80.3017	1307.2	0.20074	145.8000	0.00716	162.27	182.41	0.00078	1.26317	1.0515
32	79.7230	1310.1	0.19800	141.4923	0.00729	161.87	174.79	0.00077	1.26112	1.0307
33	79.1391	1312.0	0.19536	137.2131	0.00742	161.34	167.92	0.00075	1.25906	1.0130
34	78.5491	1312.0	0.19279	132.8393	0.00756	160.73	161.68	0.00074	1.25697	0.9979
35	77.9518	1313.8	0.19024	128.8886	0.00768	160.00	155.98	0.00073	1.25487	0.9841
36	77.3507	1314.5	0.18774	124.9214	0.00781	159.20	150.76	0.00072	1.25275	0.9726
37	76.7426	1313.7	0.18533	120.9139	0.00795	158.31	145.96	0.00071	1.25061	0.9631
38	76.1310	1310.9	0.18310	116.8606	0.00810	157.36	141.53	0.00070	1.24846	0.9553
39	75.5108	1310.0	0.18100	113.2708	0.00823	156.35	137.41	0.00069	1.24629	0.9471
40	74.8894	1309.6	0.17898	109.9089	0.00834	155.31	133.59	0.00068	1.24411	0.9395
42	73.6316	1307.3	0.17492	103.3082	0.00857	153.07	126.68	0.00067	1.23970	0.9275
44	72.3632	1304.4	0.17076	97.0078	0.00881	150.76	120.62	0.00065	1.23528	0.9189
46	71.0858	1307.0	0.16586	91.3284	0.00901	148.43	115.24	0.00064	1.23083	0.9141
48	69.8029	1308.1	0.16146	86.0786	0.00920	146.11	110.43	0.00063	1.22637	0.9091
50	68.5179	1307.9	0.15731	81.1197	0.00938	143.83	106.10	0.00062	1.22191	0.9049
52	67.2339	1308.2	0.15338	76.6029	0.00954	141.62	102.18	0.00061	1.21747	0.9006
54	65.9530	1308.3	0.14967	72.4282	0.00969	139.48	98.62	0.00060	1.21305	0.8963
56	64.6783	1308.1	0.14621	68.5652	0.00982	137.42	95.38	0.00060	1.20866	0.8917
58	63.4126	1306.8	0.14304	64.9510	0.00994	135.46	92.41	0.00059	1.20431	0.8864
60	62.1590	1306.9	0.14000	61.7071	0.01004	133.76	89.70	0.00059	1.20002	0.8796
65	59.0971	1313.9	0.13286	54.9559	0.01016	130.43	83.90	0.00060	1.18957	0.8586
70	56.1677	1328.3	0.12626	49.6016	0.01017	127.36	79.30	0.00060	1.17964	0.8412
75	53.3927	1350.4	0.12012	45.3741	0.01009	124.47	75.67	0.00062	1.17027	0.8283
80	50.7897	1381.7	0.11438	42.1081	0.00990	121.77	72.86	0.00063	1.16154	0.8186
85	48.3638	1419.0	0.10895	39.4447	0.00967	121.28	70.72	0.00066	1.15343	0.8002
90	46.1163	1465.0	0.10388	37.3989	0.00937	120.76	69.13	0.00069	1.14596	0.7863
95	44.0399	1516.4	0.09920	35.7421	0.00906	120.29	68.00	0.00072	1.13909	0.7764
100	42.1248	1573.5	0.09484	34.3878	0.00873	121.42	66.46	0.00076	1.13277	0.7520
120	35.8380	1857.7	0.07982	30.9735	0.00747	131.40	70.45	0.00095	1.11222	0.7443
140	31.1839	2171.6	0.07012	29.0286	0.00648	137.99	72.33	0.00113	1.09716	0.7372
160	27.6265	2510.4	0.06335	27.8988	0.00567	143.65	73.70	0.00131	1.08575	0.7304
180	24.8290	2860.2	0.05845	27.1517	0.00503	149.20	75.12	0.00150	1.07684	0.7239
200	22.5752	3211.3	0.05485	26.6016	0.00451	154.80	76.73	0.00170	1.06970	0.7179
220	20.7183	3561.8	0.05211	26.1833	0.00409	160.49	78.51	0.00191	1.06384	0.7124
240	19.1592	3905.8	0.05003	25.8330	0.00374	166.18	80.46	0.00214	1.05894	0.7074
260	17.8323	4243.2	0.04842	25.5464	0.00345	171.84	82.53	0.00237	1.05478	0.7030
280	16.6866	4572.0	0.04717	25.2890	0.00320	177.47	84.71	0.00261	1.05119	0.6989
300	15.6457	4882.0	0.04687	25.1355	0.00302	183.04	86.97	0.00286	1.04795	0.6997
350	13.6037	5660.4	0.04532	24.7099	0.00260	196.78	92.86	0.00354	1.04160	0.6944
400	12.0431	6416.4	0.04438	24.3855	0.00229	210.36	98.97	0.00428	1.03676	0.6907
450	10.8089	7159.4	0.04374	24.1293	0.00205	223.87	105.18	0.00509	1.03295	0.6882
500	9.8070	7901.1	0.04323	23.9216	0.00185	237.47	111.43	0.00596	1.02986	0.6866
550	8.9766	8640.2	0.04283	23.7498	0.00169	251.03	117.69	0.00689	1.02731	0.6854
600	8.2769	9382.5	0.04246	23.6056	0.00156	264.63	123.92	0.00787	1.02516	0.6846
700	7.1619	10883.8	0.04180	23.3771	0.00135	291.95	136.23	0.01002	1.02175	0.6836
800	6.3126	12423.0	0.04113	23.2044	0.00119	319.64	148.32	0.01238	1.01915	0.6833
900	5.6438	14014.5	0.04040	23.0696	0.00106	347.87	160.15	0.01495	1.01711	0.6834
1000	5.1034	15667.3	0.03961	22.9615	0.00096	440.56	171.72	0.02072	1.01546	0.5847
1200	4.2834	19138.5	0.03800	22.7992	0.00080	511.00	194.15	0.02794	1.01297	0.5841
1400	3.6905	22840.5	0.03640	22.6834	0.00069	583.16	215.68	0.03599	1.01116	0.5845
1600	3.2419	26727.4	0.03494	22.5966	0.00061	656.22	236.44	0.04486	1.00980	0.5853
1800	2.8905	30756.8	0.03367	22.5290	0.00054	729.79	256.52	0.05453	1.00874	0.5858
2000	2.6078	34966.1	0.03249	22.4743	0.00049	805.98	276.00	0.06508	1.00788	0.5854
2500	2.0938	47419.7	0.02907	22.3597	0.00039	1052.97	322.58	0.09710	1.00632	0.5712
3000	1.7421	67097.7	0.02389	22.1945	0.00033	1551.11	367.42	0.14470	1.00526	0.5247

* TWO-PHASE BOUNDARY

Table 4.12 (continued)
THERMODYNAMIC PROPERTIES OF NORMAL HYDROGEN

24.00 MPA ISOBAR

TEMPERATURE DEG. K	VOLUME CU M/KG	ISOTHERM DERIVATIVE CU M-MPA/KG	ISOCHORE DERIVATIVE MPA/K	INTERNAL ENERGY KJ/KG-K	ENTHALPY KJ/KG	ENTROPY KJ/KG-K	CV KJ / KG-K	CP KJ / KG-K	VELOCITY OF SOUND M/S
* 20.300	0.01152	2.4124	1.120	241.3	517.7	14.550	5.48	6.88	1741
21	0.01156	2.3807	1.120	245.1	522.6	14.788	5.57	7.05	1736
22	0.01163	2.3354	1.119	250.7	529.8	15.122	5.69	7.29	1729
23	0.01169	2.2899	1.117	256.6	537.2	15.451	5.81	7.52	1722
24	0.01176	2.2445	1.113	262.6	544.9	15.776	5.91	7.74	1715
25	0.01183	2.1991	1.109	268.8	552.7	16.096	6.01	7.96	1707
26	0.01190	2.1539	1.103	275.1	560.7	16.411	6.10	8.18	1700
27	0.01197	2.1089	1.096	281.6	569.0	16.724	6.18	8.39	1692
28	0.01205	2.0643	1.089	288.3	577.5	17.031	6.26	8.60	1683
29	0.01213	2.0202	1.080	295.2	586.2	17.337	6.34	8.80	1675
30	0.01221	1.9767	1.071	302.1	595.1	17.639	6.41	9.00	1666
31	0.01229	1.9338	1.061	309.3	604.2	17.937	6.47	9.20	1658
32	0.01237	1.8918	1.051	316.5	613.5	18.232	6.53	9.39	1649
33	0.01246	1.8526	1.042	324.0	623.0	18.526	6.59	9.59	1642
34	0.01255	1.8136	1.029	331.5	632.7	18.815	6.64	9.76	1633
35	0.01264	1.7746	1.016	339.2	642.6	19.101	6.69	9.94	1624
36	0.01273	1.7350	1.002	347.0	652.6	19.383	6.73	10.11	1614
37	0.01283	1.6987	0.9882	355.0	662.8	19.663	6.77	10.27	1605
38	0.01292	1.6620	0.9740	363.0	673.1	19.939	6.80	10.43	1596
39	0.01302	1.6240	0.9596	371.1	683.6	20.212	6.83	10.58	1586
40	0.01312	1.5846	0.9451	379.3	694.3	20.481	6.85	10.74	1576
42	0.01333	1.5207	0.9154	396.1	716.1	21.012	6.90	11.01	1558
44	0.01355	1.4602	0.8853	413.1	738.3	21.530	6.94	11.27	1541
46	0.01378	1.4025	0.8548	429.9	760.6	22.025	7.00	11.55	1521
48	0.01401	1.3475	0.8245	447.6	784.0	22.522	7.05	11.81	1502
50	0.01426	1.2986	0.7946	465.6	807.8	23.009	7.09	12.03	1485
52	0.01451	1.2533	0.7653	483.8	832.1	23.485	7.12	12.24	1468
54	0.01477	1.2109	0.7367	502.2	856.8	23.951	7.14	12.43	1451
56	0.01504	1.1701	0.7089	521.0	882.1	24.411	7.16	12.60	1435
58	0.01532	1.1345	0.6820	539.7	907.4	24.856	7.16	12.74	1421
60	0.01561	1.1015	0.6558	558.4	933.0	25.290	7.16	12.86	1407
65	0.01636	1.0314	0.5943	605.4	998.0	26.328	7.15	13.11	1375
70	0.01715	0.9787	0.5398	652.4	1064.0	27.307	7.16	13.28	1348
75	0.01798	0.9401	0.4919	699.3	1130.7	28.228	7.18	13.42	1326
80	0.01884	0.9126	0.4499	746.0	1198.1	29.098	7.22	13.52	1307
85	0.01972	0.8956	0.4130	792.5	1265.8	29.919	7.28	13.58	1292
90	0.02063	0.8860	0.3806	838.7	1333.8	30.697	7.36	13.62	1281
95	0.02155	0.8823	0.3522	884.8	1402.0	31.434	7.45	13.65	1272
100	0.02248	0.8836	0.3273	930.8	1470.3	32.135	7.55	13.68	1265
120	0.02626	0.9203	0.2531	1117.2	1747.4	34.660	8.12	13.88	1254
140	0.03005	0.9814	0.2057	1305.4	2026.6	36.811	8.63	14.08	1265
160	0.03382	1.0549	0.1732	1498.2	2309.9	38.702	9.07	14.27	1289
180	0.03756	1.1359	0.1494	1695.3	2596.7	40.391	9.43	14.42	1318
200	0.04125	1.2194	0.1313	1896.1	2886.1	41.915	9.72	14.53	1350
220	0.04489	1.3032	0.1171	2099.9	3177.3	43.303	9.95	14.61	1384
240	0.04849	1.3873	0.1057	2306.0	3469.8	44.576	10.11	14.66	1418
260	0.05206	1.4706	0.0963	2513.6	3762.9	45.750	10.24	14.68	1452
280	0.05559	1.5537	0.0885	2722.2	4056.4	46.837	10.32	14.68	1486
300	0.05925	1.6440	0.0828	2931.4	4353.5	47.851	10.36	14.75	1530
350	0.06806	1.8531	0.0701	3457.4	5090.8	50.125	10.43	14.73	1618
400	0.07679	2.0610	0.0609	3983.5	5826.6	52.090	10.45	14.70	1702
450	0.08549	2.2681	0.0539	4508.9	6560.6	53.819	10.45	14.66	1784
500	0.09415	2.4746	0.0483	5033.4	7293.0	55.362	10.46	14.64	1861
550	0.10280	2.6808	0.0438	5557.2	8024.5	56.756	10.46	14.62	1936
600	0.11143	2.8867	0.0401	6081.4	8755.8	58.026	10.47	14.62	2008
700	0.12868	3.2983	0.0343	7131.1	10213.3	60.285	10.51	14.65	2144
800	0.14589	3.7096	0.0300	8187.0	11688.5	62.249	10.60	14.73	2271
900	0.16310	4.1209	0.0266	9252.7	13167.2	63.988	10.72	14.85	2389
1000	0.18030	4.5323	0.0240	10331.6	14658.8	65.558	10.88	15.00	2500
1200	0.21468	5.3553	0.0200	12544.3	17696.7	68.308	11.25	15.37	2705
1400	0.24906	6.1785	0.0171	14838.4	20815.8	70.718	11.63	15.81	2891
1600	0.28343	7.0020	0.0150	17217.7	24023.0	72.864	12.13	16.24	3063
1800	0.31780	7.8257	0.0133	19682.2	27309.5	74.735	12.55	16.67	3224
2000	0.35219	8.6496	0.0120	22233.2	30685.7	76.584	12.97	17.09	3376
2500	0.43845	10.7098	0.0096	29033.7	39556.5	80.524	14.41	18.61	3718
3000	0.52673	12.7705	0.0080	36951.3	49592.3	84.139	17.46	21.99	4009

* TWO-PHASE BOUNDARY

Table 4.12 (continued)
THERMODYNAMIC PROPERTIES OF NORMAL HYDROGEN

24.00 MPA ISOBAR

TEMPERATURE DEG. K	DENSITY KG/CU M	$V(DH/DV)_P$ KJ/KG	$V(DP/DU)_V$ MPA-CU M/KJ X 10^2	$-V(DP/DV)_T$ MPA	$(DV/DT)_P/V$ 1/DEG. K	THERMAL CONDUCTIVITY W/K-M X 10^3	VISCOSITY KG/M-S X 10^7	THERMAL DIFFUSIVITY SQ M/HR	DIELECTRIC CONSTANT	PRANDTL NUMBER
* 20.300	86.8172	1285.9	0.23558	209.4394	0.00535	140.60	402.08	0.00085	1.28636	1.9669
21	86.4899	1295.6	0.23260	205.9105	0.00544	144.52	374.57	0.00085	1.28519	1.8271
22	86.0150	1307.8	0.22862	200.8776	0.00557	149.33	341.16	0.00086	1.28349	1.6649
23	85.5314	1318.2	0.22497	195.8609	0.00570	153.35	313.28	0.00086	1.28176	1.5359
24	85.0394	1327.3	0.22156	190.8687	0.00583	156.68	289.73	0.00086	1.28001	1.4318
25	84.5393	1334.9	0.21837	185.9097	0.00596	159.43	269.64	0.00085	1.27822	1.3465
26	84.0311	1341.6	0.21532	180.9929	0.00609	161.61	252.34	0.00085	1.27641	1.2766
27	83.5151	1347.6	0.21235	176.1274	0.00622	163.34	237.31	0.00084	1.27457	1.2187
28	82.9915	1352.9	0.20944	171.3226	0.00635	164.70	224.16	0.00083	1.27271	1.1702
29	82.4604	1357.4	0.20668	166.5879	0.00649	165.72	212.57	0.00082	1.27082	1.1291
30	81.9220	1360.7	0.20411	161.9329	0.00662	166.47	202.29	0.00081	1.26891	1.0938
31	81.3766	1363.8	0.20154	157.3672	0.00674	166.34	193.11	0.00080	1.26698	1.0679
32	80.8245	1366.5	0.19903	152.9007	0.00687	166.02	184.88	0.00079	1.26502	1.0459
33	80.2547	1368.6	0.19702	148.6770	0.00701	165.51	177.35	0.00077	1.26300	1.0274
34	79.6883	1371.4	0.19451	144.5215	0.00712	164.95	170.61	0.00076	1.26100	1.0099
35	79.1188	1373.4	0.19205	140.4044	0.00723	164.29	164.49	0.00075	1.25899	0.9948
36	78.5450	1374.3	0.18963	136.2770	0.00735	163.54	158.91	0.00074	1.25696	0.9822
37	77.9637	1376.3	0.18723	132.4359	0.00746	162.70	153.77	0.00073	1.25491	0.9706
38	77.3810	1376.7	0.18499	128.6050	0.00757	161.81	149.05	0.00072	1.25286	0.9605
39	76.7910	1375.2	0.18293	124.7123	0.00769	160.84	144.67	0.00071	1.25078	0.9517
40	76.2003	1371.9	0.18094	120.7474	0.00783	159.85	140.62	0.00070	1.24871	0.9447
42	75.0009	1371.8	0.17700	114.0513	0.00803	157.68	133.31	0.00069	1.24450	0.9309
44	73.7929	1372.1	0.17297	107.7547	0.00822	155.43	126.92	0.00067	1.24027	0.9204
46	72.5767	1375.8	0.16816	101.7880	0.00840	153.15	121.26	0.00066	1.23602	0.9148
48	71.3536	1377.0	0.16385	96.1475	0.00858	150.87	116.22	0.00064	1.23176	0.9096
50	70.1302	1379.2	0.15979	91.0690	0.00873	148.63	111.70	0.00063	1.22750	0.9043
52	68.9070	1381.1	0.15597	86.3606	0.00886	146.45	107.61	0.00063	1.22326	0.8994
54	67.6870	1382.5	0.15236	81.9636	0.00899	144.32	103.90	0.00062	1.21904	0.8947
56	66.4728	1382.5	0.14901	77.7819	0.00911	142.28	100.52	0.00061	1.21484	0.8902
58	65.2660	1383.1	0.14599	74.0437	0.00921	140.32	97.43	0.00061	1.21068	0.8846
60	64.0694	1384.2	0.14304	70.5749	0.00929	138.63	94.60	0.00061	1.20657	0.8778
65	61.1381	1390.6	0.13594	63.0573	0.00943	135.29	88.52	0.00061	1.19653	0.8575
70	58.3171	1404.7	0.12933	57.0761	0.00946	132.17	83.62	0.00061	1.18692	0.8405
75	55.6308	1426.4	0.12318	52.3013	0.00941	129.19	79.71	0.00062	1.17782	0.8278
80	53.0883	1455.4	0.11739	48.4474	0.00929	126.37	76.60	0.00063	1.16925	0.8193
85	50.7035	1492.8	0.11190	45.4103	0.00910	125.81	74.19	0.00066	1.16125	0.8006
90	48.4772	1537.1	0.10669	42.9526	0.00886	125.19	72.33	0.00068	1.15381	0.7871
95	46.4056	1586.9	0.10189	40.9421	0.00860	124.61	70.95	0.00071	1.14692	0.7774
100	44.4809	1642.9	0.09740	39.3019	0.00833	125.84	68.84	0.00074	1.14055	0.7483
120	38.0809	1922.6	0.08180	35.0463	0.00722	135.61	72.66	0.00092	1.11952	0.7438
140	33.2795	2235.1	0.07165	32.6598	0.00630	141.70	74.23	0.00109	1.10392	0.7376
160	29.5650	2570.4	0.06460	31.1885	0.00555	146.81	75.30	0.00125	1.09196	0.7321
180	26.6240	2919.4	0.05949	30.2420	0.00494	151.87	76.46	0.00142	1.08255	0.7261
200	24.2438	3271.7	0.05571	29.5624	0.00444	157.08	77.86	0.00161	1.07498	0.7201
220	22.2769	3622.4	0.05285	29.0318	0.00403	162.47	79.48	0.00180	1.06876	0.7147
240	20.6220	3968.1	0.05066	28.6092	0.00369	167.94	81.31	0.00200	1.06353	0.7097
260	19.2104	4306.9	0.04896	28.2515	0.00341	173.44	83.30	0.00221	1.05910	0.7050
280	17.9895	4638.6	0.04763	27.9499	0.00316	178.95	85.42	0.00244	1.05527	0.7007
300	16.8763	4941.5	0.04739	27.7448	0.00299	184.44	87.63	0.00267	1.05179	0.7009
350	14.6938	5720.3	0.04576	27.2296	0.00258	198.07	93.46	0.00329	1.04498	0.6951
400	13.0222	6476.6	0.04475	26.8391	0.00227	211.66	99.56	0.00398	1.03979	0.6913
450	11.6977	7219.7	0.04406	26.5315	0.00203	225.22	105.80	0.00473	1.03569	0.6886
500	10.6210	7961.5	0.04352	26.2827	0.00184	238.91	112.09	0.00553	1.03237	0.6868
550	9.7276	8700.6	0.04308	26.0774	0.00168	252.59	118.40	0.00639	1.02962	0.6855
600	8.9739	9443.0	0.04270	25.9051	0.00155	266.32	124.68	0.00731	1.02730	0.6846
700	7.7715	10944.3	0.04200	25.6324	0.00134	293.94	137.14	0.00929	1.02361	0.6836
800	6.8542	12483.7	0.04130	25.4266	0.00118	321.94	149.37	0.01148	1.02081	0.6833
900	6.1311	14075.5	0.04055	25.2661	0.00105	350.51	161.34	0.01386	1.01860	0.6833
1000	5.5453	15728.8	0.03974	25.1375	0.00095	440.56	173.07	0.01906	1.01681	0.5893
1200	4.6580	19201.1	0.03810	24.9447	0.00080	511.00	195.79	0.02569	1.01411	0.5890
1400	4.0151	22904.5	0.03648	24.8072	0.00069	583.16	217.61	0.03308	1.01215	0.5898
1600	3.5282	26792.6	0.03501	24.7043	0.00061	656.21	238.65	0.04122	1.01067	0.5908
1800	3.1466	30821.9	0.03373	24.6243	0.00054	729.72	259.01	0.05009	1.00951	0.5915
2000	2.8394	35027.2	0.03255	24.5597	0.00049	805.71	278.76	0.05977	1.00858	0.5913
2500	2.2808	47403.0	0.02917	24.4266	0.00039	1049.60	325.98	0.08903	1.00689	0.5779
3000	1.8985	66713.6	0.02411	24.2450	0.00033	1533.71	371.37	0.13225	1.00573	0.5325

* TWO-PHASE BOUNDARY

Table 4.12 (continued)
THERMODYNAMIC PROPERTIES OF NORMAL HYDROGEN

26.00 MPA ISOBAR

TEMPERATURE DEG. K	VOLUME CU M/KG	ISOTHERM DERIVATIVE CU M-MPA/KG	ISOCHORE DERIVATIVE MPA/K	INTERNAL ENERGY KJ/KG-K	ENTHALPY KJ/KG	ENTROPY KJ/KG-K	CV KJ / KG-K	CP KJ / KG-K	VELOCITY OF SOUND M/S
* 20.756	0.01144	2.4989	1.133	243.9	541.3	14.584	5.52	6.92	1769
21	0.01145	2.4882	1.133	245.3	543.0	14.665	5.55	6.97	1768
22	0.01151	2.4439	1.133	250.8	550.1	14.996	5.68	7.21	1762
23	0.01158	2.3994	1.131	256.5	557.5	15.321	5.79	7.43	1755
24	0.01164	2.3549	1.128	262.4	565.0	15.642	5.90	7.65	1748
25	0.01171	2.3104	1.124	268.4	572.8	15.959	5.99	7.87	1741
26	0.01177	2.2660	1.119	274.6	580.7	16.270	6.09	8.08	1734
27	0.01184	2.2218	1.113	281.0	588.9	16.579	6.17	8.28	1727
28	0.01191	2.1778	1.106	287.5	597.3	16.883	6.26	8.49	1719
29	0.01199	2.1342	1.098	294.2	605.9	17.185	6.33	8.69	1711
30	0.01206	2.0911	1.090	301.1	614.7	17.483	6.40	8.88	1703
31	0.01214	2.0485	1.080	308.0	623.6	17.777	6.47	9.07	1695
32	0.01222	2.0066	1.071	315.2	632.8	18.068	6.53	9.26	1687
33	0.01230	1.9655	1.060	322.4	642.2	18.356	6.59	9.44	1678
34	0.01238	1.9252	1.049	329.8	651.7	18.640	6.64	9.62	1670
35	0.01247	1.8904	1.039	337.3	661.5	18.924	6.69	9.80	1664
36	0.01255	1.8533	1.026	345.0	671.4	19.203	6.74	9.96	1655
37	0.01264	1.8160	1.013	352.7	681.4	19.478	6.78	10.12	1647
38	0.01273	1.7794	0.9998	360.6	691.6	19.750	6.81	10.27	1638
39	0.01282	1.7444	0.9858	368.5	702.0	20.019	6.84	10.42	1629
40	0.01292	1.7093	0.9718	376.6	712.4	20.284	6.87	10.56	1621
42	0.01311	1.6348	0.9433	392.9	733.8	20.806	6.91	10.84	1601
44	0.01331	1.5740	0.9141	409.6	755.8	21.316	6.96	11.10	1585
46	0.01352	1.5171	0.8848	426.1	777.7	21.803	7.03	11.37	1567
48	0.01374	1.4625	0.8553	443.4	800.7	22.292	7.08	11.61	1549
50	0.01396	1.4113	0.8261	461.0	824.1	22.770	7.12	11.83	1532
52	0.01420	1.3625	0.7972	478.9	848.0	23.239	7.15	12.04	1515
54	0.01444	1.3197	0.7692	496.9	872.3	23.697	7.18	12.22	1499
56	0.01468	1.2797	0.7419	515.4	897.1	24.150	7.19	12.38	1484
58	0.01494	1.2422	0.7153	533.7	922.0	24.587	7.19	12.52	1471
60	0.01520	1.2077	0.6895	552.1	947.2	25.013	7.19	12.65	1457
65	0.01588	1.1337	0.6285	598.3	1011.1	26.035	7.19	12.90	1426
70	0.01659	1.0741	0.5733	644.7	1076.1	26.998	7.20	13.09	1398
75	0.01734	1.0313	0.5243	691.0	1141.9	27.906	7.22	13.23	1375
80	0.01812	0.9976	0.4812	737.2	1208.4	28.765	7.26	13.36	1355
85	0.01892	0.9759	0.4432	783.4	1275.4	29.578	7.32	13.45	1339
90	0.01975	0.9621	0.4096	829.4	1342.8	30.349	7.40	13.52	1326
95	0.02058	0.9548	0.3798	875.4	1410.5	31.080	7.49	13.57	1315
100	0.02143	0.9522	0.3535	921.3	1478.5	31.777	7.59	13.62	1307
120	0.02488	0.9781	0.2744	1108.0	1755.0	34.297	8.16	13.88	1290
140	0.02836	1.0328	0.2233	1296.9	2034.2	36.449	8.66	14.09	1297
160	0.03183	1.1016	0.1881	1490.4	2317.8	38.341	9.10	14.30	1316
180	0.03527	1.1794	0.1623	1688.3	2605.3	40.034	9.46	14.46	1343
200	0.03867	1.2608	0.1426	1890.0	2895.5	41.563	9.75	14.57	1373
220	0.04204	1.3439	0.1271	2094.6	3187.6	42.955	9.98	14.65	1405
240	0.04536	1.4264	0.1147	2301.5	3480.9	44.232	10.15	14.70	1438
260	0.04866	1.5096	0.1045	2509.9	3774.9	45.409	10.27	14.72	1471
280	0.05191	1.5918	0.0959	2719.3	4069.1	46.499	10.35	14.72	1504
300	0.05531	1.6819	0.0898	2927.8	4366.0	47.511	10.37	14.78	1548
350	0.06344	1.8902	0.0760	3454.8	5104.3	49.788	10.44	14.75	1634
400	0.07151	2.0975	0.0660	3981.5	5840.8	51.756	10.46	14.71	1717
450	0.07954	2.3040	0.0584	4507.4	6575.4	53.485	10.46	14.67	1798
500	0.08754	2.5101	0.0523	5032.2	7308.3	55.029	10.46	14.65	1874
550	0.09552	2.7159	0.0475	5556.4	8040.0	56.424	10.47	14.63	1949
600	0.10349	2.9215	0.0435	6080.8	8771.6	57.694	10.48	14.63	2020
700	0.11941	3.3324	0.0372	7130.9	10235.5	59.954	10.52	14.66	2155
800	0.13530	3.7433	0.0325	8187.0	11704.9	61.918	10.60	14.73	2281
900	0.15118	4.1542	0.0289	9253.0	13183.7	63.658	10.72	14.85	2398
1000	0.16706	4.5652	0.0260	10332.0	14675.5	65.228	10.88	15.00	2509
1200	0.19879	5.3876	0.0216	12544.9	17713.5	67.978	11.26	15.37	2713
1400	0.23052	6.2104	0.0185	14839.1	20832.7	70.388	11.69	15.81	2898
1600	0.26225	7.0335	0.0162	17218.5	24037.0	72.534	12.13	16.24	3069
1800	0.29397	7.8569	0.0144	19683.0	27326.3	74.465	12.55	16.67	3230
2000	0.32571	8.6806	0.0130	22233.7	30702.2	76.254	12.97	17.09	3382
2500	0.40532	10.7403	0.0104	29027.8	39566.2	80.190	14.39	18.58	3724
3000	0.48673	12.8008	0.0087	36909.0	49563.9	83.842	17.34	21.85	4015

* TWO-PHASE BOUNDARY

Table 4.12 (continued)
THERMODYNAMIC PROPERTIES OF NORMAL HYDROGEN

26.00 MPA ISOBAR

TEMPERATURE DEG. K	DENSITY KG/CU M	$V(DH/DV)_P$ KJ/KG	$V(DP/DU)_V$ MPA-CU M/KJ X 10^2	$-V(DP/DV)_T$ MPA	$(DV/DT)_P/V$ 1/DEG. K	THERMAL CONDUCTIVITY W/K-M X 10^3	VISCOSITY KG/M-S X 10^7	THERMAL DIFFUSIVITY SQ M/HR	DIELECTRIC CONSTANT	PRANDTL NUMBER
* 20.756	87.4223	1333.3	0.23481	218.4609	0.00519	145.73	414.39	0.00087	1.28853	1.9666
21	87.3115	1336.9	0.23377	217.2446	0.00522	147.09	404.17	0.00087	1.28813	1.9164
22	86.8520	1350.5	0.22978	212.2542	0.00534	152.10	366.94	0.00087	1.28649	1.7390
23	86.3845	1362.3	0.22612	207.2732	0.00546	156.30	335.97	0.00088	1.28481	1.5982
24	85.9092	1372.7	0.22269	202.3095	0.00558	159.79	309.91	0.00087	1.28311	1.4846
25	85.4264	1381.6	0.21949	197.3713	0.00570	162.69	287.75	0.00087	1.28139	1.3917
26	84.9362	1389.5	0.21644	192.4670	0.00581	164.99	268.72	0.00087	1.27964	1.3156
27	84.4389	1396.6	0.21347	187.6052	0.00593	166.85	252.24	0.00086	1.27786	1.2524
28	83.9345	1403.1	0.21057	182.7946	0.00605	168.31	237.85	0.00085	1.27607	1.1996
29	83.4234	1408.7	0.20782	178.0440	0.00617	169.44	225.21	0.00084	1.27425	1.1548
30	82.9055	1413.0	0.20527	173.3623	0.00629	170.28	214.02	0.00083	1.27240	1.1162
31	82.3812	1417.0	0.20273	168.7583	0.00640	170.22	204.06	0.00082	1.27054	1.0876
32	81.8507	1420.6	0.20025	164.2410	0.00652	169.97	195.14	0.00081	1.26866	1.0632
33	81.3142	1423.5	0.19788	159.8192	0.00663	169.56	187.11	0.00079	1.26675	1.0422
34	80.7719	1426.0	0.19559	155.5019	0.00675	169.07	179.85	0.00078	1.26483	1.0237
35	80.2107	1429.7	0.19364	151.6293	0.00685	168.40	173.15	0.00077	1.26285	1.0075
36	79.6580	1432.8	0.19126	147.6268	0.00695	167.70	167.14	0.00076	1.26089	0.9928
37	79.1022	1435.0	0.18896	143.6535	0.00705	166.91	161.65	0.00075	1.25893	0.9802
38	78.5436	1436.3	0.18680	139.7629	0.00715	166.05	156.61	0.00074	1.25696	0.9690
39	77.9779	1437.3	0.18475	136.0262	0.00725	165.13	151.94	0.00073	1.25496	0.9584
40	77.4132	1437.3	0.18278	132.3252	0.00734	164.17	147.64	0.00072	1.25297	0.9493
42	76.2690	1433.2	0.17891	124.6851	0.00757	162.07	139.91	0.00071	1.24895	0.9360
44	75.1118	1435.1	0.17497	118.2239	0.00773	159.88	133.16	0.00069	1.24489	0.9242
46	73.9473	1441.4	0.17026	112.1823	0.00789	157.64	127.21	0.00068	1.24081	0.9173
48	72.7792	1445.0	0.16603	106.4422	0.00804	155.40	121.92	0.00066	1.23673	0.9110
50	71.6093	1447.8	0.16204	101.0649	0.00817	153.18	117.19	0.00065	1.23265	0.9053
52	70.4365	1449.5	0.15827	95.9713	0.00831	151.01	112.92	0.00064	1.22857	0.9004
54	69.2688	1452.4	0.15475	91.4121	0.00841	148.90	109.05	0.00063	1.22451	0.8951
56	68.1060	1454.7	0.15149	87.1538	0.00851	146.86	105.53	0.00063	1.22049	0.8898
58	66.9501	1455.9	0.14856	83.1667	0.00860	144.90	102.32	0.00062	1.21649	0.8842
60	65.8029	1457.5	0.14571	79.4700	0.00868	143.21	99.37	0.00062	1.21253	0.8775
65	62.9868	1465.4	0.13878	71.4067	0.00880	139.87	93.00	0.00062	1.20286	0.8576
70	60.2661	1478.5	0.13218	64.7324	0.00886	136.70	87.84	0.00062	1.19355	0.8413
75	57.6618	1500.7	0.12597	59.4679	0.00882	133.64	83.67	0.00063	1.18470	0.8284
80	55.1834	1528.0	0.12014	55.0509	0.00874	130.72	80.31	0.00064	1.17631	0.8206
85	52.8431	1564.5	0.11460	51.5670	0.00859	130.11	77.63	0.00066	1.16842	0.8023
90	50.6437	1607.9	0.10932	48.7231	0.00841	129.42	75.53	0.00068	1.16105	0.7888
95	48.5853	1657.0	0.10442	46.3870	0.00819	128.75	73.92	0.00070	1.15417	0.7790
100	46.6621	1711.6	0.09981	44.4301	0.00796	130.12	71.16	0.00074	1.14777	0.7448
120	40.1885	1987.8	0.08371	39.3070	0.00698	139.73	74.85	0.00090	1.12641	0.7434
140	35.2629	2298.5	0.07314	36.4196	0.00613	145.38	76.15	0.00105	1.11035	0.7381
160	31.4218	2631.9	0.06580	34.6155	0.00543	149.99	76.92	0.00120	1.09793	0.7335
180	28.3535	2979.7	0.06050	33.4409	0.00485	154.58	77.83	0.00136	1.08808	0.7280
200	25.8571	3331.7	0.05657	32.6018	0.00437	159.42	79.02	0.00152	1.08011	0.7224
220	23.7883	3684.1	0.05358	31.9687	0.00398	164.51	80.49	0.00170	1.07354	0.7169
240	22.0442	4030.1	0.05128	31.4449	0.00365	169.74	82.20	0.00189	1.06802	0.7118
260	20.5527	4371.3	0.04950	31.0273	0.00337	175.07	84.09	0.00208	1.06332	0.7070
280	19.2624	4704.2	0.04809	30.6617	0.00313	180.46	86.14	0.00229	1.05926	0.7025
300	18.0791	5001.5	0.04791	30.4077	0.00295	185.85	88.30	0.00250	1.05555	0.7021
350	15.7624	5780.6	0.04619	29.7942	0.00255	199.38	94.07	0.00309	1.04831	0.6959
400	13.9841	6537.0	0.04511	29.3312	0.00225	212.96	100.16	0.00373	1.04278	0.6918
450	12.5726	7280.3	0.04438	28.9676	0.00201	226.57	106.41	0.00442	1.03840	0.6889
500	11.4235	8022.1	0.04380	28.6740	0.00183	240.35	112.75	0.00517	1.03485	0.6870
550	10.4688	8761.2	0.04334	28.4319	0.00167	254.14	119.10	0.00597	1.03190	0.6857
600	9.6626	9503.6	0.04293	28.2290	0.00154	268.00	125.45	0.00683	1.02942	0.6847
700	8.3747	11005.0	0.04219	27.9082	0.00133	295.91	138.03	0.00868	1.02546	0.6837
800	7.3910	12544.5	0.04146	27.6665	0.00117	324.23	150.41	0.01072	1.02245	0.6833
900	6.6145	14136.5	0.04069	27.4781	0.00105	353.14	162.54	0.01295	1.02007	0.6833
1000	5.9860	15790.2	0.03986	27.3273	0.00095	440.56	174.41	0.01766	1.01815	0.5939
1200	5.0303	19263.6	0.03820	27.1013	0.00080	511.00	197.43	0.02379	1.01524	0.5940
1400	4.3380	22968.3	0.03656	26.9404	0.00069	583.16	219.56	0.03062	1.01313	0.5951
1600	3.8132	26857.6	0.03508	26.8201	0.00060	656.20	240.89	0.03814	1.01154	0.5963
1800	3.4017	30886.9	0.03379	26.7266	0.00054	729.66	261.52	0.04634	1.01029	0.5973
2000	3.0702	35088.7	0.03261	26.6512	0.00049	805.47	281.55	0.05527	1.00928	0.5973
2500	2.4672	47396.1	0.02926	26.4983	0.00039	1046.62	329.42	0.08221	1.00745	0.5847
3000	2.0545	66383.1	0.02430	26.2996	0.00033	1518.35	375.39	0.12176	1.00620	0.5402

* TWO-PHASE BOUNDARY

Table 4.12 (continued)
THERMODYNAMIC PROPERTIES OF NORMAL HYDROGEN

28.00 MPA ISOBAR

TEMPERATURE DEG. K	VOLUME CU M/KG	ISOTHERM DERIVATIVE CU M-MPA/KG	ISOCHORE DERIVATIVE MPA/K	INTERNAL ENERGY KJ/KG-K	ENTHALPY KJ/KG	ENTROPY KJ/KG-K	CV KJ /	CP KG-K	VELOCITY OF SOUND M/S
* 21.205	0.01136	2.5840	1.146	246.7	564.8	14.616	5.56	6.95	1797
22	0.01141	2.5495	1.146	251.0	570.4	14.876	5.66	7.14	1793
23	0.01147	2.5061	1.145	256.6	577.7	15.198	5.78	7.36	1787
24	0.01153	2.4625	1.142	262.3	585.1	15.516	5.88	7.57	1780
25	0.01159	2.4189	1.139	268.3	592.8	15.829	5.98	7.78	1774
26	0.01166	2.3753	1.134	274.4	600.7	16.137	6.08	7.99	1767
27	0.01172	2.3318	1.129	280.6	608.8	16.443	6.17	8.19	1760
28	0.01179	2.2884	1.122	287.0	617.1	16.743	6.25	8.39	1753
29	0.01186	2.2454	1.115	293.6	625.6	17.041	6.33	8.59	1745
30	0.01193	2.2027	1.107	300.3	634.2	17.335	6.40	8.77	1738
31	0.01200	2.1604	1.098	307.1	643.1	17.626	6.47	8.96	1730
32	0.01207	2.1188	1.089	314.1	652.2	17.913	6.53	9.14	1722
33	0.01215	2.0777	1.079	321.2	661.4	18.198	6.59	9.32	1714
34	0.01223	2.0374	1.069	328.4	670.8	18.478	6.64	9.50	1706
35	0.01231	1.9979	1.058	335.8	680.4	18.756	6.69	9.67	1698
36	0.01239	1.9594	1.047	343.3	690.1	19.031	6.74	9.83	1690
37	0.01247	1.9290	1.036	350.9	700.1	19.306	6.79	9.99	1685
38	0.01256	1.8935	1.023	358.6	710.2	19.574	6.82	10.14	1677
39	0.01264	1.8580	1.010	366.4	720.4	19.839	6.85	10.28	1669
40	0.01273	1.8243	0.9969	374.3	730.8	20.101	6.88	10.41	1662
42	0.01291	1.7570	0.9693	390.3	751.9	20.615	6.93	10.67	1645
44	0.01310	1.6857	0.9413	406.7	773.5	21.117	6.97	10.94	1626
46	0.01329	1.6273	0.9127	422.8	795.1	21.597	7.05	11.21	1609
48	0.01350	1.5737	0.8841	439.8	817.7	22.079	7.10	11.44	1592
50	0.01370	1.5221	0.8556	457.1	840.8	22.551	7.14	11.66	1576
52	0.01392	1.4731	0.8273	474.6	864.3	23.012	7.18	11.86	1560
54	0.01414	1.4282	0.7996	492.4	888.2	23.463	7.21	12.04	1545
56	0.01437	1.3862	0.7726	510.5	912.7	23.910	7.22	12.20	1530
58	0.01460	1.3465	0.7463	528.5	937.3	24.341	7.22	12.34	1516
60	0.01484	1.3108	0.7209	546.6	962.1	24.761	7.22	12.46	1504
65	0.01546	1.2344	0.6606	592.1	1025.1	25.767	7.22	12.72	1474
70	0.01612	1.1720	0.6050	637.9	1089.2	26.718	7.23	12.91	1446
75	0.01680	1.1230	0.5553	683.7	1154.1	27.614	7.26	13.07	1422
80	0.01751	1.0848	0.5112	729.5	1219.8	28.462	7.30	13.21	1401
85	0.01824	1.0576	0.4721	775.3	1286.1	29.267	7.36	13.32	1384
90	0.01900	1.0395	0.4374	821.1	1352.9	30.031	7.43	13.41	1369
95	0.01976	1.0284	0.4065	866.9	1420.2	30.758	7.52	13.48	1358
100	0.02054	1.0229	0.3790	912.7	1487.8	31.451	7.63	13.55	1348
120	0.02371	1.0378	0.2955	1099.6	1763.5	33.964	8.19	13.87	1326
140	0.02692	1.0866	0.2407	1289.0	2042.7	36.115	8.69	14.09	1328
160	0.03012	1.1497	0.2029	1483.1	2326.6	38.009	9.12	14.32	1343
180	0.03331	1.2234	0.1751	1681.8	2614.5	39.705	9.49	14.50	1367
200	0.03647	1.3033	0.1539	1884.2	2905.4	41.237	9.78	14.61	1396
220	0.03960	1.3851	0.1372	2089.6	3198.3	42.633	10.00	14.69	1426
240	0.04269	1.4673	0.1237	2297.2	3492.4	43.913	10.18	14.74	1458
260	0.04574	1.5489	0.1127	2506.3	3787.2	45.093	10.30	14.76	1490
280	0.04877	1.6308	0.1034	2716.4	4082.1	46.186	10.39	14.75	1522
300	0.05193	1.7203	0.0968	2924.5	4378.6	47.196	10.39	14.80	1565
350	0.05949	1.9276	0.0819	3452.3	5117.9	49.476	10.45	14.77	1650
400	0.06698	2.1341	0.0711	3979.7	5855.2	51.445	10.47	14.72	1732
450	0.07444	2.3401	0.0629	4506.0	6590.3	53.177	10.47	14.68	1811
500	0.08187	2.5457	0.0564	5031.2	7323.5	54.721	10.47	14.65	1888
550	0.08928	2.7511	0.0511	5555.7	8055.6	56.117	10.47	14.64	1961
600	0.09668	2.9563	0.0468	6080.3	8787.4	57.387	10.48	14.63	2032
700	0.11146	3.3666	0.0400	7130.8	10251.7	59.648	10.52	14.66	2166
800	0.12622	3.7769	0.0350	8187.2	11721.3	61.612	10.60	14.73	2291
900	0.14096	4.1873	0.0311	9253.3	13200.3	63.352	10.73	14.85	2408
1000	0.15570	4.5980	0.0279	10332.5	14692.2	64.922	10.88	15.00	2518
1200	0.18517	5.4197	0.0233	12545.6	17730.3	67.672	11.26	15.37	2720
1400	0.21463	6.2420	0.0200	14839.9	20849.6	70.082	11.69	15.81	2905
1600	0.24409	7.0648	0.0175	17219.4	24053.9	72.228	12.13	16.24	3076
1800	0.27354	7.8879	0.0155	19683.9	27343.1	74.159	12.55	16.66	3237
2000	0.30301	8.7112	0.0140	22234.4	30718.7	75.948	12.96	17.09	3388
2500	0.37692	10.7705	0.0112	29022.8	39576.7	79.882	14.36	18.55	3730
3000	0.45245	12.8306	0.0093	36871.4	49540.0	83.522	17.23	21.72	4021

* TWO-PHASE BOUNDARY

Table 4.12 (continued)
THERMODYNAMIC PROPERTIES OF NORMAL HYDROGEN

28.00 MPA ISOBAR

TEMPERATURE DEG. K	DENSITY KG/CU M	$V(DH/DV)_P$ KJ/KG	$V(DP/DU)_V$ MPA-CU M/KJ X 10^2	$-V(DP/DV)_T$ MPA	$(DV/DT)_P/V$ 1/DEG. K	THERMAL CONDUCTIVITY W/K-M X 10^3	VISCOSITY KG/M-S X 10^7	THERMAL DIFFUSIVITY SQ M/HR	DIELECTRIC CONSTANT	PRANDTL NUMBER
* 21.205	88.0083	1379.8	0.23412	227.4110	0.00504	150.73	426.76	0.00089	1.29063	1.9688
22	87.6531	1391.6	0.23094	223.4755	0.00513	154.79	394.45	0.00089	1.28936	1.8186
23	87.1999	1404.8	0.22725	218.5311	0.00524	159.16	360.09	0.00089	1.28773	1.6649
24	86.7396	1416.4	0.22379	213.5974	0.00535	162.81	331.27	0.00089	1.28608	1.5413
25	86.2722	1426.6	0.22058	208.6821	0.00546	165.85	306.83	0.00089	1.28441	1.4402
26	85.7981	1435.7	0.21752	203.7934	0.00556	168.28	285.92	0.00088	1.28272	1.3574
27	85.3174	1444.0	0.21454	198.9393	0.00567	170.25	267.85	0.00088	1.28100	1.2887
28	84.8302	1451.7	0.21164	194.1280	0.00578	171.82	252.13	0.00087	1.27926	1.2313
29	84.3368	1458.3	0.20889	189.3677	0.00589	173.05	238.34	0.00086	1.27750	1.1825
30	83.8372	1463.6	0.20634	184.6668	0.00599	173.97	226.17	0.00085	1.27572	1.1406
31	83.3317	1468.6	0.20382	180.0335	0.00610	173.98	215.36	0.00084	1.27392	1.1091
32	82.8205	1473.1	0.20136	175.4760	0.00621	173.79	205.70	0.00083	1.27210	1.0822
33	82.3037	1476.8	0.19902	171.0027	0.00631	173.44	197.03	0.00081	1.27027	1.0590
34	81.7815	1480.0	0.19675	166.6217	0.00642	172.99	189.21	0.00080	1.26841	1.0386
35	81.2543	1482.8	0.19455	162.3412	0.00652	172.43	182.11	0.00079	1.26654	1.0210
36	80.7222	1485.5	0.19237	158.1692	0.00662	171.76	175.65	0.00078	1.26466	1.0057
37	80.1708	1491.0	0.19047	154.6531	0.00670	170.95	169.63	0.00077	1.26270	0.9914
38	79.6309	1493.7	0.18835	150.7823	0.00679	170.12	164.23	0.00076	1.26080	0.9786
39	79.0888	1495.0	0.18639	146.9473	0.00687	169.24	159.27	0.00075	1.25888	0.9672
40	78.5454	1496.5	0.18450	143.2904	0.00696	168.31	154.69	0.00074	1.25696	0.9569
42	77.4471	1498.1	0.18068	136.0715	0.00712	166.27	146.49	0.00072	1.25309	0.9402
44	76.3388	1495.8	0.17683	128.6814	0.00731	164.13	139.37	0.00071	1.24919	0.9291
46	75.2198	1503.3	0.17217	122.4022	0.00746	161.92	133.10	0.00069	1.24527	0.9214
48	74.0969	1509.3	0.16802	116.6041	0.00758	159.71	127.55	0.00068	1.24133	0.9140
50	72.9730	1513.8	0.16410	111.0704	0.00770	157.51	122.59	0.00067	1.23740	0.9076
52	71.8491	1517.2	0.16039	105.8386	0.00782	155.36	118.14	0.00066	1.23348	0.9019
54	70.7270	1520.8	0.15690	101.0133	0.00792	153.25	114.11	0.00065	1.22958	0.8963
56	69.6092	1523.5	0.15370	96.4948	0.00801	151.22	110.44	0.00064	1.22569	0.8909
58	68.4956	1524.7	0.15083	92.2294	0.00809	149.24	107.09	0.00064	1.22183	0.8853
60	67.3919	1527.1	0.14806	88.3401	0.00816	147.55	104.02	0.00063	1.21802	0.8785
65	64.6769	1537.1	0.14136	79.8351	0.00827	144.20	97.38	0.00063	1.20866	0.8589
70	62.0472	1552.0	0.13480	72.7208	0.00832	141.00	91.97	0.00063	1.19964	0.8423
75	59.5203	1573.1	0.12858	66.8417	0.00831	137.87	87.56	0.00064	1.19101	0.8301
80	57.1053	1600.4	0.12268	61.9491	0.00825	134.87	83.96	0.00064	1.18281	0.8221
85	54.8122	1635.4	0.11709	57.9683	0.00814	134.22	81.05	0.00066	1.17506	0.8043
90	52.6444	1677.9	0.11176	54.7214	0.00799	133.47	78.73	0.00068	1.16776	0.7911
95	50.6046	1726.3	0.10678	52.0429	0.00781	132.73	76.89	0.00070	1.16092	0.7812
100	48.6892	1780.6	0.10208	49.8064	0.00761	134.27	73.44	0.00073	1.15452	0.7411
120	42.1725	2053.7	0.08556	43.7660	0.00675	143.77	77.00	0.00089	1.13293	0.7427
140	37.1500	2363.7	0.07460	40.3688	0.00596	149.03	78.06	0.00102	1.11648	0.7383
160	33.1965	2694.2	0.06698	38.1655	0.00532	153.16	78.56	0.00116	1.10365	0.7346
180	30.0178	3039.7	0.06149	36.7232	0.00477	157.32	79.22	0.00130	1.09341	0.7300
200	27.4181	3392.7	0.05741	35.7338	0.00431	161.78	80.21	0.00145	1.08509	0.7245
220	25.2544	3746.2	0.05429	34.9799	0.00392	166.57	81.52	0.00162	1.07819	0.7190
240	23.4260	4094.4	0.05190	34.3718	0.00360	171.58	83.10	0.00179	1.07239	0.7138
260	21.8607	4435.5	0.05003	33.8600	0.00333	176.74	84.91	0.00197	1.06744	0.7089
280	20.5035	4770.5	0.04856	33.4365	0.00309	181.99	86.88	0.00217	1.06316	0.7044
300	19.2549	5061.9	0.04842	33.1238	0.00292	187.29	89.00	0.00237	1.05924	0.7032
350	16.8102	5841.2	0.04661	32.4031	0.00253	200.70	94.69	0.00291	1.05158	0.6967
400	14.9294	6597.7	0.04548	31.8614	0.00223	214.26	100.76	0.00351	1.04572	0.6923
450	13.4340	7341.0	0.04470	31.4371	0.00200	227.92	107.03	0.00416	1.04107	0.6893
500	12.2147	8082.9	0.04408	31.0949	0.00181	241.78	113.40	0.00486	1.03729	0.6873
550	11.2005	8822.0	0.04359	30.8132	0.00166	255.68	119.81	0.00561	1.03416	0.6858
600	10.3431	9564.3	0.04315	30.5772	0.00153	269.67	126.21	0.00641	1.03152	0.6848
700	8.9719	11065.6	0.04238	30.2045	0.00132	297.87	138.93	0.00815	1.02730	0.6837
800	7.9229	12605.2	0.04162	29.9238	0.00117	326.51	151.45	0.01007	1.02408	0.6833
900	7.0941	14197.4	0.04083	29.7053	0.00105	355.76	163.72	0.01216	1.02154	0.6833
1000	6.4225	15851.5	0.03998	29.5305	0.00095	440.56	175.75	0.01646	1.01949	0.5985
1200	5.4005	19325.9	0.03830	29.2688	0.00080	511.00	199.08	0.02216	1.01637	0.5990
1400	4.6592	23031.8	0.03664	29.0827	0.00069	583.16	221.51	0.02851	1.01411	0.6004
1600	4.0969	26922.2	0.03515	28.9436	0.00060	656.19	243.14	0.03550	1.01240	0.6019
1800	3.6557	30951.8	0.03385	28.8357	0.00054	729.61	264.06	0.04311	1.01106	0.6031
2000	3.3002	35150.5	0.03266	28.7488	0.00049	805.25	284.37	0.05141	1.00998	0.6034
2500	2.6531	47397.3	0.02935	28.5748	0.00039	1043.97	332.90	0.07637	1.00802	0.5915
3000	2.2102	66096.2	0.02448	28.3582	0.00033	1504.66	379.48	0.11282	1.00668	0.5479

* TWO-PHASE BOUNDARY

Table 4.12 (continued)
THERMODYNAMIC PROPERTIES OF NORMAL HYDROGEN

30.00 MPA ISOBAR

TEMPERATURE DEG. K	VOLUME CU M/KG	ISOTHERM DERIVATIVE CU M-MPA/KG	ISOCHORE DERIVATIVE MPA/K	INTERNAL ENERGY KJ/KG-K	ENTHALPY KJ/KG	ENTROPY KJ/KG-K	CV KJ/KG-K	CP KJ/KG-K	VELOCITY OF SOUND M/S
* 21.645	0.01129	2.6677	1.159	249.4	588.1	14.647	5.60	6.99	1824
22	0.01131	2.6527	1.159	251.3	590.6	14.761	5.65	7.07	1823
23	0.01137	2.6102	1.158	256.8	597.8	15.081	5.76	7.29	1817
24	0.01142	2.5675	1.156	262.5	605.2	15.396	5.87	7.50	1811
25	0.01148	2.5247	1.153	268.3	612.8	15.706	5.97	7.71	1805
26	0.01154	2.4819	1.149	274.3	620.6	16.010	6.07	7.91	1799
27	0.01161	2.4391	1.144	280.4	628.6	16.313	6.16	8.11	1792
28	0.01167	2.3964	1.138	286.7	636.8	16.610	6.24	8.30	1785
29	0.01174	2.3539	1.131	293.1	645.2	16.905	6.32	8.49	1778
30	0.01180	2.3117	1.123	299.7	653.8	17.196	6.39	8.68	1771
31	0.01187	2.2699	1.115	306.4	662.6	17.484	6.46	8.86	1764
32	0.01194	2.2285	1.107	313.3	671.5	17.768	6.53	9.04	1756
33	0.01201	2.1876	1.097	320.2	680.6	18.049	6.59	9.21	1749
34	0.01209	2.1473	1.088	327.3	689.9	18.326	6.65	9.38	1741
35	0.01216	2.1078	1.077	334.6	699.4	18.601	6.70	9.55	1733
36	0.01224	2.0691	1.067	341.9	709.0	18.872	6.75	9.71	1726
37	0.01232	2.0313	1.056	349.4	718.8	19.141	6.79	9.87	1718
38	0.01240	1.9944	1.044	356.9	728.8	19.406	6.83	10.02	1711
39	0.01248	1.9684	1.033	364.6	739.0	19.671	6.86	10.15	1707
40	0.01256	1.9328	1.020	372.3	749.2	19.929	6.89	10.29	1699
42	0.01273	1.8695	0.9938	388.1	770.0	20.437	6.94	10.54	1685
44	0.01291	1.8048	0.9667	404.2	791.3	20.933	6.99	10.78	1669
46	0.01309	1.7369	0.9392	420.0	812.6	21.406	7.07	11.07	1649
48	0.01328	1.6806	0.9112	436.7	835.0	21.882	7.12	11.30	1633
50	0.01347	1.6300	0.8834	453.7	857.8	22.347	7.17	11.51	1618
52	0.01367	1.5811	0.8557	470.9	881.0	22.802	7.20	11.70	1603
54	0.01387	1.5340	0.8284	488.4	904.6	23.248	7.23	11.88	1588
56	0.01408	1.4913	0.8017	506.2	928.8	23.688	7.25	12.04	1574
58	0.01430	1.4507	0.7756	524.0	953.0	24.113	7.25	12.17	1560
60	0.01452	1.4139	0.7504	541.8	977.5	24.528	7.26	12.29	1548
65	0.01510	1.3324	0.6905	586.7	1039.6	25.522	7.26	12.56	1518
70	0.01570	1.2694	0.6352	631.9	1103.0	26.461	7.27	12.75	1492
75	0.01633	1.2140	0.5848	677.2	1167.2	27.347	7.29	12.93	1467
80	0.01698	1.1744	0.5399	722.6	1232.2	28.186	7.33	13.06	1446
85	0.01766	1.1411	0.4998	768.1	1297.8	28.982	7.39	13.19	1427
90	0.01835	1.1182	0.4642	813.6	1364.0	29.740	7.47	13.31	1411
95	0.01905	1.1033	0.4323	859.2	1430.8	30.461	7.56	13.40	1399
100	0.01977	1.0945	0.4038	904.9	1498.0	31.151	7.66	13.48	1388
120	0.02270	1.0997	0.3162	1091.8	1773.0	33.656	8.22	13.85	1361
140	0.02568	1.1425	0.2580	1281.7	2052.0	35.806	8.71	14.09	1359
160	0.02865	1.2008	0.2176	1476.3	2335.9	37.700	9.15	14.33	1371
180	0.03162	1.2699	0.1879	1675.5	2624.2	39.398	9.51	14.52	1392
200	0.03457	1.3471	0.1652	1878.7	2915.7	40.934	9.80	14.64	1418
220	0.03749	1.4270	0.1472	2084.8	3209.3	42.333	10.03	14.73	1447
240	0.04037	1.5078	0.1327	2293.1	3504.2	43.616	10.20	14.78	1478
260	0.04322	1.5886	0.1208	2502.9	3799.6	44.799	10.33	14.79	1508
280	0.04605	1.6701	0.1109	2713.7	4095.3	45.894	10.42	14.79	1540
300	0.04901	1.7590	0.1038	2921.2	4391.5	46.903	10.40	14.82	1583
350	0.05606	1.9653	0.0878	3449.9	5131.7	49.186	10.47	14.78	1666
400	0.06306	2.1710	0.0762	3977.9	5869.6	51.157	10.48	14.73	1747
450	0.07002	2.3764	0.0673	4504.7	6605.2	52.889	10.48	14.69	1825
500	0.07695	2.5814	0.0604	5030.3	7338.9	54.434	10.48	14.66	1900
550	0.08387	2.7863	0.0548	5555.1	8071.3	55.831	10.48	14.64	1973
600	0.09078	2.9911	0.0501	6079.9	8803.3	57.102	10.49	14.64	2043
700	0.10457	3.4007	0.0429	7130.8	10267.9	59.363	10.53	14.66	2176
800	0.11834	3.8104	0.0375	8187.4	11737.7	61.327	10.61	14.73	2300
900	0.13210	4.2204	0.0333	9253.8	13216.9	63.067	10.73	14.85	2417
1000	0.14586	4.6306	0.0299	10333.1	14708.8	64.637	10.89	15.00	2526
1200	0.17336	5.4516	0.0249	12546.4	17747.1	67.388	11.26	15.38	2728
1400	0.20085	6.2734	0.0214	14840.9	20866.4	69.798	11.69	15.81	2912
1600	0.22834	7.0957	0.0187	17220.4	24070.7	71.944	12.13	16.24	3083
1800	0.25583	7.9185	0.0166	19684.9	27359.9	73.875	12.55	16.66	3243
2000	0.28334	8.7415	0.0150	22235.1	30735.2	75.663	12.96	17.08	3394
2500	0.35231	10.8003	0.0120	29018.4	39587.7	79.595	14.34	18.52	3735
3000	0.42274	12.8600	0.0100	36837.9	49520.2	83.224	17.13	21.61	4027

* TWO-PHASE BOUNDARY

Table 4.12 (continued)
THERMODYNAMIC PROPERTIES OF NORMAL HYDROGEN

30.00 MPA ISOBAR

TEMPERATURE DEG. K	DENSITY KG/CU M	$V(\partial H/\partial V)_P$ KJ/KG	$V(\partial P/\partial U)_V$ MPA-CU M/KJ X 10^2	$-V(\partial P/\partial V)_T$ MPA	$(\partial V/\partial T)_P/V$ 1/DEG. K	THERMAL CONDUCTIVITY W/K-M X 10^3	VISCOSITY KG/M-S X 10^7	THERMAL DIFFUSIVITY SQ M/HR	DIELECTRIC CONSTANT	PRANDTL NUMBER
* 21.645	88.5767	1425.6	0.23351	236.2981	0.00490	155.60	439.19	0.00090	1.29267	1.9730
22	88.4220	1431.2	0.23210	234.5564	0.00494	157.41	423.85	0.00091	1.29212	1.9038
23	87.9818	1445.7	0.22837	229.6496	0.00504	161.95	385.75	0.00091	1.29054	1.7363
24	87.5348	1458.6	0.22489	224.7472	0.00514	165.75	353.91	0.00091	1.28893	1.6018
25	87.0814	1470.0	0.22165	219.8570	0.00524	168.92	326.99	0.00091	1.28731	1.4920
26	86.6217	1480.3	0.21857	214.9865	0.00534	171.48	304.01	0.00090	1.28566	1.4022
27	86.1559	1489.8	0.21557	210.1434	0.00544	173.55	284.22	0.00089	1.28399	1.3276
28	85.6841	1498.6	0.21266	205.3357	0.00554	175.23	267.04	0.00089	1.28231	1.2653
29	85.2066	1506.3	0.20990	200.5710	0.00564	176.54	252.01	0.00088	1.28060	1.2123
30	84.7234	1512.6	0.20736	195.8572	0.00574	177.55	238.78	0.00087	1.27888	1.1668
31	84.2347	1518.6	0.20483	191.2021	0.00583	177.62	227.05	0.00086	1.27714	1.1323
32	83.7407	1524.0	0.20238	186.6133	0.00593	177.49	216.60	0.00084	1.27538	1.1029
33	83.2416	1528.6	0.20005	182.0987	0.00603	177.19	207.24	0.00083	1.27360	1.0774
34	82.7375	1532.5	0.19781	177.6659	0.00612	176.80	198.81	0.00082	1.27181	1.0550
35	82.2286	1536.1	0.19563	173.3223	0.00622	176.28	191.18	0.00081	1.27000	1.0355
36	81.7153	1539.3	0.19348	169.0756	0.00631	175.65	184.24	0.00080	1.26818	1.0187
37	81.1976	1542.3	0.19141	164.9329	0.00640	174.93	177.92	0.00079	1.26634	1.0040
38	80.6760	1544.5	0.18947	160.9016	0.00649	174.13	172.12	0.00078	1.26449	0.9907
39	80.1347	1550.6	0.18783	157.7348	0.00655	173.20	166.68	0.00077	1.26258	0.9771
40	79.6102	1551.8	0.18598	153.8691	0.00663	172.30	161.81	0.00076	1.26072	0.9661
42	78.5503	1556.8	0.18232	146.8531	0.00677	170.30	153.10	0.00074	1.25698	0.9471
44	77.4833	1559.9	0.17853	139.8430	0.00691	168.20	145.57	0.00072	1.25322	0.9332
46	76.4083	1563.9	0.17397	132.7170	0.00708	166.03	138.97	0.00071	1.24944	0.9263
48	75.3261	1570.1	0.16985	126.5935	0.00720	163.84	133.13	0.00069	1.24564	0.9183
50	74.2419	1576.9	0.16600	121.0169	0.00730	161.65	127.93	0.00068	1.24184	0.9110
52	73.1587	1582.1	0.16235	115.6721	0.00740	159.50	123.28	0.00067	1.23805	0.9046
54	72.0777	1586.0	0.15891	110.5687	0.00749	157.41	119.07	0.00066	1.23428	0.8989
56	70.9997	1589.9	0.15573	105.8852	0.00757	155.37	115.25	0.00065	1.23053	0.8929
58	69.9274	1592.1	0.15290	101.4425	0.00765	153.39	111.77	0.00065	1.22680	0.8870
60	68.8615	1595.3	0.15018	97.3636	0.00771	151.69	108.57	0.00065	1.22310	0.8800
65	66.2364	1605.4	0.14362	88.2558	0.00782	148.33	101.67	0.00064	1.21403	0.8609
70	63.6849	1623.3	0.13722	80.8416	0.00786	145.08	96.01	0.00064	1.20525	0.8441
75	61.2321	1643.3	0.13097	74.3388	0.00787	141.90	91.38	0.00065	1.19685	0.8325
80	58.8765	1672.6	0.12505	69.1426	0.00781	138.82	87.56	0.00065	1.18882	0.8238
85	56.6333	1706.0	0.11939	64.6270	0.00773	138.15	84.44	0.00067	1.18121	0.8065
90	54.5005	1747.3	0.11401	60.9434	0.00762	137.35	81.91	0.00068	1.17400	0.7936
95	52.4837	1794.9	0.10898	57.9045	0.00747	136.55	79.87	0.00070	1.16722	0.7838
100	50.5805	1848.6	0.10421	55.3620	0.00729	138.31	75.66	0.00073	1.16084	0.7375
120	44.0441	2120.7	0.08733	48.4344	0.00653	147.72	79.13	0.00087	1.13910	0.7417
140	38.9443	2430.1	0.07603	44.4929	0.00580	152.63	79.96	0.00100	1.12234	0.7383
160	34.8991	2759.8	0.06814	41.9073	0.00519	156.33	80.21	0.00113	1.10917	0.7352
180	31.6242	3102.7	0.06246	40.1595	0.00468	160.07	80.63	0.00125	1.09858	0.7314
200	28.9273	3454.8	0.05824	38.9666	0.00424	164.17	81.42	0.00140	1.08992	0.7263
220	26.6772	3808.3	0.05501	38.0675	0.00387	168.67	82.57	0.00155	1.08272	0.7210
240	24.7712	4157.2	0.05252	37.3495	0.00355	173.45	84.03	0.00171	1.07666	0.7158
260	23.1356	4499.9	0.05056	36.7534	0.00329	178.42	85.74	0.00188	1.07147	0.7109
280	21.7153	4837.0	0.04902	36.2665	0.00306	183.55	87.65	0.00206	1.06698	0.7062
300	20.4046	5122.7	0.04892	35.8926	0.00289	188.75	89.70	0.00225	1.06285	0.7043
350	17.8377	5902.1	0.04702	35.0560	0.00250	202.03	95.33	0.00276	1.05479	0.6975
400	15.8585	6658.6	0.04583	34.4295	0.00221	215.56	101.38	0.00332	1.04861	0.6929
450	14.2821	7401.9	0.04501	33.9397	0.00198	229.26	107.66	0.00393	1.04370	0.6897
500	12.9948	8143.8	0.04435	33.5454	0.00180	243.21	114.06	0.00460	1.03971	0.6875
550	11.9229	8882.8	0.04383	33.2209	0.00165	257.21	120.51	0.00530	1.03639	0.6860
600	11.0157	9625.0	0.04338	32.9494	0.00152	271.33	126.97	0.00606	1.03359	0.6849
700	9.5629	11126.3	0.04257	32.5208	0.00132	299.82	139.82	0.00770	1.02911	0.6837
800	8.4501	12665.9	0.04178	32.1984	0.00116	328.78	152.48	0.00951	1.02569	0.6833
900	7.5698	14258.3	0.04096	31.9475	0.00104	358.37	164.91	0.01148	1.02300	0.6833
1000	6.8559	15912.7	0.04010	31.7470	0.00094	440.56	177.09	0.01542	1.02081	0.6031
1200	5.7684	19388.1	0.03839	31.4470	0.00079	511.00	200.74	0.02074	1.01749	0.6040
1400	4.9788	23095.2	0.03672	31.2338	0.00068	583.16	223.48	0.02668	1.01508	0.6057
1600	4.3794	26986.7	0.03521	31.0748	0.00060	656.19	245.41	0.03321	1.01326	0.6075
1800	3.9088	31016.4	0.03390	30.9514	0.00054	729.56	266.63	0.04032	1.01183	0.6090
2000	3.5294	35212.6	0.03272	30.8523	0.00049	805.05	287.22	0.04807	1.01067	0.6095
2500	2.8384	47405.1	0.02943	30.6557	0.00039	1041.59	336.44	0.07132	1.00858	0.5983
3000	2.3655	65845.4	0.02464	30.4206	0.00033	1492.37	383.64	0.10510	1.00715	0.5555

* TWO-PHASE BOUNDARY

Table 4.12 (continued)
THERMODYNAMIC PROPERTIES OF NORMAL HYDROGEN

35.00 MPA ISOBAR

TEMPERATURE DEG. K	VOLUME CU M/KG	ISOTHERM DERIVATIVE CU'M-MPA/KG	ISOCHORE DERIVATIVE MPA/K	INTERNAL ENERGY KJ/KG-K	ENTHALPY KJ/KG	ENTROPY KJ/KG-K	CV KJ /	CP KG-K	VELOCITY OF SOUND M/S
* 22.715	0.01112	2.8721	1.190	256.6	645.8	14.721	5.70	7.08	1889
23	0.01113	2.8606	1.190	258.1	647.8	14.809	5.73	7.14	1888
24	0.01119	2.8201	1.189	263.5	655.0	15.117	5.84	7.35	1883
25	0.01124	2.7794	1.186	269.1	662.5	15.421	5.94	7.54	1878
26	0.01129	2.7385	1.183	274.8	670.1	15.719	6.04	7.74	1873
27	0.01135	2.6975	1.179	280.7	678.0	16.015	6.14	7.93	1867
28	0.01141	2.6564	1.174	286.7	686.0	16.305	6.22	8.11	1861
29	0.01147	2.6154	1.168	292.9	694.2	16.594	6.31	8.30	1855
30	0.01152	2.5745	1.161	299.2	702.6	16.878	6.38	8.47	1848
31	0.01159	2.5338	1.154	305.6	711.1	17.159	6.46	8.64	1842
32	0.01165	2.4933	1.147	312.2	719.8	17.436	6.52	8.81	1835
33	0.01171	2.4531	1.138	318.9	728.7	17.709	6.59	8.98	1828
34	0.01177	2.4133	1.130	325.7	737.8	17.980	6.65	9.14	1822
35	0.01184	2.3740	1.121	332.6	747.0	18.247	6.70	9.30	1815
36	0.01191	2.3352	1.111	339.7	756.4	18.512	6.75	9.45	1808
37	0.01197	2.2970	1.101	346.8	765.9	18.773	6.80	9.60	1801
38	0.01204	2.2595	1.091	354.1	775.6	19.031	6.85	9.75	1794
39	0.01211	2.2228	1.080	361.4	785.4	19.286	6.88	9.88	1787
40	0.01219	2.1869	1.069	368.8	795.4	19.538	6.91	10.02	1780
42	0.01233	2.1179	1.047	384.0	815.7	20.033	6.97	10.27	1767
44	0.01249	2.0695	1.023	399.4	836.6	20.518	7.02	10.49	1759
46	0.01265	2.0085	0.9981	414.7	857.3	20.978	7.10	10.75	1744
48	0.01281	1.9548	0.9725	430.7	879.0	21.441	7.17	10.97	1730
50	0.01297	1.8968	0.9466	447.1	901.2	21.893	7.22	11.19	1715
52	0.01314	1.8424	0.9204	463.7	923.7	22.336	7.26	11.39	1700
54	0.01332	1.7924	0.8941	480.6	946.7	22.769	7.29	11.56	1686
56	0.01350	1.7490	0.8682	497.8	970.2	23.198	7.31	11.71	1673
58	0.01368	1.7066	0.8428	515.0	993.8	23.611	7.32	11.84	1661
60	0.01387	1.6653	0.8179	532.2	1017.6	24.015	7.33	11.96	1649
65	0.01435	1.5755	0.7585	575.8	1078.0	24.981	7.34	12.22	1620
70	0.01486	1.5009	0.7037	619.7	1139.8	25.896	7.35	12.45	1594
75	0.01538	1.4414	0.6533	664.0	1202.5	26.761	7.38	12.63	1571
80	0.01593	1.3958	0.6068	708.5	1266.0	27.582	7.42	12.77	1550
85	0.01649	1.3528	0.5648	753.1	1330.2	28.361	7.48	12.92	1529
90	0.01706	1.3226	0.5269	798.0	1395.2	29.104	7.55	13.05	1512
95	0.01765	1.2957	0.4930	843.1	1460.8	29.813	7.64	13.19	1496
100	0.01825	1.2775	0.4624	888.4	1527.0	30.492	7.74	13.31	1482
120	0.02071	1.2580	0.3662	1075.0	1799.7	32.976	8.30	13.78	1445
140	0.02322	1.2843	0.3004	1265.5	2078.2	35.122	8.78	14.09	1435
160	0.02574	1.3330	0.2538	1461.2	2362.1	37.016	9.21	14.34	1440
180	0.02826	1.3923	0.2195	1661.6	2650.7	38.716	9.57	14.55	1455
200	0.03077	1.4599	0.1932	1866.1	2943.1	40.257	9.87	14.71	1475
220	0.03327	1.5347	0.1722	2073.7	3238.2	41.663	10.10	14.81	1500
240	0.03574	1.6127	0.1553	2283.6	3534.6	42.953	10.27	14.86	1527
260	0.03819	1.6921	0.1413	2495.1	3831.7	44.142	10.40	14.87	1555
280	0.04061	1.7709	0.1296	2707.5	4129.0	45.243	10.49	14.87	1584
300	0.04316	1.8575	0.1213	2913.8	4424.4	46.247	10.44	14.87	1626
350	0.04921	2.0607	0.1025	3444.4	5166.6	48.536	10.50	14.82	1705
400	0.05521	2.2642	0.0889	3973.9	5906.1	50.511	10.51	14.76	1783
450	0.06117	2.4677	0.0785	4501.9	6643.0	52.247	10.50	14.71	1859
500	0.06712	2.6712	0.0704	5028.3	7377.5	53.794	10.50	14.68	1933
550	0.07305	2.8748	0.0638	5553.9	8110.6	55.192	10.49	14.65	2003
600	0.07897	3.0784	0.0584	6079.3	8843.2	56.463	10.50	14.65	2072
700	0.09078	3.4860	0.0499	7131.0	10308.5	58.726	10.54	14.67	2202
800	0.10258	3.8940	0.0436	8188.4	11778.8	60.691	10.62	14.74	2325
900	0.11438	4.3025	0.0388	9255.2	13258.3	62.431	10.74	14.85	2439
1000	0.12616	4.7115	0.0349	10334.8	14750.5	64.001	10.90	15.01	2547
1200	0.14972	5.5305	0.0291	12548.6	17789.0	66.752	11.27	15.38	2747
1400	0.17328	6.3507	0.0249	14843.5	20908.4	69.162	11.70	15.81	2929
1600	0.19684	7.1718	0.0218	17223.3	24112.7	71.308	12.14	16.24	3098
1800	0.22040	7.9936	0.0194	19687.8	27401.8	73.239	12.55	16.66	3258
2000	0.24397	8.8159	0.0174	22237.6	30776.4	75.027	12.96	17.08	3408
2500	0.30306	10.8731	0.0140	29010.1	39617.4	78.954	14.29	18.47	3748
3000	0.36333	12.9317	0.0116	36768.0	49484.4	82.559	16.92	21.37	4041

* TWO-PHASE BOUNDARY

Table 4.12 (continued)
THERMODYNAMIC PROPERTIES OF NORMAL HYDROGEN

35.00 MPA ISOBAR

TEMPERATURE DEG. K	DENSITY KG/CU M	$V(DH/DV)_P$ KJ/KG	$V(DP/DU)_V$ MPA-CU M/KJ X 10^2	$-V(DP/DV)_T$ MPA	$(DV/DT)_P/V$ 1/DEG. K	THERMAL CONDUCTIVITY W/K-M X 10^3	VISCOSITY KG/M-S X 10^7	THERMAL DIFFUSIVITY SQ M/HR	DIELECTRIC CONSTANT	PRANDTL NUMBER
* 22.715	89.9284	1537.2	0.23225	258.2851	0.00461	167.31	470.59	0.00095	1.29753	1.9915
23	89.8102	1541.8	0.23123	256.9143	0.00463	168.63	457.54	0.00095	1.29711	1.9371
24	89.3915	1558.0	0.22764	252.0961	0.00471	172.79	416.80	0.00095	1.29560	1.7717
25	88.9674	1572.5	0.22430	247.2759	0.00480	176.29	382.62	0.00095	1.29407	1.6373
26	88.5380	1585.7	0.22114	242.4608	0.00488	179.13	353.65	0.00094	1.29253	1.5275
27	88.1035	1598.2	0.21807	237.6579	0.00496	181.46	328.86	0.00094	1.29097	1.4366
28	87.6641	1609.9	0.21510	232.8743	0.00504	183.37	307.48	0.00093	1.28940	1.3606
29	87.2198	1620.3	0.21229	228.1169	0.00512	184.90	288.89	0.00092	1.28780	1.2962
30	86.7710	1629.2	0.20972	223.3927	0.00520	186.10	272.62	0.00091	1.28620	1.2408
31	86.3176	1637.6	0.20717	218.7086	0.00528	186.31	258.27	0.00090	1.28457	1.1982
32	85.8598	1645.3	0.20471	214.0714	0.00536	186.31	245.56	0.00089	1.28294	1.1616
33	85.3978	1652.1	0.20237	209.4879	0.00543	186.12	234.22	0.00087	1.28129	1.1299
34	84.9317	1658.1	0.20014	204.9647	0.00551	185.83	224.06	0.00086	1.27962	1.1020
35	84.4616	1663.5	0.19798	200.5084	0.00559	185.41	214.92	0.00085	1.27794	1.0777
36	83.9876	1668.5	0.19586	196.1255	0.00567	184.87	206.64	0.00084	1.27626	1.0566
37	83.5100	1673.0	0.19382	191.8221	0.00574	184.23	199.13	0.00083	1.27455	1.0380
38	83.0289	1676.5	0.19193	187.6046	0.00581	183.50	192.28	0.00082	1.27284	1.0215
39	82.5445	1678.9	0.19021	183.4789	0.00589	182.70	186.01	0.00081	1.27112	1.0064
40	82.0570	1681.0	0.18856	179.4508	0.00596	181.84	180.25	0.00080	1.26939	0.9930
42	81.0737	1685.2	0.18529	171.7099	0.00610	179.90	170.04	0.00078	1.26590	0.9709
44	80.0660	1699.2	0.18204	165.6944	0.00618	177.77	161.17	0.00076	1.26233	0.9514
46	79.0744	1710.9	0.17770	158.8177	0.00628	175.65	153.61	0.00074	1.25883	0.9403
48	78.0791	1722.4	0.17383	152.6322	0.00637	173.49	146.99	0.00073	1.25532	0.9298
50	77.0835	1728.7	0.17016	146.2088	0.00647	171.34	141.14	0.00072	1.25181	0.9219
52	76.0855	1734.6	0.16665	140.1801	0.00657	169.20	135.92	0.00070	1.24830	0.9149
54	75.0875	1740.6	0.16330	134.5895	0.00664	167.10	131.23	0.00069	1.24480	0.9081
56	74.0914	1747.8	0.16023	129.5832	0.00670	165.04	126.99	0.00068	1.24131	0.9010
58	73.1007	1752.4	0.15747	124.7533	0.00676	163.04	123.14	0.00068	1.23785	0.8941
60	72.1159	1756.2	0.15480	120.0965	0.00681	161.32	119.63	0.00067	1.23441	0.8869
65	69.6848	1769.3	0.14839	109.7904	0.00691	157.92	112.04	0.00067	1.22596	0.8672
70	67.3054	1787.0	0.14226	101.0215	0.00697	154.57	105.81	0.00066	1.21772	0.8521
75	65.0001	1811.6	0.13626	93.6896	0.00697	151.23	100.64	0.00066	1.20977	0.8406
80	62.7793	1844.4	0.13032	87.6305	0.00692	147.99	96.33	0.00066	1.20214	0.8313
85	60.6511	1877.7	0.12455	82.0505	0.00688	147.29	92.74	0.00068	1.19487	0.8138
90	58.6090	1920.3	0.11903	77.5149	0.00680	146.41	89.75	0.00069	1.18791	0.8002
95	56.6637	1964.4	0.11388	73.4173	0.00672	145.51	87.27	0.00070	1.18131	0.7912
100	54.8072	2016.0	0.10897	70.0140	0.00660	147.94	81.02	0.00073	1.17504	0.7291
120	48.2939	2286.4	0.09139	60.7526	0.00603	157.21	84.29	0.00085	1.15320	0.7390
140	43.0694	2593.8	0.07941	55.3153	0.00543	161.39	84.65	0.00096	1.13589	0.7388
160	38.8496	2925.1	0.07091	51.7872	0.00490	164.14	84.33	0.00106	1.12203	0.7365
180	35.3863	3265.6	0.06479	49.2697	0.00446	166.93	84.21	0.00117	1.11075	0.7339
200	32.4953	3611.8	0.06025	47.4392	0.00407	170.19	84.52	0.00128	1.10139	0.7304
220	30.0560	3965.0	0.05675	46.1257	0.00373	173.98	85.28	0.00141	1.09354	0.7258
240	27.9774	4317.0	0.05403	45.1204	0.00344	178.20	86.43	0.00154	1.08688	0.7206
260	26.1847	4664.8	0.05187	44.3060	0.00319	182.73	87.90	0.00169	1.08115	0.7154
280	24.6224	5004.1	0.05016	43.6041	0.00297	187.52	89.62	0.00184	1.07619	0.7106
300	23.1706	5275.9	0.05013	43.0404	0.00282	192.47	91.53	0.00201	1.07158	0.7070
350	20.3223	6055.0	0.04803	41.8772	0.00245	205.39	96.94	0.00246	1.06259	0.6994
400	18.1136	6811.6	0.04669	41.0123	0.00217	218.83	102.92	0.00295	1.05566	0.6942
450	16.3468	7554.7	0.04576	40.3390	0.00195	232.61	109.22	0.00348	1.05013	0.6906
500	14.8989	8296.4	0.04502	39.7984	0.00177	246.74	115.69	0.00406	1.04562	0.6881
550	13.6895	9035.2	0.04443	39.3544	0.00162	261.00	122.25	0.00468	1.04186	0.6864
600	12.6634	9777.3	0.04391	38.9832	0.00150	275.43	128.85	0.00535	1.03868	0.6852
700	11.0151	11278.2	0.04301	38.3982	0.00130	304.64	142.02	0.00679	1.03359	0.6838
800	9.7480	12817.7	0.04216	37.9589	0.00115	334.41	155.05	0.00838	1.02968	0.6833
900	8.7431	14410.5	0.04129	37.6175	0.00103	364.87	167.87	0.01012	1.02659	0.6833
1000	7.9264	16065.5	0.04039	37.3450	0.00093	440.56	180.45	0.01333	1.02409	0.6146
1200	6.6790	19542.9	0.03862	36.9381	0.00079	511.00	204.90	0.01791	1.02027	0.6165
1400	5.7709	23252.8	0.03690	36.6497	0.00068	583.16	228.44	0.02302	1.01750	0.6192
1600	5.0803	27146.9	0.03536	36.4349	0.00060	656.17	251.16	0.02862	1.01539	0.6218
1800	4.5372	31177.3	0.03404	36.2687	0.00053	729.46	273.15	0.03474	1.01374	0.6239
2000	4.0989	35368.1	0.03284	36.1356	0.00048	804.64	294.50	0.04138	1.01240	0.6250
2500	3.2996	47446.8	0.02961	35.8771	0.00039	1036.56	345.52	0.06124	1.00998	0.6156
3000	2.7523	65341.6	0.02500	35.5924	0.00033	1466.36	394.37	0.08974	1.00832	0.5748

* TWO-PHASE BOUNDARY

Table 4.12 (continued)
THERMODYNAMIC PROPERTIES OF NORMAL HYDROGEN

40.00 MPA ISOBAR

TEMPERATURE DEG. K	VOLUME CU M/KG	ISOTHERM DERIVATIVE CU M-MPA/KG	ISOCHORE DERIVATIVE MPA/K	INTERNAL ENERGY KJ/KG-K	ENTHALPY KJ/KG	ENTROPY KJ/KG-K	CV KJ /	CP KG-K	VELOCITY OF SOUND M/S
* 23.746	0.01097	3.0705	1.220	263.9	702.6	14.790	5.78	7.16	1951
24	0.01098	3.0607	1.220	265.3	704.4	14.866	5.81	7.22	1950
25	0.01103	3.0220	1.218	270.7	711.7	15.164	5.92	7.41	1945
26	0.01108	2.9829	1.215	276.2	719.2	15.457	6.02	7.60	1941
27	0.01113	2.9437	1.212	281.9	726.9	15.747	6.11	7.78	1936
28	0.01118	2.9043	1.208	287.7	734.8	16.032	6.21	7.96	1930
29	0.01123	2.8648	1.202	293.6	742.8	16.316	6.29	8.14	1925
30	0.01128	2.8253	1.197	299.7	751.0	16.594	6.37	8.31	1919
31	0.01134	2.7858	1.190	305.9	759.4	16.869	6.44	8.47	1913
32	0.01139	2.7464	1.183	312.2	768.0	17.141	6.52	8.63	1908
33	0.01145	2.7071	1.176	318.7	776.7	17.409	6.58	8.79	1901
34	0.01151	2.6681	1.168	325.3	785.6	17.674	6.64	8.95	1895
35	0.01157	2.6293	1.159	331.9	794.6	17.935	6.70	9.10	1889
36	0.01163	2.5909	1.151	338.7	803.8	18.194	6.76	9.24	1883
37	0.01169	2.5529	1.142	345.6	813.1	18.449	6.81	9.39	1876
38	0.01175	2.5154	1.132	352.6	822.5	18.702	6.85	9.53	1870
39	0.01181	2.4784	1.122	359.7	832.1	18.951	6.89	9.66	1863
40	0.01187	2.4420	1.112	366.9	841.9	19.197	6.93	9.78	1857
42	0.01200	2.3712	1.091	381.5	861.7	19.680	6.99	10.03	1845
44	0.01214	2.3036	1.070	396.4	882.0	20.152	7.05	10.27	1832
46	0.01228	2.2396	1.047	411.1	902.2	20.602	7.14	10.53	1818
48	0.01242	2.2010	1.025	426.7	923.7	21.058	7.20	10.74	1811
50	0.01257	2.1414	1.001	442.6	945.4	21.500	7.26	10.95	1798
52	0.01272	2.0954	0.9765	458.7	967.4	21.933	7.30	11.13	1787
54	0.01287	2.0472	0.9519	475.0	989.9	22.357	7.34	11.30	1775
56	0.01303	1.9978	0.9274	491.8	1012.9	22.776	7.37	11.46	1763
58	0.01319	1.9532	0.9026	508.5	1035.9	23.181	7.38	11.58	1751
60	0.01335	1.9092	0.8783	525.3	1059.2	23.575	7.39	11.71	1739
65	0.01377	1.8163	0.8195	567.7	1118.4	24.521	7.40	11.96	1713
70	0.01420	1.7308	0.7646	610.6	1178.8	25.417	7.42	12.19	1686
75	0.01466	1.6626	0.7139	653.9	1240.2	26.265	7.45	12.39	1663
80	0.01513	1.6074	0.6672	697.6	1302.6	27.070	7.50	12.56	1641
85	0.01561	1.5682	0.6242	741.6	1365.9	27.837	7.56	12.70	1623
90	0.01610	1.5299	0.5848	785.9	1429.7	28.568	7.63	12.84	1605
95	0.01660	1.4914	0.5491	830.4	1494.3	29.266	7.72	13.01	1585
100	0.01711	1.4676	0.5167	875.4	1559.7	29.936	7.82	13.14	1571
120	0.01922	1.4216	0.4137	1061.2	1829.9	32.398	8.37	13.71	1526
140	0.02139	1.4312	0.3413	1252.0	2107.5	34.536	8.85	14.06	1508
160	0.02357	1.4693	0.2892	1448.3	2391.3	36.430	9.28	14.34	1507
180	0.02576	1.5206	0.2505	1649.6	2680.0	38.130	9.63	14.56	1516
200	0.02794	1.5804	0.2207	1855.1	2972.8	39.673	9.92	14.74	1532
220	0.03012	1.6472	0.1971	2063.9	3268.7	41.082	10.16	14.86	1552
240	0.03228	1.7212	0.1778	2275.1	3566.4	42.378	10.33	14.93	1577
260	0.03442	1.7982	0.1617	2488.0	3864.9	43.573	10.47	14.95	1602
280	0.03654	1.8760	0.1482	2701.9	4163.6	44.680	10.56	14.94	1629
300	0.03877	1.9580	0.1386	2907.3	4458.2	45.678	10.48	14.91	1669
350	0.04407	2.1575	0.1171	3439.6	5202.2	47.973	10.53	14.85	1744
400	0.04932	2.3584	0.1015	3970.5	5943.1	49.952	10.53	14.78	1819
450	0.05454	2.5599	0.0897	4499.5	6681.0	51.690	10.52	14.73	1893
500	0.05974	2.7616	0.0804	5026.9	7416.4	53.239	10.52	14.69	1964
550	0.06492	2.9637	0.0729	5553.1	8150.1	54.638	10.51	14.66	2033
600	0.07010	3.1659	0.0667	6079.2	8883.2	55.911	10.52	14.65	2100
700	0.08044	3.5711	0.0570	7131.8	10349.2	58.174	10.55	14.67	2228
800	0.09076	3.9772	0.0498	8189.7	11820.0	60.139	10.63	14.74	2348
900	0.10107	4.3840	0.0442	9257.0	13299.7	61.880	10.75	14.85	2461
1000	0.11137	4.7915	0.0398	10337.1	14792.1	63.451	10.90	15.01	2568
1200	0.13198	5.6082	0.0332	12551.4	17830.8	66.202	11.28	15.38	2766
1400	0.15259	6.4267	0.0284	14846.6	20950.2	68.612	11.70	15.81	2946
1600	0.17320	7.2463	0.0249	17226.7	24154.5	70.758	12.14	16.24	3114
1800	0.19380	8.0669	0.0221	19691.2	27443.4	72.689	12.56	16.66	3272
2000	0.21442	8.8881	0.0199	22240.7	30817.5	74.477	12.96	17.07	3421
2500	0.26611	10.9435	0.0159	29004.7	39649.1	78.399	14.26	18.42	3761
3000	0.31876	13.0008	0.0133	36712.9	49463.4	81.986	16.75	21.18	4053

* TWO-PHASE BOUNDARY

Table 4.12 (continued)
THERMODYNAMIC PROPERTIES OF NORMAL HYDROGEN

40.00 MPA ISOBAR

TEMPERATURE DEG. K	DENSITY KG/CU M	$V(DH/DV)_P$ KJ/KG	$V(DP/DU)_V$ MPA-CU M/KJ X 10^2	$-V(DP/DV)_T$ MPA	$(DV/DT)_P/V$ 1/DEG. K	THERMAL CONDUCTIVITY W/K-M X 10^3	VISCOSITY KG/M-S X 10^7	THERMAL DIFFUSIVITY SQ M/HR	DIELECTRIC CONSTANT	PRANDTL NUMBER
* 23.746	91.1935	1644.6	0.23141	280.0107	0.00436	178.41	502.49	0.00098	1.30209	2.0180
24	91.0924	1649.4	0.23046	278.8086	0.00437	179.48	490.21	0.00098	1.30173	1.9708
25	90.6916	1667.0	0.22700	274.0675	0.00444	183.27	446.98	0.00098	1.30028	1.8070
26	90.2863	1683.3	0.22373	269.3192	0.00451	186.38	410.60	0.00098	1.29882	1.6736
27	89.8767	1698.7	0.22056	264.5703	0.00458	188.96	379.68	0.00097	1.29735	1.5635
28	89.4630	1713.2	0.21750	259.8275	0.00465	191.07	353.17	0.00097	1.29586	1.4716
29	89.0452	1726.4	0.21462	255.0973	0.00471	192.80	330.25	0.00096	1.29435	1.3939
30	88.6236	1737.9	0.21198	250.3862	0.00478	194.17	310.30	0.00095	1.29284	1.3272
31	88.1982	1748.8	0.20937	245.7003	0.00484	194.51	292.80	0.00094	1.29131	1.2752
32	87.7692	1759.0	0.20686	241.0461	0.00491	194.62	277.37	0.00092	1.28977	1.2305
33	87.3366	1768.0	0.20449	236.4296	0.00497	194.53	263.68	0.00091	1.28822	1.1917
34	86.9007	1776.2	0.20223	231.8569	0.00504	194.33	251.47	0.00090	1.28666	1.1576
35	86.4613	1783.7	0.20005	227.3338	0.00510	193.99	240.52	0.00089	1.28509	1.1278
36	86.0188	1790.6	0.19792	222.8662	0.00516	193.52	230.66	0.00088	1.28350	1.1019
37	85.5732	1796.8	0.19588	218.4597	0.00523	192.95	221.75	0.00086	1.28191	1.0790
38	85.1246	1801.9	0.19400	214.1198	0.00529	192.28	213.65	0.00085	1.28031	1.0586
39	84.6731	1805.8	0.19230	209.8520	0.00535	191.53	206.27	0.00084	1.27870	1.0401
40	84.2189	1809.0	0.19066	205.6614	0.00541	190.71	199.52	0.00083	1.27708	1.0236
42	83.3031	1815.2	0.18746	197.5321	0.00552	188.84	187.62	0.00081	1.27382	0.9964
44	82.3785	1821.9	0.18420	189.7700	0.00564	186.82	177.47	0.00080	1.27053	0.9754
46	81.4466	1834.6	0.18013	182.4096	0.00574	184.69	168.71	0.00078	1.26722	0.9619
48	80.4895	1856.0	0.17681	177.1544	0.00579	182.44	160.96	0.00076	1.26383	0.9473
50	79.5574	1864.4	0.17331	170.3629	0.00587	180.27	154.32	0.00074	1.26054	0.9374
52	78.6261	1878.1	0.17004	164.7520	0.00593	178.12	148.44	0.00073	1.25725	0.9276
54	77.6931	1888.1	0.16690	159.0507	0.00598	176.00	143.19	0.00072	1.25396	0.9194
56	76.7649	1894.7	0.16400	153.3596	0.00605	173.92	138.48	0.00071	1.25069	0.9124
58	75.8377	1901.1	0.16132	148.1271	0.00609	171.88	134.22	0.00070	1.24743	0.9046
60	74.9161	1906.3	0.15872	143.0324	0.00614	170.14	130.34	0.00070	1.24420	0.8968
65	72.6362	1925.0	0.15242	131.9255	0.00621	166.67	122.02	0.00069	1.23623	0.8755
70	70.4040	1943.1	0.14632	121.8556	0.00627	163.23	115.23	0.00068	1.22845	0.8607
75	68.2267	1968.9	0.14041	113.4349	0.00629	159.76	109.56	0.00068	1.22090	0.8498
80	66.1148	2001.3	0.13462	106.2738	0.00628	156.37	104.80	0.00068	1.21361	0.8421
85	64.0771	2044.3	0.12892	100.4830	0.00621	155.62	100.78	0.00069	1.20660	0.8224
90	62.1255	2087.6	0.12334	95.0470	0.00615	154.70	97.40	0.00070	1.19991	0.8087
95	60.2513	2128.9	0.11808	89.8598	0.00611	153.72	94.54	0.00071	1.19350	0.8001
100	58.4545	2182.1	0.11305	85.7859	0.00602	156.97	86.10	0.00074	1.18739	0.7209
120	52.0300	2450.7	0.09499	73.9680	0.00559	166.19	89.25	0.00084	1.16569	0.7360
140	46.7548	2757.2	0.08248	66.9139	0.00510	169.80	89.21	0.00093	1.14808	0.7389
160	42.4196	3089.8	0.07350	62.3257	0.00464	171.75	88.40	0.00102	1.13374	0.7379
180	38.8206	3431.5	0.06698	59.0296	0.00424	173.71	87.79	0.00111	1.12194	0.7359
200	35.7865	3776.1	0.06216	56.5563	0.00390	176.21	87.67	0.00120	1.11205	0.7333
220	33.2007	4124.4	0.05844	54.6870	0.00360	179.35	88.07	0.00131	1.10367	0.7298
240	30.9777	4477.0	0.05553	53.3182	0.00333	183.02	88.91	0.00143	1.09650	0.7251
260	29.0508	4828.2	0.05318	52.2382	0.00310	187.12	90.13	0.00155	1.09031	0.7199
280	27.3648	5173.8	0.05130	51.3371	0.00289	191.57	91.66	0.00169	1.08492	0.7147
300	25.7922	5430.2	0.05129	50.5022	0.00275	196.26	93.43	0.00184	1.07990	0.7096
350	22.6935	6208.9	0.04900	48.9622	0.00239	208.79	98.61	0.00223	1.07007	0.7014
400	20.2773	6965.1	0.04752	47.8225	0.00212	222.11	104.50	0.00267	1.06245	0.6956
450	18.3361	7708.0	0.04648	46.9380	0.00191	235.93	110.79	0.00315	1.05635	0.6915
500	16.7397	8449.5	0.04566	46.2291	0.00174	250.23	117.32	0.00366	1.05136	0.6888
550	15.4024	9188.1	0.04500	45.6475	0.00160	264.73	123.99	0.00422	1.04719	0.6868
600	14.2650	9929.8	0.04443	45.1617	0.00148	279.46	130.72	0.00481	1.04365	0.6855
700	12.4322	11430.3	0.04344	44.3967	0.00128	309.39	144.21	0.00611	1.03797	0.6839
800	11.0185	12969.6	0.04252	43.8227	0.00114	339.98	157.60	0.00754	1.03360	0.6833
900	9.8944	14562.4	0.04160	43.3771	0.00102	371.33	170.82	0.00910	1.03013	0.6833
1000	8.9787	16218.0	0.04066	43.0218	0.00093	440.56	183.81	0.01177	1.02732	0.6261
1200	7.5767	19697.1	0.03883	42.4922	0.00078	511.00	209.10	0.01579	1.02302	0.6292
1400	6.5536	23409.3	0.03708	42.1174	0.00068	583.16	233.48	0.02027	1.01989	0.6328
1600	5.7738	27305.7	0.03551	41.8388	0.00059	656.15	257.04	0.02519	1.01751	0.6363
1800	5.1599	31337.0	0.03416	41.6238	0.00053	729.38	279.85	0.03054	1.01563	0.6393
2000	4.6637	35523.8	0.03295	41.4520	0.00048	804.31	301.99	0.03636	1.01412	0.6411
2500	3.7578	47511.3	0.02977	41.1238	0.00039	1032.51	354.94	0.05369	1.01137	0.6334
3000	3.1371	64967.4	0.02530	40.7851	0.00033	1445.38	405.57	0.07832	1.00948	0.5943

* TWO-PHASE BOUNDARY

Table 4.12 (continued)
THERMODYNAMIC PROPERTIES OF NORMAL HYDROGEN

45.00 MPA ISOBAR

TEMPERATURE DEG. K	VOLUME CU M/KG	ISOTHERM DERIVATIVE CU M-MPA/KG	ISOCHORE DERIVATIVE MPA/K	INTERNAL ENERGY KJ/KG-K	ENTHALPY KJ/KG	ENTROPY KJ/KG-K	CV KJ /	CP KG-K	VELOCITY OF SOUND M/S
* 24.742	0.01082	3.2640	1.249	271.5	758.6	14.854	5.86	7.25	2009
25	0.01084	3.2545	1.249	272.9	760.5	14.929	5.89	7.30	2008
26	0.01088	3.2173	1.247	278.2	767.9	15.218	5.99	7.48	2004
27	0.01093	3.1798	1.244	283.7	775.5	15.504	6.09	7.66	2000
28	0.01097	3.1420	1.240	289.3	783.2	15.784	6.19	7.84	1995
29	0.01102	3.1041	1.235	295.1	791.1	16.063	6.28	8.01	1990
30	0.01107	3.0660	1.230	301.0	799.2	16.337	6.36	8.17	1985
31	0.01112	3.0278	1.224	307.0	807.5	16.607	6.43	8.33	1980
32	0.01117	2.9896	1.217	313.1	815.9	16.874	6.51	8.49	1975
33	0.01122	2.9514	1.210	319.4	824.4	17.138	6.58	8.64	1969
34	0.01127	2.9133	1.203	325.8	833.1	17.398	6.64	8.79	1963
35	0.01133	2.8753	1.195	332.2	842.0	17.655	6.70	8.93	1958
36	0.01138	2.8375	1.187	338.8	851.0	17.909	6.76	9.08	1952
37	0.01144	2.8000	1.178	345.5	860.2	18.160	6.81	9.21	1946
38	0.01149	2.7628	1.170	352.3	869.4	18.407	6.86	9.35	1940
39	0.01155	2.7260	1.160	359.2	878.8	18.651	6.90	9.47	1934
40	0.01161	2.6896	1.151	366.1	888.4	18.893	6.94	9.59	1928
42	0.01172	2.6182	1.131	380.3	907.8	19.366	7.01	9.83	1916
44	0.01184	2.5492	1.111	394.8	927.7	19.829	7.07	10.06	1904
46	0.01197	2.4829	1.090	409.0	947.5	20.269	7.16	10.31	1891
48	0.01209	2.4198	1.068	424.2	968.4	20.713	7.23	10.54	1878
50	0.01222	2.3601	1.046	439.6	989.7	21.148	7.29	10.76	1865
52	0.01236	2.3042	1.023	455.2	1011.4	21.573	7.34	10.95	1854
54	0.01250	2.2794	1.002	471.2	1033.6	21.993	7.38	11.10	1851
56	0.01264	2.2290	0.9786	487.5	1056.2	22.405	7.41	11.26	1840
58	0.01278	2.1913	0.9554	503.8	1078.9	22.803	7.43	11.37	1832
60	0.01292	2.1432	0.9323	520.2	1101.7	23.190	7.44	11.50	1820
65	0.01329	2.0456	0.8749	561.7	1159.9	24.120	7.46	11.76	1795
70	0.01368	1.9626	0.8202	603.8	1219.3	25.000	7.49	11.98	1772
75	0.01407	1.8870	0.7692	646.3	1279.7	25.833	7.52	12.18	1748
80	0.01448	1.8198	0.7219	689.2	1341.1	26.626	7.57	12.38	1725
85	0.01491	1.7672	0.6782	732.6	1403.4	27.382	7.63	12.55	1705
90	0.01534	1.7229	0.6379	776.3	1466.5	28.104	7.71	12.71	1685
95	0.01578	1.6938	0.6010	820.5	1530.4	28.794	7.79	12.83	1670
100	0.01622	1.6639	0.5671	865.0	1594.9	29.456	7.89	12.98	1654
120	0.01806	1.5871	0.4586	1049.8	1862.7	31.895	8.44	13.63	1601
140	0.01997	1.5823	0.3806	1240.5	2139.2	34.025	8.92	14.03	1578
160	0.02190	1.6082	0.3236	1437.4	2422.7	35.916	9.34	14.33	1571
180	0.02382	1.6518	0.2807	1639.3	2711.4	37.617	9.69	14.57	1576
200	0.02575	1.7050	0.2477	1845.5	3004.4	39.161	9.98	14.75	1588
220	0.02768	1.7654	0.2215	2055.2	3300.7	40.572	10.21	14.89	1605
240	0.02959	1.8336	0.2000	2267.5	3599.2	41.871	10.39	14.98	1626
260	0.03150	1.9071	0.1821	2481.6	3898.9	43.072	10.52	15.01	1649
280	0.03338	1.9838	0.1669	2696.8	4198.9	44.183	10.62	15.00	1674
300	0.03536	2.0602	0.1559	2901.6	4492.8	45.177	10.52	14.94	1711
350	0.04006	2.2557	0.1316	3435.5	5238.4	47.477	10.56	14.88	1783
400	0.04473	2.4537	0.1141	3967.6	5980.5	49.459	10.56	14.81	1855
450	0.04937	2.6528	0.1008	4497.7	6719.4	51.199	10.55	14.74	1926
500	0.05399	2.8526	0.0903	5025.9	7455.6	52.750	10.54	14.70	1995
550	0.05860	3.0529	0.0818	5552.8	8189.9	54.150	10.53	14.68	2063
600	0.06320	3.2536	0.0749	6079.4	8923.4	55.423	10.53	14.66	2128
700	0.07238	3.6562	0.0640	7133.0	10390.1	57.688	10.57	14.68	2253
800	0.08155	4.0600	0.0559	8191.6	11861.2	59.654	10.64	14.74	2372
900	0.09071	4.4649	0.0497	9259.3	13341.2	61.395	10.76	14.86	2483
1000	0.09986	4.8708	0.0447	10339.8	14833.6	62.965	10.91	15.01	2588
1200	0.11817	5.6849	0.0373	12554.7	17872.4	65.716	11.28	15.38	2783
1400	0.13648	6.5012	0.0320	14850.3	20991.9	68.126	11.71	15.81	2962
1600	0.15479	7.3191	0.0280	17230.5	24196.1	70.272	12.15	16.24	3128
1800	0.17310	8.1383	0.0249	19695.2	27484.8	72.203	12.56	16.66	3285
2000	0.19142	8.9584	0.0224	22244.5	30858.5	73.991	12.96	17.07	3434
2500	0.23735	11.0116	0.0179	29001.4	39682.2	77.910	14.23	18.39	3772
3000	0.28410	13.0674	0.0149	36668.5	49452.7	81.481	16.61	21.02	4065

* TWO-PHASE BOUNDARY

Table 4.12 (continued)
THERMODYNAMIC PROPERTIES OF NORMAL HYDROGEN

45.00 MPA ISOBAR

TEMPERATURE DEG. K	DENSITY KG/CU M	$V(DH/DV)_P$ KJ/KG	$V(DP/DU)_V$ MPA-CU M/KJ X 10^2	$-V(DP/DV)_T$ MPA	$(DV/DT)_P/V$ 1/DEG. K	THERMAL CONDUCTIVITY W/K-M X 10^3	VISCOSITY KG/M-S X 10^7	THERMAL DIFFUSIVITY SQ M/HR	DIELECTRIC CONSTANT	PRANDTL NUMBER
* 24.742	92.3840	1749.4	0.23071	301.5396	0.00414	188.97	534.95	0.00102	1.30639	2.0519
25	92.2852	1754.5	0.22980	300.3381	0.00416	189.95	521.85	0.00102	1.30603	2.0045
26	91.8995	1773.8	0.22639	295.6654	0.00422	193.31	476.31	0.00101	1.30464	1.8431
27	91.5102	1792.2	0.22310	290.9811	0.00427	196.11	437.86	0.00101	1.30323	1.7103
28	91.1173	1809.6	0.21993	286.2917	0.00433	198.43	405.09	0.00100	1.30182	1.5998
29	90.7211	1825.6	0.21695	281.6032	0.00439	200.33	376.93	0.00099	1.30039	1.5065
30	90.3216	1839.6	0.21423	276.9220	0.00444	201.86	352.54	0.00098	1.29895	1.4267
31	89.9189	1853.2	0.21155	272.2539	0.00450	202.31	331.26	0.00097	1.29750	1.3640
32	89.5132	1865.8	0.20897	267.6047	0.00455	202.52	312.59	0.00096	1.29604	1.3100
33	89.1045	1877.2	0.20654	262.9803	0.00460	202.52	296.09	0.00095	1.29457	1.2632
34	88.6930	1887.6	0.20423	258.3862	0.00466	202.40	281.45	0.00093	1.29309	1.2220
35	88.2787	1897.2	0.20201	253.8280	0.00471	202.13	268.38	0.00092	1.29160	1.1861
36	87.8617	1906.2	0.19985	249.3110	0.00476	201.72	256.65	0.00091	1.29010	1.1546
37	87.4422	1914.4	0.19778	244.8403	0.00481	201.20	246.09	0.00090	1.28860	1.1270
38	87.0201	1921.2	0.19588	240.4212	0.00486	200.57	236.54	0.00089	1.28709	1.1022
39	86.5955	1926.6	0.19416	236.0586	0.00492	199.86	227.86	0.00088	1.28557	1.0797
40	86.1687	1931.3	0.19252	231.7573	0.00497	199.08	219.96	0.00087	1.28404	1.0597
42	85.3086	1940.0	0.18932	223.3571	0.00507	197.27	206.09	0.00085	1.28097	1.0267
44	84.4405	1948.9	0.18608	215.2561	0.00516	195.28	194.34	0.00083	1.27787	1.0010
46	83.5657	1963.8	0.18205	207.4866	0.00525	193.18	184.27	0.00081	1.27475	0.9839
48	82.6853	1975.3	0.17853	200.0778	0.00534	191.00	175.55	0.00079	1.27162	0.9692
50	81.8007	1985.9	0.17525	193.0548	0.00542	188.79	167.94	0.00077	1.26848	0.9568
52	80.9135	1995.8	0.17216	186.4382	0.00549	186.58	161.22	0.00076	1.26533	0.9463
54	80.0086	2020.3	0.16959	182.3746	0.00549	184.33	155.17	0.00075	1.26213	0.9343
56	79.1301	2028.7	0.16683	176.3816	0.00555	182.20	149.90	0.00074	1.25903	0.9261
58	78.2527	2041.3	0.16436	171.4717	0.00557	180.11	145.15	0.00073	1.25593	0.9166
60	77.3824	2046.3	0.16195	165.8425	0.00562	178.33	140.87	0.00072	1.25286	0.9087
65	75.2269	2068.5	0.15584	153.8824	0.00569	174.79	131.74	0.00071	1.24529	0.8864
70	73.1119	2095.2	0.14983	143.4871	0.00572	171.25	124.33	0.00070	1.23789	0.8696
75	71.0481	2123.1	0.14392	134.0696	0.00574	167.67	118.20	0.00070	1.23069	0.8586
80	69.0373	2153.9	0.13814	125.6357	0.00575	164.15	113.03	0.00069	1.22371	0.8521
85	67.0851	2193.4	0.13247	118.5543	0.00572	163.37	108.62	0.00070	1.21696	0.8342
90	65.1997	2238.0	0.12692	112.3300	0.00568	162.39	104.86	0.00071	1.21046	0.8207
95	63.3895	2293.1	0.12164	107.3672	0.00560	161.35	101.66	0.00071	1.20424	0.8087
100	61.6472	2347.5	0.11656	102.5731	0.00553	165.49	90.94	0.00074	1.19827	0.7133
120	55.3571	2611.1	0.09815	87.8588	0.00522	174.73	94.01	0.00083	1.17690	0.7333
140	50.0760	2920.4	0.08524	79.2331	0.00480	177.86	93.63	0.00091	1.15915	0.7386
160	45.6714	3253.1	0.07588	73.4478	0.00441	179.13	92.40	0.00099	1.14449	0.7393
180	41.9732	3597.2	0.06902	69.3320	0.00405	180.36	91.35	0.00106	1.13228	0.7377
200	38.8312	3942.8	0.06393	66.2089	0.00374	182.18	90.84	0.00114	1.12197	0.7355
220	36.1318	4289.5	0.06003	63.7886	0.00347	184.71	90.88	0.00124	1.11317	0.7327
240	33.7920	4639.0	0.05697	61.9605	0.00323	187.88	91.44	0.00134	1.10558	0.7289
260	31.7500	4990.8	0.05450	60.5518	0.00301	191.56	92.42	0.00145	1.09899	0.7242
280	29.9569	5341.6	0.05246	59.4287	0.00281	195.66	93.76	0.00157	1.09322	0.7189
300	28.2816	5585.0	0.05240	58.2664	0.00268	200.09	95.37	0.00170	1.08785	0.7122
350	24.9599	6363.1	0.04994	56.3024	0.00234	212.21	100.31	0.00206	1.07726	0.7033
400	22.3557	7119.0	0.04832	54.8534	0.00208	225.39	106.09	0.00245	1.06900	0.6969
450	20.2547	7861.7	0.04718	53.7312	0.00188	239.24	112.37	0.00288	1.06238	0.6925
500	18.5211	8602.9	0.04627	52.8327	0.00171	253.67	118.95	0.00335	1.05693	0.6895
550	17.0646	9341.2	0.04554	52.0962	0.00157	268.41	125.71	0.00386	1.05237	0.6873
600	15.8228	10082.7	0.04492	51.4811	0.00145	283.42	132.56	0.00440	1.04850	0.6858
700	13.8159	11582.5	0.04384	50.5130	0.00127	314.06	146.37	0.00558	1.04225	0.6840
800	12.2628	13121.5	0.04286	49.7869	0.00112	345.48	160.13	0.00688	1.03744	0.6834
900	11.0245	14714.3	0.04190	49.2236	0.00101	377.74	173.75	0.00830	1.03362	0.6833
1000	10.0136	16370.2	0.04092	48.7748	0.00092	440.56	187.17	0.01055	1.03050	0.6376
1200	8.4622	19850.7	0.03903	48.1065	0.00077	511.00	213.33	0.01414	1.02573	0.6419
1400	7.3271	23564.9	0.03723	47.6344	0.00067	583.16	238.60	0.01813	1.02225	0.6467
1600	6.4604	27463.4	0.03564	47.2842	0.00059	656.14	263.04	0.02251	1.01960	0.6512
1800	5.7769	31495.5	0.03427	47.0144	0.00053	729.31	286.72	0.02728	1.01751	0.6550
2000	5.2241	35679.1	0.03306	46.7994	0.00048	804.04	309.72	0.03246	1.01583	0.6575
2500	4.2132	47591.6	0.02990	46.3936	0.00039	1029.15	364.72	0.04782	1.01275	0.6517
3000	3.5199	64685.0	0.02556	45.9966	0.00032	1427.99	417.28	0.06949	1.01065	0.6142

* TWO-PHASE BOUNDARY

Table 4.12 (continued)
THERMODYNAMIC PROPERTIES OF NORMAL HYDROGEN

50.00 MPA ISOBAR

TEMPERATURE DEG. K	VOLUME CU M/KG	ISOTHERM DERIVATIVE CU M-MPA/KG	ISOCHORE DERIVATIVE MPA/K	INTERNAL ENERGY KJ/KG-K	ENTHALPY KJ/KG	ENTROPY KJ/KG-K	CV KJ /	CP KG-K	VELOCITY OF SOUND M/S
* 25.708	0.01069	3.4533	1.278	279.2	814.0	14.915	5.94	7.33	2064
26	0.01071	3.4430	1.278	280.8	816.1	14.997	5.97	7.38	2063
27	0.01075	3.4072	1.275	286.1	823.6	15.279	6.07	7.56	2060
28	0.01079	3.3711	1.271	291.6	831.2	15.556	6.17	7.73	2056
29	0.01084	3.3346	1.267	297.2	839.0	15.831	6.26	7.90	2051
30	0.01088	3.2980	1.262	302.9	847.0	16.101	6.34	8.06	2047
31	0.01093	3.2612	1.256	308.7	855.1	16.368	6.42	8.21	2042
32	0.01097	3.2242	1.250	314.7	863.4	16.631	6.50	8.37	2037
33	0.01102	3.1872	1.243	320.8	871.9	16.891	6.57	8.51	2032
34	0.01107	3.1501	1.236	327.0	880.4	17.147	6.64	8.66	2027
35	0.01112	3.1131	1.229	333.3	889.2	17.400	6.70	8.80	2022
36	0.01117	3.0761	1.221	339.7	898.0	17.650	6.76	8.94	2016
37	0.01122	3.0393	1.213	346.2	907.1	17.897	6.82	9.07	2011
38	0.01127	3.0027	1.205	352.8	916.2	18.140	6.87	9.20	2005
39	0.01132	2.9663	1.196	359.5	925.4	18.381	6.91	9.32	2000
40	0.01137	2.9302	1.187	366.3	934.8	18.618	6.95	9.43	1995
42	0.01148	2.8590	1.168	380.1	953.9	19.084	7.02	9.66	1984
44	0.01159	2.7895	1.149	394.2	973.5	19.538	7.09	9.88	1972
46	0.01170	2.7220	1.129	408.1	992.9	19.971	7.19	10.13	1959
48	0.01181	2.6570	1.108	422.8	1013.4	20.407	7.26	10.36	1947
50	0.01193	2.5948	1.087	437.8	1034.4	20.834	7.33	10.57	1934
52	0.01205	2.5356	1.065	453.1	1055.7	21.252	7.38	10.76	1923
54	0.01217	2.4797	1.043	468.7	1077.4	21.662	7.43	10.94	1911
56	0.01230	2.4274	1.021	484.7	1099.7	22.068	7.46	11.10	1900
58	0.01243	2.4097	1.002	500.6	1122.2	22.464	7.47	11.21	1901
60	0.01256	2.3577	0.9796	516.7	1144.8	22.846	7.49	11.34	1890
65	0.01290	2.2638	0.9248	557.3	1202.1	23.762	7.52	11.60	1869
70	0.01324	2.1798	0.8710	598.6	1260.7	24.630	7.55	11.82	1847
75	0.01360	2.1048	0.8199	640.5	1320.2	25.452	7.59	12.01	1826
80	0.01396	2.0360	0.7723	682.8	1380.8	26.234	7.64	12.20	1804
85	0.01433	1.9734	0.7280	725.6	1442.3	26.980	7.70	12.39	1782
90	0.01472	1.9221	0.6869	768.8	1504.7	27.694	7.78	12.57	1762
95	0.01511	1.8810	0.6489	812.5	1567.9	28.377	7.87	12.72	1744
100	0.01550	1.8546	0.6139	856.7	1631.9	29.033	7.97	12.85	1730
120	0.01714	1.7591	0.5009	1040.4	1897.4	31.451	8.51	13.54	1673
140	0.01883	1.7337	0.4184	1231.0	2172.7	33.572	8.98	14.00	1644
160	0.02055	1.7491	0.3569	1428.1	2455.8	35.461	9.40	14.32	1632
180	0.02228	1.7838	0.3103	1630.5	2744.5	37.161	9.75	14.57	1633
200	0.02400	1.8323	0.2740	1837.4	3037.6	38.705	10.03	14.76	1642
220	0.02573	1.8865	0.2453	2047.7	3334.0	40.118	10.26	14.91	1655
240	0.02744	1.9485	0.2218	2260.8	3633.0	41.419	10.44	15.01	1673
260	0.02916	2.0170	0.2022	2475.8	3933.6	42.622	10.58	15.06	1695
280	0.03085	2.0909	0.1856	2692.1	4234.8	43.738	10.68	15.07	1718
300	0.03263	2.1638	0.1730	2896.6	4528.0	44.728	10.55	14.97	1752
350	0.03686	2.3550	0.1461	3431.9	5275.0	47.033	10.59	14.90	1820
400	0.04106	2.5498	0.1266	3965.3	6018.3	49.018	10.59	14.83	1890
450	0.04523	2.7463	0.1118	4496.4	6758.1	50.760	10.57	14.76	1958
500	0.04939	2.9440	0.1002	5025.4	7494.9	52.312	10.56	14.72	2026
550	0.05354	3.1424	0.0908	5553.0	8229.8	53.714	10.55	14.68	2091
600	0.05767	3.3414	0.0830	6080.1	8963.8	54.988	10.55	14.67	2155
700	0.06593	3.7411	0.0710	7134.5	10431.0	57.253	10.58	14.68	2278
800	0.07417	4.1425	0.0620	8193.8	11902.4	59.219	10.66	14.75	2394
900	0.08241	4.5453	0.0551	9262.1	13382.6	60.961	10.77	14.86	2504
1000	0.09065	4.9494	0.0496	10342.9	14875.2	62.531	10.92	15.01	2608
1200	0.10711	5.7604	0.0413	12558.4	17914.0	65.282	11.29	15.38	2801
1400	0.12358	6.5743	0.0354	14854.4	21033.4	67.693	11.72	15.81	2978
1600	0.14005	7.3904	0.0310	17234.9	24237.5	69.838	12.15	16.24	3143
1800	0.15653	8.2080	0.0276	19699.7	27526.1	71.769	12.57	16.66	3299
2000	0.17301	9.0268	0.0248	22248.8	30899.3	73.557	12.97	17.07	3447
2500	0.21433	11.0775	0.0199	28999.8	39716.2	77.473	14.20	18.36	3784
3000	0.25635	13.1315	0.0166	36632.2	49449.7	81.030	16.50	20.88	4076

* TWO-PHASE BOUNDARY

Table 4.12 (continued)
THERMODYNAMIC PROPERTIES OF NORMAL HYDROGEN

50.00 MPA ISOBAR

TEMPERATURE DEG. K	DENSITY KG/CU M	$V(DH/DV)_P$ KJ/KG	$V(DP/DU)_V$ MPA-CU M/KJ X 10^2	$-V(DP/DV)_T$ MPA	$(DV/DT)_P/V$ 1/DEG. K	THERMAL CONDUCTIVITY W/K-M X 10^3	VISCOSITY KG/M-S X 10^7	THERMAL DIFFUSIVITY SQ M/HR	DIELECTRIC CONSTANT	PRANDTL NUMBER
* 25.708	93.5095	1851.7	0.23018	322.9196	0.00396	198.99	567.99	0.00105	1.31046	2.0921
26	93.4012	1858.0	0.22917	321.5761	0.00397	199.98	552.43	0.00104	1.31007	2.0392
27	93.0286	1879.3	0.22574	316.9646	0.00402	202.99	504.76	0.00104	1.30872	1.8795
28	92.6530	1899.6	0.22244	312.3381	0.00407	205.49	464.37	0.00103	1.30736	1.7471
29	92.2745	1918.3	0.21935	307.7025	0.00412	207.56	429.87	0.00103	1.30600	1.6357
30	91.8933	1935.1	0.21652	303.0638	0.00416	209.24	400.14	0.00102	1.30462	1.5407
31	91.5094	1951.2	0.21375	298.4278	0.00421	209.79	374.34	0.00100	1.30323	1.4655
32	91.1229	1966.3	0.21109	293.8000	0.00425	210.09	351.80	0.00099	1.30184	1.4009
33	90.7340	1980.2	0.20859	289.1859	0.00430	210.16	331.98	0.00098	1.30043	1.3449
34	90.3427	1992.9	0.20621	284.5909	0.00434	210.11	314.46	0.00097	1.29902	1.2957
35	89.9491	2004.7	0.20393	280.0202	0.00439	209.90	298.88	0.00095	1.29761	1.2527
36	89.5533	2015.8	0.20172	275.4789	0.00443	209.55	284.97	0.00094	1.29618	1.2151
37	89.1553	2026.0	0.19961	270.9719	0.00448	209.07	272.48	0.00093	1.29475	1.1820
38	88.7551	2034.7	0.19767	266.5040	0.00452	208.48	261.23	0.00092	1.29331	1.1523
39	88.3530	2041.8	0.19593	262.0800	0.00456	207.81	251.05	0.00091	1.29187	1.1255
40	87.9489	2048.1	0.19426	257.7042	0.00461	207.05	241.80	0.00090	1.29042	1.1016
42	87.1352	2059.8	0.19102	249.1153	0.00469	205.28	225.65	0.00088	1.28750	1.0618
44	86.3146	2071.2	0.18777	240.7706	0.00477	203.31	212.06	0.00086	1.28456	1.0308
46	85.4880	2088.9	0.18373	232.7010	0.00485	201.22	200.49	0.00084	1.28161	1.0096
48	84.6562	2102.6	0.18022	224.9347	0.00493	199.04	190.53	0.00082	1.27864	0.9914
50	83.8203	2114.6	0.17696	217.4966	0.00500	196.83	181.87	0.00080	1.27566	0.9764
52	82.9812	2125.6	0.17390	210.4081	0.00506	194.59	174.28	0.00078	1.27267	0.9638
54	82.1403	2135.6	0.17104	203.6869	0.00512	192.36	167.58	0.00077	1.26969	0.9529
56	81.2989	2144.7	0.16841	197.3461	0.00517	190.16	161.62	0.00076	1.26670	0.9430
58	80.4310	2168.3	0.16667	193.8173	0.00517	187.90	156.13	0.00075	1.26363	0.9313
60	79.6003	2173.0	0.16434	187.6705	0.00522	186.06	151.37	0.00074	1.26069	0.9228
65	77.5440	2202.2	0.15863	175.5411	0.00527	182.42	141.30	0.00073	1.25343	0.8987
70	75.5271	2234.0	0.15279	164.6374	0.00529	178.76	133.23	0.00072	1.24634	0.8808
75	73.5548	2268.5	0.14694	154.8164	0.00530	175.06	126.58	0.00071	1.23944	0.8687
80	71.6309	2304.7	0.14118	145.8433	0.00530	171.41	121.00	0.00071	1.23272	0.8614
85	69.7620	2343.5	0.13549	137.6651	0.00529	170.64	116.25	0.00071	1.22622	0.8443
90	67.9453	2389.4	0.12991	130.5979	0.00526	169.61	112.16	0.00072	1.21993	0.8311
95	66.1884	2441.1	0.12460	124.5019	0.00521	168.50	108.64	0.00072	1.21386	0.8203
100	64.4956	2504.1	0.11948	119.6114	0.00513	173.57	95.57	0.00075	1.20803	0.7077
120	58.3468	2773.5	0.10091	102.6367	0.00488	182.89	98.60	0.00083	1.18702	0.7298
140	53.0937	3079.6	0.08772	92.0502	0.00455	185.60	97.93	0.00090	1.16927	0.7385
160	48.6518	3415.0	0.07804	85.0971	0.00419	186.27	96.31	0.00096	1.15439	0.7405
180	44.8845	3760.1	0.07090	80.0658	0.00388	186.86	94.87	0.00103	1.14188	0.7398
200	41.6582	4110.1	0.06557	76.3289	0.00359	188.05	93.99	0.00110	1.13124	0.7375
220	38.8704	4456.4	0.06149	73.3272	0.00335	190.04	93.71	0.00118	1.12210	0.7351
240	36.4373	4803.0	0.05831	70.9973	0.00312	192.72	93.99	0.00127	1.11416	0.7320
260	34.2992	5151.6	0.05574	69.1826	0.00292	196.00	94.75	0.00137	1.10722	0.7280
280	32.4108	5502.7	0.05362	67.7689	0.00274	199.78	95.89	0.00147	1.10112	0.7232
300	30.6495	5739.9	0.05347	66.3210	0.00261	203.95	97.35	0.00160	1.09544	0.7146
350	27.1292	6517.4	0.05084	63.8894	0.00229	215.66	102.04	0.00192	1.08417	0.7052
400	24.3547	7273.1	0.04909	62.0985	0.00204	228.67	107.70	0.00228	1.07534	0.6984
450	22.1071	8015.4	0.04785	60.7131	0.00184	242.52	113.95	0.00268	1.06822	0.6935
500	20.2464	8756.5	0.04686	59.6048	0.00168	257.08	120.57	0.00311	1.06235	0.6902
550	18.6789	9494.5	0.04607	58.6964	0.00155	272.03	127.41	0.00357	1.05743	0.6878
600	17.3392	10235.6	0.04539	57.9379	0.00143	287.33	134.38	0.00407	1.05323	0.6861
700	15.1678	11734.9	0.04423	56.7441	0.00125	318.67	148.50	0.00515	1.04646	0.6842
800	13.4820	13273.5	0.04319	55.8489	0.00111	350.93	162.64	0.00635	1.04122	0.6834
900	12.1343	14866.2	0.04217	55.1545	0.00100	384.12	176.67	0.00767	1.03704	0.6833
1000	11.0320	16522.3	0.04115	54.6014	0.00091	440.56	190.52	0.00958	1.03364	0.6491
1200	9.3359	20003.8	0.03921	53.7787	0.00077	511.00	217.60	0.01281	1.02842	0.6548
1400	8.0919	23719.9	0.03738	53.1985	0.00067	583.15	243.81	0.01641	1.02460	0.6608
1600	7.1402	27620.2	0.03576	52.7687	0.00059	656.13	269.18	0.02037	1.02168	0.6664
1800	6.3887	31652.9	0.03438	52.4382	0.00053	729.25	293.79	0.02467	1.01938	0.6711
2000	5.7801	35833.8	0.03315	52.1754	0.00048	803.80	317.70	0.02934	1.01752	0.6745
2500	4.6657	47683.2	0.03002	51.6845	0.00038	1026.31	374.91	0.04314	1.01413	0.6706
3000	3.9009	64470.2	0.02578	51.2248	0.00032	1413.28	429.54	0.06246	1.01180	0.6347

* TWO-PHASE BOUNDARY

Table 4.12 (continued)
THERMODYNAMIC PROPERTIES OF NORMAL HYDROGEN

55.00 MPA ISOBAR

TEMPERATURE DEG. K	VOLUME CU M/KG	ISOTHERM DERIVATIVE CU M-MPA/KG	ISOCHORE DERIVATIVE MPA/K	INTERNAL ENERGY KJ/KG-K	ENTHALPY KJ/KG	ENTROPY KJ/KG-K	CV KJ /	CP KG-K	VELOCITY OF SOUND M/S
* 26.646	0.01057	3.6392	1.307	287.1	868.6	14.972	6.01	7.41	2118
27	0.01059	3.6271	1.305	289.0	871.3	15.070	6.05	7.47	2117
28	0.01063	3.5926	1.302	294.3	878.8	15.344	6.15	7.64	2113
29	0.01067	3.5577	1.298	299.7	886.6	15.616	6.24	7.81	2109
30	0.01071	3.5225	1.293	305.3	894.4	15.883	6.33	7.96	2105
31	0.01075	3.4871	1.288	311.0	902.5	16.146	6.41	8.11	2101
32	0.01080	3.4514	1.282	316.8	910.7	16.406	6.49	8.26	2097
33	0.01084	3.4156	1.275	322.8	919.0	16.663	6.56	8.41	2092
34	0.01088	3.3796	1.268	328.8	927.5	16.916	6.63	8.55	2087
35	0.01093	3.3436	1.261	335.0	936.1	17.166	6.70	8.68	2082
36	0.01097	3.3076	1.254	341.2	944.8	17.412	6.76	8.82	2077
37	0.01102	3.2716	1.246	347.6	953.7	17.656	6.82	8.95	2072
38	0.01107	3.2357	1.238	354.1	962.7	17.896	6.87	9.07	2067
39	0.01111	3.1999	1.229	360.6	971.9	18.134	6.91	9.19	2062
40	0.01116	3.1643	1.221	367.2	981.1	18.367	6.95	9.30	2057
42	0.01126	3.0937	1.203	380.7	999.9	18.826	7.03	9.52	2047
44	0.01136	3.0244	1.184	394.4	1019.2	19.274	7.10	9.74	2036
46	0.01146	2.9566	1.165	408.0	1038.4	19.700	7.21	9.98	2023
48	0.01157	2.8906	1.145	422.4	1058.5	20.129	7.29	10.20	2011
50	0.01167	2.8269	1.125	437.1	1079.1	20.550	7.35	10.40	2000
52	0.01178	2.7656	1.104	452.1	1100.1	20.961	7.41	10.59	1988
54	0.01189	2.7071	1.083	467.3	1121.5	21.365	7.46	10.77	1977
56	0.01201	2.6515	1.061	483.0	1143.5	21.765	7.50	10.93	1966
58	0.01213	2.5990	1.040	498.5	1165.5	22.151	7.52	11.07	1956
60	0.01224	2.5499	1.018	514.3	1187.7	22.528	7.54	11.19	1946
65	0.01255	2.4754	0.9691	554.2	1244.7	23.439	7.57	11.46	1935
70	0.01287	2.3895	0.9170	594.9	1302.6	24.297	7.61	11.68	1916
75	0.01319	2.3173	0.8665	636.1	1361.6	25.111	7.65	11.87	1897
80	0.01352	2.2438	0.8187	677.9	1421.4	25.884	7.70	12.07	1875
85	0.01386	2.1803	0.7743	720.1	1482.2	26.621	7.77	12.26	1855
90	0.01420	2.1219	0.7327	762.9	1543.9	27.327	7.85	12.44	1834
95	0.01455	2.0737	0.6939	806.2	1606.6	28.004	7.94	12.61	1815
100	0.01491	2.0336	0.6579	850.0	1670.0	28.655	8.04	12.77	1797
120	0.01638	1.9310	0.5405	1032.8	1933.7	31.056	8.58	13.45	1740
140	0.01790	1.8884	0.4545	1222.9	2207.6	33.166	9.05	13.96	1707
160	0.01945	1.8907	0.3891	1420.2	2490.1	35.051	9.46	14.31	1691
180	0.02102	1.9181	0.3390	1623.0	2778.9	36.751	9.81	14.57	1688
200	0.02257	1.9596	0.2998	1830.4	3072.0	38.295	10.09	14.76	1693
220	0.02413	2.0095	0.2685	2041.3	3368.6	39.709	10.32	14.91	1704
240	0.02569	2.0648	0.2431	2255.0	3667.8	41.011	10.49	15.03	1720
260	0.02724	2.1280	0.2220	2470.7	3968.9	42.216	10.63	15.09	1738
280	0.02878	2.1975	0.2039	2687.9	4270.9	43.335	10.73	15.12	1760
300	0.03039	2.2687	0.1899	2892.3	4563.7	44.323	10.59	15.00	1792
350	0.03424	2.4553	0.1605	3428.9	5311.9	46.631	10.62	14.93	1857
400	0.03805	2.6466	0.1390	3963.4	6056.3	48.619	10.62	14.85	1924
450	0.04185	2.8404	0.1228	4495.4	6796.9	50.363	10.59	14.77	1990
500	0.04562	3.0358	0.1100	5025.2	7534.5	51.917	10.58	14.73	2056
550	0.04939	3.2322	0.0997	5553.5	8269.8	53.319	10.57	14.69	2120
600	0.05315	3.4294	0.0912	6081.2	9004.2	54.594	10.57	14.68	2182
700	0.06065	3.8259	0.0779	7136.5	10472.0	56.860	10.60	14.69	2302
800	0.06813	4.2247	0.0681	8196.4	11943.7	58.827	10.67	14.75	2416
900	0.07562	4.6252	0.0605	9265.2	13424.1	60.568	10.78	14.86	2524
1000	0.08310	5.0272	0.0545	10346.4	14916.7	62.139	10.94	15.01	2627
1200	0.09805	5.8349	0.0454	12562.5	17955.5	64.890	11.30	15.38	2818
1400	0.11302	6.6462	0.0389	14858.9	21074.8	67.300	11.73	15.80	2993
1600	0.12798	7.4601	0.0341	17239.8	24278.8	69.446	12.16	16.24	3157
1800	0.14295	8.2760	0.0303	19704.7	27567.1	71.377	12.57	16.66	3311
2000	0.15793	9.0933	0.0273	22253.7	30939.9	73.164	12.97	17.06	3459
2500	0.19548	11.1412	0.0219	28999.7	39751.0	77.077	14.19	18.33	3794
3000	0.23364	13.1933	0.0182	36602.2	49452.5	80.624	16.40	20.77	4087

* TWO-PHASE BOUNDARY

Table 4.12 (continued)
THERMODYNAMIC PROPERTIES OF NORMAL HYDROGEN

55.00 MPA ISOBAR

TEMPERATURE DEG. K	DENSITY KG/CU M	$V(DH/DV)_P$ KJ/KG	$V(DP/DU)_V$ MPA-CU M/KJ X 10^2	$-V(DP/DV)_T$ MPA	$(DV/DT)_P/V$ 1/DEG. K	THERMAL CONDUCTIVITY W/K-M X 10^3	VISCOSITY KG/M-S X 10^7	THERMAL DIFFUSIVITY SQ M/HR	DIELECTRIC CONSTANT	PRANDTL NUMBER
* 26.646	94.5778	1952.1	0.22975	344.1862	0.00380	208.55	601.68	0.00107	1.31434	2.1378
27	94.4505	1960.6	0.22849	342.5778	0.00381	209.63	581.97	0.00107	1.31388	2.0742
28	94.0893	1983.8	0.22505	338.0217	0.00385	212.31	532.32	0.00106	1.31257	1.9159
29	93.7257	2005.4	0.22183	333.4476	0.00389	214.53	490.14	0.00106	1.31125	1.7832
30	93.3597	2024.8	0.21889	328.8612	0.00393	216.35	454.00	0.00105	1.30992	1.6705
31	92.9915	2043.5	0.21602	324.2682	0.00397	217.00	422.79	0.00104	1.30859	1.5809
32	92.6212	2061.2	0.21326	319.6741	0.00401	217.38	395.65	0.00102	1.30725	1.5041
33	92.2489	2077.4	0.21068	315.0841	0.00405	217.52	371.90	0.00101	1.30590	1.4375
34	91.8746	2092.5	0.20822	310.5035	0.00409	217.53	350.98	0.00100	1.30455	1.3792
35	91.4983	2106.6	0.20587	305.9372	0.00412	217.38	332.47	0.00098	1.30319	1.3282
36	91.1202	2119.9	0.20360	301.3902	0.00416	217.07	315.99	0.00097	1.30183	1.2837
37	90.7403	2132.2	0.20143	296.8672	0.00420	216.62	301.26	0.00096	1.30046	1.2444
38	90.3586	2142.8	0.19945	292.3729	0.00423	216.07	288.03	0.00095	1.29908	1.2092
39	89.9752	2151.8	0.19766	287.9117	0.00427	215.42	276.09	0.00094	1.29770	1.1775
40	89.5903	2159.8	0.19596	283.4879	0.00431	214.68	265.29	0.00093	1.29631	1.1492
42	88.8157	2174.6	0.19266	274.7696	0.00438	212.94	246.52	0.00091	1.29353	1.1020
44	88.0353	2189.0	0.18936	266.2499	0.00445	210.99	230.82	0.00089	1.29073	1.0650
46	87.2498	2209.8	0.18529	257.9587	0.00452	208.90	217.52	0.00086	1.28791	1.0390
48	86.4597	2225.9	0.18177	249.9233	0.00458	206.71	206.14	0.00084	1.28508	1.0169
50	85.6657	2240.1	0.17850	242.1683	0.00464	204.48	196.31	0.00083	1.28224	0.9987
52	84.8687	2252.7	0.17544	234.7155	0.00470	202.22	187.73	0.00081	1.27940	0.9835
54	84.0695	2264.0	0.17259	227.5835	0.00476	199.96	180.19	0.00079	1.27655	0.9706
56	83.2691	2273.7	0.16998	220.7875	0.00481	197.72	173.51	0.00078	1.27370	0.9592
58	82.4687	2281.4	0.16767	214.3387	0.00485	195.51	167.55	0.00077	1.27085	0.9485
60	81.6692	2289.4	0.16542	208.2446	0.00489	193.60	162.21	0.00076	1.26801	0.9378
65	79.6560	2330.9	0.16072	197.1767	0.00491	189.68	150.86	0.00075	1.26088	0.9111
70	77.7140	2366.1	0.15514	185.6980	0.00494	185.88	142.00	0.00074	1.25403	0.8925
75	75.8132	2407.7	0.14946	175.6818	0.00493	182.03	134.76	0.00073	1.24735	0.8792
80	73.9677	2446.5	0.14374	165.9722	0.00493	178.27	128.76	0.00072	1.24088	0.8717
85	72.1689	2490.6	0.13811	157.3462	0.00492	177.48	123.66	0.00072	1.23460	0.8539
90	70.4197	2537.4	0.13254	149.4223	0.00490	176.44	119.28	0.00072	1.22851	0.8412
95	68.7183	2589.9	0.12717	142.5027	0.00487	175.29	115.48	0.00073	1.22261	0.8308
100	67.0733	2647.9	0.12198	136.4020	0.00482	181.23	99.99	0.00076	1.21691	0.7046
120	61.0551	2933.1	0.10322	117.8980	0.00458	190.69	103.03	0.00084	1.19625	0.7265
140	55.8554	3238.9	0.08993	105.4773	0.00431	193.06	102.11	0.00089	1.17858	0.7381
160	51.4027	3574.5	0.08000	97.1866	0.00400	193.19	100.14	0.00095	1.16359	0.7417
180	47.5849	3923.3	0.07262	91.2726	0.00371	193.19	98.34	0.00100	1.15084	0.7416
200	44.2982	4274.9	0.06708	86.8068	0.00345	193.82	97.12	0.00107	1.13994	0.7397
220	41.4370	4624.8	0.06281	83.2688	0.00322	195.29	96.54	0.00114	1.13051	0.7372
240	38.9298	4967.9	0.05952	80.3804	0.00302	197.53	96.56	0.00122	1.12229	0.7345
260	36.7130	5312.9	0.05688	78.1244	0.00284	200.44	97.09	0.00130	1.11506	0.7311
280	34.7442	5660.3	0.05471	76.3489	0.00267	203.90	98.05	0.00140	1.10866	0.7270
300	32.9061	5894.4	0.05450	74.6540	0.00254	207.82	99.36	0.00152	1.10272	0.7170
350	29.2084	6671.6	0.05170	71.7146	0.00224	219.11	103.79	0.00181	1.09082	0.7071
400	26.2794	7427.0	0.04984	69.5512	0.00200	231.94	109.32	0.00214	1.08146	0.6998
450	23.8972	8169.1	0.04849	67.8786	0.00181	245.78	115.54	0.00251	1.07388	0.6946
500	21.9189	8910.1	0.04743	66.5409	0.00165	260.45	122.19	0.00290	1.06762	0.6909
550	20.2477	9647.8	0.04657	65.4444	0.00152	275.60	129.11	0.00333	1.06236	0.6883
600	18.8162	10388.7	0.04584	64.5288	0.00141	291.17	136.19	0.00380	1.05786	0.6865
700	16.4894	11887.4	0.04460	63.0873	0.00124	323.21	150.61	0.00480	1.05058	0.6844
800	14.6771	13425.5	0.04350	62.0060	0.00110	356.31	165.13	0.00593	1.04493	0.6835
900	13.2248	15018.1	0.04243	61.1673	0.00099	390.46	179.58	0.00715	1.04042	0.6833
1000	12.0343	16674.3	0.04138	60.4995	0.00090	440.56	193.88	0.00878	1.03674	0.6605
1200	10.1984	20156.6	0.03938	59.5066	0.00076	511.00	221.91	0.01173	1.03107	0.6677
1400	8.8483	23874.1	0.03752	58.8073	0.00066	583.15	249.10	0.01501	1.02692	0.6751
1600	7.8136	27776.0	0.03587	58.2902	0.00058	656.12	275.46	0.01861	1.02374	0.6819
1800	6.9953	31809.4	0.03447	57.8931	0.00052	729.20	301.05	0.02253	1.02124	0.6877
2000	6.3319	35987.9	0.03324	57.5781	0.00047	803.60	325.94	0.02678	1.01921	0.6920
2500	5.1157	47783.0	0.03013	56.9945	0.00038	1023.86	385.52	0.03931	1.01550	0.6902
3000	4.2801	64306.6	0.02598	56.4680	0.00032	1400.61	442.39	0.05673	1.01296	0.6559

* TWO-PHASE BOUNDARY

Table 4.12 (continued)
THERMODYNAMIC PROPERTIES OF NORMAL HYDROGEN

60.00 MPA ISOBAR

TEMPERATURE	VOLUME	ISOTHERM DERIVATIVE	ISOCHORE DERIVATIVE	INTERNAL ENERGY	ENTHALPY	ENTROPY	CV	CP	VELOCITY OF SOUND
DEG. K	CU M/KG	CU M-MPA/KG	MPA/K	KJ/KG-K	KJ/KG	KJ/KG-K	KJ /	KG-K	M/S
* 27.559	0.01046	3.8220	1.334	295.1	922.7	15.027	6.09	7.49	2169
28	0.01048	3.8074	1.333	297.4	926.0	15.146	6.13	7.56	2167
29	0.01052	3.7741	1.329	302.7	933.7	15.415	6.23	7.73	2164
30	0.01056	3.7404	1.324	308.2	941.5	15.679	6.31	7.88	2161
31	0.01060	3.7063	1.319	313.7	949.4	15.940	6.40	8.03	2157
32	0.01064	3.6720	1.313	319.4	957.5	16.197	6.48	8.18	2153
33	0.01068	3.6374	1.306	325.2	965.8	16.451	6.55	8.32	2149
34	0.01072	3.6026	1.300	331.1	974.2	16.701	6.62	8.46	2144
35	0.01076	3.5677	1.293	337.1	982.7	16.948	6.69	8.59	2140
36	0.01080	3.5327	1.285	343.3	991.3	17.192	6.76	8.72	2135
37	0.01084	3.4976	1.278	349.5	1000.1	17.433	6.82	8.85	2131
38	0.01089	3.4625	1.270	355.8	1009.0	17.671	6.87	8.97	2126
39	0.01093	3.4274	1.261	362.2	1018.1	17.905	6.92	9.08	2121
40	0.01098	3.3924	1.253	368.7	1027.2	18.136	6.96	9.19	2116
42	0.01107	3.3228	1.236	381.9	1045.8	18.589	7.04	9.40	2107
44	0.01116	3.2541	1.217	395.4	1064.8	19.031	7.12	9.61	2096
46	0.01125	3.1864	1.199	408.6	1083.7	19.452	7.22	9.85	2084
48	0.01135	3.1202	1.180	422.8	1103.6	19.875	7.30	10.06	2073
50	0.01145	3.0557	1.160	437.2	1124.0	20.290	7.38	10.26	2062
52	0.01155	2.9932	1.140	451.9	1144.7	20.696	7.44	10.45	2050
54	0.01165	2.9329	1.119	466.8	1165.7	21.094	7.49	10.62	2039
56	0.01175	2.8751	1.099	482.2	1187.4	21.489	7.53	10.78	2028
58	0.01186	2.8199	1.078	497.5	1209.1	21.870	7.56	10.92	2018
60	0.01197	2.7676	1.057	513.0	1231.1	22.242	7.58	11.05	2008
65	0.01225	2.6498	1.004	552.2	1287.0	23.137	7.62	11.33	1985
70	0.01254	2.5870	0.9585	592.2	1344.8	23.993	7.66	11.57	1977
75	0.01284	2.5083	0.9089	632.9	1403.2	24.799	7.71	11.78	1958
80	0.01314	2.4474	0.8616	674.2	1462.6	25.565	7.76	11.95	1941
85	0.01345	2.3784	0.8168	716.0	1522.8	26.296	7.83	12.14	1920
90	0.01376	2.3204	0.7751	758.3	1584.0	26.996	7.92	12.33	1901
95	0.01408	2.2690	0.7361	801.3	1646.1	27.667	8.01	12.50	1882
100	0.01440	2.2209	0.6994	844.8	1709.0	28.312	8.11	12.68	1863
120	0.01574	2.1021	0.5771	1026.7	1971.1	30.699	8.65	13.36	1802
140	0.01712	2.0411	0.4886	1216.3	2243.6	32.799	9.11	13.91	1765
160	0.01854	2.0338	0.4201	1413.5	2525.7	34.680	9.52	14.29	1747
180	0.01995	2.0523	0.3668	1616.6	2814.1	36.378	9.87	14.57	1741
200	0.02138	2.0865	0.3249	1824.4	3107.3	37.923	10.15	14.77	1743
220	0.02280	2.1322	0.2912	2035.9	3404.0	39.337	10.37	14.92	1752
240	0.02422	2.1830	0.2638	2250.1	3703.4	40.640	10.54	15.03	1764
260	0.02564	2.2398	0.2411	2466.4	4004.7	41.846	10.68	15.11	1780
280	0.02705	2.3043	0.2218	2684.2	4307.3	42.967	10.78	15.15	1800
300	0.02852	2.3746	0.2067	2888.5	4599.9	43.953	10.63	15.02	1832
350	0.03205	2.5564	0.1747	3426.4	5349.2	46.264	10.66	14.95	1894
400	0.03554	2.7441	0.1514	3961.9	6094.5	48.255	10.64	14.86	1958
450	0.03902	2.9351	0.1337	4494.8	6836.0	50.001	10.62	14.79	2022
500	0.04248	3.1280	0.1197	5025.4	7574.1	51.556	10.60	14.74	2085
550	0.04593	3.3222	0.1085	5554.3	8310.0	52.959	10.59	14.70	2148
600	0.04937	3.5175	0.0992	6082.6	9044.7	54.235	10.59	14.68	2209
700	0.05624	3.9107	0.0848	7138.8	10513.0	56.501	10.61	14.69	2326
800	0.06309	4.3066	0.0741	8199.4	11985.0	58.469	10.68	14.75	2438
900	0.06995	4.7046	0.0659	9268.7	13465.5	60.211	10.80	14.86	2545
1000	0.07680	5.1045	0.0593	10350.4	14958.2	61.781	10.95	15.01	2645
1200	0.09050	5.9085	0.0494	12567.1	17997.0	64.532	11.31	15.38	2834
1400	0.10420	6.7169	0.0424	14863.9	21116.1	66.942	11.74	15.80	3007
1600	0.11791	7.5284	0.0371	17245.1	24319.9	69.088	12.17	16.24	3170
1800	0.13163	8.3423	0.0330	19710.2	27608.0	71.018	12.58	16.66	3324
2000	0.14535	9.1580	0.0297	22259.2	30980.3	72.805	12.97	17.06	3470
2500	0.17976	11.2028	0.0238	29000.8	39786.2	76.717	14.17	18.31	3804
3000	0.21471	13.2526	0.0199	36577.3	49459.7	80.253	16.31	20.66	4096

* TWO-PHASE BOUNDARY

Table 4.12 (continued)
THERMODYNAMIC PROPERTIES OF NORMAL HYDROGEN

60.00 MPA ISOBAR

TEMPERATURE DEG. K	DENSITY KG/CU M	$V(DH/DV)_P$ KJ/KG	$V(DP/DU)_V$ MPA-CU M/KJ X 10^2	$-V(DP/DV)_T$ MPA	$(DV/DT)_P/V$ 1/DEG. K	THERMAL CONDUCTIVITY W/K-M X 10^3	VISCOSITY KG/M-S X 10^7	THERMAL DIFFUSIVITY SQ M/HR	DIELECTRIC CONSTANT	PRANDTL NUMBER
* 27.559	95.5951	2051.5	0.22930	365.3665	0.00365	217.73	636.02	0.00109	1.31803	2.1885
28	95.4408	2062.7	0.22776	363.3861	0.00367	218.92	610.44	0.00109	1.31747	2.1094
29	95.0898	2087.0	0.22441	358.8802	0.00370	221.30	558.99	0.00108	1.31620	1.9517
30	94.7368	2109.1	0.22134	354.3538	0.00374	223.25	515.14	0.00108	1.31492	1.8182
31	94.3819	2130.5	0.21835	349.8125	0.00377	223.98	477.48	0.00106	1.31363	1.7119
32	94.0253	2150.7	0.21550	345.2615	0.00380	224.43	444.87	0.00105	1.31233	1.6209
33	93.6670	2169.4	0.21282	340.7062	0.00383	224.64	416.47	0.00104	1.31104	1.5422
34	93.3070	2186.9	0.21028	336.1516	0.00387	224.71	391.56	0.00103	1.30973	1.4734
35	92.9455	2203.2	0.20785	331.6027	0.00390	224.59	369.60	0.00101	1.30842	1.4133
36	92.5824	2218.8	0.20551	327.0642	0.00393	224.32	350.12	0.00100	1.30711	1.3609
37	92.2179	2233.2	0.20328	322.5406	0.00396	223.91	332.77	0.00099	1.30579	1.3146
38	91.8519	2245.9	0.20123	318.0367	0.00399	223.38	317.25	0.00098	1.30447	1.2733
39	91.4845	2256.7	0.19939	313.5566	0.00402	222.75	303.29	0.00097	1.30314	1.2361
40	91.1158	2266.5	0.19765	309.1046	0.00405	222.03	290.69	0.00095	1.30181	1.2029
42	90.3746	2284.6	0.19427	300.3011	0.00411	220.31	268.91	0.00093	1.29914	1.1474
44	89.6286	2302.2	0.19091	291.6575	0.00417	218.36	250.78	0.00091	1.29645	1.1037
46	88.8782	2326.3	0.18679	283.2026	0.00423	216.26	235.52	0.00089	1.29375	1.0724
48	88.1239	2345.4	0.18323	274.9634	0.00429	214.06	222.53	0.00087	1.29105	1.0459
50	87.3663	2362.0	0.17993	266.9642	0.00434	211.80	211.35	0.00085	1.28833	1.0240
52	86.6059	2376.7	0.17687	259.2269	0.00440	209.52	201.66	0.00083	1.28560	1.0058
54	85.8434	2389.6	0.17401	251.7707	0.00445	207.22	193.17	0.00082	1.28288	0.9904
56	85.0795	2400.6	0.17140	244.6121	0.00449	204.94	185.69	0.00080	1.28015	0.9770
58	84.3151	2409.1	0.16910	237.7642	0.00453	202.69	179.04	0.00079	1.27742	0.9647
60	83.5509	2417.6	0.16685	231.2368	0.00457	200.74	173.11	0.00078	1.27470	0.9528
65	81.6480	2441.8	0.16132	216.3540	0.00464	196.83	160.71	0.00077	1.26794	0.9251
70	79.7274	2489.8	0.15696	206.2574	0.00465	192.71	150.75	0.00075	1.26114	0.9051
75	77.8890	2531.7	0.15141	195.3693	0.00465	188.70	142.87	0.00074	1.25465	0.8917
80	76.0998	2583.7	0.14585	186.2454	0.00463	184.79	136.37	0.00073	1.24835	0.8820
85	74.3631	2629.5	0.14025	176.8644	0.00462	183.99	130.91	0.00073	1.24226	0.8640
90	72.6669	2682.1	0.13474	168.6176	0.00460	182.91	126.21	0.00074	1.23634	0.8507
95	71.0221	2737.6	0.12943	161.1474	0.00457	181.73	122.16	0.00074	1.23060	0.8405
100	69.4259	2795.3	0.12422	154.1877	0.00454	188.54	104.23	0.00077	1.22506	0.7010
120	63.5332	3091.1	0.10505	133.5510	0.00432	198.19	107.32	0.00084	1.20473	0.7233
140	58.4025	3394.0	0.09183	119.2041	0.00410	200.26	106.17	0.00089	1.18721	0.7376
160	53.9490	3733.2	0.08177	109.7236	0.00383	199.90	103.88	0.00093	1.17215	0.7428
180	50.1062	4084.0	0.07419	102.8327	0.00357	199.37	101.75	0.00098	1.15925	0.7434
200	46.7707	4436.6	0.06847	97.5868	0.00333	199.48	100.21	0.00104	1.14813	0.7420
220	43.8538	4790.4	0.06404	93.5038	0.00311	200.48	99.35	0.00110	1.13847	0.7394
240	41.2843	5135.7	0.06061	90.1241	0.00293	202.30	99.12	0.00117	1.13001	0.7366
260	39.0033	5476.2	0.05789	87.3592	0.00276	204.84	99.44	0.00125	1.12253	0.7337
280	36.9667	5819.2	0.05567	85.1829	0.00260	208.01	100.22	0.00134	1.11589	0.7301
300	35.0602	6048.3	0.05549	83.2541	0.00248	211.68	101.38	0.00145	1.10969	0.7194
350	31.2041	6825.3	0.05254	79.7700	0.00219	222.55	105.55	0.00172	1.09723	0.7089
400	28.1346	7580.6	0.05055	77.2052	0.00196	235.19	110.95	0.00202	1.08738	0.7012
450	25.6289	8322.7	0.04912	75.2225	0.00178	249.01	117.13	0.00237	1.07938	0.6956
500	23.5414	9063.6	0.04797	73.6366	0.00163	263.78	123.80	0.00274	1.07276	0.6917
550	21.7735	9801.1	0.04706	72.3365	0.00150	279.12	130.78	0.00314	1.06717	0.6889
600	20.2558	10541.9	0.04627	71.2505	0.00139	294.95	137.98	0.00357	1.06238	0.6869
700	17.7820	12040.1	0.04495	69.5398	0.00122	327.68	152.70	0.00452	1.05462	0.6846
800	15.8493	13577.7	0.04379	68.2560	0.00109	361.63	167.59	0.00557	1.04858	0.6836
900	14.2966	15170.0	0.04268	67.2599	0.00098	396.75	182.47	0.00672	1.04375	0.6834
1000	13.0213	16826.3	0.04159	66.4668	0.00089	440.56	197.24	0.00812	1.03979	0.6720
1200	11.0499	20309.2	0.03955	65.2883	0.00076	511.00	226.26	0.01083	1.03369	0.6808
1400	9.5966	24027.9	0.03765	64.4590	0.00066	583.15	254.47	0.01384	1.02922	0.6897
1600	8.4807	27931.1	0.03598	63.8466	0.00058	656.12	281.88	0.01715	1.02579	0.6978
1800	7.5971	31964.9	0.03456	63.3771	0.00052	729.16	308.52	0.02075	1.02308	0.7047
2000	6.8798	36141.4	0.03332	63.0054	0.00047	803.43	334.45	0.02464	1.02088	0.7102
2500	5.5631	47888.9	0.03022	62.3218	0.00038	1021.73	396.58	0.03612	1.01686	0.7105
3000	4.6575	64183.0	0.02615	61.7244	0.00032	1389.57	455.88	0.05198	1.01410	0.6780

* TWO-PHASE BOUNDARY

Table 4.12 (continued)
THERMODYNAMIC PROPERTIES OF NORMAL HYDROGEN

65.00 MPA ISOBAR

TEMPERATURE DEG. K	VOLUME CU M/KG	ISOTHERM DERIVATIVE CU M-MPA/KG	ISOCHORE DERIVATIVE MPA/K	INTERNAL ENERGY KJ/KG-K	ENTHALPY KJ/KG	ENTROPY KJ/KG-K	CV KJ / KG-K	CP KJ / KG-K	VELOCITY OF SOUND M/S
* 28.449	0.01036	4.0022	1.362	303.1	976.2	15.079	6.16	7.57	2218
29	0.01038	3.9846	1.359	306.0	980.4	15.226	6.21	7.66	2217
30	0.01041	3.9524	1.355	311.3	988.2	15.488	6.30	7.81	2213
31	0.01045	3.9197	1.349	316.8	996.0	15.746	6.39	7.96	2210
32	0.01049	3.8867	1.343	322.4	1004.1	16.001	6.47	8.10	2206
33	0.01053	3.8534	1.337	328.0	1012.2	16.253	6.55	8.24	2203
34	0.01056	3.8198	1.330	333.8	1020.5	16.501	6.62	8.38	2199
35	0.01060	3.7860	1.323	339.7	1029.0	16.746	6.69	8.51	2195
36	0.01064	3.7520	1.316	345.7	1037.6	16.987	6.75	8.63	2190
37	0.01068	3.7179	1.309	351.8	1046.3	17.226	6.81	8.76	2186
38	0.01072	3.6836	1.301	358.0	1055.1	17.461	6.87	8.88	2182
39	0.01076	3.6494	1.293	364.3	1064.0	17.693	6.92	8.99	2177
40	0.01081	3.6151	1.284	370.7	1073.1	17.922	6.96	9.09	2173
42	0.01089	3.5468	1.267	383.6	1091.5	18.370	7.04	9.30	2164
44	0.01098	3.4788	1.249	396.9	1110.3	18.807	7.13	9.50	2154
46	0.01106	3.4117	1.231	409.9	1129.0	19.222	7.23	9.73	2143
48	0.01115	3.3456	1.212	423.8	1148.6	19.641	7.32	9.94	2132
50	0.01124	3.2809	1.193	437.9	1168.7	20.051	7.40	10.14	2121
52	0.01134	3.2177	1.174	452.4	1189.2	20.452	7.46	10.32	2110
54	0.01143	3.1564	1.154	467.0	1210.0	20.845	7.52	10.50	2099
56	0.01153	3.0971	1.134	482.2	1231.4	21.236	7.57	10.65	2088
58	0.01162	3.0401	1.113	497.2	1252.8	21.612	7.59	10.79	2078
60	0.01172	2.9855	1.093	512.5	1274.6	21.980	7.62	10.92	2068
65	0.01198	2.8605	1.041	551.1	1329.9	22.864	7.67	11.21	2044
70	0.01225	2.7527	0.9896	590.5	1386.6	23.705	7.72	11.45	2021
75	0.01253	2.6984	0.9479	630.7	1445.1	24.512	7.76	11.68	2015
80	0.01281	2.6250	0.9006	671.5	1504.0	25.273	7.82	11.88	1996
85	0.01309	2.5726	0.8561	712.9	1563.9	25.999	7.90	12.05	1981
90	0.01338	2.5105	0.8141	754.9	1624.6	26.693	7.98	12.23	1962
95	0.01367	2.4550	0.7749	797.5	1686.2	27.359	8.07	12.42	1943
100	0.01397	2.4087	0.7382	840.7	1748.8	28.000	8.18	12.59	1926
120	0.01519	2.2696	0.6121	1021.8	2009.3	30.373	8.72	13.29	1860
140	0.01646	2.1952	0.5204	1210.8	2280.6	32.463	9.18	13.85	1820
160	0.01775	2.1754	0.4500	1407.9	2561.9	34.340	9.58	14.28	1800
180	0.01906	2.1875	0.3937	1611.3	2850.3	36.038	9.93	14.56	1791
200	0.02037	2.2138	0.3493	1819.5	3143.4	37.582	10.20	14.77	1790
220	0.02168	2.2546	0.3134	2031.3	3440.2	38.997	10.43	14.93	1797
240	0.02298	2.3018	0.2841	2246.0	3739.8	40.300	10.60	15.04	1807
260	0.02428	2.3531	0.2597	2462.9	4041.3	41.507	10.73	15.12	1821
280	0.02558	2.4115	0.2392	2681.2	4344.1	42.629	10.83	15.18	1838
300	0.02694	2.4814	0.2234	2885.3	4636.4	43.613	10.66	15.04	1871
350	0.03019	2.6583	0.1888	3424.3	5386.8	45.927	10.69	14.97	1929
400	0.03342	2.8423	0.1636	3960.8	6132.9	47.921	10.67	14.88	1991
450	0.03662	3.0302	0.1445	4494.6	6875.1	49.668	10.64	14.80	2053
500	0.03981	3.2205	0.1294	5025.9	7613.9	51.224	10.63	14.75	2114
550	0.04300	3.4125	0.1173	5555.5	8350.2	52.629	10.61	14.71	2175
600	0.04617	3.6058	0.1073	6084.3	9085.3	53.905	10.61	14.69	2234
700	0.05250	3.9954	0.0917	7141.4	10554.0	56.172	10.63	14.69	2350
800	0.05883	4.3882	0.0801	8202.7	12026.3	58.140	10.70	14.75	2460
900	0.06514	4.7836	0.0712	9272.6	13506.9	59.882	10.81	14.86	2564
1000	0.07146	5.1811	0.0641	10354.7	14999.6	61.452	10.96	15.01	2664
1200	0.08410	5.9812	0.0534	12572.0	18038.3	64.203	11.32	15.37	2850
1400	0.09674	6.7864	0.0458	14869.3	21157.3	66.613	11.75	15.80	3022
1600	0.10939	7.5953	0.0402	17250.8	24360.9	68.759	12.18	16.24	3183
1800	0.12204	8.4071	0.0357	19716.2	27648.8	70.689	12.58	16.65	3335
2000	0.13470	9.2210	0.0322	22265.1	31020.6	72.476	12.98	17.06	3481
2500	0.16644	11.2623	0.0258	29003.0	39821.8	76.385	14.16	18.29	3814
3000	0.19867	13.3096	0.0215	36556.7	49470.5	79.913	16.24	20.57	4106

* TWO-PHASE BOUNDARY

Table 4.12 (continued)
THERMODYNAMIC PROPERTIES OF NORMAL HYDROGEN

65.00 MPA ISOBAR

TEMPERATURE DEG. K	DENSITY KG/CU M	$V(DH/DV)_P$ KJ/KG	$V(DP/DU)_V$ MPA-CU M/KJ X 10^2	$-V(DP/DV)_T$ MPA	$(DV/DT)_P/V$ 1/DEG. K	THERMAL CONDUCTIVITY W/K-M X 10^3	VISCOSITY KG/M-S X 10^7	THERMAL DIFFUSIVITY SQ M/HR	DIELECTRIC CONSTANT	PRANDTL NUMBER
* 28.449	96.5666	2149.1	0.22897	386.4808	0.00352	226.54	671.06	0.00112	1.32157	2.2429
29	96.3788	2163.7	0.22708	384.0341	0.00354	227.87	637.85	0.00111	1.32089	2.1438
30	96.0369	2188.5	0.22388	379.5738	0.00357	229.95	584.76	0.00110	1.31964	1.9860
31	95.6934	2212.5	0.22077	375.0909	0.00360	230.77	539.38	0.00109	1.31839	1.8601
32	95.3485	2235.3	0.21781	370.5907	0.00362	231.29	500.29	0.00108	1.31714	1.7527
33	95.0022	2256.5	0.21502	366.0785	0.00365	231.55	466.38	0.00106	1.31588	1.6601
34	94.6545	2276.4	0.21239	361.5592	0.00368	231.67	436.78	0.00105	1.31462	1.5793
35	94.3056	2295.0	0.20988	357.0376	0.00371	231.60	410.78	0.00104	1.31335	1.5089
36	93.9554	2312.8	0.20746	352.5185	0.00373	231.36	387.81	0.00103	1.31208	1.4474
37	93.6040	2329.4	0.20516	348.0064	0.00376	230.97	367.42	0.00101	1.31081	1.3933
38	93.2515	2344.2	0.20305	343.5058	0.00379	230.47	349.23	0.00100	1.30953	1.3450
39	92.8979	2356.9	0.20115	339.0208	0.00381	229.85	332.94	0.00099	1.30825	1.3016
40	92.5432	2368.5	0.19935	334.5557	0.00384	229.14	318.27	0.00098	1.30697	1.2629
42	91.8306	2390.2	0.19589	325.7010	0.00389	227.42	293.03	0.00096	1.30439	1.1981
44	91.1142	2411.0	0.19245	316.9722	0.00394	225.47	272.14	0.00094	1.30180	1.1470
46	90.3942	2438.7	0.18826	308.3978	0.00399	223.36	254.64	0.00091	1.29921	1.1098
48	89.6710	2460.8	0.18465	300.0041	0.00404	221.14	239.82	0.00089	1.29660	1.0784
50	88.9450	2480.1	0.18132	291.8155	0.00409	218.85	227.13	0.00087	1.29399	1.0524
52	88.2165	2497.2	0.17822	283.8540	0.00413	216.53	216.17	0.00086	1.29138	1.0307
54	87.4862	2512.2	0.17535	276.1393	0.00418	214.20	206.62	0.00084	1.28876	1.0125
56	86.7546	2524.9	0.17273	268.6887	0.00422	211.88	198.24	0.00083	1.28614	0.9967
58	86.0223	2534.5	0.17042	261.5165	0.00426	209.58	190.83	0.00081	1.28352	0.9825
60	85.2899	2544.1	0.16818	254.6345	0.00429	207.59	184.23	0.00080	1.28090	0.9690
65	83.4637	2569.7	0.16264	238.7487	0.00436	203.60	170.54	0.00078	1.27439	0.9388
70	81.6539	2600.8	0.15708	224.7719	0.00440	199.51	159.83	0.00077	1.26796	0.9173
75	79.8094	2654.4	0.15300	215.3549	0.00440	195.13	150.94	0.00075	1.26143	0.9037
80	78.0755	2703.1	0.14745	204.9503	0.00439	191.08	143.92	0.00074	1.25531	0.8947
85	76.3829	2764.9	0.14195	196.5034	0.00436	190.22	138.02	0.00074	1.24935	0.8740
90	74.7374	2819.6	0.13649	187.6264	0.00434	189.10	133.01	0.00074	1.24358	0.8605
95	73.1402	2877.1	0.13125	179.5626	0.00432	187.89	128.69	0.00074	1.23799	0.8504
100	71.5815	2941.0	0.12612	172.4154	0.00428	195.52	108.30	0.00078	1.23255	0.6975
120	65.8245	3244.1	0.10665	149.3972	0.00410	205.41	111.47	0.00085	1.21261	0.7214
140	60.7642	3551.1	0.09333	133.3902	0.00390	207.23	110.13	0.00089	1.19525	0.7363
160	56.3265	3888.2	0.08335	122.5338	0.00367	206.43	107.55	0.00092	1.18017	0.7439
180	52.4598	4244.3	0.07560	114.7553	0.00343	205.40	105.10	0.00097	1.16714	0.7451
200	49.0968	4597.8	0.06973	108.6924	0.00321	205.03	103.27	0.00102	1.15588	0.7442
220	46.1349	4954.4	0.06517	104.0147	0.00301	205.58	102.14	0.00107	1.14602	0.7417
240	43.5148	5303.1	0.06161	100.1616	0.00284	207.01	101.67	0.00114	1.13736	0.7387
260	41.1807	5642.7	0.05879	96.9032	0.00268	209.22	101.79	0.00121	1.12967	0.7358
280	39.0881	5980.3	0.05652	94.2613	0.00254	212.10	102.40	0.00129	1.12281	0.7327
300	37.1199	6201.4	0.05645	92.1109	0.00243	215.52	103.42	0.00139	1.11639	0.7216
350	33.1220	5978.6	0.05335	88.0478	0.00214	225.99	107.33	0.00164	1.10341	0.7108
400	29.9249	7733.9	0.05124	85.0545	0.00192	238.43	112.59	0.00193	1.09312	0.7027
450	27.3054	8476.0	0.04972	82.7398	0.00175	252.21	118.71	0.00225	1.08473	0.6967
500	25.1167	9216.9	0.04850	80.8880	0.00160	267.07	125.40	0.00260	1.07776	0.6925
550	23.2585	9954.4	0.04752	79.3692	0.00148	282.59	132.45	0.00297	1.07186	0.6895
600	21.6597	10695.0	0.04669	78.1000	0.00137	298.67	139.75	0.00338	1.06681	0.6873
700	19.0469	12192.9	0.04529	76.0993	0.00121	332.09	154.77	0.00427	1.05858	0.6848
800	16.9995	13730.1	0.04407	74.5967	0.00107	366.90	170.04	0.00527	1.05217	0.6837
900	15.3506	15322.1	0.04291	73.4304	0.00097	403.01	185.35	0.00636	1.04703	0.6834
1000	13.9936	16978.4	0.04179	72.5016	0.00088	440.56	200.60	0.00755	1.04281	0.6834
1200	11.8910	20461.6	0.03969	71.1217	0.00075	511.00	230.64	0.01006	1.03629	0.6940
1400	10.3371	24181.4	0.03776	70.1517	0.00065	583.15	259.95	0.01285	1.03150	0.7044
1600	9.1420	28085.5	0.03607	69.4362	0.00058	656.11	288.47	0.01591	1.02782	0.7140
1800	8.1941	32119.6	0.03464	68.8884	0.00052	729.12	316.23	0.01923	1.02491	0.7223
2000	7.4239	36294.2	0.03339	68.4555	0.00047	803.28	343.27	0.02284	1.02255	0.7289
2500	6.0081	47999.6	0.03030	67.6646	0.00038	1019.85	408.13	0.03342	1.01822	0.7318
3000	5.0333	64091.2	0.02631	66.9921	0.00032	1379.81	470.07	0.04797	1.01525	0.7009

* TWO-PHASE BOUNDARY

Table 4.12 (continued)
THERMODYNAMIC PROPERTIES OF NORMAL HYDROGEN

70.00 MPA ISOBAR

TEMPERATURE DEG. K	VOLUME CU M/KG	ISOTHERM DERIVATIVE CU M-MPA/KG	ISOCHORE DERIVATIVE MPA/K	INTERNAL ENERGY KJ/KG-K	ENTHALPY KJ/KG	ENTROPY KJ/KG-K	CV KJ /	CP KG-K	VELOCITY OF SOUND M/S
* 29.318	0.01026	4.1801	1.389	311.2	1029.2	15.130	6.22	7.65	2266
30	0.01028	4.1590	1.385	314.8	1034.5	15.307	6.29	7.75	2264
31	0.01032	4.1278	1.380	320.2	1042.3	15.563	6.37	7.90	2261
32	0.01035	4.0961	1.374	325.6	1050.2	15.816	6.46	8.04	2258
33	0.01039	4.0640	1.368	331.2	1058.4	16.066	6.54	8.18	2254
34	0.01042	4.0316	1.361	336.9	1066.6	16.312	6.61	8.31	2251
35	0.01046	3.9989	1.354	342.7	1075.0	16.555	6.68	8.44	2247
36	0.01050	3.9660	1.346	348.6	1083.5	16.794	6.75	8.56	2243
37	0.01054	3.9329	1.339	354.6	1092.1	17.031	6.81	8.68	2239
38	0.01057	3.8996	1.331	360.7	1100.8	17.264	6.87	8.80	2235
39	0.01061	3.8663	1.323	366.8	1109.7	17.494	6.92	8.91	2231
40	0.01065	3.8328	1.315	373.1	1118.7	17.721	6.96	9.01	2227
42	0.01073	3.7658	1.298	385.8	1136.9	18.165	7.05	9.21	2218
44	0.01081	3.6990	1.280	398.8	1155.5	18.598	7.13	9.41	2209
46	0.01089	3.6326	1.262	411.6	1174.0	19.009	7.25	9.64	2198
48	0.01097	3.5669	1.244	425.3	1193.5	19.424	7.34	9.84	2188
50	0.01106	3.5023	1.225	439.2	1213.4	19.830	7.42	10.04	2177
52	0.01115	3.4389	1.206	453.4	1233.6	20.227	7.49	10.22	2166
54	0.01123	3.3770	1.186	467.9	1254.2	20.615	7.54	10.38	2156
56	0.01132	3.3169	1.167	482.8	1275.4	21.002	7.59	10.54	2146
58	0.01141	3.2586	1.147	497.6	1296.6	21.374	7.63	10.68	2136
60	0.01151	3.2024	1.127	512.6	1318.1	21.738	7.65	10.80	2126
65	0.01174	3.0721	1.076	550.8	1372.9	22.613	7.71	11.09	2102
70	0.01199	2.9572	1.026	589.7	1429.0	23.445	7.77	11.35	2078
75	0.01224	2.8583	0.9754	629.3	1486.3	24.236	7.82	11.56	2056
80	0.01251	2.8104	0.9375	669.7	1545.7	25.002	7.88	11.80	2051
85	0.01278	2.7389	0.8922	710.7	1605.1	25.723	7.96	11.99	2031
90	0.01304	2.6879	0.8500	752.4	1665.5	26.414	8.05	12.16	2016
95	0.01331	2.6388	0.8108	794.7	1726.7	27.075	8.14	12.33	2000
100	0.01359	2.5856	0.7738	837.6	1788.9	27.713	8.24	12.52	1982
120	0.01472	2.4419	0.6464	1018.0	2048.2	30.075	8.79	13.24	1918
140	0.01588	2.3428	0.5496	1206.6	2318.2	32.154	9.25	13.80	1870
160	0.01708	2.3086	0.4780	1403.3	2598.8	34.026	9.65	14.27	1848
180	0.01829	2.3219	0.4199	1606.9	2887.1	35.724	9.99	14.56	1840
200	0.01950	2.3430	0.3730	1815.3	3180.1	37.267	10.26	14.77	1837
220	0.02071	2.3767	0.3350	2027.6	3477.0	38.682	10.48	14.94	1840
240	0.02191	2.4198	0.3039	2242.8	3776.7	39.987	10.65	15.05	1849
260	0.02312	2.4681	0.2779	2460.1	4078.4	41.194	10.78	15.13	1861
280	0.02432	2.5196	0.2561	2678.9	4381.4	42.317	10.88	15.19	1876
300	0.02558	2.5891	0.2399	2882.6	4673.3	43.298	10.69	15.06	1909
350	0.02860	2.7609	0.2029	3422.6	5424.5	45.616	10.72	14.98	1965
400	0.03159	2.9409	0.1758	3960.1	6171.5	47.611	10.70	14.90	2023
450	0.03457	3.1257	0.1552	4494.7	6914.4	49.360	10.67	14.81	2083
500	0.03753	3.3133	0.1391	5026.8	7653.7	50.918	10.65	14.76	2143
550	0.04048	3.5030	0.1260	5557.0	8390.4	52.323	10.63	14.72	2202
600	0.04342	3.6941	0.1153	6086.3	9125.9	53.599	10.63	14.70	2260
700	0.04930	4.0800	0.0985	7144.3	10595.1	55.867	10.65	14.70	2373
800	0.05516	4.4695	0.0861	8206.3	12067.6	57.835	10.72	14.75	2481
900	0.06102	4.8621	0.0765	9276.7	13548.3	59.578	10.82	14.86	2583
1000	0.06688	5.2571	0.0689	10359.3	15041.1	61.148	10.97	15.01	2681
1200	0.07860	6.0529	0.0574	12577.4	18079.7	63.899	11.33	15.37	2865
1400	0.09033	6.8547	0.0493	14875.2	21198.4	66.309	11.76	15.80	3035
1600	0.10207	7.6609	0.0432	17257.0	24401.8	68.454	12.19	16.24	3195
1800	0.11381	8.4703	0.0384	19722.6	27689.3	70.384	12.59	16.65	3347
2000	0.12556	9.2822	0.0346	22271.6	31060.8	72.171	12.99	17.05	3491
2500	0.15502	11.3197	0.0277	29006.2	39857.6	76.078	14.15	18.27	3823
3000	0.18492	13.3644	0.0231	36539.5	49484.2	79.598	16.17	20.49	4114

* TWO-PHASE BOUNDARY

Table 4.12 (continued)
THERMODYNAMIC PROPERTIES OF NORMAL HYDROGEN

70.00 MPA ISOBAR

TEMPERATURE DEG. K	DENSITY KG/CU M	$V(DH/DV)_P$ KJ/KG	$V(DP/DU)_V$ MPA-CU M/KJ X 10^2	$-V(DP/DV)_T$ MPA	$(DV/DT)_P/V$ 1/DEG. K	THERMAL CONDUCTIVITY W/K-M X 10^3	VISCOSITY KG/M-S X 10^7	THERMAL DIFFUSIVITY SQ M/HR	DIELECTRIC CONSTANT	PRANDTL NUMBER
* 29.318	97.4967	2244.6	0.22878	407.5451	0.00341	235.03	706.81	0.00113	1.32496	2.2999
30	97.2698	2263.3	0.22652	404.5479	0.00342	236.48	664.21	0.00113	1.32413	2.1766
31	96.9362	2289.9	0.22328	400.1290	0.00345	237.38	609.63	0.00112	1.32292	2.0280
32	96.6013	2315.3	0.22019	395.6857	0.00347	237.96	562.84	0.00110	1.32170	1.9015
33	96.2654	2339.0	0.21730	391.2231	0.00350	238.28	522.44	0.00109	1.32047	1.7927
34	95.9283	2361.3	0.21457	386.7464	0.00352	238.44	487.31	0.00108	1.31925	1.6981
35	95.5903	2382.3	0.21197	382.2602	0.00354	238.41	456.58	0.00106	1.31802	1.6158
36	95.2513	2402.4	0.20946	377.7693	0.00356	238.20	429.54	0.00105	1.31678	1.5441
37	94.9113	2421.2	0.20708	373.2781	0.00359	237.84	405.62	0.00104	1.31555	1.4811
38	94.5704	2438.0	0.20490	368.7910	0.00361	237.34	384.35	0.00103	1.31431	1.4249
39	94.2287	2452.7	0.20294	364.3122	0.00363	236.74	365.36	0.00102	1.31307	1.3746
40	93.8860	2466.2	0.20109	359.8459	0.00365	236.04	348.32	0.00100	1.31183	1.3297
42	93.1984	2491.4	0.19752	350.9664	0.00370	234.32	319.11	0.00098	1.30934	1.2546
44	92.5077	2515.7	0.19399	342.1825	0.00374	232.35	295.08	0.00096	1.30684	1.1952
46	91.8141	2547.0	0.18972	333.5225	0.00378	230.22	275.04	0.00094	1.30433	1.1515
48	91.1180	2572.3	0.18605	325.0125	0.00383	227.97	258.15	0.00092	1.30182	1.1146
50	90.4196	2594.6	0.18267	316.6769	0.00387	225.65	243.76	0.00090	1.29930	1.0840
52	89.7192	2614.2	0.17954	308.5377	0.00391	223.30	231.37	0.00088	1.29678	1.0585
54	89.0173	2631.5	0.17664	300.6150	0.00395	220.93	220.63	0.00086	1.29425	1.0371
56	88.3142	2646.1	0.17400	292.9268	0.00398	218.56	211.24	0.00085	1.29173	1.0186
58	87.6105	2657.3	0.17168	285.4884	0.00402	216.21	202.97	0.00083	1.28920	1.0021
60	86.9066	2668.2	0.16943	278.3130	0.00405	214.18	195.64	0.00082	1.28668	0.9868
65	85.1501	2696.2	0.16388	261.5869	0.00411	210.11	180.52	0.00080	1.28040	0.9532
70	83.4061	2728.9	0.15830	246.6504	0.00416	205.92	168.78	0.00078	1.27418	0.9301
75	81.6851	2768.1	0.15264	233.4795	0.00418	201.63	159.41	0.00077	1.26807	0.9142
80	79.9139	2826.6	0.14884	224.5879	0.00417	197.14	151.41	0.00075	1.26180	0.9062
85	78.2693	2881.0	0.14325	214.3757	0.00416	196.25	145.08	0.00075	1.25599	0.8865
90	76.6635	2948.2	0.13782	206.0676	0.00413	195.08	139.71	0.00075	1.25033	0.8710
95	75.1048	3014.8	0.13266	198.1874	0.00409	193.81	135.10	0.00075	1.24486	0.8597
100	73.5871	3078.2	0.12758	190.2657	0.00407	202.20	112.22	0.00079	1.23955	0.6948
120	67.9484	3398.3	0.10820	165.9205	0.00390	212.38	115.50	0.00085	1.21994	0.7200
140	62.9718	3703.6	0.09441	147.5332	0.00373	213.99	114.00	0.00089	1.20280	0.7350
160	58.5562	4034.0	0.08463	135.1813	0.00354	212.77	111.15	0.00092	1.18773	0.7452
180	54.6784	4401.9	0.07689	126.9584	0.00331	211.28	108.41	0.00096	1.17460	0.7469
200	51.2917	4760.5	0.07086	120.1748	0.00310	210.46	106.29	0.00100	1.16322	0.7461
220	48.2948	5117.0	0.06619	114.7806	0.00292	210.60	104.90	0.00105	1.15320	0.7440
240	45.6339	5468.6	0.06252	110.4232	0.00275	211.66	104.21	0.00111	1.14436	0.7410
260	43.2550	5811.7	0.05960	106.7565	0.00260	213.54	104.13	0.00117	1.13650	0.7378
280	41.1164	5144.6	0.05725	103.5957	0.00247	216.15	104.58	0.00125	1.12946	0.7348
300	39.0924	5353.7	0.05737	101.2148	0.00237	219.35	105.45	0.00134	1.12283	0.7239
350	34.9676	7131.2	0.05413	96.5405	0.00210	229.41	109.11	0.00158	1.10939	0.7127
400	31.6542	7886.8	0.05191	93.0932	0.00189	241.65	114.23	0.00185	1.09868	0.7041
450	28.9300	8629.0	0.05030	90.4261	0.00172	255.39	120.30	0.00215	1.08993	0.6978
500	26.6473	9370.1	0.04901	88.2911	0.00158	270.32	126.99	0.00247	1.08263	0.6933
550	24.7046	10107.7	0.04797	86.5392	0.00146	286.01	134.10	0.00283	1.07645	0.6901
600	23.0296	10848.2	0.04709	85.0743	0.00135	302.34	141.50	0.00322	1.07114	0.6878
700	20.2852	12345.8	0.04561	82.7633	0.00119	336.43	156.81	0.00406	1.06247	0.6850
800	18.1285	13882.6	0.04433	81.0261	0.00106	372.11	172.46	0.00501	1.05570	0.6838
900	16.3874	15474.4	0.04313	79.6768	0.00096	409.24	188.22	0.00605	1.05026	0.6835
1000	14.9516	17130.6	0.04197	78.6020	0.00088	440.56	203.95	0.00707	1.04578	0.6948
1200	12.7219	20614.1	0.03983	77.0053	0.00075	511.00	235.07	0.00941	1.03886	0.7072
1400	11.0702	24334.5	0.03787	75.8835	0.00065	583.15	265.52	0.01200	1.03376	0.7195
1600	9.7974	28239.5	0.03615	75.0570	0.00058	656.10	295.22	0.01485	1.02983	0.7307
1800	8.7866	32273.7	0.03471	74.4252	0.00052	729.09	324.17	0.01794	1.02673	0.7404
2000	7.9643	36446.3	0.03345	73.9267	0.00047	803.14	352.41	0.02129	1.02420	0.7483
2500	6.4508	48113.9	0.03038	73.0211	0.00038	1018.17	420.22	0.03111	1.01957	0.7539
3000	5.4076	64025.0	0.02645	72.2696	0.00032	1371.13	485.01	0.04454	1.01639	0.7249

* TWO-PHASE BOUNDARY

Table 4.12 (continued)
THERMODYNAMIC PROPERTIES OF NORMAL HYDROGEN

80.00 MPA ISOBAR

TEMPERATURE DEG. K	VOLUME CU M/KG	ISOTHERM DERIVATIVE CU M-MPA/KG	ISOCHORE DERIVATIVE MPA/K	INTERNAL ENERGY KJ/KG-K	ENTHALPY KJ/KG	ENTROPY KJ/KG-K	CV KJ / KG-K	CP KJ / KG-K	VELOCITY OF SOUND M/S
* 30.999	0.01008	4.5298	1.441	327.6	1133.7	15.224	6.35	7.80	2358
31	0.01008	4.5298	1.441	327.6	1133.7	15.225	6.35	7.80	2358
32	0.01011	4.5007	1.435	332.9	1141.6	15.474	6.44	7.94	2355
33	0.01014	4.4712	1.428	338.3	1149.6	15.721	6.52	8.07	2352
34	0.01017	4.4412	1.421	343.8	1157.7	15.964	6.60	8.20	2349
35	0.01021	4.4109	1.414	349.4	1166.0	16.203	6.67	8.33	2346
36	0.01024	4.3802	1.406	355.1	1174.4	16.440	6.74	8.45	2343
37	0.01027	4.3491	1.399	360.9	1182.9	16.673	6.81	8.57	2339
38	0.01031	4.3178	1.391	366.8	1191.5	16.903	6.87	8.68	2336
39	0.01034	4.2863	1.382	372.8	1200.2	17.130	6.92	8.78	2332
40	0.01038	4.2545	1.374	378.9	1209.1	17.353	6.97	8.88	2328
42	0.01045	4.1906	1.357	391.2	1227.0	17.791	7.06	9.07	2321
44	0.01052	4.1263	1.339	403.8	1245.4	18.217	7.15	9.26	2313
46	0.01059	4.0620	1.322	416.3	1263.6	18.622	7.26	9.48	2303
48	0.01067	3.9979	1.303	429.5	1282.7	19.030	7.36	9.68	2293
50	0.01074	3.9342	1.285	443.1	1302.3	19.428	7.44	9.86	2283
52	0.01082	3.8712	1.266	456.9	1322.2	19.819	7.52	10.04	2273
54	0.01089	3.8092	1.247	471.0	1342.4	20.201	7.58	10.20	2263
56	0.01097	3.7484	1.228	485.5	1363.2	20.580	7.64	10.35	2254
58	0.01105	3.6888	1.209	500.0	1384.1	20.946	7.68	10.48	2244
60	0.01113	3.6308	1.190	514.7	1405.2	21.303	7.71	10.61	2235
65	0.01134	3.4936	1.141	551.9	1459.0	22.163	7.79	10.90	2212
70	0.01155	3.3690	1.092	590.1	1514.1	22.981	7.86	11.17	2188
75	0.01177	3.2580	1.043	629.0	1570.6	23.760	7.93	11.40	2164
80	0.01199	3.1609	0.9949	668.6	1628.1	24.503	8.00	11.61	2141
85	0.01222	3.0766	0.9473	708.8	1686.6	25.212	8.09	11.79	2118
90	0.01247	3.0327	0.9169	749.8	1747.7	25.911	8.17	12.05	2115
95	0.01271	2.9769	0.8751	791.5	1808.3	26.565	8.27	12.22	2097
100	0.01295	2.9212	0.8370	834.0	1869.9	27.197	8.38	12.40	2079
120	0.01393	2.7645	0.7095	1012.9	2127.2	29.540	8.93	13.17	2019
140	0.01493	2.6714	0.6055	1201.0	2395.5	31.608	9.39	13.68	1972
160	0.01596	2.5674	0.5274	1397.1	2673.8	33.464	9.78	14.19	1930
180	0.01701	2.5527	0.4682	1600.4	2961.4	35.158	10.11	14.58	1919
200	0.01807	2.5993	0.4183	1809.3	3254.9	36.704	10.38	14.77	1924
220	0.01912	2.6204	0.3764	2022.3	3551.9	38.119	10.60	14.95	1922
240	0.02017	2.6518	0.3419	2238.3	3851.8	39.424	10.76	15.07	1927
260	0.02122	2.6955	0.3131	2456.5	4153.9	40.634	10.89	15.14	1936
280	0.02226	2.7419	0.2886	2676.2	4457.2	41.757	10.98	15.20	1948
300	0.02336	2.8067	0.2723	2878.5	4747.6	42.732	10.76	15.09	1984
350	0.02600	2.9678	0.2305	3420.4	5500.5	45.054	10.78	15.02	2033
400	0.02862	3.1398	0.1999	3959.6	6249.0	47.054	10.75	14.92	2087
450	0.03122	3.3179	0.1765	4495.9	6993.1	48.806	10.72	14.84	2143
500	0.03380	3.4998	0.1581	5029.3	7733.5	50.366	10.70	14.78	2199
550	0.03638	3.6845	0.1433	5560.7	8471.1	51.772	10.68	14.73	2255
600	0.03895	3.8711	0.1310	6091.2	9207.2	53.050	10.67	14.71	2310
700	0.04408	4.2490	0.1120	7151.0	10677.3	55.320	10.69	14.70	2418
800	0.04920	4.6316	0.0979	8214.5	12150.3	57.288	10.75	14.76	2522
900	0.05431	5.0180	0.0870	9286.1	13631.1	59.030	10.86	14.86	2621
1000	0.05943	5.4076	0.0783	10369.6	15123.9	60.601	11.00	15.01	2716
1200	0.06966	6.1941	0.0653	12589.1	18162.1	63.352	11.36	15.37	2895
1400	0.07990	6.9883	0.0561	14888.0	21280.4	65.761	11.78	15.80	3062
1600	0.09016	7.7882	0.0491	17270.7	24483.1	67.906	12.20	16.24	3219
1800	0.10041	8.5924	0.0437	19736.9	27770.0	69.836	12.61	16.65	3368
2000	0.11068	9.3998	0.0394	22286.0	31140.6	71.622	13.00	17.05	3511
2500	0.13643	11.4287	0.0316	29015.2	39929.7	75.526	14.13	18.23	3840
3000	0.16255	13.4672	0.0264	36513.9	49518.2	79.032	16.06	20.36	4130

* TWO-PHASE BOUNDARY

Table 4.12 (continued)
THERMODYNAMIC PROPERTIES OF NORMAL HYDROGEN

80.00 MPA ISOBAR

TEMPERATURE DEG. K	DENSITY KG/CU M	$V(DH/DV)_P$ KJ/KG	$V(DP/DU)_V$ MPA-CU M/KJ $\times 10^2$	$-V(DP/DV)_T$ MPA	$(DV/DT)_P/V$ 1/DEG. K	THERMAL CONDUCTIVITY W/K-M $\times 10^3$	VISCOSITY KG/M-S $\times 10^7$	THERMAL DIFFUSIVITY SQ M/HR	DIELECTRIC CONSTANT	PRANDTL NUMBER
* 30.999	99.2474	2432.2	0.22854	449.5687	0.00321	250.16	780.58	0.00116	1.33135	2.4328
31	99.2472	2432.2	0.22853	449.5664	0.00321	250.16	780.54	0.00116	1.33135	2.4327
32	98.9287	2462.8	0.22519	445.2508	0.00322	250.86	713.79	0.00115	1.33019	2.2583
33	98.6096	2491.4	0.22207	440.9031	0.00324	251.28	656.67	0.00114	1.32902	2.1092
34	98.2898	2518.5	0.21912	436.5283	0.00326	251.52	607.43	0.00112	1.32785	1.9804
35	97.9695	2544.3	0.21632	432.1310	0.00327	251.55	564.71	0.00111	1.32668	1.8690
36	97.6487	2568.9	0.21362	427.7161	0.00329	251.39	527.40	0.00110	1.32551	1.7722
37	97.3273	2592.2	0.21108	423.2880	0.00330	251.06	494.63	0.00108	1.32434	1.6875
38	97.0055	2613.3	0.20874	418.8511	0.00332	250.59	465.69	0.00107	1.32317	1.6124
39	96.6832	2631.9	0.20664	414.4095	0.00334	250.00	440.01	0.00106	1.32199	1.5454
40	96.3604	2649.3	0.20466	409.9674	0.00335	249.30	417.12	0.00105	1.32082	1.4858
42	95.7136	2681.9	0.20086	401.0970	0.00338	247.56	378.19	0.00103	1.31846	1.3861
44	95.0651	2713.2	0.19713	392.2698	0.00341	245.56	346.50	0.00100	1.31611	1.3073
46	94.4152	2752.1	0.19268	383.5138	0.00345	243.37	320.34	0.00098	1.31375	1.2483
48	93.7638	2784.1	0.18886	374.8552	0.00348	241.05	298.49	0.00096	1.31139	1.1987
50	93.1112	2812.5	0.18536	366.3183	0.00351	238.65	280.02	0.00094	1.30902	1.1575
52	92.4575	2837.8	0.18212	357.9256	0.00354	236.21	264.26	0.00092	1.30666	1.1231
54	91.8029	2860.1	0.17914	349.6978	0.00357	233.74	250.70	0.00090	1.30429	1.0942
56	91.1477	2879.1	0.17644	341.6538	0.00360	231.28	238.92	0.00088	1.30193	1.0693
58	90.4922	2894.1	0.17408	333.8105	0.00362	228.83	228.63	0.00087	1.29956	1.0474
60	89.8367	2908.3	0.17179	326.1830	0.00365	226.72	219.57	0.00086	1.29720	1.0275
65	88.2001	2943.3	0.16619	308.1377	0.00370	222.47	201.10	0.00083	1.29132	0.9854
70	86.5721	2981.2	0.16059	291.6594	0.00375	218.08	186.99	0.00081	1.28548	0.9574
75	84.9602	3024.6	0.15488	276.8026	0.00377	213.57	175.89	0.00079	1.27972	0.9390
80	83.3722	3074.9	0.14908	263.5304	0.00378	209.13	166.95	0.00078	1.27406	0.9267
85	81.8151	3132.7	0.14320	251.7159	0.00376	208.18	159.60	0.00078	1.26853	0.9038
90	80.1666	3195.9	0.13998	243.1203	0.00377	206.50	152.93	0.00077	1.26269	0.8926
95	78.6848	3270.2	0.13448	234.2384	0.00374	205.16	147.72	0.00077	1.25745	0.8797
100	77.2243	3341.6	0.12939	225.5857	0.00371	214.76	119.62	0.00081	1.25231	0.6906
120	71.7961	3683.5	0.11069	198.4820	0.00357	225.62	123.22	0.00086	1.23330	0.7191
140	66.9708	4040.6	0.09627	178.9029	0.00338	226.93	121.47	0.00089	1.21656	0.7320
160	62.6633	4329.7	0.08607	160.8803	0.00328	225.00	118.14	0.00091	1.20175	0.7453
180	58.7805	4673.0	0.07881	150.0471	0.00312	222.67	114.85	0.00094	1.18850	0.7521
200	55.3429	5080.4	0.07284	143.8526	0.00291	221.02	112.22	0.00097	1.17685	0.7501
220	52.3020	5441.1	0.06793	137.0497	0.00275	220.39	110.35	0.00102	1.16661	0.7483
240	49.5830	5793.5	0.06408	131.4833	0.00260	220.77	109.24	0.00106	1.15750	0.7455
260	47.1312	6145.5	0.06101	127.0411	0.00246	222.05	108.79	0.00112	1.14933	0.7420
280	44.9199	6485.7	0.05850	123.1659	0.00234	224.14	108.91	0.00118	1.14200	0.7385
300	42.8013	6655.4	0.05913	120.1304	0.00227	226.90	109.52	0.00126	1.13500	0.7283
350	38.4605	7434.3	0.05562	114.1447	0.00202	236.19	112.68	0.00147	1.12076	0.7164
400	34.9447	8190.9	0.05319	109.7182	0.00182	248.02	117.51	0.00171	1.10932	0.7070
450	32.0350	8934.0	0.05141	106.2877	0.00166	261.65	123.45	0.00198	1.09991	0.7001
500	29.5836	9675.8	0.04998	103.5380	0.00153	276.70	130.14	0.00228	1.09202	0.6951
550	27.4879	10413.8	0.04882	101.2786	0.00141	292.71	137.36	0.00260	1.08531	0.6914
600	25.6738	11154.6	0.04784	99.3867	0.00132	309.52	144.94	0.00295	1.07953	0.6887
700	22.6868	12652.0	0.04621	96.3963	0.00116	344.94	160.82	0.00372	1.07005	0.6855
800	20.3262	14188.4	0.04482	94.1434	0.00104	382.37	177.25	0.00459	1.06260	0.6840
900	18.4117	15779.7	0.04354	92.3908	0.00094	421.57	193.92	0.00555	1.05659	0.6836
1000	16.8270	17435.7	0.04231	90.9932	0.00086	440.56	210.68	0.00628	1.05163	0.7177
1200	14.3550	20919.2	0.04007	88.9164	0.00073	511.00	244.07	0.00834	1.04393	0.7342
1400	12.5150	24640.5	0.03805	87.4585	0.00064	583.15	276.98	0.01062	1.03822	0.7504
1600	11.0919	28546.3	0.03629	86.3863	0.00057	656.09	309.25	0.01312	1.03382	0.7653
1800	9.9587	32580.2	0.03482	85.5686	0.00051	729.03	340.83	0.01583	1.03033	0.7783
2000	9.0348	36749.0	0.03355	84.9255	0.00046	802.90	371.72	0.01877	1.02749	0.7893
2500	7.3297	48350.7	0.03050	83.7689	0.00038	1015.30	446.17	0.02735	1.02226	0.8013
3000	6.1518	63952.0	0.02668	82.8475	0.00032	1356.25	517.44	0.03899	1.01866	0.7766

* TWO-PHASE BOUNDARY

Table 4.12 (continued)
THERMODYNAMIC PROPERTIES OF NORMAL HYDROGEN

90.00 MPA ISOBAR

TEMPERATURE DEG. K	VOLUME CU M/KG	ISOTHERM DERIVATIVE CU M-MPA/KG	ISOCHORE DERIVATIVE MPA/K	INTERNAL ENERGY KJ/KG-K	ENTHALPY KJ/KG	ENTROPY KJ/KG-K	CV KJ /	CP KG-K	VELOCITY OF SOUND M/S
* 32.614	0.00991	4.8726	1.492	344.2	1236.4	15.312	6.48	7.94	2444
33	0.00993	4.8621	1.489	346.2	1239.5	15.406	6.51	7.99	2443
34	0.00996	4.8345	1.482	351.6	1247.6	15.647	6.59	8.12	2441
35	0.00999	4.8065	1.474	357.0	1255.7	15.884	6.66	8.24	2438
36	0.01002	4.7780	1.466	362.5	1264.1	16.118	6.74	8.36	2435
37	0.01005	4.7491	1.458	368.2	1272.5	16.349	6.81	8.48	2432
38	0.01008	4.7198	1.450	373.9	1281.0	16.577	6.87	8.59	2429
39	0.01011	4.6902	1.441	379.8	1289.6	16.801	6.92	8.69	2426
40	0.01014	4.6602	1.432	385.6	1298.4	17.022	6.97	8.78	2423
42	0.01021	4.5996	1.415	397.7	1316.1	17.455	7.07	8.97	2416
44	0.01027	4.5382	1.397	410.0	1334.3	17.876	7.16	9.15	2409
46	0.01033	4.4764	1.379	422.1	1352.3	18.275	7.28	9.37	2400
48	0.01040	4.4143	1.360	435.1	1371.2	18.678	7.38	9.56	2391
50	0.01047	4.3522	1.342	448.3	1390.5	19.072	7.47	9.74	2382
52	0.01054	4.2904	1.323	461.8	1410.1	19.457	7.55	9.90	2373
54	0.01061	4.2290	1.304	475.6	1430.1	19.834	7.62	10.06	2363
56	0.01068	4.1684	1.286	489.8	1450.6	20.208	7.68	10.21	2354
58	0.01075	4.1087	1.267	504.0	1471.1	20.569	7.72	10.33	2345
60	0.01082	4.0500	1.248	518.3	1491.9	20.921	7.76	10.46	2337
65	0.01100	3.9093	1.201	554.9	1544.9	21.768	7.85	10.75	2314
70	0.01119	3.7786	1.153	592.3	1599.4	22.575	7.93	11.01	2291
75	0.01138	3.6597	1.105	630.6	1655.0	23.343	8.02	11.26	2267
80	0.01158	3.5532	1.058	669.7	1711.9	24.078	8.11	11.49	2244
85	0.01178	3.4592	1.011	709.5	1769.9	24.781	8.20	11.69	2220
90	0.01199	3.3767	0.9648	750.1	1828.8	25.455	8.31	11.87	2197
95	0.01219	3.3041	0.9199	791.3	1888.5	26.100	8.41	12.03	2174
100	0.01243	3.2490	0.8965	833.1	1951.6	26.746	8.51	12.33	2170
120	0.01329	3.0689	0.7615	1010.8	2207.0	29.072	9.07	13.07	2103
140	0.01418	2.9813	0.6595	1198.3	2474.8	31.135	9.53	13.64	2066
160	0.01508	2.8959	0.5722	1394.5	2751.4	32.980	9.93	14.04	2024
180	0.01599	2.7697	0.5084	1597.3	3036.3	34.658	10.24	14.54	1983
200	0.01693	2.7672	0.4587	1806.1	3329.7	36.204	10.50	14.86	1979
220	0.01787	2.8492	0.4157	2019.6	3627.7	37.624	10.71	14.97	1995
240	0.01880	2.8914	0.3780	2236.4	3928.3	38.932	10.88	15.07	2001
260	0.01972	2.9172	0.3463	2455.3	4230.4	40.142	11.00	15.16	2005
280	0.02065	2.9541	0.3196	2675.9	4534.0	41.266	11.09	15.22	2013
300	0.02163	3.0270	0.3042	2875.9	4822.6	42.233	10.82	15.11	2056
350	0.02397	3.1770	0.2578	3419.4	5576.8	44.560	10.84	15.04	2100
400	0.02529	3.3404	0.2236	3960.3	6326.7	46.563	10.81	14.95	2149
450	0.02860	3.5114	0.1975	4498.0	7072.0	48.318	10.77	14.86	2201
500	0.03090	3.6874	0.1769	5032.8	7813.4	49.879	10.74	14.80	2253
550	0.03318	3.8667	0.1603	5565.5	8551.8	51.288	10.72	14.75	2306
600	0.03546	4.0486	0.1466	6097.1	9288.6	52.567	10.71	14.72	2359
700	0.04001	4.4178	0.1253	7158.8	10759.5	54.838	10.72	14.71	2461
800	0.04455	4.7929	0.1096	8223.7	12232.9	56.807	10.78	14.76	2561
900	0.04908	5.1726	0.0974	9296.6	13713.9	58.549	10.89	14.86	2657
1000	0.05362	5.5561	0.0877	10381.2	15206.6	60.120	11.03	15.01	2749
1200	0.06269	6.3323	0.0731	12602.3	18244.4	62.870	11.38	15.37	2924
1400	0.07177	7.1179	0.0628	14902.4	21362.0	65.279	11.80	15.80	3087
1600	0.08087	7.9107	0.0550	17286.0	24564.0	67.423	12.23	16.23	3241
1800	0.08997	8.7088	0.0490	19752.8	27850.2	69.352	12.63	16.64	3388
2000	0.09909	9.5111	0.0442	22302.2	31219.9	71.138	13.01	17.04	3529
2500	0.12194	11.5300	0.0354	29027.3	40002.2	75.039	14.12	18.21	3855
3000	0.14512	13.5613	0.0296	36497.5	49558.7	78.534	15.98	20.24	4144

* TWO-PHASE BOUNDARY

Table 4.12 (continued)
THERMODYNAMIC PROPERTIES OF NORMAL HYDROGEN

90.00 MPA ISOBAR

TEMPERATURE DEG. K	DENSITY KG/CU M	$V(DH/DV)_P$ KJ/KG	$V(DP/DU)_V$ MPA-CU M/KJ X 10^2	$-V(DP/DV)_T$ MPA	$(DV/DT)_P/V$ 1/DEG. K	THERMAL CONDUCTIVITY W/K-M X 10^3	VISCOSITY KG/M-S X 10^7	THERMAL DIFFUSIVITY SQ M/HR	DIELECTRIC CONSTANT	PRANDTL NUMBER
* 32.614	100.8714	2616.2	0.22835	491.5039	0.00304	263.61	857.52	0.00118	1.33730	2.5837
33	100.7531	2628.8	0.22709	489.8700	0.00304	263.77	827.99	0.00118	1.33687	2.5088
34	100.4467	2660.7	0.22391	485.6133	0.00305	264.08	759.30	0.00117	1.33575	2.3345
35	100.1401	2691.2	0.22089	481.3228	0.00306	264.16	700.18	0.00115	1.33462	2.1847
36	99.8333	2720.4	0.21800	477.0030	0.00307	264.03	648.95	0.00114	1.33350	2.0552
37	99.5264	2748.2	0.21526	472.6586	0.00308	263.73	604.29	0.00113	1.33238	1.9423
38	99.2193	2773.5	0.21276	468.2939	0.00310	263.27	565.12	0.00111	1.33125	1.8429
39	98.9121	2796.3	0.21050	463.9132	0.00311	262.68	530.59	0.00110	1.33013	1.7545
40	98.6048	2817.6	0.20837	459.5205	0.00312	261.98	499.99	0.00109	1.32900	1.6762
42	97.9899	2857.7	0.20431	450.7152	0.00314	260.21	448.40	0.00107	1.32676	1.5458
44	97.3745	2896.3	0.20035	441.9079	0.00316	258.14	406.84	0.00104	1.32451	1.4428
46	96.7587	2942.9	0.19569	433.1268	0.00318	255.88	372.86	0.00102	1.32227	1.3650
48	96.1426	2981.9	0.19169	424.3983	0.00321	253.47	344.72	0.00099	1.32003	1.2998
50	95.5262	3016.8	0.18804	415.7470	0.00323	250.98	321.14	0.00097	1.31778	1.2457
52	94.9095	3048.0	0.18468	407.1958	0.00325	248.43	301.17	0.00095	1.31554	1.2006
54	94.2927	3075.8	0.18160	398.7661	0.00327	245.86	284.10	0.00093	1.31330	1.1626
56	93.6758	3099.8	0.17881	390.4774	0.00329	243.29	269.39	0.00092	1.31107	1.1301
58	93.0590	3119.1	0.17637	382.3478	0.00331	240.74	256.61	0.00090	1.30883	1.1016
60	92.4426	3137.1	0.17403	374.3937	0.00333	238.53	245.43	0.00089	1.30660	1.0759
65	90.9041	3180.6	0.16835	355.3682	0.00338	234.09	222.87	0.00086	1.30105	1.0230
70	89.3730	3225.6	0.16270	337.7071	0.00341	229.50	205.89	0.00084	1.29553	0.9880
75	87.8545	3275.1	0.15697	321.5186	0.00344	224.78	192.71	0.00082	1.29008	0.9654
80	86.3545	3331.0	0.15114	306.8348	0.00345	220.13	182.23	0.00080	1.28471	0.9508
85	84.8790	3394.7	0.14520	293.6117	0.00344	219.14	173.71	0.00080	1.27943	0.9266
90	83.4336	3466.3	0.13922	281.7316	0.00342	217.84	166.66	0.00079	1.27428	0.9082
95	82.0230	3542.8	0.13337	271.0114	0.00339	216.42	160.73	0.00079	1.26927	0.8931
100	80.4651	3595.2	0.13095	261.4332	0.00343	226.40	126.52	0.00082	1.26375	0.6890
120	75.2353	3964.0	0.11162	230.8886	0.00330	238.04	130.51	0.00087	1.24532	0.7168
140	70.5036	4347.3	0.09814	210.1927	0.00314	239.20	128.61	0.00090	1.22880	0.7334
160	66.3292	4712.9	0.08689	192.0795	0.00298	236.67	124.87	0.00091	1.21435	0.7407
180	62.5400	4953.2	0.07936	173.2180	0.00293	233.60	121.11	0.00092	1.20132	0.7537
200	59.0710	5294.8	0.07396	163.4613	0.00281	231.20	118.00	0.00095	1.18948	0.7584
220	55.9659	5744.0	0.06932	159.4606	0.00261	229.88	115.69	0.00099	1.17895	0.7536
240	53.1923	6131.3	0.06533	153.7984	0.00246	229.63	114.19	0.00103	1.16960	0.7494
260	50.7026	6474.3	0.06209	147.9087	0.00234	230.36	113.39	0.00108	1.16124	0.7462
280	48.4373	6813.4	0.05949	143.0900	0.00223	231.97	113.22	0.00113	1.15368	0.7427
300	46.2315	6953.3	0.06079	139.9443	0.00217	234.33	113.57	0.00121	1.14634	0.7325
350	41.7167	7734.5	0.05702	132.5354	0.00194	242.87	116.24	0.00139	1.13143	0.7200
400	38.0322	8492.9	0.05440	127.0413	0.00176	254.30	120.78	0.00161	1.11936	0.7100
450	34.9644	9237.3	0.05245	122.7753	0.00161	267.79	126.58	0.00186	1.10938	0.7024
500	32.3671	9980.5	0.05088	119.3503	0.00148	282.94	133.26	0.00213	1.10098	0.6969
550	30.1370	10719.4	0.04962	116.5315	0.00138	299.23	140.56	0.00242	1.09380	0.6928
600	28.1996	11460.9	0.04854	114.1675	0.00128	316.49	148.32	0.00274	1.08759	0.6898
700	24.9947	12958.8	0.04676	110.4230	0.00114	353.22	164.75	0.00346	1.07737	0.6861
800	22.4485	14495.1	0.04526	107.5944	0.00102	392.42	181.95	0.00426	1.06930	0.6843
900	20.3744	16086.2	0.04389	105.3896	0.00092	433.77	199.57	0.00516	1.06276	0.6837
1000	18.6513	17742.0	0.04260	103.6291	0.00085	440.56	217.40	0.00567	1.05734	0.7405
1200	15.9516	21225.1	0.04027	101.0104	0.00072	511.00	253.25	0.00750	1.04890	0.7617
1400	13.9327	24946.6	0.03819	99.1723	0.00063	583.15	288.89	0.00954	1.04262	0.7825
1600	12.3659	28852.5	0.03640	97.8224	0.00056	656.09	324.06	0.01177	1.03776	0.8017
1800	11.1146	32885.2	0.03491	96.7950	0.00051	728.98	358.65	0.01419	1.03389	0.8189
2000	10.0923	37049.9	0.03362	95.9893	0.00046	802.71	392.61	0.01680	1.03074	0.8335
2500	8.2006	48594.9	0.03058	94.5528	0.00037	1012.93	474.83	0.02443	1.02493	0.8534
3000	6.8906	63938.3	0.02687	93.4458	0.00032	1343.93	553.83	0.03469	1.02092	0.8341

* TWO-PHASE BOUNDARY

Table 4.12 (continued)
THERMODYNAMIC PROPERTIES OF NORMAL HYDROGEN

100 MPA ISOBAR

TEMPERATURE DEG. K	VOLUME CU M/KG	ISOTHERM DERIVATIVE CU M-MPA/KG	ISOCHORE DERIVATIVE MPA/K	INTERNAL ENERGY KJ/KG-K	ENTHALPY KJ/KG	ENTROPY KJ/KG-K	CV KJ /	CP KG-K	VELOCITY OF SOUND M/S
* 34.169	0.00977	5.2095	1.542	360.9	1337.6	15.394	6.59	8.08	2527
35	0.00979	5.1881	1.535	365.3	1344.3	15.590	6.66	8.18	2525
36	0.00982	5.1618	1.526	370.7	1352.6	15.822	6.73	8.30	2522
37	0.00985	5.1350	1.517	376.2	1360.9	16.051	6.80	8.41	2520
38	0.00988	5.1078	1.508	381.8	1369.4	16.277	6.87	8.52	2517
39	0.00990	5.0801	1.499	387.5	1378.0	16.500	6.92	8.62	2514
40	0.00993	5.0520	1.490	393.2	1386.6	16.719	6.98	8.71	2512
42	0.00999	4.9948	1.472	405.0	1404.2	17.148	7.07	8.89	2506
44	0.01005	4.9365	1.453	417.0	1422.2	17.566	7.17	9.07	2499
46	0.01011	4.8773	1.434	428.9	1440.0	17.961	7.30	9.28	2491
48	0.01017	4.8175	1.415	441.6	1458.8	18.360	7.40	9.46	2482
50	0.01023	4.7573	1.396	454.5	1477.8	18.750	7.49	9.64	2474
52	0.01029	4.6971	1.377	467.8	1497.3	19.131	7.57	9.80	2465
54	0.01036	4.6369	1.359	481.3	1517.0	19.504	7.65	9.95	2457
56	0.01042	4.5771	1.340	495.3	1537.3	19.874	7.71	10.09	2448
58	0.01048	4.5178	1.321	509.2	1557.6	20.231	7.75	10.22	2440
60	0.01055	4.4591	1.302	523.3	1578.2	20.579	7.80	10.34	2431
65	0.01071	4.3169	1.255	559.2	1630.6	21.416	7.89	10.62	2410
70	0.01088	4.1826	1.209	596.1	1684.4	22.214	7.99	10.89	2387
75	0.01106	4.0581	1.162	633.8	1739.4	22.973	8.09	11.14	2364
80	0.01123	3.9448	1.116	672.5	1795.8	23.700	8.19	11.38	2341
85	0.01141	3.8431	1.070	711.9	1853.2	24.397	8.30	11.60	2317
90	0.01160	3.7530	1.024	752.2	1911.7	25.067	8.42	11.80	2294
95	0.01178	3.6737	0.9796	793.2	1971.2	25.709	8.53	11.98	2271
100	0.01197	3.6035	0.9357	834.9	2031.5	26.327	8.66	12.13	2248
120	0.01276	3.3346	0.8066	1011.2	2287.3	28.656	9.22	13.03	2171
140	0.01355	3.1987	0.7030	1198.0	2553.0	30.703	9.68	13.65	2124
160	0.01437	3.1972	0.6180	1394.3	2831.2	32.560	10.07	14.02	2109
180	0.01517	3.1143	0.5439	1597.5	3114.2	34.226	10.39	14.33	2072
200	0.01597	2.9502	0.4914	1806.0	3403.4	35.749	10.64	14.82	2027
220	0.01684	2.9419	0.4488	2019.3	3703.3	37.179	10.84	15.11	2025
240	0.01768	3.0865	0.4119	2236.5	4004.7	38.491	11.00	15.12	2060
260	0.01852	3.1323	0.3779	2456.3	4307.9	39.705	11.12	15.18	2068
280	0.01934	3.1686	0.3488	2677.7	4611.7	40.831	11.21	15.23	2075
300	0.02024	3.2500	0.3354	2874.6	4898.1	41.788	10.88	15.14	2126
350	0.02234	3.3883	0.2846	3419.5	5653.5	44.118	10.89	15.07	2165
400	0.02443	3.5426	0.2470	3961.9	6404.6	46.124	10.86	14.97	2210
450	0.02650	3.7063	0.2182	4501.0	7151.0	47.882	10.82	14.88	2257
500	0.02856	3.8759	0.1955	5037.2	7893.4	49.446	10.79	14.81	2306
550	0.03061	4.0497	0.1771	5571.1	8632.6	50.855	10.77	14.76	2356
600	0.03266	4.2264	0.1620	6103.8	9369.9	52.135	10.76	14.73	2406
700	0.03674	4.5866	0.1385	7167.4	10841.7	54.407	10.76	14.72	2504
800	0.04082	4.9536	0.1210	8234.0	12315.5	56.377	10.82	14.76	2599
900	0.04488	5.3261	0.1076	9308.2	13796.6	58.120	10.92	14.86	2692
1000	0.04895	5.7030	0.0968	10393.9	15289.2	59.690	11.06	15.01	2781
1200	0.05710	6.4678	0.0808	12616.8	18326.5	62.440	11.41	15.37	2951
1400	0.06525	7.2439	0.0694	14918.2	21443.4	64.848	11.83	15.79	3110
1600	0.07342	8.0286	0.0609	17302.8	24644.6	66.992	12.25	16.23	3261
1800	0.08160	8.8200	0.0542	19770.4	27930.0	68.921	12.65	16.64	3406
2000	0.08978	9.6165	0.0489	22320.2	31298.6	70.706	13.03	17.04	3546
2500	0.11033	11.6239	0.0392	29042.2	40074.7	74.605	14.12	18.18	3868
3000	0.13115	13.6469	0.0328	36488.4	49603.8	78.089	15.91	20.14	4156

* TWO-PHASE BOUNDARY

Table 4.12 (continued)
THERMODYNAMIC PROPERTIES OF NORMAL HYDROGEN

100 MPA ISOBAR

TEMPERATURE DEG. K	DENSITY KG/CU M	$V(DH/DV)_P$ KJ/KG	$V(DP/DU)_V$ MPA-CU M/KJ X 10^2	$-V(DP/DV)_T$ MPA	$(DV/DT)_P/V$ 1/DEG. K	THERMAL CONDUCTIVITY W/K-M X 10^3	VISCOSITY KG/M-S X 10^7	THERMAL DIFFUSIVITY SQ M/HR	DIELECTRIC CONSTANT	PRANDTL NUMBER
* 34.169	102.3875	2795.8	0.22835	533.3893	0.00289	276.25	937.81	0.00120	1.34287	2.7431
35	102.1417	2824.8	0.22567	529.9203	0.00290	276.34	870.98	0.00119	1.34197	2.5784
36	101.8460	2858.7	0.22256	525.7090	0.00290	276.25	800.91	0.00118	1.34088	2.4059
37	101.5504	2890.9	0.21963	521.4625	0.00291	275.96	740.27	0.00116	1.33980	2.2564
38	101.2550	2920.6	0.21694	517.1855	0.00292	275.51	687.46	0.00115	1.33871	2.1255
39	100.9598	2947.4	0.21451	512.8820	0.00292	274.91	641.22	0.00114	1.33763	2.0097
40	100.6647	2972.7	0.21222	508.5564	0.00293	274.19	600.50	0.00113	1.33655	1.9077
42	100.0751	3020.5	0.20787	499.8546	0.00294	272.37	532.43	0.00110	1.33439	1.7385
44	99.4861	3066.5	0.20366	491.1104	0.00296	270.23	478.19	0.00108	1.33223	1.6054
46	98.8978	3120.9	0.19876	482.3525	0.00297	267.87	434.26	0.00105	1.33008	1.5043
48	98.3099	3167.1	0.19456	473.6076	0.00299	265.37	398.20	0.00103	1.32793	1.4201
50	97.7226	3208.6	0.19074	464.9006	0.00300	262.77	368.22	0.00100	1.32578	1.3505
52	97.1359	3246.1	0.18723	456.2550	0.00302	260.11	343.02	0.00098	1.32364	1.2924
54	96.5496	3279.6	0.18402	447.6927	0.00303	257.43	321.63	0.00096	1.32151	1.2435
56	95.9639	3308.8	0.18113	439.2340	0.00305	254.74	303.31	0.00095	1.31937	1.2017
58	95.3788	3332.7	0.17861	430.8979	0.00307	252.06	287.50	0.00093	1.31725	1.1653
60	94.7943	3355.1	0.17619	422.7017	0.00308	249.75	273.74	0.00092	1.31512	1.1329
65	93.3370	3408.0	0.17038	402.9244	0.00312	245.12	246.24	0.00089	1.30984	1.0668
70	91.8870	3461.0	0.16466	384.3249	0.00315	240.31	225.80	0.00086	1.30459	1.0229
75	90.4480	3517.6	0.15891	367.0497	0.00317	235.38	210.14	0.00084	1.29940	0.9945
80	89.0241	3580.0	0.15308	351.1813	0.00318	230.52	197.82	0.00082	1.29428	0.9764
85	87.6203	3650.1	0.14713	336.7366	0.00318	229.50	187.92	0.00081	1.28924	0.9498
90	86.2414	3728.6	0.14112	323.6679	0.00317	228.14	179.81	0.00081	1.28430	0.9301
95	84.8920	3813.3	0.13521	311.8664	0.00314	226.66	173.06	0.00080	1.27948	0.9145
100	83.5760	3905.6	0.12934	301.1690	0.00311	237.25	132.99	0.00084	1.27479	0.6802
120	78.3682	4220.7	0.11169	261.3250	0.00309	249.74	137.43	0.00088	1.25634	0.7169
140	73.7994	4582.8	0.09844	236.0605	0.00298	250.87	135.46	0.00090	1.24029	0.7369
160	69.5909	5047.0	0.08817	222.4929	0.00278	247.85	131.38	0.00091	1.22563	0.7431
180	65.9363	5409.2	0.07936	205.3481	0.00265	244.12	127.19	0.00093	1.21299	0.7464
200	62.6024	5569.0	0.07378	184.6887	0.00266	241.05	123.64	0.00094	1.20154	0.7601
220	59.3817	5881.1	0.06974	174.6977	0.00257	239.08	120.93	0.00096	1.19054	0.7643
240	56.5559	6408.2	0.06623	174.5602	0.00236	238.25	119.06	0.00100	1.18095	0.7556
260	54.0061	6795.8	0.06294	169.1637	0.00223	238.47	117.94	0.00105	1.17234	0.7508
280	51.7071	7152.7	0.06019	163.8406	0.00213	239.64	117.48	0.00110	1.16461	0.7464
300	49.4193	7247.8	0.06236	160.6133	0.00209	241.61	117.58	0.00116	1.15695	0.7366
350	44.7643	8031.9	0.05835	151.6740	0.00188	249.44	119.78	0.00133	1.14148	0.7236
400	40.9388	8792.6	0.05553	145.0297	0.00170	260.47	124.03	0.00153	1.12888	0.7130
450	37.7361	9539.0	0.05343	139.8609	0.00156	273.81	129.68	0.00176	1.11840	0.7048
500	35.0120	10284.1	0.05173	135.7041	0.00144	289.02	136.33	0.00201	1.10953	0.6987
550	32.6639	11024.5	0.05036	132.2775	0.00134	305.58	143.70	0.00228	1.10193	0.6942
600	30.6169	11767.1	0.04919	129.3991	0.00125	323.26	151.62	0.00258	1.09534	0.6909
700	27.2161	13266.4	0.04726	124.8291	0.00111	361.28	168.59	0.00325	1.08444	0.6867
800	24.5007	14803.2	0.04565	121.3674	0.00100	402.26	186.57	0.00400	1.07580	0.6846
900	22.2795	16394.3	0.04420	118.6628	0.00091	445.84	205.17	0.00485	1.06876	0.6839
1000	20.4277	18050.0	0.04285	116.4995	0.00083	440.56	224.14	0.00517	1.06292	0.7634
1200	17.5141	21532.6	0.04044	113.2774	0.00071	511.00	262.64	0.00684	1.05378	0.7897
1400	15.3253	25253.6	0.03830	111.0148	0.00063	583.15	301.29	0.00867	1.04695	0.8159
1600	13.6206	29158.7	0.03648	109.3547	0.00056	656.08	339.74	0.01069	1.04165	0.8403
1800	12.2556	33189.5	0.03496	108.0935	0.00050	728.94	377.78	0.01287	1.03742	0.8624
2000	11.1379	37349.7	0.03366	107.1068	0.00046	802.55	415.31	0.01523	1.03397	0.8816
2500	9.0641	48844.3	0.03064	105.3609	0.00037	1010.92	506.73	0.02208	1.02758	0.9113
3000	7.6246	63967.6	0.02702	104.0528	0.00031	1333.50	595.02	0.03126	1.02316	0.8988

* TWO-PHASE BOUNDARY

Table 4.13

THERMODYNAMIC PROPERTIES OF COEXISTING GASEOUS AND LIQUID NORMAL HYDROGEN

Temperature, K	Pressure, MPA	Isotherm derivative, m^3-MPA/kg	Isochore derivative, MPA/K	Internal energy, kJ/kg-K	Enthalpy, kJ/kg	Entropy, kJ/kg-K	CV, kJ/kg-K	CP, kJ/kg-K	Velocity of sound, m/sec	Surface tension, N/m × 10^3
13.950[a]	0.0072	1.1729	0.9234	218.2	218.3	14.079	4.67	6.36	1264	3.0033
	0.0072	0.0552	0.0005	611.4	667.4	46.635	6.21	10.52	306	
14	0.0074	1.1492	0.9234	219.5	219.6	14.173	4.72	6.47	1256	2.9950
	0.0074	0.0558	0.0006	612.5	669.3	46.301	6.21	10.54	308	
15	0.0127	1.0736	0.9064	226.3	226.4	14.640	4.92	6.91	1228	2.8290
	0.0127	0.0588	0.0009	617.9	678.2	44.763	6.24	10.67	317	
16	0.0204	1.0027	0.8893	233.5	233.8	15.104	5.12	7.36	1200	2.6635
	0.0204	0.0615	0.0014	623.0	686.7	43.418	6.27	10.85	326	
17	0.0314	0.9176	0.8750	241.1	241.6	15.568	5.30	7.88	1168	2.4987
	0.0314	0.0637	0.0021	627.8	694.7	42.227	6.32	11.07	334	
18	0.0461	0.8441	0.8612	249.2	249.9	16.032	5.47	8.42	1140	2.3346
	0.0461	0.0654	0.0029	632.2	702.1	41.158	6.37	11.34	341	
19	0.0654	0.7893	0.8465	257.8	258.8	16.498	5.62	8.93	1120	2.1712
	0.0654	0.0667	0.0040	636.3	708.8	40.188	6.43	11.66	348	
20	0.0901	0.7364	0.8300	266.9	268.3	16.966	5.75	9.45	1100	2.0085
	0.0901	0.0674	0.0054	639.8	714.8	39.299	6.49	12.04	354	
20.390[a]	0.1013	0.7093	0.8253	269.5	270.9	17.093	5.78	9.66	1089	1.9453
	0.1013	0.0675	0.0058	640.7	716.5	39.083	6.50	12.15	355	
21	0.1208	0.6619	0.8119	276.6	278.4	17.440	5.86	10.13	1070	1.8466
	0.1208	0.0675	0.0071	643.0	720.2	38.485	6.55	12.49	359	
22	0.1585	0.5993	0.7913	286.8	289.2	17.919	5.95	10.82	1044	1.6856
	0.1585	0.0670	0.0092	645.5	724.4	37.710	6.61	13.03	363	
23	0.2039	0.5264	0.7673	297.7	300.8	18.405	6.03	11.69	1010	1.5255
	0.2039	0.0658	0.0117	647.4	727.6	36.973	6.68	13.69	367	

[a] Two-phase boundary.

Table 4.13 (continued)
THERMODYNAMIC PROPERTIES OF COEXISTING GASEOUS AND LIQUID NORMAL HYDROGEN

Temperature, K	Pressure, MPA	Isotherm derivative, m^3-MPA/kg	Isochore derivative, MPA/K	Internal energy, kJ/kg-K	Enthalpy, kJ/kg	Entropy, kJ/kg-K	CV, kJ/kg-K	CP, kJ/kg-K	Velocity of sound, m/sec	Surface tension, N/m × 10^3
24	0.2579	0.4675	0.7389	309.3	313.3	18.901	6.09	12.52	980	1.3664
	0.2579	0.0639	0.0148	648.5	729.8	36.266	6.74	14.49	371	
25	0.3213	0.4108	0.7060	321.6	326.7	19.408	6.14	13.44	948	1.2085
	0.3213	0.0613	0.0186	648.9	730.7	35.579	6.81	15.52	373	
26	0.3950	0.3410	0.6676	334.8	341.2	19.929	6.20	14.80	903	1.0518
	0.3950	0.0578	0.0231	648.2	730.2	34.900	6.89	16.85	376	
27	0.4800	0.2858	0.6251	349.0	357.0	20.473	6.26	16.17	860	0.8965
	0.4800	0.0533	0.0288	646.5	728.0	34.221	6.99	18.66	377	
28	0.5770	0.2225	0.5794	364.4	374.3	21.041	6.33	18.48	806	0.7429
	0.5770	0.0479	0.0359	643.2	723.7	33.524	7.11	21.24	378	
29	0.6872	0.1624	0.5300	381.2	393.6	21.650	6.41	22.05	747	0.5912
	0.6872	0.0414	0.0449	638.1	716.6	32.795	7.28	25.19	378	
30	0.8116	0.1163	0.4759	400.2	415.4	22.315	6.51	26.59	689	0.4418
	0.8116	0.0334	0.0569	630.4	705.9	32.002	7.50	31.99	378	
31	0.9510	0.0697	0.4149	422.3	441.3	23.075	6.66	36.55	619	0.2955
	0.9510	0.0237	0.0736	618.6	689.7	31.091	7.82	46.56	375	
32	1.1068	0.0302	0.3416	450.4	474.7	24.032	6.91	65.37	534	0.1537
	1.1068	0.0129	0.0987	599.2	663.2	29.926	8.11	87.02	372	
33.180[a]	1.3130		0.1898	524.3	565.4	26.680				0.0000

[a] Two-phase boundary.

Table 4.13 (continued)
THERMODYNAMIC PROPERTIES OF COEXISTING GASEOUS AND LIQUID NORMAL HYDROGEN

Temperature, K	Density, kg/m^3	$V(DH/DV)_P$, kJ/kg	$V(DP/DU)_V$, MPA-m^3/kJ $\times 10^2$	$-V(DP/DV)_T$, MPA	$(DV/DT)_P/V$, l/K	Thermal conductivity W/K-m $\times 10^3$	Viscosity, kg/m-sec $\times 10^7$	Thermal diffusivity, m^2/hr	Dielectric constant	Prandtl number
13.950[a]	77.0235	622.6	0.25654	90.3401	0.01022	72.59	255.05	0.00053	1.25160	2.2359
	0.1254	139.8	0.06684	0.0069	0.07523	12.43	7.36	0.03395	1.00038	0.6230
14	76.8554	619.2	0.25480	88.3211	0.01046	74.62	248.01	0.00054	1.25101	2.1517
	0.1388	141.6	0.06684	0.0077	0.07441	12.54	7.48	0.03084	1.00042	0.6291
15	75.9963	621.9	0.24237	81.5880	0.01111	83.02	217.84	0.00057	1.24799	1.8128
	0.2228	150.5	0.06683	0.0131	0.07089	13.08	8.08	0.01981	1.00067	0.6591
16	75.1030	623.0	0.23130	75.3083	0.01181	88.86	194.15	0.00058	1.24486	1.6074
	0.3385	159.1	0.06680	0.0208	0.06820	13.66	8.68	0.01339	1.00102	0.6891
17	74.1710	613.1	0.22244	68.0612	0.01286	92.71	175.07	0.00057	1.24159	1.4884
	0.4922	167.1	0.06674	0.0313	0.06624	14.27	9.28	0.00943	1.00148	0.7197
18	73.1950	604.3	0.21509	61.7874	0.01394	95.43	159.34	0.00056	1.23818	1.4063
	0.6902	174.6	0.06669	0.0452	0.06495	14.97	9.88	0.00689	1.00208	0.7484
19	72.1693	601.0	0.20874	56.9619	0.01486	97.00	146.10	0.00054	1.23460	1.3451
	0.9395	181.4	0.06665	0.0626	0.06427	15.72	10.49	0.00517	1.00283	0.7783
20	71.0870	596.1	0.20314	52.3516	0.01585	98.40	134.76	0.00053	1.23083	1.2942
	1.2474	187.6	0.06666	0.0840	0.06420	16.64	11.11	0.00399	1.00376	0.8042
20.390[a]	70.7864	587.9	0.20176	50.2079	0.01644	98.92	131.98	0.00052	1.22978	1.2893
	1.3378	189.1	0.06667	0.0903	0.06424	16.94	11.29	0.00375	1.00404	0.8091
21	69.9403	577.8	0.19814	46.2935	0.01754	100.05	124.87	0.00051	1.22684	1.2648
	1.6189	193.0	0.06672	0.1093	0.06471	17.79	11.28	0.00317	1.00489	0.8241
22	68.7200	563.1	0.19347	41.1835	0.01921	100.95	116.11	0.00049	1.22261	1.2444
	2.0711	197.5	0.06686	0.1388	0.06597	19.02	12.38	0.00254	1.00625	0.8481
23	67.4149	540.5	0.18882	35.4854	0.02162	101.20	108.22	0.00046	1.21810	1.2499
	2.6119	201.2	0.06708	0.1719	0.06804	20.36	13.04	0.00205	1.00789	0.8769

[a] Two-phase boundary.

Table 4.13 (continued)
THERMODYNAMIC PROPERTIES OF COEXISTING GASEOUS AND LIQUID NORMAL HYDROGEN

Temperature, K	Density, kg/m³	$V(DH/DV)_P$, kJ/kg	$V(DP/DU)_V$, MPA-m³/kJ × 10²	$-V(DP/DV)_T$, MPA	$(DV/DT)_P/V$, 1/K	Thermal conductivity W/K-m × 10³	Viscosity, kg/m-sec × 10⁷	Thermal diffusivity, m²/hr	Dielectric constant	Prandtl number
24	66.0112	523.0	0.18378	30.8574	0.02394	100.84	101.01	0.00044	1.21325	1.2543
	3.2548	203.9	0.06739	0.2081	0.07108	21.81	13.73	0.00166	1.00984	0.9123
25	64.4917	504.2	0.17820	26.4945	0.02665	99.93	94.32	0.00042	1.20802	1.2682
	4.0171	205.8	0.06777	0.2460	0.07540	23.43	14.44	0.00135	1.01216	0.9567
26	62.8337	475.1	0.17151	21.4244	0.03116	98.43	88.03	0.00038	1.20233	1.3240
	4.9215	206.9	0.06822	0.2842	0.08144	25.24	15.19	0.00110	1.01491	1.0145
27	61.0065	451.1	0.16381	17.4346	0.03585	96.38	82.02	0.00035	1.19608	1.3764
	5.9999	207.3	0.06868	0.3201	0.09002	27.36	16.00	0.00088	1.01820	1.0916
28	58.9665	418.4	0.15535	13.1188	0.04416	93.78	76.19	0.00031	1.18913	1.5012
	7.2979	207.1	0.06912	0.3498	0.10259	29.91	16.89	0.00069	1.02216	1.1997
29	56.6460	382.6	0.14600	9.1967	0.05763	90.55	70.42	0.00026	1.18125	1.7146
	8.8866	206.2	0.06947	0.3676	0.12220	33.18	17.91	0.00053	1.02703	1.3601
30	53.9303	350.6	0.13555	6.2743	0.07585	86.57	64.57	0.00022	1.17208	1.9834
	10.8872	204.7	0.06965	0.3638	0.15631	37.87	19.17	0.00039	1.03319	1.6195
31	50.5892	310.8	0.12321	3.5282	0.11759	85.85	58.40	0.00017	1.16087	2.4864
	13.5411	202.6	0.06952	0.3205	0.22979	46.48	20.89	0.00027	1.04140	2.0930
32	45.9927	265.8	0.10745	1.3886	0.24599	91.46	51.31	0.00011	1.14555	3.6672
	17.4983	199.1	0.06953	0.2258	0.43710	66.77	23.79	0.00016	1.05373	3.1004
33.180[a]	31.4285	199.0					35.43		1.09795	

[a] Two-phase boundary.

Unit Conversions, Physical Constants, and Symbols

Chapter 5
UNIT CONVERSIONS, PHYSICAL CONSTANTS, AND SYMBOLS

A complete description of the International System of Units is given in Page, C. H. and Vigoureux, P., National Bureau of Standards, Special Publ. 330, January 1971. A good general reference for physical constants and conversion factors is Mechtly, E. A., National Aeronautics and Space Administration, SP-7012, 1964. Selected unit conversion and physical constants are given in the following table.

CONVERSION FACTORS

Density

lb/ft^3	kg/m^3	g/cm^3	mol/cm^3	Amagat
1	16.018	0.016018	7.9458×10^{-3}	178.216
0.062428	1	0.001	4.9605×10^{-4}.	11.126
62.428	1,000	1	0.49605	1.1126×10^4
125.85	2,015.9	2.0159	1	2.2428×10^4
5.6111×10^{-3}	0.089881	8.9881×10^{-5}	4.4586×10^{-5}	1

Specific Volume

ft^3/lb	m^3/kg (l/g)	cm^3/g	cm^3/mol
1	0.062428	62.428	125.85
16.018	1	1,000	2,015.9
0.016018	0.001	1	2.0159
7.9458×10^{-3}	4.9605×10^{-4}	0.49605	1

Pressure

$lb_f/in.^2$ (psi)	MPa	atm	Torr (mm Hg)	bar
1	6.8948×10^{-3}	0.068046	51.715	6.8948×10^{-2}
145.04	1	9.8692	7,500.6	10.0
14.696	0.10132	1	760.0	1.0132
0.019337	1.3332×10^{-4}	1.3158×10^{-3}	1	1.332×10^{-3}
14.504	0.1	0.98692	750.06	1

Enthalpy, Heat of Vaporization, Heat of Conversion, Specific Energies

Btu/lb	kJ/kg (J/g)	J/mol	cal/g
1	2.3244	4.6858	0.55556
0.43022	1	2.0159	0.23901
0.21341	0.49605	1	0.11856
1.8	4.1840	8.4345	1

Specific Heat, Entropy

Btu/lb-R	kJ/kg-K (J/g-K)	J/mol-K	cal/mol-K
1	4.184	8.4345	2.0159
0.23901	1	2.0159	0.48182
0.11856	0.49605	1	0.23901
0.49605	2.0755	4.184	1

Thermal Conductivity

Btu/ft-hr-R	mW/cm-K	J/s-cm-K	cal/s-cm-K
1	17.296	0.017296	0.0041338
0.057816	1	0.001	2.3901×10^{-4}
57.816	1,000	1	0.23901
241.90	4,184	4.184	1

Viscosity

lb/ft-s	kg/m-s ($N\text{-}s/m^2$)	cP (10^{-2} g/cm-s)	lb-s/ft² (slug/ft-s)
1	1.48816	1,488.16	0.031081
0.67197	1	1,000	0.020885
6.7197×10^{-4}	0.001	1	2.0885×10^{-5}
32.175	47.881	4.7881×10^{4}	1

Temperature

K	R	C	F
1	1.8	1	1.8

Velocity of Sound

ft/s	m/s
1	0.3048
3.2808	1

Surface Tension

lb_f/in.	N/m	dyne/cm
1	175.13	175.13×10^{3}
5.7102×10^{-3}	1	1,000
5.7102×10^{-3}	0.001	1

lb_f	=	pound force
lb	=	pound mass
g	=	gram
kg	=	kilogram mass
mol	=	2.01594 gram
ft	=	foot
m	=	meter
cm	=	centimeter
Btu	=	British thermal unit
J	=	joules = watt second
W	=	watt
cal	=	calorie
R	=	absolute temperature in degrees Rankine
F	=	temperature in degrees Fahrenheit
K	=	absolute temperature in Kelvins
C	=	temperature in degrees Centigrade
N	=	Newton
MPa	=	megapascal (pressure)
atm	=	atmosphere pressure
Torr	=	millimeter of mercury pressure
cP	=	centipoise
slug	=	32.174 lb
R	=	gas constant = 8.31434 J/mol-K, 8.31434×10^6 N-cm^3/m^2-mol-K
mol wt	=	2.01594

REFERENCES

1. **Roder, H. M., Weber, L. A., and Goodwin, R. D.,** National Bureau of Standards Monogr. 94, August 1965.
2. **McCarty, R. D. and Weber, L. A.,** National Bureau of Standards Tech. Note 617, April 1972.
3. **Weber, L. A.,** National Bureau of Standards, IR 74-374, 1974.
4. **Amdur, I.,** *J. Chem. Phys.,* 14(5), 339, 1946.
5. **Hermans, L. J. F.,** *Physica,* 31(10), 1567, 1965.
6. **Knaap, H. F. P. et al.,** *Discuss. Faraday Soc.,* 40, 135, 1965.
7. **Kogan, V. S. et al.,** *Ukr. Fiz. Zh.,* 14(11), 1927, 1969.
8. Lockheed Missiles and Space Co., Report No. NASA-CR-87655, LMSC-R-11-67-1, Vol. 1, March 1967.
9. **Younglove, B. A.,** *J. Chem. Phys.,* 48(9), 4181, 1968.
10. **Younglove, B. A.,** *Natl. Bur. Stand. U.S. Tech. News Bull.,* 52(7), 146, 1968.
11. **Udovidchenko, B. G. et al.,** *J. Low Temp. Phys.,* 3(4), 429, 1970.
12. **Weitzel, D. H. et al.,** National Bureau of Standards, Report No. 9796, June 1971.
13. **Stewart, J. W.,** *J. Chem. Phys.,* 40(11), 3279, 1964.
14. **Roder, H. M., Childs, G. E., McCarty, R. D., and Angerhofer, P. E.,** National Bureau of Standards Tech. Note 641, August 1973.
15. **Mason, E. A. et al.,** *Phys. Fluids,* 7(7), 1011, 1964.
16. **Waldmann, L.,** *Z. Phys.,* 124, 2, 1947.
17. **Mason, E. A. et al.,** *J. Chem. Phys.,* 36(10), 2746, 1962.
18. **Annis, B. K. et al.,** *Phys. Fluids,* 11(10), 2122, 1968.
19. **Matyash, I. V. et al.,** *Ukr. Fiz. Zh.,* 12(1), 56, 1967.
20. **Mason, E. A. et al.,** *Phys. Fluids,* 8(7), 1240, 1964.
21. **Waldmann, L.,** *Naturwissenschaften,* 32, 222, 1944.
22. **Berezhnot, A. N.,** Thermophysical Properties of Gases and Liquids, Vol. 1, Rabinovich, V. A., Ed., translated from Russian by Israel Program for Scientific Translation, Jerusalem, 1970.
23. **Sabena, S. C. et al.,** *Rev. Mod. Phys.,* 37(2), 316, 1965.
24. **Heymann, D. et al.,** *Physica,* 25, 556, 1959.
25. **Sabena, S. C. et al.,** *Rev. Mod. Phys.,* 33(2), 148, 1961.
26. **Deb, S. K.,** *Indian J. Phys.,* 41(10), 719, 1967.
27. **Nain, V. P. S. et al.,** *J. Chem. Phys.,* 51(4), 1541, 1969.
28. **Sabena, S. C. et al.,** *Mol. Phys.,* 2, 379, 1959.
29. **Rastogi, R. P. et al.,** *Trans. Faraday Soc.,* 62(528), 3325, 1966.
30. **Oost, W. A. et al.,** *Physica,* 62(3), 409, 1972.
31. **Monchick, L. et al.,** *J. Chem. Phys.,* 45(8), 3051, 1966.
32. **Farag, N. et al.,** *Z. Phys. Chem.* (Leipzig), 244(3-4), 244, 1970.
33. **Van Itterbeck, A. et al.,** *Acustica,* 7, 180, 1957.
34. **Van Itterbeck, A. et al.,** *Physica,* 13(4-5), 231, 1947.
35. **Van Itterbeck, A.,** *Physica,* 19, 525, 1953.
36. **Farag, N. et al.,** *Z. Phys. Chem.* (Leipzig), 245(3-4), 145, 1971.
37. **De Trayer, A. et al.,** *Physica,* 16, 669, 1950.
38. **Mat, P. et al.,** *Physica,* 48(4), 605, 1970.
39. **Las, J. et al.,** *Physica,* 31(8), 1346, 1965.
40. **Slieker, C. J. G.,** *Physica,* 31(9), 1388, 1965.
41. **Murphey, B. F.,** *Phys. Rev.,* 72(9), 834, 1947.
42. **Heath, H. R. et al.,** *Proc. R. Soc. London Ser. A,* 178, 380, 1941.
43. **Mason, E. A. et al.,** *J. Chem. Phys.,* 22(3), 522, 1954.
44. **Van de Ree, J. et al.,** *Physica,* 34(1), 66, 1967.
45. **Ghosh, A. K. et al.,** *J. Chem. Phys.,* 47(2), 452, 1967.
46. **Mathur, B. P. et al.,** *J. Chem. Phys.,* 41(12), 5537, 1968.
47. **Biolsi, L. et al.,** *J. Chem. Phys.,* 54(7), 3020, 1971.
48. **Van de Ree, J.,** Ph.D. thesis, Leiden University, Netherlands, 1967.
49. **Mathur, G. P. et al.,** *AIChE J.,* 11(4), 613, 1965.
50. **Diller, D. E. et al.,** *J. Chem. Phys.,* 44(7) 2604, 1966.
51. **Lund, L. M. et al.,** *J. Appl. Phys.,* 37(6), 2496, 1966.
52. **Elzinga, D. J. et al.,** *Cryogenics,* 6(4), 216, 1966.
53. **Dawson, R. et al.,** *AIChE J.,* 16(5), 725, 1970.
54. **Brezing, D.,** *AIAA J.,* 3, 1422, 1965.
55. **Grier, N. T.,** TN-D-3186, Lewis Research Center, NASA, Cleveland, 1966.
56. **Ebney, C. et al.,** *Phys. Rev. A,* 5(6), 2625, 1972.

57. Neal, W. E. J., *Proc. Phys. Soc. London*, 82, 333, 1963.
58. Paul, R. et al., *J. Chem. Phys.*, 44(5), 1622, 1965.
59. Sabena, S. C. et al., *Rev. Mod. Phys.*, 38(2), 380, 1966.
60. Sabena, S. C. et al., *Indian J. Phys.*, 36, 13, 1962.
61. Grew, K. E., *Proc. R. Soc. London Ser. A*, 178, 390, 1941.
62. Grew, K. E. et al., *J. Chem. Phys.*, 45(11), 4267, 1966.
63. Monchick, L. et al., *J. Chem. Phys.*, 49(3), 1178, 1968.
64. Mason, E. A. et al., *Chem. Phys. Lett.*, 6(6), 630, 1970.
65. Brown, N. J. et al., *J. Chem. Phys.*, 56(6), 2761, 1972.
66. Hass, W. P. A. et al., *Physica*, 26(10), 834, 1960.
67. Mathes, K. N., *Electro-Technol.*, 72(3), 72, 1963.
68. Carbonell, E., *Rev. Gen. Electr.*, 74(7-8), 611, 641, 1965.
69. Minnich, S. H. et al., *Cryogenics*, 9(3), 165, 1969.
70. Burnier, P. H. et al., *Advances in Cryogenic Engineering*, Vol. 15, Plenum Press, New York, 1970.
71. Burnier, P., *Entropie*, 28, 17, 1969.
72. Sacerdoti, G., Nota Int. No. 464, March 1970.
73. Jefferies, M. J. et al, Report No. 69-C-383, General Electric, Schenectady, New York, September 1969.
74. Lehmann, J. P., *Rev. Gen. Electr.*, 79(1), 15, 1970.
75. Burnier, P., *Bull. Soc. R. Bdge. Electr.*, 82(2), 123, 1966.
76. Cassutt, L. et al., *Advances in Cryogenic Engineering*, Vol. 7, Plenum Press, New York, 1962.
77. McCormick, N. R. et al., *Br. J. Appl. Phys.*, 5, 171, 1954.
78. Fox, G. W. et al., *J. Appl. Phys.*, 11, 592, 1940.
79. Ward, A. L. et al., *Phys. Rev.*, 122, 376, 1961.
80. Davies, D. E. et al., *Proc. 5th Int. Conf. Ionization Phenomena Gases*, Vol. 1, North-Holland, Amsterdam, 1962, 678.
81. Bobin, J. L. et al., *J. Phys.* (Paris), 26(4), 180, 1965.
82. Avivi, P. et al., *Br. J. Appl. Phys.*, 14, 819, 1963.
83. Yos, J. M., Report No. RAD-TM-63-7, Avco Corp., Wilmington, Del., March 1963.
84. Grier, N. T., Tech. Note D-3186, Lewis Research Center, NASA, Cleveland, April 1966.
85. Neusel, R. H. et al., *J. Appl. Phys.*, 36(5), 1507, 1965.
86. Heylen, A. E. D. et al., Leeds University, England, Report No. AFCRL-67-0033, P.R7, September 1966.
87. Burnier, P. et al., *Rev. Gen. Electr.*, 77(6), 600, 1968.
88. Fallou, B. et al., *Rev. Gen. Electr.*, 77(6), 594, 1968.
89. Paul, F. W. et al., Goddard Space Flight Center, NASA, Greenbelt, Md., Report No. NASA-SP-208, 1969.
90. Pearson, J. S. et al., *Br. J. Appl. Phys.*, 2(11), 1583, 1969.
91. Cookson, A. H., *Proc. Inst. Electr. Eng.*, 117(1), 269, 1970.
92. Mesnage, P., *Compt. Rend.*. 214, 702, 1942.
93. Little, A. D. et al., Final Report C-61Z92, Arthur D. Little, Inc., Cambridge, Mass., May 1961.
94. Willis, W. L., *Cryogenics*, 6(5), 279, 1966.
95. Avco Corp., Wilmington, Mass., Final Report AVSSD-0414-67-RR, 1967.
96. Neal, S., *Cryog. Eng. News*, 3(8), 30, 1968.
97. Morris, J. C. et al., *Phys. Fluids*, 13(3), 608, 1970.
98. Goodwin, R. D., *J. Res. Natl. Bur. Stand.*, Vol. 73(6), 585, 1969.
99. Roder, H. M., McCarty, R. D., and Hall, W. J., National Bureau of Standards Tech. Note 625, October 1972.
100. McCarty, R. D., National Bureau of Standards Internal Report, 74-357.
101. Roder, H. M. and McCarty, R. D., National Bureau of Standards, I.R. 75–814, 1975.
102. Michels, A., De Graaff, W., and Ten Seldom, C., *Physica*, 26(6), 393, 1960.
103. Hanley, H. J. M. and Klein, M., National Bureau of Standards Tech. Note No. 360, 1967.
104. Klein, M. and Hanley, H. J. M., *J. Chem. Phys.*, 53, 4722, 1970.
105. Hanley, H. J. M. and Klein, M., National Bureau of Standards Tech. Note No. 628, 1972.
106. Woolley, H. W., Scott, R. B., and Brickwedde, F. G., *J. Res. Natl. Bur. Stand.*, 41(5), 379, 1948.
107. IUPAC Revised Atomic Weights Values, *Chem. Eng. News*, 39, 42, 1961.
108. The International Practical Temperature Scale of 1968, *Metrologia*, 5, 35, 1969.
109. Younglove, B. H. and Diller, D. E., *Cryogenics*, 2(5), 2, 1962.
110. Diller, D. E., *J. Chem. Phys.*, 49(7), 3096, 1968.

111. **Michels, A., De Graff, W., and Wolkers, G. J.,** *Appl. Sci. Res. Sect. A,* 12(1), 9, 1963.
112. **Johnston, H., Clark, J. B., Rifkin, E. B., and Keer, E. C.,** *J. Am. Chem. Soc.,* 72, 3933, 1950.
113. **Dwyer, R. F., Cook, G. A., Shields, B. M. et al.,** *J. Chem. Phys.,* 42(11), 1965.
114. **Mullins, J. C., Ziegler, W. T., and Kirk, B. S.,** Georgia Institute of Technology, Tech. Report #1, November 1961.
115. **Clusius, K.,** *Z. Elektrochem.,* 44, 21, 1938.
116. **Dywer, R. F., Cook, G. A., and Berwaldt, O. E.,** *J. Chem. Phys.,* 43(3), 801, 1965.
117. **Mills, R. L. and Grilly, E. R.,** *Phys. Rev.,* 101(4), 1246, 1956.
118. **Goodwin, R. D.,** *Cryogenics,* 2(6), 353, 1962.
119. **Galt, J. K.,** *J. Chem. Phys.,* 16, 505, 1948.
120. **Sluijter, G. G., Knaap, H. F. P., and Beenakker, J. J. M.,** *Physica,* 30(4), 745, 1964.
121. **Corruccini, R. J.,** National Bureau of Standards Tech. Note No. 322, August 1965.
122. **Blagoi, Yu. P. and Pashkou, V. V.,** *Zh. Eksp. Teor. Fiz.,* 49(5), 1453, 1965.
123. **Roder, H. M. and Diller, D. E.,** *J. Chem. Phys.,* 52(11), 5928, 1970.
124. **Hanley, H. J. M., McCarty, R. D., and Cohen, E.,** *Physica,* 60, 322, 1972.
125. **Van Itterbeek, A., Verbeke, O. et al.,** *Physica,* 30(6), 1238, 1964.
126. **McCarty, R. D.,** National Bureau of Standards Internal Report 74-357, February 1974.
127. **Diller, D. E.,** *J. Chem. Phys.,* 42(6), 2089, 1965.
128. **Mullins, J. C. and Ziegler, W. T.,** in *International Advances in Cryogenic Engineering,* Vol. 10, Timmerhaus, K. D., Ed., Plenum Press, New York, 1963, 171.
129. **Suciu, S. W. L.,** Sikkett, Argonne National Laboratory Report No. ANL-4603, Part II, 1951, 10.
130. **Hiza, M. J., Kidnay, A. J., and Miller, R. C.,** NBS-OSRD Special Publication, to be published.
131. **Hiza, M. J.,** *Cryogenics,* 10(2), 106, 1970.
132. **Robinson, R. L. and Hiza, M. J.,** in *Advances in Cryogenic Engineering,* Vol. 20, Plenum Press, New York, to be published.
133. **Oellrich, L. R., Plöcker, U. J., and Knapp, H.,** Vapor-Liquid Equilibria, Institute of Thermodynamics, Technical University of Berlin, September 1973.
134. **Kidnay, A. J., Hiza, M. J., and Miller, R. C.,** *Cryogenics,* 13(10), 575, 1973.
135. **Wilke, C. R.,** *J. Chem. Phys.,* 18(4), 517, 1950.
136. **Mason, E. A. and Saxena, S. C.,** *Phys. Fluids,* 1(5), 361, 1958.
137. **Jones, M. C.,** National Bureau of Standards Tech. Note 390, April 1970.
138. **Mactaggart, J. W. and Hunt, J. L.,** *Can. J. Phys.,* 47(1), 65, 1969.
139. **May, A. D., Degen, V., Stryland, J. C., and Welsh, H. L.,** *Can. J. Phys.,* 39, 1769, 1951.
140. **Belozerova, V. P.,** *Sov. J. Opt. Technol.,* 34(2), 184, 1967.
141. **Forester, C. K.,** *Adv. Cryog. Eng.,* 12, 82, 1967.
142. **Bennett, M. J. and Tomkins, F. C.,** *Trans. Faraday Soc.,* 57, 185, 1957.
143. **Farkas, A.,** *Orthohydrogen, Parahydrogen, and Heavy Hydrogen,* University Press, Cambridge, 1935.
144. **Weitzel, D. H., Loebenstein, W. V., Draper, J. W., and Park, D. E.,** *J. Res. Natl. Bur. Stand.,* 60, 221, 1958.
145. **Wigner, E. and Huntington, H. B.,** *J. Chem. Phys.,* 3, 764, 1935.
146. Preliminary Report on the Thermodynamic Properties of Selected High Elements and Some Related Compounds, National Bureau of Standards, unpublished report, July 1966.
147. **Grigorev, F. V., Kormer, S. B., Mikhailov, O. L. et al.,** *JETP Lett.,* 16(5), 201, 1972.
148. **Mann, D. B., Ludtke, P. R., Sindt, C. F., and Chelton, D. B.,** *Advances in Cryogenic Engineering,* Vol. 11, Plenum Press, New York, 1966, 207.
149. **Sindt, C. F.,** *Cryogenics,* 10(5), 1970.
150. **Chelton, D. B.,** Safety in the use of liquid hydrogen, in *Technology and Uses of Liquid Hydrogen,* Scott, R. B., Denton, W. H., and Nicholls, C. M., Eds., Macmillan, New York, 1964.
151. Liquid Hydrogen, A Guide for the Safe Handling and Storage of Liquid Hydrogen at LASL Facilities, Los Alamos Scientific Laboratory, University of California, Los Alamos, New Mexico, 1962.
152. **Hord, J.,** National Bureau of Standards Report No. 10734, February 1972.
153. Hydrogen Safety Manual, NASA Tech. Memo TM-X-52454, NASA Lewis Research Center, 1968.
154. Standard for Gaseous Hydrogen Systems at Consumer Sites, Natl. Fire Protect. Assoc. Bull. NFPA No. 5DA, 1966.
155. Standard for Liquid Hydrogen Systems at Consumers Sites, Natl. Fire Protect. Assoc. Bull. NFPA No. 50B, 1971.

156. General Safety Engineering Design Criteria, Vol. 1, CPIA Publ. 194, October 1971 and Liquid Propellant Handling, Storage and Transportation, Vol. 3, CPIA Publ. 194, May 1970; prepared by JANNAF Propulsion Committee of the JANNAF Hazards Working Group, available from Chemical Propulsion Information Agency (CPIA) of the Johns Hopkins University Applied Physics Laboratory, Silver Springs, Md.
157. Corruccini, R. J., IIR-CRTBT Meet., Grenoble, France, June 8–11, 1965.
158. Jones, M. C., *J. Chem. Phys.*, 51(9), 3833, 1969.
159. Buckingham, A. D. et al., *Trans. Faraday Soc.*, 63(5), 1057, 1967.
160. Neece, G. A. et al., *J. Comput. Phys.*, 7(3), 621, 1971.
161. Bolshutkin, D. N. et al., *Din. Dislokatsii, Dokl. Soveshch. Din. Povedenujer Dislokatsii Krist.*, 1968, 683, 1969.
162. Carneiro, K. et al., *Phys. Rev. Lett.*, 30(11), 481, 1973.
163. Cuthbertson, C. et al., *Proc. R. Soc. London Ser. A*, 83, 151, 1910.
164. Corruccini, R. J., National Bureau of Standards Cryog. Eng. Lab. Note No. 64-16, August 1964.
165. Corruccini, R. J., National Bureau of Standards Cryog. Eng. Lab. Note No. 65–4, February 1965.
166. Corruccini, R. J., National Bureau of Standards Tech. Note No. 323, September 1965.
167. Russell, J. L., Jr. et al., Report No. GA-6313, General Atomic Division, General Dynamics Corp., San Diego, April 1965.
168. Victor, G. A. et al., *J. Chem. Phys.*, 50(6), 2535, 1969.
169. Zaidi, H. R., *Can. J. Phys.*, 49(3), 385, 1971.
170. Morris, J. C. et al., *Phys. Fluids*, 13(3), 608, 1970.
171. Curzon, A. E. et al., *Br. J. Appl. Phys.*, 16, 1301, 1965.
172. Bostanjoglo, O. et al., *J. Chem. Phys.*, 46(5), 2004, 1967.
173. Fiommhald, L., *Phys. Rev.*, 172(1), 118, 1968.
174. Robertson, A. G., *Aust. J. Phys.*, 24(3), 445, 1971.
175. Crompton, R. W. et al., *Aust. J. Phys.*, 24(4), 543, 1971.
176. Harrison, H. R. et al., *Chem. Phys. Lett.*, 10(4), 418, 1971.
177. Ageron, P., *Cryogenics*, 9(1), 42, 1969.
178. Lindstrom, K. et al., *Rev. Sci. Instrum.*, 41(7), 1083, 1970.
179. Unland, H. O. et al., *Advances in Cryogenic Engineering*, Vol. 17, Plenum Press, New York, 1972.
180. Kartluke, H. et al., Report RR 64-47, NASA CR-54044, Technidyne Inc., Westchester, Penn., July 1964, 213.
181. Tagashira, H. et al., *Br. J. Appl. Phys.*, 2(6), 867, 1969.
182. Halpern, B. et al., *J. Chem. Phys.*, 51(3), 1048, 1969.
183. Swamy, M. N. et al., *Br. J. Appl. Phys.*, 2(10), 1437, 1969.
184. Hiza, M. J. et al., *AIChE, J.*, 16(5), 733, 1970.
185. Kramer, H. L. et al., *J. Chem. Phys.*, 53(7), 2792, 1970.
186. Hehs, W. A. et al., Report No. NASA CR-54003, FZK-182, General Dynamics, Fort Worth, March 1964.
187. Blue, J. W. et al., NASA Tech. Natl. No. D-2591, February 1965.
188. Kogan, V. S. et al., *Sov. Phys. JETP*, 13(4), 718, 1961.
189. Mulser, P. et al., *Phys. Lett. A*, 28(10), 703, 1969.
190. Denne, D. R., *Br. J. Appl. Phys.*, 3(9), 1392, 1970.
191. Bourland, R. D. et al., *Phys. Rev. Sect. B*, 11, 3625, 1971.
192. Felsteiner, J. et al., *Phys. Rev. A*, 7(2), 826, 1973.
193. Gaines, J. R., Tech. Report No. 10, Ohio State University Research Foundation, Columbus, December 1963.
194. DeCastro, E. M. et al., *Proc. 9th Int. Conf. on Low Temperature Physics, Part B*, Plenum Press, New York, 1965.
195. Langhoff, P. W., *J. Chem. Phys.*, 57(7), 2604, 1972.
196. Mangeot, B. et al., *Compt. Rend.*, 257(21), 3134, 1963.
197. Hoylad, J. R., *J. Chem. Phys.*, 41(10), 3153, 1964.
198. Jain, D. C. et al., *Proc. Phys. Soc. London*, 89, (564), 409, 1966.
199. Saika, A. et al., *J. Chem. Phys.*, 56(7), 3312, 1972.
200. Burnier, P. et al., La Technique Moderne, September 1969.
201. Reif, F. et al., *Phys. Rev.*, 91, 631, 1953.
202. Sugawara, T. et al., *Phys. Rev.*, 95, 1355, 1954.
203. Smith, G. W. et al., *Phys. Rev.*, 117, 732, 1960.
204. Smith, G. W. et al., *Phys. Rev.*, 111, 188, 1958.
205. Lipsicas, M., *J. Chem. Phys.*, 36, 1235, 1962.

206. Dickson, S. A. Jr., Ph.D. thesis, Duke University, Durham, N.C., 1964.
207. Lipsicas, M. et al., *Can. J. Phys.*, 39, 881, 1961.
208. Harris, A. B. et al., *Phys. Rev. Lett.*, 16(19), 845, 1966.
209. Metzger, D. S. et al., *Phys. Rev.*, 147(2), 644, 1966.
210. Tward, E. et al., *Can. J. Phys.*, 46(5), 331, 1968.
211. Sung, C. C., *Phys. Rev.*, 167(2), 271, 1968.
212. Gordon, R. G. et al., *J. Chem. Phys.*, 48(6), 2655, 1968.
213. Amstutz, L. I. et al., *Phys. Rev. Lett.*, 21(16), 1175, 1968.
214. Harris, A. B. et al., *Phys. Rev.*, 175(2), 603, 1968.
215. Bloom, M. et al., *Adv. Chem. Phys.*, 12, 549, 1967.
216. Amstutz, L. I. et al., *Phys. Rev.*, 181(2), 589, 1969.
217. Homma, S., *Prog. Theor. Phys.*, 43(3), 851, 1970.
218. Harris, A. B., *Phys. Rev. Sect. B*, 2(9), 3495, 1970.
219. Constable, J. H. et al., *Solid State Commun.*, 9(2), 155, 1971.
220. Sullivan, N. S. et al., *Phys. Lett. A*, 39(1), 23, 1972.
221. Sullivan, N. S. et al., *Phys. Rev. Sect. A*, 6(3), 1102, 1972.
222. Ishimoto, H. et al., *J. Phys. Soc. Jpn.*, 35(1), 300, 1973.
223. Bloom, M., *Physica*, 23, 237, 1957.
224. Bloom, M. et al., *Can. J. Phys.*, 41(10), 1580, 1963.
225. Hardy, W. N., *Can. J. Phys.*, 44(1), 265, 1966.
226. Lipsicas, M. et al., *J. Chem. Phys.*, 44(8), 2839, 1966.
227. Miller, C. E. et al., *Phys. Rev.*, 176(1), 273, 1968.
228. Sung, C. C., *Solid State Commun.*, 6(2), 93, 1968.
229. Talita, K. et al., *Can. J. Phys.*, 47(13), 1355, 1969.
230. Amstutz, L. I. et al., *J. Phys. Chem. Solids*, 30(12), 2693, 1969.
231. Walasek, K. et al., *Phys. Lett. A*, 32(6), 399, 1970.
232. Hama, J. et al., *Prog. Theor. Phys.*, 44(2), 303, 1970.
233. Beckmann, P. A. et al., *Phys. Rev. Sect. A*, 6(4), 1684, 1972.
234. Ozerov, R. P. et al., *J. Phys. Soc. Jpn.*, 17 (Suppl. B-11), 385, 1962.
235. Bulatov, A. S. et al., *Sov. Phys. JETP*, 27(2), 210, 1968.
236. Schott, W. et al., *Phys. Lett. A*, 27(8), 566, 1968.
237. Verdan, G., *Phys. Lett. A*, 25(6), 435, 1967.
238. Schott, W., *Z. Phys.*, 231(3), 243, 1970.
239. Noolandi, J., *Solid State Commun.*, 9(21), 1809, 1971.
240. Nakamura, T., *Prog. Theor. Phys. Suppl.*, 46, 343, 1970.
241. Seiffert, W. D., Report No. EUR-4455.D, European Atomic Energy Community, ISPRA, Italy, 1970.
242. Stein, H. et al., *J. Chem. Phys.*, 57(4), 1726, 1972.
243. Sears, V. F., *Phys. Rev. Sect. A*, 7(1), 340, 1973.
244. Nielsen, M., *Phys. Rev. Sect. B*, 7(4), 1626, 1973.
245. Eder, O. J. et al., *Proc. Phys. Soc. London*, 89(566), 833, 1966.
246. Ageron, P., *Cryogenics*, 9(1), 42, 1969.
247. Schott, W., *Z. Angew. Phys.*, 30(4), 293, 1970.
248. Krascella, N. L., Report No. NASA-SP-3005, United Aircraft Corp. Research Laboratories, East Hartford, Conn., 1063.
249. Wang, C. C., *Phys. Rev. Sect. B*, 2(6), 2045, 1970.
250. Loots, V. et al., *Can. J. Phys.*, 43, 1985, 1965.
251. Leonteva, A. V. et al., *Phys. Status Solidi*, 42(2), 543, 1970.
252. Stetsenko, Y. E. et al., *Sov. Phys. Solid State*, 14(1), 149, 1972.
253. Riehl, J. W. et al., *J. Chem. Phys.*, 58(10), 4571, 1973.
254. Mash, D. I. et al., *JETP Lett.*, 2(12), 349, 1965.
255. Allin, D. J. et al., *Appl. Opt.*, 6(10), 1597, 1967.
256. Withrington, R. J., *Appl. Opt.*, 7(1), 175, 1968.
257. Baley, C. D. et al., *Can. J. Phys.*, 50(18), 2158, 1972.
258. Weinberg, M. et al., *Phys. Rev. Sect. A*, 7(4), 1413, 1973.
259. Towle, L. C., *Phys. Chem. Solids*, 26(3), 659, 1965.
260. Statsenko, Y. E. et al., *Sov. Phys. Solid State*, 12(12), 2958, 1971.
261. Bignell, N., *Cryogenics*, 13(1), 30, 1973.
262. Smith, I. E., *Progress in Cryogenics*, Vol. 3, Academic Press, New York, 1961, 77.
263. Harasima, A., *Proc. Phys. Math. Soc. Jpn.*, 23, 977, 1941.
264. Harasima, A., *Proc. Phys. Math. Soc. Jpn.*, 22, 825, 1940.
265. Harasima, A., *Proc. Phys. Math. Soc. Jpn.*, 23, 983, 1941.

266. **Good, R. J. et al.,** *Advances in Cryogenic Engineering,* Vol. 8, Plenum Press, New York, 1963, 306.
267. **Harasima, A.,** *J. Phys. Soc. Jpn.,* 22(1), 334, 1967.
268. **Ramanadham, M.,** *Indian J. Pure Appl. Phys.,* 8(4), 237, 1970.
269. **Avery, D. A.,** *Phys. Rev.,* 186(1), 108, 1969.
270. **Scott, G. G. et al.,** *Phys. Rev. Sect. A,* 2(5), 2080, 1970.
271. **Kogan, V. S. et al.,** *Sov. Phys. JETP,* 10(3), 485, 1960.
272. **Bulatova, R. F. et al.,** *Sov. Phys. JETP,* 21(1), 89, 1965.
273. **Mills, R. L. et al.,** *Phys. Rev. Lett.,* 15(18), 722, 1965.
274. **Mills, R. L. et al.,** *Phys. Rev. Lett.,* 17(22), 1131, 1966.
275. **Schuch, A. F. et al.,** *Phys. Rev.,* 165(3), 1032, 1968.
276. **Hord, J. et al.,** Report NBS IR 75-803, National Bureau of Standards, 1975.
277. **Stewart, R. B. et al.,** in *Technology and Uses of Liquid Hydrogen,* Scott, R. B., Denton, W. H., and Nicholls, C. M., Eds., Macmillan, New York, 1964, chap. 11.
278. **Roder, H. M. et al.,** National Bureau of Standards Tech. Note 130, December 1961.
279. **Johnson, V. J.,** A Compendium of the Properties of Materials as Low Temperatures – Phase 1: Part I, Properties of Fluids, National Bureau of Standards Cryogenic Engineering Laboratory WADD Tech. Report 60-56, Part I, Contr. No. AF 33(616)58-4, July 1960.
280. **Angerhofer, P. E. et al.,** Report No. 19700, National Bureau of Standards, Boulder, Colo., August 1971.
281. **Hilsenrath, J.,** Tables of Thermal Properties of Gases, Circ. No. 564, National Bureau of Standards, 1955.
282. **Ho, C.Y., Powell, R. W., and Liley, P. E.,** Standard Reference Data on the Thermal Conductivity of Selected Materials, Part 3, Final Report NBS-NSRDS Contr. CST-1346, Purdue University, T.P.R.C., West Lafayette, Ind., September 1968.
283. **Corruccini, R. J.,** Dielectric Constant of Liquid Parahydrogen, National Bureau of Standards Tech. Note 144, April 1962.
284. **Dean, J. W.,** National Bureau of Standards Tech. Note 120, 1961.
285. **Hust, J. G. and Stewart, R. B.,** National Bureau of Standards No. 8812, 1965.
286. **Roder, H. M. et al.,** Report No. NBS-IR 75-123, National Bureau of Standards, Cryogenics Division, Boulder, Col., 1975.
287. **Hendricks, R. C. et al.,** *Progress in Refrigeration Science and Technology,* Vol. 1, Int. Congr. Refrigeration, Avi Publishing, Westport, Conn., 1973, 695.

Index

INDEX

CRC PUBLICATIONS OF RELATED INTEREST